Seeing

Seeing

The Computational Approach to Biological Vision

Second Edition

John P. Frisby and James V. Stone

The MIT Press
Cambridge, Massachusetts
London, England

MIT Press books may be purchased at special quantity discounts for business or sales promotional use. For information, please email special_sales@mitpress.mit.edu or write to Special Sales Department, The MIT Press, 55 Hayward Street, Cambridge, MA 02142.

This book was set in Adobe InDesign CS3 by John P. Frisby. Printed and bound in the United States of America.

Library of Congress Cataloging-in-Publication Data

Seeing: the computational approach to biological vision / John P. Frisby and James V. Stone—2nd ed.

   p.   cm.

Includes bibliographical references and index.

ISBN 978-0-262-51427-9 (pbk.: alk. paper)

1. Visual perception. 2. Optical illusions. 3. Psychophysiology. I. Stone, James V. II. Title.

BF241.F74   2010

152.14—dc22   2009031989

Dedications

John Frisby: to Jo, without whose support this project would probably never have been started, and without which it most definitely would never have been finished.

James Stone: to Nikki, Sebastian, and Teleri, without whom this book would have been finished sooner, but with fewer beautiful photographs.

In seeing eye to eye there's more than meets
The eye. And motes and beams within the eye
Of mind are harder lost to sight than all
My eye and Betty Martin could have seen.
We are all eyes. Even the blind have more
Than half an eye, and someone is their apple,
Or they cast those of sheep or they are sight
For sore, or have a good one for or find
Some happy way of being all. The eyes
Will have it, seeing how we may. For now
We sharply see it; close them; now we do.
We're ever learning from the pupilled eye.

Trevor Dickinson
April 26, 1980

Written after discussing *Seeing, First Edition,*
with John Frisby.

Contents

Preface to Second Edition

Writing an introductory text to the vast topic of seeing is rather like painting a picture: both entail a great deal of selection. Not everything can be included, even in brief, and so decisions have to be made about what to put in. In this book we have used as our guide the computational approach to vision. The intellectual foundations of this approach have a long pedigree, extending back at least as far as the work of the pioneers in the field of cybernetics that emerged in the 1950s and 60s. Cybernetics is usually defined as the interdisciplinary study of the structure of regulatory systems, and is hence closely related to control theory and systems theory. One of its hallmarks was a concern with stating clearly the problems that need to be solved. This feature is also the cornerstone of the computational approach to vision, best known from the work of David Marr and his colleagues.

Marr proposed three distinct levels of analysis for the study of seeing: first, the *computational level*, specifying the precise nature of the vision problem at hand and identifying constraints that can be used to solve the problem in principle; second, the *algorithmic level*, giving details of a definite method (an algorithm) for exploiting the these constraints; and third, the *hardware implementation level*, which specifies how to execute the method in a physical system (e.g., the neurons mediating seeing). Note that the computational approach has nothing to do with computers, nor is it to be confused with the field of study called computer vision.

It is a curious fact that the authors of some textbooks on vision and cognitive psychology begin by introducing Marr's three-levels approach, but then largely ignore it in the rest of the book. We also describe Marr's framework early on, but we follow this up by trying to use it systematically in tackling a wide range of vision problems. In each chapter, this approach is applied to a different topic in vision by examining problems that the visual system encounters in interpreting retinal images; the constraints available to solve these problems; the algorithms that can realize the solution; and the implementation of these algorithms in neurons.

Both authors have had considerable experience of using this approach in teaching vision courses to university students. It has led them to dispense with the usual kind of introductory lecture, because abstract ideas (such as *levels of analysis for understanding complex information systems*) tend to "fly right over the heads" of students with no hands-on experience of the problems of vision. Instead, we "plunge in" with some specific topics, and then use these to extract lessons about how the computational approach works out in practice. Hence, one feature of our teaching is that our "introductory" lecture is the final one in the course. We adopt this approach here, in that our final chapter, *Seeing Summarized*, evaluates Marr's approach and discusses it in the context of other approaches. We hope that by the time readers get to that chapter they are in a good position to engage with the debates it contains. If you are a reader who wants that context first, then consider skim-reading it at the outset.

Despite delaying our introduction of general contextual and abstract material to the final chapter, in the first chapter we try hard to make accessible the topic of seeing to the general reader and to students who have not previously attended a vision course. We hope that such readers will find that this book gives them a good general understanding of seeing, despite the fact that much of it is well suited to the needs of advanced undergraduates and postgraduates. In short, this book needs no special background or prior familiarity with basic concepts in vision. A touchstone of our approach is to try to *teach* the subject. For us, this means we cover at length, using many illustrations, topics that we know from long experience can be hard to grasp, (an example is the convolution of a receptive field with an image). We provide at the end of Ch 1 a guide as to what each chapter contains, and all chapters (except the first and last) have a short note at the outset entitled *Why read this chapter?* These are intended to help readers to find a path through the book that suits their expertise and interests.

The upshot of all the foregoing is that we have not set out to write a handbook, or to cover all the topics to which many other texts devote whole chapters. There is no such thing as the "perfect" textbook on seeing and we certainly make no claim in that regard. Readers will, indeed should, consult other texts, and, of course, pursue their particular interests using Web searches (see advice below and our sections at the end of each chapter on *Further Reading*). What we do hope we have achieved is to provide enough basic coverage of the subject to give readers a sound understanding of core topics, so that they are well equipped to tackle the vast vision literature. And of course, we hope that readers become persuaded that the computational approach has strong merits as a way of thinking about seeing.

Finally, how does this second edition differ from the first? They have much in common, for example, a general emphasis on the computational approach, many illustrations, and an attempt to combine accessibility with rigor. (See Preface to First Edition, below.) That said, this second edition is much more explicit in teaching the computational approach, and it is much longer (roughly three times as many pages). Much of this increase is down to the introduction of new topics, which include mathematical material aimed at advanced students. Of course, we have also brought this second edition up to date. However, although a great deal has happened in the vision literature since the first edition was published in 1979, we have been surprised to realize that much of the first edition remains core material today. Moreover, we believe the general tenor of the computational approach espoused in 1979 in the first edition of *Seeing* still holds valuable lessons for students and vision researchers in the twenty first century.

Acknowledgements, second edition
We are greatly indebted to Neville Moray and Tom Stafford, who read and commented on all chapters. David Buckley, Helen Davis, and John Elliott gave detailed helpful advice on numerous chapters. John Mollon read several chapters as part of a publisher's review and his comments and suggestions were extremely useful, as were those from our students. Any errors remaining are of course all our own work. Emily Wilkinson gave us much-needed advice on navigating our way through the perilous seas of publishing. We thank Robert Prior of MIT Press for his support, and for the advice of MIT press staff on various aspects of layout.

JVS would like to thank Raymond Lister for taking the time to introduce to a (then) young psychologist about complexity theory, and Stephen Isard who taught an even younger psychologist how to think in straight lines.

John P. Frisby and James V. Stone   *October 2009*

Additional Note It has been a pleasure to work with Jim Stone on this second edition. His expertise, particularly using Bayesian methods, has been crucial to updating. We have exchanged 11,053 emails during the 4+ years it has taken to complete the project. And we are still friends.   *John Frisby*

Preface to the First Edition

Novelists sometimes say that their characters take over even the best laid plot and determine an ending different from the one originally envisaged. Something of the sort happened in the writing of this book, even though it is not a work of fiction, or at least is not meant to be. I started out to write *The General Reader's Fun Book of Visual Illusions*, with extended captions to give some explanation of why the illusions were as they were. But it proved impossible to say much of interest about them until the scene had been set with some introductory chapters on the fundamental nature of seeing and the basic neural machinery of the visual system.

When I had done this, the natural format for the book turned out to be not just a series of illusory figures, but instead a series of chapters dealing with selected topics from the standpoint of how the visual system performs certain jobs. To be sure, pictures of illusions abound—they provide valuable clues about visual mechanisms—but now they complement and illustrate a "story", rather than performing the central guiding role as originally planned. This change makes the book suitable not only for a general readership, but also for an introductory course on vision, especially one in which the teacher wishes to emphasize a combined psychological, physiological, and computational approach. But no attempt has been made to write a conventional textbook. The coverage is anything but encyclopedic, and a course tutor will find many holes that need plugging. But that seems to me no bad thing: why else have a course tutor? More important in my view is to have a text which communicates some fundamental ideas in visual science, as it is being practised today, so laying a foundation upon which the teacher can build.

But despite this change from my expectations and plans, I hope the book is still "fun." It is certainly still intended for the general reader, in that no special knowledge is demanded and all terms are defined as the book proceeds. Even so, understanding visual mechanisms is intrinsically tricky, and although I have made every attempt to reduce each problem area to its simplest essentials, the general reader will sometimes need a fairly large dose of motivation to grapple with all the book has to say. But whenever the text becomes a bit too intricate for to cope with, the general reader is encouraged to flick over the pages and treat the book as it was originally meant to be written—as a pleasure book of visual illusions. There are plenty of them and newcomers to the area will enjoy peering at the illustrations, often in disbelief that their visual systems can possibly get it all so wrong. Visual illusions are, quite simply, fascinating.

Acknowledgements, first edition

I owe a special debt to David Marr, whose writings on seeing are for me the work of genius, and whose influence can be felt throughout this book. I am particularly grateful to John Mayhew for our frequent discussions about vision, which have helped illuminate for me so many different aspects of the subject. He read most of the manuscript in draft, made useful comments, and provided all the computer-drawn figures. Kevin Connolly, my Head of Department, provided me with excellent facilities, good advice, and encouragement throughout. Without his willingness to arrange my teaching and administrative commitments in a way which still left some time over for peace and quiet, the book would not be written yet. I am very appreciative of these things. Henry Hardy—of the Roxby Press when the project began, now of the Oxford University Press—my editor, and the person who invited me to write the book in the first place, has provided patient and careful guidance at all stages, always helping toward greater clarity of presentation and also offering much useful advice from his own considerable knowledge of perception. Hugh Elwes of the Roxby Press is to be thanked, now the book is finally written, for bullying me into continuing with it when my energies were flagging and I was near to quitting. David Warner and his team of artists have made a first-rate job of designing the book and contributed much inventiveness and skill in transforming my figure sketches into proper illustrations. Numerous other people have contributed in many other ways, some perhaps not realizing just how helpful they have been, even if indirectly, and I would particularly like to mention Bela Julesz, Richard Gregory, Chris Brown, and Paul Dean. Finally, my family deserves and gets my warm thanks for having put up so gamely with so much while the book was being written They have amply earned the book's dedication.

John P. Frisby   *Sheffield, May 1978*

Figures and Cover Design

Unless stated otherwise, figures were created by the authors, or by Robin Farr whom we thank for for assistance in recreating many figures from the first edition of *Seeing*. We thank Len Hethrington for help with photographs. Most technical figures (e.g., most random dot stereograms, gratings, and convolutions) were created by J.V. Stone using MatLab. The copyright for all these figures is held by the authors. Figure permissions for other figures, where needed, are given in the legend for each figure. Cover based on an idea by J.V. Stone (photograph by Andrew Dunn). We welcome any comments on figure citations, and particularly any omissions or errors. Corrections will be posted on http://mitpress.mit.edu/seeing.

Further Reading

References for sources cited in each chapter are listed at the end of each chapter, often with a comment added. Suggestions for *Further Reading* are also given in those lists, again usually with comments.

Also, conducting Web searches is nowadays standard practice to follow-up particular topics.

There are many fine books on vision and it is hazardous to pick out a few that we think have particular merits that complement this book, but here goes, in alphabetical (not priority) order:

Bruce V, Green PR, and Georgeson MA (2003)) *Visual perception: physiology, psychology, and ecology.* Psychology Press. *Comment* A wide-ranging text with much advanced materia!. Its account of spatial and temporal filtering is a fine extension of the material given in this book.

Marr D (1982) *Vision: A computational investigation into the human representation and processing of visual information* San Francisco, CA: Freeman. *Comment* A seminal work that provides the inspiration of much of this book. Soon to be republished by MIT Press. Not an introductory text by any means but its opening and closing sections set out and debate Marr's approach in an accessible way. We particularly recommend the *Epilogue* which gives an imaginary dialogue based on Marr's conversations with Francis Crick and Tomaso Poggio. It seems to us that its arguments for Marr's approach stand up well to the test of time.

Mather G (2008) *Foundations of Perception Second Ed.* Psychology Press. *Comment* Provides excellent tutorial sections on many technical topics, such as the photometric system used for measuring light including luminance.

Snowden R, Thompson P and Troscianko T (2006) *Basic Vision* Oxford University Press.
Comment An engaging and amusingly written text with fine accounts of many visual phenomena.

Quinlan P and Dyson B (2008) *Cognitive Psychology.* Pearson Education Limited. *Comment* Takes as its theme Marr's approach, applying it in the field of cognitive psychology, and hence a good complement to the present book whose emphasis is on the early stages of visual processing.

Regan D (2000) *Human Perception of Objects.* Sinauer Associates, Inc. Publishers: Sunderland, Mass. See pp.116–120. *Comment:* An excellent book that we recommend strongly for readers who want to pursue various topics in human and animal vision at an advanced level.

Wade NJ and Swanston MT (2001) *Visual Perception: an Introduction,* 2nd Edition. Psychology Press. *Comment* A short and admirably clear introduction, which is particularly strong on the history of the subject.

CVonline: The Evolving, Distributed, Non-Proprietary, On-Line Compendium of Computer Vision

http://homepages.inf.ed.ac.uk/rbf/CVonline/
Comment A remarkable, evolving, and free Web resource for anyone wishing to explore computer vision techniques, including edge detection. Its creation has been heroically masterminded by Bob Fisher at the University of Edinburgh.

http://webvision.med.utah.edu/. *Comment* An excellent account of retinal structure and physiology with many figures of details going well beyond the material in this book.

Reference Style

For the novice, it may not be obvious how to interpret the data about references given at the end of each chapter. The format of our references follows a fairly standard convention, and we use the following as an example of a journal paper, as this is the most common type of publication cited.

Swindale NV, Matsubara JA, and Cynader MS (1987) Surface organization of orientation and direction selectivity in cat area 18. *Journal of Neuroscience* **7**(5) 1414–1427.

This reference refers to a research paper, and has several key components:

1) *Authors*: Swindale NV, Matsubara JA and Cynader MS. This part is self-evident.

2) *Date*: (1987) The year of publication.

3) *Title*: Surface organization of orientation and direction selectivity in cat area 18. Title of paper.

4) *Journal name*: By convention, the journal name is printed in italics, often obscurely abbreviated. For example, *Journal of Neuroscience* is often written as *J. Neuro*. To remove ambiguity we give full journal names in this book.

5) *Volume of journal*: **7**. By convention, the volume is usually printed in bold.

6) *Number*: (5). Where relevant, the part of the volume is given in brackets, and printed just after the volume number.

7) *Page numbers*: 1414–1427. The final part of the reference is the page numbers of the paper. Within the main body of the text of this book a citation of this reference would appear as:

Swindale, Matsubara, and Cynader (1987)
or, if we are trying to save space when citing references that have many authors, simply as

Swindale et al. (1987)
The words et al. are an abbreviation of the Latin et alii, which means "and others." Some journals and texts set this in italics and without a full stop.

Web Site for Seeing, Second Edition

Readers might find it useful to consult The MIT Press Web site for this book: http://mitpress.mit.edu/seeing. They will find there Matlab code for demonstrating how convolution works, and also the anaglyphs in Ch 18, which give better depth effects when viewed on screen. Also, any errata drawn to our attention will be posted there.

1

Seeing: What Is It?

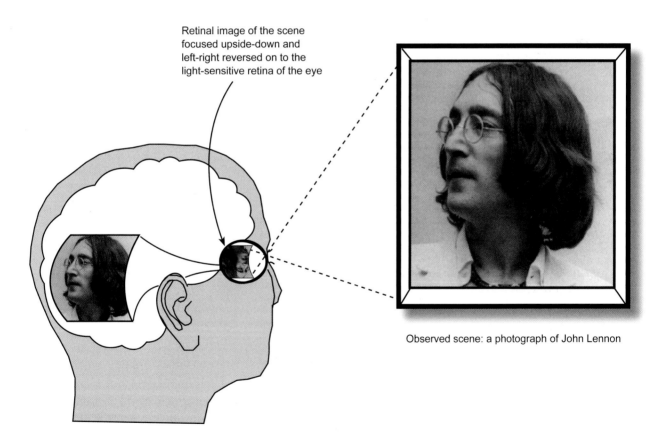

Retinal image of the scene focused upside-down and left-right reversed on to the light-sensitive retina of the eye

Observed scene: a photograph of John Lennon

1.1 An "inner screen" theory of seeing
One theory of this kind proposes that there is a set of brain cells whose level of activity represents the brightness of points in the scene. This theory therefore suggests that seeing is akin to photography. Note that the image of Lennon is inverted in the eye, due to the optics of the eye, but it is shown upright in the brain to match our perceptions of the world—see page 8. Lennon photograph courtesy Associated Newspapers Archive.

What goes on inside our heads when we see? Most people take seeing so much for granted that few will ever have considered this question seriously. But if pressed to speculate, the ordinary person who is not an expert on the subject might suggest:

> Could perhaps there be an "inner screen" of some sort in our heads, rather like a cinema screen except that it is made out of brain tissue? The eyes transmit an image of the outside world onto this screen, and this is the image of which we are conscious?

The idea that seeing is akin to photography, illustrated in **1.1**, is commonplace, but it has fundamental shortcomings. We discuss them in this opening chapter and we introduce a very different concept about seeing.

The photographic metaphor for seeing has its foundation in the observation that our eyes are in many respects like cameras. Both camera and eye have a lens; and where the camera has a light-sensitive film or an array of light-sensitive electronic components, the eye has a light-sensitive retina, **1.2**, a network of tiny light-sensitive receptors arranged in a layer toward the back of the eyeball (Latin *rete*—net). The job of the lens is to focus an image of the outside world—the retinal image—on to these receptors. This image stimulates them so that each receptor encodes the intensity of the

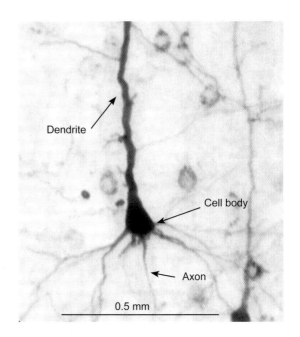

1.3 Pyramidal brain cell
Microscopic enlargement of a slice of rat brain stained to show a large neuron called a ***pyramidal cell.*** The long thick fiber is a dendrite that collects messages from other cells. The axon is the output fiber. Brain neurons are highly interconnected: it has been estimated that there are more connections in the human brain than there are stars in the Milky Way. Courtesy P. Redgrave.

small point of light in the image that lands on it. Messages about these point by point intensities are conveyed from the eye along fibers in the optic nerve to the brain. The brain is composed of millions of tiny components, brain cells called ***neurons***, **1.3**.

The core idea of the "inner screen" theory illustrated in **1.1** is that certain brain cells specialize in vision and are arranged in the form of a sheet—the "inner screen." Each cell in the screen can at any moment be either active, inactive, or somewhere in between, **1.4**. If a cell is very active, it is signaling the presence of a bright spot at that particular point on the "inner screen"—and hence at the associated point in the outside world. Equally, if a cell is only moderately active, it is signaling an intermediate shade of gray. Completely inactive cells signal black spots. Cells in the "inner screen" as a whole take on a pattern of activity whose overall shape mirrors the shape of the retinal image received by the eye. For example, if a photograph is being observed, as in **1.1**, then the pattern of activ-

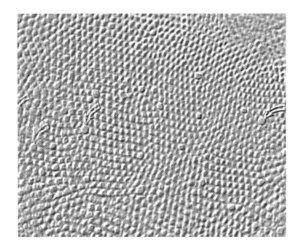

1.2 The receptor mozaic
Microphotograph of cells in the center of the human retina (the fovea) that deals with straight ahead vision. Magnification roughly x 1200. Courtesy Webvision (http://webvision.med.utah.edu/sretina.html#central).

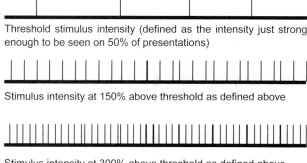

Threshold stimulus intensity (defined as the intensity just strong enough to be seen on 50% of presentations)

Stimulus intensity at 150% above threshold as defined above

Stimulus intensity at 300% above threshold as defined above

→ Time

1.4 Stimulus intensity and firing frequency
Most neurons send messages along their **axons** to other neurons. The messages are encoded in **action potentials** (each one is a brief pulse of voltage change across the neuron's outer "skin" or membrane). In the schematic recordings above, the time scale left-to-right is set so that the action potentials show up as single vertical lines or **spikes**. The height of the spikes remains constant: this is the **all-or-none law**—the spike is either present or absent. On the other hand, the **frequency of firing** can vary, which allows the neuron to use an **activity code**, here for representing stimulus intensity. Firing rates for brain cells can vary from 0 to several hundred spikes per second.

ity on the "inner screen" resembles the photograph. The "inner screen" theory proposes that as soon as this pattern is set up on the screen of cells, the observer has the experience of seeing.

This "inner screen" theory of seeing is easy to understand and is intuitively appealing. After all, our visual experiences do seem to "match" the outside world: so it is natural to suppose that there are mechanisms for vision in the brain which provide the simplest possible type of match—a physically similar or "photographic" one. Indeed, the "inner screen" theory of seeing can also be likened to television, an image-transmission system which is also photographic in this sense. The eyes are equivalent to TV cameras, and the image finally appearing on a TV screen connected to the cameras is roughly equivalent to the proposed image on the "inner screen" of which we are conscious. The only important difference is that whereas the TV-screen image is composed of more or less brightly glowing dots, our visual image is composed of more or less active brain cells.

Seeing and Scene Representations

The first thing to be said about the "inner screen" theory of seeing is that it proposes a certain kind of *representation* as the basis of seeing. In this respect it is like almost all other theories of seeing, but to describe it in this way requires some explanation.

In this book we use the term representation for anything that stands for something other than itself. Words are representations in this sense. For example, the word "chair" stands for a particular kind of sitting support—the word is not the support itself. Many other kinds of representations exist, of course, apart from words. A red traffic light stands for the command "Stop!", the Stars and Stripes stands for the United States of America, and so on. A moment's reflection shows that there must be representations inside our heads which are unlike the things they represent. The world is "out there," whereas the perceptual world is the result of processes going on inside the pink blancmange-like mass of brain cells that is our brain. It is an inescapable conclusion that there must be a representation of the outside world in the brain. This representation can be said to serve as a *description* that encodes the various aspects of the world of which sight makes us aware.

In fact, when we began by asking "What goes on inside our heads when we see?" we could as well have stated this question as: "When we see, what is the nature of the representation inside our heads that stands for things in the outside world?" The answer given by the "inner screen" theory is that each brain cell in the hypothetical screen is describing (representing) the brightness of one particular spot in the world in terms of an *activity code*, 1.4. The code is a simple one: the more active the cell, the lighter or more brightly illuminated the point in the world.

It can come as something of a shock to realize that somehow the whole of our perceived visual world is tucked away in our skulls as an inner representation which stands for the outside world. It is difficult and unnatural to disentangle the "perception of a scene" from the "scene itself." Nevertheless, they must be clearly distinguished if seeing is to be understood. When the difference between a perception and the thing perceived is fully grasped, the conclusion that seeing must involve a representation of the viewed scene sitting somewhere inside our heads becomes easier to accept. Moreover, the problem of seeing can be clearly stated: what is the nature of the brain's representation of the visual

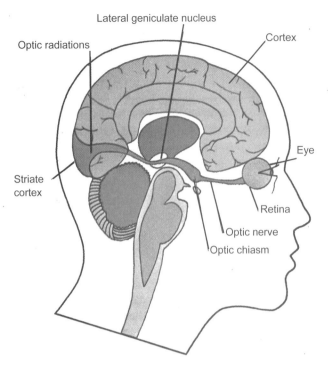

Lateral geniculate nucleus

Optic radiations

Cortex

Striate cortex

Eye

Retina

Optic nerve

Optic chiasm

1.5 Diagrammatic section through the head
This shows principal features of the major visual pathway that links the eyes to the cortex.

world, and how is it obtained? It is this problem which provides the subject of this book.

Perception, Consciousness, and Brain Cells

One reason why it might feel strange to regard visual experience as being encoded in brain cells is that they may seem quite insufficient for the task. The "inner screen" theory posits a direct relationship between conscious visual experiences and activity in certain brain cells. That is, activity in certain cells is somehow accompanied by conscious experience. Proposing this kind of parallelism between brain-cell activity and visual experience is characteristic of many theories of perceptual brain mechanisms. But is there more to it than this? Can the richness of visual experience really be identified with activity in a few million, or even a few trillion, brain cells? Are brain cells the right kind of entities to provide conscious perceptual experience? We return to these questions in Ch 22. For the moment, we simply note that most vision scientists

get on with the job of studying seeing without concerning themselves much with the issue of consciousness.

Pictures in the Brain

You might reasonably ask at this point: has neuroscience has anything to say directly about the "inner screen" theory? Is there any evidence from studies of the brain as to whether such a screen or anything like it exists?

The major visual pathway carrying the messages from the eyes to the brain is shown in broad outline in **1.5**. Fuller details are shown in **1.6** in which the eyes are shown inspecting a person, and the locations of the various parts of this scene "in" the visual system are shown with the help of numbers.

The first thing to notice is that the eyes do not receive identical images. The left eye sees rather more of the scene to the left of the central line of sight (regions 1 and 2), and vice versa for the right eye (regions 8 and 9). There are other differences between the left and right eyes' images in the case of 3D scenes: these are described fully in Ch 18.

Next, notice the *optic nerves* leaving the eyes. The fibers within each optic nerve are the *axons* of certain retinal cells, and they carry messages from the retina to the brain. The left and right optic nerves meet at the *optic chiasm*, **1.6** and **9.9**, where the optic nerve bundle from each eye splits in two. Half of the fibers from each eye cross to the *opposite* side of the brain, whereas the other half stay on the same side of the brain throughout.

The net result of this partial crossing-over of fibers (technically called *partial decussation)* is that messages dealing with any given region of the field of view arrive at a common destination in the cortex, regardless of which eye they come from. In other words, left- and right-eye views of any given feature of a scene are analyzed in the same physical location in the *striate cortex*. This is the major receiving area in the cortex for messages sent along nerve fibers in the optic nerves.

Fibers from the optic chiasm enter the left and right *lateral geniculate nuclei*. These are the first "relay stations" of the fibers from the eyes on their way to the striate cortex. That is, axons from the retina terminate here on the dendrites of new neurons, and it is the axons of the latter cells that then proceed to the cortex. A good deal of mystery still

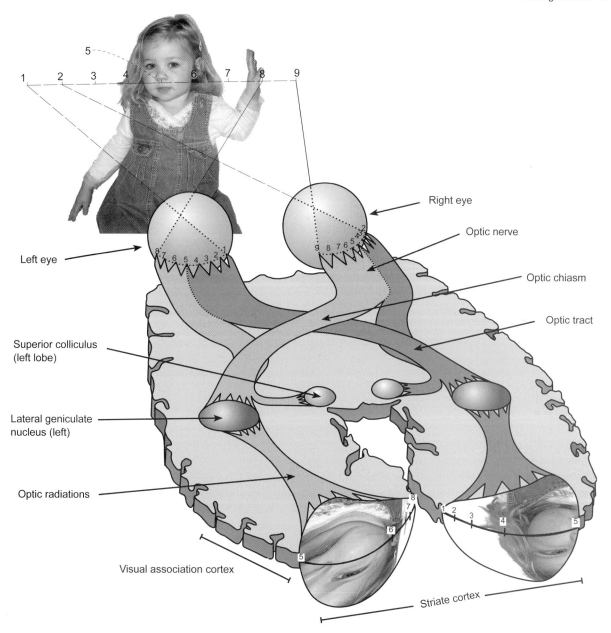

1.6 Schematic illustration of two important visual pathways
One pathway goes from the eyes to the striate cortex and one from the eyes to each superior colliculus. The distortion in the brain mapping in the striate cortex reflects the emphasis given to analysing the central region of the retinal image, so much so that the tiny representation of the child's hand can hardly be seen in this figure. See **1.7** for details.

surrounds the question of what cells in the lateral geniculate nuclei do. They receive inputs not only from the eyes but also from other sense organs, so some think that they might be involved in filtering messages from the eyes according to what is happening in other senses. The lateral geniculate nuclei also receive a lot of fibers sending messages from the cortex. Hence there is an intriguing

two-way up-down traffic going on in this visual pathway and we discuss its possible functions in later chapters.

Before we go on to discuss the way fiber terminations are laid out in the striate cortex, note that the optic nerves provide visual information to two other structures shown in **1.6**—the left and right halves of the ***superior colliculus***. This is a

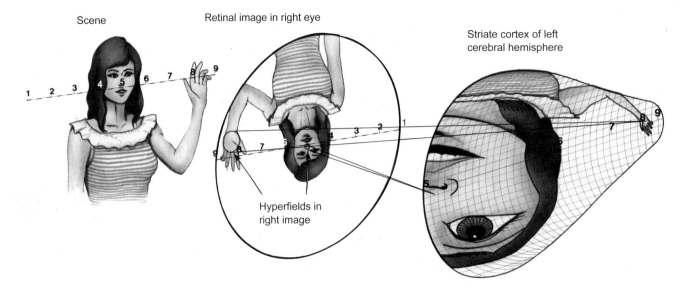

Scene　　　　Retinal image in right eye

Striate cortex of left cerebral hemisphere

Hyperfields in right image

1.7 Mapping of the retinal image in the striate cortex (schematic)
Turn the book upside-down for a better appreciation of the distortion of the scene in cortex. The **hyperfields** are regions of the retinal image that project to hypothetical structures called **hypercolumns** (denoted as graph-paper squares in the part of the striate cortex map shown here, which derives from the left hand sides of the left and right retinal images; more details in Chs 9 and10). Hyperfields are much smaller in central than in peripheral vision, so relatively more cells are devoted to central vision. Hyperfields have receptive fields in both images but here two are shown for the right image only.

brain structure which lies underneath the **cortex, 1.5**, so it is said to be *sub-cortical*. Its function is different from that performed by regions of the cortex devoted to vision. The weight of evidence at present suggests that the superior colliculus is concerned with guiding visual attention. For example, if an object suddenly appears in the field of view, mechanisms within the superior colliculus detect its presence, work out its location, and then guide eye movements so that the novel object can be observed directly.

It is important to realize that other visual pathways exist apart from the two main ones shown in **1.6**. In fact, in monkeys and most probably also in man, optic nerve fibers directly feed at least six different brain sites. This is testimony to the enormously important role of vision for ourselves and similar species. Indeed, it has been estimated that roughly 60% of the brain is involved in vision in one way or another.

Returning now to the issue of pictures-in-the-brain, the striking thing in **1.6** is the orderly, albeit curious, layout of fiber terminations in the striate cortex.

First, note that a face is shown mapped out on the cortical surface (*cortical* means "of the cortex"). This is the face that the eyes are inspecting.

Second, the representation is upside-down. The retinal images (not shown in **1.6**) are also upside-down due to the way the optics of the eyes work, **1.1**. Notice that the sketch of the "inner screen" in **1.1** showed a right-way-up image, so it is different in that respect from the mapping found in the striate cortex.

Third, the mapping is such that the representation of the scene is cut right down the middle, and each half of the cortex (technically, each **cerebral hemisphere**) deals initially with just one half of the scene.

Fourth, and perhaps most oddly, the cut in the representation places adjacent regions of the central part of the scene farthest apart in the brain!

Fifth, the mapping is spatially distorted in that a greater area of cortex is devoted to central vision than to peripheral: hence the relatively swollen face and the diminutive arm and hand, **1.7**. This doesn't mean of course that we actually *see* people in this distorted way—obviously we don't. But it reveals that a much larger area in our brain is assigned to *foveal vision* (i.e., analyzing what we are directly looking at) than is devoted to *peripheral vision*. This dedication of most cortical tissue to foveal vision is why we are much better at seeing details in the region of the scene we are looking at than we are at seeing details which fall out toward the edges of our field of view.

All in all, the cortical mapping of incoming visual fibers is curious but orderly. That is, adjacent regions of cortex deal with adjacent regions of the scene (with the exception of the mid-line cut). The technical term for this sort of mapping is *topographic*. In this instance it is called *retinotopic* as the mapping preserves the neighborhood relationships that exist between cells in the retina (except for the split down the middle). The general orderliness of the mapping is reminiscent of the "inner screen" proposed in **1.1**. But the oddities of the mapping should give any "inner screen" theorist pause for thought. The first "screen," if such it is, we meet in the brain is a very strange one indeed.

The striate cortex is not the only region of cortex to be concerned with vision—far from it. Fibers travel from the striate cortex to adjacent regions, called the *pre-striate cortex* because they lie just in front of the striate region. These fibers preserve the orderliness of the mapping found in the striate region. There are in fact topographically organized visual regions in the pre-striate zone and we describe these *maps* in Ch 10. For the present, we just note that each one seems to be specialized for a particular kind of visual analysis, such as color, motion, etc. One big mystery is how the visual world can appear to us as such a well-integrated whole if its analysis is actually conducted at very many different sites, each one serving a different analytic function.

To summarize this section, brain maps exist which bear some resemblance to the kind of "inner screen" idea hesitantly advanced by our fictional "ordinary person" who was pressed to hazard a guess at what goes on the brain when we see. However, the map shown in **1.6-1.7** is not much like the one envisaged in **1.1**, being both distorted, upside-down and cut into two.

These oddities seriously undermine the photographic metaphor for seeing. But it is timely to change tack now from looking inside the brain for an "inner screen" and to examine in detail serious logical problems with the "inner screen" idea as a theory of seeing. We begin this task by considering man-made systems for seeing.

Machines for Seeing

A great deal of research has been done on building *computer vision* systems that can do visual tasks. These take in images of a scene as input, analyze the visual information in these images, and then use that information for some purpose or other, such as guiding a robot hand or stating what objects the scene contains and where they are. In our terminology, a machine of this type is deriving a scene description from input images.

Whether one should call such a device a "perceiver," a "seeing machine," or more humbly an "image processor" or "pattern recognizer," is a moot point which may hinge on whether consciousness can ever be associated with non-biological brains. In any event, scientists who work on the problem of devising automatic image-processing machines would call the activity appearing on the "inner screen" of **1.1** a kind of *gray level description* of the painting. This is because the "inner screen" is a representation signaling the various shades of gray all over the picture, **1.8**. (We ignore color in the present discussion, and also many intricacies in the perception of gray: see Ch 16). Each individual brain cell in the screen is describing the gray level at one particular point of the picture in terms of an *activity code*. The code is simple: the lighter or more brightly illuminated the point in the painting, the more active the cell.

The familiar desktop image scanner is an example of a human-made device that delivers gray level descriptions. Its optical sensor sweeps over the image laid face down on its glass surface, thereby measuring gray levels directly rather than from a lens-focused image. Their scanning is technically described as a *serial* operation as it deals with different regions of the image in sequence.

Digital cameras measure the point by point intensities of images focused on their light sensitive surfaces, so in this regard they are similar to biological eyes. They are said to operate in *parallel* because they take their intensity measurements everywhere over the image at the same instant. Hence they can deliver their gray levels quickly.

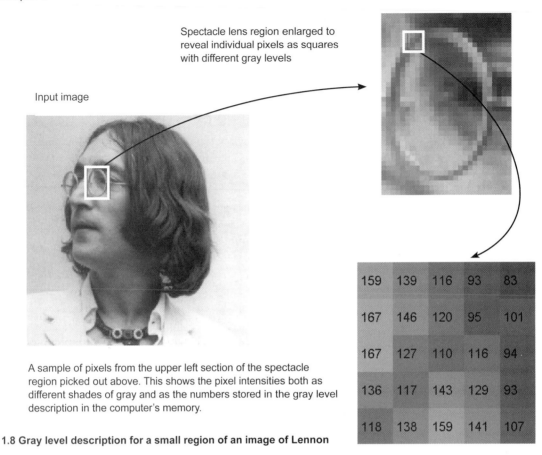

Spectacle lens region enlarged to reveal individual pixels as squares with different gray levels

Input image

A sample of pixels from the upper left section of the spectacle region picked out above. This shows the pixel intensities both as different shades of gray and as the numbers stored in the gray level description in the computer's memory.

1.8 Gray level description for a small region of an image of Lennon

The intensity measurements taken by both scanners and digital cameras are recorded as numbers stored in a digital memory. To call this collection of numbers a "gray level description" is apt because this is exactly what the numbers are providing, as in **1.8**.

The term "gray level" arises from the black-and-white nature of the system, with black being regarded as a very dark gray (and recorded with a small number) and white as a very light gray (and recorded with a large number).

The numbers are a description in the sense defined earlier: they make *explicit* the gray levels in the input image. That is, they make these gray levels immediately usable (which means there is no need for further processing to recover them) by subsequent stages of image processing.

Retinal images are upside-down, due to the optics of the eyes (Ch 2) and many people are worried by this. "Why doesn't the world therefore appear upside down?", they ask.

The answer is simple: as long as there is a systematic correspondence between the outside scene and the retinal image, the processes of image interpretation can rely on this correspondence, and build up the required scene description accordingly. Upside-down in the image is simply interpreted as right-way-up in the world, and that's all there is to it.

If an observer is equipped with special spectacles which optically invert the retinal images so that they become the "right-way-up," then the world appears upside-down until the observer learns to cope with the new correspondence between image and scene. This adjustment process can take weeks, but it is possible. The exact nature of the adjustment process is not yet clear: does the upside-down world really begin to "look" right-way-up again, or is it simply that the observer learns new patterns of adjusted movement to cope with the strange new perceptual world he finds himself in?

Try squinting to blur your vision while looking at the "block portrait" versions. You will find that Lennon magically appears more visible. See pages 128–131.

An ordinary domestic black-and-white TV set also produces an image that is an array of dots. The individual dots are so tiny that they cannot be readily distinguished (unless a TV screen is observed from quite close).

Representations and Descriptions

It is easy to see why the computer's gray level description illustrated in 1.8 is a similar sort of representation to the hypothetical "inner screen" shown in 1.1. In the latter, brain cells adopt different levels of activity to represent (or *code*) different pixel intensities. In the former, the computer holds different numbers in its memory registers to do exactly the same job. So both systems provide a representation of the gray level description of their input image, even though the physical stuff carrying this description (computer

1.9 Gray level images
The images differ in pixel size from small to large.

Gray Level Resolution

The number of **pixels** (shorthand for **pic**ture **ele**ment*s*) in a computer's gray level description varies according to the capabilities of the computer (e.g., the size of its memory) and the needs of the user. For example, a dense array of pixels requires a large memory and produces a gray level description that picks up very fine details. This is now familiar to many people due to the availability of digital cameras that capture high resolution images using millions of pixels. When these are output as full-tone printouts, the images are difficult to discriminate from film-based photographs.

If fewer pixels are used, so that each pixel represents the average intensity over quite a large area of the input image, then a full-tone printout of the same size takes on a block-like appearance. That is, these images are said to show **quantization** effects. These possibilities are illustrated in **1.9**, where the same input image is represented by four different gray level images, with pixel arrays ranging from high to low resolution.

hardware vs. brainware) is different in the two cases. This distinction between the functional or design status of a representation (the job it performs) and the physical embodiment of the representation (different in man or machine of course) is an extremely important one which deserves further elaboration.

Consider, for example, the physical layout of the hypothetical "inner screen" of brain cells. This is an anatomically neat one, with the various pixel cells arranged in a format which physically matches the arrangement of the corresponding image points.

In sharp contrast to this, the computer registers that perform the same job as the hypothetical brain cells would not be arranged in the computer in a way which physically matches the input image. That is, the "anatomical" locations of the registers in the computer memory would not necessarily be arranged as the hypothetical brain cells are, in a grid-like topographical form that preserves the neighbor-to-neighbor spatial relationships of the image points.

Instead, the computer registers might be arranged in a variety of different ways, depending on many different factors, some of them to do with how the memory was manufactured, others stemming from the way the computer was programmed to store information. The computer keeps track of each pixel measurement in a very precise manner by using a system of labels (technically, *pointers*) for each of its registers, to show which part of the image each one encodes. The details of how this is done do not concern us: it is sufficient to note that the labels ensure that each pixel value can be retrieved for later processing as and when required. Consequently, it is true to say that the hypothetical brain cells of **1.1** and the receptors of **1.2** are serving the same *representational function* as the computer memory registers of **1.8**—recording the gray level of each pixel—even though the *physical nature* of the representation in each case differs radically. It differs in both the nature of the pixel code (cell activity versus size of stored number) and the anatomical arrangement of the entities that represent the pixels.

The idea that different physical entities can mediate the same information processing tasks is the fundamental assumption underlying the field of ***artificial intelligence***, which can be defined as the enterprise of making computers do things which would be described as intelligent if done by humans.

Before leaving this topic we note another major difference between the putative brain cells coding the gray level description and computer memory registers. Computers are built with an extremely precise organization of their components. As stated above, each memory register has a label and its contents can be set to represent different things according to the program being run on the computer. One moment the register might be holding a number within a spreadsheet, a few moments later it might be holding the code for a letter in document being edited using a word processor, or whatever. Indeed, the capacity for the *arbitrary* assignment of computer registers, to hold different contents that mean different things at different times according to the particular computation being run, is held by some to be the true hallmark of ***symbolic computation***.

But this capacity for arbitrary and changing assignment is quite unlike the brain cells supporting vision, which, as far as we presently understand things, are more or less permanently committed to serve a *particular* visual function (but see the caveat below on learning). That is, if a brain cell is used to represent a scene property, such as the orientation of an edge, then that is the job that cell always does. It isn't quickly reassigned to represent, say, a dog, or a sound, etc., under the control of other brain processes.

It may be that other brain regions do contain cells whose functional role changes from moment to moment (perhaps cells supporting language?). If so, they would satisfy the arbitrary assignment definition of a symbolic computational device given above. However, some have doubted whether the brain's wiring really can support the highly accurate cell-to-cell connections that this would require. In any event, visual neurons do not appear to have this property and we will not use this definition of symbolic computation in this book.

A caveat that needs to be posted here is to do with various phenomena in ***perceptual learning***: we get better at various visual tasks as we practice them, and this must reflect changes in vision brain cells. Also, ***plasticity*** exists in brain cell circuits in early development, Ch 4, and parts of the brain to do with vision may even be taken over for other functions following blindness caused by losing the eyes (or vice versa: the visual brain may encroach on other brain areas).

But this caveat is about slowly acting forms of learning and plasticity. It does not alter the basic point being made here. When we say the visual brain is a symbolic processor we are *not* saying that its brain cells serve as symbols in the way that programmable computer components serve symbolic functions using different symbolic assignments from moment to moment.

Levels of Understanding Complex Information Processing Systems

Why have we dwelt on the point that certain brain cells and computer memory registers could serve the same task (in this case representing point-by-point image intensities) despite huge differences in their physical characteristics?

The answer is that it is a good way of introducing the linking theme running through this book. This is that we need to keep clearly distinct *different levels of discourse* when trying to understand vision.

This general point has had many advocates. For example, Richard Gregory, a distinguished scientist well known for his work on vision, pointed out long ago that understanding a device such as a radio needs an understanding of its individual components (transistors, resistors, etc.) and also an understanding of the design used to connect these components so that they work as a radio.

This point may seem self-evident to many readers but when it comes to studying the brain some scientists in practice neglect it, believing that the explanation must lie in the "brainware." Obviously, we need to study brain structures to understand the brain. But equally, we cannot be said to understand the brain unless we understand, among other things, *the principles underlying its design.*

Theories of the design principles underlying seeing system are often called **computational theories.** This term fits analyzing seeing as an information processing task, for which the inputs are the images captured by the eyes and the outputs are various representations of the scene.

Often it is useful to have a level of analysis of seeing intermediate between the computational theory level and the hardware level. This level is concerned with devising good procedures or **algorithms** for implementing the design specified by the computational theory. We will delay specifying what this level tries to do until we give specific examples in later chapters.

What each level of analysis tries to achieve will become clear from the numerous examples in this book. We hope that by the time you have finished reading it we will have convinced you of the importance of the computational theory level for understanding vision. Moreover, we hope we will

have given a number of sufficiently well-worked out examples to illustrate its importance when it comes to understanding vision. For the moment, we leave this issue with a famous quotation from an influential computational theorist, David Marr, whose approach to studying vision provides the linking theme for this book:

> Trying to understand vision by studying only neurons is like trying to understand bird flight by studying only feathers: it just cannot be done. (Marr, 1982)

Representing Objects

The "inner screen" of **1.1** can, then, be described as a particular kind of symbolic scene representation. The activities of the cells which compose the "screen" describe in a symbolic form the intensities of corresponding points in the retinal image of the scene being viewed. Hence, the theory proposes, these cell activities represent the lightnesses of the corresponding points in the scene. We are now in a position to see one reason why this is such an inadequate theory of seeing: it gives us such a woefully impoverished scene description!

The scene description which exists inside our heads is *not* confined simply to the lightnesses of individual points in the scene before us. It tells us an enormous amount more than this. Leaving aside the already noted limitation of not having anything to say about color, the "inner screen" description does not help us understand how we know what *objects* we are looking at, or how we are able to describe their various features—shape, texture, movement, size—or their spatial relationships one to another. Such abilities are basic to seeing—they are what we have a visual system for, so that sight can guide our actions and thoughts. Yet the "inner screen" theory leaves them out altogether.

You might feel tempted to reply at this point: "I don't really understand the need to propose anything more than an "inner screen" in order to explain seeing. Surely, once this kind of symbolic description has been built up, isn't that enough? Are not all the other things you mention—recognizing objects and so forth—an immediately given consequence of having the photographic type of representation provided by the "inner screen"?"

One reply to this question is that the visual system is so good at telling us what is in the world around us that we are understandably misled into taking its effortless scene descriptions for granted. Perhaps it is because vision is so effortless for us that is tempting to suppose that the scene we are looking at is "immediately given" by a photographic type of representation. But the truth is the exact opposite. Arriving at a scene description as good as that provided by the visual system is an immensely complicated process requiring a great deal of interpretation of the often limited information contained in gray level images. This will become clear as we proceed through the book. Achieving a gray level description of images formed in our eyes is only the first and easiest task. It is served by the very first stage of the visual pathway, the light-sensitive receptors in the eyes, **1.2**.

This point is so important it is worth reiterating. The intuitive appeal of the "inner screen" theory lies in its proposal that the visual system builds up a photographic-type of brain picture of the observed scene, and its suggestion that this brain picture is the basis of our conscious visual experience. The main trouble with the theory is that although it proposes a symbolic basis for vision, the symbols it offers correspond to points in the scene. Everything else in the scene is left unanalyzed and not represented *explicitly*.

It is not much good having the visual system build a photographic-type copy of the scene if, when that task is done, the system is no nearer to using information in the retinal image to decide what is present in the scene and to act appropriately, e.g., avoiding obstacles or picking up objects. After all, we started the business of seeing in the retina with a kind of picture, the activity in the receptor mosaic. It doesn't take us any further toward using vision to guide action to propose a brain picture more or less mirroring the retinal one. The inner screen theory is thus vulnerable to what philosophers call an ***infinite regress***: the problems with the theory cannot be solved by positing another picture, and so on *ad infinitum*.

So the main point being argued here is that the inner screen theory shown in **1.1** totally fails to explain how we can recognize the various objects and properties of objects in the visual scene. And the ability to achieve such recognition is anything but an immediate consequence of having a photographic representation. A television set has pixel

images but it is precisely because it cannot decide what is in the scene from whence the images came that we would not call it a "visual perceiver." Devising a seeing machine that can receive a light image of a scene and use it to describe what is in the scene is much more complicated, a problem which is as yet unsolved for complex natural scenes.

The conclusion is inescapable: whatever the correct theory of seeing turns out to be, it must include processes quite different from the simple mirroring of the input image by simple point-by-point brain pictures. Mere physical resemblance to an input image is not an adequate basis for the brain's powers of symbolic visual scene description. This point is sometimes emphasized by saying that whereas the task of the eyes is forming images of a scene, the task of vision is the opposite: ***image inversion***. This means getting a description of the scene from images.

Images Are Not Static

For simplicity, our discussion so far has assumed that the eye is stationary and that it is viewing a stationary scene. This has been a convenient simplification but in fact nothing could be further from the truth for normal viewing. Our eyes are constantly shifted around as we move them within their sockets, and as we move our heads and bodies. And very often things in the scene are moving. Hence vision really begins with a stream of time-varying images.

Indeed, it is interesting to ask what happens if the eyes are presented with an unchanging image. This has been studied by projecting an image from a small mirror mounted on a tightly fitting contact lens so that whatever eye movement is made, the image remains stationary on the retina. When this is done, normal vision fades away: the scene disappears into something rather like a fog. Most visual processes just seem to stop working if they are not fed with moving images.

In fact, some visual scientists have claimed that vision is really the study of motion perception; all else is secondary. This is a useful slogan (even if an exaggeration) to bear in mind, particularly as we will generally consider, as a simplifying strategy for our debate, only single-shot stationary images.

Why do we perceive a stable visual world despite our eyes being constantly shifted around?

As anyone who has used a hand-held video camera will know, the visual scenes thus recorded appear anything but stable if the camera is jittered around. Why does not the same sort of thing happen to our perceptions of the visual world as we move our eyes, heads and bodies? This is an intriguing and much studied question. One short answer is that the movement signals implicit in the streams of retinal images are indeed encoded but then interpreted in the light of information about self-movements, so that retinal image changes due to the latter are cancelled out.

Visual Illusions and Seeing

The idea that visual experience is somehow akin to photography is so widespread and so deeply rooted that many readers will probably not be convinced by the above arguments against the "inner screen" theory. They know that the eye does indeed operate as a kind of camera, in that it focuses an image of the world upon its light-sensitive retina.

An empirical way of breaking down confidence in continuing with this analogy past the eye and into the visual processes of the brain is to consider visual illusions. These phenomena of seeing draw attention to the fact that what we see often differs dramatically from what is actually before our eyes. In short, the non-photographic quality of visual experience is borne out by the large number and variety of visual illusions.

Many illusions are illustrated in this book because they can offer valuable clues about the existence of perceptual mechanisms devoted to building up an explicit scene description. These mechanisms operate well enough in most circumstances, but occasionally they are misled by an unusual stimulus, or one which falls outside their "design specification," and a visual illusion results. Richard Gregory is a major current day champion of this view (Gregory, 2009).

Look, for example, at **1.10**, which shows an illusion called ***Fraser's spiral***. The amazing truth is that there is no spiral there at all! Convince yourself of this by tracing the path of the apparent spiral with your finger. You will find that you return to your starting point. At least, you will if you are careful: the illusion is so powerful that it can even induce incorrect finger-tracing. But with due care, tracing shows that the picture is really

1.10 Fraser's spiral (above)
This illusion was first described by the British psychologist James Fraser in 1908. Try tracing the spiral with your finger and you will find that there is no spiral! Rather, there are concentric circles composed of segments angled toward the center (left).

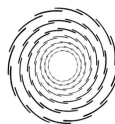

made up of concentric circles. The spiral exists only in your head. Somehow the picture fools the visual system, which mistakenly provides a scene description incorporating a spiral even though no spiral is present. A process which takes concentric circles as input and produces a spiral as output can hardly be thought of as "photographic."

Another dramatic illusion is shown in **1.11**, which shows a pair of rectangles and a pair of ellipses. The two members of each pair have seemingly different shapes and sizes. But if you measure them with a ruler or trace them out, you will find they are the same. Incredible but true.

You might be wondering at this point: are such dramatic illusions representative of our everyday perceptions, or are they just unusual trick figures dreamt up by psychologists or artists? These illusions may surprise and delight us but are they really helpful in telling us what normally goes on inside our heads when we see the world? Some distinguished researchers of vision, for example, James Gibson, whose ***ecological optics*** approach to vision is described Ch 2, have argued that illusions are very misleading indeed.

But probably a majority of visual scientists would nowadays answer this question with a defi-

1.11 Size illusion
The rectangular table tops appear to have different dimensions, as do the elliptical ones. If you do not believe this then try measuring them with a ruler. Based on figure A2 in Shepard (1990).

nite "yes." Visual illusions *can* provide important clues in trying to understand perceptual processes, both when these processes produce reasonably accurate perceptions, and when they are fooled into generating illusions. We will see how this strategy works out as we proceed through this book.

At this point we need to be a bit clearer about what we mean by a "visual illusion." We are using illusions here to undermine any remaining confidence you might have in the "inner screen" photographic-style theory of seeing. That is, illusions show that our perceptions often depart radically from predictions gained from applying rulers or other measuring devices to photographs.

But often visual illusions make eminently good sense if we regard the visual system as using retinal images to create representations of what really is "out there." In this sense, the perceptions are not illusions at all—they are faithful to "scene reality." A case in point is shown in **1.12**, in which a checkerboard of light and dark squares is cast in shadow. Unbelievably, the two squares picked out have the same intensity on-the-page in this computer graphic but they appear hugely different in lightness. This is best regarded not as an "illusion,"

1.12 Adelson's figure
The squares labeled A and B have roughly the same gray printed on the page but they are perceived very differently. You can check their ink-on-the-page similarity by viewing them through small holes cut in a piece of paper. Is this a brightness illusion or is it the visual system delivering a faithful account of the scene as it is in reality? The different perceived brightnesses of the A and B squares could be due to the visual system allowing for the fact that one of them is seen in shadow. If so, the perceived outcomes are best thought of as being "truthful," not "illusory." See text. Courtesy E. H. Adelson.

1.13 A real-life version of the kind of situation depicted in the computer graphic in 1.12
Remarkably, the square in shadow labeled A has a lower intensity than the square labeled B, as shown by the copies at the side of the board. The visual system makes allowance for the shadow and to see what is "really there." [Black has due cause to appear distressed. After: 36 Bf5 Rxf5 37 Rxc8+ Kh7 38 Rh1, Black resigned. Following the forced exchange of queens that comes next, White wins easily with his passed pawn. Topalov vs. Adams, San Luis 2005.] Photograph by Len Hetherington.

strange though it might seem at first sight. Rather, it is an example of the visual system making due allowance for the shading to report on the true state of affairs (veridical).

The outcome in **1.12** is not some quirk of computer graphics, as illustrated in **1.13** in which the same thing happens from a photograph of a shaded scene.

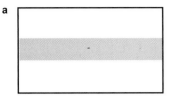

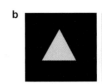

1.14 Brightness contrast illusions
a The two gray stripes have the same intensity along their lengths. However, the right hand stripe appears brighter at the end which is bordered by a dark ground, darker when adjacent to a light ground.
b The small inset gray triangles all have the same physical intensity, but their apparent brightnesses vary according to the darkness/lightness of their backgrounds. We discuss brightness illusions in Ch 16.

Other brightness "illusions" are shown in **1.14**. These also illustrate the slippery nature of what is to be understood by a "visual illusion." Figure **1.14a** could well be a case of making allowance for shading but **1.14b** doesn't fit that kind of interpretation because we do not see these figures as lying in shade. However, **1.14b** might be a case of the visual system applying, unconsciously, a strategy that copes with shading in natural scenes but when applied to certain sorts of pictures produces an outcome that surprises us because we don't see a shaded scene. This is not a case of special pleading because most visual processes are unconscious, so why not this one? On this argument, the varied perceptions of the identical-in-the-image gray triangles is "illusory" only if we are expecting the visual system to report the intensities in the image.

But, when you think about it, that doesn't make sense. The visual system isn't interested in reporting on the nature of the retinal image. Its task is to use retinal images to deliver a representation of what is *out there in the world*. The idea that vision is about seeing what is in retinal images of the world rather than in the world itself is at the root of the delusion that seeing is somehow akin to photography.

That said, if illusions are defined as the visual system getting it seriously wrong when judged against physically measured scene realities, then human vision is certainly prone to some illusions in this sense. These can arise for ordinary scenes, but go unnoticed by the casual observer. The teacup illusion shown in **1.15a** is an example. The photograph is of a perfectly normal teacup, together with a normal saucer and spoon. Try judging which mark on the spoon would be level with the rim of the teacup if the spoon was stood upright in the cup.

Now turn the page and look at **1.15b** (p. 18). The illusory difference in the apparent lengths of the two spoons, one lying horizontally in the saucer and one standing vertically in the cup, is remarkable. Convince yourself that this percep-

1.15a Teacup illusion
Imagine the spoon stood upright in the cup. Which mark on the spoon handle would then be level with the cup's rim? Check your decision by inspecting **1.15b** on p.18.

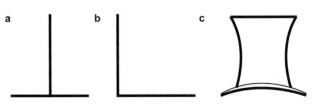

1.16 The vertical-horizontal illusion
The vertical and horizontal extents are the same (check with a ruler). This effect occurs in drawings of objects, as in **c**, where the vertical and horizontal curves are the same length.

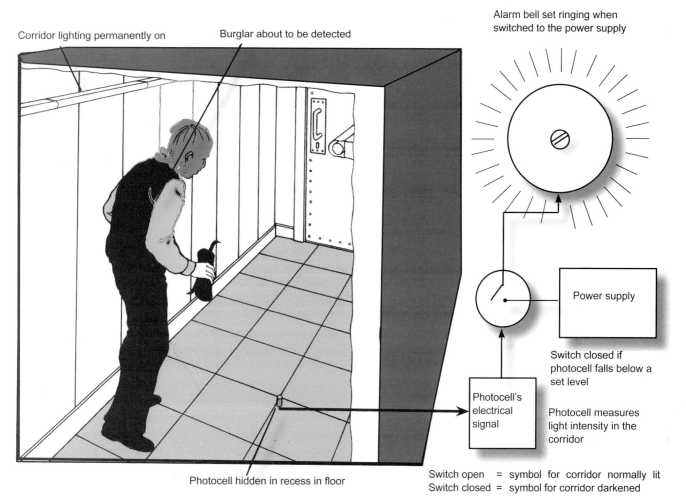

1.17 A simple burglar alarm system operated by a photocell

tual effect is not a trick dependent on some subtle photography by investigating it in a real-life setting with a real teacup and spoon. It works just as well there as in the photograph. Real-world illusions like these are more commonplace than is often realized. Artists and craftsmen know this fact well, and learn in their apprenticeships, often the hard way by trial and error, that the eye is by no means always to be trusted. Seeing is *not* always believing—or shouldn't be.

This teacup illusion nicely illustrates the usual general definition of visual illusions—as perceptions which depart from measurements obtained with such devices as rulers, protractors, and light meters (the latter are called *veridical* measurements). Specifically, this illusion demonstrates that we tend to over-estimate vertical extents in comparison with horizontal ones, particularly if

the vertical element bisects the horizontal one.

The simplest version of this effect, illustrated in **1.16a**, is known as the vertical-horizontal illusion. The effect is weaker if the vertical line does not bisect the horizontal as in **1.16b** but it is still present. It is easy to draw many realistic pictures containing the basic effect. The brim in **1.16c** is as wide as the hat is tall, but it does not appear that way. The perceptual mechanisms responsible for the vertical-horizontal illusion are not understood, though various theories have been proposed since its first published report in 1851 by A. Fick.

The illusions just considered are instances of ***spatial distortions***: vertical extents can be stretched, horizontal ones shortened, and so on. They are eloquent testimony to the fact that perceptions cannot be thought of as "photographic copies" of the world, even when it comes to a visual experience as apparently simple as that of seeing the length of a line.

Scene Descriptions Must Be Explicit

Explanations of various illusions will be offered in due course as this book proceeds. For the present, we will return to the theme of *seeing is representation*, and articulate in a little more detail what this means.

The essential property of a scene representation is that it makes some property of the scene *explicit* in a code of symbols. In the "inner screen" theory of **1.1**, the various brain cells make explicit the various shades of gray at all points in the image. That is, they signal the intensity of these grays in a way that is sufficiently clear for subsequent processes to be able to use them for some purpose or other, without first having to engage in more analysis. (When we say in this book a representation makes something explicit we mean: immediately available for use by subsequent processes, no further processing is necessary.)

A scene representation then, is the result of processing an image of the scene in order to make attributes of the scene explicit. The simplest example we can think of that illustrates this kind of system in action is shown in **1.17**, perhaps the most primitive artificial "seeing system" conceivable—a burglar alarm operated by a photocell. The corridor is permanently illuminated, and when the intruder's shadow falls over the photocell detector hidden in the floor, an alarm bell is set ringing. Viewed in our terms, what the photocell-triggered alarm system is doing is:

1.   Collecting light from a part of the corridor using a lens.

2.   Measuring the intensity of the light collected—the job of the photocell;

3.   Using the intensity measurement to build an explicit representation of the illumination in the corridor—*switch open* symbolizes *corridor normally lit* and *switch closed* symbolizes *corridor darkened*.

4.   Using the symbolic scene description coded by the state of the switch as a basis for action—either ringing the alarm bell or leaving it quiet.

1.15b Teacup illusion (cont.)
The vertical spoon seems much longer than the horizontal one. Both are the same size with the same markings.

Step 3 requires some **threshold** level of photocell activity to be set as an **operational** definition of "corridor darkened." Technically, setting a threshold of this sort is called a **non-linear** process, as it transforms the linear output of the photocell (more light, bigger output) into a YES/NO **category decision**.

Step 4 depends on the assumption that a darkened corridor implies "intruder." It would suffer from an "intruder illusion" if this assumption was misplaced, as might happen if a power cut stopped the light working.

The switch in the burglar alarm system serves as a symbol for "burglar present/absent" only in the context of the entire system in which it is embedded. This simple switch could be used in a different circuit for a quite different function. The same thing seems to be true of nerve cells. Most seem to share fundamentally similar properties in the way they become active, **1.3**, but they convey very different messages (code for different things, represent different things) according to the circuits of which they are a part. This type of coding is thus called **place coding**, or sometimes **value coding**, and we will see in later chapters how the visual brain uses it.

A primitive seeing system with similar attributes to this burglar detector is present in mosquito larvae: try creating a shadow by passing your hand over them while they are at the surface of a pond and you will find they submerge rapidly, presumably for safety using the shadow as warning of a predator.

More Visual Tricks

The effortless fluency with which our visual system delivers its explicit scene representation is so beguiling that the skeptical reader might still doubt that building visual representations is what seeing is all about. It can be helpful to overcome this skepticism by showing various trick figures that catch the visual system out in some way, and reveal something of the scene representation process at work.

Consider, for example, the picture shown in **1.18**. It seems like a perfectly normal case of an inverted photograph of a head. Now turn it upside-down. Its visual appearance changes dramatically—it is still a head but what a different one.

These sorts of upside-down pictures demonstrate the visual system at work building up scene descriptions which best fit the available evidence. Inversion subtly changes the nature of the evidence in the retinal image about what is present in the scene, and the visual system reports accordingly. Notice too that the two alternative "seeings" of the photograph actually *look* different. It is not that we attach different verbal labels to the picture upon inversion. Rather, we actually *see* different attributes of the eyes and mouth in the two cases. The pattern of ink on the page stays the same, apart from the inversion, but the experience it induces is made radically different simply by turning the picture upside-down.

The "inner screen" theory has a hard time trying to account for the different perceptions produced by inverting **1.18.** The "inner screen" theorist wishes to reserve for his screen the job of represent-

1.18 Peter Thompson's inverted face phenomenon
Turn the book upside-down but be ready for a shock.

ing the contents of visual experience. Fundamentally different experiences emerge upon inversion; therefore, fundamentally different contents must be recorded on the screen in each case. But it is not at all clear how this could be done. The "inner screen" way of thinking would predict that inversion should simply have produced a perception of the same picture, but upside-down. This is not what happens in **1.18** although it is what happens for pictures that lack some form of carefully constructed changes.

1.19 Interpreting shadows
The picture on the right is is an inverted copy of the one on the left. Try inverting the book and you will see that the crater becomes a hill and the hill becomes a crater. The brain assumes that light comes from above, then it interprets the shadows to build up radically different scene descriptions (perceptions) of the two images. Courtesy NASA.

a

b

1.20 Ambiguous figures
a Duck or rabbit? This figure has a long history but it seems it was first introduced into the psychological literature by Jastrow in 1899.
b Vase or faces? From www.wpclipart.com.

Another example of the way inversion of a picture can show the visual system producing radically different scene descriptions of the same image is given in **1.19**. This shows a scene with a crater alongside one with a gently rounded hill. It is difficult to believe that they are one and the same picture, but turning the book upside-down proves the point. Why does this happen? It illustrates that the visual system uses an assumption that light normally comes from above, and given this starting point, the ambiguous data in the image are inter-

preted accordingly—bumps become hollows and vice versa on inversion.

This is a fine example of how a visual effect can reveal a design feature of the visual system, that is, a principle it uses, an assumption it makes, in interpreting images to recover explicit scene descriptions. Such principles or assumptions are technically often called ***constraints***. Identifying the constraints used by human vision is a critically important goal of visual science and we will have much to say about them in later chapters.

Figure/Ground Effects

Another trick for displaying the scene-description abilities of the visual system is to provide it with an ambiguous input that enables it to arrive at different descriptions alternately. **1.20** shows two classic ambiguous figures, The significance of ambiguous figures is that they demonstrate how different scene representations come into force at different times. The image remains constant, but the way we experience it changes radically. In **1.20a** picture parts on the left swap between being seen as ears or beak. In **1.20b** sometimes we see a vase as ***figure*** against its ***ground***, and then at other times what was ground becomes articulated as a pair of faces—new figures.

Some aspects of the scene description do remain constant throughout—certain small features for instance—but the overall look of the picture changes as each possibility comes into being. The scene representation adopted thus determines the figure/ground relationships that we see. Just as with the upside-down face, it is not simply a case of different verbal labels being attached at different times. Indeed, the total scene description, including both features and the overall figure/ground interpretation, quite simply *is* the visual experience each time.

One last trick technique for demonstrating the talent of our visual apparatus for scene description is to slow down the process by making it more difficult. Consider **1.21** for example. What do you see there? At first, you will probably see little more than a mass of black blobs on a white ground. The perfectly familiar object it contains may come to light with persistent scrutiny but if you need help, turn to the end of this chapter to find out what the blobs portray.

1.21 What do the blobs portray?
Courtesy Len Hetherington.

Once the hidden figure has been found (or, in our new terminology, we could say represented, described, or made explicit), the whole appearance of the pattern changes. In **1.21** the visual system's normally fluent performance has been slowed down, and this gives us an opportunity to observe the difference between the "photographic" representation postulated by the "inner screen" theory, and the scene description that occurs when we see things. The latter requires active interpretation of the available data. It is not "immediately given" and it is not a passive process.

One interesting property of **1.21** is that once the correct scene description has been achieved, it is difficult to lose it, perhaps even impossible. One cannot easily return to the naive state, and experience the pictures as first seen.

Another example of a hidden-object figure is shown in **1.22.** This is not an artificially degraded image like **1.21** but an example of animal camouflage. Again, many readers will need the benefit of being told what is in the scene before they can find the hidden figure (see last page of this chapter for correct answers).

The use of prior knowledge about a specific object is called *concept driven* or *top down processing*. If such help is not available, or not used, then the style of visual processing is said to be *data*

1.22 Animal camouflage
There are *two* creatures here. Can you find them? Photograph by Len Hetherington.

1.23 Can you spot the error?
Thanks to S. Stone for pointing this out.

driven or ***bottom up***. An example of the way expectations embedded in concept driven processing can sometimes render us oblivious to what is "really out there" is shown by how hard it is to spot the unexpected error in **1.23**. For the answer, see p. 28.

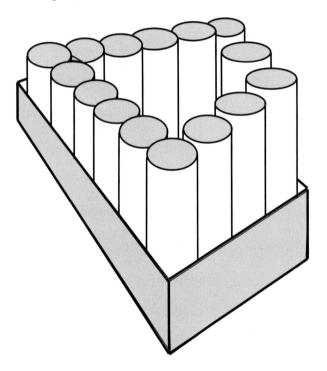

1.24 Impossible pallisade
Imagine stepping around the columns, as though on a staircase. You would never get to the top (or the bottom). By J.P. Frisby, based on a drawing by L. Penrose and R. Penrose.

Three-Dimensional Scene Descriptions

So far we have confined our discussion of explicit scene descriptions to the problems of extracting information about objects from two-dimensional (2D) pictures. The visual system, however, is usually confronted with a scene in three dimensions (3D). It deals with this challenge magnificently and provides an explicit description of where the various objects in the scene, and their different parts, lie in space.

The "inner screen" theory cannot cope with the 3D character of visual perception: its representation is inherently flat. An attempt might be made to extend the theory in a logically consistent manner by proposing that the "inner screen" is really a 3D structure, a solid mass of brain cells, which represent the brightness of individual points in the scene at all distances. A kind of a brainware stage set, if you will.

It is doubtful whether complex 3D scenes could be re-created in brain tissue in a direct physical way. But even if this was physically feasible for 3D scenes, what happens when we see the "impossible pallisade" in **1.24?** (You may be familiar with the drawings of M. C. Escher, who is famous for having used impossible objects of this type as a basis for many technically intricate drawings.)

If this pallisade staircase is physically impossible, how then could we ever build in our brains a 3D physical replica of it? The conclusion is inescapable: we must look elsewhere for a possible basis for the brain's representation of depth (***depth*** is the term usually used by psychologists to refer to the distance from the observer to items in the scene being viewed, or to the different distances between objects or parts of objects).

What do impossible figures tell us about the brain's representation of depth? Essentially, they tell us that small details are used to build up an explicit depth description for *local* parts of the scene, and that finding a consistent representation of the entire scene is not treated as mandatory.

Just how the local parts of an *impossible triangle* make sense individually is shown in **1.25**, which gives an exploded view of the figure. The brain interprets the information about depth in each local part, but loses track of the overall description it is building up. Of course, it does not entirely lose

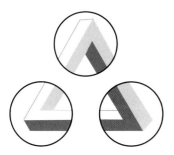

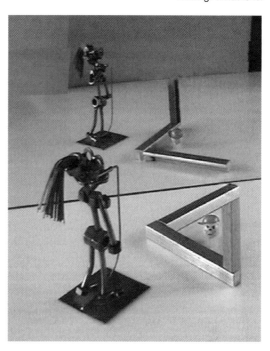

1.25 Impossible triangle
The triangle you see in the foreground in the photograph on the right is physically impossible. It appears to be a triangle only from the precise position from which the photograph was taken. The true structure of the photographed object is seen in the reflection in the mirror. The figure is included to help reveal the role of the mirror. This is a case in which the visual system prefers to make sense of local parts (the corners highlighted in the figure shown above), rather than making sense of the figure as a whole. Gregory (1971) invented an object of this sort. To enjoy diverse explorations of impossible objects, see Ernst (1996).

track of this global aspect; otherwise, we would never notice that impossible figures are indeed impossible. But the overall impossibility is a rather "cognitive" effect—a realization in thought rather than in experience that the figures do not "make sense."

If the visual system insisted on the global aspect as "making sense" then it could in principle have dealt with the figures differently. For example, it could have "broken up" one corner of the impossible triangle and led us to see part of it as coming out toward us and part of it as receding. This is illustrated by the object at the left of **1.25**.

But the visual system emphatically does not do this, not from a line drawing nor from a physical embodiment of the impossible triangle devised by Gregory. He made a 3D model of **1.25**, left. When viewed from just the right position, so that it presents the same retinal image as the line-drawing, then our visual apparatus still gets it wrong, and delivers a scene description which is impossible globally, albeit sensible locally. Viewing this "real" impossible triangle has to be one-eyed; otherwise, other clues to depth come into play and produce the physically correct global perception. (Two-eyed depth processing is discussed in detail in Ch 18.)

One interesting game that can be played with the trick model of the impossible triangle is to pass

another object, such as one's arm, through the gap while an observer is viewing the model correctly aligned, and so seeing the impossible arrangement. As the arm passes through the gap, it seems to the observer that it slices through a solid object!

An important point illustrated by **1.25** is the inherent ambiguity of flat illustrations of 3D scenes. The real object drawn in **1.25**, left, has two limbs at very different depths: but viewing with one eye from the correct position can make this real object cast just the same image on the retina as one in which the two limbs meet in space at the same point.

This inherent ambiguity, difficult to comprehend fully because we are so accustomed to interpreting the 2D retinal image in just one way, is revealed clearly in a set famous demonstrations by Ames, shown in **1.26**. The observer peers with one eye through a peephole into a dark room and sees a chair, **1.26a**. However, when the observer is shown the room from above it becomes apparent that the real object in the room is *not* the chair seen through the peephole. In the example shown in **1.26b**, the object is a distorted chair suspended in space by invisible wires, and in **1.26c** the room contains an odd assemblage of luminous lines, also suspended in space by wires. The collection of parts is cunningly arranged in each case to produce

a

b

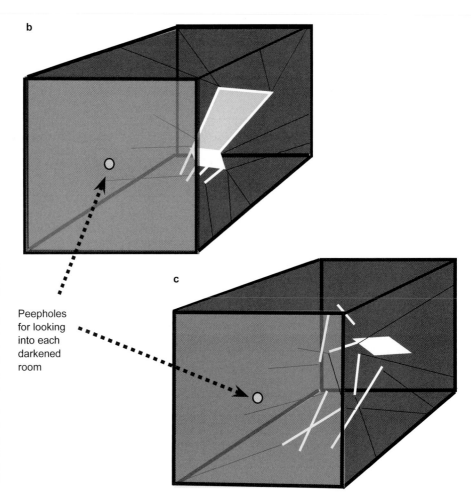

1.26 Ames's chair demonstration
a What the observer sees when he looks into the rooms in **b** and **c** through their respective peepholes.
b Distorted chair whose parts are held in space by thin invisible wires. The chair is positioned in space such that it is seen as a normal, undistorted, chair through the peephole, without distortion.
c Scattered parts of a chair that still look like a normal chair through the peephole, due to the clever way that Ames arranged the distorted parts in space so that they cast the same retinal image as **a.**

Peepholes for looking into each darkened room

c

a retinal image which mimics that produced by the chair when viewed from the intended vantage point. In the most dramatic example, the lines are not formed into a single distorted object, but lie in space in quite different locations—and the "chair seat" is white patch painted on the wall.

The point is that the two rooms have things within them which result in a chair-like retinal image being cast in the eye. The fact that we see them as the same—as chairs—is because the visual system's design exploits the assumption that is "reasonable" to interpret retinal information in the way which normally yields perceptions that would be valid from diverse viewpoints. It is "blind" to other possibilities, but that should not deceive us—those possibilities do in fact exist.

Another way of putting this is to say that the Ames's chair demonstrations reveal that the visual

system (along with a typical computer vision system) utilizes what is called the ***general viewpoint constraint.*** A normal chair appears as a chair from all viewpoints, whereas the special cases used by Ames can be seen as chairs from just one special vantage point. The general viewpoint constraint justifies visual processes that would yield stable structural interpretations as vantage point changes.

The general viewpoint constraint can be embedded in bottom up processing. It is not necessary to invoke top down processing in explaining the Ames chair demonstrations—that is, knowing the shape of normal chairs, and using this knowledge to guide the interpretation of the retinal image.

Normal scenes are usually interpreted in one way and one way only, despite the retinal image information ambiguity just referred to. But it is possible to catch the visual system arriving at

Initial perception

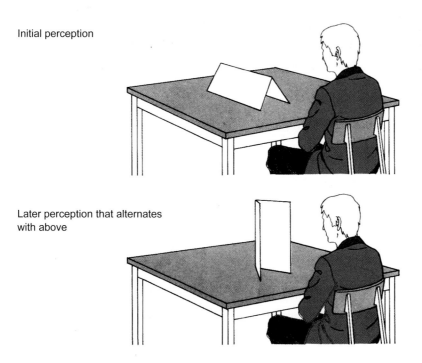

Later perception that alternates with above

1.27 Mach's illusion
With one eye closed, try staring at a piece of folded paper resting on a table (upper). After a while it suddenly appears not as a tent but as a raised corner (lower).

different descriptions of an ambiguous 3D scene in the following way. Fold a piece of paper along its mid-line and lay it on a table, as in **1.27**. Stare at a point about mid-way along its length, using just one eye. Keep looking and you will suddenly find that the paper ceases to look like a tent as it "should" do, and instead looks like a corner viewed from the inside. The effect is remarkable and well worth trying to obtain.

The point is that both "tent" and "corner" cast identical images on the retina, and the visual system sometimes chooses one interpretation, sometimes another. It could have chosen many more of course, and the fact that it confines itself to these two alternatives is itself interesting.

Another famous example of the same sort of alternation, but from a 2D drawing rather than from a 3D scene, is the Necker cube, **1.28**.

Conclusions

Perhaps enough has been said by now to convince even the most committed "inner screen" theorist that his photographic conception of seeing is wholly inadequate. Granted then that seeing is the business of arriving at explicit scene representations, the problem becomes: how can this be done?

It turns out that understanding how to extract explicit descriptions of scenes from retinal images is an extraordinarily baffling problem, which is one reason why we find it so interesting. The problem is at the forefront of much scientific and technological research at the present time, but it still remains largely intractable. Seeing has puzzled philosophers and scientists for centuries, and it continues to do so. To be sure, notable advances have been made in recent years on several fronts

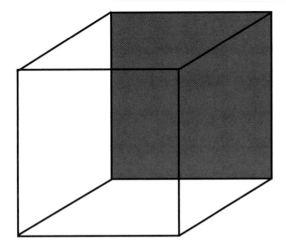

1.28 Necker cube
Prolonged inspection results in alternating perceptions in which the shaded side is sometimes seen nearer, sometimes farther.

within psychology, neuroscience, and machine image-processing, and many samples of this progress will be reviewed in this book. But we are still a long way from being able to build a machine that can match the human ability to read handwriting, let alone one capable of analyzing and describing complex natural scenes.

This is so despite multi-million dollar investments in the problem because of the immense industrial potential for good processing systems. Think of all the handwritten forms, letters, etc. that still have to be read by humans even though their contents are routine and mundane, and all the equally mundane object handling operations in industry and retailing.

Whether we will witness a successful outcome to the quest to build a highly competent visual robot in the current century is debatable, as is the question of whether a solution would impress the ordinary person.

A curious fact that highlights both the difficulties inherent in understanding seeing and the way we take it so much for granted is that computers can already be made which are sufficiently "clever" to beat the human world champion at chess. But computers cannot yet be programmed to match the visual capacities even of quite primitive animals. Moves are fed into chess playing computers in non-visual ways. A computer vision system has not yet been made that can "see" the chessboard, from differing angles in variable lighting conditions for differing kinds of chess pieces—even though the computer can be made to play chess brilliantly.

Even so, most people would probably be more impressed with a world-class chess-playing computer than they would be with a good image-processor, despite the fact that the former has been realized whereas the latter remains elusive. It is one of our prime objectives to bring home to you why the problem of seeing remains so baffling. Perhaps by the end of the book you will have a greater respect for your magnificent visual apparatus.

Meanwhile, we have said enough in this opening chapter to make abundantly clear that any attempt to explain seeing by building representations which simply mirror the outside world by some sort of physical equivalence akin to photography is bound to be insufficient. We do not see our retinal images. We use them, together with prior knowledge, to build the visual world that is our representation of what is "out there." We can now finally dispatch the "inner screen" theory to its grave and concentrate henceforth on theories which make *explicit scene representations* their objective.

In tackling this task, the underlying theme of this book will be the need to keep clearly distinct three different levels of analysis of seeing. Ch 2 explains what they are and subsequent chapters will illustrate their nature using numerous examples. We hope that by the time you have finished the book that we will have convinced you of the importance of distinguishing between these levels when studying seeing, and that you will have a good grasp of many fundamental attributes of human and, to a lesser extent animal, vision.

Navigating Your Way through This Book

This book is organized to suit two kinds of readers: students or general scientific readers with little or no prior knowledge of vision research, and more advanced students who want an introduction to technical details presented in an accessible style. Accordingly, we now flag up certain chapters and parts of chapters that delve more deeply into technical issues, and these can be missed out by beginners.

The underlying theme of this book is the claim, first articulated clearly by David Marr, that it is important to keep separate three different levels of discussion when analyzing seeing:

computational theory specifying how a vision task can be solved;

algorithm giving precise details of how the theory can be implemented, and

hardware (which means neurons in the case of brains) for realizing the algorithm in a vision system.

Ch 2 explains what these levels are and it illustrates the first two by examining the task of using texture cues to build a representation of 3D shape.

Ch 3 then begins our account of how neurons work by introducing a key concept, the receptive field. This involves a description of the kinds of visual stimuli that excite or inhibit neurons in the visual system. It also describes how the outputs of populations of neurons can be integrated to detect the orientation of an line in the retinal image.

Ch 4 describes how psychologists have studied neural mechanisms indirectly using visual phenomena called aftereffects.

Chs 5-8 are designed to be read as a block culminating in a discussion of object recognition.

Ch 5 describes how the task of edge detection can be tackled.

Ch 6 expands this account by presenting ideas on how certain neurons in the retina have properties suggesting that they implement a specific theory of edge detection. This account provides a particularly clear example of how it is important to keep separate the three levels of analysis: computational theory, algorithm, and hardware.

In Ch 7, we discuss how detected edges can be grouped into visual structures, and how this may be used to solve the figure/ground problem.

In Ch 8, we then present various ways in which the separated "figures" can be recognized as arising from particular objects.

Chs 9 and 10 are best read as a pair. Together, they describe the main properties of brain cells involved in seeing, and the way many are arranged in the brain to form maps.

Ch 11 is a technical chapter on *complexity theory.* It introduces this by exploring the idea of recognizing objects using receptive fields as templates. This idea is found to make unrealistic demands on how the brain might work, and a range of possible solutions are explored.

Ch 12 explains the intricacies of *psychophysical methods* for measuring the phenomena of perception. This is a field of great practical importance but it can be skipped by beginners. However, this chapter explains certain technical concepts, such as probability density functions, that are needed for some parts of subsequent chapters.

Ch 13 presents an account of seeing as *inference.* This explains the basics of the Bayesian approach to seeing, which is currently attracting a great deal of interest. It provides technical details that may be difficult for beginners who may therefore wish just to skim through of the opening sections to get the basic ideas.

Ch 14 introduces the basics of motion perception, and Ch 15 continues on that topic giving more technical details.

Chs 16 and 17 are best read as a pair. Ch 16 considers the problem of how our perceptions of black, gray, and white surfaces remain fairly constant despite variations in the prevailing illumination. Ch 17 tackles the same kind of problem for colored surfaces.

Chs 18 and 19 deal with stereoscopic vision, with the second giving technical details omitted from the first. Ch 18 contains many stereograms called anaglyphs that are designed to be viewed with the red/green filters mounted in a cardboard frame that is stored in a sleeve on the inside of the back cover of the book.

Ch 20 considers a topic that has been much studied in recent years using the Bayesian approach: how do we combine information from many different cues to 3D scene structure?

Ch 21 describes in detail a particular figure/ground task that was much studied in the early days of the field artificial intelligence: deciding which edges in images of a jumble of toy blocks belong to which block. This culminates in a list of hard-won general lessons from the **blocks world** on how to study seeing. We hope the beginner will find here valuable pointers about how research on seeing should and should not be studied.

Ch 22 discusses the vexed topic of seeing and consciousness. This is currently much-debated, but our conclusion is that little can be said with any certainty. Visual awareness is as mysterious today as it always has been. We hope this chapter reveals just why making progress in understanding consciousness is so hard.

Ch 23 ends the book with a review of our linking theme, the computational approach to seeing. Amongst other topics, it sets this review within a discussion of the strengths and hazards of different empirical approaches to studying seeing: computer modelling, neuroscience methods, and psychophysics.

We hope that by the time you have finished the book you will have a firm conceptual basis for tackling the vast literature on seeing. We have necessarily presented only a small part of that literature.

We also hope that this book will have convinced you of the importance of distinguishing between the computational theory, algorithm, and hardware levels when studying seeing. Most of all, we hope that this book leaves you with a sense of the profound achievement of the brain in making seeing possible.

Further Reading

We include these references as sources but they are not recommended for reading at this stage of the book. If you want to follow up any topic in more depth then try using a Web search engine. Google Scholar is suitable for academic sources.

Ernst B (1996) *Adventures with Impossible Objects.* Taschen America Inc. English translation edition.

Gibson JJ (1979) *The Ecological Approach to Visual Perception.* Boston, MA: Houghton Mifflin. *Comment* See commentary in Ch 2.

Gregory RL (1961) The brain as an engineering problem. In WH Thorpe and OL Zangwill (Eds) *Current Problems in Animal Behaviour.* Cambridge: Cambridge University Press 335–344.

Gregory RL (1980) Perceptions as hypotheses. *Philosophical Transactions of the Royal Society of London: Series B:* **290** 181–197.

Gregory RL (1971) *The Intelligent Eye.* Wiedenfeld and Nicolson.

Gregory RL (2007) *Eye and Brain: The Psychology of Seeing.* Oxford University Press, Oxford.

Gregory RL (2009) *Seeing Through Illusions: Making Sense of the Senses.* Oxford University Press.

Marr D (1982) *Vision: A computational investigation into the human representation and processing of visual information.* San Francisco, CA: Freeman.

Shepard RN (1990) *Mindsights.* W.H.Freeman and Co.: New York. *Comment* A lovely book with many amazing illusions.

Thompson P (1980) Margaret Thatcher: a new illusion. *Perception* **9** 483–484. *Comment* See *Perception* (2009) **38** 921-932 for commentaries on this classic paper.

Collections of Illusions, Well Worth Visiting

http://www.michaelbach.de/ot/

http://www.grand-illusions.com/opticalillusions

www.telegraph.co.uk/news/ newstopics/howaboutthat/3520448/ Optical-Illusions---the-top-20.html

www.interestingillusions.com/en/subcat/ color-illusions/

http://lite.bu.edu/vision-flash10/applets/lite/lite/ lite.html

http://illusioncontest.neuralcorrelate.com

http://www. viperlib.york.ac.uk/browse-illusions. asp

Hidden Figures

1.21 contains the dalmatian, left.

1.22 has a frog (easy to see) and a snake (harder).

1.23 The word *that* is printed twice. It is remarkable that this error was not picked up in the production process.

2

Seeing Shape from Texture

2.1 Using the shape of image texture elements to find shape from texture
The hat is decorated with a texture of circular discs. Each disc projects to an ellipse in the image, where the degree of "ellipticalness" is systematically related to the local surface orientation. These image ellipses were used as the input for a shape-from-texture computer program working on similar principles to those described toward the end of this chapter. Each needle sticks out at a right angle from the surface, and indicates the local surface orientation, as estimated by the shape-from-texture program. These local estimates of 3D orientation are a good starting point for building up a representation of the shape of the whole hat. Courtesy A. Loh (2004).

Look at the painting by Pieter de Hooch, 2.2. Your visual system causes you to see the tiled floor extending into the distance. Yet the painting itself is flat. Clearly, there must be information in the image of the tiled floor for you to be able to see the depth in the painted scene. The first thing we need to do is examine carefully what this information might be.

Texture Density Gradients

The striking thing about the image of the tiled floor is that the amount of texture varies from place to place within the image. Precisely how texture varies within the image depends on the ***slant*** of the surface. For a surface with zero slant (such as a vertical wall, viewed "head on"), an image of that surface has the same amount of texture everywhere. Conversely, for a surface with a large slant (e.g., the tiled floor in **2.2**), image texture in the background is much more densely packed than texture in the foreground. In this case, ***texture density*** increases as we move from the foreground to the background of the image. More importantly, the precise manner in which texture density changes across an image is *systematically* related to surface slant.

Below we describe how to exploit this systematic relationship between texture density and surface orientation. But the first and vital step is simply recognizing that the relationship between image *texture* and surface *slant* is systematic, because this allows us to gain a firm foothold on the problem of how to derive ***shape from texture***.

You may have noticed that we have been discussing how to obtain surface *slant*, and not surface *shape* as indicated by the title of this chapter. This is because we have implicitly assumed that obtaining the orientation of small patches on a curved

Why read this chapter?

The problem of how to obtain shape from texture is an accessible example of a vision problem that can be attacked using Marr's computational framework. We therefore use this chapter to 'kill two birds with one stone': explaining a classic topic in vision research, how to obtain shape from texture, and explaining how Marr's framework helps in formulating solutions to vision problems. Marr's framework is a unifying theme of this book, so we present this chapter both as a general introduction to the framework and as an example of how it can be applied to a specific problem.

2.2 Pieter de Hooch, 1629–1684
Woman drinking with two men, with maidservant
Probably 1658. Oil on canvas. 73.7 x 64.6 cm. The floor provides an example of a *texture gradient* that yields a vivid sense of depth. The superimposed dashed lines reveal the vanishing point used by de Hooch. Such points help artists create accurate perspective projections and hence vivid depth effects. © The National Gallery, London, UK.

surface (**2.1**) is a good start toward the goal of representing its overall 3D shape.

Before we continue, we need to define some terms introduced above. First, the term ***slant*** is used in everyday language to mean slope, a definition which captures its technical meaning quite well: slant is simply the *amount of slope* of a surface (details later). Second, we assume that texture consists of discrete, geometric texture elements, or ***texels*** (derived from the words ***tex***ture ***el***ements).

Thus, texels can derive from diverse scene entities, such as discs (**2.1**), tiles (**2.2**), and pebbles (**2.3**). The assumption that texture consists of discrete texels is made mainly for the purposes of exposition, and does not apply in general. Third,

Relatively high texture density

Relatively low texture density

2.3 Texture density variation in an image as a cue to the slope of a surface, here a beach
The two identical trapezoids in the image mark out different rectangular areas on the beach but they project to trapezoids with identical areas in the image (even though the top one appears larger).

when we refer to the "amount of texture" we mean the amount of texture per unit area, known as **_texture density_**. We have already observed that texture density varies from place to place in the image; for example, in **2.3**, there may be 6 pebbles per square centimeter in the background region of the image, say, but only 2 pebbles per square centimeter in the foreground.

Perspective Projection

In order to proceed, we need to explain the concept of **_perspective projection_**. Consider the case of a pinhole camera, **2.4a,b**. All the light rays from the scene pass through the pinhole. Consider now two surface regions of the same size. Each contains the same amount of texture, but the surface region in the background gets squeezed by the optics of perspective projection into a smaller region in the image than does the region in the foreground. Thus, a slanted surface that has the same texture density everywhere defines an image in which the texture density changes from place to place. As

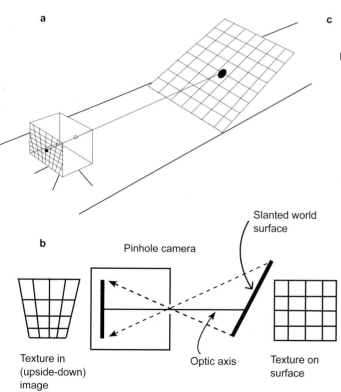

a

b

Pinhole camera

Slanted world surface

Optic axis

Texture in (upside-down) image

Texture on surface

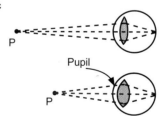

c

The lens brings light rays from point *P* entering the pupil to a focus on the retina.

Pupil

The thicker lens bends light rays to a greater extent than in the above diagram, as required because *P* is nearer.

2.4 Perspective projection
a Pinhole camera, producing an inverted image
b Cross-section of pinhole camera forming an image of a plane slanted with respect to the optic axis of the camera, as in **a**. Notice that the plane has an homogenous texture because it has equal numbers of texture elements (the outline squares) in same-sized areas on the plane surface. Not drawn to scale.
c Role of the lens in focusing light rays
These cross-sectional diagrams of the eye do not show the important role of the *cornea* (Ch 6) in focusing light rays. They illustrate only the role of the lens (shaded), whose shape and hence power can be adjusted, in young people, in a process called *accommodation*. With aging the lens gradually becomes inelastic leading to an inability, called *presbyopia,* to focus on (technically, accommodate to) near objects—hence the need for many elderly people to use reading glasses.

b Observer holding a device which allows a painting (P) of the Baptistry to be seen through a central peephole (shown by central black blob in the painting) by reflection from the mirror (M). By removing/replacing the mirror the observer can see alternately the Baptistry and the painting, and hence compare them.

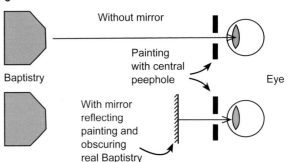

2.5 Perspective and painting

Brunelleschi is generally regarded as the first artist to have correctly used linear perspective in paintings.

a Baptistry of St John in Florence. Brunelleschi's painting of this building, created around 1413, is lost.

b Brunelleschi created a peephole in his painting through which he invited observers to view the real Baptistry. He also provided a mirror that the observer could place in front of the painting so that the observer would then see a reflection of the painting. By alternately viewing with and without the mirror through the peephole the observer could compare seeing the real Baptistry with Brunelleschi's painting of it. Apparently observers found it difficult to tell the difference between them, thus demonstrating the accuracy of his use of perspective.

c Plan views explaining the role of the mirror. See Kubovy (1986).

neighboring surface regions project to neighboring image regions, the amount of "squeezing" varies smoothly as we move across the image. There is thus a smooth change in image texture density. The resultant gradual change in texture density across the image defines a *texture density gradient*.

Imaging systems such as the eye and cameras can be thought of as sophisticated versions of the pin-hole camera shown in **2.4a**. The line which connects the pin-hole to the center of the image is the *optic axis*, and the distance along the optic axis from the image to the pin-hole is the *focus length*.

A pin-hole camera makes a good, but very dim, image. Making the pin-hole larger makes the image brighter because more light is let in. However, light rays from each point in a scene get spread over a larger image region, resulting in a more fuzzy image. In order to overcome this problem, our eyes make use of a lens to focus the light rays arriving through the *pupil*, a much larger aperture

than a pin-hole. The main function of the lens is to focus the light rays arriving from any given scene point to a single point in the image, **2.4c**. This is why the lens-focused image is brighter than the image formed in a pinhole camera. We will not delve deeper into the optics of the eye, but will proceed using the simplifying assumption that a pinhole camera is a good model of the optics of the eye for our purposes.

It is important to note that perspective projection is a mapping operation from a three-dimensional (3D) world to a two-dimensional (2D) image. Consequently, only certain physical world properties are accurately preserved in the image. For example, straight lines in the world project to straight lines in the image, so that "straight-ness" or *colinearity* is preserved by perspective projection. In contrast, size is not preserved by perspective projection because the image of a distant object is smaller than that of the same object when it is nearby. Moreover, the fact that size is not preserved with perspective projection ensures that regions of the same size on a slanted plane project to regions of different sizes in the image.

It is a surprising fact that a deep understanding of perspective projection was not achieved until early in the 15th century, **2.5**. Thereafter it was widely used by artists to provide realistic impressions of depth. Indeed, the fact that the eye forms an image at all was generally accepted only after the famous astronomer Kepler (1571–1630) published an account of the formation of the retinal image in the eye in 1604. In 1619 Christoph Scheiner scraped away the tissue at the back of a ox eye until it became like a translucent projector screen. Scheiner then observed on this screen an image of what the bull would have been looking at (if the eye were still in the bull's head of course).

The fact that the retinal image is inverted was interpreted by others as a major objection to these findings on the grounds that we do not perceive an upside-down world. N-C Fabri de Peiresc (1580–1637) and Pierre Gassendi (1592–1655) even proposed the existence of a retinal mirror to make the retinal image appear upright. This provides an insight into how easy it is to mistake seeing for the process of image formation. Many people even to-

2.6 Inhomogeneous vs. homogeneous textures
The floor tiles in **2.2** and the pebbles in **2.3** are examples of *homogeneous* surface textures, meaning that they have the same texture density everywhere. This property is not possessed by the Mars scene shown here, an example of an *inhomogeneous surface texture*. Courtesy NASA.

day are perplexed by the fact that the retinal image is inverted, as discussed in Ch 1.

If the amount of texture per unit area is about the same everywhere on a surface then it is said to be **homogeneous**; otherwise, it is called **inhomogeneous**, **2.6**. This can be used as a constraint in solving the problem of estimating the orienta-

a

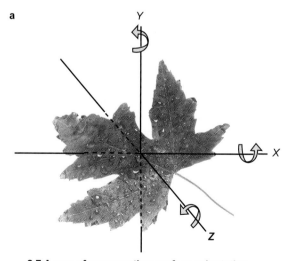

b

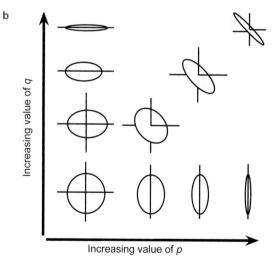

Increasing value of q

Increasing value of p

2.7 A way of representing surface orientation
a Falling leaf with three perpendicular axes X (left-right), Y (vertical), and Z (depth). The leaf's orientation can be defined in terms of the amount of rotation around any two axes. By convention, these are the X and Y axes; rotation around the X-axis is represented by the parameter q, and rotation around the Y-axis is represented by parameter p.
b If we think of the leaf as a planar surface and we draw it as a circular shape for simplicity then its orientation can be described with the two parameters, p and q. Another illustration explaining the *p,q representation* is **21.10**, in Ch 21.

tion of that surface. Armed with the **homogeneity constraint**, we are almost ready to describe a method for finding the orientation of a textured surface plane. But first, we need to specify a way of representing surface orientation.

Representing Surface Orientation

We will describe two ways of representing the 3D orientation of a planar surface.

First, imagine a perfectly flat leaf as it falls from a tree, gliding from side to side at is slides through the air. Now imagine that the leaf is pierced at its center by two lines: a vertical line, and a horizontal line which extends from left to right from your point of view, **2.7a**.

Clearly, the leaf can rotate around each of these lines or **axes** as it falls through the air. More importantly, any orientation that the leaf can adopt is completely specified by its rotation around these two axes. In essence, these axes can be used to define the leaf's orientation in terms of two numbers, or **parameters**, giving rotation p around the vertical axis and rotation q around the horizontal axis, **2.7b**. For example, the surface shown in the beach scene in **2.3** is rotated around the horizontal axis by about $60°$, and is rotated around the vertical axis by about $10°$. Thus, in this case, we say that (p, q) equals $(10, 60)$. (In fact, usually p and q are defined as the tangents of the rotation angles, but this is of little consequence for our purposes).

To understand a second way of representing surface orientation, imagine looking down on the top of an opened umbrella, **2.8**. Each rib of the umbrella has a different angle "around the clock," as it were. This angle is called the **tilt**, and varies between zero and $360°$. Now, as we move out along each rib of the umbrella, the *amount of slope* or **slant** on each patch of the umbrella surface increases. On moving from the umbrella's tip, say out along a rib, the patch changes its slant angle from zero at the tip through to $90°$ of slant when the orientation of the patch is parallel to the axis of the umbrella's shaft.

So here we have a way of representing surface orientation, not with the parameters (p, q), but with a different pair of parameters, (*slant, tilt*). Equations exist for translating one pair of parameters into the other (p. 52). This fact illustrates that these two representations are alternatives.

Notice that in each case, surface orientation needs exactly two numbers to be specified. This reflects the underlying physical fact that a plane's orientation has what are technically called two **degrees of freedom**. What this means is that any one of all the different possible orientations of a plane can be described with a particular combination of two parameters, which can be either slant and tilt, or p and q. Each parameter (e.g., slant) is free to vary independently of the other parameter (e.g., tilt), and hence the phrase two degrees of freedom.

In what follows, we will use the (p, q) parameters, and ask: how can (p, q) be estimated from measurements of image texture density? The answer depends on knowing how image texture density constrains possible values of (p, q). We will show how this works for p and q separately.

Tilt angle changes around axis of the umbrella handle

Slant angle changes from umbrella tip to rim

2.8 Slant and tilt illustrated with an umbrella
An alternative to representing surface orientation by p and q, as in **2.7**, is using *slant* and *tilt* angles. The small panels of the umbrella show surface patches with different orientations in space. Slant increases as the patches move away from the umbrella tip. In contrast, the direction of slant, called its tilt, changes as we move around the umbrella tip. Slant and tilt are defined with respect to a reference axis, which is the umbrella stalk in this figure. More usually, this axis would pass through the center of the camera/eye.

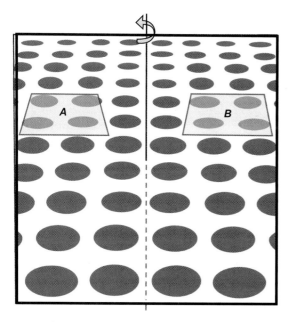

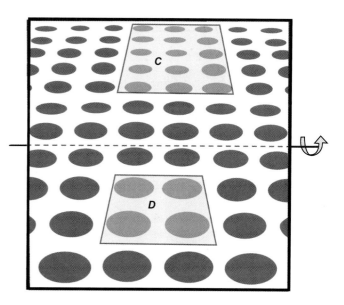

2.9 Finding _p_, rotation around a vertical axis
The value of _p_ here is zero, as implied by the fact that regions _A_ and _B_ have the same texture density. The arrow indicates anti-clockwise rotation around the vertical axis, which increases _p_.

2.10 Finding _q_, rotation around a horizontal axis
The different amount of texture in regions _C_ and _D_ implies that _q_ is not zero. The arrow indicates rotation direction around the horizontal axis required to increase _q_. (_C_ and _D_ are the same size on the page but look very different. This is an example of the Ponzo illusion, **2.23**.

Finding _p_

If we measure the texture density in two image regions _A_ and _B_ which lie on any horizontal line in the image then we can ascertain _p_, the rotation of the surface plane around its vertical axis. For example, in **2.9** the density in image region _A_ is given by the number of texels in _A_ divided by the area of the image region _A_, which is about 5 square centimeters (cm²). Therefore, the image texture density in _A_ is 4 (texels) divided by 5cm² or about 0.8 texels/cm². A similar calculation yields an image texture density of 0.8 texels/cm² in region _B_.

In general, if texture density is the same in any two image regions lying along a horizontal line then we can conclude that the two corresponding surface regions must be equidistant from the camera. If these surface regions were at different distances then the texture densities in the image regions _A_ and _B_ would be different, and we would conclude that the surface was rotated around the vertical axis to yield a non-zero value for _p_. The equal texture densities in _A_ and _B_ in **2.9** therefore imply that the amount of rotation _p_ around the vertical axis is zero.

But what, you may ask, if the texture densities in image regions _A_ and _B_ are not equal? This is dealt with using an example of finding _q_ in the next section. Finding _p_ and finding _q_ involve the same type of exercise, because each simply defines the amount of rotation around one axis.

Finding _q_

If we measure the texture density in two image regions which lie on any vertical line in the image then we can ascertain _q_, the rotation of the plane around its horizontal axis. This is illustrated for regions _C_ and _D_ in **2.10**. We can divide the density in region _C_ by the density in region _D_ to obtain a ratio:

$$R_{CD} = \frac{\text{Texture density in image region } C}{\text{Texture density in image region } D}$$

Clearly, the magnitude of the ratio R_{CD} increases as the surface slope increases. If we know the image positions of the two image regions _C_ and _D_ (which we do), and if we know the texture density in both image regions (which we do), then we can use

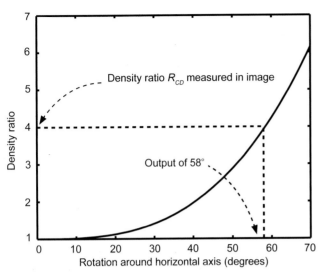

2.11 Ratio of image texture densities in *C* and *D* (from 2.10) plotted against *q*
This ratio of densities R_{CD} can be used to read off the precise value of rotation q around the horizontal axis. In this example, the output value of q is 58°.

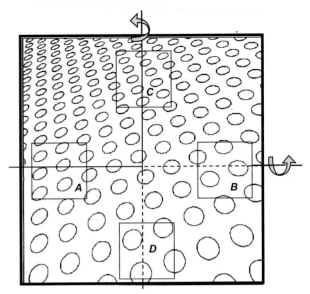

2.12 Surface oriented such that neither *p* nor *q* is zero
This is revealed by the different texture densities in the outlined areas. In general, only the texture density in four image regions is required to obtain p and q. Rotation p around the vertical axis is obtained from the ratio of densities in regions *A* and *B*, whereas rotation q around horizontal axis is obtained from the ratio of densities in regions *C* and *D*. The graph shown in **2.11** can then be used to look up the values of both p and q.

some standard mathematics to construct a graph (**2.10**) of image texture density ratio versus rotation q around the horizontal axis. We can then use this graph as if it were a ***look-up table***.

To summarize, we measure the texture density in image regions *C* and *D*, find the ratio R_{CD} and then read off q from the graph, **2.11**. At this point, we have a value for q, the amount of plane rotation around the horizontal axis, and as we already know for this example that $p = 0$ (see above), we have therefore solved the problem of recovering (p, q) for our simple textured plane surface, i.e. $(p, q) = (0°, 58°)$.

Above, we used simple logic to show that $p = 0°$ in **2.9**. If we had a graph of image texture density ratio versus p then we could use this to find p. Such a graph would be identical to that shown in **2.11**, and we so we can use the same graph to find values of p and q. Specifically, we know that the texture densities in image regions *A* and *B* are the same (0.8 texels/cm²), and therefore that the ratio of densities is $R_{AB} = 0.8/0.8 = 1$. If we now look up this value in **2.11** then we see that a value of $R_{AB} = 1$ corresponds to a rotation (q in this case) of zero degrees; a result we anticipated above.

This technique can be applied to planes with any orientation, **2.12**. As above, we require two

pairs of image regions. Thus, the ratio of texture densities in *A* and *B* can be used to find p, whereas the texture densities in *C* and *D* can be used to find q.

We have used a very simple example in order to demonstrate how surface orientation can be obtained from texture information *in principle*. In practice, many of the simplifying assumptions can be discarded, and the basic principle can be made to work under realistic conditions.

In fact, perspective projection gives rise to many different texture gradients, as shown in **2.13**. Moreover, most of these can be used to estimate surface orientation based on the principles outlined above for texture density.

What have we learned? Shape from texture provides a clear and accessible example of how to use a specific visual cue to recover surface orientation from a flat image. But even more importantly, it nicely illustrates important features of a general approach to tackling seeing problems, as we explain next.

Surface (plan view of part of railway line)

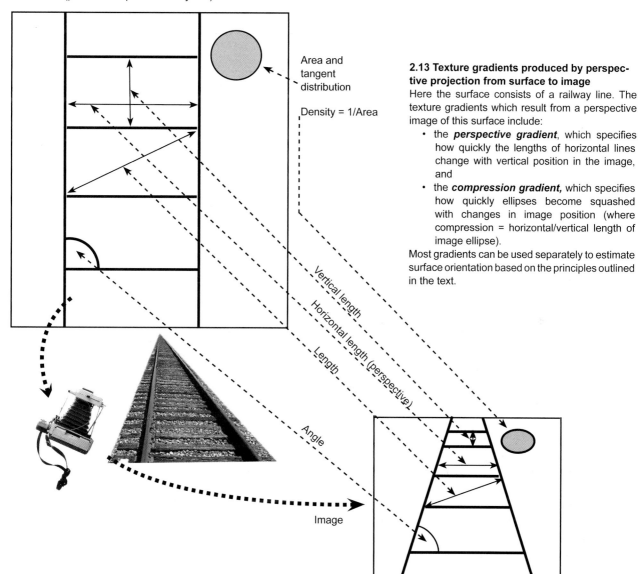

Area and
tangent
distribution

Density = 1/Area

Vertical length

Horizontal length (perspective)

Length

Angle

Image

2.13 Texture gradients produced by perspective projection from surface to image
Here the surface consists of a railway line. The texture gradients which result from a perspective image of this surface include:
- the *perspective gradient*, which specifies how quickly the lengths of horizontal lines change with vertical position in the image, and
- the *compression gradient,* which specifies how quickly ellipses become squashed with changes in image position (where compression = horizontal/vertical length of image ellipse).

Most gradients can be used separately to estimate surface orientation based on the principles outlined in the text.

Marr's Computational Framework

One particularly fruitful theoretical approach for solving the problem of seeing is the *computational framework,* introduced briefly in the previous chapter. In fact, without saying so, we have already employed in this chapter the computational framework for solving a specific vision problem, shape from texture. This framework was developed and popularized by David Marr (with various significant antecedents in the vision literature, Chs 1, 23). It provides a strategy for tackling a wide range of problems in vision. The framework consists of three different *levels of analysis*, **2.14**, which we will examine in the context of shape from texture.

First, the *computational level* specifies the general nature of the problem to be solved. In this chapter, the problem to be solved consists of finding the three-dimensional orientation of a textured surface. This level also identifies specific *constraints* that might be useful in solving the problem under consideration, and these are usually derived from examining very carefully the information available. In our case, we observed

a

Computational theory What is the nature of the problem to solved, what is the goal of the computation, why is it appropriate, and what is the logic of the strategy by which it can be carried out?

Representation and algorithm How can this computational theory be implemented? In particular, what is the representation for the input and output, and what is the algorithm for the transformation from input to output?

Hardware implementation How can the representation and algorithm be realised physically?

b

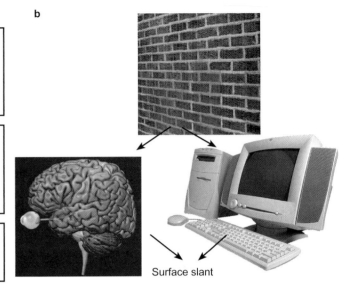

Surface slant

2.14 Marr's three levels
a Table based on Figure 1-4 in Marr's seminal book *Vision* (1982), in which he argued that these were "three levels at which any machine carrying out an information-processing task must be understood."
b Under Marr's framework, it does not matter, as far as the computational theory level is concerned, if shape from texture is implemented by the neurons in a brain or by the electrical components in a computer. Brain photo courtesy Mark Dow.

that the amount of surface texture per unit area is about the same in all surface regions (i.e., surface texture is homogeneous), and we then used this as a constraint in solving the problem. The solution amounted to measuring and then comparing texture density in a different image regions. Notice that this is not the only theory of how to compute surface orientation from texture—far from it, as we will describe in due course other computational theories of extracting shape from texture.

Second, the *representation* and *algorithmic level* specifies three entities: the nature of the input data, a particular method or *algorithm*, and the nature of the output (the representation) produced by the algorithm. The algorithm is based on the computational theory used to process input data, and the precise nature of the output required. In our case, the input data is in the form of image texture, and the output consists of the two parameters (p, q), which together specify the orientation of a surface plane. The algorithm we developed consists of measuring the image texture density in two pairs of image regions: one pair positioned along a horizontal image axis, and one pair along a vertical image axis. We found the ratio of texture density in each pair of regions, and then used a graph as

a look up table to ascertain the values (p, q) which were consistent with the measured texture density ratios. The use of these look-up tables could only be justified by the constraint, identified at the computational level, that the surface texture is homogeneous.

Third, the *hardware level* describes how the algorithm specified above is executed in hardware. In other words, the hardware level states the nature of the physical device used to execute the operations required by the algorithm. Two obvious choices are the neurons in a brain or the electronic components in a computer. A less obvious choice, and one adopted for our simple example of shape from texture, is a pencil and paper, a calculator, and, most importantly, a person to mark out image regions and to count the number of texels in each one.

Notice that the output (p, q) does not depend on what hardware is chosen, provided the chosen hardware executes the specified algorithm, **2.14b**. In other words, there is a degree of *modularity* within the three levels of analysis framework. This is a crucial fact. The inner workings of each level are to a considerable extent independent of the inner workings of the levels above and below it. The algorithm does not "care" whether surface orientation is obtained by neurons firing in a brain, or by shuffling bits around a computer, or even by shunting cupcakes around a baker's shop in order to keep track of texel numbers; provided all of these different physical hardware implementations carry out the same underlying calculations as specified at the algorithmic level.

Similarly, the computational level specifies which problem is to be solved, but it does not place limits on which algorithm could be used to solve that problem. We chose to exploit the constraint of texture homogeneity with an algorithm based on texture density in a set of image regions. However, we could have chosen a different algorithm for exploiting this constraint.

The point is that, just as the algorithmic level does not care which hardware implementation is used, so the computational level does not care which algorithm is used. What really matters is that the hardware implements the algorithm specified at the algorithmic level, and that the algorithm is sufficient to solve the problem specified at the computational level.

There is thus a hierarchical cascade of influence, from the computational, to the algorithmic, and finally, to the hardware level. Each level constrains, but does not specify precisely, how each subsidiary level operates. This hierarchical structure provides a degree of modularity between levels, and this, in turn, provides massive flexibility in terms of designing (in the case of artificial systems) and in terms of evolving (in the case of biological systems) solutions to the problems of vision.

Notice that Marr's framework constitutes a hierarchy in which the computational theory level is placed at the top because of its *logical* priority. However, it does not necessarily have *methodological* priority. That is, the computational level theory does not have to be constructed before trying out some preliminary ideas at the algorithmic level. These experiments can lead to discovery through failures where shortcomings exist in understanding the way to solve a visual task. Also, as we will see as we go through this book, good ideas for computational theories have emerged by getting clues from studies of the visual mechanisms in eyes and brains, or from experiments on human vision. The goal is indeed to obtain a complete account of

a seeing problem with solid proposals at all three levels: computational theory, algorithmic, and hardware. But the process of constructing such an account is often incremental, and requires the exploitation of research results from many different sources. Such is the nature of scientific enquiry.

We conclude this section by summarizing, using Marr's framework, the main features of estimating surface orientation using homogeneity.

Computational Theory

The computational level specified the *homogeneity constraint* (equal density of texels over the surface), and the way that the density gradient in an image increases as the slant of the surface increases (the graph in **2.11**).

Representation and Algorithm

This specified a number of steps:

1. *Choose an input representation.* We have used texels, assuming that a prior process has made this representation available.

2. *Choose a surface orientation representation.* We chose to use the (p, q) parameters. It is important to note that it is possible to build an algorithm using the texture homogeneity constraint, and hence the same computational theory, using the (*slant, tilt*) rather than (p, q) parameters. So choosing which parameter pair to use, (p, q) or (slant, tilt), is made at the algorithm level in Marr's three-levels framework. It is not a matter of necessarily having a different computational theory of the shape-from-texture task for the chosen representation.

3. *Find p, rotation around the vertical axis*:

a. *Image Regions.* Draw two regions A and B centered on the same horizontal line in the image.

b. *Image Region Area.* Measure the area of image regions A and B as A_A and A_B, respectively.

c. *Texel Number.* Count the number of texels in image regions A and B to give N_A and N_B, respectively.

d. *Texture Density.* Divide the number of texels in each region by the area of that region to give the texture density in A and B as $\rho_A = N_A/A_A$ and $\rho_B = N_B/A_B$, respectively.

e. *Ratio of Texture Densities.* Divide ρ_A by ρ_B to give R_{AB}, the ratio of texture densities in image regions A and B.

f. *Look Up.* Use the graph in **2.11** to look up the value of p consistent with the observed ratio R_{AB}.

2.15 Size constancy and depth cues
The image of the man marked by an arrow has been copied to three other locations in the image. It is difficult to believe that these images are the same size on the page because they appear so different. This phenomenon is usually interpreted as a consequence of the visual system using depth cues to create size perceptions that take into account how far away the object is. This process normally operates to ensure that size perceptions remain the same when an object moves away from or toward us, hence its name—*size constancy*, p. 41. The photograph is of a piece of street art by Kurt Wenner reproduced with his permission. The illusory but very convincing perception of a hole in the pavement that has been created by moving slabs to one side is a testament to the skill of the artist in using depth cues to create vivid 3D effects.

4. *Find q.* Repeat steps in 2 using regions C and D centered on the same vertical line in the image to find q.

Hardware

In this case, the calculations specified could be done with pencil and paper.

A cautionary note. As we will see as we proceed through this book, in some cases the distinction between computational theory and algorithm can become a bit blurred. That said, it is still a very good discipline, by way of assisting clarity, to be guided by Marr's framework. If a computer vision experiment isn't working out very well then it is important to ask whether the problem lies in a flawed theory (e.g., an ill-specified constraint) or a poor algorithm (e.g., a procedure that doesn't put a decent constraint to work in an effective way).

These conceptual issues also crop up in studying biological vision systems, as we will see.

An Optical Size Illusion

An intriguing feature of the image regions C and D shown in **2.10** is that they do not appear to be the same size. This is a version of a classic optical illusion, the **Ponzo illusion**. Regions C and D are in fact the same size on the page (try measuring them with a ruler to convince yourself). However, because they are overlaid on an image of a surface, your visual system assumes that they are actually on that surface. And if these two regions really were on the surface then they would have different sizes. We can see this if we describe the situation a little more precisely.

Let's use the labels C' and D' to refer to the surface regions that appear as regions C and D in the image. As can be seen from **2.9**, the surface region C' that corresponds to the image region C contains more texture than the surface region D'. Using our assumption that the surface contains the same amount of texture everywhere, this implies that region C' must be bigger than surface region D'. This is the key to understanding the illusion: surface region C' is bigger than surface region D', even though the corresponding image regions C and D are identical in size. Your visual system simply, and quite sensibly, chooses to perceive the two regions not as two identical image regions C and D with identical image areas, but as two *surface* regions C' and D' with different surface areas. In other words, your visual system forces you to perceive the size of each region as it would appear *out in the world*, and not at the size it actually is in the image.

The reason we have elaborated this at some length is that it is an example of a general rule: it is almost always the case that what we see reflects the visual system's function of telling us what is out there in the world. We do not see our retinal images. This is a deep truth worth pondering.

A different, but less helpful, way of saying this is that our perceptions tend to remain *constant* despite changes in the retinal image. For example, the image of an object enlarges or shrinks as it moves toward or away from us but the perception of the object's size tends to remain the same. This phenomenon is usually described in the vision literature as **size constancy, 2.15**. However, this label neglects the fact that same-sized objects do in fact appear to some considerable degree smaller when they move farther away, albeit not as small as would be predicted from the corresponding change in retinal image size. This is sometimes described as a failure of size constancy.

We think that a clearer way of regarding the kind of situation portrayed in **2.15** is that the the visual system's goal is to build a representation of "things out there in the world." It represents the sizes of objects in such a way that as an object moves farther away it looks *both* smaller *and* of the correct size *given its position in the scene*. In other words, the representation of size in human vision meets two requirements. It gives information about the physical sizes of objects, which is why we can judge reasonably well the relative sizes of objects at

different distances. It also places those objects in a representation of the world which necessarily has to cope with the huge amount of space in the scene before us—hence the consequence that farther objects look "smaller" and at the same time farther away. We therefore suggest it would be helpful to dispense with the confusing term *size constancy* and instead concentrate on the issue of the nature and functions of the size representations that are built by our visual system.

Size constancy is just one of many **perceptual constancies**. For example, **lightness constancy** is the tendency for a surface to appear to have the same surface reflectance even when its illumination changes so that its retinal image changes, see Chs 16 and 17.

Skip to Summary Remarks?

So far, we have considered how to estimate slant by comparing texture density in different image regions. Equivalent, and in some respects more intuitively obvious, methods exist, but these require a more detailed understanding of perspective projection. The remainder of this chapter deals with those methods but some readers may wish to postpone considering them. We have now said enough by way of introducing the main theme of this book— the computational approach to seeing—to allow readers if they so wish to move to the *Summary Remarks* (p. 52) and then on to the next chapter.

Finding Surface Orientation Using Back-projection

Consider what would happen if we could freeze time at the instant a photograph of a textured surface is taken, and simply reverse the direction of each light ray, **2.16**. Light would spring from the photographic image, through the pin-hole and onto the textured surface. This is **back-projection**. If we now replace the textured surface with a white card then the image formed on that surface would naturally be identical to the texture on the original world surface. Critically, this is true only if the card and the world surface used to create the image are at exactly the same orientation. In order to avoid confusion, let's call this white card the **model surface**, and the original textured world surface as the **world surface**.

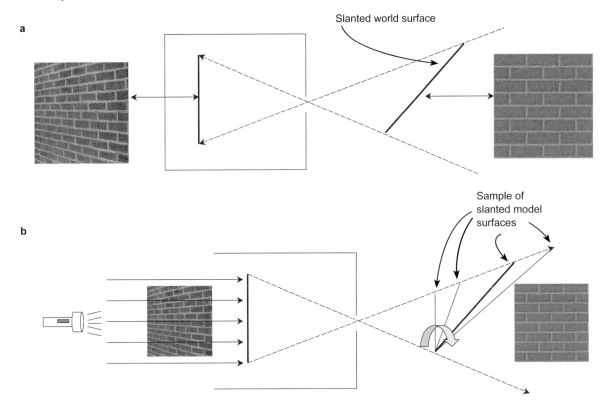

2.16 Using back-projection to find the slant of a plane
a Side view of the perspective projection of a brick wall, with a homogeneous texture, that is slanted with respect to the camera's optic axis.
b Side view of the back-projection of the image captured in **a**. The slant of the model surface on which the back projected image falls is adjusted until the back projected image has a homogeneous texture.

The model surface will be used as a hypothesis for the orientation of the world surface. We are thus projecting the photographic image back out into the world, and onto the model surface, which is why this process is called back-projection. All we are really doing is using the camera as if it were a projector (which is straightforward if we think of a projector as a camera in which the light is going in the "wrong" direction).

In terms of image formation, the light does not care in which direction it travels. Back-projection can therefore be achieved by replacing the back of a camera with a transparent photograph and then shining a light through the photograph, **2.16**.

More generally (that is, without the need to freeze time), if we take a transparent photograph of a textured plane and put it in a camera which is being used as a projector then light rays exit through the pin-hole and can be made to form an image on our model surface. However, we now know that,

in order to reproduce the texture as it was on the original world surface, this model surface would have to be held at exactly the same orientation as the world surface. But how would we know when the model surface is at the correct orientation?

The answer is implicit in the assumption that the original surface texture is homogeneous. Suppose we notice that the back-projected image texture on our model surface is homogeneous when we hold the model surface at a specific orientation. Moreover, we find that any other orientation produces an image of a non-homogeneous texture on the model surface. What might we conclude from this?

Recall that in our back-projection setup light does not care in which direction it travels. It is therefore obliged to form an image at the back of the camera if light travels from a world surface and through the pin-hole. The light is equally obliged to form an image if the light travels through a

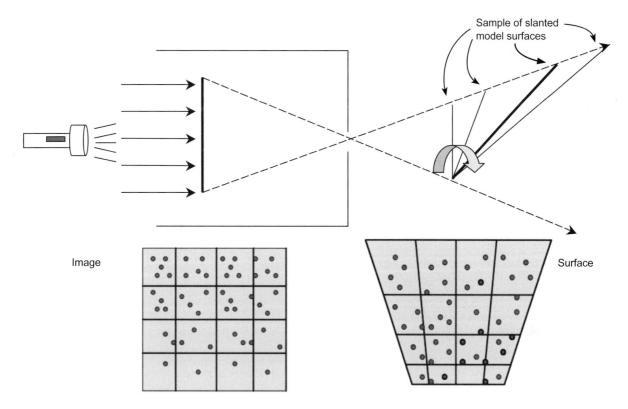

2.17 Measuring texture density inhomogeneity by comparing 16 model surface regions
If the model surface has the correct orientation then the 16 back-projected image regions would have the same texture density. The homogeneous texture on the world surface projects to an inhomogeneous image texture, as indicated by the different amount of texture in each image region. The image texture is inhomogeneous because the world surface is slanted. The back-projection of the image texture on to the model surface will be homogeneous only if the model surface and the world surface have identical orientations.

transparent photograph at the back of the camera, then through the pin-hole, and onto our model surface.

In other words, if the back-projected image on the model surface is identical to the texture on the world surface then it follows that this world surface and the model surface must have the same orientation.

In essence, this is the core of the algorithm. Given an image of a textured world surface, we place a transparent photograph of that image in a projector. We then adjust the orientation of a model surface (a large piece of white card) until the back-projected texture on that model surface appears to be homogeneous. This adjustment involves a degree of "hunting" for the best orientation, but it is not difficult in practice. The important point is that, once we have finished adjusting the model

surface, the orientation of the world surface and our model surface are identical, **2.16**.

Measuring Homogeneity

Rather than waving a piece of card about in front of a projector and guessing when the texture looks homogeneous, we can exploit the basic idea to provide a well-defined algorithm for obtaining the orientation of a textured surface.

First, we divide our image into, say, 16 equal areas, **2.17**. Each image region back-projects to a trapezoidal region on the model surface. Crucially, we know that if the model surface has the correct orientation then the texture density of all of these 16 regions is identical, or almost identical (almost identical because, in practice, textures are usually only approximately homogeneous). Therefore, we need to define a measure of homogeneity that we

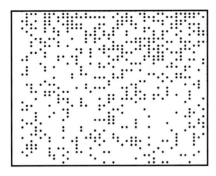

2.18 Image of a slanted plane textured with dots
This does not create a vivid perception of a slanted plane despite there being a strong density gradient cue.

can apply to the back-projected texture on our model surface. In practice, it is easier to work with measures of inhomogeneity, as we will see.

One way to think about inhomogeneity is as the variability in the density of back-projected texture in the 16 regions on our model surface. Clearly, this variability is about zero at the correct orientation because the density of back-projected texture is the same in all 16 regions. At any other orientation, the texture density in the 16 back-projected regions is not the same, and the total variability across these 16 regions is therefore greater than zero.

In order to construct an algorithm we need a formal measure of the variability in texture density. This takes the form of a quantity known as **variance**. The details of how variance is computed need not concern us here; suffice it to say that variance provides a formal measure of inhomogeneity. In essence, as the inhomogeneity of the back-projected texture increases, so the variance of the density of the 16 back-projected regions also increases.

Armed with a formal measure of inhomogeneity, we no longer have to guess "by eye" when the back-projected image on the model surface is homogeneous for any given model surface orientation. However, we still have to wave a card (our model surface) about in front of a projector in order to adjust the model surface orientation. This is clearly unsatisfactory in the age of computers. Fortunately, it is a simple matter, at least in principle, to translate the physical act of card-waving and

back-projection into a computer program that will happily wave its virtual card in front of a virtual projector (for that is exactly what it does) until the virtual cows come home.

Now we have a photograph, a virtual projector, a virtual model surface, and an algorithm for finding the best orientation of this model surface. As it is likely that our photograph is from a digital camera, it is no hardship to place the photograph in the virtual projector. We then press the "go" button on the computer, which starts the program waving its virtual card about. Every time the card's orientation is changed, the variance of the back-projected texture is measured across the 16 regions on the virtual model surface. When the model surface orientation has been adjusted until this variance cannot be reduced any further, the computer beeps and displays two numbers, p and q, on the screen. These numbers define the orientation of the model surface, which represents the best estimate of the original world surface.

We can summarize this procedure within Marr's framework as follows:

Computational Theory
This is the *homogeneity constraint* (equal density of texels over the surface) used previously.

Representation and Algorithm
This specified a number of steps:

1. *Choose an input representation* We again use texels, assuming that a prior process has made this representation available.

2. *Choose a surface orientation representation* We have again chosen to use the (p, q) parameters.

3. *Find p, q using back-projection based on homogeneity*

   a. Place transparent image of a textured world surface in a projector (e.g., at back of pin-hole camera).

   b. Shine light through the transparency, and measure texel density in a number of regions as back-projected on to the model surface.

   c. Adjust orientation of model surface to find the minimum variance in texel densities.

   d. Measure orientation (p, q) of model surface; this is the estimate of the original world surface orientation.

Hardware
Back-projector and a calculator or a computer for calculating variance.

The key point to observe here is the exploitation of the same constraint, *homogeneity*, in a very different algorithm. This illustrates Marr's point about keeping his three levels of discourse distinct.

Human Vision Uses Texel Shape

Having seen how the problem of obtaining surface orientation from texture can be solved in principle, we naturally want to know if the method we have developed bears any relation to the method used by ourselves when we look at a textured surface.

The techniques used to answer this question involve viewing surfaces with carefully controlled textures at different orientations and measuring how much surface slope is seen. To do this properly involves a field of research called ***psychophysics***, Ch 12. There have been many such studies on human texture perception, and we will have more to say about them later. For the moment, we can obtain an indication of the answer to our question using certain demonstration figures with textured surfaces created using computer graphics.

An image of a textured plane is shown in **2.18** where each texel is a dot. The dots on the surface

are homogeneous. If the human visual system makes use of the assumption that surface texture is homogeneous and if it measures texture density in the way described above then we should perceive this figure as a perfectly flat plane with a surface orientation (p, q) of $(0°, 70°)$. In fact, the surface appears to slope much less than this. Indeed, the impression of any 3D slope is very weak indeed.

This simple demonstration stimulus is sufficient to raise serious doubts about whether human vision implements the computational theory we have set out. This is intriguing indeed. Either we don't depend on an assumption of texture homogeneity or dots aren't used as texels for the shape from texture computation—which would be an odd and puzzling limitation. So what strategy might human vision use for the convincing perceptions of slope seen in **2.1, 2.2, 2.3**? Just what texture information is it using?

Figure **2.19a** shows circles on a surface plane. This surface texture is homogeneous (i.e., it has the same amount of texture everywhere). Figure **2.19b** shows ellipses on a plane with the same surface orientation as in **2.19a**. This surface texture

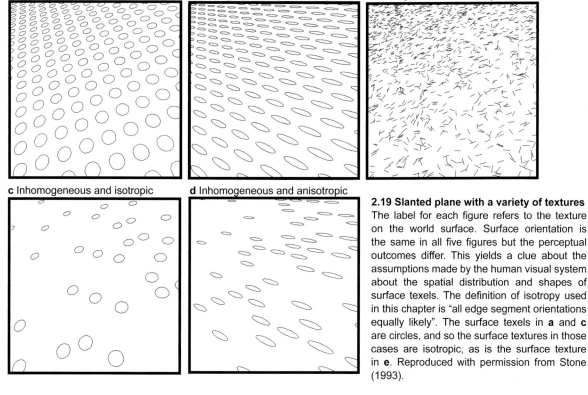

a Homogeneous and isotropic

b Homogeneous and anisotropic

e Homogeneous and isotropic texture

c Inhomogeneous and isotropic

d Inhomogeneous and anisotropic

2.19 Slanted plane with a variety of textures
The label for each figure refers to the texture on the world surface. Surface orientation is the same in all five figures but the perceptual outcomes differ. This yields a clue about the assumptions made by the human visual system about the spatial distribution and shapes of surface texels. The definition of isotropy used in this chapter is "all edge segment orientations equally likely". The surface texels in **a** and **c** are circles, and so the surface textures in those cases are isotropic, as is the surface texture in **e**. Reproduced with permission from Stone (1993).

2.20 Surface covered with circular texels (right)
It is easy to perceive the way the surface orientation varies.

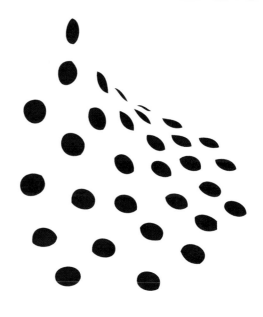

2.21 3D object covered with dots (below)
Its shape is difficult to discern.

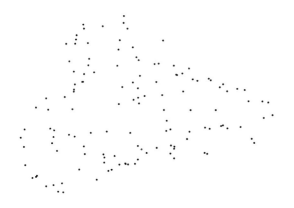

is also homogeneous. Do you see the same surface orientation in these two figures? We do not. This is interesting because they do in fact portray the same surface orientation. This implies that the change in image texture density (the density gradient) within both images is the same in both images. Despite this, the impression of a plane receding in depth is much more compelling in **2.19a** than it is in **2.19b**.

Indeed, the latter does not even appear to be an image of a plane but rather of a surface gently curved in 3D. If our visual systems depend only on homogeneity then both figures would appear as plane surfaces with the same orientation. The fact that both images seem to depict different surface orientations suggests that homogeneity is not the only factor in determining shape from texture. This reinforces the conclusion drawn from **2.18.**

If we do not depend heavily on homogeneity to obtain shape from texture then how do we do it? The answer is that there are several cues based on texture that can, in principle, provide surface orientation. One of these involves the *shape* of texels. Basically, our visual system seems to assume that texels are roughly circular.

A pertinent observation here is that it is quite difficult to look at an ellipse without perceiving it as a circle on a slanted surface, as you can check by

examining the various figures in this chapter containing ellipses. In other words, our visual system seems to assume any ellipse in an image is actually a circle on an oriented surface. It seems probable that the reason we tend to perceive an ellipse as a slanted circle is because our visual systems assume that *image* ellipses correspond to slanted *surface* circles.

The importance for human vision of utilizing information in the shape of texels can explain why we see so little slant in **2.18**. It is because the shape of a dot appears to remain the same irrespective of surface orientation.

As another example on this theme, consider the object shown in **2.20**, which is covered with circular texels. It appears much more solid than the object in **2.21**, which is covered in dot texels. The change in image texel shape resulting from local changes in surface orientation of **2.20** gives a compelling impression of three-dimensional shape. In contrast, local changes in surface orientation have little perceptual effect on the shape of dots in **2.21** so that the only remaining cue to surface orientation is image texel density.

We conclude this chapter by summarizing a different computational theory related to texel shapes that can make use of a back-projection algorithm similar to the foregoing demonstrations. Our main

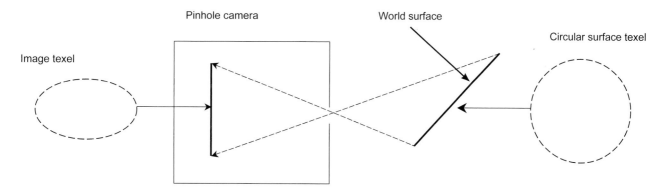

Place transparency of photograph taken above here and adjust slant of model surface until back-projected texels are circles

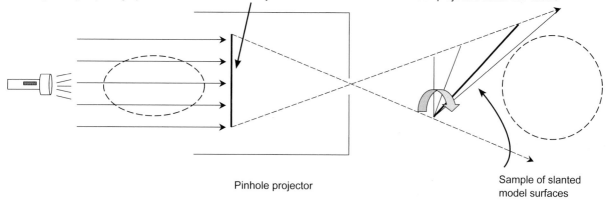

2.22 Using back-projection to find the orientation of a surface with circular texels

goal is to reinforce the main features of the computational framework for studying vision.

This new computational theory is based on the assumption of ***isotropy***, which can be used as a constraint in estimating shape from texture. If a texture is isotropic then it contains equal proportions of all edge orientations. This means that a histogram of the orientations of edge segments within that texture is flat, and is said to have a uniform distribution.

Isotropy is quite a general concept, and we will discuss two examples of it here in terms of surface textures that have: a) circular texels, and b) ***uniform edge orientation histograms*** (also called ***tangent distributions***).

Just as the assumption of homogeneity can be used as a constraint in devising algorithms, so too can the assumption of isotropy.

Shape from Texture Using Circular Texels

This algorithm is analogous to the one developed for homogeneous textures insofar as it depends on back-projection. Given a transparent photograph of a textured world surface, where each texel is a circle, we back-project that photographic image on to a model surface. Using a similar line of reasoning to that developed for homogenous textures, if the orientation of the model surface is identical to the original world surface then the back-projected texture on the model surface also consists of circles, **2.22**. At any other orientation, the back-projected texture on the model surface consists of ellipses, where the amount of squashing of ellipses increases as we move away from the correct orientation.

What might we conclude from this? If the orientation of the model surface is adjusted until the

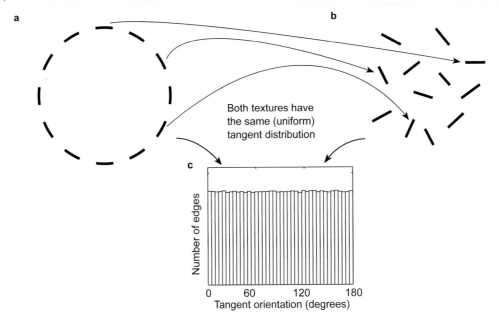

Both textures have
the same (uniform)
tangent distribution

2.23 Tangent distributions of isotropic textures
a The edge elements of a circular texel define a uniform tangent distribution, shown in the histogram in **c**, and this corresponds to an isotropic texture.
b If the edge elements in **a** are placed randomly on the surface (whilst preserving the orientation of each edge) then they still yield the uniform tangent distribution shown in **c**, and also correspond to an isotropic texture. Thus, an isotropic texture can consist either of texels that are circles or texels that are scattered edge elements which occur at all edge orientations with equal probability.

back-projected texture consists of circles then the orientation of the model surface and the orientation of the original world surface must the same.

In terms of Marr's framework, the computational problem consists in finding surface orientation from texture. However, the constraint now being exploited depends on texel shape and not on texel density gradients. Thus, instead of adjusting a model surface until the back-projected texture has homogeneous density, the new algorithm adjusts the model surface until the back-projected texels are circular, **2.22**, which corresponds to an isotropic texture. As above, the algorithm could be implemented physically by waving a card about in front of a projector, or by a computer program.

Developing this algorithm in sufficient detail to allow it to be implemented as a computer program requires specifying how to measure whether a back-projected surface texel is circular. There are various ways this could be done. For example, one could locate a texel's center (there are image processing methods for doing this) and then measure the lengths of a sample of lines of varying

orientations from the center to the texel's boundary. If they are all the same length then they can be assumed to be the radii of a circle. Hence this algorithm would vary the slant of the model surface until an orientation is found at which the lengths of these hypothesized radii of texels are all similar. We will not go further into the details of this possible algorithm because it turns out that there is a more generally applicable algorithm for exploiting the isotropy constraint.

Shape from Isotropic Surface Texels

It is unlikely to have escaped your notice that surfaces are not always, or even often, marked with roughly circular texels. More typically, surface textures consist of imprecise patterns, **2.3**, or no pattern at all, **2.6**.

However, as Andrew Witkin showed in 1981 (see also Blake and Marinos, 1990), we can still make use of the isotropy constraint for such textures because the assumption that a texture consists of circular texels is actually a specific version of a more general assumption. If we consider a circular

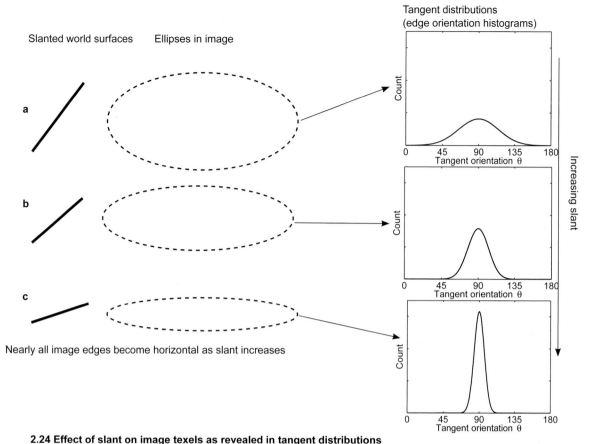

Slanted world surfaces Ellipses in image

Tangent distributions
(edge orientation histograms)

a

b

c

Increasing slant

Nearly all image edges become horizontal as slant increases

2.24 Effect of slant on image texels as revealed in tangent distributions
As slant increases (left column) circular surface texels project to increasingly "squashed" image ellipses (middle column). For such ellipses, the distribution of tangent orientations becomes narrower (less spread out) as a surface slant is increased (right column).

surface texel as made up of a series of discrete line elements or edges, **2.23a**, then all edge orientations are equally represented in the texel. This is revealed by plotting a histogram of edge (or more properly *tangent*) orientations, **2.23c**. It is uniform (flat) because all tangents are equally represented in a circular texel.

Taking this idea further, we can consider what would happen if we begin with circular texels but we dismantle each texel into a set of oriented edges. We then scatter these edge segments on the surface (taking care to preserve the orientation of each edge element), **2.23b**. The resultant histogram, or *tangent distribution* as it is called, of these scattered edges remains the same, **2.23c**. This is neat because it shows we do not have to rely on circular texels being present on the surface. All we need to assume is that the surface texture has an

isotropic distribution (i.e., a uniform edge tangent distribution) regardless of whether the texels are circles or edges. Thus, circular texels are just one very specific (and physically improbable) way of having an isotropic texture.

Now, we know that a circular texel projects to an elliptical image texel, and that a circular texel has a uniform tangent distribution. In contrast, if a circular texel is squashed from top to bottom to become an ellipse then there is a bias toward horizontal edge orientations. Accordingly, if we plot, as in **2.24**, a histogram of the edge orientations of an elliptical image texel then it has a bump in the middle because of the bias in the distribution of edge orientations, **2.24a**. As the surface slope increases, so each image texel becomes progressively more squashed, and its edge orientations become more biased, **2.24b,c**. Crucially, **2.24** shows that

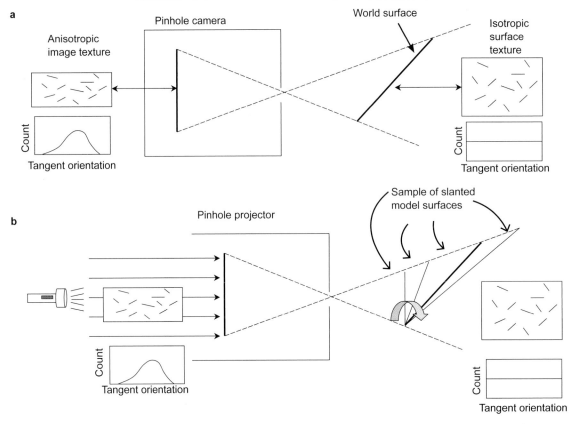

2.25 Searching for a model surface slant that yields a uniform distribution of tangents on the model surface
a Capturing an image of surface with an isotropic texture and associated uniform tangent distribution.
b Using back-projection to find a model surface slant with a uniform tangent orientation distribution, which implies an isotropic back-projected texture.

the amount of bias increases systematically as the surface slope increases. It is therefore possible to use the amount of bias in image edge orientation to estimate surface orientation.

We can exploit edge orientation histograms to measure isotropy, or its converse, ***anisotropy***, in a back-projection algorithm. As before, we begin by capturing an image of an isotropic textured surface at an unknown orientation, **2.25a**. We then place a transparency of this image in our pin-hole projector, **2.25b**. Next, we adjust the orientation of our model surface until the back-projected texture on that model surface is isotropic, at which point we conclude that the orientation of the world surface and our model surface are identical.

Rather than estimating when the texture on the model surface is isotropic "by eye" we can define a formal measure of isotropy, and then use this as an objective measure. Given that an isotropic

texture has a uniform edge orientation histogram, we could use any one of several formal measures of isotropy. As discussed above, a uniform histogram implies an isotropic texture. Accordingly, a measure of histogram uniformity would provide a measure of isotropy. One such measure is known as ***entropy***, which is largest in the case of a uniform histogram (i.e., an isotropic texture). The details of how entropy is defined need not concern us here, suffice it to say that a reliable measure of the uniformity of histograms exists. In a sense, because we cannot measure isotropy directly, we are using entropy as a proxy for isotropy. Using proxy measures is "standard practice" within the vision literature, although of course a good argument is needed to justify them in any particular case.

Finally, we can translate this rather cumbersome physical analogue computer into an algorithm that can be run as a digital computer program.

Given a digital image of a textured surface, this program will automatically search for an orientation of its virtual model surface that produces an isotropic texture. Specifically, at each orientation, the image edges that are back-projected onto the model surface are used to form a histogram, and the isotropy of the back-projected texture is measured as the entropy of this histogram. The model surface orientation is adjusted until the entropy of the back-projected texture cannot be increased any more. At this point, the texture on the model surface is maximally isotropic; the program stops and prints out the orientation of the model surface as two parameters (p, q) on the computer screen.

Summarizing this using Marr's levels, we get:

Computational Theory

This is the *isotropy constraint,* the assumption that the surface texture has a uniform edge tangent distribution.

Representation and Algorithm

1. *Choose an input representation.* Measures of the orientations of edge segments of surface markings.

2. *Choose a surface orientation representation.* We have again chosen to use the (p, q) parameters.

3. *Find (p, q) using back-projection based on isotropy*

a. Make a histogram of edge segment orientations.

b. Adjust a model representation of surface orientation in a computer until the histogram of edge orientations is flat (highest entropy). This is back-projection modelled on a computer and is equivalent to shining light through a transparency of an image of the world surface.

c. Accept (p, q) of the highest entropy model surface as an estimate of the orientation of the world surface.

Hardware

A computer for modelling back-projection and measuring flatness of edge histogram.

A cautionary note. We should point out one apparently fatal, but in practice unimportant, problem for methods based on texel shape. If the image is formed on a plane (as it is in the back of a typical camera) then these methods work only if texels are close to the image center unless appropriate allowance is made for a flat image plane. This is because forming an image on a flat image plane introduces shape distortions that become marked in the peripheral image regions which lie at an appreciably different distance from the lens than the central regions. As far as human vision is concerned, we tend to look directly at surfaces. This ensures that surface texels project to the center of the retinal image. Additionally, the retinal surface is approximately spherical, which offsets the image distortion factor because all parts of the retina are roughly equidistant from the center of the lens.

Isotropy Considered within Marr's Framework

In terms of Marr's framework, the isotropy method addresses the same computational problem as the homogeneity method: this problem is finding surface orientation from texture. However, the constraint identified at the computational theory level is that textures are usually isotropic; that is, the distribution of edge orientations on the world surface is roughly flat.

At the algorithmic level, the new procedure we have described exploits this constraint. Just as for the homogeneity method, it could use either the (p, q) representation for surface orientation or (slant, tilt).

We have outlined two different isotropy algorithms that feed off the same constraint identified at the computational theory. The first was restricted to circular texels. The second was of much wider scope because it is defined isotropy in terms of tangent distributions, not in terms of the shape of surface texels. Thus, instead of adjusting a model surface until the back-projected texels are circles, the second algorithm adjusts the model surface until the back-projected image edges form a uniform (flat) tangent distribution on the model surface.

How about Human Seeing?

A study by Rosenholtz and Malik in 1997 investigated whether our visual systems make use of texture gradients or texture shape as a cue to surface orientation. In their experiment, a texture gradient was defined in terms of texture homogeneity, whereas texture shape was defined in terms of isotropy. Without going into details, the authors found that for humans perceived orientation of textured surfaces depends on both homogeneity and isotropy. Thus, the brain depends on both gradients of texture as well as texture shape (in the image) in estimating surface orientation.

It is perhaps not surprising that human vision, during its evolution over millions of years, has found a variety of ways of exploiting image texture cues for recovering surface orientation. What we have learned in this chapter is the value of understanding what human vision is doing by examining very carefully the image information available, thereby identifying valid constraints, and then thinking how those constraints can be exploited in suitable algorithms.

It would be nice to round off this chapter with some account of the brainware mediating shape from texture. Sadly, nothing much is known about this at present. We can however look forward to future studies that will reveal what neuronal structures and circuits implement the methods tested psychophysically by Rosenholtz and Malik, and the many others who have studied the use of texture cues in human seeing.

Translating (*p*,*q*) to (slant, tilt)

For simplicity, we have been using p to refer to the angle of rotation α (alpha) around the Y-axis, but this is more commonly defined as $p = \tan(\alpha)$. Similarly, we have been using q to refer to the angle of rotation β (beta) around the X-axis, but this is more commonly defined as $q = \tan(\beta)$. Using these definitions, the equations for translating from (p, q) to (slant, tilt) are slant $= \operatorname{atan}(\sqrt{(p^2 + q^2)})$ and tilt $= \operatorname{atan}(p/q)$. In fact, if we set p to zero (which can be done by rotating the image around the optic axis) then the complicated relation between (p, q) and (slant, tilt) simplifies a great deal. Thus, setting $p = 0$ yields slant $= q$ and tilt $= 0$. Hence if $p = 0$ then tilt $= 0$ and slant is given by the angle of rotation around the X-axis. For this reason, we have used planes with $p = 0$ in this book.

Shape from Texture, Not Depth from Texture

You may be struck by the fact that the title of this chapter is "shape from texture", and not "depth from texture." This is a subtle, but important, distinction. Shape is defined by the changing local orientation of points on an object's surface, as in **2.16**. In contrast, depth refers to the distance of points on an object from the eye (as measured

along optic axis of the eye). Both texel density and texel shape can be used to find the orientation of a plane, and texel shape can also be used to find surface shape. However, neither can be used, in principle, to find surface depth. For example, the orientation of the surface shown in **2.19a** could be the picture of a carpet with large patterns, or a sketch of hand-held bubble wrap; you have no way of knowing which it is. If it is carpet then the surface has large texels that are far away, whereas if it is bubble wrap then it has small texels that are close to the eye.

Summary Remarks

This chapter has been concerned with a particular computational problem: recovering 3D surface orientation from texture. We have treated this as a problem that can be usefully considered on its own. That is, we have not found it necessary to delve into color perception or motion perception, or object recognition. And we have arrived at a solution that can be regarded as a ***vision module***.

Regarding vision as being carried out by a series of computations within the visual system, each one performed by an independent vision module, is an important aspect of Marr's approach. Each module needs to be based on a good computational theory. It needs to be precise about the algorithmic requirements of clear specifications for input and output representations. It also needs to be clear about the processing steps undertaken to get from the input data to the output.

The job of the computational theory is to provide a well-founded method for solving a computational problem, using knowledge derived from the nature of the task and/or the nature of the world being viewed. This knowledge leads to the identification of useful assumptions or ***constraints***.

It is important to recognize that the above scheme for studying vision makes no mention of the physical structures, such as parts of the human brain or the components of a computer that might actually carry out the tasks of any given module. These structures are the hardware, or neurons in the case of biological vision systems. It we want to understand any particular vision system completely, be it man-made or biological, then finding out how its hardware implements a given computational theory is essential. But the central tenet of the computational approach is that studying hardware is necessary but not sufficient. Studying only hardware neglects the crucially important requirement of understanding the nature of the task that the hardware is carrying out. Discovering or devising well-founded methods for solving visual processing tasks is the key feature of the computational approach.

It is worth repeating here that the three levels of conceptual analysis of a vision task—computational theory, algorithm, and hardware—are linked, but often only loosely. Confusion can await those who mix up the concerns of each level.

Also, it is worth noting that the term *computational* has the unfortunate drawback that it misleads some people into thinking this approach has something uniquely to do with computers. To be sure, computers are typically used to check the viability of any proposed solution to a seeing problem. Doing so has the merit that it exposes, ruthlessly, flaws in the logic of a proposed solution or critical limitations deriving from the constraints being exploited. But we emphasize again the remarks made earlier that computational theories and algorithms based on them can in principle be implemented using paper and pencil.

This book will pursue the computational theme in tackling diverse aspects of seeing. Indeed, we have chosen from the vast literature on vision a set of topics to which this approach has been applied, while trying not to neglect basic facts about seeing. The main goal of this chapter has been to illustrate the main features of the computational approach in order to lay a firm foundation for what is to come.

Ecological Optics and the Work of J. J. Gibson

The first person to examine in great detail the fact that texture gradients comprise a rich source of visual information about 3D scene structure was James Gibson, an American psychologist working in the middle of the last century (1904–1979). He coined the term **ecological optics** for the enterprise of studying the information contained in the **optic array** about the visual world in which we live. This is perhaps his most enduring contribution. He initially defined the term optic array as the collection of light rays that enter the eyes. He later changed this to a nested set of solid angles, each solid angle

corresponding to one of the large or small facets of the environment.

Some eminent vision scientists regard Gibson as a genius. However, what he failed to do was to develop computational theories arising from the information he identified in textures.

One of the terms coined by Gibson was **direct perception**, which was the term he used for the claim that the visual system simply *resonated* to the information in texture gradients. To many, including ourselves, this is an obscure notion which is best regarded as a piece of the history of vision science, rather than a useful idea on which to build current research.

In the age of computers and computer vision it is possible to be very precise about how useful information can be extracted from images, and it is the computational approach that we pursue in this book.

But a cautionary remark is in order: the computational approach is regarded by some as a false turning. For some (not Gibson, incidentally) the "truth" about seeing will be found by studying the brain structures that carry it out. We are very interested in the brain structures that support seeing, as many chapters of this book make clear. But we agree with Marr—understanding brain structures must include an understanding the nature of the computational problems entailed in seeing. The strong claim is: only when we understand how those problems can be solved *in principle* will we fully grasp why brain structures have the properties that they do.

Further Reading

There is a large literature on the psychophysics of human texture perception but rather little of it is directly relevant to the topics we discuss in this chapter.

Aloimonos J (1988) Shape from texture. *Biological Cybernetics* **58** 345–360. *Comment* Explains how to find shape from Texture based on the assumption of homogeneity.

Blake A and Marinos C (1990) Shape from texture: Estimation, isotropy and moments. *Artificial Intelligence* **45** 323–380. *Comment* A highly technical account of shape from texture based on isotropy.

Gibson JJ (1960) *The Perception of the Visual World.* Boston: Houghton Mifflin. *Comment* A classic work mainly of historical interest, which appears controversial and/or dated from a modern, computational perspective.

Kubovy M (1986) *The Psychology of Perspective and Renaissance Art.* Cambridge University Press. *Comment* A delightful book. Out of print but available on Kubovy's website: http://people.virginia.edu/~mk9y/home.html

Loh A (2004) Available at http://www.csse.uwa.edu.au/~angie/

Marr D (1982) *Vision: A Computational Investigation into the Human Representation and Processing of Visual Information.* New York: Freeman. *Comment* This book has had a massive impact over the past two decades on vision research. Its main contribution has been to expound the computational framework for studying vision. This makes it one of the few books on seeing that goes beyond descriptions of the phenomenology of visual perception and/or brain mechanisms for seeing. However, Marr's work was cut short prematurely (he died at the age of 35 from leukemia). Parts to concentrate on regarding the content of this chapter concern his exposition of the three-levels framework for studying seeing, pp. 19–31. The *Epilogue* is also well worth reading as it provides an illuminating imaginary conversation between Marr and an interviewer who asks for clarifications and who puts forward criticisms of Marr's approach for his reply.

Rosenholtz R and Malik J (1997) Surface orientation from texture: Isotropy or Homogeneity (or Both?). *Vision Research* 37 2283–2293. http://www.cs.berkeley.edu/~malik/papers/rosen-malik97.pdf *Comment* One of the few papers to investigate the relative roles of isotropy and homogeneity in human perception of surface orientation. The authors conclude that both cues are used in human perception.

Stone JV and Isard S (1995) Adaptive scale filtering: a general method for obtaining shape from texture. *IEEE Transactions Pattern Analysis and Machine Intelligence* 17(7) 713–718. *Comment* Technically demanding, but one of the few papers on how to identify texels which vary in size across an image of a textured surface in order to obtain shape from texture based on homogeneity.

Stone JV (1993) Shape from local and global analysis of texture. *Transactions Royal Society London (B)*, **339**(1287) 53–65. *Comment* Discusses the difference between methods which rely on image–wide (global) measures of texture such as homogeneity, and those that rely on localized measures of texture, such texel shape and isotropy.

Witkin AP (1981) Recovering surface shape and orientation from texture. *Artificial Intelligence* **17** 17–45. *Comment* One of the first papers to formulate a vision problem in terms of Bayesian estimation, an approach to vision problems that we discuss in detail in later chapters. A technically demanding, but highly readable, paper of shape from texture based on isotropy.

3

Seeing with Receptive Fields

3.1 Jumping spider
Notice the large forward-pointing eyes. These are used in hunting prey and finding mates—see overleaf.
Courtesy Thomas Shahan (http://www.flickr.com/photos/7539598@N04/).

Jumping spiders (Family *Salticidae*) have remarkable visual abilities. They have no fewer than four pairs of eyes, the two biggest being forward-pointing at the front of the head, **3.1**. These eyes all have single lenses like mammalian eyes, unlike the insects that have compound eyes in which each receptor has its own separate lens.

Jumping spiders have been called the tigers among spiders. They hunt their prey rather than using webs. They scan the scene using body movements and by moving the retinae of their forward-pointing eyes under the lenses which are themselves fixed in position in the spider's carapace, a rigid material which acts as an external skeleton. These retinal movements are roughly equivalent to our own eye movements. Insect prey can be detected from about 30-40 cm. If prey is detected the spider slowly advances. When near enough, it jumps and grabs the prey with its jaws.

Jumping spiders are thus experts at recognizing prey using vision. But how do they distinguish between prey and fellow spiders with whom to mate?

Drees explored this issue in 1952 by testing spiders with a variety of patterns, **3.2**. Michael Land (1969b) notes that the spiders will attack any small dark object that has just moved unless the object possesses a pair of oblique lines ("legs") on either side of a central spot (the "body"). Such an object is treated as another jumping spider, and will elicit a courtship display from a male spider (see Land and Nilsson, 2001).

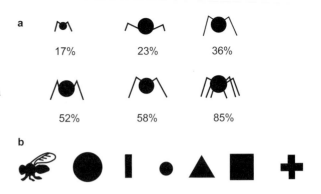

3.2 Pattern recognition by jumping spiders
(*Epiblemum scenicum*)
a Potential mates The numbers are the percentage of trials on which courtship was evoked.
b Potential prey These stimuli evoked prey capture.
Redrawn with permission from Drees (1952).

This leads naturally to the idea that spiders use certain distinguishing *features* to discriminate between potential mates and potential prey. For example, it seems they can detect *bar-shaped features* similar to the legs of fellow spiders. But how do they do this? And how do we do it, because human vision is also certainly good at detecting bar features? In this chapter, we discuss problems to be solved in designing a decent bar detector. This is a good topic to explain many core ideas in the task of edge detection, which is pursued in detail in Chs 5 and 6.

Why read this chapter?
The concept of a *receptive field* is explained by considering the task of trying to build a *template detector* for a bar-shaped image feature. The orientation tuning of the receptive fields of *simple cells* found in the striate cortex is then described. The contrast vs. orientation ambiguity in the responses of these cells is examined and resolved in a model that computes edge orientation from a range of 18 or so *coarsely tuned* populations of simple cells called **channels**. The model utilizes a *weighted-average* algorithm. The model can account for where peaks are found in human sensitivity to changes in stimulus orientation. The combinatorial explosion that arises in the number of cells required for template-based pattern recognition is briefly introduced, leaving Ch 11 on **complexity theory** to deal with the details.

3.3 Jumping spider retinae overlaid on spider image
The white dots in each boomerang-shaped retina symbolize a set of receptors arranged in a bar-shaped pattern. The spider aligns its two retinae with leg-shaped bar features in the input. Adapted from Drees (1952) using photo in **3.1**.

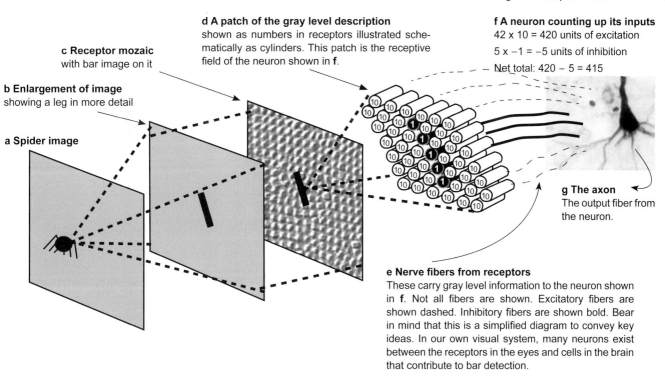

d A patch of the gray level description shown as numbers in receptors illustrated schematically as cylinders. This patch is the receptive field of the neuron shown in **f**.

c Receptor mozaic with bar image on it

b Enlargement of image showing a leg in more detail

a Spider image

f A neuron counting up its inputs
42 x 10 = 420 units of excitation
5 x −1 = −5 units of inhibition
Net total: 420 − 5 = 415

g The axon The output fiber from the neuron.

e Nerve fibers from receptors
These carry gray level information to the neuron shown in **f**. Not all fibers are shown. Excitatory fibers are shown dashed. Inhibitory fibers are shown bold. Bear in mind that this is a simplified diagram to convey key ideas. In our own visual system, many neurons exist between the receptors in the eyes and cells in the brain that contribute to bar detection.

3.4 Designing a bar detector using a template implemented as a receptive field of the off-center type

Recognition using Templates

As stated above, the jumping spider has eight eyes, each of which is similar in design to the human eye. Six of these eyes are at the sides of the head and they have fish-eye like lenses, giving them a wide field of view. The two forward-looking eyes have very high resolution (almost as good as ours). The jumping spider uses its low-resolution, wide field-of-view eyes to detect the presence of moving objects and prey, and then it quickly orients its body to point its high-resolution eyes toward the object of interest. This is akin to humans foveating an object.

The spider also shifts the position of each retina of these eyes under the fixed lenses. Michael Land has shown that these scanning movements align the boomerang-shaped retinae with the images of leg-shaped bar features, **3.3**. This retinal alignment, coupled with the odd shapes of these retinae, suggest that perhaps the jumping spider uses a form of image processing known as ***template matching*** to recognize bar-shaped features. We will examine this using the simple case of the image of a dark bar on

a lighter ground focused on the kind of receptor mosaic found in mammalian retinae, **1.2**.

The dark bar image in **3.4b** leads to a gray level description (Ch 1) of the kind illustrated schematically by the patch of receptors in **3.4c**. Each receptor is represented in **3.4d** as a cylinder, and the number in each one shows its activity level, its ***gray level***. The lighter background is coded by larger numbers (e.g., 10 in **3.4**) than those of the dark bar (e.g., 1 in **3.4**).

Now imagine nerve fibers connecting all the receptors in this patch of retina to a neuron, as in **3.4e**. (Take another look at **1.3** if you have forgotten what neurons look like.) These connections are of two sorts. The ***excitatory*** ones (shown dashed in **3.4e**) pass messages from receptors that try to activate the receiving neuron. That is, they try to make the neuron fire a stream of ***action potentials*** down its output fiber, the ***axon*** (**1.3**). Another way of saying this is that the signals sent via the excitatory fibers are passed on with a ***positive weighting***. This is a good term because if the gray level value is high then more excitatory effect is generated than if it is low. The excitatory effect is caused by the terminations of these fibers on the receiving neuron releasing **excitatory chemical transmitters**.

Inhibitory connections (shown bold in **3.4e**) do the opposite. Their terminals deliver inhibitory chemical transmitters that try to switch the neuron off, so their messages have a **negative weighting**.

All this implies that the neuron is rather like a ballot box. It receives votes (some excitatory, some inhibitory) and counts them up to arrive at a total. If the total exceeds a certain **threshold** then the neuron is switched on. The threshold defines a limit that, if exceeded, leads to the neuron sending action potentials down its axon.

The patch of receptors sending messages to the neuron is called its **receptive field**. This is an appropriate name because it nicely captures the idea of the neuron having a "field of view" from which it receives messages. For illustration in **3.4d**, the receptive field is shown as a patch of 47 receptors within the entire receptor mosaic. Five of these receptors are linked to where the bar feature falls in the image. From these receptors the neuron receives 5 messages of the inhibitory kind, yielding $5 \times 1 = 5$ inhibitory votes. The 1s reflect the low gray levels coded in these receptors for the dark-bar image shown in **3.4**. These receptors have low gray levels because the dark bar in the scene reflects little light. As negative votes implicitly carry a negative weighting, the inhibitory total is –5 votes.

The remaining 42 receptors in the receptive field are in the excitatory region surrounding the central bar-shaped inhibitory region. The excitatory votes (sent along the dashed fibers in **3.4e**) give a total of $42 \times 10 = +420$. The gray levels of 10 code the relatively bright part of the image surrounding the dark bar.

Adding up the two sorts of votes, we get $+420 + (–5) = +415$ votes. If we set the threshold for the neuron becoming active at, say, +400 then this neuron will become excited because its total input of +415 exceeds the threshold. The neurons will therefore generate a stream of impulses to the brain, indicating *bar present*. The general idea is much like the burglar detector described in Ch 1, but here the neuron is operating as a switch-like mechanism for *bar present* or *bar absent*. Notice

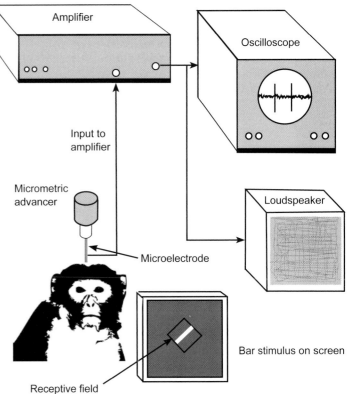

Amplifier

Oscilloscope

Input to amplifier

Micrometric advancer

Loudspeaker

Microelectrode

Bar stimulus on screen

Receptive field

3.5 Single cell recording
The primate looks alert in the picture, and indeed fully conscious animals are sometimes used (the brain has no pain receptors). However, the animal is usually anaesthetized to achieve complete immobilization. This helps control accurately where the eyes are looking.

that the neuron is coding a decision regarding presence/absence of a *pattern*, and not just an image darkening as for the burglar detector when it was triggered by a shadow.

What happens if there isn't a bar image present and the whole of the receptive field receives light of equally bright intensity? The five receptors in the bar-shaped inhibitory region will then respond with levels of 10 just like those in the surround of **3.4**. This leads to an inhibitory total of $–5 \times 10 = –50$ being sent to the neuron. Subtracting these from the excitatory votes from the surround we get $+420 – 50 = +370$. This total falls below the threshold level of +400, so the neuron remains quiet. Hey presto! We now have a neuron that seems to be a bar detector, which is what we set out to design. Just how good this bar detector is will be examined carefully in this chapter.

This interplay of excitation and inhibition is a neat trick. Initially you may have wondered why the inhibition was necessary but you should now appreciate that without it the bar detector would have fired saying "bar present" for just a uniform bright image patch. That would of course be hopeless for a *pattern* detector.

This example is of a form of image processing called template matching. The key idea is to encode in the seeing system a pattern (the template) whose shape directly matches the input pattern to be detected. In the example we have considered, the encoded pattern is made up of excitatory and inhibitory regions in the receptive field of the neuron. Hence,the latter acts as kind of on/off switch signaling *pattern present* or *pattern absent* in the receptive field. In the spider's case, if a bar is detected then this signal would be used as a trigger for a courtship display, not an attack. That is, the activated bar detecting neuron in a male spider would be used as a signal for potential mate, not potential meal.

Receptive Fields in Monkeys and Man

Do "bar detecting" neurons of the kind shown in **3.4** actually exist in biological visual systems? Neurons with this kind of receptive field have indeed been found using a neurophysiological technique called *single cell recording*, **3.5**. It has been used for many creatures—crabs, fish, frogs, birds, rats, cats as well as monkeys. It has also been used with human patients being operated on for brain abnormalities. However, whether the neurons that have been found with bar-shaped receptive fields really are "bar detectors" is a crucial issue that we discuss in detail after having described their properties.

In **3.5** a monkey is looking at a screen on which a variety of stimuli can be displayed. A delicate probe called a *microelectrode* is inserted with great precision into a chosen part of the visual system. This electrode, usually a thin wire insulated except at its tip, picks up the tiny electrical changes which take place around active nerve cells. These minute voltage changes (of about 100 millivolts) are amplified and displayed on the screen of a computer system. As explained in **1.4**, an active neuron emits a series of pulses called action potentials. These appear on the screen as vertical *spikes*. If a cell is responding strongly (i.e., excitation outweighs inhibition) then a fast and prolonged series of spikes is seen on the screen. If the cell's excitatory inputs

are balanced by inhibitory inputs then just a few spikes are seen—this is the neuron's *maintained rate* of firing (sometimes also called its *baseline* or *resting discharge rate*). And if the cell is strongly inhibited then no spikes whatever appear on the screen. The pluses from the amplifier are also sent to a loudspeaker so that the neurophysiologist can literally listen to the brain's chattering language of action potentials. What is heard from an active neuron is a series of clicks, each click caused by an individual action potential. The firing rate of a neuron is usually measured as *impulses per second*.

The general strategy underlying the technique of single cell recording is to advance the microelectrode very gradually until activity from a neuron is detected. Having impulses made audible via the loudspeaker is very useful because it frees the experimenter from constantly having to observe the screen. Once having picked up a neuron's activity with the microelectrode probe, the experimenter then proceeds to try out a range of visual stimuli on the stimulus screen in an attempt to discover what images excite the cell and what images inhibit it.

Recording from Single Cells in Striate Cortex

We now describe the properties of cells in the *striate cortex* discovered using the single cell recording technique The striate cortex is a part of the brains of mammals where many fibers from the retina arrive, **1.30**.

The first important property of striate cells is that each is concerned with a limited patch of retina, just as for the bar detector in **3.4**. So (as described above), each cell has a receptive field: the patch of retina that constitutes its own "window on the world." In the experiment shown in **3.5**, the receptive field of the cell being recorded can be drawn on the screen because the monkey's eyes are immobilized. If the animal's eyes were allowed to move about (as is permitted in some experiments) then the receptive field of the cell would be swept successively over different regions of the screen.

Many different kinds of cells have been found in the striate cortex. They are classified using the characteristics of their receptive fields. We deal in this chapter only with so-called *simple cells*. This term was invented by the discoverers of these cells' properties, David Hubel and Torsten Wiesel, who won a Nobel Prize in 1981 for their work on information processing in the visual system. They chose the label *simple* because the receptive fields of these cells can be mapped fairly simply into *excitatory* and *inhibitory* sub-regions using stationary spots of light flashed on the receptive field. This was not true for other types of cells, which they therefore called **complex cells** (we describe these in Ch 9).

[*Terminological note*: The definition of receptive field is the patch of retina that influences the output of the cell in question. Strictly speaking, the receptive field is not defined in terms of its pattern of weights. Nevertheless, sometimes (as in this book) the term receptive field is used interchangeably with a pattern of weights. When used to refer to a pattern of weights, the receptive field concept is similar to *template, operator, mask,* or *weighting function* used in the computer vision literature.]

In some areas of a simple cell's receptive field a stationary spot causes the cell to become excited and emit a small burst of impulses. Conversely, if the spot is flashed in other regions then the cell becomes inhibited, i.e., it stops emitting impulses, or at least its firing rate falls below its resting discharge rate. In practice, many simple cells have very low, even zero, resting discharge rates. Thus, if we wish to demonstrate the presence of an inhibitory region in the receptive field then it may be necessary first to excite the cell by flashing a stimulus on the excitatory zone of its field, and then flash an additional spot in the inhibitory zone. The latter flash reveals its inhibitory effect by reducing the ongoing rate of firing caused by the excitatory stimulus.

The question now is: what are the shapes of the excitatory and inhibitory regions of the receptive fields of simple cells? This can be answered by flashing a small spot of light at different locations within the receptive field. A record is kept of the excitatory or inhibitory effect of the spot at each location by marking plus signs and minus signs, respectively, on the screen. Some typical receptive field maps of simple cells are shown in **3.6a**. Observe that each field is divided into two sub-regions, one excitatory and one inhibitory. This is shown by the plus and minus symbols, a plus symbol indicating where a spot of light caused the cell to become active, and a minus symbol where the spot caused inhibition.

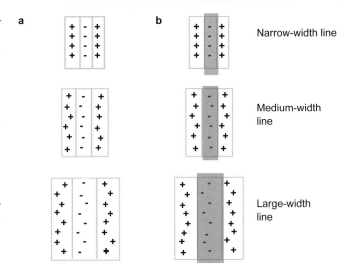

3.6 Receptive fields of line detector simple cells
(Drawn schematically). These are of a type sometimes called (misleadingly, see text) **line detectors**.
a Fields plotted using spots of light. The symbol + shows a place where the spot excited the cell, and the symbol - shows where the spot inhibited the cell.
b The optimal stimulus for these fields is a dark bar on a light ground as this ensures bright light falls on the excitatory zones and relatively little light falls on the inhibitory zones.

Moving on from single spots, we can now ask: what *pattern* of light would **optimally** excite these cells? The answer is: bar-shaped patterns, as shown in **3.6b**. Indeed, the receptive fields in **3.6**

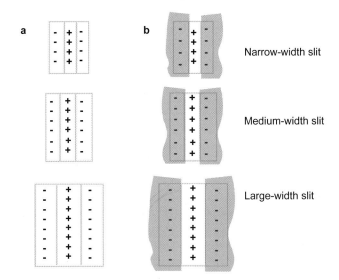

3.7 Simple cells whose optimal stimuli are "slits"
a Receptive fields plotted with spots of light.
b The optimal stimuli are suitably positioned slits that ensure the excitatory (+) zone is stimulated with light and the inhibitory (-) zone receives little light.

Receptive fields plotted with small spots of light

Optimal stimuli

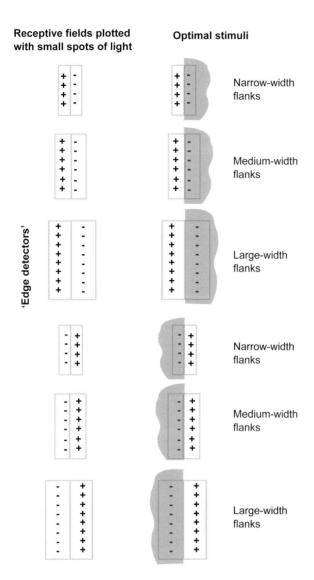

Narrow-width flanks

Medium-width flanks

Large-width flanks

Narrow-width flanks

Medium-width flanks

Large-width flanks

3.8 Simple cells whose preferred stimuli are vertical edges of various sizes

resemble quite closely the kind of receptive fields we invented when trying to design a template for a bar detector. To be sure, the receptive fields illustrated in **3.6** are maximally stimulated by vertical bar-shaped patterns of light rather than by the oblique bar in **3.4**. But other simple cells with obliquely oriented receptive fields are commonplace, so this is an unimportant difference. Note that it is the orientation of the boundaries between the excitatory and inhibitory regions that defines the *orientation tuning* (see below) of the cell in question.

Different Types of Simple Cell

We will explain orientation tuning more fully shortly, but first notice from **3.7** another way in which simple cells differ from each other. While some respond best to a dark bar on a light background, **3.6**, others respond best to a *light* bar on a *dark* surround, **3.7**. This is because the excitatory and inhibitory regions in **3.7** are swapped with respect to those in **3.6**.

Hubel and Wiesel called these cells *slit detectors*. Yet other types of simple cells are tuned neither to bars nor slits but to dark/light edges or light/dark edges as shown in **3.8**.

Hubel and Wiesel came upon the discovery that lines could be very effective stimuli for some simple cells by accident, as the following quotation from Hubel (1988) explains:

The position of the microelectrode tip was unusually stable, so that we were able to listen to one cell for about nine hours. We tried everything short of standing on our heads to get it to fire. After some hours we began to have a vague feeling that shining light on one particular part of the retina was evoking some response, so we tried concentrating our efforts there. To stimulate, we used mostly white circular spots and black spots. For black spots we would take a 1 by 2 inch glass laboratory slide, onto which we glued a black spot, and shove it into a slot in an optical instrument to project onto the retina. For white spots, we used a slide made of brass with a hole drilled in it. After about five hours of struggle, we suddenly had the impression that the glass with the dot was occasionally producing a response, but the response seemed to have little to do with the dot. Eventually, we caught on: it was the sharp but faint shadow cast by the glass as we slid it into the slot that was doing the trick. Most amazing was the contrast between the machine-gun discharge when the orientation of the stimulus was just right, and the utter lack of a response if we changed the orientation or simply shined a bright flashlight into the cat's eyes.

So Hubel and Wiesel's landmark discovery was that each simple cell is highly selective for the orientation of a stimulus edge or line falling in its receptive field. Even though such a cell can be weakly excited by a single spot, it is maximally excited when presented with a suitably oriented edge.

This is an example of serendipity in science: pursuing lucky events from unplanned experiments.

Of course, the investigators must have the acumen and open-mindedness to realize the importance of the chance results that have come their way. This is also a good example of how "*Chance favours the prepared mind*," as noted by Louis Pasteur (1822–1895): "*La chance ne sourit qu'aux esprits bien préparés*." He famously took advantage of serendipity in laying the scientific foundations for the first vaccine for rabies.

Orientation Tuning of Simple Cells

We will now explain in a bit more detail the orientation tuning of the receptive fields of simple cells. Consider the vertically tuned "***edge detector***" cell shown in **3.9**. (We use the apostrophes to indicate that calling these cells *edge detectors* is probably very misleading, Ch 5). At the top left of **3.9** is the optimal stimulus for exciting this cell. This stimulus causes the vigorous discharge of impulses shown alongside it as a burst of impulses. How this comes about is shown in the upper right diagram in **3.9,** where light is shown falling on the plus (+) symbols, thereby exciting the cell. The dark gray region represents a low light level stimulating the inhibitory zone. As little light falls on that zone, little or no inhibition is generated, and hence the cell bursts into life when the stimulus appears because the excitatory inputs exceed the inhibitory inputs.

Now look in **3.9** at what happens if a *non*-vertical stimulus falls on this vertically tuned receptive field. The response is reduced according to the amount by which the non-vertical stimulus diverges from the vertical. This is because now darkness falls on a part of the excitatory zone (so reducing the excitatory input to the cell). At the same time, the change in stimulus orientation causes light to fall on a part of the inhibitory zone, increasing the inhibitory input and adding to the reduction caused by the weaker excitation.

When the stimulus edge is rotated all the way to horizontal, the cell receives as much inhibition as excitation and so its firing rate is then reduced to the resting discharge level. The balance between excitation and inhibition in this case is shown by the equal numbers of pluses and minuses triggered by the horizontal stimulus. Smaller shifts of orientation from vertical, either clockwise or anticlockwise, produce smaller reductions in firing rate than a full 90° twist because they do not take so much light off the excitatory zone and they do not place so much on the inhibitory zone. The conclusion from all this is: the vertical edge detector of **3.9** is vertically tuned because the total excitation-minus-inhibition count is greatest with a vertical stimulus. The overall relationship between orientation of input edge and output is shown the graph at the bottom of **3.9**. This sort of graph is often called a ***tuning curve***.

In the case of the cell in **3.9** the excitatory and inhibitory zones of each receptive field are shown as carrying equal weight overall. That is, if *uniform* illumination covers the receptive field of this cell (bottom stimulus of **3.9**), then the cell does not "notice" it at all because the pluses are canceled by the minuses: there is a roughly equal balance between them. This is different from what was shown in **3.4,** but in fact roughly equal total weights from the inhibitory and excitatory regions of the receptive field are quite common in simple cells. This means that such simple cells ignore the illumination level of a uniform field because if the intensity of the illumination rises then both the excitatory and inhibitory responses increase by the same factor, and they continue to cancel. The upshot is that simple cells (and many other visual neurons) respond best to patterns that generate luminance *differences*, e.g., edges, and they ignore overall illumination level.

What has just been said about the optimal stimulus for the edge detecting cell in **3.9** applies equally well to all the other types of cells shown in **3.6** and **3.7**.

In every case, the ***optimal*** or ***preferred stimulus*** is one which provides maximum excitation and minimum inhibition. And because the excitatory and inhibitory regions are mapped out in an oriented way, they show the same kind of orientation tuning illustrated in **3.9**.

Different Cells for Different Orientations

We have told this story about simple cells using vertically tuned cells as examples. What about other orientations? Hubel and Wiesel found in 1962 that different orientations were dealt with

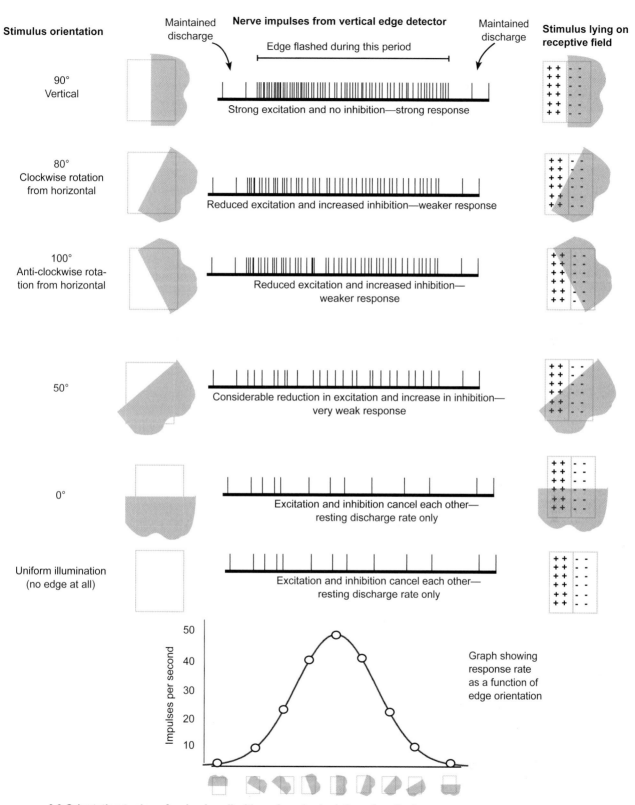

3.9 Orientation tuning of a simple cell with preferred orientation of vertical

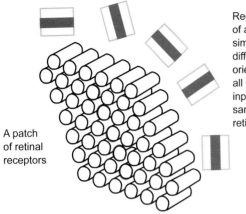

Receptive fields of a sample of simple cells with differing preferred orientations but all receiving their inputs from the same patch of retinal receptors

A patch of retinal receptors

3.10 Many simple cells analyze each patch of retina
Beware: this is a highly schematic diagram—it does not show the cells intervening between the receptors in the retina and the simple cells in the brain. The properties of these intervening cells are described in Ch 6.

by different cells. That is, any given patch of retina is "looked at" by a whole range of cells, each cell dealing with one or other orientation. This is illustrated in **3.10** for cells with "bar detecting" fields. The angle to which each cell is tuned is determined by the pattern of its excitatory and inhibitory regions.

"Slit" and "edge" simple cells are also found for a full range of orientations. Cells of those types are also "looking at" any given patch of the retina and hence the associated region of the input image. Thus, a wide range of cell types is found analysing what is present in each patch of retina. These cells all share the same receptors and hence work from the same gray level description. But the fibers going from the receptors to each cell are different and the patterns of excitatory and inhibitory connections are thereby different.

Different Cells for Different Places

So far we have considered analyzing the image in just one patch of retina. But the mammalian retina consists of millions of receptors and each patch of receptors needs to be analyzed. **Convolution** is the technical term for applying a given receptive field type all over an image. We will explain that concept in detail in Ch 5. For the present we note that a convolution is implemented in biological visual systems by having many cells of the same type, each one devoted to its patch of retina.

Given that cells with all the receptive field types shown in **3.6–3.7**, each for a full range of orientations, are needed for each of the myriad retinal patches, one can see that the striate cortex is a very elaborate biological image processing device indeed. We will examine its structure and function in Ch 9.

Are Simple Cells Feature Detectors?

If you look at the optimal stimuli shown in **3.6–3.7** then you can easily see why these cells have often been dubbed **feature detector**s. That is, it has commonly been assumed that if a cell has a specific edge feature as its optimal stimulus then that cell must be a template recognition mechanism for signaling the presence of an edge feature in a given position on the retina.

But computer experiments that make use of simple cell models to perform image processing have shown this simple-minded "feature detector" idea must be wrong. The reason is that the output of any given simple cell is far too ambiguous to be regarded as a reliable feature detector on its own.

For example, consider a cell with a preferred orientation of 90° (vertical) with the tuning curve shown in **3.11a**. Such a cell would by definition respond most strongly when a high contrast vertical black edge falls in the appropriate position in its receptive field. If the **contrast** of the edge is reduced, by making the black zone a dark gray and the white zone a light grey, then the cell still responds but less strongly, **3.11b**. The response is lowered because less excitation and more inhibition is being fed to the cell than before.

Now consider the cell's response to a high contrast edge rotated by, say, ±20° or so away from the optimal vertical orientation, **3.11c**. The cell would respond just as well to this non-optimally oriented stimulus as it would to the optimally oriented one of lower contrast. If activity in this cell is taken simply and directly as the neural representation of a vertical edge then we would be susceptible to some very awkward illusions. We would confuse faint vertical edges with high contrast just-off-vertical ones, a quite unsatisfactory state of affairs which fortunately does not arise. The brain has mechanisms for solving this problem, and we describe these next.

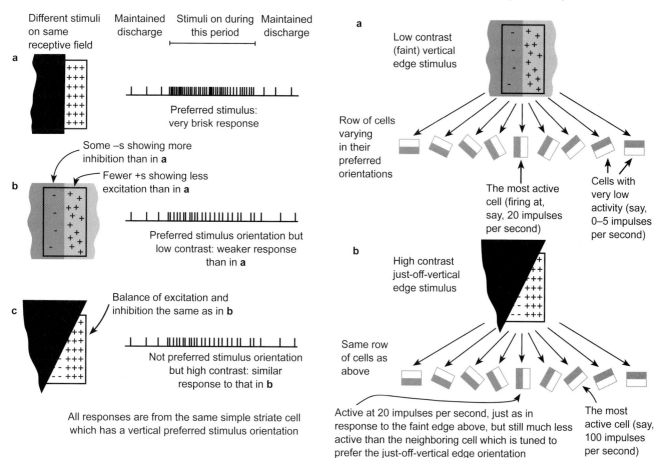

Left figure (3.11):

Different stimuli on same receptive field | Maintained discharge | Stimuli on during this period | Maintained discharge

a
+++ +++ +++ +++ +++ +++

Preferred stimulus: very brisk response

Some −s showing more inhibition than in **a**

Fewer +s showing less excitation than in **a**

b
− + +
+ +
+ +
+
+ +
− + +

Preferred stimulus orientation but low contrast: weaker response than in **a**

Balance of excitation and inhibition the same as in **b**

c
+
+ +
+ + +
+ + +
− + + +
− − + + +

Not preferred stimulus orientation but high contrast: similar response to that in **b**

All responses are from the same simple striate cell which has a vertical preferred stimulus orientation

3.11 Simple cell ambiguity

Right figure (3.12):

a

Low contrast (faint) vertical edge stimulus

− + +
+ +
+ +
+
+
+ +
− + +

Row of cells varying in their preferred orientations

The most active cell (firing at, say, 20 impulses per second)

Cells with very low activity (say, 0–5 impulses per second)

b

High contrast just-off-vertical edge stimulus

+
+ +
+ + +
+ + +
− + + +
− − + + +

Same row of cells as above

Active at 20 impulses per second, just as in response to the faint edge above, but still much less active than the neighboring cell which is tuned to prefer the just-off-vertical edge orientation

The most active cell (say, 100 impulses per second)

3.12 Interpreting simple cell responses in context

Many Cells Make Light Work

The general answer to the ambiguity problem just introduced is to consider each cell's output, not on its own, but in the *context* of the activities of other cells examining the same patch of retina.

For example, suppose mechanisms were present that identified the *most active* neuron when assessing the orientation of an input feature. (A better scheme will be described shortly but this example is a helpful starting point.) In this case, for the faint vertical edge in **3.11b,** the most active cell would still be vertically tuned, **3.12a.** This cell would not be firing as briskly as it would have been if the edge were a high contrast vertical one. Nevertheless, it would still "come out on top" against the opposition from other, even less active cells.

Hence, the hypothetical pattern decision mechanism would note that this cell was the most active one, and register the input feature as having a vertical orientation.

Now consider responses to a high contrast edge oriented about 20° from vertical, **3.12b.** This non-vertical stimulus stimulates the *vertically* oriented receptive field just as well as the vertical but faint edge shown in **3.12a.** But in **3.12b** the cell tuned to 20° from vertical fires at an even greater rate than the vertically tuned one because 20° from vertical is its preferred orientation.

Consequently, the mechanism which detects the most active neuron would register the input feature *not* as a vertical one but at its true orientation of 20° from vertical.

In this way, by taking advantage of the context in which any given cell's activity occurs, the brain need not be fooled by the intrinsically ambiguous responses of individual simple cells. (If cell outputs are free of any jitter or noise, the outputs of only two cells are sufficient to correctly resolve this ambiguity; Ch 11.)

This orientation-contrast example illustrates very nicely the idea of regarding each simple cell as delivering *image measurements for interpretation,* and not as an "edge detectors" pure and simple. They can't serve as feature detectors because their individual responses are far too ambiguous.

The example of ambiguity just discussed involves two stimulus dimensions, orientation and contrast. We examine in detail in Ch 11 how to resolve ambiguities of that sort. We now consider how to compute the orientation of a stimulus when its contrast is held constant. Our goal is to show that there is a much better way of using the responses of a range of cells with different preferred orientations than simply to seek the most active one.

Computing Stimulus Orientation

Single cell recordings show that there are simple cells for only 18-20 different preferred orientations within each patch of retina. Hence, if cell responses were interpreted as suggested in **3.12**, with the *most active* cell signalling the orientation of the input feature, then we could only see 18-20 different orientations. This implies that we would be limited to discriminating between lines differing in orientation only if their orientations differed by about 10°. That is, the 180° of orientation would be shared between, say, 18 orientation detectors with the peaks of their tuning curves (**3.13**, **3.14**) separated by 180°/18 =10°. But our perceptual capabilities are much better than this: we can manage discriminations of less than 0.25°. Clearly, there is a need for some method of *interpolation* between neighboring orientation measurements. By this we mean there must be a way for the brain to estimate stimulus orientations lying in between the 18 preferred orientations encoded by simple cells.

For example, compare the two situations illustrated schematically in **3.13** and **3.14**. The input feature in **3.13** is vertical and this causes a symmetric distribution of simple cell firing rates with peak firing shown for the vertically-tuned cell. Either side of this peak response, simple cells whose preferred stimuli are bars with orientations of 80° and 100° are shown firing quite briskly—but not as fast as the vertically (90°) tuned cell. Cells with optimal

orientations of 70° and 110° also fire, but relatively weakly, and the responses from cells tuned to 60° and 120° are very small indeed. For cells with optimal orientations further away from vertical than about ±30° the firing rates quickly reduce to resting discharge levels. This symmetric pattern of firing rates centered on the vertically tuned cell can be regarded as a "signature" profile of activity for the feature representation *vertical bar.* It would be the signature noted by the interpretive mechanisms whose task it is to decide what the simple cell measurements convey about the input.

Now study **3.14**, which illustrates firing rates that arise for an input bar whose orientation is 92°. As you can see, the distribution is slightly skewed, not symmetric. The vertically tuned cell is still firing fastest but it has almost been caught up by the 100° cell, which is now firing well above the level of the 80° cell. That wasn't the case for the 90° input. Equally, the 110° cell is firing more briskly than the 70° cell. This skewed activity profile is the signature profile for a 92° bar.

This is a much more sophisticated approach to the interpretation of simple cell measurements (outputs) than simply finding the most active neuron and assuming its preferred orientation is the orientation of the input feature. What we now have is a system in which simple cells can be said to **sample the stimulus dimension of orientation,** and infer what is going on in the image from the pattern of simple cell outputs.

One advantage of this scheme for finding edge orientation is that it is a very neat way of avoiding more simple cells than are necessary to do the job. This makes it economical in terms of number of cells required. Another clever brain trick.

Integrating Channel Outputs: Weighted Means

A *channel* is defined as a population of cells that all have the same preferred value of a particular stimulus property. For example, if all the cells in a given population have the same preferred orientation then together they constitute a single *orientation channel*. Note that the cells in such a channel are not necessarily located close to one another. In fact, they can be distributed widely across the striate cortex (Ch 9). The examples shown in **3.13** and **3.14** illustrate orientation channels, but show only one single cell taken from each channel with

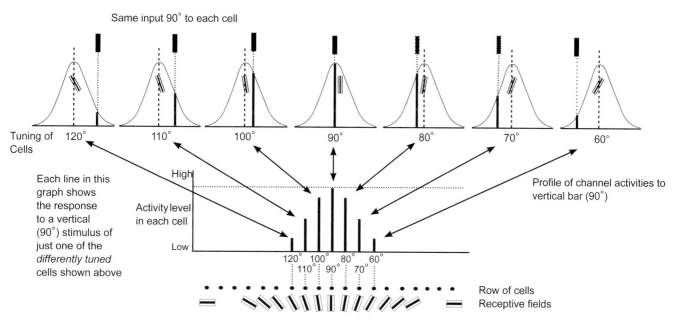

Same input 90° to each cell

Tuning of Cells   120°   110°   100°   90°   80°   70°   60°

Each line in this graph shows the response to a vertical (90°) stimulus of just one of the *differently tuned* cells shown above

High

Activity level in each cell

Low

120° 100° 80° 60°
110° 90° 70°

Profile of channel activities to vertical bar (90°)

Row of cells
Receptive fields

3.13 Activities of simple cells to a vertically oriented stimulus

The upper graphs show the tuning functions of cells with preferred orientations from 60° to 120°. The striped bars under each tuning curve represent the receptive field of the cell: their orientations show the preferred orientations of each cell. Above each tuning curve is shown a vertical bar as input. Following down the dotted lines from the input bars leads to the firing rate, shown by the thin vertical line under each tuning curve, for each cell to the vertical bar stimulus. The vertically tuned (90°) channel fires most vigorously. Cells with preferred orientations close to vertical also become activated by the vertical input bar though to a lesser extent. The overall pattern of activity is symmetrical around vertical. (Drawn schematically.)

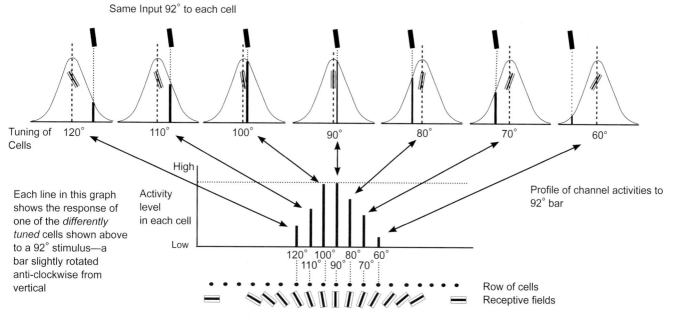

Same Input 92° to each cell

Tuning of Cells   120°   110°   100°   90°   80°   70°   60°

Each line in this graph shows the response of one of the *differently tuned* cells shown above to a 92° stimulus—a bar slightly rotated anti-clockwise from vertical

High

Activity level in each cell

Low

120° 100° 80° 60°
110° 90° 70°

Profile of channel activities to 92° bar

Row of cells
Receptive fields

3.14 Activities of simple cell to a stimulus oriented at 92°

As in **3.13**, the vertically tuned (90°) cell fires most vigorously but to a slightly lesser extent. However, the cell tuned to 100° fires much more strongly than in **3.13**, indeed almost as strongly as the 90° cell. Also, the cell tuned to 80° fires less briskly than it did when the stimulus was a vertical bar (compare **3.14**). The overall pattern of activity is thus asymmetrical, being skewed toward the cells tuned to orientations over 90°. (Drawn schematically to emphasize main points.)

a particular orientation tuning (90°, 80°, 70°, and so on).

We now explore further the idea of interpreting simple cell outputs by giving an example of one particularly simple way of doing it. This is to regard all the activities of the cells as a mass of data and work out the **weighted mean** of all these data. Skip these details if you prefer and go to the next section.

To keep things arithmetically straightforward, let each impulse in a given time interval of, say, 1 second be regarded as one data item. Also, we will consider computing a weighted mean from just three cells, with preferred orientations of 80°, 90° and 100°. The basic idea is to let each cell contribute to the computation according to its firing rate (output) in comparison with the other cells.

Let's first take the situation in which these cells are responding to a vertical bar (90°), and let's suppose their firing rates are as shown in the second column of **3.15**. The total number of impulses in 1 second is $35 + 50 + 35 = 120$ impulses per second. We now *weight* the contribution of each cell taking into account how much each cell is firing in comparison with the other cells. Thus,, we weight the output of the 80° cell by the fraction $35/120 = 0.29$, the 90° cell by $50/120 = 0.42$ and the 100° cell by $35/120 = 0.29$. You can think of this weighting as reflecting how much influence is to be given to each of the cells in computing the stimulus orientation they are dealing with.

The final column of the table multiplies the preferred orientation of each cell by its weighted firing rate, which sum to give the weighted mean of 90°. This is exactly as it should be, of course, for a 90° input—this what we want the feature code to be saying when this particular symmetrical "signature tune" is "playing" in the orientation channels.

But now consider **3.16** which shows *asymmetrical* firing rates in the same three channels for a stimulus just-off-vertical, 92°. The weighted mean is now 92°. This output can be regarded as the result of *interpolating* between the preferred orientations of the three orientation channels to find the orientation of the stimulus. Progress.

This strategy of using weighted outputs from a set of channels has the advantage that it averages out the effects of noise in responses. Obviously, this will be better if more channels are used, for

example, the full range shown in **3.13** and **3.14**. We say more about the problems of noise in Ch 5.

[This simple scheme for calculating a weighted mean is used as an example to explain the basic idea. However, it would need to be refined in a practical system, as it collapses arithmetically for any orientation coded as 0°. This problem can be fixed by expressing angles trigonometrically, but we will not go into detail here.]

But you might say: the brain doesn't have a calculator for doing arithmetic, so how might it use its neurons to implement this type of interpolation calculation? The general answer is: the brain can "do" arithmetic using the processes of excitation and inhibition in combination.

Suppose the brain did do something along the lines of a weighted mean calculation, and then used just one neuron to encode each discriminable orientation. Given that we can distinguish orientations as little as 0.25° apart then that would entail having a few hundred neurons to encode the orientation of every edge feature in each patch of retina. Each such neuron would then be said to be a **local code** for just one particular orientation.

Alternatively, perhaps the brain doesn't do a weighted means calculation to decide which one of a set of neurons should become activated as the code for a given orientation. Perhaps instead the patterns of simple cell responses shown in **3.13** and **3.14** are used as a **population code** for the feature representation *vertical bar present*.

After all, our simple weighted mean calculation has demonstrated that the population of simple cells taken together has the orientation of the bar encoded in its activity pattern. Perhaps this **distributed representation** is sufficient as it stands for the uses the brain has for orientation data. If so, why bother going a step further and making the bar orientation *explicit* in a local code? This question raises some fundamental issues in trying to understand seeing and the brain. We return to them in detail in later chapters.

Coarsely Tuned Channels Are a Good Idea

We said above that simple cells can be viewed as *channels sampling the stimulus dimension of orientation*. It turns out that the basic principle underlying this sort of sampling scheme applies generally in vision.

Preferred Orientation of Cells	Cell firing rates (impulses per sec)	Weighted firing rates	Orientations Weighted (degrees)
80°	35	35/120 = 0.29	80°×0.29 = 23.43°
90°	50	50/120 = 0.42	90°×0.42 = 37.50°
100°	35	35/120 = 0.29	100°×0.29 = 29.17°
	Total = 120		Total = 90.0° This is the Weighted Mean

3.15 Calculating a weighted mean from the responses of three channels to a vertical stimulus (90°)

Preferred Orientation of Cells	Cell firing rates (impulses per sec)	Weighted firing rates	Orientations Weighted (degrees)
80°	24	24/118 = 0.20	80°×0.20 = 16.27°
90°	46	46/118 = 0.39	90°×0.39 = 35.08°
100°	48	48/118 = 0.40	100°×0.40 = 40.68°
	Total = 118		Total = 92.0° This is the Weighted Mean

3.16 Calculating a weighted mean from the responses of three channels to a 92° stimulus

Take color vision for example. Have you ever had to replace colored inks in a color printer and wondered: how can just three inks (cyan, magenta, and yellow) be sufficient to create all the colors that you see on the printed page?

The short answer is: evolution has "discovered" that having just three types of color receptors (often referred to as red, green, and blue cones; Ch 17) is a *sufficient* sampling of the stimulus dimension of wavelength of light. That is, just three measurements obtained using red, green, and blue cones are sufficient to infer all the colors that matter to us for survival. The three colored inks of a printer are sufficient to trigger these three receptor types appropriately for almost all the colors we can see. The result is a clever economy in stimulus creation (the printer) and in stimulus sampling (the eye and brain).

One consequence of having a limited number of samples of a stimulus dimension is that the cells taking each measurement need to be **broadly tuned.** This is why using such channels is called **coarse coding.** The sampling idea would not work for color vision if each retinal cone were exquisitely sensitive to just one wavelength.

The reason is that as soon as a wavelength appeared that was outside its narrowly tuned range it would fall silent. It would literally have nothing to say about that wavelength. Thus, we would either have to be blind to that wavelength, or we would need myriad sharply tuned cones, each specializing in just one wavelength. That is unworkable. A massive number of cones would be needed for each small patch of the fovea (the central region of the retina that mediates highest resolution vision). How could this myriad be packed in without unacceptable loss of spatial resolution for fine details?

Broad tuning is, then, an important requirement for exploiting the economy offered by using channels. However, broad tuning means that the coarsely coded cells in question cannot serve as a **local code** for a feature assertion. As stated above, the output of each broadly tuned cell is far too ambiguous to serve that role. Either the activities in the whole set of channels need to be used as a population code, or a process of *interpretation* needs to be performed to create a local code with the required resolution.

Whichever of these coding schemes, local or population, turns out to be used by the brain, it is clear that we need to break away entirely from the idea that simple cells are feature detectors, pure and simple. It is this realization which lay behind our earlier remarks about the inappropriateness of calling such cells detectors: "slit detectors," "bar detectors," and so on.

This conclusion was drawn by Marr and others from computer-based image processing experiments that modelled simple cells. The ambiguity in the responses of their models of simple cells meant that these cells must be regarded as image-measuring devices which provide useful data about features of the input image but either this

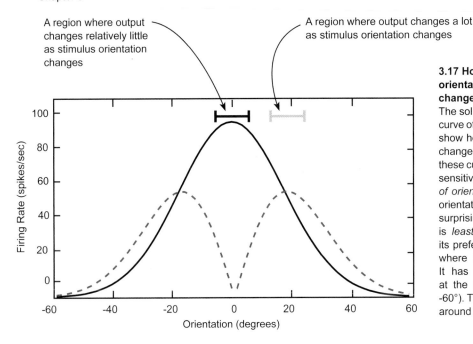

A region where output changes relatively little as stimulus orientation changes

A region where output changes a lot as stimulus orientation changes

3.17 How a change in stimulus orientation is related to the change in cell response
The solid curve depicts the tuning curve of a cell. The dashed curves show how the tuning curve *slope* changes with stimulus orientation; these curves therefore depicts the sensitivity of the cell to *changes of orientation* at points along the orientation continuum. Note that, surprisingly at first sight, the cell is *least* sensitive to *changes* at its preferred orientation (here 0°) where its firing rate is highest. It has similarly poor sensitivity at the extremes (i.e., +60° and -60°). The cell is most sensitive at around +20° and -20°.

data needs to be interpreted before a proper feature description can be asserted, or their responses have to be treated as a population code.

One method for reducing the effects of noise is to *average* responses from many cells. The underlying assumption here is that the noise in any given cell will be independent of that affecting other cells. This means that the noise variations will tend to cancel out when an average is taken.

This is one reason why it may be a good idea for the brain to consider the responses of entire populations of cells when attempting to recover the parameters of the stimulus that caused those cells to respond. Using averaging to get around the problem of noise is explored in detail Ch 5 in connection with the task of edge detection from noisy images.

Problem of Parameter Resolution

A question that we discuss in Ch 11 is: how few channels are needed to resolve the ambiguity in simple cells responses? It turns out that, in principle, only two, very broadly tuned. However, if we had only two cells to span the entire range of 180° then each cell would be very insensitive to most orientation changes. This is because a large part of each cell's orientation range would be on regions of the tuning curve that change very slowly with changes in input orientation, so that changes in

stimulus orientation would cause very little change in output, **3.17**. The output of such a very broadly tuned cell changes very little unless retinal line orientation falls on the flanks of the tuning curve. These flanks are where the slope of the tuning curve is greatest, and therefore where the change in cell output per degree of change in line orientation is greatest.

One way to ensure that most retinal line orientations coincide with this ′sensitive″ part of a tuning curve is to use a large number of cells with tuning curves which, taken together, tile the space of all possible orientations to yield adequate resolution, as in **3.18**. This represents another reason for using many cells to encode orientation, which is independent of the noise problem above.

A curious side effect of this way of extracting orientation is that it predicts peaks and troughs in the system's sensitivity to orientation. A peak of high sensitivity should be found in the region where the two response curves are changing sharply. Troughs should arise in the regions covered by the top of each cell's response curve. This is because at the tops the *change* in response as retinal orientation changes is not as great.

This is paradoxical. It would, at first sight, be natural to expect *maximum* sensitivity for orientations falling on the highest point of each cell's tuning curve. But on careful examination, each cell is

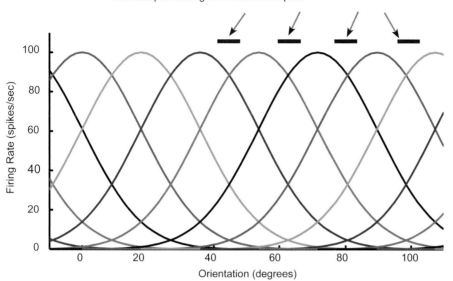

Examples of regions of peak sensitivity to stimulus orientation changes because these are regions with steepest changes in channel outputs

3.18 Using many simple cells with different preferred orientations to "tile" the full stimulus orientation range

most sensitive to *changes* in orientation about half way down from the top.

So it is reasonable to ask: does human vision show these predicted peaks and troughs in orientation sensitivity? The answer is yes. Regan and Beverley reported experiments in 1985 on human orientation discrimination which confirmed the prediction (see Regan, 2000).

Can Templates Ever Work as Recognizers?

We started this chapter by considering the use of bar templates to detect stimulus bars. We discovered that the problem of response ambiguity bedevils their use, but this led us to a general principle: ambiguities can be resolved by drawing inferences from the outputs of many templates (channels).

Even so, simple templates can be made to work well as pattern recognition devices in some special limited contexts. Consider, for example, a bank check number recognition system, **3.19**. One way that the numbers can be recognized is to build into the number-recognizing machine a set of templates, one for each numeral. Then the task is be to note which template best fits the number on the check being analyzed.

Using a template in this way is similar to the template bar detector in **3.4**, except that the

number template has a much more complex pattern than the simple bar feature. Moreover, bank check numbers are made more readily distinguishable one from another by using specially designed numerals with lines of different thicknesses to facilitate recognition.

But that trick alone would not be enough to get such templates to serve as pattern recognizers. The crucial added ingredient is using a special check scanning device which prevents complications arising from large variations in the input images in terms of the brightness, contrast, shape, size, and orientation of numerals. This permits a template recognition system that works well for the task it tackles.

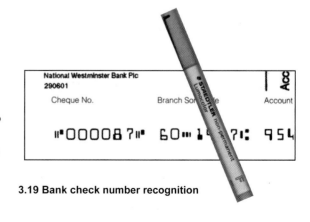

3.19 Bank check number recognition

However, that task is a very simple one by the standards of biological vision systems. They have to cope with all manner of variations in the way objects appear in retinal images, variations over which they have no control. How human vision copes with some of these variations is an issue that we will address in later chapters, particularly Ch 8, *Seeing Objects*. But we pursue here the topic of template recognition a bit further by way of introducing some basic facts that illuminate why the seeing problem is so hard.

Templates and the Combinatorial Explosion

You might be wondering: could a template recognition system be made to work by having a range of different templates, each tuned to deal with one or other source of image variation?

The way we coped with the problem of variable image bar orientation illustrates this idea: we found it a good idea to have 18 or so differently oriented bar templates, and to use these coarsely coded measurements of orientation to work out bar orientation using weighted means. What if this approach was extended for other sources of image variation, such as color and size?

Each such variable, often called a *parameter* as already noted, would need its own set of coarsely coded templates, each one dealing with a limited range of the parameter's possible values.

The trouble with this idea is that it immediately hits a major snag, called the ***combinatorial explosion***. If we need 18 templates for orientation then for each one of these we will need a suitable range of templates for size. Let's say for simplicity that this would also be 18. Hence, 18 × 18 templates would be needed to deal with all combinations of orientation and size.

Now consider adding another variable, such as contrast, and suppose that too needed 18 templates. This takes us to 18 × 18 × 18 = 5,832 templates for one numeral.

Well, the brain has a lot of neurons so perhaps that isn't so bad. But things rapidly get worse when other sources of image variation are brought into the equation.

Take object position on the retina for example. Again to keep things simple, let's suppose the parameter of vertical position needs 18 templates and similarly for the horizontal position parameter.

We now have 18 × 18 × 18 × 18 × 18 = 1,889,568 templates.

And, once again, this is for just one object, for example just one of the numerals that our bank check number template recognizer would have to deal with if it was stripped of careful control of variations in the input image.

Imagine needing this huge number of templates for all the different objects that we so readily recognize—numerals, letters, birds, trees, chairs, people, and so on.

There is a general formula for working out the number of combinations of parameters involved in this combinatorial explosion: the total number equals N^k where N is the number of templates per parameter (18 in our example), and the exponent k is the number of parameters. Don't worry if this ***exponential*** formula seems a bit opaque: if you want to know more about it then read Ch 11 where we discuss its implications at length.

The combinatorial explosion reveals just how hard the problem vision is. Any attempt to solve it using simple-minded templates doesn't work: the brain just doesn't have enough neurons.

Binding Problem

You might think at this point this is silly; surely there is no need to have a cell for every combination of parameter values? Why not simply have one population of cells that encodes only one parameter exclusively, for all objects.

For example, one population could encode only color, and another could encode only orientation. Using this scheme we would require k populations to encode k parameters. If each population consisted of one million cells then no more than k million cells would be required in total. The brain may do something along these lines (as we will see), but this raises another fundamental vision problem: how do we attach parameter values to objects? This is known as the ***binding problem*** and we discuss it in Ch 11.

Back to the Jumping Spider

We can now see that a spider would have a very hard time using templates as a means of deciding the question: *is the object over there mate or prey?* Such a spider would also be subject to the combinatorial explosion (further details in Ch 11).

However, it may be that the jumping spider has, as suggested earlier, evolved some special-purpose visual mechanisms that are quite unlike our own. Land has suggested that these spiders use scanning movements of their boomerang-shaped retinae to align them with the orientations of leg-like bars in the input image. This trick might allow the spider to avoid the 18 or so different orientation tuned channels that monkeys and humans seem to possess.

These retinal scanning movements might also solve the problem of variable object position in the image. They would do that if they ensured that the object's image falls on exactly the right spot for the spider's limited number of templates to be able to recognize *Mate* or *Prey*. Perhaps these and other special-purpose adaptations give the spider a working template-based recognition system.

The general idea here is that perhaps the spider has *not* evolved a general-purpose vision system, such as our own, but one specialized for its particular ecological niche. It can survive if it can capture insect prey and find mates. Perhaps it has a visual system set up to do those tasks and very little else. In this respect it may be a bit like the simple-minded but highly specialised bank check recognition system described above.

Concluding Remarks

This chapter has explored the task of building a bar detector using templates to discover what problems have to be overcome. Along the way we defined some essential technical terms, many to do with basic facts about the "hardware" of biological visual systems. A key concept has been that of a *receptive field* with *excitatory* and *inhibitory* regions.

Linked to this is the idea of receptive fields of various types, organised as *channels* analysing each patch of retina and providing measurements about a *stimulus dimension* (such as orientation) from which the brain can work out which features are present in the input image. In the next chapter, Ch 4, we use these ideas in showing how certain illusions called **aftereffects** have revealed orientation channels in the human visual system without need for invasive single cell recordings.

This chapter has also introduced a fundamental concept in vision research: the *combinatorial explo-*

sion arising from the exponential N^k formula. We examine that problem in considerable depth in Ch 11 to explain in more detail why vision is such a hard problem.

Finally, armed with core ideas introduced in this chapter, we are ready to have a much closer look in Ch 5 at the task of edge detection. We consider there and in Ch 6 the tasks of how to recover feature properties other than orientation, such as bar width, and whether an edge is a sharp or fuzzy.

Ch 5, *Seeing Edges,* will also remedy a nagging irritation that may have formed in your mind. We emphasized in Chs 1 and 2 the need to be very clear about the *computational theory, algorithmic,* and *hardware levels of task analysis* when studying vision. But we have blatantly ignored our own advice in this chapter. We have jumped straight into considering a particular sort of algorithm, applying templates, without any guidance from a computational theory as to the design of those templates. You might feel a bit cheated. But we have done this simply to introduce a wide range of basic concepts and terms that it is best to get out of the way first.

In any event, if you do feel a bit cheated then you have drawn the right conclusion because this chapter illustrates just how unsatisfactory it can be to start addressing an image processing task without a clear computational theory of that task. Template matching is a species of algorithm. The design and use of templates demands the clarity afforded by a decent theory of the task. We investigate in Ch 5 how we can get a much better understanding of what the receptive fields of simple cells are doing by thinking a lot harder about the task of feature detection. That will be seen to be the moral of the story of simple cells told here.

Further Reading

This main function of this early chapter has been to explore some core ideas needed to understand seeing, such as receptive fields, channels, coarse and fine coding. We do not recommend much further reading at this stage on this material. We will be making suggestions for further reading for later chapters that use the core concepts dealt with here.

Hence, it is not suggested you consult the sources overleaf at this juncture but we provide them so that you can follow them up if you wish.

Barlow HB (1972) Single units and sensation: A neuron doctrine for perceptual psychology, *Perception* **1** 371–394 *Comment* Classic paper that discusses the relationship between the firing of single neurons in sensory systems and subjectively experienced sensations. Commentaries celebrating this landmark paper are in *Perception* **38** 795–807. It is probably best tackled after reading Ch 11.

Drees O (1952) Untersuchungen über die angeborrenen Verhaltensweisen bei Springspinnen (Salticidae). *Z. Tierpsychol.* **9** 169–209.

Hubel DH and Wiesel TN (1962) Receptive fields, binocular interaction and functional architecture in the cat's striate cortex. *Journal of Physiology* **160** 106–154.

Hubel DH (1988) *Eye, Brain and Vision.* Scientific American Library. New York. WH Freeman. *Comment:* This gives a highly readable overview of Hubel and Wiesel's Nobel Prize–winning research. It is the source of the quotation from Hubel given in this chapter. However, it is best left until after reading Chs 8 and 9.

Jonsson E (2008) *Channel–Coded Feature Maps for Computer Vision and Machine Learning.* Linkoping Studies in Science and Technology, Linkoping University, Sweden. ISBN 978–91–7393–988–1. *Comment:* An advanced mathematical treatment of theory underlying the use of channels in computer vision. We cite it because it is up-to-date and of possible interest to students of computer vision.

Land M (1969a) Structure of the retinae of the principal eyes of jumping spiders (*Salticidae: Dendryphantinae*) in relation to visual optics. *Journal of Experimental Biology* **51** 443–470.

Land M (1969b) Movements of the retina of jumping spiders (*Salticidae: Dendryphantinae*) in response to visual stimuli. *Journal of Experimental Biology* **51** 471–493.

Land MF and Nilsson D-E (2001) *Animal Eyes.* Oxford University Press. *Comment:* A short, erudite, and fascinating account of animal eyes in all their diversity. Subsumes papers by Land cited above.

Regan D (2000) *Human Perception of Objects.* Sinauer Associates, Inc. Publishers: Sunderland, Mass. See pp.116–120. *Comment:* An excellent book that we recommend strongly for readers who want to pursue various topics in human and animal vision at an advanced level, and specifically the work of Regan and Beverley on peaks and troughs in orientation sensitivity.

4

Seeing Aftereffects: The Psychologist's Microelectrode

a

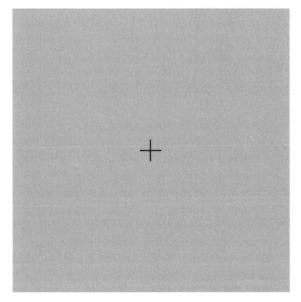

b

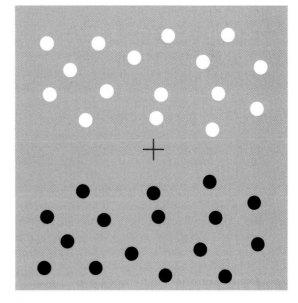

4.1 Obtaining negative after-images
a Test Stimulus Look briefly at the cross and note it is set in a uniform gray field.
b Adapting Stimulus Next, stare steadily at the cross for 15 seconds or so, and then quickly transfer your gaze back to the cross in **a**. You will see **negative after-images** of the dots. They are called *negative* because the black dots give light after-images, and vice versa for the white dots.

Neurophysiology has told us a great deal about the visual machinery of the brain. In Ch 3 we introduced the Nobel Prize–winning work of Hubel and Wiesel on orientation-tuned cells in the striate cortex. In the present chapter we ask: is there any *psychophysical* evidence that reveals the operations of these orientation-tuned cells? The work to be reported tests the claim made in Ch1 that visual illusions can afford interesting clues about the mechanisms of vision?

Aftereffects

One large group of illusions used to relate neurophysiology to psychology relies upon the fact that certain curious illusory phenomena are experienced after prolonged or intense constant stimulation. Such illusions are called *aftereffects*. They occur widely in sensory systems.

Perhaps the most commonly experienced aftereffect comes from accidental observations of bright lights: the sun, for example, or a naked light bulb. Following such unwanted exposures, it is usual to find that an *after-image* of the light source remains apparent for some time afterward, super-

Why read this chapter?

In this chapter we begin by describing how aftereffects have been used to study neurons indirectly in the human visual system—which is why aftereffects have been dubbed the *psychologist's microelectrode*. A core concept in vision research is described—the *spatial frequency tuned channel*. Allied to this is the *contrast sensitivity function*, which describes how well different spatial frequencies are detected. This function is interpreted as the envelope of a range of spatial frequency tuned channels. The *tilt aftereffect* is explained using the ideas introduced in Ch 3 on how orientation can be computed from coarsely tuned channels with overlapping sensitivities to orientation. The various topics in the chapter are dealt with using a blend of three approaches to studying seeing: computational modeling, psychophysics, and neuroscience. The chapter ends with a technical section that can be skipped on first reading. In this section we explain the *modulation transfer function* and *bandpass filtering* of images. We also analyze a theoretical account of a curious kind of aftereffect, called the *contingent aftereffect*, in terms of calibration processes in human vision. This leads back to the tilt aftereffect and its interpretation in terms of processes designed for *maximizing information* transfer.

imposed upon whatever we see in the scene we next happen to observe. (Beware looking directly at the sun: doing so can cause permanent damage to retinal receptors.)

Bright lights are not essential for obtaining after-images, however. For example, **4.1** explains how to obtain *negative after-images* from black and white dots. They are called *negative* because the after-images have the opposite brightness to the discs which induced them. The after-images last only a few seconds. This contrasts with after-images from very bright lights which can last several minutes. The persistence of the after-images from **4.1** can, however, be increased by lengthening the time spent staring at the cross during *adaptation*. The illusory after-images are almost certainly due to some kind of bleaching of photosensitive pigments in the receptors in the retina caused by the prolonged exposure to the unvarying stimulation.

Some examples of non-visual aftereffects are as follows. If you run a finger to and fro along a curved edge for a few minutes with your eyes closed, a straight edge will subsequently seem to be curved in the opposite direction, as first reported by James Gibson. If you spend a long while (say 90 minutes or so) in an atmosphere containing an excess of carbon dioxide then fresh air can subsequently give an illusory smell of ammonia. If you listen repeatedly to a tone that increases in intensity then a subsequently presented tone of constant intensity is likely to sound as though it is decreasing in intensity.

Aftereffects have long been studied by psychologists but they became the subject of particularly intensive research in the 1960s and 1970s. The reason was that they gave the psychologist a tool for probing the workings of sensory mechanisms discovered by neurophysiologists around that time. They are still widely used as a way of getting at these sensory mechanisms indirectly. Indeed, aftereffects have been dubbed the *psychologist's microelectrode* because of the neat relationship which often exists between psychological findings from aftereffects and neurophysiological discoveries made using real microelectrodes. And in some ways indirect probing with aftereffects is more powerful because it can reveal facts about a whole population of cells working together, rather than about one cell or few cells measured individually using microelectrodes.

Test-Adaptation-Test Cycle

It will be helpful at this point to outline the typical procedure used in studies of aftereffects, and to introduce some associated terminology.

First, the observer looks at a ***test stimulus*** (e.g., the gray field of **4.1a**) and notes some property of its appearance, such as its brightness, its color, and so on. In experiments, as distinct from demonstrations, the perceptual attribute of interest would be measured using psychophysical techniques. ***Psychophysics*** is the business of measuring the relationship between the perceived and the physical attributes of stimuli (Greek *psyche:* mind; *physic:* of nature.)

Second, the observer stares at an ***adapting stimulus*** for a prolonged ***adaptation period***.

Third, the observer reverts to the ***original test stimulus*** and notes its appearance once more, or in formal experiments the perceptual property of interest is re-measured.

This sequence is called a ***test-adaptation-test cycle***. The effect of the adaptation is noted (or measured) by comparing the appearance of the test stimulus in its pre- and post-adaptation presentations.

Grating Stimuli

The type of stimulus most frequently used during the last 40 years or so of visual psychophysics has been the ***grating***. This is a repeating collection of bars that can vary in their orientation, width, the sharpness of their edges, their contrast, and their color.

To take ***contrast*** first, if a grating has very dark bars and very light white ones (**4.2**, top) then it is said to have a high contrast. Equally, if the "black" bars are not very dark and reflect light with almost equal intensity to the not-very-light "white" bars (**4.2**, bottom) then the grating is said to have a low contrast. Of course, intermediate contrasts are also possible (**4.2**, middle). Each grating in the series has a ***luminance profile*** associated with it. This is a graph showing how luminance, which is defined as the intensity of light reflected from a surface, varies across the surface of the grating. These profiles help clarify what is meant by contrast.

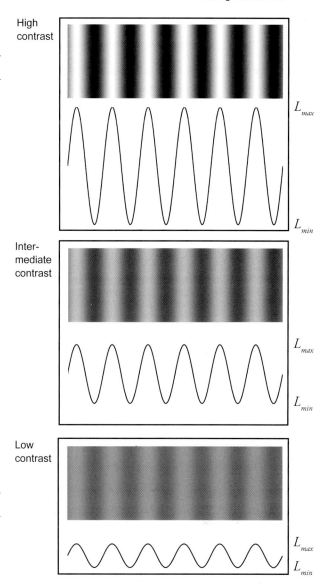

4.2 Sine wave gratings of varying contrast
The graphs below the gratings show the variation in luminance across each one. They reveal the sinusoidal shape of the luminance variation.

Note that although contrast varies from low to high in **4.2**, the total amount of light reflected from each grating as a whole remains the same. Imagine if all the light from the top grating were collected somehow and its intensity measured with a photocell. If this measurement were compared with a similar measurement taken from the grating with the lower contrast, the two measurements would be identical. So a higher contrast is not obtained by increasing the total amount of light reflected from

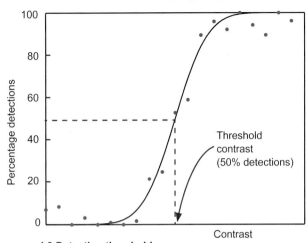

4.3 Detection threshold
The dots are synthetic data to which the curve has been fitted. The *ogive* shape of this graph is the typical shape of the *psychometric function*. The detection threshold is often defined as the stimulus which yields detection on 50% of trials although sometimes a 75% cut-off is used.

the page. Rather, contrast is increased by "packaging" the same total quantity of light to create "whiter" regions, at the expense of also creating "blacker" regions. Thus, the *average luminance* of all gratings in **4.2** is identical, even though the contrast difference between the upper and lower gratings is enormous.

Grating contrast is usually defined formally using the equation:

$$\text{Contrast} = \frac{L_{max} - L_{min}}{L_{max} + L_{min}} \ ,$$

where L_{max} is the maximum luminance in the display, and L_{min} is the minimum (see middle grating in **4.2**). The top line (numerator) of this equation captures the idea that contrast rises if the difference between the maximum and minimum luminance increases. The bottom line (denominator) expresses this difference as a fraction of a measure of the overall luminance. This makes sense because a small difference between L_{max} and L_{min} will appear to have greater contrast if the overall luminance is low than if it is high.

Contrast Thresholds for Gratings

The gratings shown in **4.2** vary in their contrast and hence in their visibility. If the series from high to low contrast were extended further at the low end (a difficult thing to do in a printed illustration:

hence the limited range in **4.2**) then the gratings would become harder to detect, and finally a point would be reached when a very low contrast grating would not be seen at all. That is, the very low contrast grating would appear no different from a field of uniform grey. Technically, it is then said to fall below *contrast threshold*.

Threshold is a technical term in psychophysics. Initially it was conceived as a stimulus value at which there is a transition from seeing to not seeing the stimulus, or vice versa. However, it soon became evident that such a point does not exist because of variability in the responsiveness of receptors and neurons. Hence a detection threshold is usually defined as the stimulus which yields detection on 50% of trials.

In the case of contrast thresholds, there is no contrast which defines for all occasions the boundary between seeing and not seeing. Rather, an observer's ability to detect a near-threshold grating varies from moment to moment: it is a case of "sometimes you see it, sometimes you don't." This variability happens even if the contrast of a near-threshold grating is held fixed. It is caused by fluctuations in the sensitivity of the visual system. The way in which the same physical stimulus can sometimes lead to seeing and at other times to not seeing contrasts is evident in the *psychometric function*, **4.3**.

Psychophysicists have to take account of this variability when measuring thresholds, Ch 12. Meanwhile, note that contrasts just above the threshold range produce the perception of a grating of low contrast: a grating of washed-out gray stripes which can only just be seen. Contrasts well above the threshold range produce clearly visible gratings appearing as a collection of black and white stripes.

Psychophysical Methods

There are a variety of specialized *psychophysical methods* that allow for the variability in detecting stimuli close to threshold. The simplest is the *method of adjustment* which finds an estimate of the threshold in a "quick and dirty" fashion. Here, the observer is simply asked to adjust the stimulus value to find a point at which they can "just see it." This is repeated a few times and the mean value from these trials is taken as the threshold.

In the ***method of constant stimuli*** a set of stimulus values ranging from very weak ('rarely if ever seen') to reasonably strong ('seen on almost all occasions') is chosen. The members of this "constant" set are then presented again and again in a random order over a long series of presentations. Each presentation is called a ***trial***. The resulting data are then plotted as a psychometric function, **4.3**. This is the S-shaped curve (technically, an ***ogive***) that best fits the data points. The stimulus value on the ogive that is seen on 50% of trials is read off as the threshold.

The methods of adjustment and of constant stimuli both have technical disadvantages. The main one concerns distinguishing the genuine sensitivity of the observer's visual system from the observer's inevitable idiosyncratic bias in choosing a criterion for the level of visibility required before saying, "Yes, I see the stimulus." The preferred method for avoiding this hazard is ***two alternative forced choice*** (**2AFC**), which requires the observer to say which of two stimuli has the target-to-be-detected. We describe the details of 2AFC in Ch 12 along with ***signal detection theory***, which underlies the design of modern psychophysical methods.

Orientation-Specific Contrast Threshold Elevation

We now discuss an aftereffect that can be obtained from the stimuli in **4.4** using the Test-Adapt-Test procedure. First, look at the column of low-contrast test gratings shown in **4.4** from a distance of about 1 meter. Note that they are only just visible and that their orientations vary between vertical and horizontal (this is the initial Test phase). If you cannot discriminate the gratings from this viewing distance (i.e., if they fall below your contrast threshold) then move the book a little closer until you can just see them.

Precautionary note If you suffer from epileptic seizures or migraines that you think might be triggered by staring at stripes then do not do these experiments.

4.4 Orientation-specific contrast threshold elevation caused by adaptation
The high contrast gratings are for adaptation; the low contrast gratings are for testing the effects of adaptation.

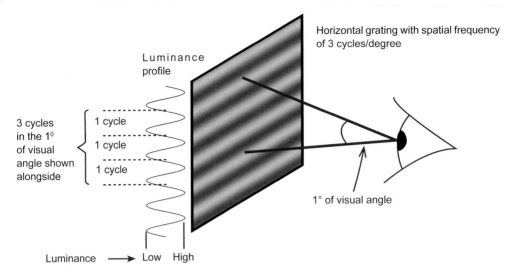

Luminance profile

Horizontal grating with spatial frequency of 3 cycles/degree

3 cycles in the 1⁰ of visual angle shown alongside

1 cycle

1 cycle

1 cycle

1° of visual angle

Luminance ⟶ Low High

4.5 Spatial frequency measured as cycles/degree
The distance between two retinal points is usually measured in terms of degree of visual angle. Here an observer is shown viewing a horizontal grating that is projected into the eye with a spatial frequency of 3 cycles/degree. This means that 3 complete cycles of sinusoidal luminance occur in each degree of visual angle. Each cycle is one "wave" or "oscillation" of luminance change that goes from mid-grey through light gray to mid-grey again, then to dark grey, and then back to mid-grey. A grating of 1 cycle/degree would have just one such cycle of luminance in a single degree of visual angle. To give you a feel for visual angle, note that a line of length 1 cm viewed from 57.3 cm projects a visual angle of 1°. So, roughly, the width of a finger nail viewed at arms length projects 1°. The dark bar in a 1 cycle/degree grating viewed from 57.3 cm is 0.5 cm in width, as the dark bar is one half of the luminance cycle.

Next, adapt your visual system to vertical bars by gazing at the high contrast vertical adaptation grating (top left) for about one minute. Use the same viewing distance as you used for inspecting the test gratings. Allow your gaze to wander within the circle drawn on the adaptation grating during your one-minute inspection of it. This is an important precaution which prevents the build-up of a conventional after-image, **4.1**.

Finally, when the full minute of adaptation has expired, quickly transfer your gaze to the column of test gratings. This is the second Test phase. You will observe (if all has gone well) that the *vertical* test grating (top right) is no longer visible. All you will see in that test stimulus after the adaptation is uniform grey.

On the other hand, observe that the horizontal test grating (bottom of right hand column) *does* remain visible after the period of adaptation. That is, it is as detectable post-adaptation as it was pre-adaptation. Look also at the test gratings of intermediate orientations and try to decide whether they remain visible post-adaptation or not. Be careful to note that the state of adaptation fades after a few seconds. Therefore, keep returning to the vertical

adaptation grating for "top up" adaptation periods of at least 15 seconds each between short inspections of the test stimuli.

The phenomenon illustrated in **4.4** provides a demonstration of a species of aftereffect from adaptation called ***orientation specific contrast threshold elevation*** The *non-vertical* gratings in **4.4** could be seen just as well post-adaptation to vertical as they could pre-adaptation.

However, a caveat is in order. You may find that the visibility of the 67.5° grating (second grating down in **4.4**) might become temporarily invisible post-adaptation. This is because the adaptation effect spreads out around the adapted orientation (vertical). This is a property that provides a revealing clue about the ***orientation tuning*** of the cells thought to be involved, as we will see shortly. But this refinement doesn't alter the basic result that the adaptation is orientation-specific, albeit quite broadly tuned.

You should convince yourself that the orientation specificity of this aftereffect holds just as well for other adaptation orientations, besides vertical. For instance, adapt to the 45° high contrast grating shown in **4.4** and note that the 45° low contrast test grating becomes momentarily invisible post-adaptation, whereas the other test gratings are relatively unaffected. Similarly, adapt to the horizontal high contrast grating and observe that now the aftereffect is restricted to the horizontal test grating. Allow a few minutes of rest between changes of adaptation stimulus, so that the effects of adaptation to one orientation have time to dissipate before you proceed to test the adaptation effects from another.

Spatial Frequency

Gratings can be made to vary in contrast or orientation as described above. They can also vary in the width of their bars and in the fuzziness of the edges of their bars.

Look back at the gratings of **4.2** and you will see that the profiles underneath each one have a wavy shape. Their bars do not have sharp edges but instead the luminance gradually changes from light-to-dark. The wavy shape used in **4.2** is that of a *sine wave*, also known as a *sinusoid*. We explain later how sine waves are useful for describing some overall properties of the visual system using the concept of the *modulation transfer function*. For the moment, simply note that sine wave gratings are usually described not in terms of bar width but instead as having a *spatial frequency* of so many *cycles per degree* of *visual angle*, **4.5**, which is abbreviated to cycles/degree.

In order to explain what this means, let's begin first with the concept of spatial frequency. The term *frequency* is used because the luminance can be thought of as *oscillating* up and down across the grating.

You will probably be familiar with the concept of oscillations in connection with sound. The frequency of oscillations in sound pressure level on your ear drums is heard as *pitch*. A high frequency of oscillation is heard as a high pitched tone, a low frequency a low pitched tone. For hearing, frequency refers to the number of cycles per second, or *Hertz* (abbreviated to *Hz*): each cycle is defined as one complete oscillation in variation of sound pressure from high to low and back to high again.

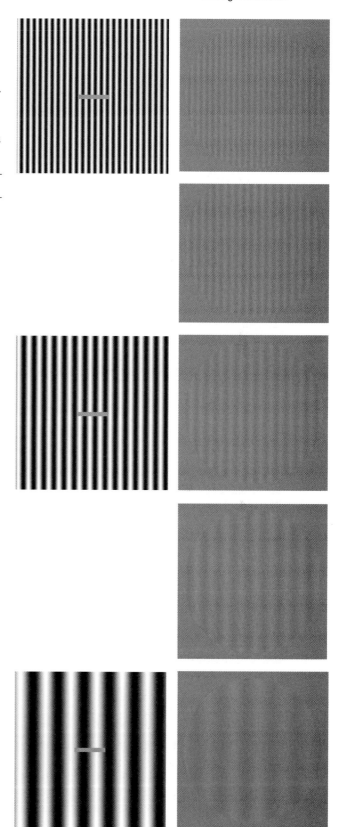

4.6 Spatial frequency-specific elevation of contrast thresholds
Adaptation gratings on the left, test gratings on the right.

81

The concept of frequency is also used in vision but with two key differences. The first is that the oscillations in question are variations in luminance, not sound pressure. The second is that the oscillations occur not over time but over space, that is, across a distance in the image. Hence the term *spatial* frequency.

We now need to explain the concept of visual angle. This is a convenient way of expressing the sizes of entities in images. For example, a line 1 cm long viewed from 57.3 cm *project*s an image of a line that is 1° visual angle. A line of 0.5 cm viewed from half that distance (28.7 cm) also projects an image of 1° of visual angle. Using visual angle to express the sizes of entities in images avoids needing to state the sizes of their associated entities in the scene, or the distance from which they are viewed.

Figure **4.6** illustrates these ideas by comparing gratings with spatial frequencies of 4 cycles/degree and 1 cycle/degree. A grating of 4 cycles/degree has 4 complete sinusoidal cycles in luminance (light through to dark cycles) for every degree of visual angle, whereas a grating of 1 cycle/degree has just one cycle of luminance variation per degree visual angle.

Spatial Frequency-Specific Contrast Threshold Elevation

The aftereffect described above for orientation-specific adaptation can be matched by a similar demonstration of spatial frequency-specific adaptation. Look at the column of low contrast test gratings in **4.6** and you will see that they vary in the spatial frequency of the bars composing each grating. Inspect these gratings from about 1 meter as before and use this viewing distance throughout.

Next, adapt your visual system to the grating of thin bars shown at the top left of **4.6**. Again, allow your gaze to wander to and fro along the gray bar at the center of this grating to avoid an afterimage. After a minute's adaptation, quickly look at the test grating alongside with stripes of similar spatial frequency, and note that they are invisible for a few moments.

Renew your state of adaptation by re-inspecting the high contrast grating for a further 15 seconds and then look at the other test gratings. These gratings are not affected by adaptation to the thin-bar

grating, their own bars remaining visible post-adaptation. Here then is another specific aftereffect, this time one specific to the spatial frequency of the bars of the grating.

Check that this specificity holds for other adaptation spatial frequencies by repeating the test-adaptation-test cycle for the other adaptation gratings shown in **4.6**. Follow throughout the precautions, described for **4.4,** about viewing distance, length of adaptation, topping up the adaptation, etc. You should find that the lowered sensitivity caused by adaptation is restricted to stimuli of the same spatial frequency as those used for adaptation.

Adaptation: The Recalibration Hypothesis

What is the nature of the changes in the visual system caused by viewing the adaptation stimulus? One intriguing idea is that adaptation reveals an unnoticed but ever-present process of ***recalibration***. If the striate cortex, Ch 3, could speak then it might explain this in the following way:

> My vertically tuned cells have been telling me for some minutes that there are a lot of high contrast vertical edges out there. Assuming this is not some crazy temporary hyper-excitability in these cells, it means that I am not using my available processing resources to encode edges with maximal efficiency. I need to make adjustments to the tuning of my cells to allow me to deal with the challenge of a world in which vertical edges predominate.

The striate cortex is here facing up to a problem well known to electrical engineers. This is the problem of developing a computational theory of self-calibrating systems and algorithms for implementing the theory. The task of the theory is to specify how to design systems that can re-tune themselves when changes occur in the inputs they have to process.

Self-calibration processes also have to cope with variations in the performance of components. Processes in the striate cortex must be able to make adjustments if it turns out that some of the cells really have gone crazily hyperactive rather than in response to a sustained change in the contents of the visual world.

We discuss in some detail later how self-calibration might be implemented.

Adaptation: The Fatigue Hypothesis

Another possible explanation of why adaptation to gratings causes raised contrast thresholds is that fatigue is involved. That is, could it be that cells stimulated during adaptation became tired out from firing away at a high rate for a few minutes in response to their preferred stimulus? The idea here is that the cells become so fatigued that when a weak (low contrast) stimulus is presented they do not respond to it as they would normally. This is a much simpler explanation than self-calibration because it is a theory couched purely in terms of events going on at the hardware level. We will show how this idea can explain certain aftereffects shortly.

Of course, the self-calibration and fatigue explanations need not be mutually exclusive. Perhaps both are involved. However, it turns out that an aftereffect can be "stored" for several minutes by a period of darkness in between adaptation and test. It seems reasonable to think that the dark period would give neurons a chance to recover from fatigue, which suggests that that fatigue cannot be the whole story.

Aftereffects: The Psychologist's Microelectrode

You have probably guessed by now why grating stimuli have figured so prominently in this chapter. Gratings provide a convenient way of presenting bars differing in orientation and width. And these two stimulus characteristics are, of course, amongst those which are very important for determining the preferred stimuli for certain visual neurons, Ch 3. You will doubtless remember that some striate neurons were most excited by vertical thin bars, others by horizontal thin bars, yet others by oblique thick bars, and so on, in a large range of different permutations (refer back to Ch 3 for a reminder on this point if you are in any doubt).

Any given grating may fall on the receptive fields of many cells. Not all cells will be stimulated by the grating but those that are will all be of the same type in terms of tuning to orientation and width. Each grating is, therefore, a powerful probe stimulus for exciting cells of a specific kind. This fact has been exploited in literally thousands of experiments over the past four decades or so. We will illustrate some of the main findings as this chapter proceeds.

Each one has tried to contribute some new twist to the story of early visual processing, by adding, for example, details on how the tuning of striate cells for orientation relates to the orientation tuning of a perceived aftereffect.

The overall body of knowledge provided by this work reflects a remarkably intimate relationship between physiology and psychology. In many cases, the dovetailing of psychological and physiological findings is so good that few doubt that both kinds of study are tapping the same mechanisms. Of course, care is necessary in any given instance of interpretation. Finding a similarity between physiological and psychological results is not conclusive proof that the latter stem directly from the mechanisms revealed by the former. But when all is said and done, the use of aftereffects as psychophysical probes to match the microelectrode probes of the neurophysiologist is one of the great achievements in psychophysics.

Contrast Sensitivity and Orientation

The availability of cheap and fast computers means that gratings for psychophysical experiments can be conveniently generated on a computer screen, and their contrast, orientation, and spatial frequency can be easily adjusted by the experimenter.

The graph in **4.7** gives results from an imaginary experiment conducted using such stimuli. The data are based on findings from real experiments simplified for our purposes. They show how sensitivity to contrast varies according to the orientation of the

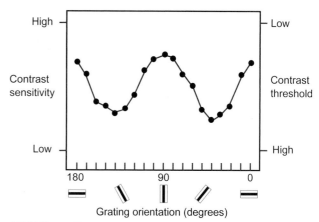

4.7 Oblique effect
Contrast sensitivity is worst for oblique gratings, best for vertical and horizontal gratings. Schematic data based on typical findings.

grating stimuli. High sensitivity (left-hand vertical axis) is the result of low contrast thresholds (right-hand vertical axis), and low sensitivity is the result of high thresholds. For instance, if the subject had a very low contrast threshold for a grating, this would mean that it was seen with a very weak contrast, so it is sensible to say that the subject was very sensitive to it.

The striking feature of **4.7** is the ***oblique effect***: the observer in question was least sensitive to oblique gratings and most sensitive to vertical and horizontal gratings. In other words, more contrast was needed to see an oblique grating rather than to detect a vertical or horizontal one. The possible causes of this intriguing oblique effect are discussed later.

Measuring Orientation-Specific Contrast Elevation

What would the graph in **4.7** look like if the data for it had been collected after a period of adaptation to a high contrast grating of, say, vertical orientation?

This experiment could be performed by placing a second computer screen showing a high contrast vertical grating alongside the one shown in **4.7**. The observer could adapt to this grating and then immediately transfer gaze to the other computer screen showing the test stimuli for the purpose of collecting post-adaptation threshold measurements.

The kind of results that would be expected are shown in **4.8**. This includes the pre-adaptation sen-sitivity curve of **4.7** as a baseline against which the effects of adaptation to vertical can be judged. The key point is that sensitivity is hardly affected at all by adaptation exposure to gratings more than 20-30° away from vertical: this is shown by the way the pre- and post-adaptation curves intertwine for all orientations other than those around vertical. Here then we have a psychophysical demonstration of ***orientation specific adaptation*** which matches the informal demonstration you experienced for yourself when inspecting **4.4**.

Suppose the subject had been adapted not to vertical but instead to an oblique orientation of, say, 140°. The data would then have looked rather like those in **4.9**. Again, the pre-adaptation sen-sitivity curve is shown for comparison, and again there is an adaptation notch cut out from the post-adaptation curve. This time the notch is centered on 140° as one would expect because this was the adaptation orientation but it has roughly the same width. So once again we have orientation-specific adaptation, but now to a different orientation.

The advantage of graphs like those of **4.8** and **4.9** is that the effects of adaptation can be seen in detail for a wide range of test-grating orientations. Thus, a measurement can be made of the ***tuning*** of the adaptation effect, i.e., the range of test-grating orientations affected. The tuning is revealed by the width of the notch in the post-adaptation graph. A narrow notch indicates sharp tuning, a wide notch indicates broad tuning.

It turns out that the nature of this psychophysi-

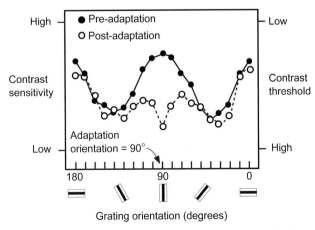

4.8 Contrast sensitivity to gratings before and after adaptation to a vertical grating

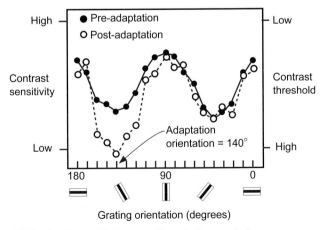

4.9 Contrast sensitivity to gratings before and after adaptation to an oblique grating of 140°

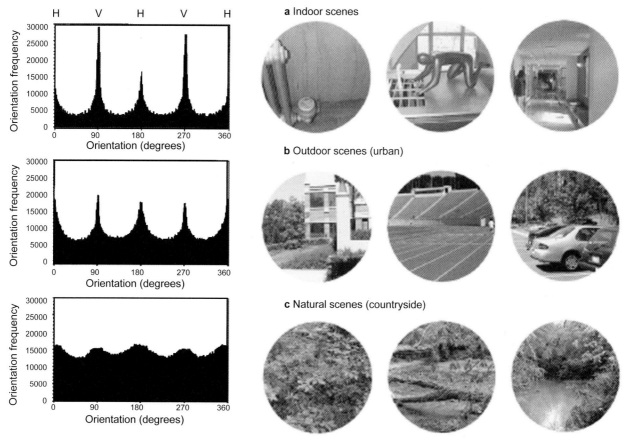

4.10 Orientation frequencies in different kinds of images
Each graph shows the averages of analyses of 50 representative scenes of the type shown to the right of each graph. Vertical contours (V) are 90° and 270°, and horizontal contours (H) are 0° and 180°. These orientations predominate, even in natural scenes **c**, but particularly in indoor **a** and outdoor **b** scenes. With permission from Coppola, Purves, McCoy and Purves (1998).

cally observed orientation tuning fits in remarkably well with the neurophysiologically observed orientation tuning of cells in the striate cortex. It is this kind of intimate interplay between data from the two approaches which has caused such excitement and a vast body of interrelated work over the past few decades. When data like those in **4.8** and **4.9** are compared with the orientation tuning graph for a simple cell shown in Ch 3, **3.9**, it is no wonder that, as we noted earlier, adaptation aftereffects have been called the *psychologist's microelectrode.*

The graphs of **4.8** and **4.9** have been discussed in terms of *lowered contrast sensitivity* caused by adaptation. An alternative way of saying the same thing is that the adaptation has caused *elevated contrast thresholds.* These two quantities are related by the equation:

contrast sensitivity = 1/(contrast threshold).

Try turning the book upside-down and you will see peaks of raised contrast thresholds, again centered on each adaptation orientation. These peaks are of course identical to the notches cut in the sensitivity curves. As explained for **4.7**, whether one talks in terms of contrast sensitivity or contrast thresholds is a matter of convenience or suitability for the particular experiment.

Orientation Tuning and the Environment

We referred earlier to the oblique effect which is the finding that oblique stimuli have higher thresholds than vertical and horizontal stimuli. This phenomenon has been observed in many different types of psychophysical experiments. Various hypotheses have been proposed about why this should be so, ranging from optical factors to do with the lens system of the eye (an improbable idea, in fact, at least for explaining all oblique effect data) to structural neurophysiological causes. An example of a neurophysiological hypothesis is that the striate cortex is particularly specialized for dealing with vertical and horizontal orientations, either by having more cells for those orientations and/or those cells being more finely tuned. Which of these possibilities is the case, or indeed whether both are true, is a current research issue.

The visual system's preference for vertical and horizontal could be a result of the kind of self-calibration processes mentioned below as an explanation of the tilt aftereffect. The idea here is that self-calibration could lead to the striate cortex becoming optimally suited to our environment. Studies of the orientation of edges in images of a wide variety of scenes have shown that there is a predominance of vertical and horizontal edges, **4.10**. This isn't surprising for indoor and urban scenes given the nature of western built environments. You can easily confirm it informally for yourself just by looking around. But curiously, it is also true of country scenes, **4.10c**, albeit to a lesser extent. Could it be that somehow the visual system takes note of this environmental fact and becomes specialized accordingly?

Thus, perhaps changes in threshold contrast sensitivity caused by adaptation to vertical edges and the oblique effect might reflect the same underlying self-calibration neural processes. The oblique effect might differ from the reduced sensitivity seen in adaptation experiments to oriented stimuli only in arising from *everyday* exposure to a preponderance of vertical (and horizontal) edges.

A recently developed theory of perception developed by Dale Purves and R. Beau Lotto in 2003 fits in well with the results shown in **4.10**. They suggest that the visual system exploits what they call a **radical empiricist strategy** in resolving image ambiguities. The idea is that these ambiguities are resolved by *associations* learned from experience of the relative frequency of occurrence of the diverse possible sources of any given ambiguous retinal stimulus. That is, we learn to see the most probable scene entities that could "explain" retinal images and we ignore possible but rare interpretations of those retinal images. The associations are gleaned from the success or failure of behavioral responses that have been made to the same or similar

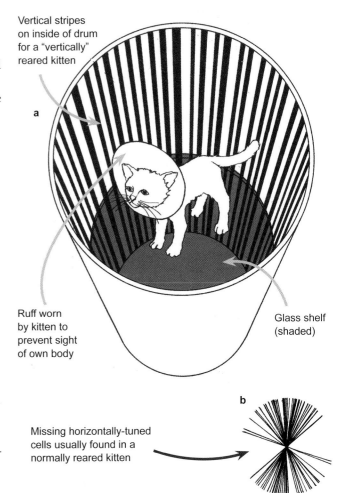

Vertical stripes on inside of drum for a "vertically" reared kitten

a

Ruff worn by kitten to prevent sight of own body

Glass shelf (shaded)

b

Missing horizontally-tuned cells usually found in a normally reared kitten

4.11 Blakemore and Cooper's experiment
a Drum apparatus
b Each line represents the preferred orientation of a striate cell found in the kitten's cortex after rearing with vertical stripes on the inside of the drum. The lack of horizontally-tuned cells is clear. Kittens raised with horizontal stripes in the drum stripes showed a loss of vertically-tuned cells. Redrawn with permission from Blakemore and Cooper (1970).

stimuli in the past. For example, Purves and Lotto have argued that the perception of angles reflects the relative frequencies with which different angles in a scene project into retinal images.

Purves and Lotto have contrasted their empiricist approach to vision with explanations cast in terms of orientation tuned channels in the striate cortex. Their core argument is that because any element of a visual stimulus could have arisen from infinitely many different objects "the output of any detector to the rest of the visual system is necessarily as ambiguous as the stimulus it presumably encodes" (see p. 5 of Purves and Lotto, 2003). But the flaw in that argument, in our view, is that the business of feature detection, and indeed vision generally, is precisely to resolve the inherent ambiguity in the stimulus *using constraints derived from a computational analysis of the problem.*

Evidence for environmentally driven tuning of the cells in the visual cortex has come from an experiment conducted by Colin Blakemore and Grahame Cooper in 1970. They showed that early visual experience can play an important part in the development of the striate cortex in cats. They raised kittens in complete darkness except for certain limited exposures each day to a special visual environment. Some kittens were exposed only to vertical stripes, others only to horizontal ones. This was achieved by placing the kittens in a drum with internal surfaces decorated to give the required stimulation, **4.11a**. The kittens wore a ruff to ensure that they could not even see their own bodies, and to restrict visual experience to either vertical or horizontal bars and nothing else. Subsequent neurophysiological recordings from the kittens" brains showed that the "vertically reared" kittens lacked horizontally tuned striate neurons, **4.11b**, and that the "horizontally reared" kittens lacked vertically tuned ones, **4.11**. In other words, the strange visual environments in which the kittens were reared had a dramatic effect on the development of their visual cortex.

Subsequent work has shown that this environmental tuning effect is not always easy to obtain. The reasons for this are not yet clear. The central question engaging much current research is: to what extent is the visual system *pre-wired* by the unfolding of genetically determined growth processes in early life, and to what extent are genetically determined structures *tuned* to deal with the particular visual environment they encounter?

Different research workers hold different views on this question of *visual plasticity*, as it is called, and the answers will be different for different animals. Some creatures are born with their eyes closed (e.g., kittens) and have a relatively gradual visual development which seems to be most active several weeks after birth (at around the fifth week

of life for kittens). Other animals (e.g., lambs, wildebeest) are born in a high state of readiness to deal with their environments, both by running away from predators and, possibly, by seeing them in the first place.

Where humans fit into this range is controversial but there is no doubt that babies improve their visual capacities in early life and that if they suffer from certain eye abnormalities then this can affect them for life. For example, a congenital eye deviation (squint; technically a strabismus) which is uncorrected for several years after birth is almost certainly followed by an inability to see stereoscopic (two-eyed) depth for life. This is because there appears to be a special **critical period** around 3 months of age for the normal human infant, which is when stereo vision develops. We discuss stereovision in Chs 18 and 19.

Discoveries about the role of early visual experience in animals have had important influences on the way clinicians treat children with visual disorders such as squints. The general impetus given by the animal work is to try early treatment, hopefully while the visual system is still in a flexible or "plastic" state of development.

However, some surgeons take the view that operating early in life to correct a squint can do more harm than good as it can be difficult to get the desired alignment of the eyes when operating on young babies. This is a highly controversial issue in current clinical research on treating squints in young children.

Before leaving the question of why vertical and horizontal edges are treated preferentially by the visual system, we must point out that not all races seem to show this effect equally. Some studies have suggested, for example, that Asians show little difference between vertical/horizontal and obliques when compared with Caucasians—even Asians brought up in a westernized environment which emphasized vertical and horizontal features. Consequently, one cannot rule out a genetic explanation for the effect, in which tuning for the environment was achieved by natural selection.

Once again, we note that the issue of how the visual system might tune itself during its lifetime rather than over generations to deal efficiently with its visual environment is taken up later in this chapter.

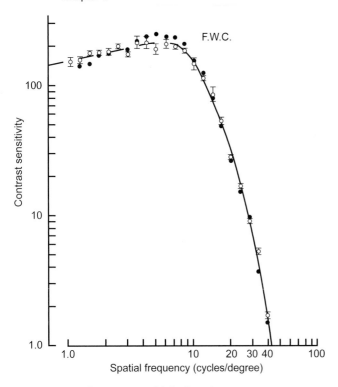

4.12 Contrast sensitivity function
Data for gratings of different spatial frequency (equivalently, bar width). These data come from Fergus Campbell, whose initials are F.W.C. and who was instrumental in early work using aftereffects as probes of neurophysiological mechanisms. Redrawn with permission from Blakemore and Campbell (1969).

Spatial Frequency Contrast Sensitivity Function

The kind of experiment described above for orientation-specific adaptation can be conducted for gratings which vary in *spatial frequency*. Here the experiment would begin by measuring the observer's contrast sensitivity for a wide range of different spatial frequencies. Then the observer would be adapted to a grating of just one spatial frequency and contrast sensitivity for the full range of spatial frequencies re-measured. In this experiment, orientation would be kept constant throughout, as the variable stimulus property of interest is simply spatial frequency.

This time we can look at actual data taken from a classic paper by Colin Blakemore and Fergus Campbell published in 1969. Thus, **4.12** shows Campbell's pre-adaptation contrast sensitivity graph for gratings of varying bar width. This graph is called a ***contrast sensitivity function***.

The term function is used by mathematicians for a relationship between two variables. Here the variables are contrast sensitivity plotted on the vertical axis and spatial frequency plotted on the horizontal axis. Recollect that contrast sensitivity is the reciprocal of the contrast threshold, that is, it equals 1/(contrast threshold), and that spatial frequency is the technical term used for the widths of bars in sinusoidal luminance profiles of the kind shown in **4.5**.

The axes in **4.12** use ***logarithmic*** steps as a convenient way of squeezing a wide range of values into a graph. Thus, the main steps in the contrast sensitivity axes are 1, 10, 100, so the main steps in this graph refer to ten-fold changes.

The most prominent point illustrated in **4.12** by Campbell's contrast sensitivity function is that spatial frequencies higher than about 10 cycles/degree needed more contrast to be seen than spatial frequencies lower than 10 cycles/degree (equivalently, bar widths smaller than about 0.5 mm viewed from 57.3 cm, **4.5**). In other words, Campbell showed a marked *decrease* in contrast sensitivity for spatial frequencies above about 10 cycles/degree. The fall-off in contrast sensitivity becomes very steep in the 20-30 cycles/degree

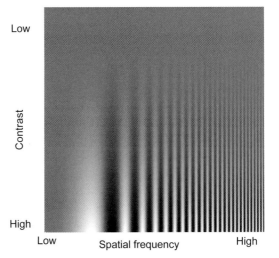

4.13 The contrast sensitivity function
Gratings vary from low to high spatial frequency from left to right, and from high to low contrast from bottom to top. If you view this figure from about 1 meter you will see a "hump" of lines centered just to the left of the middle of the horizontal axis. This is a demonstration that the mid-range of the spatial frequencies depicted can be seen with lower contrasts than the high or low spatial frequencies. You are thereby seeing the shape of your contrast sensitivity function, **4.12**.

region. The limit of this decrease is reached for spatial frequencies of 40-50 cycles/degree (bar widths of 0.125-0.1 mm viewed from 57.3 cm). Beyond this, Campbell was unable to see a grating no matter how high its contrast. That is, his visual system could no longer pick out the grating bars—they had become invisible to him.

The contrast sensitivity function also shows a decrease in sensitivity to very low spatial frequencies, although the range used in **4.12** does not show this very well.

To help you get a clear idea of what the contrast sensitivity function is all about, **4.13** shows a picture that allows you to see your own contrast sensitivity function, with reduction in sensitivity for both low and high spatial frequencies.

The thinnest width of bars revealed by the upper visibility cut-off in **4.12** is called *visual acuity*, the measurement that optometrists are interested in when they assess whether a person needs glasses. However, they usually go about discovering it using charts composed of letters or symbols with *high* contrasts, **4.14**.

Effect of Adaptation on the Contrast Sensitivity Function

What happens to Campbell's contrast sensitivity function after adaptation to a grating of, say, 7 cycles/degree? The results obtained from an experiment of this kind are shown in **4.15**. The pre-adaptation curve is plotted as a continuous line: it is in fact a line fitted to the data points of **4.12.** The notch of decreased sensitivity caused by the adaptation is clearly evident in the post-adaptation data. The decreased sensitivity means that gratings in the region of this notch needed more contrast to be seen. This notch is centered on the adaptation spatial frequency. There is some spread of the adaptation effect, just as there was in the orientation case, but spatial frequencies well away from the one used for adaptation show no decrease in sensitivity. So once again we see the *specific* nature of the adaptation, this time spatial frequency (or bar width) specific adaptation.

Contrast sensitivity function as an envelope enclosing many spatial frequency channels

The *specific* nature of the spatial frequency adaptation, just described, has implications for how we

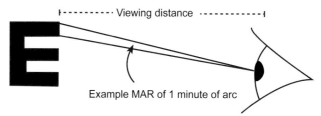

4.14 Visual acuity
This is usually defined as the *minimum angle resolution*, **MAR**. This varies for different stimuli. It is usually measured by optometrists using letter charts: if the letters are so small that they fall below the MAR then the patient can no longer identify the letter. The MAR illustrated here is 1 minute of arc visual angle (1/60 of a degree). This is the angle projected by a letter stroke of just under 2 mm width viewed from 6 m, which is the viewing distance typically used by optometrists for testing *distance visual acuity*. People who have the same small MAR for both near and far viewing distances are said to have **emmetropic** vision: they don't need glasses for either distance or near vision. Visual angle is unaffected by viewing distance but, of course, the size of object required to project a given visual angle does vary with distance. See Jackson and Bailey (2004) for a tutorial.

should think about the contrast sensitivity function of the human visual system. It suggests that this function is an "envelope" marking the boundary of individual sensitivities of a set of component *channels*, each one tuned to a narrow range of spatial frequencies, **4.16a**. Each individual channel is a population of visual neurons within the visual system tuned to a limited range of spatial frequen-

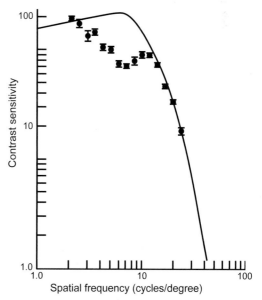

4.15 Contrast sensitivity function following adaptation to a grating of 7 cycles/degree
With permission from Blakemore and Campbell (1969).

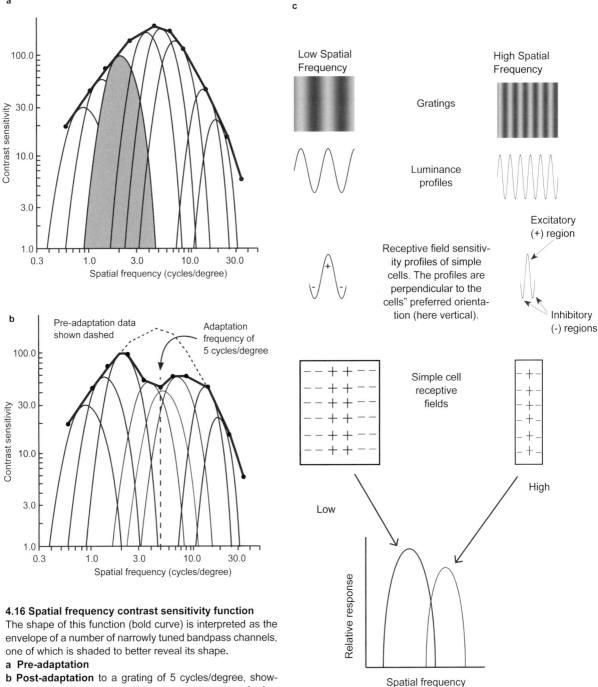

a

c

Low Spatial Frequency

High Spatial Frequency

Gratings

Luminance profiles

Receptive field sensitivity profiles of simple cells. The profiles are perpendicular to the cells" preferred orientation (here vertical).

Excitatory (+) region

Inhibitory (-) regions

b

Pre-adaptation data shown dashed

Adaptation frequency of 5 cycles/degree

Simple cell receptive fields

Low

High

Relative response

Spatial frequency (cycles/degree)

4.16 Spatial frequency contrast sensitivity function

The shape of this function (bold curve) is interpreted as the envelope of a number of narrowly tuned bandpass channels, one of which is shaded to better reveal its shape.

a Pre-adaptation

b Post-adaptation to a grating of 5 cycles/degree, showing a notch of lowered sensitivity due to adaptation of a few channels on and close to the adapting frequency. Drawn schematically guided by typical findings.

c Bandpass filters

Each channel can be described in two complementary ways. First, as a ***bandpass filter***: cells in each channel can be thought of as filters "passing" ('letting through') the limited range of spatial frequencies to which they are sensitive. Second, as devices comprising cells with receptive fields of the simple cell type (Ch 3). Thus, the match of each receptive field to a grating can be thought of in terms of a "key and lock" mechanism. The cells in a channel tuned to low spatial frequencies would have large receptive fields that would "fit" a low spatial frequency grating. Conversely, cells in a channel tuned to high spatial frequencies would have small receptive fields.

cies. The curve that fits the peaks of the individual channels is the envelope.

The ***tuning*** of each channel can be measured neurophysiologically by using microelectrodes to measure the responses of neurons to gratings of different spatial frequencies. Thus, we have two complementary ways of thinking about channels, as explained in the legend to **4.16c.**

Alternatively and non-invasively, psychophysics can be used to measure channel tuning by noting the width of the notch cut into the contrast sensitivity function by adaptation to a grating of a particular spatial frequency, **4.16b.** The idea here is that adaptation reduces the sensitivity of only those channels activated by the adaptation grating. Hence the wider the notch in the kind of plot shown in **4.15,** the broader the tuning of the spatial frequency channels. Equally, a narrow notch would indicate narrowly tuned channels. This is a fine example of the kind of interlocking neurophysiological and psychological studies that have absorbed so much research effort over the past few decades.

Spatial Frequency Contrast Sensitivity Function for Above-Threshold Contrasts

The data described so far have been derived from experiments using gratings with contrasts close to the threshold for seeing them. These low contrasts have told a revealing story about multiple spatial frequency tuned channels. But what happens when gratings are used that have stronger contrasts?

This question was asked by Mark Georgeson and Geoff Sullivan in 1975. They did a contrast matching experiment and found that if contrast was increased sufficiently then frequencies with the *same physical* contrast generated the *same apparent* contrast, **4.17.** This was true of the full spatial frequency range and it accords with what we see when we look around a well-lit scene: high spatial frequency stimuli appear pretty much as "contrasty" as low spatial frequency stimuli, other things being equal.

The results in **4.17** don't mean, however, that contrast threshold data are misleading. Rather, it is best to regard them as telling us an interesting story

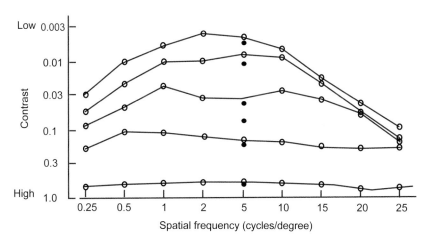

4.17 Contrast matching
The central column of filled dots represents the contrasts of various gratings of 5 cycles/degree, called the *reference gratings*. One of these was shown on any given experimental presentation, along with another *comparison grating* with a spatial frequency in the range 0.25 to 25 cycles/degree. The observer's task was to adjust the contrast of the comparison grating so that it matched the apparent contrast of the reference grating. The matched contrasts are shown by the unfilled dots. Data from the top graph are from presentations on which the reference grating had a low contrast: it has the characteristic shape found for detection contrast thresholds—the contrast sensitivity function in **4.12.** This bowed shape reveals that gratings with spatial frequencies at either end of the frequency range required more physical contrast to appear as "contrasty" as the 5 cycles/degree reference grating when the latter was set to a low contrast. As the contrast of the reference grating was increased the comparison gratings appeared, regardless of their spatial frequency, more and more to have the same apparent contrast as the reference grating. For the bottom graph, all gratings were perceived with about the same contrast. Based on data from Georgeson and Sullivan (1975).

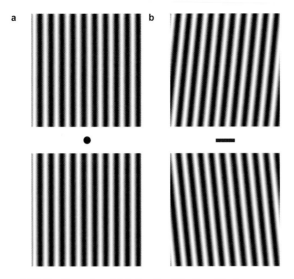

4.18 Gratings for obtaining the tilt aftereffect
a Vertical test gratings
b Tilted adaptation gratings. See text for how to view.

about the channels in the early stages of the visual system. The above-threshold contrast matching data are a reminder that later stages *interpret* the data delivered by those channels in generating our perceptions. Indeed, Georgeson and Sullivan suggested that their results indicated that spatial frequency channels in the visual cortex are organized to compensate for earlier attenuation, and that this achieves "deblurring" of the image.

Tilt Aftereffect

So far we have considered only aftereffects of contrast threshold elevation caused by adaptation to grating stimuli but there are other revealing consequences of such adaptation.

Look at **4.18a.** Hold the book vertically while doing this and fixate (i.e., fix your gaze on) the dot between the two gratings. Check that these gratings are composed of vertically oriented bars.

Then, after you have finished reading this paragraph, transfer your gaze to fixate the short horizontal bar between the two gratings of tilted bars in **4.18b.** Keep looking at this short bar for about a minute, but as you do so move your fixation to and fro along it, to avoid an after-image. (Another precaution is that you shouldn't do this experiment if you suffer from epilepsy or migraines as staring at the gratings can trigger attacks.)

After the 1 minute *adaptation period*, quickly transfer your gaze back to the left-hand test grat-

ings, again fixating the spot between them. You should now see that the test gratings momentarily appear tilted from their true orientation of vertical. The illusory tilt for each grating is in a direction *away from* the corresponding adaptation grating, which is why this is an example of a negative aftereffect. Thus, whereas the adaptation gratings together look like a chevron pointing to the right, the vertical test gratings look, post-adaptation, like a chevron pointing to the left.

This illusion, discovered by James Gibson in 1931 and known as the ***tilt aftereffect***, shows that adaptation to bars of one orientation can subsequently affect the perceived orientation of bars of another orientation. The tilt aftereffect finds a ready explanation in terms of the activities of simple cells, as follows.

Consider **4.19.** At the top is shown a row of bell-shaped graphs. Each one describes the orientation tuning curve of a simple cell (compare with **3.9** and **3.12** in Ch 3). The number below each graph gives the *preferred orientation* of the cell, that is, the orientation that creates the highest rate of firing of the cell. This orientation is illustrated by the small gratings set under the peak of each bell-shaped curve. Only cells with preferred orientations of 120° through 90° (vertical) to 60° in steps of 10° are shown, for simplicity.

Above each tuning curve is a black vertical bar which represents one of the vertical bars in **4.18a.** Consider this bar as an input stimulus to the cell when **4.18a** is viewed. The dotted line descending from each of these vertical input bar points to the activity that would be generated in each cell by this vertical bar. These levels of activity are represented by the line drawn on each graph connecting the tuning curve to the horizontal axis.

The set of seven neuronal responses to the vertical bar are brought together and plotted in the graph labeled *Pre-adaptation Profile of Channel Activities to Vertical Bar* in **4.19a.** As can be seen, the profile is symmetric and is centered on the cell with a vertical preferred orientation (90°). This profile can be thought of as a "signature tune" for a vertical stimulus. In essence, this profile of activities is characteristic of a vertical bar.

Now consider **4.19b** (the lower figure). This depicts the neuronal response profile induced by the same vertical grating as above, but follow-

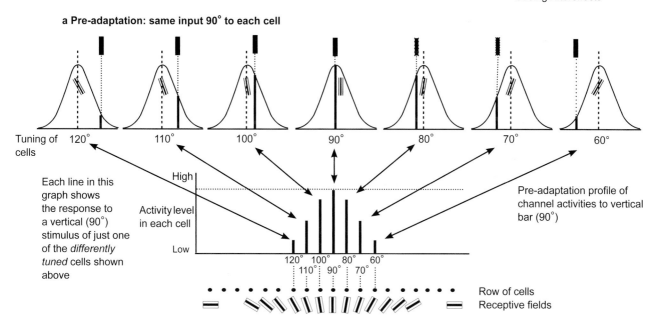

a Pre-adaptation: same input 90° to each cell

Tuning of cells: 120° 110° 100° 90° 80° 70° 60°

Each line in this graph shows the response to a vertical (90°) stimulus of just one of the *differently tuned* cells shown above

Pre-adaptation profile of channel activities to vertical bar (90°)

High

Activity level in each cell

Low

120° 100° 80° 60°
110° 90° 70°

Row of cells
Receptive fields

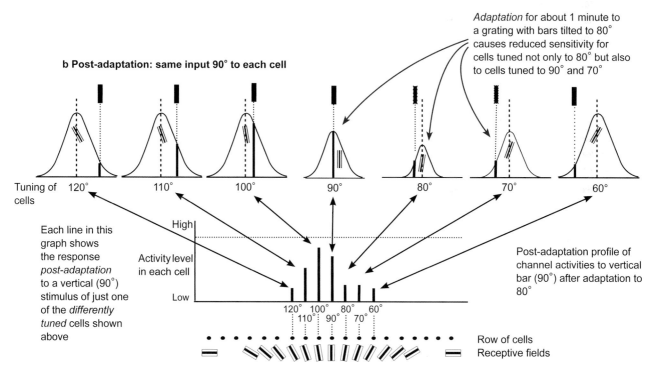

Adaptation for about 1 minute to a grating with bars tilted to 80° causes reduced sensitivity for cells tuned not only to 80° but also to cells tuned to 90° and 70°

b Post-adaptation: same input 90° to each cell

Tuning of cells: 120° 110° 100° 90° 80° 70° 60°

Each line in this graph shows the response *post-adaptation* to a vertical (90°) stimulus of just one of the *differently tuned* cells shown above

Post-adaptation profile of channel activities to vertical bar (90°) after adaptation to 80°

High

Activity level in each cell

Low

120° 100° 80° 60°
110° 90° 70°

Row of cells
Receptive fields

4.19 Explaining the tilt aftereffect in terms of adaptation in orientation tuned channels

The adaptation causes a change in the sensitivity of a sub-set of orientation tuned channels. The upper figure **a** shows the symmetric *pre-adaptation activity profile* generated by a vertical bar (90°), as in **3.13**. The lower figure **b** shows the activity profile to the same vertical bar after about 1 minute of scanning a grating oriented at 80°. The adaptation causes a reduction in the sensitivity of cells tuned to 80° and also those tuned to 90° and 70°. These changes in sensitivity cause the vertical bar to generate a skewed *post-adaptation activity profile.* This skewed pattern is similar to that caused normally by a bar tilted a few degrees anti-clockwise from vertical. (Compare the skewed pattern generated by the 92° bar shown in **3.14**.) Hence, post-adaptation we see a bar tilted a few degrees away from the adaptation orientation—this is the tilt aftereffect. Drawn schematically and approximately to emphasize main points.

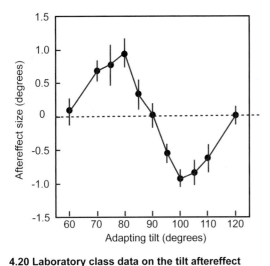

4.20 Laboratory class data on the tilt aftereffect
Each point is the mean of data from 12 students attending a course in the University of Sheffield, who each saw only one adaptation condition. The vertical lines through each point show the range within which 68% of the data points used to calculate the mean fell. The graph shows the *distance paradox*. Kindly supplied by David Buckley.

ing adaptation to a grating of 80°. This adaptation reduces the sensitivity of cells with preferred orientations around 80°, as shown by the lower amplitude (peak height) of the tuning curve for the 80° cell (ie the cell with preferred orientation of 80°), and the somewhat lower amplitudes for the 70° and 90° cells.

The reduction in sensitivity in these channels dictates that, when the vertical grating is shown post-adaptation, the response profile is no longer symmetric, but skewed, and (more importantly) its peak is no longer at 90°, **4.19b**. In fact, this response profile is a "signature tune", not for a bar at 90°, but for a bar at 92°. (Refer back to Ch 3, **3.13** for a reminder on this point.) Therefore a 90° bar is perceived as a 92° degree bar. In summary, the channels hypothesis for explaining the tilt aftereffect is that the post-adaptation response profile induced by a 90° bar (for example) is consistent with a 92° bar. Thus, we have a neat explanation of the tilt aftereffect in terms of changes in the responses of cells in orientation tuned channels.

Notice that this explanation works whether one supposes that adaptation reduces sensitivity by recalibration, by fatigue, or a combination of both (although there might be subtle differences depending on the details of the various theories; this is a current research issue).

It is intriguing that the principles of the explanation exhibited in **4.19** were devised in 1929 by Georg von Békésy, a Nobel Prize laureate for his work on the ear, long before simple cells were discovered around 1960.

There are several properties of the tilt aftereffect consistent with this channels explanation. For example, adapting stimuli with tilts further from vertical than +/-30° do *not* affect the percept of vertical post-adaptation. This fits the theory when cast in terms of simple cells with the tuning curves set out in **4.19**. This is because adaptations far away from vertical will not alter the sensitivities of cells tuned to vertical and near-vertical orientations because these cells are not responsive to far-from-vertical orientations—such non-vertical orientations fall outside their tuning curves.

A second property that fits the theory is that adapting to a *vertical* grating does not alter the post-adaptation percept of the orientation of vertical stimuli. The reason is that, although the vertical adaptation reduces sensitivity to vertical and to near-to-vertical orientations, the dip in the post-adaptation profile still yields a *symmetric* activity profile in the test phase to vertical. This is the key property of the "signature tune" for vertical. Very neat.

The ***distance paradox*** is the name given to two related facts: (1) adaptation has no effect when the test stimulus has the *same* orientation as the adaptation stimulus and (2) adaptation has an effect only when the test stimulus orientation lies nearby and *to one or other side* of the adaptation orientation. These facts are evident in the tilt aftereffect data shown in **4.20**.

So all's well for the channels theory so far. But one tricky problem remains: adaptation to horizontal gratings *does* have an effect on the percept of vertical gratings. This was noted by James Gibson when he discovered the tilt aftereffect in 1931. This is odd, as cells with the tuning curves shown in **4.19** are insensitive to horizontal and so they should have no part to play in the percept of vertical according to the theory. Is this a killer for the theory? The answer is *yes* for the fatigue theory but a version of the recalibration theory can cope with this result.

The recalibration theory of Clifford et al. (2000) holds that the recalibration process has two parts.

One part has the result that adaptation orientations close to the test orientation produce the standard tilt aftereffect. A second part has the result that adaptation orientations far from the test orientation also produce an aftereffect. The details are provided later in this chapter (pages 108-109).

The distance paradox is a characteristic of feature attributes which are said to be ***place coded***. In the present case, the different simple cells for each orientation constitute the different "places." Of course, the difference in physical position in the brain is tiny: the key point is that the difference in stimulus attribute is represented by activity in differently located but physically similar components. (Perhaps a better technical term for place coding might be "coding by connections." It is also sometimes called ***value coding***, Ch 11.)

Neuroanatomy has shown us that all the cells comprising a given orientation channel are structurally identical, and neurophysiology has found that they have the same type of nerve impulses as all brain cells. Thus, all the cells sketched schematically in **4.19** are, it seems, physically alike. What serves to distinguish them is the connections they receive, so that some are "turned on" by certain feature orientations, others by other feature orientations, and so on. This is a kind of coding much used in computers: each transistor can be identical in electrical design but each can serve a vastly different function because of its place in the circuits of which it is a part.

The fact that the tilt aftereffect can be explained neatly in terms of simple cells is another example of why aftereffects have been dubbed the *psychologist's microelectrode*.

Size Aftereffect

If the neurophysiological explanation of the tilt aftereffect is true, then this effect is based on changes in orientation-specific cells caused by adaptation. We know that these detectors are also fussy about spatial frequency (bar width, or size), as well as bar orientation, and we also know that spatial frequency-specific adaptation can be demonstrated using the elevation-of-contrast-thresholds adaptation technique. Consequently, it should be possible to generate a size aftereffect, similar to the tilt aftereffect, but with an illusory shift in perceived *bar width* following adaptation.

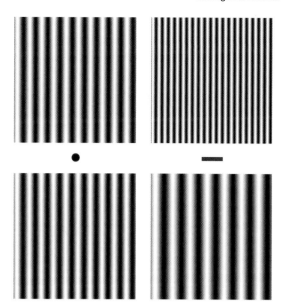

4.21 Gratings for obtaining the size aftereffect
Test gratings are on the left, adaptation gratings on the right.

This is how Colin Blakemore and Peter Sutton reasoned, and they discovered just such an effect in 1969. You can experience this for yourself using the gratings in **4.21**.

First, fixate the spot between the two vertical gratings on the left and check that they appear to be composed of bars with the same width (as indeed they are). Next, adapt your visual system, using the gratings on the right of the figure. Allow your gaze to run along the horizontal bar between the upper (thick) and lower (thin) bars. Continue the adaptation for about a minute or so. Even longer can be helpful in seeing this aftereffect, which is not quite as noticeable as the tilt aftereffect. Following the adaptation period, quickly look back to the spot between the two test gratings. If all has gone well, you should notice that these gratings briefly look different in width: the upper grating seems to be composed of thicker bars than the lower grating. Do not expect a big size effect but the size difference should be clear enough. Note that gratings are not needed to get this effect, as demonstrated in **4.22**. We explain later in this chapter why spatial frequency tuned channels are sensitive to the sizes of stimuli other than bars.

It is easy to apply exactly the same type of physiological reasoning to explain Blakemore and Sutton's size aftereffect as used for the tilt aftereffect this patch of texture is in an intermediate sized font and is suitable for use as the test pattern for seeing the size aftereffect caused by adaptation to the

this patch of texture is in a small font and so it is best picked up by high spatial frequency tuned channels this patch of texture is in a small font and so it is best picked up by the high spatial frequency tuned channels this patch of

●

▬

this patch of texture is in an intermediate sized font and is suitable for use as the test pattern for seeing the size aftereffect caused by adaptation to the smaller and larger

this patch of texture is in a larger font and so it is best picked up by the low spatial frequency tuned channels this patch of texture

4.22 Text samples for obtaining the size aftereffect
Test stimuli on the left, adaptation stimuli are on the right.

in **4.16**. Thus, the adaptation causes changes in the sensitivities of cells tuned to the spatial frequencies of the adaptation bars. Later on, when the test gratings are re-inspected, the profile of activity in spatial frequency tuned cells is changed so that the apparent width of the bars in each test grating is "pushed" away from the width of the adaptation grating. Thus, the upper test grating appears thicker and the lower one thinner in its bar composition. If the labeling of the horizontal axes in **4.20** for orientation is replaced by labeling for spatial frequency then the graphs apply just as well to the size aftereffect as they do to the tilt aftereffect.

The size and tilt aftereffects have been much studied. The tuning of these effects—the degree and extent of their influence—can be established using psychophysical techniques, and inferences drawn about the probable tuning of orientation and/or spatial frequency tuned channels in the brain.

Motion Aftereffect

The emphasis of this chapter has so far been on orientation-specific and size-specific aftereffects because these relate directly to the simple cells described in Ch 3. But the family of aftereffects is a large one. In Ch 14, we introduce a much studied and quite different example of the genre, the *motion aftereffect*.

Simultaneous versus Successive Illusions

Aftereffects are examples of *successive illusions*, ones in which stimulation at one moment in time has consequences for the perception of subsequent events. There are also *simultaneous illusions*, in which two or more stimuli presented together interact to cause an illusory effect. It turns out that there are very many illusions which stem from the presentation of several bar stimuli at one and the same time. Some of these illusions appear highly comparable to the tilt and width aftereffects, and have therefore invited neurophysiological speculations about their causes comparable to the explanation of aftereffects set out above.

Consider, for example, **4.23a**. The two central gratings of circular shape are in fact composed of vertical bars, but the bars look tilted away from vertical in the opposite direction to the angle of the bars that surround them. Are the surrounding bars somehow inducing a change in the activity profile dealing with the central bars, a change of shape comparable to that described for the tilt aftereffect? Blakemore, Carpenter, and Georgeson proposed in 1970 that this might be so, and further suggested that *inhibition* between bar detectors might play the same role in this context that adaptation played in the equivalent successive illusion, i.e., the tilt aftereffect.

The underlying idea here is that orientation-tuned cells might inhibit one another, with each cell attempting to reduce the activity in its "orientation-neighbors," so to speak. This inhibition would disturb the activity profiles produced in response to the vertical bars of **4.23a**, and produce profiles characteristic of non-vertical bars, hence the illusion. The *Zollner illusion,* **4.23b**, provides another example of a vivid orientation illusion which can be explained in the same general way.

What might be the computational purpose served by such inhibition? Why introduce such a curious process capable of generating distortions? In Ch 3, we showed that inhibition (combined with excitation) did the job of giving certain brain cells pattern-sensitivity, of making them possess preferred stimuli for activation. This proved very useful in creating detectors as a step on the way to building up an edge feature representation. But here inhibition is given a rather different role.

a

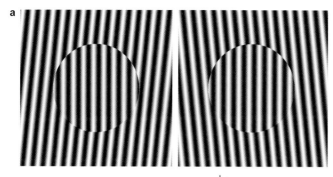

b

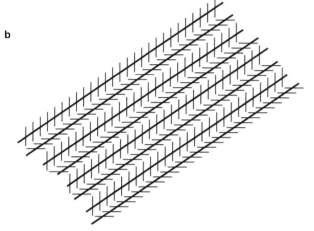

4.23 Simultaneous tilt illusions
a The lines in the discs are all physically but not perceptually parallel to the edges of the page.
b Zollner illusion: the long oblique lines are physically but not perceptually parallel.

Blakemore, Carpenter, and Georgeson, like previous authors who have considered the role of inhibition in sensory systems, suggest that it might in certain circumstances serve to improve the "sharpness" of the tuning profiles of orientation detectors. This might facilitate the interpretation of the activity profiles shown in **4.19** in determining the perceived orientation of the bar features in question. We discussed in Ch 3 the issue of how resolution in orientation detection relates to tuning curves. Related ideas are explored in Ch 15 in connection with *maximum likelihood estimation* for computing motion direction from a set of broadly tuned motion detectors.

An illusion which rounds off very nicely the main empirical content of this chapter so far is **4.24a**. This shows a simultaneous width-of-bar illusion invented by Donald MacKay. The horizontal bars in the inset central region are of the same

width throughout but they look strikingly different in the left and right halves of this region. That is, the region placed against the thin-bar background appears to contain thicker bars than does the region placed against the thick-bar background. This is a simultaneous version of the bar-size aftereffect, **4.21.** An explanation in terms of inhibition between width-specific detectors can be offered for it, just as inhibition between orientation-specific cells was posited for the simultaneous tilt illusion of **4.23**.

Interestingly, MacKay's simultaneous size-contrast illusion is abolished, or at least substantially reduced, if the bars in the central region are set at 90° to those in the surround: compare **4.24a,b**. This suggests that orientation-tuned processes are involved. Moreover, if this effect is caused by inhibition, then this inhibition takes place only between cells with the same preferred orientation. In terms of cells in the striate cortex, this suggests that

a

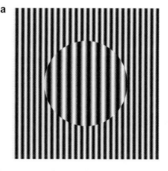

b

4.24 Size-contrast effects
a The lines in the small inset discs are of the same width in both halves of the figure but where surrounded by thick lines (right) they appear narrower than where surrounded by thin lines (left).
b The size-contrast effect is much reduced if the inducing lines in the surround are set at right angles to the central lines. This shows that the effect is at least in part orientation-dependent.

the inhibition is restricted to cells with the same or similar orientation tuning.

Three Approaches to the Problem of Seeing

This chapter has shown how the psychological approach to the study of seeing blends superbly with the neurophysiological and computational approaches described in earlier chapters.

The logic of the *psychological approach* is: let us treat visual phenomena as the output of a device (the visual system) which takes as its input the retinal image, and ask what mechanisms must be present in the device in order to explain the output, given the input. Often, the phenomena chosen for study are illusions, in the hope that when a system breaks down (i.e., when an illusion occurs), this is a good opportunity to gather clues about how the system normally works. The key assumption here is that illusions can be used as probes for studying normal functioning. They are not to be regarded as useless or misleading simply because they are (often) untypical of normal seeing. This psychological approach to visual mechanisms is the natural complement to the computational and neurophysiological approaches.

The *neurophysiological approach* studies perceptual mechanisms directly by physiological and anatomical studies of the visual system, for example, Hubel and Wiesel's work on simple cells, Ch 3.

The *computational approach* attempts to tackle the problem of seeing by investigating the nature of the visual tasks being undertaken. The theories developed are often tested by taking advantage of *computer* modeling. This uses the powerful resources of modern computers to build a model to check if a task theory is adequate when faced with images of various kinds, both artificial and arising from natural scenes.

However, it is important to note that computer modeling is an approach that is not the same thing as the computational approach, as we use the latter term. The hallmark of the latter is discovering good *constraints* that can be exploited in a suitable *algorithm* to recover the desired *representation,* Ch 2. Computer modeling may or may not be used with these specific goals in mind.

All three approaches have become integrated in recent years, so that it is sometimes difficult to

know who is the computational scientist, who is the psychologist, and who is the neurophysiologist. It is increasingly common for any one visual scientist to wear any one of these hats at different times, depending on the particular problem under study. However, because of the different skills involved, it is more usual to find multidisciplinary research teams attacking a problem with their complementary approaches.

But it is prudent to end this chapter on a cautionary note. The richness of visual phenomena and neuroscience findings is such that there is temptation for psychologists and neurophysiologists to neglect the importance of asking: what is the task being solved by the visual system and how might it be solved? Those who subscribe to the computational way of thinking believe that ignoring this question can lead to psychologists and neurophysiologists missing out on devising studies guided by theories of seeing that could point the way to new and revealing experiments.

Skip from Here to Summary Remarks?

We have now explained a number of core concepts, particularly spatial frequency tuned channels, in sufficient detail for the reader to proceed to *Summary Remarks* on p. 109, and then to move on to Ch 5. The remainder of this chapter introduces some related topics using more technical detail than has been done so far.

Modulation Transfer Function

We introduced earlier in this chapter the concept of *spatial frequency tuned channels.* Here we explain further how those channels can be thought of as *bandpass filters.*

Images of natural scenes usually have intensity variations of different kinds. Some edges in images are very sharp, e.g., those arising from many object boundaries, whereas others are quite fuzzy, e.g., those arising from some kinds of shadows. Technically, this is described as images having intensity variations at different *scales*. We discuss this issue in detail in Ch 5. For the moment, we show how the concept of spatial frequencies is well suited to describing information in images at different scales. We will explain how a *coarse* image scale is said to deal with low spatial frequencies, and a *fine* image scale with high spatial frequencies, **4.25.**

It is possible to process an image to select a particular range of spatial frequencies. It can be done using receptive fields of suitable size and type. For example, a simple cell with a receptive field composed of large regions of excitation and inhibition selects a band of low spatial frequencies, whereas a receptive field with small regions picks out a band of high spatial frequencies, **4.16c**. This kind of image processing is called spatial frequency *filtering*. To explain this idea further it is helpful to draw a parallel with hearing.

Consider a device used to amplify music signals stored on, say, a compact disc. The performance of the amplifier can be described in terms of a graph showing how well it amplifies each sound frequency. This graph is called a ***modulation transfer function***, **4.26**. This is an excellent name

because it captures exactly what the amplifier is doing: "transferring" different sound modulations (frequencies) from the storage medium (e.g., a CD) to the speakers and thus into the sound waves that enter the listener's ears.

Notice that the modulation transfer function shown in **4.26** transfers best frequencies in a middle range. It is said not to ***attenuate*** those frequencies very much. On the other hand, it transfers high and low frequencies less well—it attenuates them. Indeed, in **4.26** the high frequency component isn't passed at all, so it doesn't appear in the output waveform, which is why the latter has lost the spiky character of the input. For these reasons this modulation transfer function is said to be that of a ***bandpass filter*** because it "passes" or "filters" only a limited band of frequencies. It blocks all others.

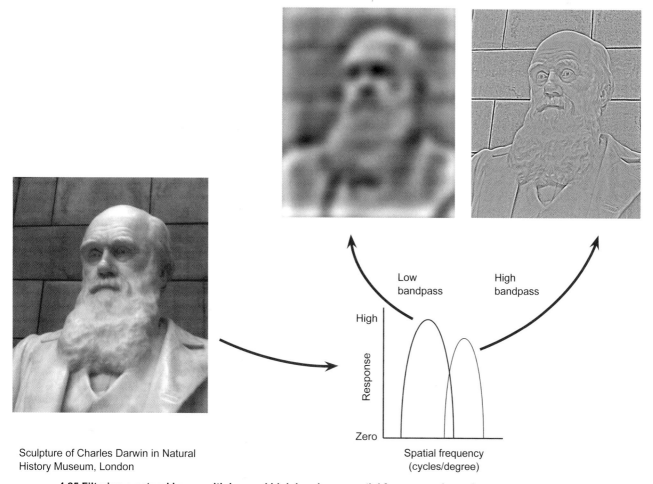

Sculpture of Charles Darwin in Natural History Museum, London

4.25 Filtering a natural image with low and high bandpass spatial frequency channels

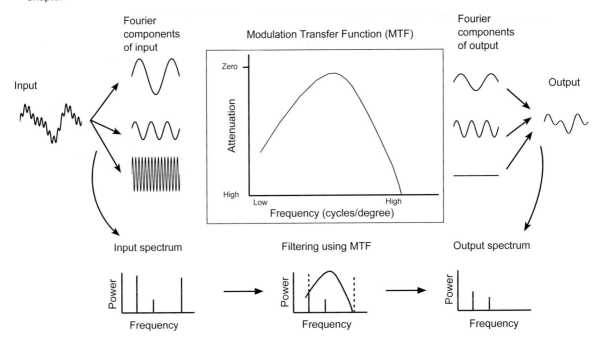

4.26 Modulation transfer function
The input waveform has three main sine wave (Fourier) components. Their relative strengths (that is, their amplitudes, or power) are shown by the vertical bars in the input spectrum graph. If this input is fed into a device with the modulation transfer function (MTF) shown then the output is not a faithful copy. Instead, the strengths of the components are changed. In this case, the component with the high frequency is not passed by the MTF at all: it is said to be "filtered out." The low frequency component is also altered, by reducing its power. The middle frequency component is unchanged. The output spectrum graph summarizes these changes. When the changed components are added together the output waveform is as shown on the right. Notice how the sharp, spiky (high frequency) features of the input do not appear in the output as they have not been passed by the MTF. The coarse structure of the input is, however, maintained in the output.

The modulation transfer function is a convenient and succinct way of describing a great deal of useful information about the amplifier's performance. Obviously, you would not want to pay much for an amplifier if you were told that its modulation transfer function had a very narrow bandpass characteristic, because then you would miss musical sounds outside this narrow band of frequencies (e.g., perhaps little or no bass or treble).

Spatial Frequency Contrast Sensitivity Function as a Modulation Transfer Function

We now apply to image processing the ideas just explained in the context of audio amplifiers. How well does the visual system "transfer" (that is, "see") different spatial frequencies? For stimuli at or close to contrast threshold, this information is given by the contrast sensitivity function, **4.12**. This graph shows how the human visual system can be regarded (at contrast threshold) as a bandpass sys-

tem, dealing with spatial frequencies over a limited range, with an upper cut-off of about 30-40 cycles per degree for most people.

We showed earlier how the contrast sensitivity function can be interpreted as the "envelope" of a set of component filters, each one bandpass-tuned to a narrow range of spatial frequencies, **4.16a**. Each individual filter, or channel, is a population of visual neurons which share the same tuning characteristics.

The tuning of each channel can be measured neurophysiologically by studying the responses of neurons to gratings of different spatial frequencies using microelectrodes. Alternatively and non-invasively, psychophysics can be used to measure channel tuning by noting the size of the notch cut into the contrast sensitivity function by adaptation to a grating of a particular spatial frequency, as was shown in **4.16b**.

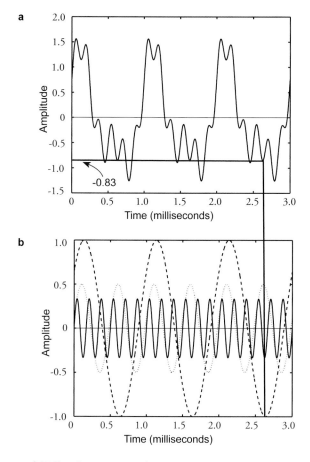

4.27 Fourier components
a A complex waveform.
b The result of analyzing the complex waveform into its sine wave (or fourier) components. This means that if the components were added together, the complex waveform would be recreated exactly. To help see how this works on a point for point basis, a vertical line is shown straddling the two figures: the values of the Fourier components at this point are 0.5, -0.33, and -1.0, which when summed give the total of -0.83 picked out in the upper figure.

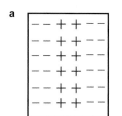
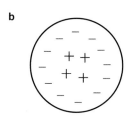

4.28 Oriented and circular receptive fields
If light falls on the regions shown with + symbols the cell becomes excited. Conversely, light falling on regions marked with – symbols inhibits the cell. Fields are for:
a Receptive field of simple cell found in the striate cortex.
b Receptive field of retinal bipolar cell.

Fourier Analysis

The theoretical underpinning of the modulation transfer function is a theorem proved by the great French mathematician, Jean Baptiste Joseph Fourier (1768–1830). This theorem states that any **waveform** can be decomposed into a set of component sine waves, each with a particular **amplitude** and **phase** for the waveform in question, **4.27**. The amplitude says how much power there is in each component sine wave, that is, how large the difference is between the peaks and the troughs of the sine wave. The phase specifies the left-right position in which each sine wave has to be placed when adding all the sine waves together to yield the original waveform.

The set of component sine waves is known as the set of **fourier components** of the waveform. So using this terminology, we can say that the modulation transfer function shows how well the system in question deals with each fourier component in the input.

Spatial Frequency Filtering of Natural Images

We have already shown an example of what bandpass filtered images look like, **4.25**. They were created by analyzing an image with spatial frequency bandpass receptive fields. That can be done with a range of oriented receptive fields of the general kind shown in **4.28a**. However, that would mean showing the results of using many receptive fields, each one dealing with a particular combination of orientation and spatial frequency tuning. We do that in Ch 9, but for those reading this chapter immediately after Ch 3, we now consider the simpler process of using **circularly symmetric** receptive fields, **4.28b**. Certain cells in the retina have receptive fields of this type, and these fields respond to intensity variations in the image at all orientations.

The bandpass images of Darwin in **4.25** were obtained using two different bandpass circular receptive fields. The large and small receptive fields "picked up," or "selected," the low and high spatial frequency components respectively. To understand **4.25**, bear in mind that the gray level of each pixel in the bandpass images represents the result of placing the receptive field in the corresponding place in the original image. By "result," we mean the answer obtained by taking the excitation created by light

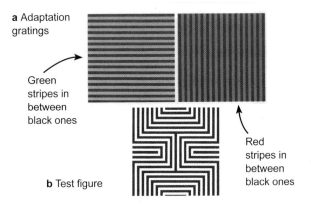

a Adaptation gratings

Green stripes in between black ones

b Test figure

Red stripes in between black ones

4.29 McCollough effect
See p. 418 in Ch 17 for the proper colored version of this figure.

landing on the excitatory region of the receptive field, and subtracting the inhibition generated by light landing on the inhibitory region. This simple subtraction yields a number and mid-grey is used to show when that number is zero. If the number is positive (excitation exceeding inhibition) then it is shown in **4.25** as somewhere between mid-grey and white, depending on how big this number is. Similarly, if the number is negative (excitation less than inhibition) then it appears in **4.25** as somewhere between mid-grey and black.

The main fact illustrated by **4.25** is that low spatial frequency filtering picks up coarse scales of the changes in image intensities. It misses the details. The converse happens for high spatial frequency filtering. It captures (equivalently, "transfers" or "passes") only the fine scale details. Ch 5 will delve into further details on this.

Contingent Aftereffects

In 1965, Celeste McCollough (pronounced McCulloch) discovered a particularly odd kind of aftereffect which has attracted a great deal of research, with hundreds of published reports on it. You may see the aftereffect for yourself using the proper colored version of **4.29** on p. 418.

First, look at the test stimulus **4.29b**, which is simply a pattern of black and white bars, some vertical and some horizontal. Be sure to confirm that this stimulus appears quite colorless to you. Next, adapt your visual system using **4.29a**. Look for 10 seconds at the red-vertical pattern and then

for 10 seconds at the green-horizontal one, then go back to the red-vertical for a further 10 seconds, and so on. Alternate between the two adapting stimuli for a period of at least 3 minutes. If you can bear it and want to obtain a good effect, do this for 10 minutes. When the adaptation period is up, transfer your gaze back to **4.29b** and note what you see.

After adaptation the black and white test stimulus appears colored; this is the **McCollough effect**. However, these color illusions are not color after-images of the usual sort, **17.17**, p. 408. If you look carefully you will see that the vertical bars in the test stimulus appear faintly greenish (the vertical adaptation bars are red) and the horizontal test bars appear faintly reddish (the horizontal adaptation bars are green).

However, recollect that all parts of the retina were equally exposed to red and green during the adaptation phase so the illusory colors cannot be due to the same cause (bleaching of just one set of color receptors) that is thought to underlie the normal negative colored after-images of **17.17**. Somehow, the negative color aftereffects have become "tied to" the orientation of the inducing stimulus.

Try turning **4.29** through 90°. You should find that the illusory colors in the test stimulus change place, showing that "vertical" and "horizontal" are defined with respect to the retina, not the world, for this effect.

The McCollough effect was the first of several such aftereffects discovered in the decade after its first report. As a group they are called **contingent aftereffects**, because the appearance of an aftereffect of some kind is dependent, or "contingent," upon some other stimulus characteristic (for the McCollough effect, it is contingent on the vertical and horizontal contours).

Similar color aftereffects can be made contingent on direction of movement (e.g., leftward versus rightward movement), or on spatial frequency (thick versus thin stripes), or on curvature (convex versus concave bars). Moreover, it is also possible to do things the other way around and, for example, make a movement aftereffect contingent upon the color of display. All kinds of such combinations can be employed, hence the many research studies on these effects.

But there have been reasons other than sheer curiosity behind the flurry of research into contingent aftereffects. The hope has been that contingent aftereffects would provide a new tool for probing the workings of the visual system, a new kind of microelectrode if you like. Thus, various people have speculated that these aftereffects might reflect the presence in the visual system of "doubly-tuned" neurons, ones which require that a stimulus should have two (or more) characteristics before the neuron will fire.

For example, there may be neurons which require their preferred stimulus to be not only of the right orientation but also of the right color. Examples might be cells tuned to vertical-and-red, oblique-and-red, horizontal-and-red, vertical-and-green, oblique-and-green, and so on. Now if such cells were coupled in an opponent-process manner, much as we supposed to be the case when explaining the movement aftereffect, then a neural basis exists for explaining the aftereffects. The effects of the period of adaptation may be to recalibrate and/or fatigue one half of each opponent pairing, with the aftereffect then emerging as the post-adaptation consequence. Encouragement for this type of explanation comes from neurophysiological reports that some orientation detectors are indeed also color-specific.

But there are problems for explanations couched in terms of doubly-tuned neurons. First, given the large number of contingent aftereffects, this explanation implies vast numbers of doubly-tuned cells, enough to cover all the possible combinations of stimulus attributes now known to give contingent aftereffects. This seems rather wasteful of neural machinery. On the other hand, there are indeed vast numbers of cells in the visual cortex. We discuss this problem at length in Ch 11 when we introduce **complexity theory**.

Second, contingent aftereffects can be very durable. If you managed to last out the full 10 minutes of adaptation to **4.29a**, your contingent aftereffect is probably still present. So try looking again at the test stimulus **4.29b**. Did you see a fleeting reappearance of the illusory colors? If so, your experience is in keeping with many reports of the long-lasting nature of these effects. They can last for weeks if the initial adaptation is long enough (e.g., half an hour or so). This longevity suggests a

mechanism quite different from the usual short-term adaptation of neurons posited for normal aftereffects. It favors an explanation involving some kind of learning. As John Mollon succinctly put it in 1974, it suggests that if we do not have neurons jointly specific to color and orientation before we adapt to **4.29a**, perhaps we do when we have finished. In other words, perhaps adaptation can *create* doubly-tuned cells.

Adaptation: The Recalibration Hypothesis

What is the nature of the changes in the visual system caused by viewing the adaptation stimulus? We described in brief earlier in this chapter the fatigue and recalibration hypotheses. We now delve deeper into processes of recalibration. Despite the plausible account that recalibration provides, it should be borne in mind that this remains a working hypothesis.

Recollect that we said earlier that the striate cortex has to face up to a problem well known to electrical engineers. This is the problem of designing self-calibrating systems that can re-tune themselves when certain types of changes occur in their inputs and/or their components.

The next few sections contain technical details that you may choose to skip on first reading. If so, proceed to the *Summary Remarks*, p. 109.

Correlation and the Contingent Aftereffect

In statistical terms, if two neurons, say *a* and *b*, provide similar information then their outputs, Ψ_A and Ψ_B, (respectively) are said to be **correlated**. This means that if Ψ_A is known then we can deduce something about Ψ_B. The higher this correlation is, the more we can deduce about Ψ_B from Ψ_A, and vice versa.

As the correlation between Ψ_A and Ψ_B increases they are said to become increasingly **redundant**. This term is used because, in the limiting case, when Ψ_A and Ψ_B are identical, then knowing Ψ_A makes knowing Ψ_B redundant. In this case, there is a correlation of $r = 1.0$, where r is the conventional symbol used to denote correlation. Conversely, if the output Ψ_A is independent of the output Ψ_B then this implies a correlation of $r = 0$.

Thus, one way for the brain to ensure an efficient use of its neurons is to minimize the amount of redundancy, which implies that *a* and *b* have outputs

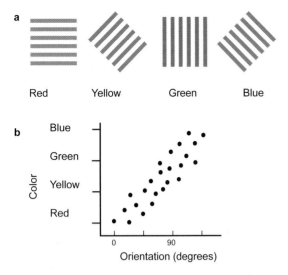

4.30 Generalized contingent aftereffect
a Instead of associating two colors with two orientations, a spectrum of colors is associated with the full range of orientations (only four displayed here).
b A graph of orientation versus color for this set of stimuli shows a large correlation between orientation and color. See p. 254 for the colored version of this figure.

that are independent of each other, and that this is true of every pair of neurons. This perspective has become increasingly popular, and is associated primarily with the work of Horace Barlow; it is known under various names, examples being **decorrelation** and **redundancy reduction**. Our account builds on Barlow's work.

Let's consider a generalization of the normal contingent aftereffect in which red stripes are vertical and green stripes are horizontal. This generalization consists of a range of orientations and stripes, such that each orientation is usually presented in a specific color, as shown in **4.30a**. This introduces a correlation between color and orientation, where the amount of correlation can be varied systematically between zero and unity.

For example, if line orientation is made to vary systematically, so that horizontal lines are usually red, lines at 45° are usually orange, lines at 90° are usually yellow (and so on), then this introduces a correlation, where the amount of correlation is determined by how reliably each color is associated with a specific orientation. If we plot a graph of orientation versus color for the stimuli in **4.30a** then each color-orientation stimulus defines a point, as shown in **4.30b**. For example, if a color-orientation stimulus is colored red with an orien-

tation of 45° then this defines a point low down on the vertical (color) axis and mid-way along the horizontal (orientation) axis.

If we reduce the correlation between color and orientation to zero then we obtain color-orientation pairs which correspond to our everyday experience, in which every orientation occurs in all colors with equal probability, **4.31a**. In order to proceed, we need to explain the two columns of graphs in **4.31**.

The left-hand column contains graphs (labeled **a**, **c**, **e**, and **g**) which represent the physical color-orientation values within a set of stimuli, and each dot represents a stimulus with a specific color, which we denote as A, and orientation, which we denote as B. If all combinations of color-orientation values A and B are equally probable then the dots are scattered roughly uniformly, producing a **uniform joint distribution** of stimulus values, as in **4.31a**. Note that this uniform joint distribution can exist only if the following three conditions are true: 1) all *values A* of color are equally probable, 2) all *values B* of orientation are equally probable, and, 3) all *combinations* of color A and orientation B values are equally probable.

As a counter-example, consider **4.31c**, where color and orientation are correlated. Here, all values of color are equally probable and all values of orientation are equally probable, but some *combinations* of color and orientation values are more probable than others, giving rise to a non-uniform joint distribution.

The right-hand column contains graphs (**b**, **d**, **f**, and **h**) which represent the outputs Ψ_A and Ψ_B of two hypothetical neurons a and b. Specifically, each graph represents the outputs Ψ_A and Ψ_B when the neurons a and b are presented with stimuli in the corresponding graph in the left hand column. For example, the axis label Ψ_A refers to the output of a color-sensitive neuron, and Ψ_B refers to the output of an orientation-sensitive neuron. Thus, a point in **4.31a** which represents a pair A and B of physical values (color and orientation) gives rise to a corresponding point in **4.31b**, which represents the outputs Ψ_A and Ψ_B of a pair of neurons. As **4.31a** represents a uniform distribution of color-orientation stimuli, the resultant set of neuron output pairs is also a uniform distribution, as shown in **4.31b**.

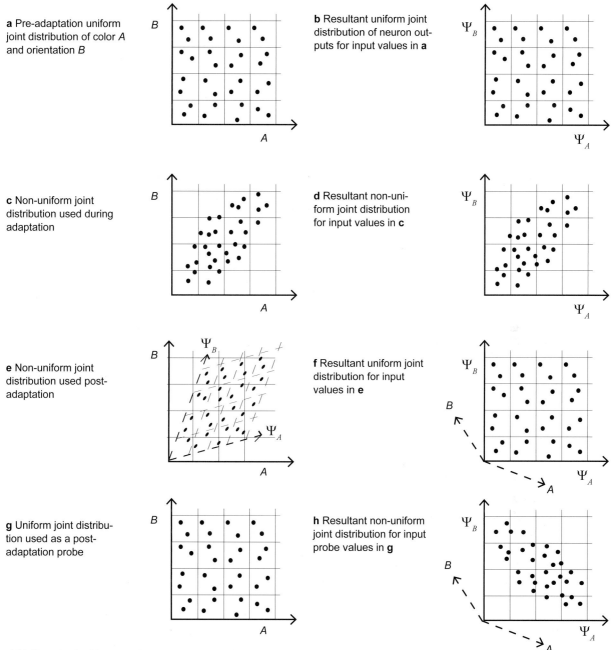

a Pre-adaptation uniform joint distribution of color *A* and orientation *B*

b Resultant uniform joint distribution of neuron outputs for input values in **a**

c Non-uniform joint distribution used during adaptation

d Resultant non-uniform joint distribution for input values in **c**

e Non-uniform joint distribution used post-adaptation

f Resultant uniform joint distribution for input values in **e**

g Uniform joint distribution used as a post-adaptation probe

h Resultant non-uniform joint distribution for input probe values in **g**

4.31 Hypothetical input and output values before, during, and after recalibration
Redrawn with permission from Barlow (1997).

For the uniform distribution of stimuli (i.e., any orientation can be paired with any color), all the squares in the grid in **4.31a** contain roughly equal numbers of dots, which implies that every possible combination of the values *A* and *B* occur with equal probability. The graph in **4.31b** implies that the resultant combinations of neuronal outputs Ψ_A

and Ψ_B also occur with equal probability. It is as if the world presents stimuli at all possible combinations of color and orientation, so that every square in **4.31a** contains roughly equal number of points.

If, after observing the world for many days, a single square contained no dots, as in **4.31d**, then this would imply that certain combinations of orientation and color did not occur, as in **4.31c**. More importantly, it would imply a waste of neuronal coding ability, because the outputs Ψ_A and Ψ_B, which correspond to this empty square, would never get used.

More precisely, if every square in this graph did not contain about the same number of dots then the full range of output values of each neuron is not being used, and this represents an inefficient use of neurons.

This is exactly what happens when the brain is presented with the ***non-uniform*** or ***biased*** distribution of color-orientation stimuli used in the generalization of the contingent aftereffect described above and in **4.29**. As the adaptation phase introduces a correlation between color and orientation, squares in the top left and bottom right of in **4.31d** are empty, and only squares along the diagonal contain any dots at all. This follows from the fact that, during adaptation, most combinations of color and orientation never co-occur, which implies that most combinations of neuronal output values Ψ_A and Ψ_B never co-occur.

One way for the brain to remedy this state of affairs is to recalibrate each neuron such that every possible combination of Ψ_A and Ψ_B values occurs with equal probability, as in **4.31f**. This can be achieved if each of the two neurons is modified as follows.

Before recalibration (i.e., before adaptation), neuron a responds exclusively to the physical parameter A (color), and neuron b responds exclusively to parameter B (orientation). After recalibration, neuron a is still excited by A, but now B also has a small *inhibitory* effect on neuron a. Similarly, after adaptation, neuron b is still excited by B, but now A also has a small *inhibitory* effect on neuron b.

These changes can be interpreted to mean that a and b now respond to a ***linear combination*** or, equivalently, a ***weighted average*** of both parameters A and B, as indicated by the skewed (dashed) axes Ψ_A and Ψ_B overlaid on **4.31e**. These axes have been added to show how the axes Ψ_A and Ψ_B in **4.31f** are related to the physical parameters A and B.

For comparison, we have also overlaid a set of (dashed) axes to **4.31f** to show how the physi-

cal quantities A and B in **4.31e** are related to the neuron output values Ψ_A and Ψ_B. Recalibration effectively compresses each square associated with Ψ_A and Ψ_B in **4.31e** into a rhombus. Note that the number of dots in all rhombuses is about the same, which implies that all combinations of Ψ_A and Ψ_B values are, once again, equally probable. This can be seen more clearly in **4.31f** where we have "de-skewed" the Ψ_A and Ψ_B axes. Note that this "de-skewing" does not imply any difference between the response characteristics of the neurons shown in **4.31e** and **4.31f**.

Thus, after recalibration, the neuronal output values Ψ_A and Ψ_B have a uniform joint distribution even though the joint distribution of colors and orientations (A and B) that give rise to those output values is non-uniform.

For the mathematically inclined reader, it can be seen that the lines labeled Ψ_A and Ψ_B in **4.31f** are defined by the equations $\Psi_A = A - pB$ and $\Psi_B = B - qA$, where p and q are constants. These equations indicate that the output Ψ_A (for example) increases with A, and *decreases* with B, so that A excites the neuron a, whereas B *inhibits* it. In other words, neuron a's preferred orientation still excites a, but now the adapting stimulus color inhibits neuron a.

Of course, when we de-skew **4.31e**, we also change the axes associated with the physical values of color and orientation A and B so they are no longer perpendicular, as shown by the dashed axes in **4.31f** labeled A and B. Thus, in a world where color and orientation are correlated, the outputs of two neurons can only be uncorrelated if each neuron responds to a combination of color and orientation.

At this point the recalibrated system is said to be fully ***adapted*** to the stimuli. Thus, after being exposed to this biased world for some time, this recalibration theory of the contingent aftereffect suggests that neurons adjust themselves to reflect the biased distribution of color-orientation stimuli. More precisely, when presented with stimuli with properties that do not have a uniform joint distribution (as in **4.31e**), neurons recalibrate so that the joint distribution of their outputs is uniform, as shown in **4.31f**.

With regard to the contingent aftereffect, the key question is: what does the distribution of neuron outputs look like when presented with a uniform distribution of parameter values? The answer can be obtained by plotting the neuronal outputs Ψ_A and Ψ_B that result from a uniform distribution of stimuli, as shown in **4.31g**. As shown in **4.31h**, the resultant cloud of points lies along the downward sloping diagonal, which implies that Ψ_A and Ψ_B are now *negatively* correlated. This means that if Ψ_A is large then Ψ_B is usually small, and vice versa. Bear in mind that these responses are induced by a uniform (*un*biased) post-adaptation distribution of color-orientation stimuli, which before adaptation, yielded a uniform distribution of dots, as in **4.31b**.

In terms of perception, let's examine the consequences of This negative correlation when the pair of neurons has been adapted to a black and green horizontal grating, and is then presented with a horizontal black and white grating.

Before recalibration, let us suppose that neuron *b* responds to a line at its preferred (horizontal) orientation but is unaffected by the color (green) of that line. After recalibration, this neuron still responds to horizontal lines, but it is now inhibited by the color green.

Similarly, before recalibration, another neuron *a* responds to a line with its preferred color (green) but is unaffected by the orientation of that line. After recalibration, this neuron still responds to the color green, but it is now inhibited by horizontal lines. This implies that, post-adaptation, when such a color-sensitive neuron *a* is exposed to a black and white horizontal grating, it is inhibited to some extent, simply because it is now inhibited by any horizontal grating. What are the perceptual consequences of this inhibition?

In order to explore this, consider a whole population of color-sensitive neurons. If neurons which responded to green before adaptation are now inhibited by the presence of horizontal lines then this yields a population of neurons in which those neurons that respond to green are inhibited.

Now, as far as the brain is concerned, the perceived color of any image region is determined by the *relative* amounts of red and green detected in that region (see Ch 17). Accordingly, if any part of the image contains "negative amounts of green" (which is what inhibition amounts to) then this green deficiency is interpreted as an excess of green's complementary color, *red*. Therefore, the

white stripes in a horizontal black and white grating appear to have a red-ish hue after adaptation.

In summary, when presented with a set of stimuli defined by two or more physical parameters, the recalibration hypothesis posits that the brain organizes neurons in such a way that the joint distribution of their outputs is uniform. If these physical parameters are correlated, as in the contingent aftereffect adaptation stimuli, then maximizing uniformity has the effect of removing correlations in the outputs of neurons which respond to different parameters. After such recalibration, a "neutral" stimulus, such as a horizontal black and white grating, is perceived to have red stripes, at least until the brain recalibrates itself once again to a visual world in which orientation and color are no longer correlated. However, the process of recalibration to a "normal" world may take several weeks, because that is the amount of time it can take for contingent aftereffects to disappear.

Implementation: A Cautionary Note

The observant reader would have noticed that the output of neuron *a* increases **monotonically** with color, and that the output of neuron *b* increases monotonically with orientation. Some neurons do provide this form of **variable coding** (Ch 11), but neurons sensitive to color and orientation usually have bell-shaped non-monotonic tuning curves with a specific preferred color or orientation. In reality, therefore, the outputs of a color- and orientation-sensitive neuron would not define the uniform joint distribution in **4.31a**.

We can remedy this by positing a different pair of neurons for each square in **4.31a**, where one neuron in each pair is sensitive to a small range of colors, and the other is sensitive to a small range of orientations. Using these **value coding** neurons (Ch 11), the logic described above still applies, and makes the same predictions regarding the contingent aftereffect.

Information and the Tilt Aftereffect

We have argued above that neurons transmit information efficiently if their outputs are independent. But this argument is based on an unstated assumption, namely, that each neuron is a *perfect* transmitter of information. However, we know that neurons are to some degree noisy devices. Thus, for every input a neuron receives, its output reflects this input *plus some noise*. So neurons are not perfect transmitters.

For example, if two neurons have the same inputs then the noisiness of each neuron ensures that their outputs are not perfectly correlated. One way to estimate the output *that would have been obtained in the absence of noise* is to take the average output of a number of noisy neurons, as discussed with respect to "noise cleaning" for edge detection, Ch 5. Thus, rather than regarding the complete removal of redundancy between the outputs of neurons as a desirable goal, the inevitable hazard of noise suggests that a certain amount of redundancy is actually a good thing.

Notice that this contradicts our earlier point that a neuronal population with correlated outputs is a waste of neurons. That argument rested on the implicit assumption that neurons are noiseless devices. Because neurons are noisy, it makes sense for different neurons to signal the same event. This "muddies the waters" to some extent because it implies that the amount of information transmitted about a given patch of retina (e.g., an oriented line) increases with the number of neurons that have their receptive fields centered on that patch.

However, if we add neurons one by one then the amount of information transmitted increases rapidly at first, but then tails off (with the square root of the number of neurons). That is, there is a process of "diminishing returns," with additional neurons adding progressively less and less information as the number of neurons increases. Given that neurons are metabolically expensive, there comes a point where the (diminishing) informational benefits of adding each neuron is outweighed by the energy cost it incurs.

Maximizing Information and the Tilt Aftereffect

Taking account of noise forces us to reconsider the argument that neuron outputs should exhibit a uniform joint distribution by de-correlating those outputs. It turns out that a more general strategy consists of *maximizing the amount of information transmitted* by neurons. This does not conflict with the idea of achieving a uniform joint distribution but subsumes it under a more general framework.

The idea of *maximizing information* was tested on the tilt aftereffect in a paper by Martin

Wainwright published in 1999. He argued that neurons in a given population should adjust their outputs so as to maximize the amount of information they transmit. Rather than considering the tuning function of each individual neuron, Wainwright considered the composite *envelope function* which results from the overall effect of many individual tuning functions. The concept here is similar to the envelope of channels shown in **4.16** for the case of spatial frequencies. Just imagine an envelope of a similar kind in which each point on the envelope is the sum of all neuronal responses to a given stimulus orientation.

Given a biased input distribution of line orientations, as happens in the tilt aftereffect, the problem consists of finding that particular envelope function which maximizes the information transmitted about that distribution. In other words, which composite envelope function transmits most information about a given input distribution of line orientations?

A crucial factor in this calculation is the amount of noise present. Wainwright showed that, using plausible levels of noise, maximizing information resulted in a unique envelope function. We refer the interested reader to his paper for further details.

Of course, the "litmus test" of this research is the extent to which Wainright's model exhibits the tilt aftereffect. After the "maximum information" envelope function had been found for a biased distribution of line orientations, the response to other line orientations was obtained. The model did indeed show the general characteristics of the tilt aftereffect shown in **4.20**.

Wainright's model also predicts that the discrimination of orientation around the adaptation orientation should be increased after adaptation. This is because, according to his theory, the receptive fields of neurons tuned to the adapted orientation become more densely packed and more narrowly tuned. We will not go into the details of the theory in this regard, but we note that psychophysical data collected by K Beverley and David Regan in 1975 confirms this prediction even though their experiment was not carried out to test Wainright's model.

In 2000, Clifford, Wenderoth, and Spehar presented a more mechanistic model, which was nevertheless inspired by ideas from information theory. Within their model, presenting neurons with an adapting stimulus not only reduces the magnitude of their responses, it also increases the width of their tuning curves. We have considered the consequences of reducing response magnitudes above. In line with the objective of maximizing information transmission, an increase in tuning curve width decorrelates responses of different neurons over the distribution of stimulus orientations that includes the adapting stimulus.

One impressive aspect of Clifford et al.'s model is that it provides an account of psychophysical data derived from aftereffects not only of orientation (including the strange fact that adaptation to horizontal causes an aftereffect on orientations around vertical; p. 94), but also of motion direction, and hue. A comprehensive review of recent work in this area can be found in Schwartz, Hsu, and Dayan (2007).

Conclusions on Recalibration

In summary, both the tilt aftereffect and the contingent aftereffect may be a consequence of the brain trying to maximize the efficiency of its neurons by constantly recalibrating them to suit characteristics of the prevailing visual world. In both cases this can be achieved by adapting neurons so that their outputs jointly transmit as much information as possible about their inputs.

In all the foregoing, it is important to understand that the pursuit of efficiency takes place in the context of the particular visual environment of the perceiver. This takes us back to the oblique effect, **4.7**, in which edges around vertical and horizontal have lower contrast thresholds for detection and whose orientations are perceived more accurately. This has been linked to the predominance of vertical and horizontal edges in our visual worlds, **4.10**. It thus seems highly likely that the kind of recalibration processes just discussed can also explain the oblique effect.

Having visual mechanisms that optimally adapt their owners to their visual worlds is just as it should be, of course, as that will assist survival. Indeed, from a Darwinian perspective one could say that the matching of a visual system to its visual environment is the overriding consideration.

Of course, encoding efficiency is also desirable but perhaps it is best thought of as a secondary benefit, with its true significance being that it provides a way of ensuring the advantage of matching the visual system to its owner's visual niche, thereby conferring an advantage in terms of enhanced Darwinian fitness.

Summary Remarks

This chapter has introduced some key concepts that need to be understood if you decide to tackle the literature on current visual research.

In particular, the idea of a spatial frequency and orientation tuned channel is fundamental. The chapter has also laid a foundation for understanding research on the task of edge detection described next in Ch 5.

Although this chapter has thus been concerned with using aftereffects to study what are called *low level* processes, we end by noting that aftereffects have been used increasingly in recent years for studying *high level* processes involved in face recognition. For example, Jaquet et al. reported in 2007 aftereffects relating to Caucasian and Chinese faces that cannot be explained by adaptation in low level visual mechanisms.

Further Reading

Barlow HB (1997) The knowledge used in vision and where it comes from. *Philosophical Transactions of the Royal Socitey of London* (B) 352 1141–1147. *Comment* Discusses the contingent aftereffects in terms of their possible function, and suggests a computational mechanism for how they arise. An insightful and thought–provoking paper.

Beverley KI and Regan D (1975) The relation between discrimination and sensitivity in the perception of motion in depth. *Physiology* **24**9(2) 387–398.

Blakemore C (1973) *The Baffled Brain*. In: Gregory RL and Gombrich EH (1973) *Illusion in Nature and Art*. 9–46 Duckworth. *Comment* A fine account for the general reader to early work linking illusions to neurophysiological mechanisms.

Blakemore C and Campbell FW (1969) On the existence of neurons in the human visual system selectively sensitive to the orientation and size of retinal images. *Journal of Physiology* **203** 237–260. *Comment* One of the classic papers on aftereffects. It is the source of the data shown in **4.12**, **4.15**.

Blakemore C and Cooper GF (1970) Development of the brain depends on the visual environment. *Nature* **228** 477–478. *Comment* Reports the kitten-in-the-drum experiment illustrated in **4.11**.

Blakemore C and Sutton P (1969) Size adaptation: a new aftereffect. *Science* **166** 245–247. *Comment* This describes the size aftereffect in **4.21**.

Blakemore C, Carpenter RHS and Georgeson MA (1970) Lateral inhibition between orientation detectors in the human visual system. *Nature* **228** 37–9.

Campbell FW and Maffei L (1974) Contrast and spatial frequency. *Scientific American* **231** 106–13.

Clifford CWG, Wenderoth P and Spehar B (2000) A functional angle on some after-effects in cortical vision. *Proceeding Royal Society of London* (B) **267** 1705–1710.

Clifford CWG (2002) Perceptual adaptation: motion parallels orientation. *Trends in Cognitive Sciences,* **6**(3) 136–143. *Comment* This review paper proposes that orientation and motion share a common mechanism, and compares Clifford's model with that of Wainwright et al. (1999).

Coppola DM, Purves HR, McCoy A and Purves D (1998) The distribution of oriented contours in the real world. *Proceedings National Academy of Sciences USA* **95** 4002–4006. *Comment* The source of data and pictures in **4.10**.

De Valois RL and De Valois KK (1988) *Spatial Vision*. Oxford University Press. *Comment* An advanced text giving a thorough account of early work on orientation and spatial frequency tuned channels.

Georgeson MA and Sullivan GD (1975) Contrast constancy: deblurring in human vision by spatial frequency channels. *Journal of Physiology* **252** 627–656.

Jackson AJ and Bailey IL (2004) Visual acuity. *Optometry in Practice* **5** 53–70 *Comment* An excellent tutorial review on the theory and practice of measuring visual acuity in clinical settings.

Jaquet E, Rhodes G and Hayward WG (2007) Opposite aftereffects for Chinese and Caucasion faces are selective for social category information and not just physical face differences. *Quarterly Journal of Experimental Psychology* **60**(11) 1457–1467. *Comment* A starting point for searching the literature on aftereffects and face processing.

MacKay D See http://www.asa3.org/ASA/articles/MacKay_bib.html *Comment* Various simultaneous size contrast illusions.

McCollough C (1965) Adaptation of edge-detectors in the human visual system. *Science* **149** 1115–1116.

Mather G, Verstraten F and Anstis S (1998) *The Motion Aftereffect: A Modern Perspective*. MIT Press. *Comment* A excellent overview provided by 7 chapters written by 19 authors. This provides excellent accounts of studies of the motion aftereffect, both ancient and modern, and also a good place to seek out references on after effects generally.

Nundy S, Lotto B, Coppola D, Shimpi A and Purves D (2000) Why are angles misperceived? *Proceedings National Acadamy of Sciences USA* **97** 5592–7. *Comment* This describes the application of Purves and Lotto's radical empiricist theory of vision to some angle illusions.

Purves D and Lotto RB (2003) *Why we see what we do: an empirical theory of vision*. Sinauer Associates, Inc.: Sunderland, Mass., USA. *Comment* A wide-ranging overview of their empiricist theory of vision. It contains many stunning computer graphics illustrating various illusions.

Schwartz O, Hsu A and Dayan P (2007) Space and time in visual context. *Nature Reviews Neuroscience* **8** 522–535. *Comment* A review paper attempting to reconcile computational (mainly Bayesian) functional models with mechanistic models and with physiological findings. A source of references and an excellent overview of the field.

Wainwright MJ (1999) Visual adaptation as optimal information transmission. *Vision Research* **39** 3960–3974.

5

Seeing Edges

a

b

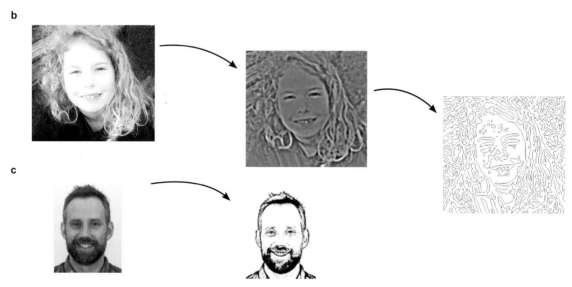

c

5.1 Edges in art and in computer vision
a Sketch of some of the main contours used in a self-portrait by Albrecht Dürer, illustrating how artists use lines to mark the boundaries of objects and their parts, such as here the eyes, nose, mouth, etc. The visual system seems to create similar edge-based representations, marking significant entities in retinal images.
b A processing sequence from image (left), via convolution using a circularly symmetric receptive field (center), to a representation of many of the edges in the image (right). This is the processing sequence described in this chapter.
c From image to sketch, by the computer vision system of Bruce Gooch, Erick Reinhard, and Amy Gooch (2004), reproduced with permission. The early stage of their system starts with a similar sequence to that shown in **b**, but then adds other processing stages.

Take a look around the scene before you and you will have no trouble seeing *edges* formed by the boundaries of objects, of surfaces, of surface markings, and of shadows. This ability to see edges must reflect the fact that there are edge feature representations somewhere inside our heads.

Artists often use lines to pick out object edges, **5.1**, and they have known about the importance of edges in seeing since the beginning of art, as cave drawings testify, **5.2**.

It is of great theoretical interest to the science of seeing that line drawings of objects do not need to be complete for us to be able to recognize objects, **5.3** (see also Ch 8). In its search for object boundaries, human vision has ways to get by with partial evidence. The term *illusory contour* is used for an edge that is seen where none exists in the image (Ch 16). The importance of being able to cope with breaks in edges soon became apparent in the early days of computer vision. It was found, to the astonishment of investigators, that sometimes

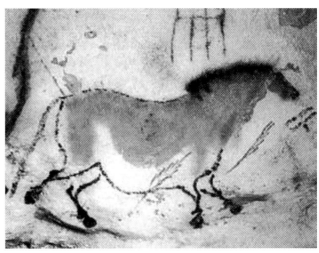

5.2 Cave painting
From Lascaux, France. PD-Art.

the pixel values in their grey level images simply did not have changes marking edges which they themselves could see when looking at the original images. This shocking realization led some of the pioneers of computer vision to postpone working from natural images and instead use hand-made edge maps for exploring various visual processes.

Brain representations for edges cannot be all there is to seeing, of course, because we are able to describe scene characteristics much more complicated than edges. Even so, edge feature representations might be immediately useful for guiding a grasping action around an object. They can also serve as an important first step on the way to more complex perceptual tasks, such as object recognition or depth perception, as we will see.

How should edge detection be tackled? Bars are a species of edge feature, and we considered in Ch 3 the task of designing a bar detector. This task proved surprisingly tricky but we showed how bar orientation could be computed from the outputs

Why read this chapter?

If we are to understand human vision it will be vital to understand the early stages of image processing in the visual pathway. We introduce this topic with Marr and Hildreth's computational theory of edge detection. A ubiquitous problem in finding edges is that images are imperfect; they contain glitches that are referred to as *noise*. The theory tackles the problem of noise by blurring images, using a process called *convolution*. In essence, this consists of applying a particular kind of *operator* (cf. receptive field in biological vision) all over the image. This operator has a gaussian profile because the computational theory specifies that this is the shape that optimally combines smoothing away noise with not disturbing too greatly where edges are to be found in the convolved image. The next step is to find regions in the image where there are abrupt changes in image intensities, because this is where the edges are located. This requires measuring intensity *gradients* and/or *changes in gradients*. These are called the *first* and *second derivatives* respectively. Various biologically plausible algorithms are described for implementing the theory. These algorithms involve operators that are remarkably similar to the receptive fields possessed by cells in the retina and the striate cortex. Using the task of edge detection, the chapter illustrates the value of distinguishing between the computational theory, algorithm, and hardware levels when trying to understand complex information processing systems.

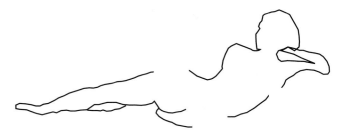

5.3 Objection recognition from incomplete contours
The shape of the figure is well captured by the line drawing in which some parts of the body boundaries are missing.

of a group of simple cells whose preferred stimuli were bars of different orientations. But there are many types of features other than bars and much more to the edge detection problem than just computing edge orientation. It is time to have a closer look at the seeing problem of edge detection.

Edges in Images

In studying seeing, it is always important to distinguish clearly between scenes and the images of those scenes. So the first question to ask in analyzing the problem of edge detection is: how do edges in scenes show up in images?

An image of John Lennon is shown in **5.4a**. The pixel intensities of this image are shown as a hilly landscape in **5.4c**, in which height is used as a way of representing intensity. You may be surprised how broken up and cluttered this landscape appears. Can our clear percepts of the scene in front of the eyes really start from such a basis? The answer is *Yes*. Natural images usually are as messy as this landscape suggests.

Part of the messiness in **5.4c** is caused by the many edges from the hairs on Lennon's head. But look at the background in the original image and in the landscape. The former appears a pretty flat grey to our eyes whereas the latter is surprisingly "bubbly." Some of this "bubblyness" is caused by images having been captured with electronic light detectors that are a little "noisy." You can think of this *noise* as similar to the "snow" seen on a poorly tuned television. Receptors in the retina also suf-

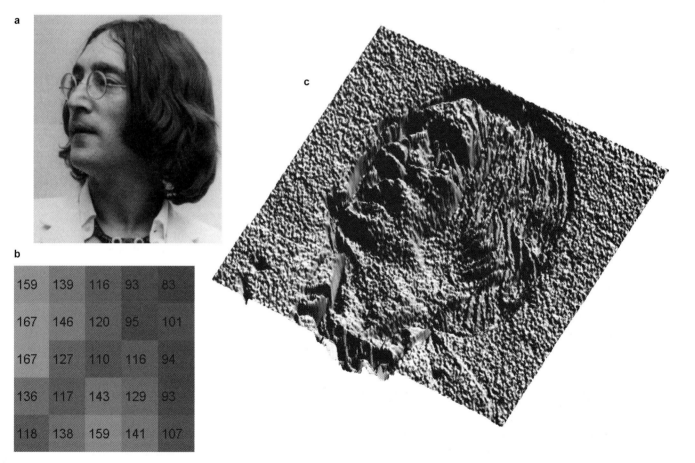

5.4 Pixel Intensities as a landscape
a Photograph of John Lennon.
b Sample of pixels from a greatly magnified image. Each square represents one pixel whose intensity (gray level) is given as a number.
c Pixel intensities plotted as a 3D graph, which shows them as a hilly landscape in which the valleys indicate low intensities and the peaks high intensities.

a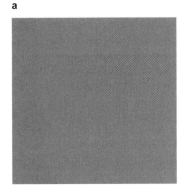
b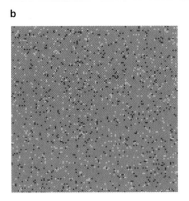
c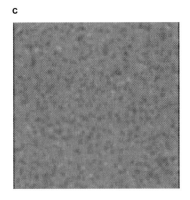

5.5 Image noise cleaning
a Original test image. **b** Original image with added "salt and pepper" random noise. **c** The result of smoothing **b** by local 3x3 neighborhood averaging.

fer from random noise fluctuations, so noise is a problem for human as well as computer vision. Our first problem to be solved is thus clear. We need to find a principled way to eliminate image noise, or at least reduce it, prior to seeking edges in images. This ensures that any edges found can be relied upon to reflect scene edges and not spurious noise-created image intensity changes.

Computational Theory: Getting Rid of Image Noise

Analyzing this noise cleaning task properly would mean developing a deep understanding of where the noise comes from. This would be needed to develop a principled way of getting rid of it or sidestepping it in some way. In short, and as usual in studying seeing, a full *task analysis* would be needed to devise a good computational theory for getting rid of noise.

We will not explore noise sources in any detail, as that would lead us into some complex details about various visual neurons. Instead, we will consider just noise in receptors and we will proceed on the basis of a very simple idea. We will assume that the main sources of receptor noise are spatially random. What this means is that the noise in neighboring receptors is assumed to be independent. Think of this as each receptor being subject to its own "private" noise, so that noise fluctuations in one receptor cannot be predicted from those in its neighbors.

This assumption (another example of what is technically called a *constraint*) allows a neat trick

for getting rid of noise: exploit the "law of averages" to cancel out a lot of the noise. That is, take the average of the activities in neighboring receptors and high noise in one will tend, due to the noise independence assumption, to be canceled out by low noise in another. Thus the total receptor noise will tend to average out to zero. [This scheme is not aimed at noise elsewhere in the visual system, particularly low spatial frequency noise.]

The price to be paid for using this constraint is that taking neighborhood averages blurs the grey level description a little, as we will see. But it does lead to smoother, less noisy, images and the penalty of some blur turns out to be worth paying. Indeed, we will find that using a range of images blurred to different extents is an important aspect of edge detection, because it is part of the process of finding edges of different types (from sharp through to fuzzy).

You might be thinking that the averaging scheme just proposed is a bit extreme. Wouldn't it be better to give most weight to the central receptor under consideration than to weight it equally with its neighbors? That is, instead of substituting the intensity recorded in each receptor with the average value of that receptor and its neighbors, how about using a scheme in which the neighbors somehow contribute relatively less weight to the end result? It turns out that this is exactly the right thing to do. But before explaining this important refinement, we will describe the equal-weighting idea first because it is simpler for conveying some core ideas.

An Algorithm for Image Noise Cleaning

How can neighborhood averaging be implemented? A possible computer vision *algorithm (or procedure)* is as follows:

Step 1 For each pixel, add up the pixel intensities of its immediate neighbors, and then add this total to the intensity of the pixel in question. Call this the neighborhood total.

Step 2 For each pixel, divide its neighborhood total by the number of pixels contributing to that total, to calculate a mean (average) intensity. Substitute this mean for the original pixel intensity.

In short, add up all the pixel intensities in a region and divide by the number of pixels in that region.

The region can be of various sizes but for the present we will use 3×3 patches of pixels (that is, each pixel plus its 8 closest neighbors).

Does this kind of neighborhood averaging work in getting rid of noise? The idea can be tested by taking a uniform grey image, **5.5a**, and adding noise randomly to each pixel, **5.5b**. To highlight the basic ideas, much more noise has been added than would be expected in either artificial or biological images. Running the 3×3 local averaging algorithm on this very noisy image shows that it reduces the noise quite a lot, **5.5c**, although it does not get rid of it all. Nevertheless, this test demonstrates that averaging can help reduce noise for independent receptor noise sources.

But what about local averaging on natural images? It produces a smoother image for the John Lennon picture, as can be seen in **5.6**.

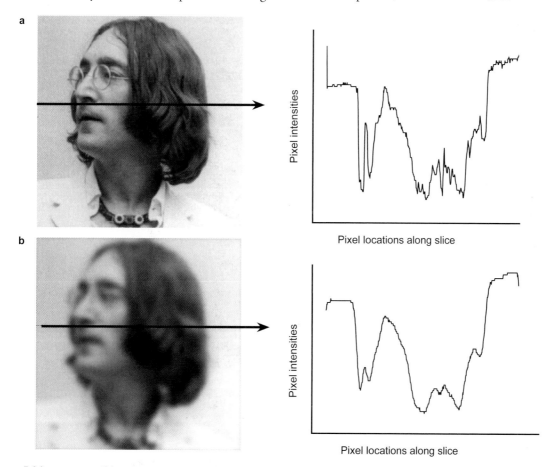

5.6 Image smoothing
a Original Lennon image with alongside it the intensity profile of a one-dimensional slice indicated by the horizontal line.
b The same after the image has been smoothed using 3x3 local neighborhood averaging. Notice that both profiles show image intensity variations at different scales from steep to shallow.

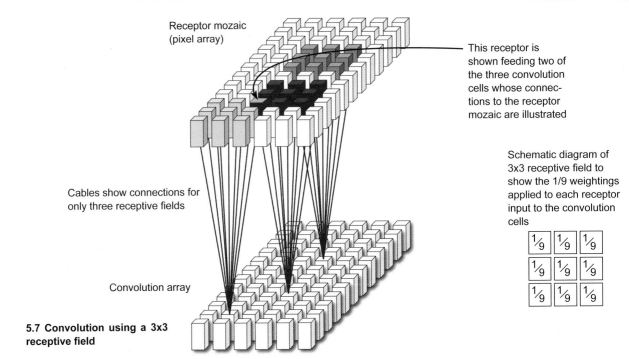

Receptor mozaic
(pixel array)

This receptor is
shown feeding two of
the three convolution
cells whose connec-
tions to the receptor
mozaic are illustrated

Schematic diagram of
3x3 receptive field to
show the 1/9 weightings
applied to each receptor
input to the convolution
cells

Cables show connections for
only three receptive fields

$\frac{1}{9}$	$\frac{1}{9}$	$\frac{1}{9}$
$\frac{1}{9}$	$\frac{1}{9}$	$\frac{1}{9}$
$\frac{1}{9}$	$\frac{1}{9}$	$\frac{1}{9}$

Convolution array

5.7 Convolution using a 3x3 receptive field

Convolution

The procedure just outlined is all very well for computers, as they are good at arithmetical operations, but how might the brain do the averaging job with only neurons at its disposal? Can they be made to "perform" the required arithmetic using a method that fits how they work? The answer is yes, but to explain it entails introducing a generally useful procedure for computers as well as brains, called *convolution*.

Convolution is illustrated in **5.7**. At the top is shown schematically a *receptor mosaic* (roughly equivalent to an array of pixels in computer vision, Ch 1) whose activities encode image grey levels. This receptor mozaic can be thought of as a highly simplified model of a small region of the receptor mosaic of a human eye. The receptors send their outputs into a second array of elements. We will call these *convolution cells,* and they are arranged in a *convolution array*. The latter array carry out a function similar to that performed by the *bipolar cells* in the retina, as we will see in Ch 6.

Each convolution cell in **5.7** receives inputs from 9 receptors, arranged in a 3×3 square centered on one receptor. This is illustrated with two sample sets of fibers connecting the receptors to the convolution array. The receptor clusters for

the two samples are picked out with shading to help illustrate what each one is "looking at" in the receptor mosaic. Each receptor cluster defines the *receptive field* of the associated convolution cell.

[*Terminological note repeated from Ch 3*: The standard definition of receptive field is the patch of retina that influences the output of the cell in question. Strictly speaking, it is not defined in terms of the pattern of weights associated with that area, but in practice the term receptive field is often used to refer to that pattern, as we do in this book. For the latter usage, the receptive field concept is similar to *template, operator, mask,* or *weighting function* used in the computer vision literature.]

The two sheets of cells, receptor mosaic and convolution array, are shown in **5.7** neatly lined up one over the other. That is, the central cell in each receptor cluster feeds a convolution cell whose position in the convolution array has a matching spatial location. This helps us keep track of what is going on in the figure but the physical layout would not be critical in a vision system. The key property is where the connecting fibers come from and go to, a point that was made in Chs 3 and 4 in connection with *place coding*.

For simplicity, only the fibers to three convolution cells from the receptor mozaic are shown in in **5.7**, but in fact every convolution cell would be connected to the receptor mosaic in a similar fashion. The key difference between convolution cells is that the set of inputs defining their receptive fields come from receptor clusters in slightly different positions in the receptor mosaic. There would be hundreds of fibers linking the two arrays. In **5.7** there are only 9×45=405 fibers but there would be very many more in a realistically sized vision system.

Because of these multiple connections, each receptor has to feed many different convolution cells. Again for reasons of simplicity, **5.7** shows only one instance of this kind—this is the receptor picked out with the arrow+label. In other words, the two convolution cells in question have wiring connections such that they share one receptor in common.

The small inset in **5.7** shows an example of a possible set of *receptive field weightings,* which are the same for each cell in the convolution array. For example, if a receptor is signaling 18 units of activity (this number is its code for the associated image intensity) then the influence transmitted to the convolution array is much smaller than this, in fact only 1/9th. Why 1/9th? Because there are 9 inputs in this particular 3×3 receptive field, and so multiplying each one by 1/9th comes to the same thing as adding up all the 9 inputs and dividing by 9 to get the mean. The conclusion is that weighting inputs is a neat trick for doing the averaging arithmetic we require.

[To see why this works, consider if all the 9 receptors in a receptive field had the same activity level, say 18 units. The neighborhood total for all nine will come to 9 × 18 = 162. Obviously, dividing this by 9 to get the mean will give 18, which is the answer we want because all receptors in this example were registering 18 units of activity. Now consider doing the same thing by weighting. Each input fiber delivers 18 × 1/9 = 2 units of activity, and adding up all 9 inputs gives 9 × 2 = 18.]

This is good news because weighting is convenient if neurons are all you have to do the job. They work by transmitting excitation or inhibition, Ch 3. In **5.7** all the fibers from the receptors pass on excitation, with the precise amount of excitation adjusted to suit the weighting required for the computation in question.

To summarize so far: convolution using suitably weighted connections is one way to implement a local neighborhood averaging noise cleaning algorithm. This is said to be ***biologically plausible*** because it embodies a procedure that lends itself to being readily implemented in neurons.

Convolving an input image with a receptive field is a widely used technique for processing images, in both man-made and biological visual systems. It is essential to understand convolution to understand vision.

In the case we have been considering, convolution replaces the original grey level description with a smoothed grey level description, which is said to be the ***convolved image***. However, many other sorts of receptive fields can be used and for them the convolved image will not be a more or less close replica of the original grey level description. Rather, the activity levels in the convolved image will vary according to the "goodness of fit" of the receptive field at each location in the input image. The "bar-tuned" receptive fields we investigated in Ch 3 are a case in point, and we will see other examples in due course. For the moment, remember that each element in the convolved image has a position which signifies the point in the input image on which the receptive field was centered when the receptive field's goodness of fit with the relevant point of the image was calculated.

To reiterate the key terminology, convolution can be defined as applying a receptive field all over an image.

Equivalent Algorithms

Summarizing the simple weighting trick for implementing noise cleaning by taking a local neighborhood average gives us:

Step 1 For each pixel *P*, multiply its intensity and that of its neighbors by 1/*N*, where *N* is the number of elements in the receptive field centered on pixel *P*.

Step 2 Add up the values so obtained, and put the result into the convolution cell corresponding to the pixel *P*.

The point of giving this second procedure in this form is to emphasize that a computational theory of a task can usually be implemented with one of a number of algorithms. Each one ***implements*** the

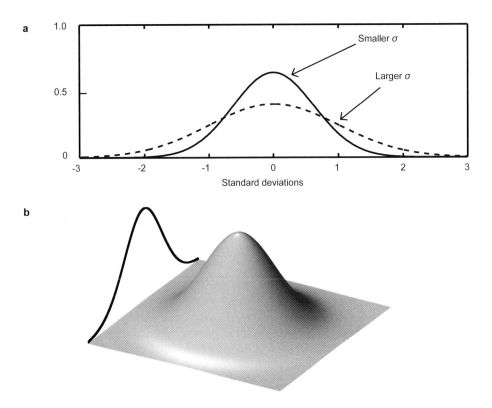

5.8 Gaussian distributions
a Cross-sections of two gaussians of different widths or "spreads." The parameter that determines the size of the spread is the standard deviation (σ, pronounced sigma). Weight strength is plotted on the vertical axis.
b 3D picture of a two-dimensional gaussian, with its profile shown as the graph on the left.

same theory and achieves an equivalent output but each does the job in a different way. Which one is chosen will be determined in part by the hardware or brainware available for running the procedure. Our interest in biologically plausible procedures leads us naturally to the kind of weighting just set out. We will see weighting of this kind in the receptive fields of the bipolar cells of the retina, in Ch 6.

The noise cleaning receptive field used in **5.7** raises some interesting questions. For example, how big should the receptive field be? Is it sufficient to use in general just the closest 8 neighbors for each pixel to smooth the noise away? Or should a larger size than 3×3 be used?

It turns out that biological vision systems have a range of receptive field sizes. Before explaining why, we need to delve a bit deeper into the question of now best to average out noise.

Gaussian Smoothing

We said earlier that the 3×3 averaging procedure was a very simple scheme for noise cleaning, chosen to introduce some core ideas. It turns out on closer inspection that instead of all the pixels having equal weights when feeding into the convolution array those near to the center of the receptive field should be weighted more highly and those further way less so.

The particular shape of the weighting distribution has to optimize two conflicting goals: smoothing away noise and not disturbing too greatly where edges are to be found in the convolved image. There is a theorem (see Marr and Hildreth, 1980) stating that the shape of weighting distribution that achieves an optimal trade-off between the two goals is bell-shaped. Examples are shown in **5.8a**.

The technical name for a distribution with this shape is a ***gaussian***, after the mathematical genius Karl Friedrich Gauss (1777–1855) who discovered it. As can be seen in **5.8a**, the pixels nearest the center of the receptive field have largest weights and thus have most influence. The weights then reduce, smoothly and gradually, to zero at the boundary of the field. This pattern of weights is shown in **5.8b** as a three-dimensional view of a receptive field, with weight strength again plotted on the vertical axis.

An example of the benefit of gaussian smoothing over smoothing with a rectangularly shaped field is shown in **5.9**. The former but not the latter uncovers the vertical grating to which a weaker higher spatial frequency (Ch 4) horizontal grating has been added as noise.

Image Edges have Different Scales

A close look at the image intensity landscape of the John Lennon image in **5.4** reveals that its variations in intensity range over different ***spatial scales***. That is, some intensity changes occur over small image regions and are thus quite sudden, e.g., the steep, sharp edges from the spectacles. Others occur more gradually over larger image regions, e.g., the shallow, fuzzy edges from shadows around the chin. The profiles of two slices of this image shown in **5.6** nicely illustrate this fact.

The different scales reflect different sorts of scene entities. Sharp edges often arise from object boundaries or texture markings, whereas shallow ones usually arise from shadows. It is important to find a way of dealing with this fundamental fact of scale in developing a theory of edge detection. (Small and large scales of image intensity changes are often referred to as conveying high and low ***spatial frequencies*** respectively, for reasons made clear in Ch 4.)

One way of finding edges at different scales is to use a range of receptive field types, each one "tuned" to a limited range of edge scales. Each

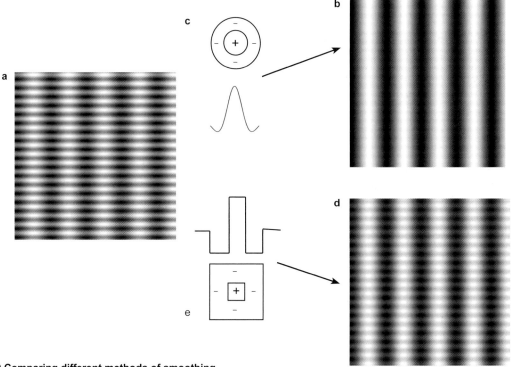

5.9 Comparing different methods of smoothing
a Vertical grating to which has been added a horizontal grating of equal contrast.
b Smoothing **a** with a circular symmetric gaussian filter with the receptive field of the kind shown in plan view and in profile in **c**. This preferentially attenuates the narrow (horizontal) grating.
d Smoothing **a** with a rectangular-shaped filter with a profile of the kind shown in **e**. This rectangular filter does not attenuate the narrow (horizontal) grating as much as the gaussian filter.

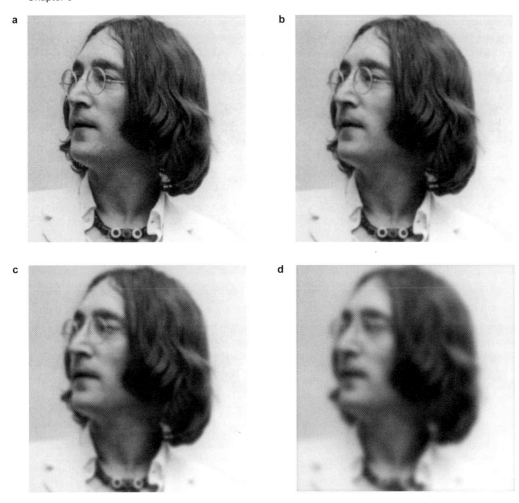

5.10 Convolving the Lennon image with gaussians of different standard deviations, σ
From **a** to **d**: σ = 1, 2, 4, and 8 pixels of the receptive field used in the convolution.

such receptive field is said to operate as a ***filter*** because it "lets through" only edges at a particular scale. But what are the principles for designing scale-tuned receptive field filters? This question, as usual, forces us to find a decent computational theory of the task.

Finding Edges at Different Scales

It was David Marr and Ellen Hildreth who first clearly stated, in 1980, edge detection as an optimization problem. Specifically, they stated that edge detection required an optimal trade-off between two conflicting tasks: selecting edges at a given scale and accurately localizing the positions of edges in the image at that scale. This led them to use gaussian receptive fields spanning a range of sizes, thereby exploiting the theorem referred to above in connection with noise cleaning. This is a nice example of a computational theory, because it is a mathematically well-founded solution to a clearly stated problem.

So, filtering for different scales is achieved by having a range of different convolution arrays, each one using a differently sized receptive field, but all having weightings of the basic bell shape, **5.8**. The ***parameter*** that determines the size, or equivalently the spread, of a gaussian is its ***standard deviation***, σ (pronounced sigma). Profiles of gaussians with large and small σ are illustrated in **5.8a**.

Results from using the Marr/Hildreth scale filtering scheme on the Lennon image are shown in **5.10** using four gaussians, each with a different σ.

As can be seen, the smallest (σ =1 pixel) preserves the sharpest image intensity variations. The image is increasingly blurred as σ is enlarged—compare the results for σ=1, 2, and 4 pixels. The largest bell-shaped receptive field completely removes the fine variations in intensity, both those due to noise and to sharp scene edges.

Computational Theory for the Task of Measuring Image Gradients

We started out by saying that image intensities can be thought of as a hilly landscape, **5.4**. This makes it natural to think of edge detection as the task of measuring *image intensity gradients*. So we now need a task theory telling us how to measure these gradients in our scale-filtered images. A simplified account of this is possible using the metaphor of the input image intensities forming a hilly landscape because the task then becomes one of measuring the gradients of hills. Technically, this is called finding the *first derivative* of the image landscape.

A road sign saying 14%, or 1 in 7, warns you that you will shortly go up or down 1 meter for every 7 meters travelled along the road. The numbers $1/7 = 0.14$, or 14% , are called *gradients*. This example shows that measuring a gradient means working out a ratio: the change of height for a given horizontal extent. In our case, this translates to finding edges by measuring how much intensity goes up or down over a given region of the image. That is as much gradient measuring theory as we need in this introductory account.

Algorithm for Measuring Image Gradients

The task theory tells us that we should measure image gradients and what this means. We now need a biologically plausible procedure for doing it. A simple algorithm is:

Step 1 Measure the difference in intensity (height) between image points.

Step 2 Divide this difference by the distance between the points, to obtain the gradient.

As an example, consider a case of the image intensity gradient between neighboring pixels, whose intensities happen to be 10 and 5 units. The gradient is then simply $(10 – 5)/1 = 5$ units of intensity per unit of distance. The latter is measured in pixel widths, which in this example is 1 as it concerns neighboring pixels.

Can this be procedure be realized in neurons? An equivalent but more suitable scheme turns out to be:

Step 1 For each pair of neighboring pixels, weight the inputs from one as positive, the other as negative, by multiplying by +1 and –1 respectively.

Step 2 Add together the weighted inputs.

This procedure is arithmetically equivalent to the previous one. It is biologically plausible because it is easy to implement with neurons.

Measuring Gradients in a Slice of an Image Intensities Landscape

To illustrate this procedure it is helpful to begin with just a single slice of an image intensity landscape, and then return later to considering the whole image. The slices shown in **5.6** are a bit complicated for showing the basic ideas and so an artificially generated slice is illustrated in **5.11**. It shows a cross-section in which, working from the left side, intensities rise and then fall, reflecting the presence of a bright ridge. Note that the gradient at Q is steeper than at S: walking up the hill at Q increases your height considerably more than walking at S for the same lateral distance.

The weighting of neighboring pixels in **5.11** is achieved by using a receptive field with –1 and +1 weights. For example, the pixel value on one side of the point labeled P is multiplied by –1 and the pixel value on the other side by +1, and then the two quantities so obtained are added together. In the case of P, this gives $(–5) + (+5)$, which equals 0. Zero makes sense in this case as P is on a horizontal part of the image intensity profile. The zero result is entered in the graph labeled convolution profile, **5.11b**.

The same procedure is also applied in **5.11** for the points Q, R, S, T, and U. But of course, as the gradient of the whole profile is required, it is necessary to apply the $(–1, +1)$ receptive field all along the image intensity profile shown in **5.11a**, rather than just at the chosen points Q, R, S, T, and U. Applying a receptive field all over an image is called convolution, as stated above. The image is said to have been *convolved with* the receptive field.

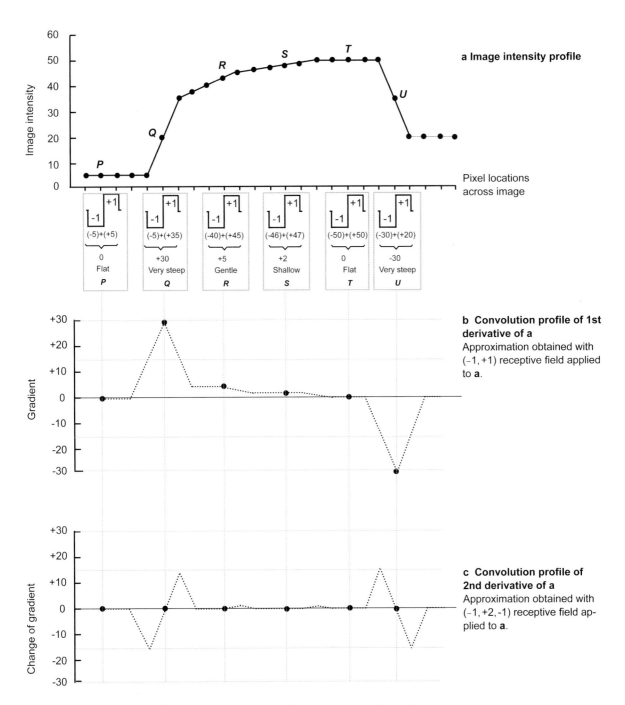

a Image intensity profile

Pixel locations
across image

**b Convolution profile of 1st
derivative of a**
Approximation obtained with
(−1, +1) receptive field applied
to **a**.

**c Convolution profile of
2nd derivative of a**
Approximation obtained with
(−1, +2, −1) receptive field applied to **a**.

5.11 Measuring gradients and changes of gradients using receptive fields
The units of intensity plotted in **a** are arbitrary but chosen to convey the main ideas.

To sum up so far, **5.11b** is the convolution profile derived from **5.11a**, using the (–1, +1) receptive field. This profile illustrates what is called the *first derivative*. It shows that when the intensity profile is flat, the output given by this receptive field is 0, as required. On the steepest parts in this example, the output is +30 or –30 (which means 30 units of intensity change for a shift across the image of 1 pixel). The plus sign indicates an up gradient, and the minus sign indicates a down gradient. The slope of the shallowest non-zero gradient is +1, e.g., at point S in **5.11a**.

Measuring Changes in Gradients

Measuring gradients (finding the first derivative) is a useful thing to do but it is also desirable to extract information about *changes in gradients*. Doing this is called finding the **second derivative**. It may seem a bit odd at first to think of measuring changes in changes. However, changes in intensity gradients are significant because they are usually caused by things in the world that the visual system wants to know about, such as illumination changes, surface orientation changes, and changes in surface reflectance (e.g., the edges or objects or surface markings).

One way in which a gradient change can be located is to search for a peak in the first derivative profile. The changes in slope in **5.11a** either side of Q produce a sharp peak in its first derivative, **5.11b**. Note that a trough (a "negative peak") occurs for the point U. This point is similar to Q in that it has sharp changes of slope on either side. However, a trough occurs in the convolution output rather than a peak simply due to the signs of the (–1, +1) receptive field coupled with the downward direction of the slope at U. If a (+1, –1) receptive field had been used then the trough at U would have been a peak, and the peak associated with Q would have become a trough.

To summarize, the first derivative measures gradients and the second derivative measures changes in gradients. The latter can be found by measuring gradients in the first derivative.

The second derivative of the intensity profile in **5.11a** is shown in **5.11c**. Note that at points where a peak or trough is located in the first derivative, **5.11b**, the second derivative passes from positive to negative. Such a point is called, natural-

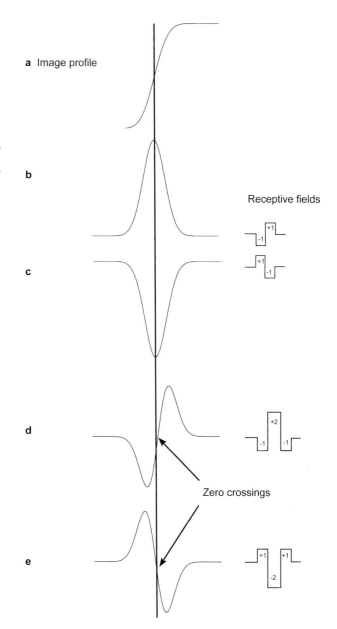

a Image profile

b

c

d

e

Receptive fields

Zero crossings

5.12 Convolving receptive fields with an edge
The intensity change in **a** gives rise to a peak or a trough in the first derivatives **b** and **c**. It gives rise to a zero crossing in its second derivatives **d** and **e**.

ly enough, a *zero crossing*, **5.12**. Hence, there are two ways of locating changes in intensity gradient: either find zero crossings in the second derivative or find peaks or troughs in the first derivative.

Single-Step Way of Measuring Changes in Gradients

We have said that the second derivative can be found by applying the $(-1, +1)$ receptive field twice over. That is, use the following algorithm:

Step 1 Measure gradients using receptive fields of the type $(-1, +1)$ or $(+1, -1)$.

Step 2 Measure gradients in the output of Step 1, again using fields of either $(-1, +1)$ or $(+1, -1)$.

It turns out that there is a way of measuring changes in gradients using a single processing step. The receptive fields for doing this have profiles of either $(-1, +2, -1)$ or $(+1, -2, +1)$. We now explain why this works.

At a point where the gradient in the image intensity profile changes, the gradient to the right of the point is different from the gradient to the left. This is illustrated in **5.13** for a point labeled *V*. So if we measure the gradient to the left and subtract from it the gradient to the right we obtain a measure of their difference and hence a measure of the second derivative.

An easy way to do this in a single-step convolution is to use a receptive field built up in the following way:

	V		
			Point *V* is where the 2nd derivative is being measured
−1	+1		Gradient to the left of *V*
	−1	+1	Gradient to the right of *V* for subtraction
−1	+2	−1	Resulting receptive field

Beware a possible confusion at this point. In the middle column of the table the subtraction can be written as +1 minus (−1) = +2. Recollect that a rule of arithmetic is that "two minuses give a plus." That is, doing the minus operation twice over yields a plus. Thus, minus (−1) becomes +1, and this is why the answer is +2.

The result of applying this receptive field to the intensity profile in **5.11a** is shown in **5.11c**, in

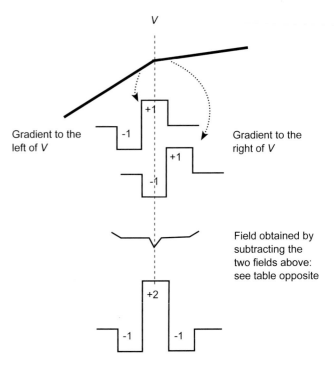

5.13 Receptive field of (−1, +2, −1) type measures changes of gradient
This figure should be studied in conjunction with the associated table shown opposite.

which the zero crossings mark points of gradient change in the image. We will have more to say about zero crossings as edge location markers in Ch 6.

The same simple arithmetic can be used to derive the $(+1, -2, +1)$ weighting profile by subtracting neighboring pairs of $(+1, -1)$ fields.

The conclusion we have thus come to is: convolving with a $(-1, +2, -1)$ or $(+1, -2, +1)$ receptive field is a neat way of getting the second derivative in a single processing step. This could explain why many cells in biological vision systems have these sorts of fields.

Two-Dimensional Convolutions

For purposes of exposition, only a one–dimensional slice of an image intensity profile has been considered so far. However, a visual image is usually a two–dimensional array of intensities. So how can the receptive field profiles just described can be extended to cope with two dimensional images?

A straightforward way to do this is to keep the receptive field functionally a 1D device by making it sensitive to gradients in one particular orienta-

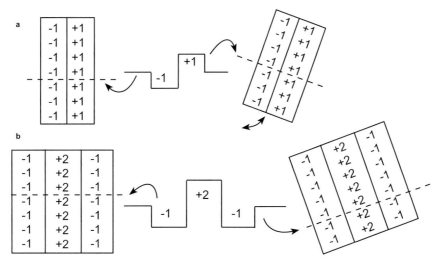

5.14 Oriented receptive fields for measuring directional derivatives
Fields with different orientations are shown together with the profiles across them in the directions shown with the dashed lines that are perpendicular to field orientation. **a** 1st derivatives, **b** 2nd derivatives.
Note the similarity of these fields to those of the simple cells described in Ch 3.

tion only. This is done by the simple expedient of elongating each receptive field, **5.14**. In this case the operators are said to provide ***directional de-rivatives*** because the derivative is tied to a particular orientation.

Having read Ch 3, you may recognize how these extended one–dimensional receptive fields bear a strong resemblance to the receptive fields of some of the *simple cells* found in the striate cortex of cats and monkeys by Hubel and Wiesel. This similarity makes it tempting to suggest that simple cells really are devices for delivering directional derivatives. If this is so, gradient measurements would need to be sampled in a fairly large number of different orientations at each point in the image. This may not be a great handicap for the brain because it has lots of orientation tuned cells (more details on these in Ch 9).

The idea that the receptive fields of simple cells can be interpreted as providing oriented second derivatives shows how far we have come from Ch 3. There we discussed simple cells as candidates for "bar detectors." Careful examination of the task of edge detection has hugely refined our understanding of what these cells might be doing. This is a fine example of the benefits that can come from task analysis at the computational theory level.

That said, it is also important to note that whether simple cells really are best thought of as

delivering directional derivatives is still an open question. We know too little about brain mechanisms to be sure at present. Even so, some believe it to be the current "best bet" for the functional role of at least some types of simple cells.

Measuring the Image Gradients Using an Isotropic Receptive Field

Is it possible to measure the first and second derivatives (i.e., image intensity gradients and changes in image gradients, respectively) in two–dimensional images *without* using oriented receptive fields? That is, can two-dimensional receptive fields be constructed for measuring derivatives that are sensitive to image intensity changes irrespective of their orientations? The technical term for describing any process that is the same in all directions is ***isotropic*** (*iso* = same; *tropia* = direction).

It turns out that the first derivative cannot be measured with an isotropic receptive field but this can be done for the second derivative using a ***laplacian*** receptive field, **5.15a**. You can think of this receptive field as being created by spinning a $(-1, +2, -1)$ set of weights around its center. We will see in Ch 6 that this type of receptive field is similar to those possessed by the certain cells in the retina, **5.16**. This will prove important when we consider what certain retinal cells seem to be doing and how they feed into brain cells.

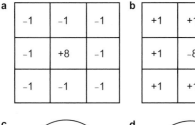

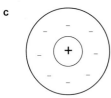

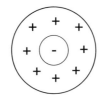

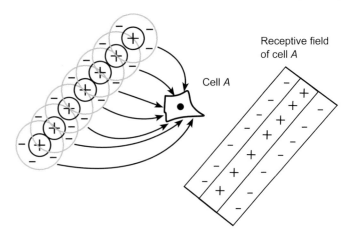

Receptive field of cell A

Cell A

5.15 Laplacian receptive field

a A 3x3 array of weights that can be thought of as a combination of four $(-1, +2, -1)$ fields, one each for vertical, horizontal and the two obliques directions.

b A similar combination of four $(+1, -2, +1)$ fields, producing a receptive field with opposite signs to those shown in **a**.

c and **d** Respectively, on-center and off-center circularly symmetric receptive fields of the antagonistic cente--surround kind found in many biological visual systems. Their qualitative resemblance to the laplacian fields in **a** and **b** is striking. We discuss in detail receptive fields of this type in Ch 6, on the retina.

5.17 Combining a set of isotropic gradient measurements to create a cell signalling an orientation-tuned second derivative

The circularly symmetric fields measure changes of gradient in slightly different positions on the image. When their outputs are fed into *cell A* then this cell has a oriented receptive field of the kind possessed by some simple cells. *cell A* can thus be regarded as an *operator* implementing a **figural grouping** *process* that links edge points: see Ch 7 for details. This is different way of interpreting what simple cells may be doing.

It is easy to compute directional derivatives from isotropic gradient measurements. This is done simply by combining the outputs of circularly symmetric (isotropic) receptive fields dealing with suitably located nearby image points. An example is shown in **5.17**. One way of looking at this is as a *figural grouping operation* which links together edge points that form an edge with a certain orientation (details in Ch 7).

Skip from Here to Summary Remarks?

Readers who do not want technical details on how to combine blurring with measuring image gradients can move on to the Summary of this chapter on p. 132, and then on to Ch 6.

5.16 Circularly symmetric receptive field

This field is responsive to edges of all orientations, as demonstrated in detail in Ch 6.

a 3D picture of the response profile of an on-center receptive field. Fields of this kind are informally called *Mexican hat* receptive fields because of their similarity to the cross-sectional shape of a classic form of Mexican head wear.

b Explanation of the profiles shown in **a**. The graph rises above the horizontal "zero response" line when excitatory influences arising from light falling on to the field's center exceeds inhibitory ones arising from light falling on the surround.

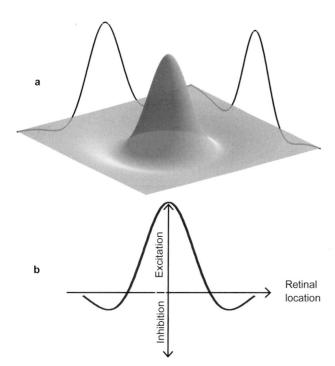

a

b

Excitation

Inhibition

Retinal location

Combining Image Blurring and Gradient Measuring

Recollect that we began by explaining how receptor noise could be reduced by taking the average of local neighborhoods of receptors. Having done that, we turned to the task of measuring image gradients in the "blurred" or "smoothed" images. In each case we showed how this could be done with convolutions using suitable receptive field types.

What we now do is explain how, by exploiting a clever bit of mathematics, it is possible to combine blurring and gradient measuring in a single receptive field. We won't go into the details of the math but we can convey the basic idea quite simply.

The type of receptive field that we introduced for noise cleaning was our old friend, the gaussian. Now, imagine applying *to this gaussian* the receptive field for measuring changes of gradient, $(-1, +2, -1)$.

If this is done it creates a new kind of receptive field, called D^2G for short (pronounced *delta squared G*). This can be used as a receptive field in its own right to deliver image blurring and gradient measuring in a single step, **5.18**. Very neat.

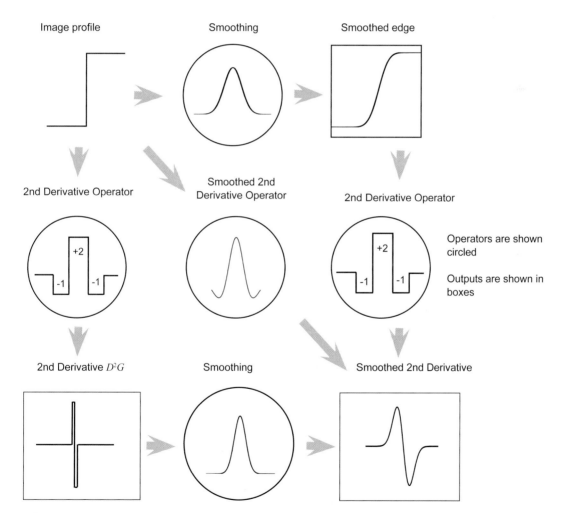

Image profile — **Smoothing** — **Smoothed edge**

2nd Derivative Operator — **Smoothed 2nd Derivative Operator** — **2nd Derivative Operator**

Operators are shown circled

Outputs are shown in boxes

2nd Derivative D^2G — **Smoothing** — **Smoothed 2nd Derivative**

5.18 Convolution is commutative
This means that the same outcome is achieved regardless of the order in which operations are carried out. Hence, starting from the same image profile (upper left), the sequences shown by the green arrows and the blue arrows produce the same outcome (lower right). The sequence shown by the arrows uses an operator (center) that produces that same output in a single stage.

The reason this works is because the arithmetic doesn't care in which order these operations are performed. That is, the following two sequences of operations produce the same result:

Step 1 Convolve the grey level image with a gaussian receptive field to give a gaussian smoothed image.

Step 2 Convolve the gaussian smoothed image with the laplacian receptive field to obtain the second derivative of that image.
or
Step 1 Convolve the laplacian receptive field with the gaussian smoothing receptive field to create the D^2G receptive field.

Step 2 Convolve the D^2G receptive with the grey level image to obtain the second derivative of the smoothed image

It may surprise you that these two procedures yield the same results. Technically, mathematicians say that it works because convolution is *commutative*. To help understand this, note that multiplication is also commutative. Thus $2 \times 3 \times 4 = 24$, and $2 \times 4 \times 3 = 24$. The order of doing the multiplications doesn't matter.

What does the receptive field look like? It is illustrated in profile in the middle of **5.18** and as a 2D shape in **5.18**. It is the so-called ***Mexican hat receptive field*** shown in **5.16**.

Mexican hat (or D^2G) receptive fields come in different sizes according to the scale of image edge that they have been designed to deal with. That is, convolving a laplacian with a narrow gaussian (small standard deviation) produces a smaller D^2G

receptive field than does convolving a laplacian with a wide gaussian (large standard deviation). This is illustrated with the Mexican hat profiles shown in **5.19**

To round off this account, **5.20** illustrates and summarizes the task of finding an edge in a noisy image.

Block Portraits and Cross-Channel Combination

What do you see in the block pattern in **5.21**? Most people report nothing more than a mass of blocks of varying greys. Having read Ch 1 you might choose to describe it as a grey level description with a very coarse pixel size, so that each block has a grey value which is the average grey level for quite a sizeable area of the input image. But that doesn't help you see what the original input image was.

Now try looking at **5.21** with your eyes screwed up so that your vision is blurred. If you wear spectacles you might find that you can blur your input image satisfactorily by looking at the picture without spectacles, perhaps from a few feet away. In any event, if you are successful in blurring your vision, you will almost certainly now be able to recognize that **5.21** contains **5.10a**. So we have here a curious state of affairs. Normally we are used to sharp vision giving us the best hope of recognizing objects, not the worst. Millions of people wear spectacles for this very reason. What is going on?

First of all, we pointed out early in this chapter that we can recognize blurred pictures, and so that particular aspect of this demonstration is not mysterious. The problem to be tackled is: how is it that the object is "hidden" in **5.21** when the image is sharp, and "released" when the image is blurred?

It is necessary at this point to understand that the block portrait **5.21** can be considered as a composite of two pictures added together. The first of these is a coarse picture of John Lennon in which only low spatial frequencies are present (re–read Ch 4 if you need a reminder on spatial frequencies). This is what appears when the blocks image is blurred as the low spatial frequencies survive the blurring.

The second composite picture in **5.21** portrays the sharp edges of the blocks. This high spatial frequency information is removed by blurring. Thus, one explanation of why we don't see the object in

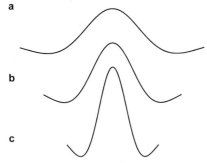

5.19 Profiles of receptive fields for measuring second derivatives of image intensities at different scales
a Coarse scale. **b** Medium scale. **c** Fine scale.

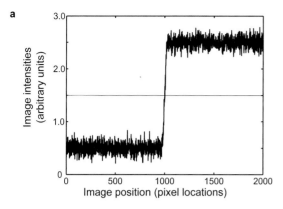

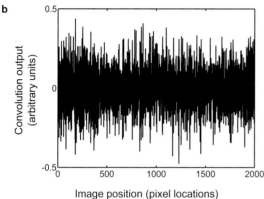

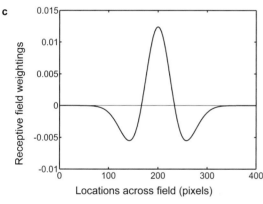

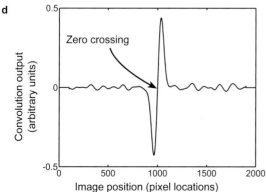

5.20 Summarizing edge detection using the Marr-Hildreth algorithm (left)

a One–dimensional profile of a step edge in an image to which has been added low amplitude noise.

b The results of finding the 1st derivative of **a** without image smoothing. The edge is lost in a forest of edges, most of which come from the noise.

c A receptive field that combines smoothing and finding the second derivative (note scale on vertical axis). This is a profile of the laplacian (i.e. second derivative) of a gaussian, or the D^2G receptive field.

d The results of convolving the D^2G receptive field with the noisy edge in **a**. This both blurs away the noise and produces an output which allows the edge to be located at the zero crossing.

5.21 unless it is blurred is that the high spatial frequencies are "masking" the information in the low frequencies. But this account begs the question, which can be posed as: *why* should inclusion of the high frequencies mask the information carried by the low frequencies?

We agree with Marr's answer to the conundrum posed by **5.21**. When both the coarse portrait and the sharp edges are present, we build up a representation of what is "really" there—a set of blocks of various shades of grey. When the information about the sharp edges is stripped away by blurring, the blocks disappear and we see what the low spatial frequencies on their own contain: Lennon.

Putting this another way, we can say that our representation of the *blocks* in the block portrait simply does not match our stored visual representation of John Lennon. Consequently, we do not

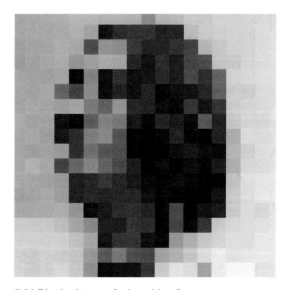

5.21 Block picture of what object?

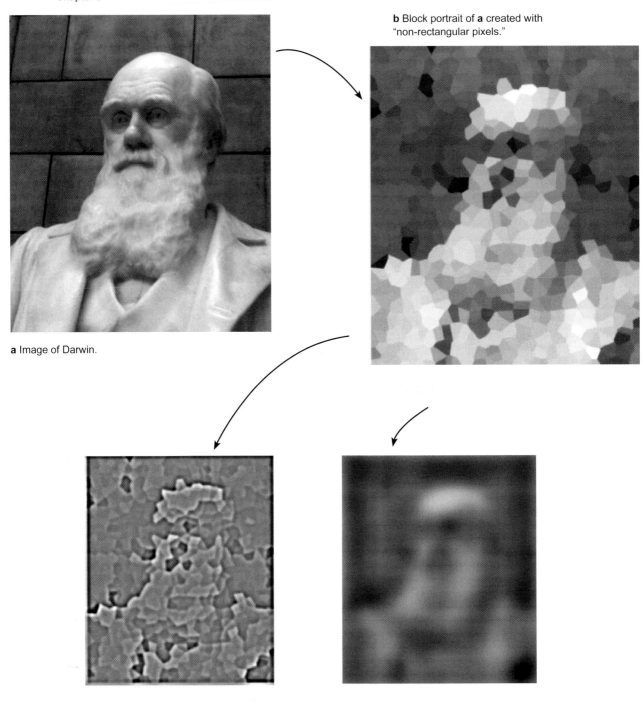

b Block portrait of **a** created with "non-rectangular pixels."

a Image of Darwin.

c Filtered version of **b** revealing its high spatial frequency components, i.e., the sharp edges of the "pixels."

d Filtered version of **b** revealing the low spatial frequency components. This creates a species of blurring that allows recognition of the portrait hidden in **b**.

5.23 An explanation of why blurring a block portrait allows recognition of the hidden face
Blurring the pixelated version of **a** strips away the sharp edges of the pixels (themselves shown in **c**) and allows the coarse structure to be recognized. The fact that recognition is possible from such blurring is testimony to the power of human object recognition, but that is another story discussed in Ch 8.

recognize him in **5.21** when we view it well–focused. On the other hand, when the information about the sharp edges is stripped away by blurring, the structural description for John Lennon *is* now built up, and recognition duly occurs. The mystery is resolved. (We discuss the visual representation of faces in Ch 8.)

The joke picture in **5.22** (p. 204) that has been widely circulated on the Web (unattributed: creator please contact us) can be interpreted in the same way.

You might be wondering: why is a block portrait a composite? The answer is that it is a fact of physics that when a block picture is made, by averaging light-intensity over sub-regions of an image to obtain the grey levels for the various blocks, this averaging does not disturb very greatly the coarse structure of the original image (as long as the blocks are not too big). However, it destroys the sharp detail of the original and replaces it with new sharp detail—the edges of the blocks. The two components in a block portrait, the edges of the edges and the coarse structure, are illustrated in **5.23** for a block portrait of Darwin.

Block portraits raise a critically important issue: the combination of information in different spatial frequency channels. This is the problem of ***cross-channel combination***. Often the "signature tune" for a sharp edge in a natural image is a particular pattern of activity in several channels, each one tuned to a limited range of spatial frequencies (Ch 4). This pattern of activity in several channels has to be recognized somehow to build up accurate representations of features in images. Marr and Hildreth (1980) discuss this problem (see also Georgeson, May, Freeman, and Hesse, 2007).

One interesting conclusion we can draw from block portraits is that we do not have conscious awareness of information carried in the different spatial frequency tuned channels described in Ch 4. Rather, what we see is the result of combining information from *all* channels, as appropriate.

A second conclusion is that block portraits provide evidence against the idea that low spatial frequency channels play a special and direct role in mediating object recognition, as has been thought by some. If that proposal were true then we should surely see Lennon in **5.21**, or Darwin in **5.23b**, despite these being well–focused.

Finally, we find it interesting that a block picture of a very ordinary original scene can have considerable aesthetic appeal. Or at least, so it seems to us. A very mundane visual scene indeed is shown in **5.24a**—a house chimney. It is reproduced as a block picture in **5.24b**. The aesthetic improvement effected by this transformation is, we think, remarkable, and reminiscent of certain styles in twentieth–century art, for example, various paintings by Braque, Mondrian, and Picasso. The next time you see some cubist art, try blurring your vision and see if that helps reveal the object that has been "block rendered" by the cubist style.

It is tempting to speculate on the basis of these various demonstrations that one strand in modern art has been, perhaps unwittingly, to stretch to the limit the visual system's capability for recognizing easily what a painting contains. Perhaps it is when the visual system is hovering on the brink of successful recognition, but not quite able to achieve it, that certain aesthetic experiences become manifest.

a

b

5.24 Block pictures as art?
a Mundane house image. **b** Blocked version of **a**.

Summary Remarks

Marr and Hildreth's edge detection theory described in this chapter has served well our goal of introducing core ideas such as convolution and derivatives. It has been replaced in computer vision by mathematically more refined techniques (see Gonzalez and Woods, 1987) but we have chosen to describe it in detail for three reasons.

First, it has allowed us to introduce some basic image processing ideas in an accessible way for non-mathematical readers. Second, their theory has considerable relevance to the study biological vision systems. And third, their theory has an important place in the history of seeing research over the past three decades. Understanding how edge detection can be regarded as the task of measuring image gradients and changes of image gradients is fundamental to anyone wanting to tackle the modern vision literature.

That said, it is important to note also that the kind of zero crossing map in **5.1b** (other examples in Chs 6 and 7) can be at best only a useful first step to the edge representations used by human vision. The sketch built up by the system of Gooch et al. in **5.1c** shows that computer vision studies using further processing can yield edge maps much closer to the kind of edges drawn by artists in their sketches, **5.1c**.

Also, it is worth noting that psychophysical studies have cast doubt on whether human vision uses zero crossings at all. Perhaps it extracts other edge markers from its convolutions (see Morgan and Watt, 1997).

The main general lesson to be taken from Marr and Hildreth's work is that it is essential to develop a sound computational theory of a seeing task. The task in question has been: how can edges be detected in noisy images? This led to their theory which treats edge detection as an exercise in measuring image gradients in images smoothed with a gaussian because that optimally combines smoothing with preservation of edge locations. The theory yielded several algorithms each with a clear rationale. Those algorithms provide possible interpretations of the receptive fields of certain types of cells found in biological visual systems. Hence we see here the value of the computational approach to seeing.

Further Reading

Georgeson MA, May KA, Freeman TCA, and Hesse GS (2007) From filters to features: scale-space analysis of edge and blur coding in human vision. *Journal of Vision* **7**(13) 1–21. *Comment* A fine starting point to get into the scale space literature.

Gonzalez RC and Woods RE (1987) *Digital Image Processing* 2nd Ed Prentice-Hall:New Jersey. *Comment* A textbook on image processing that provides a broad review for anyone wanting to explore the wide range of approaches to edge detection.

Gooch B, Reinhard E, and Gooch A (2004) Human facial illustrations: creation and psychophysics. *ACM Transactions on Graphics* **23** 27–44.

Marr D and Hildreth E (1980) Theory of edge detection. *Proceedings of the Royal Society of London B* **207** 187–217. *Comment* This describes the technical details of the general approach to edge detection described here. See also Marr D (1982) *Vision* WH Freeman and Company. However, it is worth noting that Marr and Hildreth considered and rejected the idea of simple cells as delivering directional first derivatives. Rather, they preferred to see them as performing a grouping operation of the kind shown in **5.17** and considered in more detail in Ch 7.

Morgan MJ and Watt RJ (1997) The combination of filters in early spatial vision: A retrospective analysis of the MIRAGE model. *Perception* **26** 1073–1088.

CVonline: The Evolving, Distributed, Non-Proprietary, On-Line Compendium of Computer Vision http://homepages.inf.ed.ac.uk/rbf/CVonline/ *Comment* A remarkable, evolving, and free Web resource for anyone wishing to explore computer vision techniques, including edge detection. Its creation has been heroically masterminded by Bob Fisher at the University of Edinburgh.

6

Seeing and the Retina

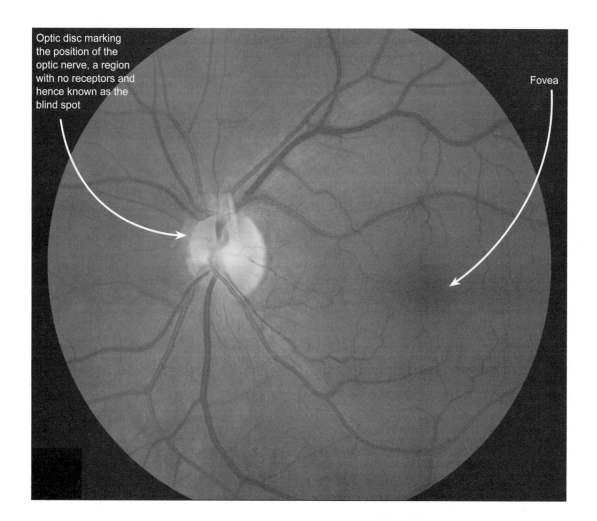

Optic disc marking the position of the optic nerve, a region with no receptors and hence known as the blind spot

Fovea

6.1 Looking into a human eye

This is a view seen using an ophthalmoscope, a device that allows the retina to be illuminated and observed at the same time. The branching structures are arteries and veins lying close to the surface of the retina. Courtesy R Farr. Visit http://webvision.med.utah.edu to see a colored version of this sort of picture and also various Webvision figures used in this chapter.

The retina is a 0.5 mm thick layer of cells located on the inner surface of the eyeball, **6.1**. It contains the light-sensitive *receptors*, **6.2**, which initiate the whole business of building scene representations. The lens system of the eye focuses an image of the observed scene on to the *receptor mosaic*, which contains the 126 million or so receptors that respond when light shines on them. They send messages to various nerve cells in the retina, and the retina's output is sent to the brain via a stream of pulses (*action potentials*) in the roughly 1.2 million fibers in the *optic nerve*.

Anatomy of the Retina

The most important retinal cells and their inter-connections are illustrated schematically in **6.3** and **6.4b**. The *receptor → bipolar cell → ganglion cell* pathway is sometimes referred to as the *vertical organization* of the retina, to distinguish it from the *horizontal organization*, which consists

Why read this chapter?

The chapter begins with the basics of retinal anatomy. The receptive fields of *bipolar* and *retinal ganglion* cells are then described. We explain how the combination of antagonistic excitatory and inhibitory signals from the central and surrounding regions of these receptive fields can be modeled as a *difference of gaussians* (DOG) to yield the characteristic *Mexican hat response profile*. We then discuss how bipolar cells could perform the convolution operation specified by the Marr/Hildreth edge detection theory described in Ch 5. The existence of two different types of bipolar cell receptive fields (on-center/off-surround and off-center/on-surround) is interpreted as a means of coping with both the positive and negative parts of the Marr/Hildreth convolution output. The problem of detecting edges is modeled as the task of finding zero crossings in the convolution output, although psychophysical results casting doubt on this being a satisfactory model for human vision are noted. The classic explanation of the *Hermann grid illusion* as the outcome of the center/surround convolution is described, but this explanation is rejected because it fails to predict correctly what is seen in a number of figures. The chapter ends on a cautionary note: the retina is an immensely complicated structure containing a wide variety of cell types, and can therefore be considered as a "mini-brain" whose operations are, as yet, not well understood.

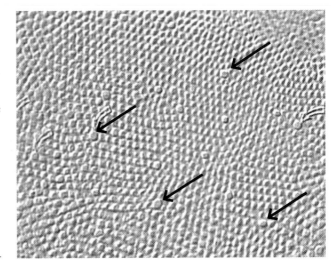

6.2 Receptor mosaic: plan view
This is a tangential section through the fovea in the receptor layer. The arrows mark examples of the larger and scarcer *S*-cones that are sensitive to short wavelengths (blue light). Curiously, there are no *S*-cones in the central fovea and they comprise only about 4–6% of the receptor population. See Ch 17. Courtesy Webvision.

of cells whose fibers have sideways connections. The latter cells are mainly of two types: *horizontal cells*, which carry lateral messages at the level of the receptor → bipolar junctions, and *amacrine cells*, which carry lateral messages at the bipolar → ganglion cell junctions, **6.4b**.

The first and most curious point about retinal anatomy is that the receptors are tucked away inside the retina. They are not on the surface, which receives the light. This is an odd arrangement because the light has first to pass through many (albeit near-transparent) layers of cells as well as blood vessels, before it can activate the receptors and thus set going the processes of seeing.

Squids and octopuses have evolved eyes which are similar to our own except that their receptor layer is on the surface of the retina, where light can strike receptors directly. That seems, at first sight, a much more sensible arrangement. But perhaps evolution discovered that having the receptors at the back of the retina provides a way of giving mammalian eyes "sun block" protection. The central region of the retina, called the *macula* has a yellowish pigment called the *macula lutea* thought to play the valuable role of absorbing ultraviolet light. Squids and octopuses don't have this sunlight protection problem in their aquatic world.

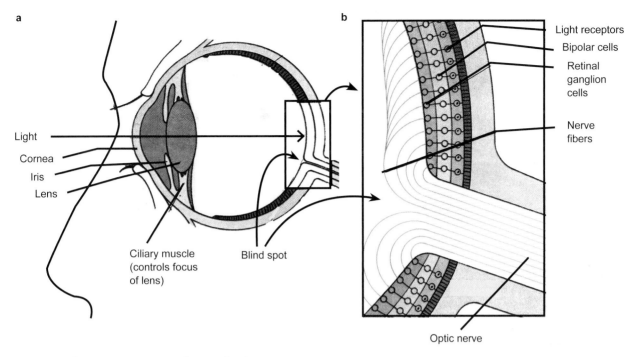

a

Light

Cornea

Iris

Lens

Ciliary muscle
(controls focus
of lens)

Blind spot

b

Light receptors

Bipolar cells

Retinal
ganglion
cells

Nerve
fibers

Optic nerve

6.3 Schematic diagrams of eye and retina
a Vertical cross-section of eye.
b Expansion of inset in **a** to show the 3 main layers of cells: receptors, bipolars, and ganglion cells.

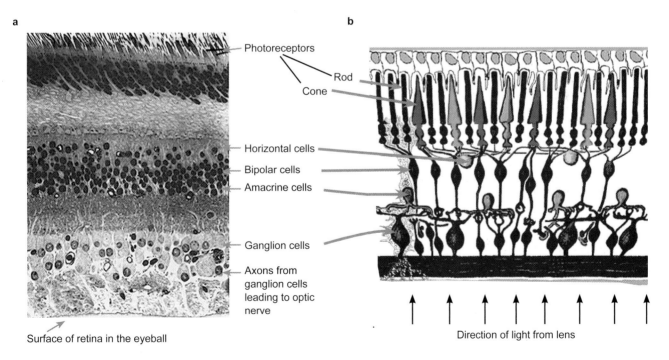

a

b

Photoreceptors

Rod

Cone

Horizontal cells

Bipolar cells

Amacrine cells

Ganglion cells

Axons from
ganglion cells
leading to optic
nerve

Surface of retina in the eyeball

Direction of light from lens

6.4 Cross-sections through the human retina
a Light microscopic section. **b** Diagram showing the main types of retinal cells and their arrangements in layers.
Courtesy Webvision, whose website (http://webvision.med.utah.edu) is a good starting point if you wish to find out more
about retinal anatomy.

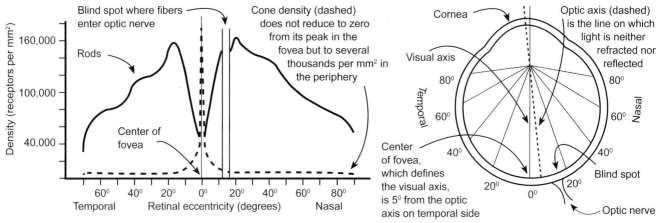

6.5 Rod and cone densities across the retina
Approximate distribution of cones and rods showing how their densities vary with retinal eccentricity.

6.6 Concept of retinal eccentricity (schematic)
Horizontal slice through left eye showing how retinal position is measured in terms of angular difference from the visual axis.

Rods and Cones

The receptors are of two types, called *rods* and *cones* on the basis of their shapes. They contain different types of *photosensitive pigment*, differentially sensitive to different wavelengths of light, which enable them to specialize in different tasks. We have about 120 million rods and about 6 million cones. They can be thought of as two distinct light-sensitive systems packaged together into a single "camera"—the eye. The distribution of rods and cones varies across the retina, with the cones most dense centrally, **6.5**, **6.6.**

The rods have a pigment called *visual purple*, or *rhodopsin*, which is about 500 times more sensitive to light than the pigments in the cones. The rods are thus well suited to low-level "night" vision (*scotopic vision*). They are thought to play no role in color vision, their job being to signal light intensities. Rods are most sensitive to green light—rhodopsin absorbs green light relatively well. It reflects red and blue light relatively well, which is why it appears purple. The importance of the rods for night vision is recognized by the requirement for computer displays in the bridges (control rooms) of ships sailing at night to be covered with a red filter. The rods are less sensitive to red light and so the filters preserve their rhodopsin for responding to the dim light coming from potential hazards, such as other ships.

Cones contain other photosensitive molecules—the pigments that make color vision possible. They require more light to work than the rods and hence support "daylight" or *photopic vision*. There are three types of cones. They absorb light maximally at wavelengths corresponding roughly to long (red-yellow), medium (green-blue), and short (blue) wavelengths of light. Hence they are referred to as *L, M,* and *S* cones. We discuss them further in Ch 17.

At the center of the fovea is the *macula* which contains the *foveal pit*, **6.7**, which has only cones, packed in at the high density of about 150,000 cones per square millimetere. The very central region of the foveal pit, where the cones are smallest and packed tightest, is very small—only about 1/10,000 of a millimeter across. This region subtends a visual angle of about 20 seconds arc or 0.3°, so it would be filled by an image of a disc measuring about 0.3 cm viewed from about 57 cm.

[Visual angle is a convenient way of describing the size of a retinal image irrespective of the particular size or distance of the object which gave rise to it, **4.5**. A handy fact is that a finger nail (which is roughly 1 cm across) viewed at arm's length (which is roughly 57 cm) subtends about 1°.]

The foveal pit probably contains only about 2000 cones in all, but even within this densely packed region of retina, further variations in the size of the cones can be detected with careful scrutiny. At the very center of our gaze, served by the very smallest cones packed most tightly together, seeing may depend on a total of only a couple of dozen receptors. It is a startling thought that when we fixate something very carefully in an effort to maximize our ability to detect fine details, we

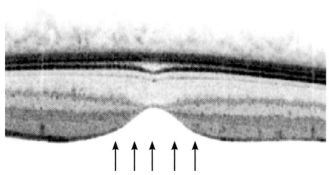

↑ ↑ ↑ ↑ ↑
Direction of light from lens

6.7 Cross-section of retina showing foveal pit
Optical coherence tomography, courtesy Tom Evans and
John Burke.

might be relying on just 20–30 cones, each with a diameter of about 1 *micron* (one millionth of a meter).

Retina and Image Compression

Each receptor is not served by a single cable to the brain. This is evident from the fact that we have about 126 million receptors (roughly equivalent to pixels in digital cameras) in each eye, but only about 1.2 million nerve fibers in the optic nerve of each eye leading to the brain. The average ratio of receptors to optic nerve fibers is thus about 100:1. In the fovea, the ratio is much closer to 1 to 1, but even there it is misleading to think of each foveal receptor having its own "private telephone line" to the brain. This is because in between each receptor and the output fibers of the retina are many cells, **6.4**, which combine receptor signals in various ways.

The 100:1 ratio of receptors to optic nerve fibers implies that information encoded by the receptors has to be sent in an abbreviated form to the brain. That is, the retina has to encode and send on to the brain only the important things about the image in "informationally compact" messages. This is called an *image compression* task and it needs a lot of processing power. Biologically, that means lots of nerve cells. It is thus no surprise that the retina is embryologically a part of the brain in humans—it grows in the fetus from the same lump of cells as the brain. It has many neurons because it is a mini-brain engaged in a complex image processing task.

Interestingly, squid and octopus eyes, so similar to our own except for having their receptor mosaics close to the surface of the retina, do *not* grow from embryological brain tissue. This is one line of evidence for this eye having evolved independently of the mammalian eye, although we probably share some eye-related genes with them that arose in very ancient common ancestors.

What would be a good strategy for solving the image compression task performed by the retina? Many important things in our visual world, such as object edges, usually show up as image intensity gradients that change rapidly, Ch 5. We might therefore expect to find neurons in the retina that measure changes in image gradients. Armed with this expectation, we will turn shortly to describing the image processing properties of the cells in the retina.

But first a word of caution: different animals have different sorts of retinas. Animals with well-developed visual brain mechanisms, such as humans, monkeys, and cats, have relatively simple (but still complex) retinas. Other animals, without the same extensive central brain machinery for vision, such as frogs and birds, rely rather more upon their retinas for visual analysis, so their retinas have evolved to be more complicated. In the neurophysiological account which follows, we describe only the typical workings of the more straightforward mammalian retinas.

Before doing so, we discuss briefly a question that may have occurred to you: why didn't evolution simply make the optic nerve bigger and hence avoid the retina having to deal with an image compression task?

One possible answer could be keeping the brain down to a manageable, indeed portable, size. Suppose the brain analyzed every part of the field of view with the same high resolution it uses for foveal vision. Calculation shows that the brain would then have to be huge. Not a very practical proposition for a hunter-gatherer species. Instead, evolution came up with moveable eyes to scan the scene using the high resolution foveal pathway *sequentially* over the field of view. Our perceptions of an apparently sharply defined visual world are in fact composed from a series of glances, rather like shining a torchlight around a scene and building up a representation of what comes into view.

But having a fovea and mobile eyes comes at a price: finding a solution to the problem of getting messages from the eyes to the brain via an optic nerve that is small and light enough not to inhibit eye movements. So perhaps this is a second reason why the optic nerve has fewer fibers than there are receptors.

Note in **6.4b** that each receptor can serve more than one bipolar cell and that ganglion cells can collect messages from more than one bipolar cell. This is why the idea that receptors have private lines to the brain is deeply misleading.

It must be emphasized that **6.4b**, though adequate for our purposes, is a highly schematic outline of retinal anatomy. Subdivisions can be distinguished within the various cell classes, just as the receptors can be divided into rods and cones. To illustrate another facet of retinal complexities, about 1–3% of retinal ganglion cells have the photo-pigment *melanopsin* that makes them intrinsically photosensitive. This pigment responds more slowly to light than the pigments in rods and cones and is thought to subserve the control of daily bodily *circadian* rhythms and the control of pupil size.

Retinal anatomy is a well developed field of knowledge. What is less well understood are the details of how the retina functions and what computational tasks it performs.

Neurophysiology of the Retina

Some important clues about retinal function have come from neurophysiological studies using the technique of single cell recording described for striate cells in Ch 3. The general idea is to insert a *microelectrode* into a retinal cell and record from it while stimulating the retina with a pattern of light, **3.4**. The pattern of light is manipulated in various ways, and the processing job performed by the cell is then inferred from the nature of its responses and types of stimuli to which it responds.

The creature that contributed most to our understanding of retinal function in the early pioneering work is the mudpuppy (*Necturus maculosus*). This curious amphibian, a type of salamander, lives in the muddy depths of silt-laden rivers. Ironically, the environment of the mudpuppy is such that it can hardly depend much on vision for its survival. And yet this animal proved a boon for neurophysiologists interested in the retina because it is blessed with relatively huge retinal cells. John Dowling and Frank Werblin (1969) exploited that fact to insert the fine tips of microelectrodes into every type of cell, even receptors. Such penetrations could not then be made in most animals without causing significant damage. Once the electrode was inserted, it was possible to record from the cell as it responded to stimuli flashed on a screen viewed by the animal.

Receptors Deliver a Gray Level Description

With such techniques, Dowling and Werblin found that the receptors respond to the luminance of the input image in just the way required of a gray level description. The receptor response is a change in voltage across their cell membrane (not a code in terms of action potentials) that increases with the intensity of incident light, and lasts as long as the stimulus lasts. The change in voltage level across a receptor's membrane can thus be thought of as functionally equivalent to the pixel numbers captured by digital cameras.

But remember that the concept of "gray level" is used very broadly in this book. It includes a cone responding selectively to light of a particular color. (Or more properly, light of a particular wavelength, because light is not itself colored. Color is an attribute of our perceptions. However, it is common practice to refer to the colors of light as a short-hand.)

This might prove confusing to some readers who might prefer to think in terms of cones as coding "blue light levels," "red light levels" or "green light levels," as the case may be. We tackle color perception in Ch 17.

Bipolar Cells: Center-Surround Receptive Fields

The receptors then serve the role of delivering a point by point image intensity description akin to the gray level description in a computer vision system. What about the cells that the receptors send their outputs to? These are the **bipolar cells**. Their receptive fields have also been studied with single cell recording techniques.

Diagrams of the spatial layouts of the two types of bipolar cell receptive field are shown in **6.8**. A spot of light shone on the central region of the on-center field type causes the cell to become

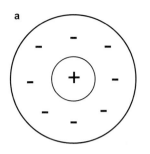 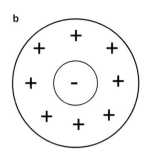

6.8 Circularly symmetric receptive fields
a On-center/off-surround   b Off-center/on-surround.
Cross-sections of the sensitivity profile for **a** are shown in **6.11**.

excited. This is indicated with the plus signs in the central region of **6.8a**. The opposite happens in the surround, where the spot causes the cell to be inhibited. This why this type of receptive field is called *on-center/off-surround*.

The other sort of center/surround field has the opposite arrangement and is shown in **6.8b**. Naturally enough, this sort is called an *off-center/ on-surround*. We discuss later why two types of receptive fields are needed in biological visual systems but not in computer vision systems. The current best bet is that having the two kinds solves the problem of representing negative numbers using nerve cells that cannot have negative firing rates.

Bipolars respond to the net sum of their excitatory and inhibitory inputs with a graded potential change across their cell membranes, just like the receptors, even though the bipolars encode a very different sort of information. Graded potentials are a viable code for receptors and bipolars because they are physically very close and so do not have to send messages over long distances during which a membrane potential code might become degraded. Retinal ganglion cells are close to the bipolars and so they can receive their graded potential messages quite safely. However, the ganglion cells have the job of sending message to the brain down the optic nerve and for that task action potentials serve as a suitably robust code.

The wiring diagram for bipolars is shown in highly schematic and abbreviated form in **6.9**. This unit receives a message from a central receptor via a junction called a *synapse* (Ch 1). Other receptors surrounding this central receptor also feed messages to the bipolar, but indirectly via horizontal

cells whose inputs come from receptors surrounding the central receptor. This layout is an excellent way of providing inhibitory or excitatory messages from the surround receptors because it solves the hardware problem: how can a graded potential change in a receptor be used to excite one bipolar and at the same moment inhibit another? The answer: horizontal cells can do the task of changing the effect of the receptor signal, from excitatory to inhibitory or vice versa as appropriate. Bear in mind that the receptors feeding the surround in **6.9** also feed excitation to other on-center bipolars: receptors are connected to very many bipolars.

For the sake of simplicity lots of interesting details are left out in **6.9**. For example, several receptors in the center of the receptive field might feed signals into the bipolar cells, and the surrounds can have very many receptors also. In short, the receptive fields of some bipolars are small, others are large. Their different sizes seem to be mediating analysis of the image at different *spatial scales* (Ch 5). We will come back to this issue shortly.

Another interesting complication which is not made evident in **6.9** is that, in addition to the receptor → horizontal → bipolar connections, there are receptor → horizontal → receptor linkages. Dowling and Werblin suggested that these links enable receptors to be adjusted in their sensitivity according to the general prevailing illumination of a scene.

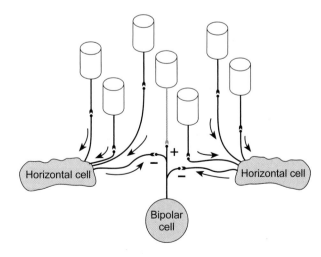

6.9 Wiring diagram linking receptors, bipolars, and horizontal cells

We emphasize the very important point that receptors can feed two or more bipolars simultaneously. Some of these bipolars might be on-center units, others off-center ones. About equal numbers of bipolars of each type are found and this makes good sense because the two types of bipolars carry different but complementary components of the image gradient change measurements.

Both off-center and on-center bipolars have the same basic wiring diagram shown in **6.9**. The difference is that the location of excitatory and inhibitory influences is reversed. For the off-center bipolars the central receptor feeds inhibition and the associated horizontal cells feed excitation; vice versa for the on-center bipolars.

Receptive Fields of Retinal Ganglion Cells

The bipolars send their outputs to the *ganglion cells*. This is done in a way which preserves the center-surround receptive fields of the bipolars. Thus, most ganglion cells also have the sort of *antagonistic center/surround receptive fields* shown in **6.8**. One major difference though is that, when ganglion cells are excited, they generate action potentials along their axons which comprise the optic nerve to the brain.

The center-surround receptive field organization of many ganglion cell has been known since the very early days of single cell recording in the retinas of such diverse creatures as cats, frogs, and rabbits. Typical recordings obtained from a cat are shown in **6.10**. For this particular ganglion cell, a spot of light was shone on the screen in such a position that it landed on the center of the receptive field caused the cell to emit a vigorous burst of impulses, **6.10a**. On the other hand, a spot shone anywhere in the periphery of the receptive field (**6.10b, c**) caused inhibition, shown by elimination of the resting (sometimes called "maintained") discharge rate. Clearly, such a cell is an on-center/off-surround unit.

Note that when the spot is switched off in the surround, the cell then emits a burst of impulses. It is as though the inhibition produced by the spot had "held the cell down" in some way, only for it to "bounce back" briefly as soon as this inhibition ended with the spot being switched off. It was this kind of response which gave rise in the first place to the terms off-surround and (where appropriate)

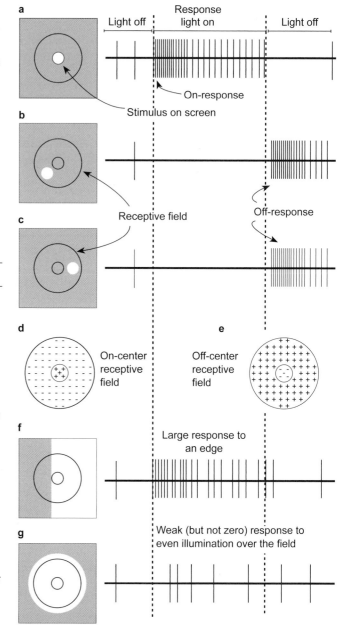

6.10 Single unit recording to reveal receptive field properties of retinal ganglion cells
a-d, f on-center, e off-centre field, g illumination over whole field causes weak transient response for both on- and off-center fields.

off-center, with the post-stimulus burst of impulses being called an off-response.

Also observe that the center and surround regions of ganglion cells are concentric. These regions can be plotted out by exploring with a spot of light to give the receptive field diagrams shown in **6.10d**. Compare these with the similar fields shown for bipolars in **6.8**.

Retinal ganglion cells respond very well to suitably positioned edges. For example, in **6.10f** a light/dark edge is shown falling on the receptive field in such a position that the excitatory center receives a high light intensity all over it, whereas the inhibitory surround is only partly covered. This stimulus arrangement produces a lively response because insufficient inhibition is generated from the surround to cancel the excitation produced from the center.

What happens if a patch of light is used which is large enough to cover the whole of the receptive field? Most ganglion cells seem much less responsive in these circumstances, **6.10g**. But it is not true that such a stimulus produces no response at all, at least, not true for all types of retinal ganglion cells. Many of them are quite responsive to even illumination, albeit not so responsive as to a suitably positioned edge.

What about the ***amacrine cells***? They make extensive contact with the ganglion cells but what they do to shape the messages sent by the ganglion cells to the brain is not yet clear. One line of thinking is that they have a role in the temporal modulation of the signals, that is, they play a role in the detection and encoding of motion signals (Ch 14).

Before leaving the topic of retinal neurophysiology, mention must be made of a curious property of some retinal ganglion cells. This is the ***periphery effect***, reported in 1964 by James McIlwain. Suppose a weak stimulus is used that is just insufficient to excite a ganglion cell (such a stimulus is said to be ***sub-threshold***). McIlwain found that if an edge is then presented well to one side of the receptive field of the ganglion cell, then the cell can become active and "notice" the *sub-threshold* input which previously went undetected. This is odd because here we have a stimulus falling *outside* the conventionally-plotted receptive field (using small spots of light) which is nevertheless capable of influencing the cell. This effect is a warning that there is much about the retina, and indeed cells in the visual pathway generally, that we do not as yet understand.

So far we have described the responses of just one kind of retinal ganglion cell. There are in fact four main types of ganglion cell and their outputs are sent to different parts of the brain structures that receive messages from the retina via the optic

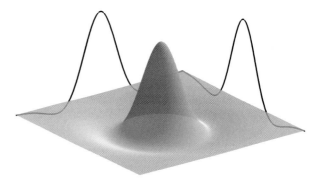

6.11 Response sensitivity of a retinal ganglion cell
Three-dimensional plot showing the sensitivity to light spots placed in different regions of the receptive fields of cells found in the retina. The spike in the middle shows where light excites the cell. The trough around this spike shows where light inhibits the cell. This shape explains why these cells are sometimes said to have Mexican hat receptive fields.

nerve. It is best to review them in the context of those brain structures and so we delay describing them until Ch 9.

The diagrams sketched in **6.9** and **6.10** of the receptive fields of bipolar cells and ganglion cells show the zones of excitation and inhibition. But what of the quantitative aspects: how *much* excitation and inhibition is triggered within these zones?

Figure **6.11** shows schematically the results of using a spot of light flashed successively in different places over the receptive field of a ganglion cell. It reveals why these fields are often referred to as of the *Mexican hat* type, introduced briefly in Ch 5.

Modeling Center-Surround Cells as DOGs

The Mexican hat type of receptive field, **6.11**, can be modeled quite accurately as a ***difference of gaussians***, or DOGs as they are known for short, **6.12**. One gaussian describes the way the excitatory effects vary over the field; the other describes the inhibitory effects. The response of the cell is predicted by subtracting the inhibitory gaussian from one from the excitatory one. Each gaussian has a different "spread" or width (technically, a different standard deviation, Ch 5). The net result is the Mexican hat profile.

Edge Detection Theory and Receptive Fields

What computational theory might underlie the design of the circularly-symmetric on/off receptive fields we have just described?

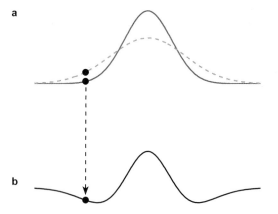

6.12 Modeling an on-center Mexican hat receptive field as a difference of gaussians, or DOG

a The solid line shows how the excitatory effect of light varies across the receptive field, producing most excitation at the center. The dotted line shows how the inhibitory effect of light varies—its graph has a smaller peak and is more spread out (it has a larger standard deviation).

b Subtracting the inhibitory effect at each point in the field from the excitatory effect at the same location produces the Mexican hat profile. This is shown for a point in the periphery of the field where the inhibition exceeds the excitation.

In Ch 5, we explained how it is possible to measure in two-dimensional images intensity gradients and changes in image gradients (technically, the first and second derivatives respectively) using oriented receptive fields. We then asked the question: can these derivatives be measured with fields sensitive to *all* edge orientations? We noted briefly that this can't be done for the first derivative but the second derivative can indeed be measured *isotropically* using circularly-symmetric fields. (*Iso* = same; *tropia* = direction; so here isotropic means doing the same job for all orientations.) And the circularly-symmetric fields for doing this are the ones we have described above possessed by bipolar cells and ganglion cells.

We will not delve here into the theory associated with the circularly-symmetric field type: go to Ch 5 if you want to know about that. But what we will do is show these fields at work analyzing an image.

Convolution in the Retina

So far we have described individual cells and their receptive field properties. It is time to get a feel for how such cells work as an image processing system.

Much was said about **convolution** in Ch 5. To remind you of the general idea, **6.13** shows in schematic form a receptor mosaic coupled by various sets of connecting fibers to a **convolution array** of cells. Each convolution cell extracts from the receptor mosaic a particular type of information according to the design of the excitatory and inhibitory connections which feed into it. In any given convolution array, all the cells are tuned to the same type of information, but they all "look" at different parts of the input image.

In the example shown in **6.13**, each convolution cell receives inputs from an approximately circular cluster of receptors. The clusters for just four convolution cells are picked out with heavy outlines to help illustrate what each such cell is "looking at." In Ch 3, when introducing the idea of templates, the receptor cluster was or bar-shaped, and in Ch 5 we used 3×3 clusters for local image averaging. We are now using the same basic ideas here but our goal is to model what bipolar cells are doing.

Note that in **6.13** the central receptor in each cluster feeds a convolution cell whose position in the convolution array exactly matches that of this central receptor. The two sheets of cells, receptor mosaic and convolution array, are thus neatly lined up, one over the other. In **6.13**, only the fibers connecting up a few convolution cells are shown, for simplicity, but in fact every convolution cell would be connected to the receptor mosaic in a similar fashion, each with a set of inputs coming from a receptor clusters in a slightly different position in the receptor mosaic. Because of these multiple connections, each receptor has to feed many different convolution cells. But, again for reasons of simplicity, **6.13** shows only one small overlap— that between the receptive fields of the two central convolution cells whose wiring diagram is shown in full. These share one receptor in common.

As in our previous convolutions, the receptor connections carry either excitatory or inhibitory influences to the cells in the convolution array, which then proceed to count up the sum of these two types of inputs. In **6.13**, the central receptor in each cluster feeds excitation to its convolution cell, whereas those in the surround feed inhibition. These two types of influence are marked on the receptor mosaic with + or – respectively. Because center and surround connections are thus antagonistic, the convolution cells model the *on-center/*

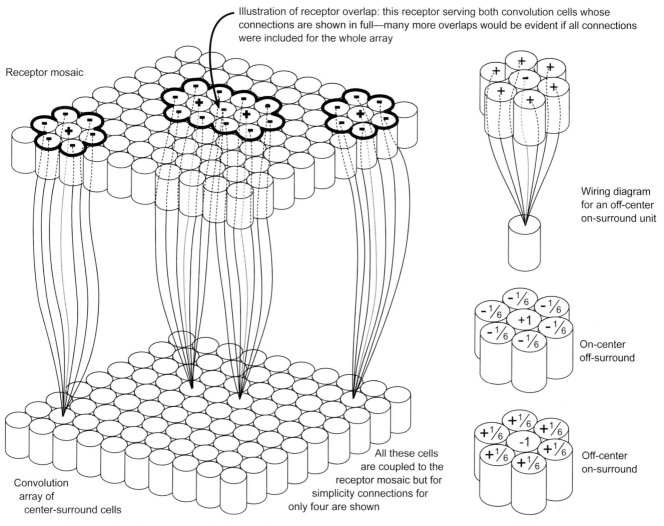

Illustration of receptor overlap: this receptor serving both convolution cells whose connections are shown in full—many more overlaps would be evident if all connections were included for the whole array

Receptor mosaic

Wiring diagram for an off-center on-surround unit

	$-\frac{1}{6}$	
$-\frac{1}{6}$	$-\frac{1}{6}$	$-\frac{1}{6}$
$-\frac{1}{6}$	$+1$	$-\frac{1}{6}$

On-center off-surround

	$+\frac{1}{6}$	
$+\frac{1}{6}$	$+\frac{1}{6}$	$+\frac{1}{6}$
$+\frac{1}{6}$	-1	$+\frac{1}{6}$

Off-center on-surround

Convolution array of center-surround cells

All these cells are coupled to the receptor mosaic but for simplicity connections for only four are shown

6.13 Measuring changes of gradient using the on-center receptive fields of bipolar cells

6.14 Bipolar cell receptive fields and their weightings

off-surround bipolar cells described above. It is possible, **6.14**, to have cells wired up with the reverse arrangement to model *off-center/on-surround* units,

Bear in mind that **6.13**, as a model for the receptors → bipolars linkages in the retina, is highly simplified. For example, it leaves out the horizontal cells that pass on to the bipolars the surround influences.

To keep track of what we are doing in this section, note that we are explaining how convolution machinery using circularly-symmetric fields can measure the second derivative of image intensity gradients isotropically, as the first step in detecting edges. (Recollect that the second derivative measures changes in image gradients.) Therefore, it is desirable to make the excitatory and inhibitory

influences on a cell add up to zero if no edge is resting on a receptor cluster serving the cell.

This can be done by giving each receptor in the cluster a certain weighting in its influence, so that if all the receptors are active to the same extent because they are stimulated by an area of even illumination, then the net influence of all the receptors is zero.

Suitable weightings for achieving this with our simple center/surround clusters are +1 or –1 for the centers and +1/6 or –1/6 for the surrounds, depending on the type of on/off units in question. Weighting diagrams designed on this basis are shown in **6.14** for the two types of unit.

You can readily appreciate that the +1 of the on-center is canceled by the six receptors of the

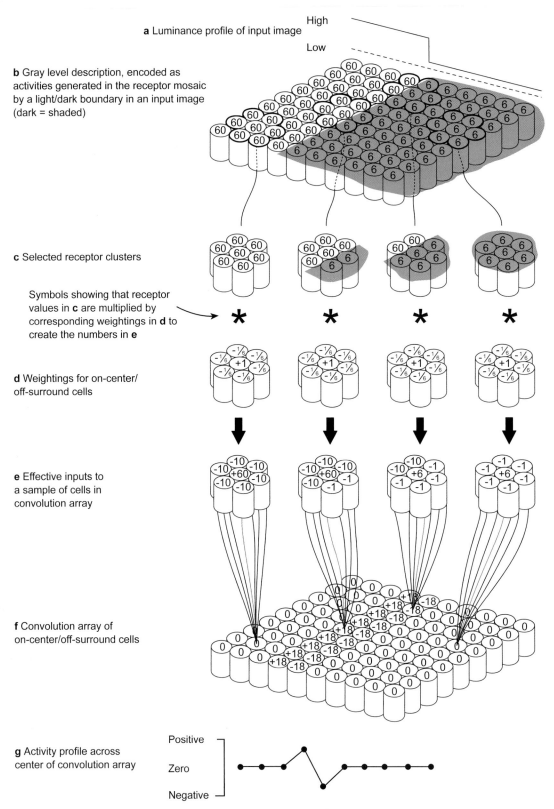

a Luminance profile of input image

b Gray level description, encoded as activities generated in the receptor mosaic by a light/dark boundary in an input image (dark = shaded)

c Selected receptor clusters

Symbols showing that receptor values in **c** are multiplied by corresponding weightings in **d** to create the numbers in **e**

d Weightings for on-center/ off-surround cells

e Effective inputs to a sample of cells in convolution array

f Convolution array of on-center/off-surround cells

g Activity profile across center of convolution array

6.15 On-center bipolar cells at work on a steep image edge

off-surround, each feeding –1/6 of their activity as inhibition. Thus, (6 x –1/6) = –1, and then 1 +(–1) = 0. Equally, the –1 of the off-center is canceled by the six "doses" of surround excitation, each weighted +1/6. Thus, for each type of unit, the set of weightings given to center and surround receptor activities ensures that a region of uniform illumination gives zero output, as required.

Let's now see how things work out if this convolution machinery is faced with the type of input it is supposed to detect—an edge in the image.

Bipolar Cells at Work on an Image

An input image containing a steep luminance profile is shown resting on a receptor mosaic in **6.15a**, with the dark portion of the edge shaded as gray and the light portion left as white. The dark region sets up only weak receptor activity, shown as the number 6 in each receptor to indicate "6 units of activity." Because receptors respond with a change in potential across their membranes, think of this number 6 as illustrating a number of millivolts. But of course, we have chosen the number 6 for convenience of illustration and to keep the arithmetic simple. It is not intended as a realistic model of receptor response in millivolts of change.

The light side of the edge in **6.15b** sets up receptor activity of 60 units. The whole pattern of numbers throughout the entire receptor mosaic constitutes a gray level description of the input image. Having obtained a gray level description via the receptor mosaic, the convolution array of **6.15f** proceeds to "look" for changes in image intensity gradients. Each convolution cell "inspects" one particular region of the receptor mosaic, as described for **6.15**. It counts up the excitatory and inhibitory influences coming from this region.

Receptor clusters feeding four convolution cells are shown picked out with a heavy border in **6.15b**, and then again as isolated clusters in **6.15c**. Each receptor's activity has to be weighted in its influence, and weighting diagrams for on-center/off-surround units are shown in **6.15d**. Each receptor's activity is multiplied by the appropriate weighting, an operation symbolized by the asterisks between **6.15c,d**. The results of the multiplications are shown in **6.15e**, which is the set of effective inputs (i.e., post-weighting inputs) fed through to the convolution cells for counting.

The counts obtained are shown in **6.15f**, where you can see that most of the convolution array is inactive (the zeros), but that high positive and high negative counts are present along the two sides of the edge boundary. These edge counts are +18 and –18 respectively. An "activity profile" through the convolution array is shown in **6.15g**. This emphasizes the fact that the array has succeeded in locating the edge as a change in image intensity gradients. These gradients are zero either side of the edge, i.e., on neither side is there an intensity change as the gray level description is all 60s on one side of the edge and all 6s on the other.

We have said that the circularly-symmetric fields of bipolars are sensitive to edges isotropically, i.e., they do the same job regardless of edge orientation. This is illustrated in **6.16**. This figure is similar to **6.15** but notice that the edge in the image has a different orientation. This is useful because it means that this very first stage of the edge computation can proceed without need for many different convolution arrays, each serving a different orientation.

Notice that the edge in **6.16** has less contrast than the one in **6.15**—the edge appears in the gray level description in the receptor mosaic as a boundary between 60s and 54s rather than 60s and 6s. This is to illustrate how the output in the convolution array varies with edge contrast.

As a final difference from **6.15**, the bipolars modeled in **6.16** have off-center/on-surround fields, to show that the machinery works in exactly the same way—except that the signs are reversed. You should have no difficulty in following through the arithmetic for yourself, and understanding how the edge comes to be represented in the convolution array as a strip of cells responding at levels of activity of +2 and –2.

That is, the edge now appears in the convolution array as a –2/+2 boundary (instead of the +18/–18 boundary in **6.15**). The key thing to grasp is, therefore, that off-center bipolars work in just the same way as the on-center ones but they do the convolution arithmetic the other way round, as it were. This means they encode as positive those counts which appear as negative in **6.15f**.

This raises the question: why have both sorts of bipolars? The best-bet answer at present is: this

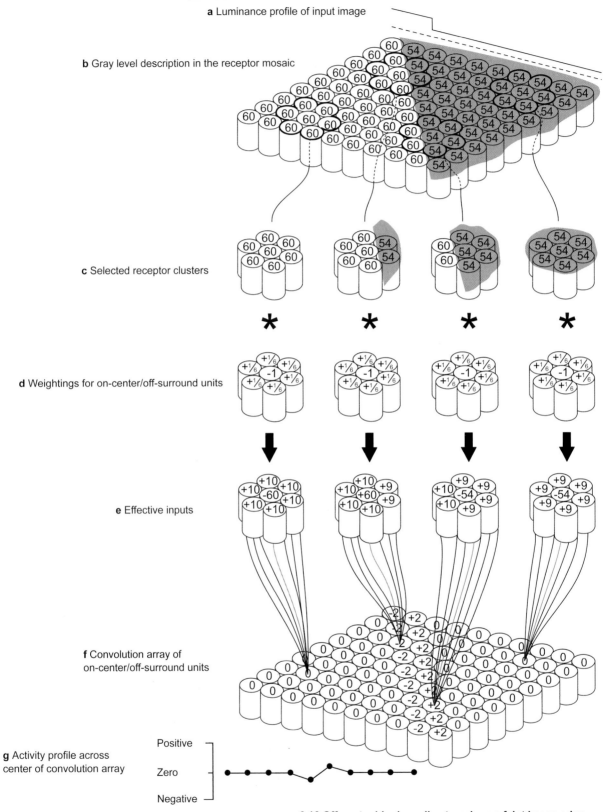

a Luminance profile of input image

b Gray level description in the receptor mosaic

c Selected receptor clusters

d Weightings for on-center/off-surround units

e Effective inputs

f Convolution array of on-center/off-surround units

g Activity profile across center of convolution array

Positive

Zero

Negative

6.16 Off-center bipolar cells at work on a faint image edge

is a convenient way of keeping track of negative numbers in biological systems. Neurons such as ganglion cells, whose outputs are trains of action potentials sent up the optic nerve, cannot have negative firing rates; they cannot fire more slowly than zero, obviously. The outputs of the bipolars are not action potentials but graded potentials. Perhaps it might be possible in principle to have bipolars so set up that their resting level is a mid-value potential (signaling zero) and voltages higher than this code positive numbers and lower ones code negative. However, that isn't how things seem to be. Instead, the two type of bipolars seem to work as pairs with the on-center ones coding the positive outputs in **6.15g** and the off-center ones coding the negative outputs in **6.15g**. The details get a bit tricky, and are anyway speculative, so we will come back to them later rather than risk disrupting the main story now.

For ease of exposition, the particular examples of edges chosen for **6.15** and **6.16** lie neatly on the boundaries between receptors. Other edge orientations (as well as the same orientation, but simply shifted slightly sideways) could cause the minor problem of deciding when a partially-covered receptor was or was not deemed to be active. But this is a detail we can ignore, and assume that in a real receptor mosaic of very fine resolution (such as the human retina contains) the problem is immaterial.

Making Edge Locations Explicit

The edge in **6.15f** shows up clearly as the +18/–18 boundary. But how can that boundary be made *explicit*? By this we mean represented in such a way that subsequent processes can use it without themselves needing to do any further work.

The +18/–18 boundary is an example of what was introduced in Ch 5 as a *zero crossing*. This is a good label because it draws attention to the most salient property of the boundary: it is where the convolution output passes through zero in changing from positive to negative.

A simple way of building a "zero crossing detector" is shown in **6.17**. Its design is a "logical AND gate." This is the technical term for a device that becomes active if and only if all its inputs are present. In the case shown in **6.17**, the AND gate is set up so that it fires if *input A* is positive AND *input B* is negative. It is easy to envisage this

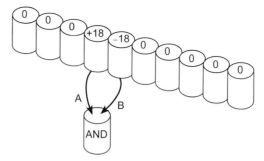

6.17 AND detector for finding a zero crossing
The detector becomes active if and only if *input A* is positive AND *input B* is negative.

AND detector being implemented with a neuron. However, whether neurons actually exist that signal zero crossings is still an open question, as far as we are aware. This is remarkable given all the neurophysiological research on the visual system since this scheme was discussed in the 1980s by Marr and Hildreth. There has been much psychophysical research trying testing whether zero crossings are used by the human visual system as explicit codes for edges. The present answer seems to be: not sure but probably not—see Ch 5. It could be that biological vision systems implement the final part of the theory not with AND gates finding zero crossing but in some other way. Or perhaps they implement a different edge detection scheme altogether. Seeing edges is still mysterious.

However, it has been worthwhile taking you through the zero crossings story as it introduces basic ideas and it has informed much research.

Zero Crossings in Natural Images

Figure **6.18** shows zero crossings found in the John Lennon convolution arrays of Ch 5. Zero crossings are shown at three different scales, from fine to coarse. At last we have achieved what we have been after for so long: explicitly encoded representations of the locations of edges in images

Admittedly, you might be a bit surprised, even disappointed, that they appear so messy. Sorting out the sense from this mass of *edge points* for each image is a topic we discuss in Ch 7.

Dealing with Negatives

A problem that cropped up earlier was how the Marr/Hildreth edge detection scheme could be realized in biological visual systems because nerve cells cannot encode negative numbers. The underlying problem is that where inhibition "wins out" over excitation in the bipolar cell convolution, the bipolar response goes to zero (i.e., the resting membrane potential). The bipolars do not have a way of encoding negative values.

This worry draws attention to the general problem of keeping track of negative quantities in biological convolutions. In a computer system negative quantities can easily be dealt with by the computer's arithmetical machinery, but in a biological system composed of nerve cells, negative quantities pose problems for representation. A cell can either be active or inactive: it cannot be "negatively active."

Of course, it would be possible in principle to have zero represented not by inactivity but by a "medium" level of activity, and then code "positive" quantities by activity above this medium level and "negative" ones by activity below it.

This could be done using as a "medium" level either a particular value of graded potential, in the case of bipolars, or a particular firing rate, in the case of retinal ganglion cells. Biological brains sometimes seem to use this sort of option. However, in the case of the computation of edges it seems likely that positive quantities are encoded by one channel of cells (the on-center cells) and negative ones by another channel (the off-center cells). We set out a possible scheme here.

Combining Signals from Off-Center and On-Center Bipolars

A luminance profile with a shallow gradient on which is superimposed a step edge is shown in **6.19a**. The gray level description is shown as levels of activity in the receptor mosaic in **6.19c** , with the profile of one slice shown in **6.19b**. The numbers in **6.19b,c** show that the shallow intensity gradient appears as a series of small changes: 60/57, 57/54, and 24/21. On the other hand, the step edge creates the much greater change of 54/24.

The next thing to note is that the gray level description is convolved with center-surround units to produce the outputs in **6.19d**. In fact, to keep things simple, just a central strip of convolution cells from a convolution array is shown in **6.19d**. The convolution outputs results arise from receptive fields of the kind shown in **6.15d** and **6.16d**

which have receptive field center weightings of either +1 or −1, with their surround units weighted −1/6th or +1/6th respectively. For these convolutions, negatives were allowed, and so some negative outputs and so some convolution cells have negative outputs.

Try working out for yourself why the numbers are as they are, referring back to **6.15** and **6.16** for guidance.

Having obtained the on-center and off-center convolution outputs, the next step, shown in **6.19e**, is to apply a **threshold** of 2. This means that in, only convolution values greater than or equal to +2 are retained; all others are set to zero, as in **6.19e**.

The idea behind applying this threshold is to get rid of the shallow gradient. The assumption being made here is that this shallow gradient can be safely ignored as being due to irrelevant causes, such as prevailing illumination conditions. Of course, what we are really after are gradients caused by scene entities such as object boundaries and texture markings. Just what level of threshold should be set for any given image for this purpose is an important matter. We tackle some aspects of it in Ch 16 when discussing the computation of surface lightness.

The result of applying the +2 threshold to each array in **6.19d** is that, as required, it leaves just the large gradient change measurements of +10 and −10 as the only ones appearing in **6.19e**. Bear in mind that **6.19d,e** show convolution outputs with "non-biological" units, as negatives are allowed.

The next step is to find the transitions (zero crossings) between +10/−10 using AND gates, as shown in **6.19f**. The active AND gates here serve as "edge symbols."

The problem we now tackle is this: how can this sort of edge detection computation be carried out with biologically plausible convolution cells for which negatives are not allowed (e.g., a neuron cannot fire with a negative firing rate). A way of doing this is shown in **6.19g,h**, in which convolution outputs that would have been negative are set to zero. The threshold of 2 has been applied as before. The convolutions with on-center and off-center units are illustrated separately, again with the results shown from just one strip across a full convolution array.

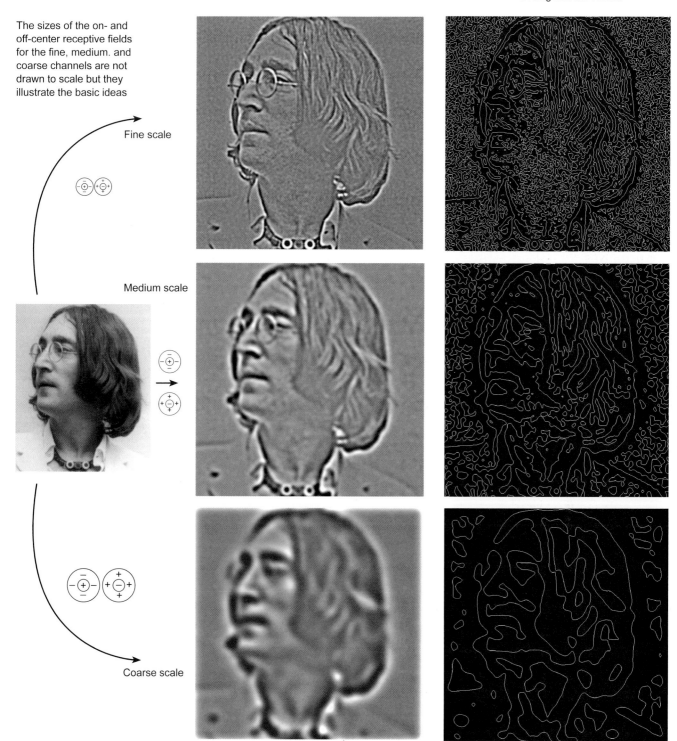

The sizes of the on- and off-center receptive fields for the fine, medium. and coarse channels are not drawn to scale but they illustrate the basic ideas

Fine scale

Medium scale

Coarse scale

6.18 Zero crossing edge maps of the image of John Lennon

Left column: convolution images in which all pixels lighter than mid-gray indicate output levels that would be obtained from cells with on-center receptive fields, whereas pixels darker than mid-gray indicate output levels of cells with off-center receptive fields (with darker pixels indicating increasing off-centre output). Each convolution image was calculated at a different scale, from fine to coarse.

Right column: edge maps. It is evident that many more zero crossing edge tokens are found at the fine scale than at the coarse scale.

See Ch 5, pp. 128–131 for a discussion of the problem of combining channel outputs at different scales.

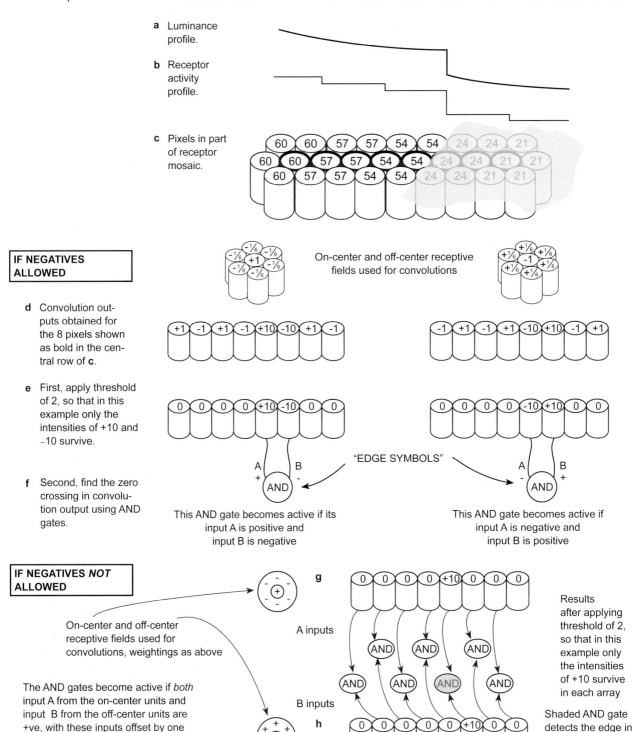

a Luminance profile.

b Receptor activity profile.

c Pixels in part of receptor mosaic.

On-center and off-center receptive fields used for convolutions

IF NEGATIVES ALLOWED

d Convolution outputs obtained for the 8 pixels shown as bold in the central row of **c**.

e First, apply threshold of 2, so that in this example only the intensities of +10 and −10 survive.

f Second, find the zero crossing in convolution output using AND gates.

"EDGE SYMBOLS"

This AND gate becomes active if its input A is positive and input B is negative

This AND gate becomes active if input A is negative and input B is positive

IF NEGATIVES *NOT* ALLOWED

On-center and off-center receptive fields used for convolutions, weightings as above

The AND gates become active if *both* input A from the on-center units and input B from the off-center units are +ve, with these inputs offset by one pixel in this example

g

A inputs

B inputs

h

Results after applying threshold of 2, so that in this example only the intensities of +10 survive in each array

Shaded AND gate detects the edge in this example

6.19 Dealing with negative numbers using neurons

Refer back to **6.15** and **6.16** for an explanation of the way the convolution outputs in **d** are obtained by applying the 2D pattern of weights in the receptive fields shown above **d** to the gray levels in **c**.

As can be seen, only +10 appears in each strip, but where they appear is different in each case. Thus, the –10 value in **6.19e** for the on-center units appears as +10 in the outputs of the off-center units, **6.19h**. This might sound complicated, but the complexity is more apparent than real, as you will discover if you study carefully the various steps set out in **6.19**.

The two "biological" post-threshold convolution outputs are aligned one under the other in **6.19g**. Of course, this is just for exposition: it doesn't really matter where the convolution output cells are placed, it is the connections they make that matter.

The task now is to make the location of the edge explicit by coding it as an active edge detector cell. Discarding negatives by using both on-center and off-center bipolars means that there is no longer a zero crossing to be hunted down. Rather, what we need is to find a positive signal in *each type of bipolar* dealing with each side of the image edge. So the AND detectors shown between **6.19g,h** now have as their trigger requirement: *active if and only if input A* AND *input B are both positive*. A cluster of such detectors is shown but only one becomes active—the one (shown gray) picking up the +10/+10 pair of responses in the bipolar arrays. All the others remain inactive because their inputs are zero.

So here we have a biological implementation of the edge detection problem which has the nice feature of avoiding any problems with negatives.

Of course, **6.19** shows a very simplified scheme indeed. It purpose is just to show the general idea of how the two kinds of receptive field, on-center/ off-surround and off-center/on-surround, could operate as pairs to solve the problem of how to encode the negative values in convolution outputs.

The zero crossings shown in the Lennon images in **6.18** could have been produced either by bipolar-type arrays feeding +/+ AND detectors or, if negatives were allowed, using +/– AND detectors. The computational theory is identical in each case. The difference is at the algorithm and hardware levels only.

Perceptive Fields and Hermann Grid Illusion

A famous contrast illusion is shown in **6.20**, an illustration taken from a book written by John Tyndall in 1869 on sound. While reading this book, Hermann noticed that dark shadowy dots appeared at the intersections of the component figures. Thus, it was that Tyndall's book on sound came to make an indirect contribution to vision.

Hermann reported his observation in 1870, and thenceforth the so-called Hermann grid illusion has been much studied and debated by visual scientists. A classic, but now generally discredited, explanation treats the illusion as an unwanted side effect of center-surround analysis. The general idea is shown in **6.21**, where it can be seen that an on-center unit dealing with an intersection receives four "doses" of inhibition from the surrounding arms of the grating, whereas a similar unit dealing with a zone in between intersections would receive only two "doses" of inhibition. The net result would be that a weaker center/surround signal would be generated at the intersections than in between them. Hence, the theory claims, this is why the intersection appears gray in comparison with the regions in between which appear white.

The proposal here is that a signal set up in bipolar cells in the very first step of visual analysis somehow survives as a "flaw" right the way through to the brain sites dealing with the perception of surface lightness, see Ch 16. This is an extraordinary claim, which we find hugely implausible. It amounts to saying: this is an early penalty of center-surround edge detecting hardware that never gets put right.

There are, however, some things which this theory can count in its favor. One interesting property of the Hermann grid is that the gray spots do not

6.20 Matrix of Chladni figures
This is the type of pattern in which Hermann first observed illusory dark dots at unfixated intersections.

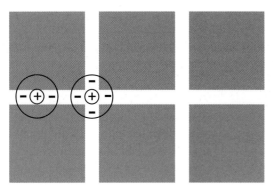

6.21 Explanation of Hermann grid Illusion in terms of lateral inhibition and on-center receptive fields
Greater inhibition from the surround would be generated for a receptive field dealing with an intersection (shown by four minus signs) than for one dealing with a zone between intersections (shown by only two minus signs).

appear at the intersection which is being viewed directly, but only at those intersections which fall on peripheral retina.

At least, this is so in **6.20**, but one can obtain a Hermann grid effect in central vision if the size of the lines is reduced sufficiently, **6.22**. For example, you should be able to see small gray spots in **6.22**, where the lines are thin, even with central (or at least near-central) fixation of the intersections. The reason for this could be related to the fact that the size of center-surround receptive fields varies across the retina, with small ones found in central regions and progressively larger ones out toward

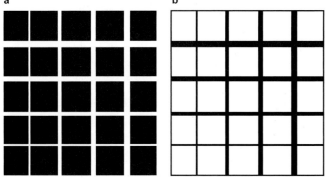

6.22 Hermann grids with intersections of varying sizes
Try looking just to one side of the intersections and you will find that you need to look much further away from thick-line intersections to obtain the illusory dark spots than you do from thin-line intersections. Note also that whereas **a** produces darker-than-white spots at the intersections, **b** produces whiter-than-black. The latter effects can be explained using the idea illustrated in **6.21** and substituting off-center for on-center fields.

the periphery. This obviously fits in well with the explanation of the illusion in terms of center/surround analysis, because this explanation relies on a fairly "neat" fit of size of intersection to size of on-center, **6.21**.

Given this explanation, one can turn the tables on the Hermann grid, as it were, and instead of regarding it as (only) an illusion to be explained, one can regard it as a device for measuring center-surround receptive field sizes in human observers, for whom microelectrode penetrations are obviously inappropriate. This has been termed measuring *perceptive fields*. It can be done by finding the size of intersection which is necessary to give an illusory gray spot for a whole range of different eccentricities of fixation (i.e., distances from central gaze of the intersection in question). A plot of "size of intersection required" against "eccentricity of viewing" can then be used to estimate the sizes of field centers in different retinal locations. This used to be the very first visual psychophysics experiment performed by students attending our course on seeing in the University of Sheffield, not least because it gave rock-solid results even for naïve observers.

The receptive field/lateral inhibition style of theorizing just described for the Hermann grid has been tried for other contrast illusions. So why is it generally discredited? The answer is because many figures have been devised that should provide a Hermann grid-type effect given the receptive field theory, but they don't.

An example is shown in **6.23b**, which is a version of the Hermann grid in which the corners of the squares have been rounded off. Do you see illusory shadows at the intersections in **6.23a** for the normal squares but not for the rounded ones? That is what we see. And yet there is no obvious reason why the rounded squares should fail to produce the illusion, as receptive fields could be "fitted" to this figure in the same way as shown in **6.21**.

We conclude (along with others: see Ch 16 for more examples, discussion and references) that the receptive field explanation of the Hermann grid illusion is incorrect. This is a pity, because it is such a nice way of linking neurophysiological findings to perceptions. It seems we have here an occurrence that is commonplace in science generally: harsh facts killing off an attractive theory.

a

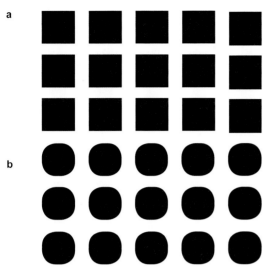

b

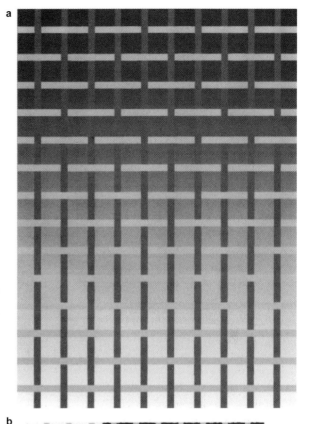

6.23 Hermann grid type patterns
a Standard Hermann grid in which illusory dark spots appear at non-fixated intersections.
b A grid of rounded squares in which we do not see illusory dark spots at any intersections.

Before leaving this topic, we note that the advent of cheap computer graphics has led to an explosion of studies exploring variants of classic illusions. A prize-winning contribution of this kind has argued that the Hermann grid is really a variant of a wider class of figures, consisting of intertwined bars called *weaves*, **6.24**. Try making your own new illusions, and if you come up with a cracker then you could win a prize.

Concluding Remarks

This chapter has introduced the anatomy and neurophysiology of the retina. In keeping with the theme of this book, which is to consider what tasks have to be solved by the visual system, we have concentrated on viewing the retina as implementing a solution to an image compression problem.

We explained why, within this framework, it makes sense for the retina to begin the task of finding edges because these will be where particularly significant information is to be found in retinal images. Finding edges can be treated as the task of measuring changes of image intensity gradients.

The receptive field properties of bipolar cells fit nicely with the requirements for doing just this, regardless of edge orientation. In technical language, we explained that bipolars might be delivering isotropic measurements of changes in image intensity

b

6.24 Weave patterns
In 2007 Kai Hamburger and Arthur Shapiro created patterns of intertwined bars that they call *weaves*. These stimuli generate Hermann grid-like illusory light or dark "smudges" at points where the bars overlap.
a Smudges do not appear in the middle section which has intermediate background luminance. This is not as expected given the standard receptive field explanation of the Hermann grid illusion.
b Weaves are resistant to many different sorts of distortion that usually disrupt the Hermann grid illusion. An example shown here is a weave made of jagged bars in which the smudges still appear, whereas jagged edges in Hermann grids tend to destroy the illusion.

gradients using circularly-symmetric antagonistic on/off Mexican hat receptive fields. The problem of measuring intensity gradients at different scales was solved by using fields of this kind covering a range

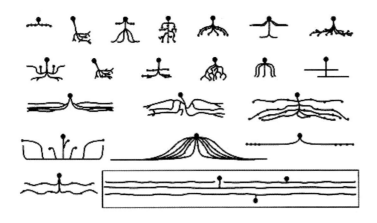

6.25 Diverse kinds of amacrine cells in rabbit retina
It is generally thought that each type of amacrine cell has a different function, arising from the different sizes and shapes of their fibers and synapses. The type enclosed in the box are characterized by extremely long dendritric fibers spreading across the retina, and they may be involved in local motion detection according to Olveczky, Baccus, and Meister (2003). Reproduced with permission from Masland (2003).

of sizes. Theoretical details explaining why circularly symmetric fields deliver isotropically measure of changes in image gradients can be found in Ch 5.

We end this chapter by emphasizing that our overview of the retina has necessarily been brief and many complexities have not been dealt with. For example, there are very many different types of each retinal cell, **6.25**. And the retina is involved with much more than just edge detection. For example, it contains neural circuits that adjust the retina's sensitivity to match the ambient illumination and contrast. Also, a recent investigation by Bruce Olveczy, Stephen Baccus and Marcus Meister suggests that the retina has mechanisms that suppress global movement signals deriving from eye movements that would otherwise complicate the detection of local movements arising from objects. The retina is embryologically a part of the brain and its complexities justify thinking of it as a truly impressive processor of visual information.

Further Reading

Dowling JE and Werblin FS (1969) Organization of retina of the mudpuppy *Necturus maculosus*. I. Synaptic structure. *Journal of Neurophysiology* **32** 315–338. *Comment* A classic paper.

Hamburger K and Shapiro A (2007) *Weaves and the Hermann Grid* See http://ilusioncontest.neuralcorrelate.com/index.php?module=pagemasterandPAGE_user_op=view_pageandPAGE_id=112.

McIlwain JT (1964) Receptive fields of optic tract axons and lateral geniculate cells: peripheral extent and barbiturate sensitivity. *J Neurophysiology* **27** 1154–1173.

Marr D and Hildreth E (1980) Theory of edge detection. *Proceedings Royal Society of London (*B) **207** 187–217. *Comment* A classic paper.

Masland RH (2003) The retina's fancy tricks. *Nature* **423** 387–389.

Ninio J (2001) Flashing lines. *Perception* **30** 253–257. *Comment* A new lightness illusion. See Ch 16 for more discussion and references.

Ninio J and Stevens KA (2000) Variations on the Hermann grid: an extinction illusion. *Perception* **29** 1209–1217. *Comment* A range of figures that cast doubt on the receptive field explanation of the Hermann grid.

Olveczy BP, Baccus SA, and Meister M (2003) Segrgation of object and background motion in the retina. *Nature* **423** 401–408.

Wässle H (2004) Parallel processing in the mammalian retina. *Nature Reviews Neuroscience* **5** 747–757. *Comment* Draws attention to the fact that there are at least 10–15 different morphological types of ganglion cell in any mammalian retina and that there are at least 10 different types of bipolar cell. Argues that the retina is engaged in many visual processing tasks that have hitherto been thought to be carried out in the brain.

http://webvision.med.utah.edu/. *Comment* An excellent account of retinal structure and physiology with many figures of details.

7

a

Seeing
Figure from Ground

b

7.1 Edges found in an image of the office of J V Stone
a Original image.
b Enlarged copy of this image in which pixels have been replaced by "edge points" found using the output of a center-surround filter of the type described in Chs 5 and 6.

We are amazingly good at finding visual structures within a mass of irrelevant details. Consider for example **7.1a** which shows an image of a room and **7.1b**, in which each mark shows the location of an *edge feature* found in **7.1a** by a computer using edge detection procedures. You have no trouble picking out the edge features related to the chair—but how did your visual system know which ones were from the chair and which ones from the desk, the books, and so on?

This is an instance of the *figure/ground problem* introduced briefly in Ch 1. Trying to understand how to solve the problem of finding a figure as a distinct entity separated from the other things around is a hugely important and fundamental aspect of vision. One reason for its importance is that if structures made up of related-features can be found then this is a starting point for finding objects in images. We will examine the figure/ground problem using the computational approach to seeing set out in Ch 2.

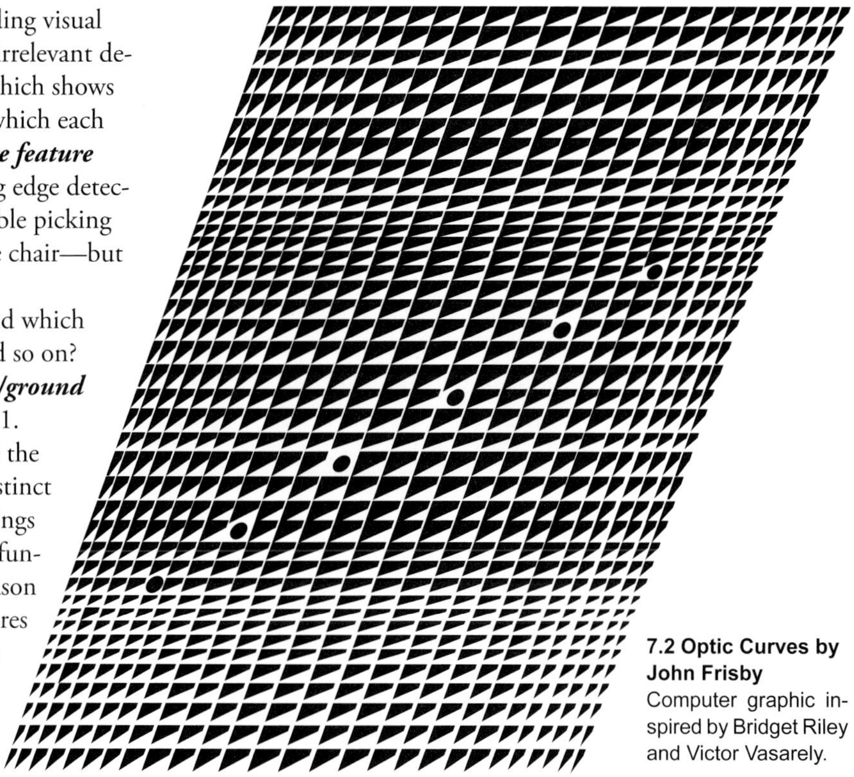

Each black mark in **7.1b** shows where the computer, running an edge detection procedure, found an *edge point* at the corresponding pixel in the original image, **7.1a**. An edge point is defined as a place where there is an *intensity change* straddling neighboring pixels in the image. In some areas, such as the path, there are rather few intensity changes that the computer deemed to be edge points; in others there are very many, such as the regions of foliage. For each edge point marked in **7.1b** the

7.2 Optic Curves by John Frisby
Computer graphic inspired by Bridget Riley and Victor Vasarely.

computer will have stored in its memory a list of edge properties, such as location in the image, edge orientation and edge fuzziness. These edge points can be thought of as similar to the explicit edge markers identified using biological receptive fields described in Chs 5 and 6.

The trick in finding related features is to use edge point properties to link together those features that come from a given figure. That is easily said but we need some principles for guiding how this can be done. In the terminology introduced in Ch 2, this means finding *constraints* for resolving all the many ambiguities in linking together the edge points in **7.1b**.

Why read this chapter?

We have examined in previous chapters the idea that the first stages of image processing in the human visual system are concerned with the detection of edges. Modeling this produces a mass of small edge segments. This chapter explores solutions to the *figure/ground problem*: which edge segments should be grouped together if they come from the same *figure*? The result of this grouping process yields what Marr called the *primal sketch*. Various *figural grouping rules* are described, and justified in terms of *constraints* arising from the typical properties of object boundaries.

Gestalt Grouping Principles

It turns out that some classic phenomena of human vision provide a wealth of useful constraints for finding structures in images made up of related features. These phenomena were first emphasized and studied as important clues about perceptual mechanisms by the Gestalt psychologists, working in the 1920s and 1930s. The German word *Gestalt* is not readily translated but roughly it means *form* or *configuration*.

The Gestalt psychologists noticed that many perceptions exhibit the grouping of elements together. As Max Wertheimer wrote in 1938:

> The fundamental formula of Gestalt theory might be expressed this way: There are wholes, the behavior of which is not determined by that of their individual elements, but where the part-process are themselves determined by the intrinsic nature of the whole. It is the hope of Gestalt theory to determine the nature of such wholes.

From this insight they developed their famous saying:

> *The perceptual whole is more than the sum of its parts.*

That is, parts are *not* treated as separate and isolated entities in perception. Rather, parts interact to produce a Gestalt which can differ markedly from what would be expected if parts did not affect one another.

Gestalt **grouping principles** have been exploited in so-called *Op Art*, which is why paintings of that genre form a nice way of illustrating some of the main grouping principles. For example, look at the computer graphic in **7.2** inspired by similar works by Bridget Riley and Victor Vasarely. Your eyes are immediately caught by certain structures within it. This is so even though the picture contains just numerous individual elements of varying shape in a semi-regular pattern. For example, you instantly see the round elements as a distinct pattern standing out from the straight-edged ones. Also, a group of colored shapes on the left stand out as a cluster, contrasting with the other shapes surrounding them. These are instances of **perceptual groupings**. Each group or whole "brings together" into a perceptual structure certain elements which share a common property of some kind. In the examples

just given, the common property is similar shape, either roundness or angularity, or color.

Another grouping principle at work in **7.2** is that of **similar size** which operates to "pull out" some rows as separate from others.

Yet another principle of grouping operating in **7.2** is that of **continuity**. All the elements other than the discs have straight sides. The curves are the result of your visual system grouping together edges of roughly similar orientation into higher-level perceptual structures, so producing the apparent presence of curved lines sweeping across the pattern. The curves are "put in" by your visual system, which links individual elements together on the grounds of *continuity in their orientations*. The power of this grouping principle is well attested by Fraser's spiral, **1.10**, one of the most dramatic of all visual illusions. The illusory spiral is seen because the visual system finds that the best continuity between individual elements is that given by a spiral, rather than by the concentric circles which are "really" present.

A very simple principle of grouping, and yet one of the most important, is that of **proximity**. In **7.3a**, for example, we see the row of discs broken up into three pairs, each one a result of the proximity of pair members. In **7.3b** all the discs are equally spaced one from another and we can therefore see rows of discs or columns of discs with equal ease. If the matrix of dots is flattened, as in **7.3c**, then columns become perceptually dominant because the flattening has brought certain discs into closer proximity. The visual system picks up this fact and "reports" (that is, sees) columns. A dominance of rows is equally easy to produce by an opposite alteration of the shape of the matrix of discs, **7.3d**, again showing the proximity principle at work.

7.3 Grouping principle of proximity
a Pairs of dots are seen.
b Rows and columns are seen with equal ease.
c Columns dominate due to proximity on vertical axes.
d Rows dominate due to proximity on the horizontal axes.
The various Gestalt grouping principles interact in complex ways: see Quinlan and Wilton (1998).

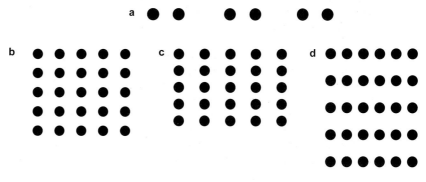

7.4 Grouping principle of closure

The grouping principle of **closure** is illustrated in **7.4**. The upper row of this figure shows a set of brackets which can be grouped in various ways. In the lower row, lines have been drawn between the tips of the brackets to give shapes with a "closed" contour. In the left half of this lower row, the closed-contour shapes are the outlines of television screens, in the right half they are columns. In each case, the principle of closure, of seeing elements which produce a completed form as a single structure, dictates how the ambiguous vertical brackets should be grouped to give enclosed entities.

Other grouping principles exploit similarities in **color** and **motion**.

Nick Wade has pointed out that these various grouping principles were well known to the ancients, **7.5**, being used by artists more than 2000 years ago. However, it is the distinctive contribution of the Gestalt psychologists to draw enduring conclusions from these various effects about fundamental aspects of the nature of human vision.

7.5 Mosaic from Rome, second century
The whorls anticipate the kind of patterns explored by the Gestalt psychologists and Op Artists. Reproduced with kind permission of Nick Wade.

We have reviewed a sample of grouping principles used by the visual system: *similar size* or *shape, continuity, proximity, closure*. But why can the visual system rely on these Gestalt grouping principles for finding the "right" structures in typical scenes?

The answer lies in the nature of the entities important to our survival in our visual environment. The objects we need to find and deal with can, crudely put, be said to fill "continuous lumps of space." And the boundaries of these objects thereby tend to show such properties as closure and continuity.

Moreover, their textures often show regularities which justify the grouping principles of proximity and similarity. It is these properties of our visual worlds and our ecological niche within it that are the constraints on which our visual system has fed in evolving reliable Gestalt grouping principles. This is an argument at the computational theory level, as it concerns finding and justifying good constraints for solving a visual task.

Seeing Symmetry

Another powerful grouping principle that was emphasized by the Gestalt psychologists is that of *symmetry*. If a property of an object remains the same (technically, *invariant*) over a transformation then the original and transformed patterns are said to be symmetric (Pani, 1996). Thus, the concept in science is a very broad one, extending well beyond the use of the term symmetry in ordinary language. Various examples of the scientific meaning of of symmetry are shown in **7.6**, some of which contain combinations of transformations.

Symmetry has long been an important principle in both physics and art. Its use as an attention-grabbing device in paintings has been exploited for millennia, **7.7**. It is classed as a grouping principle in visual psychology because the human visual system is highly adept at using symmetry for "bringing together" parts of an image into a perceptual structure.

Symmetry detection has been much studied, both in visual psychology and in computer vision (Tyler, 1996, provides an excellent starting point for searching the literature). Figure **7.9** shows an example of the kind of *random dot* computer-generated stimuli introduced by Bela Julesz into psychophysical studies of symmetry detection.

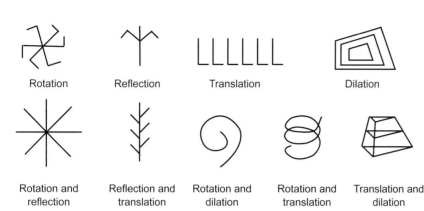

Rotation Reflection Translation Dilation

Rotation and reflection | Reflection and translation | Rotation and dilation | Rotation and translation | Translation and dilation

7.6 Various types of symmetry
The technical definition of symmetry illustrated here is described in the text. Redrawn from Pani (1996). See also Barrow (2005) for an analysis of symmetries used in patterns in mosaics, friezes, etc.

Patterns of this sort have revealed that we see some sorts of symmetries with much greater ease than others. Specifically, **7.8** shows that we see reflective symmetry very easily. Why might this be so?

The most likely answer is because this kind of mirror symmetry is a property of many objects of huge importance to us. For example, Chris Tyler and others have suggested that mirror symmetry betrays the presence of living organisms. Inanimate objects, such as rocks, do not in general possess any types of regular symmetry (unless viewed at microscopic scales). In sharp and striking contrast, animals that move linearly through the environment, notes Tyler, are formed with an axis of mirror symmetry aligned with their usual direction of movement, **7.9** . This kind of structure has the desirable property that it avoids asymmetries that might cause a bias from the direction of movement.

7.7 Angor Wat, Cambodia: temple wall illustration
The soldiers are an example of early artistic use of repetition (translational) symmetry. Courtesy Tyler (1996).

7.8 Random dot patterns exploring human symmetry perception
a Base pattern shown in the other three panels with different kinds of repeats.
b Translational symmetry.
c Rotation symmetry.
d Reflective, or "mirror," symmetry.
The relative ease of seeing the reflective symmetry is evident.

7.9 Insects showing symmetry
From top: antlion, longhorn beetle, cicada.

Whatever the truth of this argument regarding the origins of symmetry in animals, using reflective symmetry to pick out animals is obviously a highly adaptive visual strategy. And it is why military camouflage often tries to disrupt the perception of symmetry. So, just as for the other grouping principles, the use of a symmetry for grouping is justified by it being a constraint observed by important objects in our visual worlds.

Tyler also points out, as have others, that plants also show reflective symmetry, for example in their stems, as it is a natural consequence of their growth processes. However, plant structures also show forms of symmetry other than reflective, **7.10**.

We will meet symmetry detection again in Ch 10 when we consider *generalized cones*, 7.7 bottom right, as candidates for representing the parts of objects (legs, tree trunks, etc.).

7.10 Fern frond showing both repetition and reflective symmetry

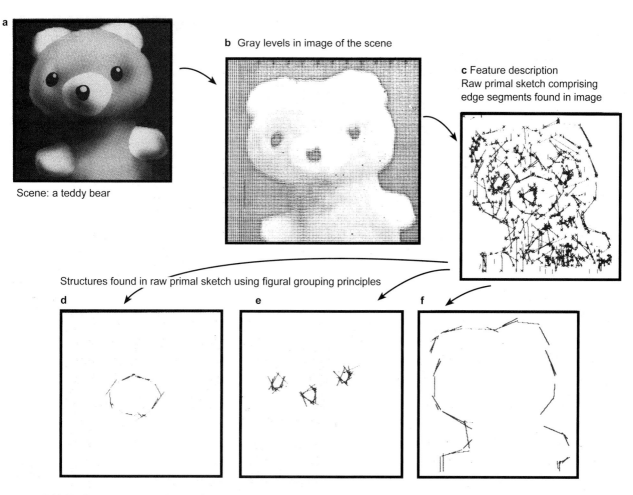

a

Scene: a teddy bear

b Gray levels in image of the scene

c Feature description
Raw primal sketch comprising
edge segments found in image

Structures found in raw primal sketch using figural grouping principles

d   **e**   **f**

7.11 Finding structures within a feature description
Reproduced with permission from Marr (1976) courtesy of the Royal Society, London.

Putting Grouping Principles to Work

The problem of finding structures within a mass of edge features is illustrated by **7.11**. The scene being observed contains a teddy bear, **7.11a**, and a grey level description of an image of this object is shown in **7.11b**.

The next item in the series is an ***edge representation***, shown in **7.11c** as myriad small line segments. These were found all over the image by a computer vision system devised by David Marr using the edge detection theory described in Chs 5 and 6. This theory was heavily based on findings about how our own visual system works. This suggests that perhaps feature representations rather like those shown in **7.11c** are created at an early stage of the visual analysis that goes on in our brains.

The edge features in **7.11c** are abbreviated versions of the edge data held in the computer. Just as for the office image, **7.1**, each edge point is stored along with data on its location in the image, its orientation, and its fuzziness. The picture would be impossibly complicated if all this information was somehow included in **7.1c**. However, we must think of it being present, just as it is in our own brains when we view the teddy bear image. This means that each edge point shown in **7.11c** represents a much more richly specified edge feature than each printed mark can possibly suggest.

The first thing to notice about the edge representation, even in this simplified display, is its messy and confused state. Edge features appear almost everywhere. This is perhaps not so surprising when you look at the teddy bear itself and note that your own visual system also detects many features with varying properties in lots of different places. But the profusion of these different features is not usually immediately apparent to us because our visual system can readily *group together* clusters of features. In this way, it can see structures within the otherwise unwieldy mass of individual edge features. The awareness of these larger structures pushes the awareness of their component features into the background.

Three edge clusters separated out by Marr's computer program are shown below the whole set of edges. Each is shown as a structure in its own separate picture. These structures—*perceptual groupings* if you like—are the overall outline of the teddy bear, **7.11d**, the teddy bear's eyes and nose, **7.11e**, and the teddy bear's muzzle, **7.11f**. Each structure was found "hidden" within the mass of confused **low-level** edge point data in **7.11c**.

The outline-of-the-bear structure, **7.11f**, was achieved primarily on the basis of the grouping principle of continuity of orientation. The eyes and nose demonstrate a grouping of features primarily on the basis of proximity and similarity, and the muzzle shows the principle of closure at work, features being brought together which make a completed oval shape. In fact, obtaining each structure did not depend upon any one principle of grouping because all principles were used at once. Their application in unison combined to give the end results shown in **7.11d,e,f**.

The manner of the discovery of these groupings represented a considerable achievement in the mid-1970s when this work was done. It was a significant step toward solving a fundamental difficulty for computer vision systems called the **segmentation problem**. This is synonymous with the *figure/ground problem* studied in the psychology vision literature. Computer vision has found it a very difficult job to take a mass of low-level edge features and find out what regions go together and hence form structures.

Marr's demonstration of how the grouping principles used by human vision can be exploited to solve a computer vision task was particularly interesting because it showed that they can be applied without any information on the specific nature of the scene from which the edge features came. This is an important point. Marr's computer

vision system found the structures shown in the lower part of **7.11** using a host of Gestalt grouping principles without "knowing" that it was looking for a teddy bear or parts of a teddy bear. It is thus said to be an example of a **bottom-up** or **data driven** processing.

The converse approach can also be taken, as we will soon see. That is, sometimes structures can be found with the help of knowledge about the object being sought. When **high level** knowledge about specific objects is used, the processing is said to be **top-down** or **concept driven.** Our own visual system can use both types of process, fluently integrating them as appropriate to the scene.

Early Grouping: The Raw Primal Sketch

We delved quite deeply in Chs 5 and 6 into the edge processes used to create the representation in **7.11c**. One aspect of that story we didn't deal with was the grouping procedures employed in finding the edges shown in **7.11c**, *before* using grouping principles to find the eyes, muzzle, and outline.

This early grouping procedure is illustrated in **7.7**. Notice that we are now moving on from the computational theory of grouping (finding constraints) to discussing grouping at the algorithmic and hardware levels. Each AND gate in **7.12** is the explicit code for a single edge point and their outputs are shown being fed to another unit which becomes active *if* the AND gates are active. The nature of this edge grouping operation is very conservative. This is because the locations of edge points found by the AND gates are specified in such a way that the grouping unit links together neighboring edge points only if they come from a very limited range of image edge orientations—see legend to **7.7**. This is why the edges in **7.11c** show up as small line segments, rather than as the edge points in **7.1b**.

Using an initial very tightly circumscribed grouping stage as a first step is an example of a rule of thumb that has been found valuable in computer vision: the **principle of least commitment**. This states that nothing should be done until we can be fairly sure that it is the correct thing to do, because it is usually impossible to unpick mistakes later on.

Marr used a special term for an initial edge representation produced using only very safe grouping decisions. He called it the **raw primal sketch**.

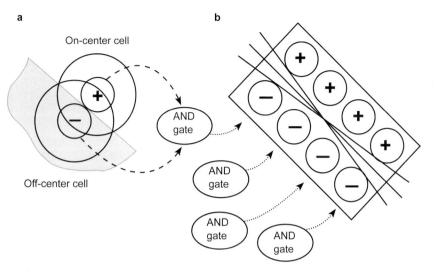

a

On-center cell

Off-center cell

AND gate

AND gate

AND gate

AND gate

+

−

b

+

−

+

−

−

+

−

7.12 Marr and Hildreth's scheme for detecting oriented edge segments
a Receptive fields of a pair of on-center and off-center cells whose positions are adjacent. The shaded area represents the location of an edge in an image. An AND gate is shown that will become active if both cells are active. Refer back to Chs 5 and 6 if you need a reminder on these concepts
b A receptive field created by linking a series of AND gates of the kind shown in **a**. Each gate derives from cells dealing with closely adjacent areas of the image. Thus, this field can be thought of as implementing a grouping operation. Activity in a cell with this field would be an explicit code for an oriented edge segment for the narrow range of orientations shown by the thin lines in **b**. Adapted with permission from Marr (1976).

He suggested it is a representation of the intensity changes in the input image at a level of complexity roughly similar to that of the low-level features in the visual image of which we are conscious. He proposed that all subsequent processes of analysis and interpretation use the primal sketch for their supply of information. That is, they do *not* have access to the preceding data on which this sketch is based. Thus, for Marr, the raw primal sketch was the point at which visual awareness begins. It was in his system the sole input to later visual processes, such as those of object recognition.

This notion of the primal sketch as an early "solid" database computed without any top-down help (Ch 1) has been challenged. We will see in Ch 9 that there exist many *back projection fibres* from higher levels of the visual pathway to lower levels. It seems almost certain therefore that early processes are influenced at a very early stage of analysis by later processes.

It easy to envisage the AND gate grouping procedure in **7.12** implemented in a neuron. The receptive field in **7.12b** is strongly reminiscent of the kind of field possessed by some simple cells in the striate cortex—refer back to **3.7**.

Later Grouping: The Full Primal Sketch

The next stage in Marr's scheme was to use the "raw" features shown in **7.11c** to build a data structure called the *full primal sketch*. The entities represented there are such things as boundaries around regions in the image that share some common property. They are thus at a "higher level" than the simple edge representations of the raw primal sketch. Nevertheless, the full primal sketch is still itself quite low level in the sense that it does not make explicit anything about what entities in the scene "caused" the image regions, such as particular objects.

The structures in **7.11d–f** are shown as a collection of their constituent features. However, in Marr's computer program they would be attached to what he called an *assertion* representing each one.

This is a good example of another rule of thumb articulated by Marr as defining good practice in the field of artificial intelligence: the *principle of explicit naming.* This states that if you are building a data structure and you need to manipulate its components then you have to give each one a name so that it can be accessed.

Thus, in the teddy bear example, each of the structures in **7.11d–f** would be labeled in some way. Not with the labels "eye" or "muzzle" or "outline" because such semantic labels are not the business of Gestalt grouping principles and the primal sketch. Rather, some convenient neutral labels would be employed, capable of being referred to by subsequent processes. In the computer, such labels might be Gestalt *No. 42,* or whatever.

This is so important that it is worth going over it again. Each structure in **7.11d–f** would be represented in Marr's computer program with a label. The label would have associated with it a

Scene

Feature description: the raw primal sketch

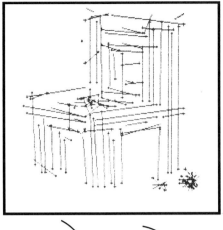

Grouped clusters of edge segments found using figural grouping processes

list of its constituent features, their locations, their properties, etc. The label (the name) would be the unifying factor giving the cluster of features its structural identity. Thus, the proposal made here by Marr is that we can roughly equate the "naming" by his program of a discovered structure with our visual system "noticing" one of the structures in, say, the Vasarely painting, **7.7**. Putting this the other way around, our "noticing the structure" can be regarded as an act of explicit naming. There must be some sort of code inside our heads which quite simply *is* this explicit naming.

This is not an easy concept to grasp, but we hope enough was said in Ch 1 about seeing being the business of building explicit scene representations for you to understand the central point being made. In any event, we give next another example to help make things clearer.

Another Example of Grouping Processes

In the scene illustrated in **7.13** there is a chair and Marr's program has produced its customary "busy" raw primal sketch, a mass of edge features. At the next stage, clusters of features are separated out as belonging together in different regions of the scene. These are shown in **7.13** printed out in separate pictures. The clusters were found mainly by grouping together edge features with similar orientation. Each cluster has been "made explicit," and given a label (not shown). Also, the computer will have stored data attached to each label specifying the cluster's position, orientation, and extent. If

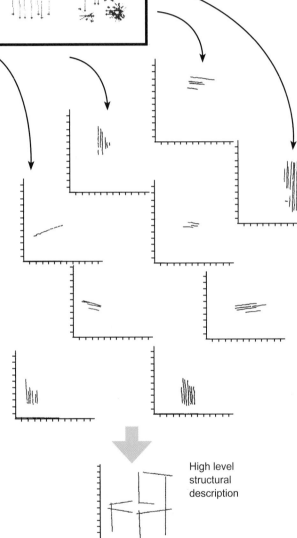

High level structural description

7.13 Grouping edge segments to build a stick figure representation
Reproduced with permission from Marr (1976) courtesy of the Royal Society, London.

164

the "things" represented by these labels are printed out, then a skeleton outline of the chair appears, **7.13** bottom. Each "thing" is a perceptual unit, a Gestalt if you like, and Marr has shown us how they can be extracted (i.e., explicitly named) from the input image.

You will doubtless notice the similarity between the skeleton chair in **7.13** and the kind of sketch which a cartoonist might make of the actual chair in question. Could it be, therefore, that the reason we are so good at recognizing cartoons is that they are similar in crucial ways to at least one of the kinds of representations of objects which our visual system normally uses? Marr was very struck by this idea and we will explore it in detail when we tackle the problems of object recognition, Ch 8.

Figure from Ground in Ambiguous Figures

When we look at **7.14,** we see either a white cross standing out against a black surround, or a black cross on a white surround. This shows that the visual system can come up with different answers at different times when seeking Gestalts. We could as well describe this as the visual system "segmenting" **7.14** differently at different times, or "explicitly naming" different entities at different times.

The grouping principle used for segmentation in **7.14** is brightness. Either the white elements are brought together, leaving the black ones as ground, or vice versa. This is a very straightforward basis for grouping but a valuable one. Color can also form a good basis for grouping. Indeed, color vision may have initially evolved for this very reason, for example, helping our ancestors find fruit as differently colored Gestalts amongst the surrounding green foliage.

The ambiguous figures in Ch 1, **1.20,** illustrate figure/ground phenomena very nicely and, once again, effects of this sort were known to the ancients. For example, **7.15** is based on a mosaic of the second century. As you look at it, various structures become dominant and then disappear as others take their place.

Whenever figure and ground oscillate, as in these ambiguous pictures, what is probably happening is that the application of the various grouping principles does not produce a single "solution." Instead, competing perceptual structures are

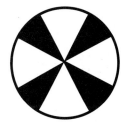

7.14 Ambiguous figure/ground
We can see either a white cross on a black background, or a black cross on on white background.

discovered at different times. In an ordinary scene it is almost always the case that a unique answer is found to the problem of segmenting the various elements of the feature representation. This is the "true" answer in the sense that the structures found are those that correspond to reality as judged by our interactions with the scene—such as walking in it, or picking things up.

Figure/Ground & Games

Figure/ground phenomena are nowhere better illustrated than in a familiar kind of child's doodling pad. The geometric line patterns provided on each page of these pads are capable of endlessly different

7.15 Figure/ground reversals
Circles, squares, and crosses appear and disappear. Based on second century mosaic from Antioch.

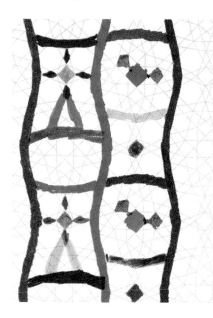

7.16 Seeing different structures
Two young children were given similar line patterns for coloring in. They saw very different figure/ground structures, as their choices of colored-in patterns demonstrate.

interpretations. This is shown by the two examples provided in **7.16** by one of the author's daughters, done when they were young. As one looks at the patterns, a large variety of different perceptual organisations emerge as the grouping principles incorporated in our visual system constantly produce new inventive figure/ground solutions to the hugely ambiguous input. What pattern is "really" there? There is no answer to this question. And indeed, even in normal scenes, where it is quite clear which pattern it is "sensible" to see, there can be other theoretical possibilities which go unnoticed.

The automatic character of figure/ground separation tends to make us take it for granted. But another kind of puzzle-pad provides its fascination by taking away grouping clues, leaving a task of finding hidden words **7.17**. Normally we group together letters belonging to a word on the principle of proximity. But in **7.17** this clue is removed because all the constituent letters are equally spaced. This forces us to use high-level knowledge about words to group individual letters. This is surprisingly difficult to do. When our Gestalt grouping principles are disarmed, visual life can become tough.

Figure/Ground and Texture Segmentation

Figure/ground separation is sometimes studied using patterns with carefully controlled visual textures. For example, look at **7.18a** and you should be

able to see fairly easily within it a "hidden" square area. The square is segmented as a region separate from the rest of the figure by its slightly different texture. The difference in this instance is that the small U shapes within the square region are upright (U), whereas in the surround they are lying on their sides. Our visual system detects the difference and we see the square separated out as a discrete perceptual entity.

A common strategy in this kind of research is find instances of texture differences which we *cannot* discriminate. These failures serve as clues

```
W M T E R D D O U B T F U L
F E X T R A Z U J E A A G E
B I D J H R O T L E E M A N
T L R R L T F U L L P E O T
T O N E E Y O O K I M E E T
Q I H P Y S O R D I D P E K
P U F Y L P D R O O T E X T
A O I L Y I P A R T Y E L F
L Y O T E B S V Y O Z R G T
E D I R E A L I S T I C W I
R E A L L Y I L I E L O A K
A R K B O Y M T U R N E S N
```

7.17 How many words can you discover?
Search horizontally, vertically, and diagonally.

a 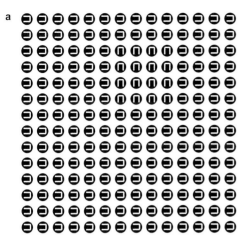 b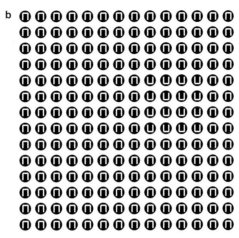

7.18 Texture discrimination task
Can you find the hidden square of elements in each array? Most people find it easier to do this in **a** than in **b**.

in working out what kinds of texture information the visual system does use to identify those regions. Consider **7.18b**, for example. Can you see the hidden square this time? Probably not, without looking very carefully and deliberately, using high-level cognitive processes (e.g., memorizing some elements and then checking others against them). In other words, the square in **7.18b** does not immediately "pop out" as figure against ground, as does the square in **7.18a**, and this gives a clue about mechanisms of immediate texture vision.

In particular, it suggests that differences in orientation can be used (because such differences exist in **7.18a**), whereas simple shifts in the position of certain texture elements are not so helpful. Thus, the square in **7.18b** is made of upside-down Us, whereas the surround is made of normal Us, so the only difference between them is in the position of the "cross piece" of the component shapes. The

conclusion has to be that the relative positions of edges within micropatterns, considered independently of any other properties such as orientation, are not always used by human vision for texture segmentation.

And yet orientation differences in the textures of two regions can be quite marked and still go unnoticed. Consider, for example, **7.19a**. It is easy to see a square standing out as figure because its "woven texture" has a different orientation from that of the surround. But can you readily see the odd-quadrant-out in **7.19b**? Almost certainly not without scrutiny. And yet the *only* difference between these two figures is that in the latter the boundaries between sub-regions of the picture have been obscured with a black line. Look carefully and check this surprising fact for yourself. The conclusion seems to be that segmentation by analysis of visual texture may sometimes depend crucially

a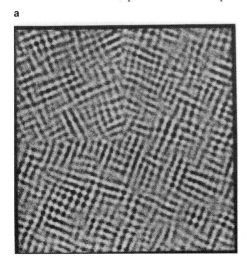

b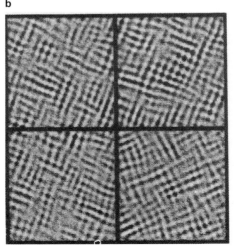

7.19 Texture boundaries
a The odd-quadrant-out is easy to discern by virtue of its components being twisted a little with respect to those of its surround.
b Which is the odd quadrant? The patterns are identical to those in **a** except for the figure as a whole being rotated to a new position, and for the inclusion of black lines masking the texture boundaries.
Created by Mayhew and Frisby (1978).

on detecting differences at the *boundaries* between regions. When these are masked, even by quite a thin line, figure/ground perception via texture segmentation may fail.

This is surprising because by far the largest part of the texture regions in **7.19b** are unaffected by the black borders. One would have thought that region-finding should have been able to proceed quite happily on this basis. A computer programmed to seek orientation differences pure and simple would have no trouble whatever with **7.19b**. But we do. Once again, we see here a case of the human visual system being interested in edges. The edges in question here, the texture boundary edges, are technically called **second order**, as the average point-by-point image intensities on either side of them are identical. The term **first order** is used for simple intensity differences, such as those used to find the Maltese cross in **7.15**. There the human visual system uses average intensity, be it black figure on white surround, or vice versa, to find figure and ground.

In **7.19** the odd-square-out cannot be detected in terms of a first order difference, as none exists. That is, if the image intensities for all the points in each square were to be added up and the mean value worked out, all four squares would be found to have the same mean intensity. This is why their orientation differences are said to be second order.

If you look back at the chair example, **7.13**, the way Marr's program brought together edges with similar orientations can be thought of as a instance of finding structures on the basis of second order texture differences.

Mayhew and Frisby pointed out in 1978 that the difficulty of finding the odd-quadrant-out in **7.19b** is difficult to reconcile with any theory of human visual texture discrimination based on immediate access to the outputs of oriented spatial frequency tuned channels of the kind described in Chs 4 and 5. If such access was available then we should be able to find the target quadrant in **7.19b** just as readily as we can in **7.19a.** Mayhew and Frisby concluded that their results supported Marr's proposal that texture discrimination operates on the basis of primal sketch entities derived from spatial frequency channel outputs. If you wish to follow up on this particular debate, Quinlan's (2003) review paper provides a good starting point for considering the large literature on visual search.

Camouflage and Constraints

Principles of grouping take advantage of certain typical properties of objects in the world. That is, it is usually the case that objects or sub-regions of objects actually *are* made up of similarly textured elements, or that their boundaries *are* defined by features with continuity of orientation. These are the common properties of everyday objects. The grouping principles used so successfully in computer vision take advantage of this fact. As we said above, putting this in technical language, one can say that grouping principles exploit certain *constraints* on feature relations shown by typical objects.

But not all objects show these constraints, camouflaged ones being good examples to the contrary. In **7.20** the aircraft is difficult to spot because deliberate attempts have been made to make its boundaries conflict with normal rules. For example, the principle of continuity leads to the grouping together of the edges of the camouflage blobs painted on the aircraft. These then become dominant at the expense of the boundary of the object. Many examples of animal camouflage can be interpreted in the same general way, i.e., as creating boundaries within and across the animal which conflict with the normal constraints on how features from objects cluster together. The examples given in Ch 1, **1.20–21**, further illustrate the point.

7.20 An example of military camouflage
Can you find the jet fighter plane?

Figure/Ground in Line Drawings

Consider the image of a jumble of toy building blocks in **7.21**, which shows an example of a **blocks world scene.** Segmenting images of this sort of scene was much studied in the early days of computer vision. The task offered a tough challenge but one that seemed tractable enough to hold out hope for an early solution.

We can easily segment **7.21** into the constituent blocks. We have no difficulty at all in seeing the arch, the triangular prism, the wedge resting on cubes, as separate entities. But how do we actually come to see adjacent regions as belonging to one block or another? Seeing the blocks is easy, understanding how we do this seeing is a serious challenge. Various approaches have been tried in finding a solution. The various twists and turns in this story are instructive because they helped discover

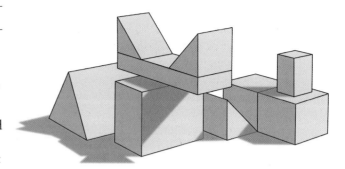

7.21 Blocks world scene

what should count as "a deep understanding of a visual task" as distinct from an "*ad hoc* even if quite effective algorithm." The story of research into the blocks world is told in Ch 21. We examine there in some detail the *algorithm* level for implementing constraints for solving the figure/ground problem.

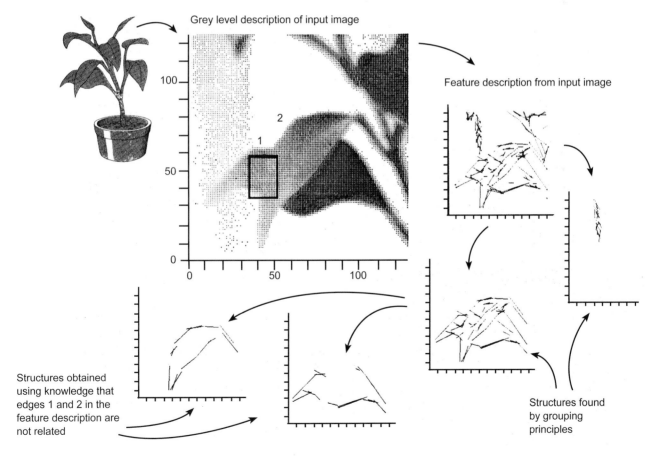

Grey level description of input image

Feature description from input image

Structures obtained using knowledge that edges 1 and 2 in the feature description are not related

Structures found by grouping principles

7.22 Segmenting a feature description when grouping principles fail
Reproduced with permission from Marr (1976) courtesy of the Royal Society, London.

Segmentation via Object Recognition

Marr's full primal sketch program was based on the idea of segmenting an edge representation without help from object recognition processes, that is in a wholly bottom-up fashion. The teddy bear outline in **7.11** was found as a perceptual structure without the program being told to "look for" a teddy-bear shape. Nor was the success of the program dependent on its discovering halfway through that a teddy bear was present, and then using this information to search out eyes, nose, and outline. All these structures were found without the program "knowing" what it was doing in terms of finding recognized objects. There are occasions, however, when ambiguities in segmentation crop up. Then the use of information from higher-level representations of object recognition can be very helpful.

An example of this is shown in **7.22**. The scene is a plant, and the part of the grey level representation concerned with the leaves is shown as the part being segmented into constituent structures. The feature representation shows the by now familiar mass of features requiring interpretation. The structure in the upper left area is easily discovered using rules of proximity and similarity. But the leaves themselves remain stubbornly grouped together. They are not separated out as two distinct objects, which they obviously are to our visual system. The problem is that in the area of overlap of the two leaves (shown as an inset rectangle in the grey level representation), the initial edge detecting processes cannot find enough edge features to demarcate the two leaves clearly. The grey levels are just too closely matched in those areas.

Marr found that he could get his program to solve this problem *if* he told it that the line segments deriving from regions labeled 1 and 2 in the grey level representation come from different objects. Given this clue, the program then had no further difficulty in separating out the two leaves, as **7.22** illustrates.

In this instance, the information about regions 1 and 2 came from Marr himself, who told his program what to do about them. In principle, however, one can imagine this information coming from higher-level object knowledge added on to Marr's program. These processes might have managed to recognize, from such segmentation as

had already been achieved, that a plant was present in the scene, and therefore that leaves could be expected in various locations. Knowledge about the likely *shapes* of leaves might then be fed down to the lower-level segmentation processes and provide the essential clue about regions 1 and 7. This would then be an instance of top-down or ***conceptually driven processing*** (see earlier definitions).

There may be, however, other top-down ways of dealing with the problem in **7.22** without resorting to high level knowledge about object shapes. For example, when human vision discovers that it is having no success in segmenting a large feature cluster, it might well tell the eyes to have a "closer look." Thus, eye movements might be directed to this region to maximize the chances of picking up helpful new features. Or perhaps the scene might be looked at from a new vantage point altogether, with the observer moving his head or body position to achieve this. Alternatively, attention might be directed to other types of clues, for example distance information, or color.

These are various ways in which help can come from ***downward flowing information***. This is a good term because it clearly brings out the key attribute of the information, namely that higher-level processes are influencing the operation of lower-level processes. If the downward flowing information provides help deriving from knowledge about objects (e.g., the probable shapes of leaves in the above example), then the term reserved for this is conceptually driven processing. That is, a concept (e.g., "leafness") drives interpretation of information extracted by the input image.

On the other hand, if the downward flowing information merely guides the new application of low-level processes (e.g., generates a head movement to bring new information into play), without any reliance on object concepts or knowledge about objects, then it is exerting a ***control function***. In the latter case, all processing is said to be data driven, rather than conceptually driven, because high level knowledge about objects plays no part, even though there are high level control processes "noticing" how well the low-level processes are getting on, and guiding their operations if any "difficulties" arise. The latter could include failure in object recognition. Such events might happen for example in driving when "unexplained" entities

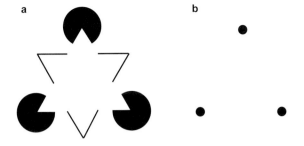

7.23 Inferring objects from incomplete data
a Kanizsa's triangle, showing enhanced brightness and illusory contours.
b A triangle of dots with inferred boundaries but no illusory brightness and no illusory contours. The illusory triangle in a can be perceptually enhanced by placing tracing paper over the display, and/or viewing it from a distance or with eyes half closed to create blurred images.

grab attention and cause a momentary change in head/eye positions.

A current research topic is determining when and how knowledge about the world in terms of object concepts takes over from, or joins in, with vision processes relying on general-purpose low-level constraints. Ch 9 reviews evidence that suggests the human visual system first performs a "quick and dirty" process to arrive at a decision as to what objects are in the scene, or what the nature of the scene is (street, beach, etc.). This decision then drives subsequent processes, even perhaps those Marr believed to be involved in the primal sketch, in a top-down manner, using *back projection fibres* from higher levels of the visual pathway to lower levels.

Illusory Edges

An example of the controversy as to when and how high-level information interacts with low-level image processing concerns an illusion called Kanizsa's triangle, **7.23a**. This figure consists of three discs with sectors cut out of them, plus some lines forming corners, lying on a ground of uniform light-intensity. What we see, however, is an illusory bright triangle, appearing in front of and obscuring the other pattern elements. It is tempting to suggest, as many have done, that the illusory bright contour is "created" by our visual system to "make sense" of the input image. The various black pattern elements with sectors missing are evidence for the object concept "triangle." So to are the interrupted lines. This information is enough, runs the argument, for bright edges to be "invented" to give this triangle a complete boundary contour.

If this explanation is correct, then a high-level object concept ('triangle present') has contributed not only to segmentation of the triangular figure but also to the creation of edges in the feature rep-

resentation. Thus, this theory goes beyond simply saying that the various bits and pieces of Kanizsa's figure have been grouped together, and then a representation of "triangle" arrived at. This latter type of process happens in the three dots, **7.23b**, but there it is *not* associated with illusory bright edges.

But *why* is the enhanced brightness missing in **7.23b**? if identification of a triangle is held to be a *sufficient* basis for the bright triangle seen in Kanizsa's triangle? We conclude that the illusory bright edges in Kanizsa's triangle require parts of the whole edges to be present in the image as an intensity difference (e.g., those provided by the sectored discs). This conclusion seems forced even if high level object knowledge to do with the concept of a triangle also plays a part. We return to this question in more detail when we examine lightness and brightness perception in Ch 16. We note in Chs 9–10 the brain regions that have been identified as mediating the creation of illusory contours.

More on High Level Knowledge in Vision

That high-level knowledge can play a part at some stage in vision is, of course, not in doubt. Remember, for example, **1.21**. Until you were told that a Dalmatian dog was present, you probably had great difficulty in seeing this figure as anything other than a set of blobs. A similar effect was present in animal camouflage figure, **1.21**. These are clear-cut instances of high-level object knowledge being exploited in seeing. The controversy is thus not a matter of either/or. ***Bottom-up*** and ***top-down*** processes are not mutually exclusive possibilities. The current debate is altogether more interesting. It is about the details of what knowledge is used when and how in seeing. The back projection pathways in biological vision systems are a tantalizing glimpse of something very important going on. But what is it? We take this up in Ch 9.

Summing Up

This chapter has continued the theme of this book set out in Ch 2. That is, a seeing problem is identified (here the figure/from ground problem), and possible solutions considered (analysis at the computational theory level). This led to the identification of various constraints, mainly those revealed by figural grouping principles. These principles are an example of how knowledge of the typical structures in our visual world can be used to identify constraints. These can be used as generally reliable assumptions when deciding which features are to be linked (grouped) together.

In addition, we found that knowledge about the properties of specific objects can be valuable in some circumstances. This high-level knowledge about specific objects is to be contrasted with the low-level general-purpose knowledge revealed by the figural grouping principles.

We have only touched on the algorithm and hardware levels, while considering the computation of the raw primal sketch. In Ch 21 we describe some algorithms that have been developed for exploiting constraints in the bocks world.

Also, we have only mentioned in passing eye and/or brain mechanisms that might be biological implementations of grouping algorithms. We will return to that topic after we have introduced some facts about eye/brain structures in Ch 9.

In the present chapter, we have concentrated on pressing home the importance of analyzing the *computational problems* that any seeing device, be it man-made or biological, has to confront and overcome. And the general lesson is clear: solving seeing problems requires finding *constraints* that are well-founded on a thorough analysis of a seeing task. These constraints can then serve as reliable assumptions upon which visual processes can be built. This is a core theme of this book.

Further Reading

Barrow JD (2005) *The Artful Universe Expanded.* Oxford University Press. *Comment* A delightful book, which includes an interesting account of the use of different symmetries in mosaics and friezes.

Julesz B (1970) *Foundations of Cyclopean Perception.* Chicago University Press: Chicago. *Comment* Classic monograph summarizing research in diverse fields, including symmetry perception, using random dot stimuli soon after their introduction into vision research by Julesz.

Marr D (1976) Early processing of visual information. *Philosophical Transactions Royal Society London (B)* **275** 483–524. *Comment* Marr's classic paper on the computation of the primal sketch.

Mayhew JEW and Frisby JP (1978) Texture discrimination and Fourier analysis in human vision. *Nature* **275**(5679) 438–439. *Comment* This paper presents research using patterns of the kind used in **7.19b**.

Pani JR (1996) The generalised cone in human spatial organisation. Pages 383–393 in Tyler (1996) cited below.

Quinlan P (2003) Visual feature integration theory: past, present and future. *Psychological Bulletin* **129**(5) 643–673. *Comment* A good starting point for searching the large literature on visual search and visual attention.

Quinlan P and Wilton RN (1998) Grouping by proximity or similarity? Competition between the Gestalt principles in vision. *Perception* **27**(4) 417–430. *Comment* A nice experimental investigation concluding that it appears necessary to consider processes that operate within and between groups of elements that are initially identified as *Gestalts* on the basis of proximity. Argues that whether such groups survive further analysis depends critically on the feature content of the constituent elements.

Tyler CW (1996) Ed *Human Symmetry Perception and Its Computational Analysis.* VSP: Utrecht *Comment* An excellent collection of papers and a good starting point for finding references on symmetry. Tyler's introductory chapter sets the scene admirably clearly and has inspired much of the content here on symmetry.

Wade N (2004) Good figures. *Perception* **33** 127–134. *Comment* A fascinating review of grouping principles in the art of the ancients.

Wertheimer M (1938) *Gestalt Theory.* Translation in *A Source Book of Gestalt Psychology.* Ed. Ellis WD, Harcourt, Brace and Co.: New York, 1938.

8

Seeing Objects

8.1 Cartoon/Photograph Pair of John Lennon
Notice how the cartoon caricatures Lennon by exaggerating some of his already distinctive features, such as sharp nose, and prominent chin. It is possible to "doctor" photographs using computer graphics to create a caricature-styled "photograph." People judge such images to be more like the person than the photograph from which they were made. This suggests that human vision may build representations of faces of individuals in terms of their distinctive features. To follow up that topic, the fine book by Vicki Bruce and Andy Young (1998) reviewing the face recognition literature is a good starting point. Lennon photograph courtesy Associated Newspapers Archive.

We are amazingly good at recognizing objects by sight. Have you ever pondered, for example, our extraordinary ability to recognize cartoons of public figures? A few well-chosen strokes of the artist's pen capture a "likeness" of the person, **8.1**. We can see the two "objects," person and cartoon, as alike in some fundamental respects even though they differ enormously in their details.

At first sight, our success at recognizing cartoons suggests that the information crucially required for recognition consists of the sharply defined contours of an input image, all else not being used for the job. But this is not so because we are remarkably good at recognizing blurred photographs, **8.2**.

Another feat of recognition is the ease with which we read different typefaces. We can easily see all the patterns in **8.3** as upper case Ts, despite wide variations in the nature of their constituent features. Our ability to cope with very different specimens of handwriting is another remarkable achievement of the human capacity for visual recognition, **8.3**.

The visual system's fluent and usually automatic ability to recognize objects can mislead us into thinking that the task is a simple one. In fact, its true complexity is so great that understanding how it is done at the brain's level of sophistication has so far defeated all those who have ever studied it, be they psychologists, engineers, neurophysiologists, mathematicians, or computer scientists This failure is all the more noteworthy when it is realized that very large sums have been spent on investigating the ***object recognition problem***, as it is sometimes called, because of the immense industrial and military potential which successful understanding would bring.

8.2 Recognition can survive severe distortions

It is easy to recognize John Lennon even when his photograph in **8.1** is blurred (above) or reproduced as an edge map (right; Chs 4 and 5).

Why read this chapter?

Images of objects display enormous *appearance variability* and yet object recognition is fluently and speedily performed by the human visual system. This feat is far from being matched by computer vision systems. Various approaches have been tried for overcoming appearance variability. The main ones are reviewed and evaluated using the framework of constraints proposed by Marr and Nishihara. Object-based and view-based schemes are described, together with psychophysical studies that have tried to tease out the object recognition mechanisms of human vision. The Hough transform is introduced as a biologically plausible algorithm for matching representations of "figures" found in images to stored collections of object models.

8.3 Feats of recognition
The capital letter T in different typefaces (above) and highly variable handwritten numerals (opposite page).

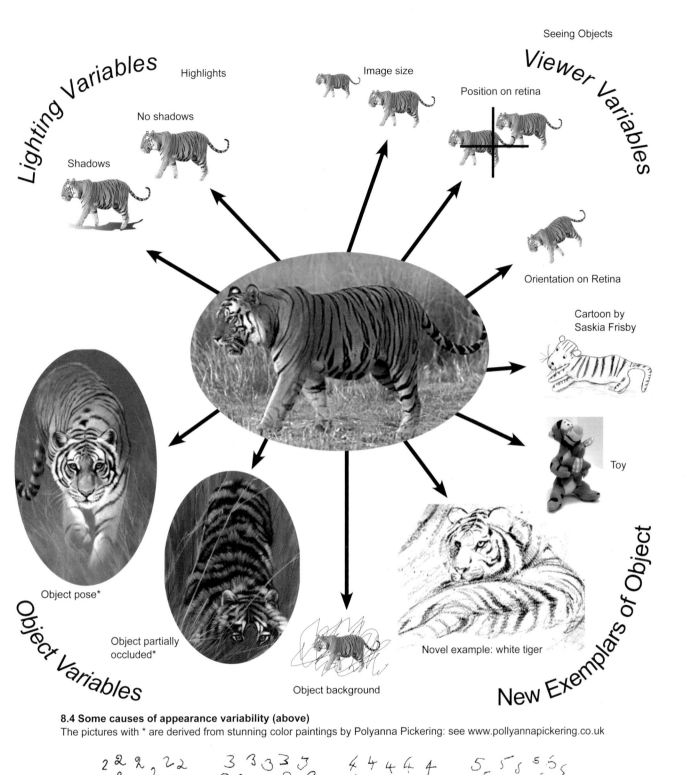

Lighting Variables

Highlights

No shadows

Shadows

Image size

Seeing Objects

Viewer Variables

Position on retina

Orientation on Retina

Cartoon by
Saskia Frisby

Toy

Object pose*

Object partially
occluded*

Object Variables

Object background

Novel example: white tiger

New Exemplars of Object

8.4 Some causes of appearance variability (above)
The pictures with * are derived from stunning color paintings by Polyanna Pickering: see www.pollyannapickering.co.uk

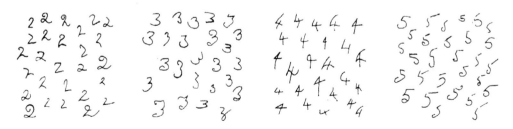

Competent visual robots, if competence is measured on anything like the human scale of achievement, are a long way off. Nonetheless, advances in recent decades have been sufficient to enable us at least to see more clearly the nature of the problems to be tackled, and to sketch in a conceptual framework within which research can be profitably directed. The main aim of this chapter is to elucidate this framework and to summarize the main approaches.

Appearance Variability

The central problem to be overcome in recognizing objects is that their appearance can vary dramatically on different occasions. We described in Ch 3 a simple system for bank check recognition which worked only because strictly controlled industrial imaging more or less eliminated image variability. But in everyday seeing, the same object will project radically different images at different times. We summarise some of the main causes of *appearance variability* in **8.4** as a guide for exploring the problems that have to be overcome.

One major source of variability is that images of a given object vary according to the particular *viewing position* of the observer. Thus, the object's image can be smaller or larger depending on the distance from which it is viewed. This is the problem of *size* or *scale*. Also, the image of an object will not always fall on exactly the same position on the retina. This is the problem of *image position* or *image translation*, although looking directly at an object removes most of the variability in position because this ensures an object's image always lands on the fovea. Or the observer's head position may change leading to variation in image *orientation*. Another example of a viewer-dependent variable is the optical quality of the retinal image: is it *sharply focused* or *blurred*?

A major factor causing considerable appearance variability is *lighting*. There may different *shadows* or *highlights* on different occasions of viewing, or neither. Also, *color* can vary greatly on different occasions of viewing.

Yet another factor is the *background* against which the object is seen. This leads to the *figure from ground* problem explored in Ch 7.

Many objects do not remain a constant shape. For example, people stand up, sit down, walk, run, raise their arms, and so on. These *articulations* between object parts are examples of **object instabilities** that can cause huge variations in images of the same object. Again, this variability causes little or no interference in our ability to recognize people quickly and easily, either as particular instances of the class "people" or as specific individuals. Indeed, the particular way people move can help their recognition, as we will describe later.

Another example of object instability that causes huge variations in the appearance of rigid objects is *object orientation* (or *pose*), *object position*, and *object size*. For example, we can still recognize a coffee mug whether it is the right way up or upside-down, and regardless of its position over a wide range of distances and sizes.

Finally, we are often confronted by a new example of a given object class whose image differs vastly from anything we have seen before and yet we can still recognize the object. Caricatures are a case in point. Despite such large differences we usually glide smoothly over them and see the similarity to the original character.

The diverse nature of appearance variability renders forlorn any attempt to recognize objects simply and solely in terms of image matching of the simple template-of-pixels kind discussed in Ch 3. Moreover, the diversity of appearance is so great that seeking a single all-embracing theory of human object recognition may be unrealistic. Perhaps the brain has diverse recognition techniques to match the diverse inputs with which it has to deal.

What is the Object Recognition Task?

Why bother with object recognition? This is a harder question than it appears. Indeed, defining precisely what we mean by object recognition is itself quite tricky. Shimon Ullman suggested in 1996 that the benefit of object recognition is that it enables us to access useful information stored in our brains about the object before us, information that is not itself visible, such as: Is the object good to eat? Should we run away from it? Should we try to court it? How can we pick it up? And what could we expect to see in hidden parts of the object?

Those are valuable things to know but object recognition is far from being the be-all and end-all of seeing. Ch 2 introduced the computational

Lines and blobs found in an image by edge detection
These are the kinds of entities represented in the first stages of building the primal sketch, Ch 7. The long lines would at this stage be simply unlinked strings of edge points.

Figural grouping using the principles of similarity and proximity "peels off" the dots from the lines, yielding a new set of primal sketch tokens, each representing one *blob cluster*. Note that this process of *figural abstraction* would find the blob clusters even if the clusters had been composed of quite different elements, as long as they were similar and close to one another.

approach to vision using the example of achieving a three-dimensional representation of the scene using texture gradients. This kind of representation has obvious value in guiding movements (see also Ch 18 on stereo vision).

Consideration of the task of object recognition is constrained by what else the visual system in question can do. For example, any recognition system will be limited by the range of features that it can compute, and the resolution that it can deliver in representing those features.

Figural grouping of the blob cluster tokens using the principle of colinearity yields two new tokens, each representing one *line* of blob clusters. The parameters of *line length* and *line orientation* would be attached to these line tokens.

Structural Descriptions

As a starting point for discussing how to solve the task of object recognition, note the following simple observation: human vision is excellent at seeing the *parts* of which objects are composed. Look around and you will have no problem in seeing, say, the legs of a chair or of a person as components of the object in question. This does not force the conclusion that this ability is used in object recognition, but it has often been argued that a good way of talking about the recognition of visual similarity is in terms of ***structural descriptions*** of objects made up of parts.

Input to matching processes for recognition

8.5 Levels of figural abstraction
Finding structures following edge detection. See Ch 7 for details of figural grouping.

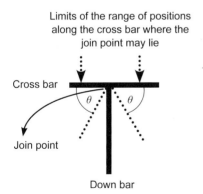

Limits of the range of positions along the cross bar where the join point may lie

Cross bar

θ θ

Join point

Down bar

8.6 Structural description of a capital T as two bars
Allowable ranges are shown for just two of the parameters defining the spatial arrangement of the bars with respect to one another, for the bars to be deemed a T. These example parameters are: (1) the range of positions of the join point along the cross bar, and (2) the angle θ (pronounced *theta*) between the down bar and the cross bar.

All the patterns in **8.3** are Ts because they all share a common structure which can be defined abstractly as: a T is a *down bar* which is joined at its upper end to the middle of a *cross bar*. As long as the terms down bar and cross bar are themselves defined at a sufficiently high *level of abstraction*, **8.5,** then all the Ts in **8.3** fit the description as stated, and therein lies their T-ness. Thus, the structural description for a T is a general formula for a T, **8.6**.

Structural description theories of object recognition claim that what first happens when we recognize any one of the patterns in **8.3** as a T is that our visual system breaks up the input pattern

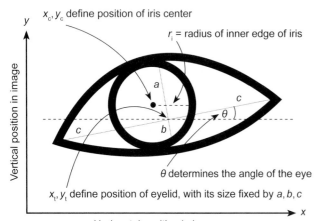

8.7 Capital B or numeral 13?
The middle character in the two boxes are physically identical.

into parts. This can be thought of as a process of abstraction in which details are lost, yielding a kind of summary of the pattern as a set of a set of constituent parts, technically called ***primitives***. Certain shape properties are attached to each primitive. For example, in the case of Ts the primitives are *bars* whose relative length and thickness have to lie within limits specified in the details of the structural description for a T, **8.6**. These primitives might be extracted as shown in **8.5** using a cascade of the various grouping processes set out in Ch 7 for separating *figure from ground* (see **7.13** for a reminder on how a stick figure chair structure was found from an image of a chair).

Note that in **8.6** the bars comprising the T are specified as highly abstract entities; indeed, they can be thought of as "mini-objects" in their own right. Using such high level abstractions helps solve some aspects of the appearance variability problem.

An important part of building a structural description is making explicit the spatial relationships between the primitives. This is shown in **8.6** for a capital T using the relationship *joined to*, with ***parameters*** defining (a) the range of locations

where the join point can fall, and (b) the allowable range of angles between the bars where the join happens. If a structure is found which is made up of two joined bars but with parameters outside the allowable ranges then it won't be matched as ('seen as') a T.

An important caveat here is that if a particular context strongly favours the interpretation of T for a two-bar structure whose elements are outside the allowable parameters then a T might well result in a T being seen. An example of context at work is shown in **8.7**, in which the central pattern tends to be seen either as a B or as 13 depending on the neighbouring characters that provide ***top-down,*** or equivalently ***concept driven***, processing. We discuss in Ch 13 the theoretical background provided by Bayes' theorem for using *prior information* for various seeing tasks, including object recognition.

The relationship *joined to* is just one of a set of spatial relationships that might be used for specifying structural descriptions. Others are *below, above,* and *crossing*. Also, complex sets of parameters can be defined for structures that are much more intricate than letters. For example, **8.8a** shows a

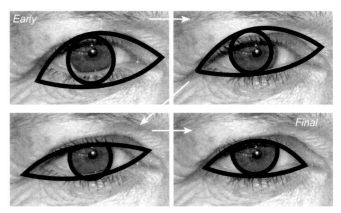

8.8 Deformable eye model
a A sample of the parameters used by Yuille and Halliman (1992). Their recognition system adjusted the values of these and other parameters in searching for a good fit between image and eye template This sample is chosen to illustrate the principle of using multiple parameters in defining an abstract structure. Redrawn with permission from MIT Press.
b Various stages in fitting over many iterations the deformable model to an image of a face as the parameter values are adjusted. Schematic plots (not real data) drawn to illustrate the work described in Yuille and Halliman (1992).

parameterized structural description for an eye. There are numerous parameters needed for this highly complex entity. We will not explain their details. We use this example simply to illustrate that many parameters may be needed to specify an object model.

Finding a Match between Input and Model

Having a structural description for an object model is all very well, but how is this model to be matched to the input? Even if the processes of abstraction have produced a representation of the input as a structure cast in the same language as the object models held in store, the input might differ from the correct model in its details. So the **matching problem** is finding which one of a store of object models fits the representation of the input, **8.8**.

We will consider the matching problem in more detail later but to give you an idea of what is involved, the eye model of **8.8a** is shown being fitted to a face image in **8.8b**. The computer program that did this gradually adjusted the values of the parameters for the eye model within allowable ranges during a search to see if it could find a good fit between input and model. If a criterion for goodness of fit was met, as in this illustrated case, then the eye template was deemed to be matched and recognition thereby achieved.

Of course, whatever the processes of recognition might be in the brain, their operations are not something of which we are conscious. When we look at **8.3** and see all its patterns as Ts, we do not deliberately and consciously scrutinize each pattern to break it up into its parts and then check that these parts and their relationships do indeed fit what we were told long ago constitutes a T. Rather, the process is automatic and unconscious, at least in fluent adult readers dealing with reasonably legible text.

But although we are unaware of what is going on, a stored representation of a T must exist somewhere inside our brains, and the act of recognizing a T must involve matching a representation obtained from the visual input against this stored representation. The stored description for a T is possibly in the inferotemporal lobe (Ch 10), laid down as a memory built up when we learned to read.

Object Recognition: Overview

It is time to take stock. The general scheme for object recognition we have been introducing is set out in the flow chart in **8.9**. To expand on this scheme a little, let's take another example, that of recognizing the cartoon of Lennon in **8.1**.

The structural description approach (and there are others, see later) suggests that we first build up a structural description from this cartoon, **8.1**. We then find that it matches a stored structural description for "Lennon" which we created as a memory when we first took note of him. This description might include such items as: "Pointed nose, round spectacles, long hair, long chin," and so on. But these particular components would need to be tied somehow to the structure of a face, which might itself be described as: "Oval outline in which is set a horizontal bar [mouth] in the lower region,

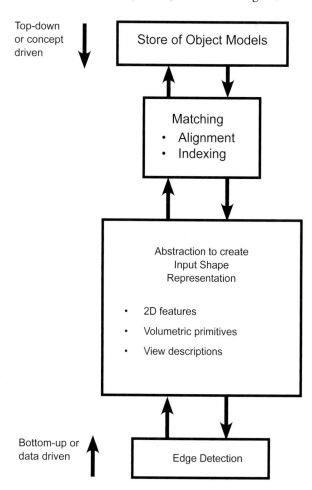

8.9 Flow chart of object recognition

a vertical bar [nose] in the central region, and two blobs [eyes] in the middle-to-upper region." When we recognize a cartoon of other public figures, the structural description for "face" will be the same, but with a different set of details attached to it, each with different parameters of size and relative positions, leading finally to the recognition of the person depicted. This face description is two-dimensional but a three-dimensional version could be developed based upon the structure of the head.

Explaining structural descriptions in terms of words should not be taken to mean that the object representations in our heads are themselves word-like. They are not verbal symbols as such, although some find it helpful to use the language analogy and call these representations **visual symbols**. But beware the dangers of using the language analogy: words deal with many concepts as well as visual objects, and many animals are expert at object recognition and yet they lack language.

Possible brain codes serving the role of visual object symbols were discussed in Ch 3. It can be strange to think of seeing as a matter of building up a set of symbols (a set of representations, if you prefer) for entities in a scene. But we have to get used to the idea that when we look at a T, or a picture of Lennon, myriad visual symbols become activated inside our brains, and that among this multitude is one standing for T-ness if a T is present, or one standing for Lennon if a suitable picture of him is present. These symbols are regarded as "high-level" as they derive from the activation of the relatively "low-level" ones representing features such as bars. However, a bar as defined in **8.6** is much higher in the abstraction hierarchy than, say, an edge—and far above that of a pixel.

You might also find it a strange thought that, as you look around at the scene before you, everything you "see," be it a simple feature, a cluster of features, or a battered armchair, the "seeing" must have been mediated in each case by some kind of symbol becoming selected or activated inside your brain. This symbol (equivalently, *representation*) stands for the visual feature or object and constitutes its explicit description in the brain. According to this view, seeing objects occurs when the descriptions built up for entities in the viewed scene have been matched to one or other of a vast number of visual symbols, most of which lie

dormant inside our heads most of the time. The idea here is that the collection of activated symbols quite simply *is*, somehow, the substrate of our visual experience of the objects seen at any given moment. It is a surprising, even fantastic, realization, but it seems an inescapable one.

But talking about object recognition as building up and matching descriptions of an input image to one of a set of stored of object models is far from being a full-blown computational theory of recognition. It is more a way of stating in broad outline what some believe a successful theory must accomplish. First, it must explain how suitably abstract descriptions of the input image are built up in a language of visual primitives and their spatial relationships. Second, it must tell us how such descriptions can be stored. And third, it must tell us how any given description obtained by looking at an input image can be matched to the "right one" of all the huge number of stored descriptions which we possess—and how this can be done, usually, in a fraction of a second. These are tall orders for any theory, and perhaps you begin to appreciate rather more why the object-recognition problem continues to be such a baffling one.

More on Terminology and Basic Concepts

Before going into details about theories of object recognition, we need to clarify certain technical terms and distinctions.

First, the recognition of a *specific instance* of an object ("That's my coffee mug you have taken, I recognize its handle.") is different from recognizing an object as a member of a *class* of objects ("That's a coffee mug, not a tea cup."). The latter task is called **object classification** and instances of objects within a class are called **exemplars**. Lions, tigers, and house cats are all exemplars of the class *cat*. All the Ts in **8.3** are exemplars of the class T-shape.

Second, we will mainly be tackling how recognition might be achieved on the basis of **shape**, which can be either *two-dimensional* (such as hand written or printed characters) or *three-dimensional* (objects such as people and chairs). However, we need to bear in mind that some objects can be recognized using properties other than shape. For instance, the surfaces of some objects have distinctive properties, such as *surface texture*. We can recognize zebras on the basis of their characteristic stripes.

We can also use *color* to recognize, for example, robins on the basis of their red breast. Bird watchers also often recognize different species by their characteristic flight. Indeed, birders have developed a special skill, called recognition by *jizz*. This is the ability to absorb a wide range of information about a bird in the few seconds before it dives behind a rock or up a tree. Jizz entails taking account of a bird's color, size, sound, and movement—and also but not only its shape. Jizz refers to the "feel" of the bird. The word may derive from US Air Force slang for "general impression and shape," or "gis."

Third, by way of recapitulation: we have said that recognition is defined as *matching* the representation of an *input* to one of a stored collection of *object representations*. Both input and stored representations are often called *object models*. Once recognition has been attained, information stored about the object, such as the nature of its surfaces that are not visible from the current viewpoint, can be used to guide further visual processing and for planning suitable actions.

Recognition by Motion

The jizz used by bird watchers includes sensitivity to motion patterns, and human vision also uses movements to aid recognition, as noted by Shakespeare: "'Tis Cinna, I do know him by his gait" (from *Julius Caesar*).

In 1973, Gunnar Johansson showed how remarkably adept the human visual system is in recognizing patterns of moving dots depicting biological motion. When shown as a sequence, images of dot patterns like those in **8.10** are known as a ***Johansson movie***. It is surprising how much information human vision can extract from such sequences. For example, it is possible to disentangle the dots depicting two people ballroom dancing. Cells selectively responsive to biological motion have been found in a region of the temporal lobe (see Chs 8 and 9).

George Mather and Linda Murdoch used Johansson movies to demonstrate that we can identify the sex of the figure, based purely on its

characteristic motion. In essence, this was achieved by creating "hybrid Johansson movies," which combined the motion characteristics of a man with the physical structure of a woman (or vice versa) and then asking subjects to judge the sex of the depicted figure. Here's what they did in a little more detail.

Each observer was shown a series of Johansson movies, and was asked to judge the sex of the figure in each movie. Unknown to the observer, each figure was defined in terms of both its structure and its dynamics. If the figure had a male structure then it had wide shoulders and narrow hips, but if it had female structure then it had narrow shoulders and wide hips. If the figure had male dynamics then it had small lateral (side to side) hip motion, but if it had female dynamics then

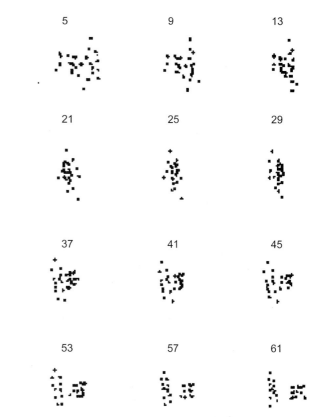

8.10 Johansson movie
Each dot marks the position of a limb joint at a particular moment in time. It is not clear what is depicted in each frame of the movie but when seen in sequence the percept of a person walking a dog is immediately obvious. Many fascinating movies of this kind are at http://astro.temple.edu/~tshipley/mocap/dotMovie.html. From Rashid (1980) with permission ©1980 IEEE.

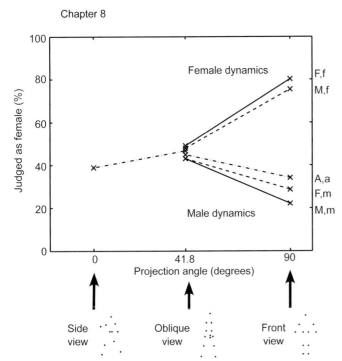

8.11 Effect of motion dynamics on perceived sex: data from Mather and Murdoch

The effect of male or female dynamics had almost no effect for side-on and oblique views (i.e., 0 and 41.8 degrees). However, for frontal views, where lateral hip motion is most apparent, the male/female dynamics of the figure determined whether the figure was perceived as a male or a female. Specifically, figures which had female dynamics were perceived as female, whether they had female structure (labelled as F,f—walks like a woman and built like a woman), or male structure (M,f—walks like a woman but built like a man), and these movies yielded responses which were not significantly different from each other. Similarly, figures which had male dynamics were perceived as male, whether they had male structure (M,m—walks like a man and built like a man), or female structure (F,m— walks like a man but built like a woman). It Thus, appears that dynamics rather than structure determines perceived sex of a moving figure in Johansson movies. Adapted with permission from Mather and Murdoch (1994).

it had the large lateral hip motion more typical of females. Each movie consisted of a figure with one of five possible combinations of structure and dynamics: (1) male structure and male dynamics (denoted as M, m), (2) male structure and female dynamics (M, m), (3) female structure and male dynamics (F, m), (4) female structure and female dynamics (F, f), and, (5) androgenous structure and androgenous dynamics (A, a), where the androgenous figure had a structure and dynamics that was between that of a male and a female. In addition, each figure was shown from one of three viewpoints: (a) 0 degrees: this was a side view, and all stimuli appeared identical from this view be-

cause lateral motion was invisible from this angle, (b) 41.8 degrees: this was an oblique view, (c) 90 degrees: this was a frontal view, in which lateral motion of hips was most apparent.

The results from these various conditions are summarized in **8.11**. The conclusion to be drawn from this elegant experiment is that dynamics rather than structure seems to determine the perceived sex of a moving figure in Johansson movies. This makes it likely that motion also plays an important role in sex identification in normal viewing.

While it may seem plausible that articulated objects such as human bodies can be recognized from their motion, it is less obvious that this would also apply to rigid objects. However, the ethologist Tinbergen showed that chicks of ground-dwelling birds (e.g., grouse) react very differently to the silhouette of a bird in the sky, depending purely on

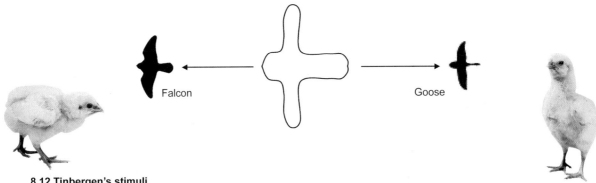

8.12 Tinbergen's stimuli
Chicks respond to the same single shape (center) as a harmless goose when travelling "pointed end first," but as a bird of prey when travelling "blunt end first'."

a

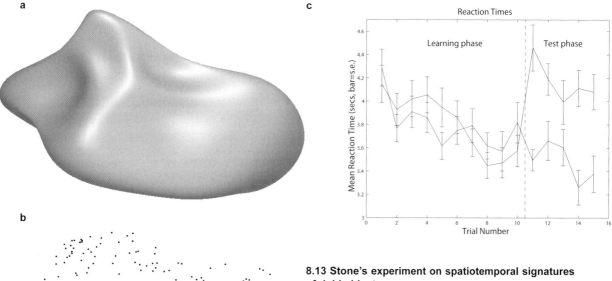

b

c

8.13 Stone's experiment on spatiotemporal signatures of rigid objects

a SIngle frame from a movie of a tumbling ameboid solid object.
b Single frame from Johansson movie of a similar object.
c Mean reaction times during learning (trials 1–10) and test (trials 11–15), for control (dashed line) and experimental (solid line) conditions. Rotation-reversal of target objects increased reaction time relative to non-reversed rotations of control condition. Bars indicate standard errors. Reproduced with permission from Stone (1998).

its direction of travel, **8.12.** The key lesson from this classic experiment is that both the shape and motion of an object define its identity.

More recently, Stone proposed that object recognition can be based on ***directed view-sequences*** (or equivalently ***spatiotemporal signatures***) as well as specific viewpoint. He used movies of solid ameboid objects rotating in a tumbling motion (objects rotated around an axis which changed over time), **8.13a**, and this was later repeated using point-light Johansson movies of similar objects, **8.13b**. In each trial, the observer was presented with the movie of an object: a movie which was the same each time a specific object was shown. The observer's task was simply to indicate whether or not each object had been seen before. Naturally, the learned "target" objects were interspersed with "distractor" objects, each of which was seen only once.

There were four targets, and after 10 presentations of each target's movie, observers could recognize the targets with about 80% accuracy. During this *learning phase* of the experiment, reac-

tions times fell steadily from 2.2 to 1.8 secs, **8.13c**. Then, without any change from the subject's point of view, two of the four target movies were played backwards throughout the rest of the experiment, in relation the movie's direction during the learning phase.

Now, as each movie depicted the tumbling motion of an object, running a movie backwards should have little impact on performance. More importantly, the 3D shape information conveyed by the movie is *exactly* the same whether a movie is presented forwards or backwards. So if a subject's performance changed after a learned target had its movie reversed then this should not be due to a change in the 3D interpretation of the object's shape.

In fact, subjects' performance did get worse, but only for those *targets* which had their movies *reversed*. For the remaining two control objects, which did not have their movies reversed, performance remained the same, or continued to improve as in the learning phase. The performance changes

8.14 When is a face not a face?
Features in the cluster on the right are the same as those on the left but shifted to different positions.

for target objects were reflected both in terms of accuracy and reaction times, but only reaction times are shown in **8.13c** (because these showed a more clear-cut effect).

The empirical conclusion from all these studies of recognition from motion is that, once again, the human visual system has remarkable abilities to extract information from visual images. What is now needed is for these studies to stimulate computational theories of the task of extracting recognition decisions from motion sequences. We discuss next theoretical issues in object recognition generally.

Shape Primitives and their Relationships

Two key decisions have to be made in specifying a theory of object recognition. They concern the nature of the representations of the input and of the stored models. These two decisions entail (a) specifying the *primitives* to be used to describe shapes and (b) a scheme for representing the spatial relationships between primitives.

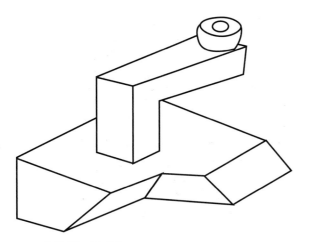

8.15 What is it?
We can readily see volumetric components in novel objects that we cannot recognize.

We have already illustrated these concepts in relation to the Ts in **8.3** and the eye model in **8.8**. In these cases the primitives were *image features*, such as the bars, regions, edges, and corners (these are the entities comprising the primal sketch representation of the image, Chs 5, 6 and 7). Such features have the advantage of solving some aspects of the appearance variability problem. This is because changes in pixel intensities due to variations in, say, lighting, that are unhelpful for object recognition, can be removed during the creation of, for example, edge representations.

But beware. The concept of a feature is a slippery one. An eye can be considered as a feature of a face but it is also a sophisticated object-concept in its own right, as revealed in **8.8**.

Of course, in some instances, such as some cartoons or sketches, an "eye feature" might be far from complex. It could be depicted as just a simple blob, dot, or line, but interpreted as an eye in the context of a pattern of such entities comprising a face. This issue relates to the concept of levels of abstraction illustrated in **8.5**.

An important issue here is keeping track of the positions of features, and to do this requires specifying a coordinate framework of some kind. In the case of the pixel primitives used for the simple bar detector in Ch 3, the spatial relationships of pixels were coded using the x, y coordinates defining each pixel position in the input image.

This is a *retinotopic coordinate system* (also described in Ch 9, page 208), and it can also be used for feature primitives, as shown in **8.8** for the eye model. In that case, the x, y coordinates are specified in the stored eye model as a range of "allowable" x, y values. This gives a desirable degree of fuzziness in the feature relationships required for recognition, which is as a further step towards solving the appearance variability problem.

For example, the exact positions of eyes, nose, and mouth in a face vary across individuals; hence it makes good sense to use a range of allowable positions. Of course, if the allowable ranges are exceeded then recognition will not follow, which is presumably why we would not see the jumble of features shown in **8.14** as a face.

The idea of allowable parameter ranges was also used for the model of a T, **8.6**, but in that case the allowable range for the join point between

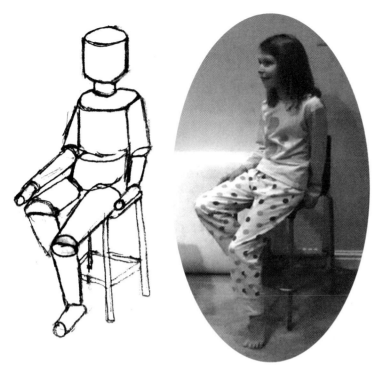

Given the ability of the visual system to find the **volumetric primitives** comprising an object, it is natural to enquire whether such primitives play any part in object recognition. Many theorists believe that they do.

One way of representing a volumetric primitive is as a stick specifying the axis of a cylinder, **8.17**. Indeed, the fact that the human visual system is so adept at recognizing an object depicted with such an impoverished stimulus as a stick figure was one the factors that led David Marr and Keith Nishihara in 1976 to take seriously the idea that our brains do actually compute axis-based volumetric shape representations. Their argument runs: the reason we can "see" a stick figure as depicting an object is that the human visual system generates stick-figure-like representations. So, when a stick figure is presented as a stimulus, it provides a shortcut as it were to the three dimensional representation that is *normally* computed from low-level representations such as the primal sketch.

8.16 Drawing the human form
Artists often start by sketching cylinders to represent components (left). Paul Cézanne famously remarked: "*...treat nature by means of the cylinder, the sphere, the cone...*"

the cross and down bars was not expressed in *x*, *y* retinotopic coordinates. Rather, the range was specified as a range *on the cross bar*. This is an example of using an *object-centered coordinate system* because the position of the join point is specified on a component of the object. Hence this position is independent of the viewpoint of the observer, or where the T falls on the retina. This is a key distinction and we will have more to say later about coordinate systems for representing relationships between primitives.

Volumetric Primitives

A more complicated type of primitive than a 2D bar is one representing a *volume* of 3D space. The human visual system is extremely good at seeing objects as made up of "lumps of stuff" filling up a bit of the three-dimensional world, even if the viewed object has never been seen before and is unrecognized, **8.15**. Indeed, one technique taught to artists is to compose a figure from a set of cylinders and then, quite literally, to flesh these out into the human form, **8.16**.

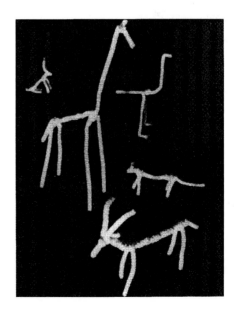

8.17 Stick figure axis-based representations
Reproduced with permission from Marr and Nishihara (1978).

8.18 Recognizing figures whose components are joined flexibly
Sculpture of Kate Moss by Marc Quinn. Photo courtesy Tom Roundell, Sotheby's Collection, Chatsworth 2007.

Just as for the Ts, one way of representing the spatial relationships of stick primitives is to specify the points where the various sticks join. This is an example of an object-centered description because it is rooted on the object itself. Again, using a range of allowable positions for primitives meeting at any given point can be a way of solving some aspects of the appearance variability problem. We illustrated this for the join in the T-structure in **8.3**. In addition, specifying a range of allowable angles between stick primitives that join is one way of tackling the problem of recognizing articulated objects, **8.18**. Marr and Nishihara called the parameters defining the allowable joins between stick primitives the *adjunct relations*.

More formally, an object-centered coordinate system has its *origin* located within the object itself. In this case, the location of each point or feature on the object is specified as a distance along the X, Y, and Z axes which pass through the origin, which is located somewhere in the object, say at the object's center. The origin has, by definition, coordinates $(X, Y, Z) = (0, 0, 0)$. For example, if a feature on the object is 3cm to the right, 2 cm above, and 1cm behind the origin then its coordinates can be written as $(X, Y, Z) = (3, 2, 1)$.

An object-centered representation can be contrasted with a *viewer-centered representation*. In this case, the origin of the coordinate system is attached to the viewer. It may be at the centre of the eye, or sometimes it is a point midway between the eyes, called the *cyclopean eye*, **8.18**.

For example, if an object is 30 cm the right, and 20 cm above, and 10 cm in front of the viewer then its coordinates are $(X, Y, Z) = (30, 20, 10)$.

Note that the point described above which has coordinates $(X, Y, Z) = (3, 2, 1)$ in the object-centered coordinate frame, has coordinates $(X, Y, Z) = (33, 22, 11)$ in the viewer-centered coordinate frame.

Another theorist who believes that volumetric primitives play an important role in object recognition is Irving Biederman. He suggested in 1987 using a limited vocabulary of 3D shapes (36 in **8.20**), that he dubbed *geons* (an abbreviation of *ge*ometric i*on*). These vary in simple qualitative ways along just a few dimensions. Sticks and/or geons can be thought of as "visual letters" used to build up "visual words," that is, objects.

Biederman shared with Marr & Nishihara the view that the spatial relationships between geons should be specified in an object-centered system, although he did not label his scheme in that way. He used a set of simple terms such as *above*, *below*, rather than the numerical values of Marr and Nishihara's adjunct relations. Biederman argued that the speed and fluency of human object recognition made it likely that it relied on simple qualitative distinctions rather than quantitative details as the latter might take a lot of time to compute and match.

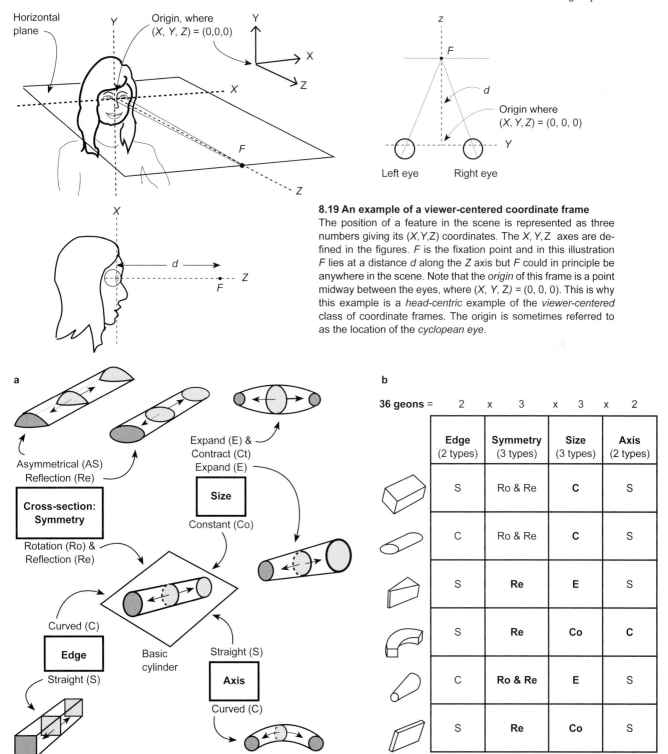

8.19 An example of a viewer-centered coordinate frame

The position of a feature in the scene is represented as three numbers giving its (X, Y, Z) coordinates. The X, Y, Z axes are defined in the figures. F is the fixation point and in this illustration F lies at a distance d along the Z axis but F could in principle be anywhere in the scene. Note that the *origin* of this frame is a point midway between the eyes, where $(X, Y, Z) = (0, 0, 0)$. This is why this example is a *head-centric* example of the *viewer-centered* class of coordinate frames. The origin is sometimes referred to as the location of the *cyclopean eye*.

8.20 Definitions of the volumetric primitives comprising Biederman's set of geons

a Each geon is defined as a qualitatively different variation from the basic cylinder (shown in the diagonal box) in terms of: *Axis* (straight/curved), *Size* (expand+contract/expand/constant), Edge (straight/curved), Cross-section symmetry. These qualitative contrasts provide 36 combinations (2 × 3 × 3 × 2), and hence 36 geons in all.

b Examples of geons, using the coding symbols defined in **a**. Adapted from Biederman (1987).

a

Human

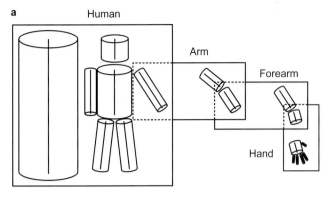

Arm

Forearm

Hand

b

Cylinder

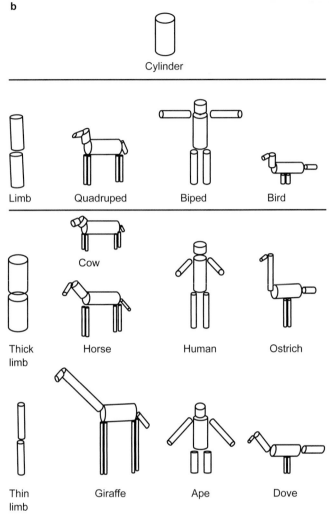

Limb   Quadruped   Biped   Bird

Cow

Thick
limb   Horse   Human   Ostrich

Thin
limb   Giraffe   Ape   Dove

8.21 Marr and Nishihara's scheme for satisfying the classification and identification constraints

a Representing a human form as a series of models, from coarse to fine. Marr and Nishihara explained their scheme as follows:

> First the overall form—the "body'—is given an axis. This yields an object-centered ordinate system which can then be used to specify the arrangement of the "arms," "legs," "torso," and "head." The position of each of these is speci-fied by an axis of its own, which in turn serves to define a coordinate system for specifying the ar-rangement of further subsidiary parts. This gives us a hierarchy of 3D models: we show it extend-ing downward as far as the fingers. The shapes in the figure are drawn as it they were cylindrical. but that is purely for illustrative convenience: it is the axes alone that stand for the volumetric qualities of the shape, much as pipe-cleaner models can serve to describe various animals.

b Store of object models representing objects both as members of a class and as specific exemplars of that class. Redrawn with permission from Marr and Nishihara (1978).

Criteria for a "Good" Representation of Shape

How should we judge theoretical proposals regard-ing primitives and indeed other aspects of object recognition theories? Marr and Nishihara set out three criteria that must be met by a "good" repre-sentation of a three-dimensional shape:

1 *Accessibility*
Can the representation be computed easily?

2 *Scope and Uniqueness*
Is the representation suitable for the class of shapes with which it has to deal (*scope*), and do the shapes in that class have one or more canonical represen-tations *(uniqueness)?* A **canonical representation** is one in which a particular shape has a *single* descrip-tion in the representation.

For example, if the same shape description was arrived at for all the different appearances of a tiger in **8.4** then this description would be said to be canonical. The representation of a T as two bars joining together is also an example of a canonical representation.

Unique representations are advantageous when it comes to storing and matching object models, because they reduce the possibilities that have to be considered. However, it may not be feasible to compute canonical representations for many types of objects, in which case taking that path for them will fail the accessibility criterion.

A *non*-canonical representation scheme is one in which the object model is stored as a number of different representations which collectively serve as its code. An example of this kind that we will discuss at length later is ***view-based recognition.***

A scheme of this type for a tin of beans might contain a list of representations for different viewpoints, such as "flat circular patch" when the view is along the central axis, and "convex cylindrical patch" when the tin is viewed side on. This type of scheme may lead to the need for complicated processes for solving the problem of matching a representation of an input image to one or other of the multiple views stored as the object model.

3 *Classification and Identification*

Can the representation be used to capture both the *similarities* (for deciding class membership) and the *differences* (for identifying specific instances) between the objects with which it has to deal? For example, we can readily see that a horse and a donkey share some properties in common as well having some differences. That is, they are both exemplars of the class *quadruped*, but they possess important differences. Marr and Nishihara used the terms ***stability and sensitivity*** rather than the now more commonly used terms *classification* and *identification*.

A good object representation scheme must lend itself naturally to delivering the opposing but complementary requirements of classification and identification. This demands a separation of the invariant information characterizing a particular category of shapes from information about the subtle differences between shapes in the same category.

These three criteria are useful for discussing the pros and cons of different object recognition theories. Specifically, they are useful for guiding the search for solutions to three *design problems* that, Marr and Nishihara argued, have to be faced in building a object representations. These problems are: What *primitives*? In what *coordinate system* are the spatial relationships between primitives to be located? What is an appropriate *organization* of the primitives?

Marr and Nishihara's Theory

We said above that Marr and Nishihara's theory was based on axis-type volumetric primitives (stick figures). Their spatial relationships were specified in an object-centered coordinate system making explicit where the sticks joined and the allowable variations in their positions and angles.

This scheme scores well in terms of the scope criterion as it is suited to many objects occupying a volume of space—but not all: a crumpled newspaper was Marr and Nishihara's illustration of a clearly unsuitable object because it could be described with so many alternative axes.

Their scheme also does well on the uniqueness criterion because axis-based primitives allow an object to be represented by a single description of the spatial relations of the object's parts, that is, a description arrived at from whatever angle the object is viewed. However, an axis-based canonical representation may do less well on the accessibility criterion. Computing them appears to be beyond the current state of the art of computer vision.

Marr and Nishihara proposed a way of using axis-based representations that meets the criterion of supporting both classification and identification. They suggested that object representations were multi-level entities ranging from coarse to fine. At each level, the model comprised two sub-levels: one representing the overall axis of the object or object component and another representing a set of joined components at that level, **8.21a**. Many different objects would be able to be matched at the coarsest level and this would satisfy the classification criterion—matches here bring out the similarities between objects, **8.21b**. Descending through the hierarchy of objects brings into play finer discriminations as finding matches at the level of fine details discriminates between different albeit similar objects, hence satisfying the identification criterion.

This proposal is an attempt to deal with an often neglected issue in object recognition theories: how should the object models be organized in order to facilitate the recognition? Should there be a "flat" organizational structure, in which all models have an equal status? Or should there be a *hierarchical* organization, **8.21b**, in which primitives in models at the highest level convey coarse information about the shape of an object, with lower levels giving the details? A hierarchical scheme satisfies the classification/identification criterion. This is because similarities at the highest level can deliver the competence to "see" the sameness between a horse and donkey, or even a horse and tortoise, with lower levels in the hierarchy dealing with their differences.

Marr and Nishihara's ideas have yet to be fully implemented as a computer vision program although ways have been explored for extracting axis-based representations from images. Marr died at the age of 35 from leukemia in 1981, sadly cutting short his distinguished and influential life. Whether their theory will ever be implemented is a moot point, as its complexities may fail fatally on the accessibility criterion.

Biederman's Theory of Object Recognition

We have already outlined some of the key features of Biederman's theory, specifically his proposal of a limited set of geons as volumetric primitives and a simple set of spatial relationships between them. His emphasis on using simple qualitatively

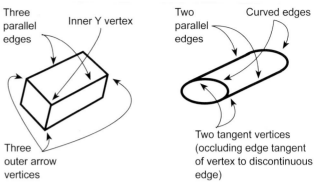

8.22 Using non-accidental properties to identify geons in edge representations
Adapted from Biederman (1987).

defined primitives suggests his theory will score well on the accessibility criterion, and a version has been implemented in a computer vision program (Hummel and Stankiewicz, 1996). His theory, like that of Marr and Nishihara, which preceded it by a decade, fares similarly on the scope and uniqueness criteria. It originally contained no explicit provisions for meeting the classification criterion but

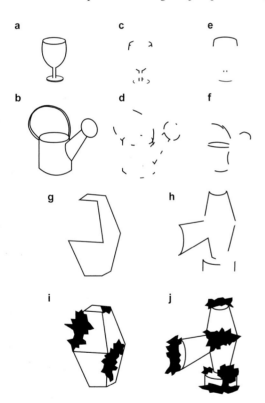

8.23 Are junctions specially important?
Recognition of objects **a** and **b** is greatly hindered in **e** and **f** in which junctions defining non-accidental properties (NAPs) are deleted, but hardly at all in **c** and **d** in which contour deletions preserve these junctions. (Reproduced with permission from Biederman, 1987.) However, whereas **g** and **h** appear as 2D patterns, if evidence of occlusion is provided, as in **i** and **j**, then 3D volumetric objects are readily seen behind the black patches, thus, revealing important subtleties in the way NAPs are utilized. (Reproduced with permission from Moore and Cavanagh, 1998.)

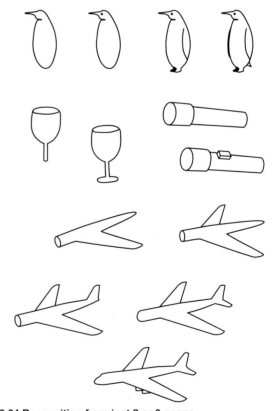

8.24 Recognition from just 2 or 3 geons
With permission from Biederman (1987).

since them attempts have been made to tackle this limitation.

One interesting point of detail about Biederman's theory is that he proposes that geons can be identified in edge representations, such as those in the primal sketch, using "signatures" for each type of geon cast in a language of ***non-accidental properties***, **8.22**. The idea here is get around some of the problems of appearance variability by finding patterns of edge features that are valid for a given geon type regardless of the pose of the object or the position of the viewer.

Biederman has spent a lot of time studying human object recognition and finds evidence favouring his ideas. For example, if those parts of the edge description that contain the non-accidental properties are deleted then recognition is much more severely impaired than if similar lengths of edges are deleted that spare them, **8.23**. Also, Biederman has shown that objects can be recognized if only a few geons are visible, which is consistent with his idea of representing objects as simply as possible. Hence he argues for the use of stored models made up of just a few geons, **8.24**.

View-Based Object Recognition

We seem to see most 3D objects as comprising volumetric components, and can describe their relationships verbally using an object-centered representation. However, it does not follow that human object recognition relies critically, or even at all, on that ability. Over the last two decades or so, evidence has steadily accumulated suggesting that human vision may recognize objects on the

basis of a *set of views* stored in memory. This could be in addition to object-centered representations—the two methods are not mutually exclusive.

Using a set of views clearly fails the uniqueness criterion because each object representation is not a single canonical entity but instead a collection of views, each one of which has to be recognized separately. At first sight, this sounds a fatal weakness because it seems to propose storing a prodigious amount of information about different views. But there are ways of making this sort of theory viable by restricting the infinitely large number of possible views to just a few typical or ***characteristic views*** of each object. For example, you rarely need to recognize a car from underneath, unless you are a car mechanic of course. This kind of thinking has a long history in the psychological literature on object recognition.

One important idea here is that of the ***view potential*** of objects, introduced by Jan Koenderink

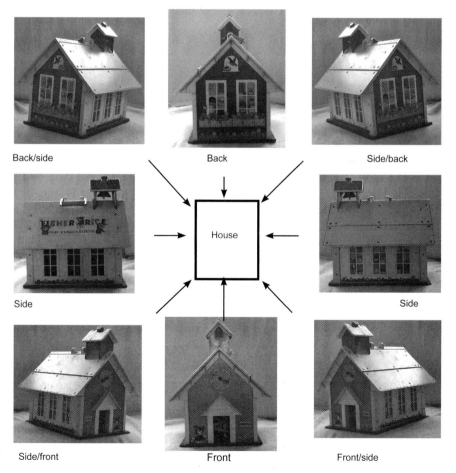

Back/side   Back   Side/back

Side   House   Side

Side/front   Front   Front/side

8.25 View potential of a house
This comprises the eight qualitatively different views available from viewing positions on the ground plane.

and Andrea van Doorn in 1977. They observed that the *qualitative* character of many views of an object is often the same, in that they share the same main features even if these features show differences in quantitative details (e.g., in shape due to perspective effects arising from a different viewpoint).

For example, imagine walking around a detached house, starting at the front, **8.25**. You would initially see the *front* view, which would include such features as the front door, front windows, eaves, perhaps a gable end. As you walked around from that starting point, no features would be lost and no new features added until a side elevation suddenly comes into sight, whereupon a new *front/side* view would arise. This would comprise all the previous front view features plus, say, a set of side windows. This change in the list of features that can be seen is what is meant by *qualitative* differences between views.

Further walking round the house would bring you to a point where the front view features were lost, leaving the side view features on their own. This point marks a further boundary between qualitatively different views contain different sets of features, not just different positions of features.

Given these qualitative jumps, the infinity of possible views seen from the ground plane can (in this simple case) thus be grouped into only eight classes: front, front/side, side, side/back, back, back/side, side, side/front, as shown in **8.25**. Each one is canonical in the sense that a single description of it is computed from the infinity of possible viewpoints which can give rise to it. Thus, this *view-based scheme* can be thought of as representing the object as a set of "sub-objects," each one a canonical characteristic view.

Koenderink and van Doorn's term *view potential* is apt because it draws attention to the fact that an object has a limited number of qualitatively different views. While it may seem more difficult to specify a limited number of such views for less "regular" objects than houses, there is intriguing evidence suggesting that view-based representations are used in human vision.

For example, Heinrich Bulthöff and Shimon Edelman found in 1992 that if observers are trained to recognize objects from two viewpoints separated by 75 degrees then recognition was best when the objects were shown from the same views as were used during training. Performance was not so good for views which were between the specific views used during training and worst for views that fell outside the range of views used during training, **8.26**. These results were repeated in a series of experiments which used both the paperclip-like objects in **8.26** and amoeboid objects similar to those used in Stone's experiment, **8.13**. They argued that these results are hard to explain if recognition is mediated by object-centered 3D structural descriptions. They concluded that view-specific representations mediated recognition because if a 3D structural description had been built up during training then all test views should have been equally easy to recognize.

View-based schemes have proved viable in computer vision. For example, Shimon Ullman has demonstrated a recognition scheme in which a series of views of an object are stored. Recognition is then achieved if an input image of a non-stored view can be matched to a view created as an interpolation between stored views. This scheme is explained in the legends to **8.27** and **8.28**.

A classic paper that has a bearing on the issue of viewpoints and recognition performance is that of Shepard and Metzler on **mental rotation**. Observers were given a pair of rotated 3D objects and asked to judge whether they were identical, **8.29**. Observers responded *yes* or *no* and reaction times were recorded. Stimuli could be rotated either within the picture plane or in depth. The results showed a remarkably orderly relationship between time required for mental rotation and the difference in 3D rotation between the objects. Thus, Shepard and Metzler's results suggest we do have access to the 3D structure of objects, but Bulthöff and Edelman's results indicate that we do not use this in object recognition, and instead use specific 2D views in order to perform object recognition.

Of considerable interest for the present discussion are some of the findings of Dave Perrett, Ed Rolls, and others who have found cells in the inferotemporal cortex of the macaque monkey responsive to faces. Some of these cells show a selective response to an individual face over a wide range of viewpoints. However, their selectivity does not always cover the whole range from profile view to full-face view. Instead, the response is often

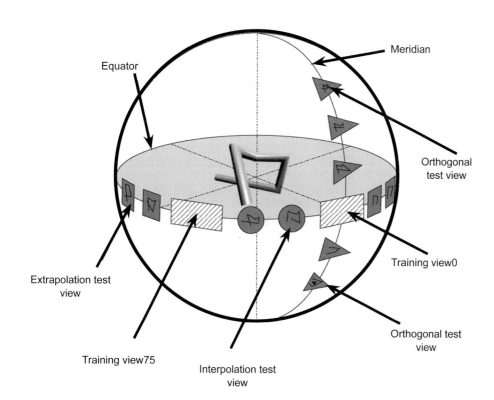

a The viewsphere
This provides a way of representing all possible views of an object. Imagine crawling around over the surface of the sphere and looking at the object at the center of the sphere. You would see a different view depending on where you are on the viewsphere. The extrapolation and interpolation views used during testing in Bulthöff and Edelman's experiment were located on the equator, whereas orthogonal views fell on the meridian. See legend below. Adapted with permission from Bulthöff and Wallis (2002).

8.26 Bulthöff and Edelman's (1992) experiment on view-based recognition
This experiment consisted of a training phase followed by a test phase.

1) *Training phase*: observers learned to recognize a set of bent paperclip-like target objects presented from two training viewpoints (labeled Training View0 and Training View75) in **a**. These views were separated by 75 degrees around the equator of the viewsphere.

2) *Test phase*: on each trial, the observer was presented with a pair of objects, and had to indicate which one was a target (the other was a "distractor" paperclip object). The viewpoints used in the test phase (see **a**) were:

  (i) *Interpolation views*: these views were on the equator of the viewsphere, and were *in between* the two views (View0 and View75) used in the training phase.

  (ii) *Extrapolation views*: these views were on the equator of the viewsphere, but were *outside* the range of the View0 and View75 training views.

  (iii) *Orthogonal views:* these views were on the meridian of the viewsphere, which is orthogonal (perpendicular) to the equator.

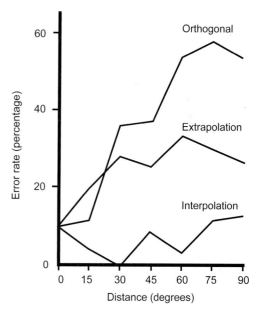

b Error rates for the three conditions Interpolation, Extrapolation, and Orthogonal. The key finding is that, for a given learned object, more accurate recognition performance is obtained for test views that are closer to training views. This was especially true for the orthogonal test views that lie along the meridian of the viewsphere.
Redrawn with permission from Bulthöff and Edelman (1992).

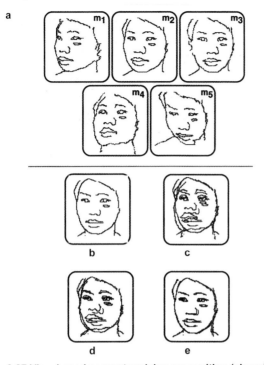

optimal for a particular view orientation, and then decreases steadily as the viewpoint changes. Thus, these cells could perhaps be mediating a view-point-based recognition scheme.

We describe in Ch 10, pages 250–251, further research on regions of the brain that seem specially concerned with object recognition.

Evaluating View-Based Object Recognition

It is time to ask how view-based representations fare when judged against Marr and Nishihara's criteria for evaluating object representations.

The Accessibility Criterion
Ullman's computer implementation of his view-based recognition system demonstrates that at least some systems of this sort meet the accessibility criterion. This is because he showed it was feasible to compute in reasonable time views created as interpolations between stored views, and then to use these interpolations for matching to input views.

The Scope and Uniqueness Criteria
Viewpoint-based schemes also seem to do well on scope as they are adaptable to a wide range of object types. And the uniqueness criterion seems to have been neatly finessed by not storing huge numbers of views of each object, but just a selected set.

The Classification and Identification Criteria
A limitation of view-based schemes is that it is not easy to see how they can be extended naturally to recognize classes of objects, rather than just specific objects. It has been suggested that just as an intermediate view can be computed between two stored views (e.g., as in **8.27**) so interpolation could be performed between different exemplars within a class. However, whether this is a viable salvation for view-based classification is a moot point. Could it really cope with the often quite pronounced differences in the way objects in a class appear, despite having the same underlying 3D structures?

Of course, it could be that one could achieve a decent level of object classification simply by as-

8.27 View-based computer vision recognition (above)
a In this example, the object model consists of the top five pictures m_1 to m_5 that show edge maps for five views of a face.
b The input to be recognized: a novel face image that is not in the model set.
c The input image superimposed on m_4. There is a poor match.
d The matching system's first attempt to improve the input-to-model match by calculating a new view from the model views that is an interpolation between m_4 and another model view. This improves the match but it is still not good.
e The result after 10 attempts to find a better interpolated view from the model set. Each attempt is guided by a measure of the size of the errors between input and previous interpolation-between-model-views. As can be seen, the final fit is remarkably good and would justify the conclusion of *Recognition Achieved.*

8.28 Another example of view-based recognition (left)
a-d As shown in **8.27** but here the model in **a** is matched to a face **b** from a different person.
e A good match is not attained after 10 attempts.
f The interpolated view used in **d**.
Both **8.27** and **8.28** are courtesy Ullman (1998).

sociating together a long list of those objects which are deemed to be exemplars of a class. This could certainly be done in industrial settings, where the computer vision designer has the power to make such assignments when programming the system. But that approach seems to miss the point when modelling human object recognition: dogs and cats and cows *look* alike in certain fundamental ways. It is not just a case of assigning them arbitrarily to a class of four-legged creatures. This suggests that ideally an object recognition scheme should deliver the capacity to recognize the exemplars in a class of objects as similar, despite their differences, as a natural consequence of the way it is built. This is what Marr and Nishihara were striving to achieve when they proposed their hierarchical axis-based coarse-to-fine recognition scheme.

We can also ask: are view-based representations in fact just a special case of structural descriptions or are they quite different?

The answer hinges on what is represented in each view. View-based adherents in the psychological literature have not always been clear on this critical issue. For the house example, **8.25**, we noted that each view can be considered as a "sub-object" capable of being represented as a structural description in which the required set of features for that particular view are listed, together with allowable ranges for their spatial relationships. This is very reminiscent of the structural descriptions described for Ts and eyes, **8.6** and **8.8**. Hence the spatial relationships of the parts seen in each view could be based on object-centered coordinates.

It is worth noting in this connection that the data structures used to represent the features of the faces in Ullman's view interpolation scheme, **8.27**, stored spatial information about the locations of those features in a coordinate frame that applied to all views. These various considerations suggest that the view-based vs. object-centered dichotomy may be more apparent than real. What view-based schemes do not do, or any rate none yet seems to have done, is to compute volumetric primitives in each view. But one may ask: could this be worth trying?

Indexing and the Matching Problem

When matching an input object representation to a collection of stored models, the question arises

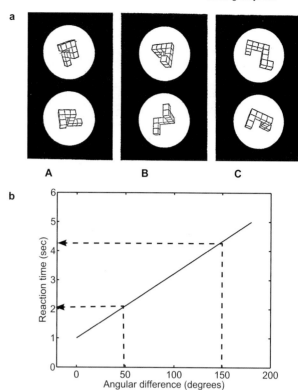

8.29 Shepard and Metzler's experiment on mental rotation

a Examples of pairs of stimuli used in same/different task (see text). **A** Same pair, which differs by a rotation of 80 degrees in the picture plane. **B** Another same pair, which differs by an 80° rotation in depth. **C** A different pair.

b Graph showing that reaction times can be accurately predicted from the angular difference between the pair of test objects. The data suggest that observers were rotating representations of the objects at a constant rate. The dashed lines show that the reaction time for two objects that differed in orientation by 50 degrees was 2 seconds, and a difference of 150 degrees took just over 4 seconds. This graph is for picture-plane (i.e., non-depth) rotations, and a slightly shallower slope was obtained for 3D (in-depth) rotations. Redrawn with permission from Shepard and Metzler (1971).

as to how to search the collection of stored models for a good fit. It would be time-consuming to work through *all* the object models, one after the other, trying to find a good alignment match. If the task was recognizing a tiger one might well be eaten before a match was obtained.

To facilitate matching speed, one technique which has a long history in computer vision is to search the input representation for a few distinctive features (say those with non-accidental properties, **8.22**) and then use them to *index* into the store of

**F-table storing properties of the
features of the stored model**

Feature	Length L	Orientation θ (theta)
1	40	0
2	50	80
3	20	0
4	50	100
5	10	0
6	12	0
7	20	90
8	15	0

Stored model of a face mask

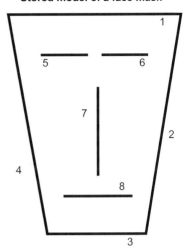

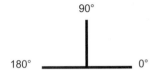

Convention used here for the angle of the rotation parameter r. Thus, a horizontal line feature has $r = 0°$.

A-table for counting votes for distortions of scale and rotation (s, r)

s \ r	0°	- - -	40°	- - -	90°	- - -	130°	- - -
- - -								
0.25								
- - -								
0.5								
- - -								
0.75								
- - -								
1.0								
- - -								
- - -								
- - -								
etc								

The matching problem

Is the input below a distorted instance of the stored model face mask (above)?

Test by calculating for each input feature in turn the (s, r) distortion between input and model.

Record the distortions found in the appropriate cell in the A-table.

Examples shown are for Model Feature 1 as a match for Input Features A and B:

Input Feature A as a candidate match yields $s = 20/40 = 0.5$ and $r = 40° - 0° = 40°$, a correct match.

Input Feature B as a candidate match yields $s = 10/40 = 0.25$ and $r = 130° - 0° = 130°$, an incorrect match.

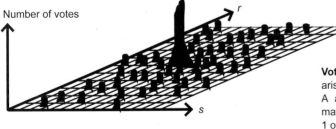

Voting for distortions arising from the parts A and B if they are matched with Feature 1 of the model.

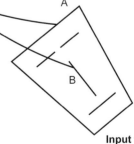

Input to be recognized
It is half the size of the stored model and it is rotated by 40°, so the accumulator cell with most entries is $s = 0.5$ and $r = 40°$ in this case.

Three-dimensional histogram of the cell counts in the A-table after all votes have been cast. Height codes frequency. One (s, r) cell has many more votes than all others. A clear spike of this sort is good evidence that the input is a distorted version of the model, justifying the conclusion of *Object Present*—in this example, that the input contains a distorted version of the stored face mask.

8.30 Hough transform used for matching

object models. This is much as one might use an individual letter to index into a word dictionary. Having found a limited set of candidate models in this way, then matching proceeds by seeing if the model can be "stretched," technically *aligned*, to fit other features in the input.

We describe this kind of indexing approach in Ch 21 where we set out the work of Roberts using object recognition as a way of solving the figure-ground problem in the blocks world. He showed that, having found a particular line junction that was a good cue for a particular kind of block, say a cube, then the search could guided to look for other cues by way of testing this "cube hypothesis." The idea of regarding *perception as hypothesis testing* has a long history in the study of perception. It is particularly associated with the great German physicist Hermann von Helmoltz (1821–1894), who used the phrase *unconscious inference* for this way of thinking about perception. Ch 13 provides details and discusses modern work using Bayes' theorem to implement this approach

Hough Transform for Overcoming Distortions

Let's begin by defining some terms. The input model is the representation created by the visual system from the retinal image. The task is to find out what this input model is by finding out which one, if any, of a collection of stored models it matches. Suppose the collection of stored models does contain the input model. Usually the input model differs from the correct stored model by virtue of one or more distortions. An example of a distortion would be an orientation difference between an input model of a T-structure and a stored model of a T. Size and position differences between input and stored models are other examples of distortions.

Suppose an orientation distortion existed because the observer viewed the T-in-the-scene with head held at an angle. We now make a critical observation. In a case such as this, both bars of the input model have the same distortion (here orientation difference) from the bars of the stored model T. That is, both bars have consistent distortions. This turns out to be a very useful constraint for solving the matching problem. We will describe how the consistent distortion constraint can be exploited by considering the example of matching a

input model of a face mask to a stored model face mask, **8.30**. The matching algorithm we describe is called the ***Hough transform***. It takes its name from its inventor Paul Hough, who patented it in 1962, but the generalized version we explain here is based on the work of Dana Ballard (1987). The term transform is appropriate because the key idea is to find out if the input model is a transformed version of one or other stored model. (Transform here is used synonymously with distortion.)

The first step in using the Hough transform is to define the stored model in a data structure we will call the ***F-table***, because in our example it is a list of features. Think of this table as a representation of the object to be recognized, whose primitives are the constituent features. Each feature is given a number to keep track of it. In **8.30**, the F-table specifies the orientation and size of various features of the face mask. To keep things simple, this mask is composed only of straight lines. The ***feature dimensions*** of the F-table in this example are those of orientation (θ) and size (here, line length L).

The next step is to create another data structure called the **accumulator table**, or ***A-table***. This is used to record the goodness of fit between an input model and the stored model coded in the F-table. An example of an A-table that suits the present example is shown in **8.30**. It has *distortion dimensions* of rotation angle r and size scaling s.

Each cell in the A-table represents a specific rotation and scaling transformation (s and r) which is applied to the input image. In essence, it is assumed that there exists exactly one pair of rotation and scaling transformation values which transforms the stored model into the input image. It is also assumed that exactly one of the cells in the A-table corresponds to this specific transformation. The job of the Hough transform method is to find this cell, and thereby, to find the transformation parameter values that map the stored model to the input image. In more detail, the A-table is used as follows.

Imagine that a line of certain length and orientation is found in the input image (say, as a result of processes computing the primal sketch, Chs 5, 6 and 7). This input line could be a distorted (transformed) instance of any one of the lines recorded in the F-table. That is, this input line could be rotated from the orientation of any given line logged in the F-table, and/or it could be a longer or shorter version of one of those lines. Such distortions often occur as we move our eyes and heads, even if an object stays stationary in the world.

What happens next can be described informally as follows: "I have been told the length of a particular feature in the primal sketch derived from the input image—for this feature its length L is 20 (arbitrary units). I am considering whether this feature could be a distorted version of a particular feature in the F-table, let's say Feature No. 1 in **8.30**, whose length is 40 units. Clearly, if this is a correct match, then the input is a shrunken version of the model. Specifically, the input feature is only half the length of Feature No.1 in the stored model. Hence to achieve a match, Feature No.1 has to be scaled down by dividing its length by 2, or equivalently multiplied by 0.5. That is, the size distortion s between input and stored models implied by this potential match is 0.5."

A more economical description is expressed in the following equations for the value of s that maps L_{input} to L_{stored}:

$$L_{input} = s \times L_{stored}$$

and so

$$s = L_{input} / L_{stored}$$

If there is no distortion present then $s = 1$, because in that case

$$L_{input} = L_{stored}.$$

Similarly, if the rotation value that maps the stored line orientation θ_{stored} to the input line orientation θ_{input} is r then

$$\theta_{input} = \theta_{stored} + r,$$

and so

$$r = \theta_{input} - \theta_{stored}.$$

In these equations, s and r are the size scaling and rotational distortion parameters, respectively, that would need to be applied to achieve a match between the image model feature line and the stored model feature in the F-table.

We have now arrived at the following position. We have evidence, from the hypothesized match between one line in the input image and one line in the F-table, that the input pattern we are trying to recognize could indeed be in the image but if so it has been scaled in size and rotated, with estimated values of the distortion parameters s and r.

We need to keep a record of this evidence. Accordingly, the appropriate (s, r) cell in the A-table is incremented by one, i.e., 1 is added to the number this cell contains. Counting up instances of estimated possible (s, r) distortions is why this data structure is called the accumulator table.

Another way of describing all this is to say that a "vote" has been logged in the A-table. The vote supports the claim that the stored model is present in the image but in a distorted version defined by the calculated values of the parameters s and r.

This process of logging votes in the accumulator has to be repeated many times, taking all possible feature matches into account. That is, every line recorded in the F-table for the face mask has to be considered as a possible match for the input line being considered. This is because, by definition, it is not known in advance which hypothesized match between input line and F-table line is correct, if indeed any is correct. And having voted for a specific pair (s, r) of distortion parameter values for the first input image line, the whole process has to be repeated for all the other input line features.

What happens next? After all the votes have been recorded, the A-table is inspected to see if any cell is a clear winner. That is, does any cell have a much bigger number in it than any other?

If the target pattern is present, the "peak" cell in the A-table will have at least as many votes as the number of feature matches considered. This is because the "winning" cell will receive a vote from every input line, as they will all have been distorted consistently if the input image is a version of the stored model.

If one cell clearly does have more votes than any other, the conclusion is drawn that there is a distorted version of the stored model in the input, and that it is distorted according to the parameters (s, r) associated with the winning cell.

On the other hand, if there is no clear winner then the opposite conclusion is drawn: pattern absent. The latter decision would be justified if the votes were to be "smeared out" over all the cells more or less equally, with no clear winning cell.

To illustrate all this for our very simple example, note that the input model in **8.30** contains 8 line features and there are also 8 entries stored in the F-table. As each input line is considered as a possible match for every F-table line, there will be $8 \times 8 = 64$ votes cast in all. Suppose a resolution for the A-table was chosen that allowed for 10 size scaling values and 36 rotation values. In this case, each possible match would be approximated to one of the $10 \times 36 = 360$ (s, r) combinations.

Only one of these (s, r) combinations would encode the correct distortion. The cell coding that combination would have 8 votes cast for it. This is because there are 8 input lines, each one of which would cast a correct vote.

Each input line would also generate 7 incorrect votes, as it is considered as a possible match for all the other lines. But these incorrect votes ('noise') would be smeared over the A-table in a random fashion, so that cells encoding incorrect distortions would get very few votes. Given that in this example 64 votes are cast over the 360 cells, the probability of any cell getting a vote at random is 64/360, less than one in five. Hence the 8 votes cast for the correct cell will make that cell stand out clearly from all the rest, **8.30**. If a clear-cut "spike" of this sort emerges, then it is safe to conclude: *Object Present*. If there is no such spike then the conclusion is: *Object Absent*.

In a typical computer system the large number of hypothesized feature matches would be considered serially (i.e., one after another). But in a biological brain, it is plausible to suppose that the matches are done in parallel. For example, suppose that there is one neuron serving as the "neurophysiological implementation" of each cell in the A-table. Votes to these cells could be implemented by excitation delivered by axons from other brain cells whose function is to calculate the (s, r) parameter values. We will not go into the details of neural circuits that might be designed to make this work. We simply draw attention to the way the Hough transform lends itself in principle to biologically plausible parallel implementations. This is why the Hough transform is very attractive for cognitive neuroscientists interested in what are called neural network or (equivalently) connectionist architectures.

The neat trick exploited in this use of the Hough transform algorithm is that the input model is recognized if and only if evidence can be found for all its visible features having been distorted consistently from a stored model. So in this matching algorithm, distortions are not treated as something to be got rid of, as in the approach of "normalizing" the input image which is sometimes done for template recognition schemes, Ch 3. Rather, the distortions are seen as something to be exploited. Clever.

Issues Arising from Using Hough for Matching

Readers who are uninterested in issues of detail arising from the above can skip this section.

(1) The customary use of the Hough transform is as an algorithm for parameter estimation where there is a need to integrate information from many sources of noisy data. In that standard usage, the actual values of the estimated parameters are required. We describe using the Hough transform in this way in Ch 19 for the purposes of working out where the two eyes are looking, which is an important sub-problem in stereoscopic vision. The Hough transform can also be used to implement the intersection-of-constraints algorithm for solving the motion aperture problem, Ch 14. The unusual but clever feature of using the Hough transform as an object recognition matching algorithm is that the estimated parameter values are ignored! All that is needed for making the decision *Object Present/Absent* is to find out whether there exists a set of parameters that show that all the features of the stored model have been distorted in the same way. The specific values of those parameters are irrelevant for the matching task.

(2) In the face mask example described above, only two feature dimensions were considered, size (here represented as line length) and orientation. But of course, many other kinds of distortions might have been present, due to the appearance variability problem, **8.3**. Each additional possible type of distortion could in theory be accommodated by extending the A-table to have an additional dimension, with one dimension for each distortion, and with one cell for each *combination* of distortion values.

The trouble with this approach is that it leads to the *combinatorial explosion* that we met in Ch 3 when considering receptive fields as templates for recognition. We set out there the general formula for finding the number of combinations of parameters (and hence cells needed) arising from this explosion: the total number $M = N^k$ where N is the number of cells per parameter, and the exponent k is the number of parameters. We examine this exponential formula in Ch 11 where we point out that as extra parameters are added it is quite easy to find that more neurons are required than are available in the brain. In the present case, a new parameter will arise for each feature dimension included in the F-table.

The use of the Hough transform for object matching cannot escape the combinatorial explosion. However, it does provide a convenient way of side-stepping this explosion by choosing the number of dimensions in the F-table and keeping this number as small as feasible. But ignoring a dimension comes at a price—the matching will be made less general. Whether a particular feature dimension matters in any given case will depend on the needs of the specific object recognition task in hand.

Our face mask example provides a case in point. We pruned away most feature dimensions, to illustrate the principles, and ended up with only two, size and orientation. But this very simple two-dimensional F-table would make this matching system prey to the illusion of concluding Mask Present if the features in the input were consistently distorted in length or orientation but randomly shuffled in position.

In short, this system would "see" both the mask-to-be-recognized and some sorts of "jumbled" masks as one and the same thing, **8.14**.

This illusion can be avoided by encoding the spatial relationships between the features in the F-table. But to do this would entail using extra feature dimensions, raising again the spectre of the combinatorial explosion. So the question is: just how many dimensions are sufficient for any specific system of object recognition? Again, we refer you to Ch 11 for a detailed examination of this question.

(3) In Ch 13, we also explain the role of Bayes' theorem in studying inferential processes in vision. John Porrill has demonstrated mathematically how the Hough transform can be interpreted as

*It's hard work digging clay
Save it for a rainy day.*

8.31 What is the last word in each sentence?
This word is written identically in both cases.

implementing Bayesian estimation of parameters (Porrill, Frisby, Adams, and Buckley, 1999).

(4) Note that the Hough transform is neutral with regard to the object recognition theory it is used to implement. It is "general purpose" in character. For example, it can be "tuned" to any dimensions of the primitives deemed critical to the theory of object recognition being implemented, be these primitives 3D generalized cylinders or 2D view-based feature lists. Thus, the Hough transform resides very clearly at Marr's algorithm level.

(5) You might be wondering at this point: can the Hough transform cope with cases where the input pattern is partially occluded? The answer is yes—as long as there are enough features visible for a clear peak to emerge in the A-table. That is all that is needed—there is no requirement that all features be present in the input.

(6) What about the figure/ground problem: does the Hough transform need to follow a stage where the pattern to be recognized is first isolated from its surround? The answer is no. Again, as long as there are enough features visible for a clear peak to emerge in the A-table then the process works. Indeed, using the Hough transform for pattern recognition is sometimes described as achieving segmentation-by-recognition. (Image segmentation is a term used in the computer vision literature for solving the figure/ground problem, Ch 7.)

(7) It is worth emphasizing again that one of the major attractions of the Hough transform is that it lends itself readily to implementation on parallel hardware, with huge benefits for processing speed. This fact about the Hough transform is particularly important to those who are interested in "natural computation," because a highly distinctive characteristic of biological visual systems is that they exploit immense parallelism. Presumably one reason they do so is to overcome the handicap of having to use slow computing elements. Neurons typically send impulses at rates well under

1000 impulses per second. The equivalent figure for today's solid state computer circuitry is many orders of magnitude faster (an order of magnitude is a technical way of saying ×10). Even cheap PCs now have cycle speeds measured in gigahertz (1G = a thousand million).

(8) The scope of the Hough transform is so wide that it must be emphasized that it is only within the context of a particular scheme that it can be evaluated. For example, it would be possible to design the F-table in **8.30** to allow variations within limits of the relative positions of the primitives in the image, or even subsets of them. Also, the process of incrementing the accumulator is amenable to all sorts of special-purpose weighting-of-evidence strategies that seem sensible in a particular application.

However, such choices should be guided by a principled task theory justifying them. That is, in Marr's terms, they should be guided by a satisfactory computational theory of the task in hand, lest the exercise becomes one of hacking up ad hoc "solutions" that sometimes work and sometimes not for no very clear reasons. This is why the Hough transform is, in Marr's recommended framework for studying complex information processing systems such as the human visual system, quite clearly an algorithmic tool. Its twin attractions are that it is capable of being tailored to suit the requirements of a potentially very wide range of computational theories, and that it is a biologically plausible candidate for modeling brain systems.

Bottom-Up and Top-Down

We have discussed this issue in several previous chapters. For example, we described how Marr's use of figural grouping principles sometimes got stuck in ways which could be relieved by using stored information in a top-down manner, Ch 7. We have also discussed the massive nerve fiber projections from higher visual brain centers to lower ones, such as the lateral geniculate nucleus, Ch 9. And we have also described Simon Thorpe's finding that we can recognize object categories very quickly, which then get elaborated with more processing, also in Ch 9.

We mentioned earlier in this chapter the important influence of *context* on recognition. On a theoretical note, we have made mention of the cur-

8.32 Can you spot the printing error?

rent unifying principle set out in Bayes' theorem, for using prior expectations to guide the interpretation of image data (explored in detail in Ch 13). All these issues point the same way: human vision is a system in which higher and lower levels are intimately engaged in a joint enterprise.

The role of conceptually driven processing in seeing is very evident in reading. Whether we see "clay" or "day" in **8.31** depends very much on the context set by the sentences within which the word appears.

Just how "locked in" to one given interpretation of text we can become is illustrated in the pamphlet cover of **8.32**. We leave you to discover for yourself the printing error which this cover contains. It is a quite genuine cover, proofread at various stages in its production by the Printing Unit of our University. Yet the error was missed by all who saw it until it was too late and it was widely distributed.

Indeed, the new co-author of this second edition of *Seeing* (Jim Stone) could not find the error without help from the author of the original edition (John Frisby).

Here we have a fine example of conceptually driven processing at work, dictating what is "seen," in this case at the expense of what is really there.

Concluding Remarks on Seeing Objects

The capacities of the human visual system to recognize an enormously wide range of different objects despite the massive problems of appearance variability is staggering. Current computer vision is nowhere near its level of competence. This chapter has set out the main themes in current research but there is a very long way to go in understanding how humans, and indeed many other animals, are so good at object recognition.

We do not yet even have an answer to the question: does human vision rely on view-based or object-centered volumetric representations? This question is the subject of contemporary lively controversy. It is perfectly possible that in grappling with the massive problems of appearance variability, human vision utilizes a whole set of processes, some using volumetric primitives in an object-centered coordinate system, others using a set of views comprised of 2D features of one or other kind. The future may tell us the answer. But it may not, because vision could be just too hard to be understood by the comparatively paltry non-visual human intelligence.

Further Reading

Ballard DH (1987) Cortical connections and parallel processing: structure and function in vision. In Arbib MA and Hanson AR (Eds) *Brain and Cooperative Computation* 563–621. *Comment* A clear and wide-ranging account of how the generalized Hough transform can be used to model neural computations in vision.

Biederman I (1987) Recognition-by-components: a theory of human image understanding. *Psychological Review* **94** 115–147. *Comment* Describes Biederman's highly cited, but by some highly criticised, recognition-by-components (RBC) structural description theory.

Bruce V and Young A (1998) *In the Eye of the Beholder: The Science of Face Perception.* Oxford University Press. *Comment* A fine popular science book that is an excellent place to start accessing the huge amount of research over the past three decades on that very special class of stimuli faces. It was written to accompany an exhibition on *The Science of the Face* at the Scottish National Portrait Gallery. Its copious illustrations include many of paintings and sculptures.

Bulthöff HH and Edelman S (1992) Psychophysical support for a two-dimensional view interpolation theory of object recognition. *Proceedings of the National Academy of Sciences of the USA* **89** 60–64. *Comment* A widely cited paper promoting a viewpoint-based theory of human object recognition

Bulthöff HH and Wallis G (2002) Learning to recognize objects. In Fahle M and Poggio T (Eds), *Perceptual Learning* MIT Press.

Hummel JE and Biederman I (1992) Dynamic binding in a neural network for shape recognition. *Psychological Review* **99** 480–517.

Hummel JE and Stankiewicz BJ (1996) An architecture for rapid hierarchical structural description. In Inui T and McClelland J (Eds) *Attention and Performance* XVI MIT Press: Cambridge MA 92–121. *Comment* This paper and the previous one report computational experiments on implementing Biederman's RBC theory.

Johansson G (1973) Visual perception of biological motion and a model for its analysis. *Perception and Psychophysics* **14** 201–211.

Koenderink JJ and van Doorn AJ (1977) How an ambulant observer can construct a model of the environment from the geometrical structure of the visual inflow. In Hauske G and Butenandt E (Eds.) *Kybernetik* 224–247 Oldenburg. *Comment* A seminal early paper introducing the concept of the view potential.

Marr D and Nishihara HK (1978) Representation and recognition of the spatial organization of three-dimensional shapes. *Proceedings of the Royal Society of London B* 200 269–294. *Comment* A classic paper that has had a major impact. Introduced a set of clear criteria for judging object recognition theories.

Mather G and Murdoch L (1994) Gender discrimination in biological motion displays based on dynamic cues. *Proceedings of the Royal Society of London (B)* **258** 273–278.

Porrill J, Frisby JP, Adams WJ, and Buckley D (1999) Robust and optimal use of information in stereo vision. *Nature* **397**(6714) 63–66 *Comment* Explains how the Hough transform can implement a Bayesian solution to a vision problem but it is hard going for non-mathematicians. The gist is that each cell in the A-table could be interpreted as the probability that the input matches the model given the transformation parameter values associated with that cell. Using this perspective the A-table is an approximation to the likelihood function and the cell which receives most votes corresponds to the maximum likelihood estimate (MLE) of the true transformation parameter values. See Ch 13 for further details on the Bayesian approach and its relationship to MLE.

Rashid RF (1980) Towards a system for the interpretation of moving light displays. *IEEE Transactions on Pattern Analysis and Machine Intelligence* **6** 574–581.

Quinlan P and Dyson B (2008) *Cognitive Psychology*. Pearson Education Limited. *Comment* Takes as its theme Marr's approach, applying it in the field of cognitive psychology and hence a good complement to the present book.

Shepard RN and Metzler J (1971) Mental rotation of three-dimensional objects. *Science* **171** (3972) 701–703.

Stone JV (1998) Object recognition using spatiotemporal signatures. *Vision Research* **38** 947–951.

Tarr MJ and Bulthöff HH (1998) Image-based object recognition. *Cognition* **67** 1–20. *Comment* A wide-ranging review of the strengths and weaknesses of the image/viewpoint-based theories of object recognition. A good starting point for grappling with the issues involved in trying to make image-based schemes cope with the requirement to recognize object classes.

Tinbergen N (1948) Social releasers and the experimental method required for their study. *Wilson Bulletin* **60** 6–52.

Ullman S (1996) *High Level Vision* MIT Press Cambridge MA. *Comment* An excellent survey. Particularly strong on computational theories.

Wallis G, Backus BT, Langer M, Huebner G, and Bülthoff H (2009) Learning illumination- and orientation-invariant representations of objects through temporal association. *Journal of Vision*, **9**(7):6, 1–8.

Yuille A and Halliman P (1992) Deformable templates. In Blake A and Yuille A Eds (1992) *Active Vision* MIT Press. Pages 21–38.

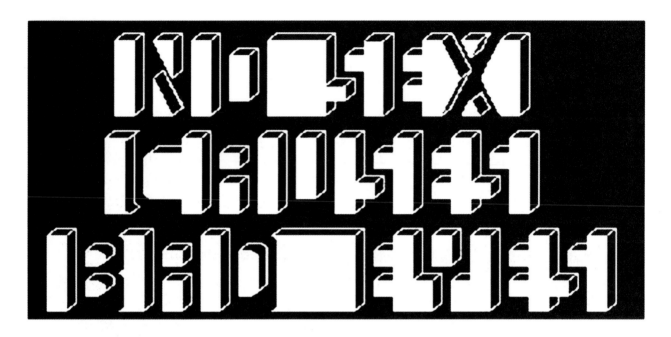

5.22 What is the hidden message?
Find out by screening up your eyes to achieve blurred
vision, and/or view from a few metres. Alternatively, see
the blurred version on p. 228. This figure is discussed
in Ch 5 on p. 131. We have been unable to trace the
creator of this figure, which has been widely circulated
unattributed on the Web. We will be pleased to cite the
source on this book's Web site if its origin becomes known
to us.

Seeing with Brain Cells

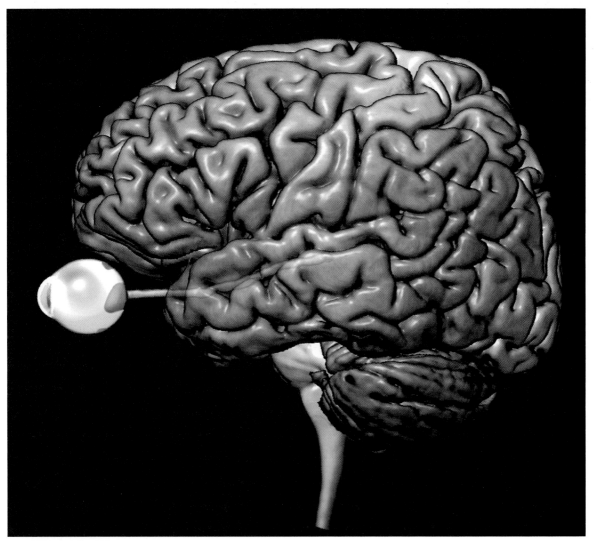

9.1 Eye and brain
Each eye contains about 126 million receptors. Each optic nerve contains only 1 million or so fibers carrying messages from the eye to the brain, which is why the retina can be thought of as engaged in an image compression task (see Ch 6). The brain's analysis of the information from the eyes entails a huge expansion of the number of cells involved. The brain contains about 100 billion neurons and it has been estimated that about 40% of the primate brain is involved in mediating seeing. This suggests that there are about 40 million brain cells involved in vision for each optic fiber.
Figure reproduced courtesy of Mark Dow.

In this chapter we begin with a brief overview of brain anatomy, concentrating on the parts devoted to vision. We then describe in some detail the structure and functions of *low level* brain regions that come first in the chain of processing sites that support seeing. We discuss in considerable detail in the next chapter the nature of the *sensory maps* that are widespread in the brain. We also deal there with brain regions that contain *high level* object representations.

Only a simplified account can be given in this book of how the brain is built and how it seems to work, and even this is given only for the parts of the brain most directly concerned with vision. The aim is to give a glimpse of what recent research findings have to say about how the visual world might be represented in brain tissue, and to discuss some of the clues these findings have provided about how to solve image-processing problems in general.

Overview of the Anatomy of the Brain

The human brain is an immensely complicated structure which sits tightly packed inside its protective casing, the skull. The brain's overall appearance is similar to that of a gray ish pink blancmange with a wrinkled skin, and its consistency is about that of a stiff blancmange, or of tofu. Somehow this superficially unimpressive blob, this pudding, mediates such complex processes as thinking, remembering, feeling, talking, walking, and seeing. The cells in the brain that carry out the information processing tasks required for these activities are called **neurons**. There are about 10^{11}

Why read this chapter?

We begin with an outline of the anatomy of the *visual pathway* from eye to brain. Evidence is described showing how the *simple* and *complex* cells found within the striate cortex are organized into columns. Each column is arranged perpendicular to the cortical surface, and the simple and complex cells within each column are optimally tuned to the same stimulus orientation. The so-called ice cube model of *hypercolumns* is explained, along with the results of a computer model of imaging processing by hypercolumns. The concerns of this chapter are largely pitched at the *hardware level* within Marr's framework for understanding complex information processing systems.

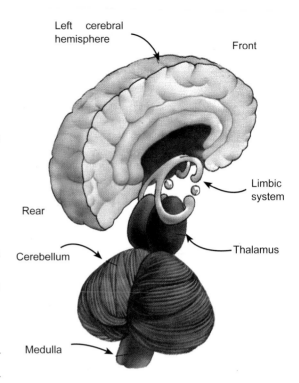

9.2 Exploded section of some main brain structures
The large structure at the top is the left cerebral hemisphere. The limbic system is concerned with emotions; see **9.3** for the functions of the other parts.

neurons in the brain (10^{11} = 100,000,000,000 = 100 billion; unpacking this number in this way may help you appreciate the very large number of neurons in the brain). As might be expected, the mysteries of brain function far outweigh the items of certain knowledge at the present time, even with regard to something as basic as knowing what task different brain components are doing.

Cerebral Hemispheres

The most prominent part of the brain, and the part you would see if you lifted off the top of the skull as though it were a cap, is the top of the left and right **cerebral hemispheres** (Latin: *cerebrum* = brain). These structures sit on top of many other brain structures, **9.1–3**. They contain the bulk of the brain's machinery for vision, although there are other important sites, as we shall see.

Communication between the cerebral hemispheres occurs via a neural information highway called the **corpus callosum**, which connects them.

The surface layer of each hemisphere is called the ***cerebral cortex*** (Latin *cortex:* "bark'). The cortex is about 3–4mm thick in man and its natural color is a gray ish pink but in the brain section of **9.4** the cortex has been specially stained so that it shows up as blue ribbon.

To visualize the structure of the cortex, imagine a soccer ball which has been deflated and crumpled up to fit inside the skull. If you now think of the crumpled skin of the soccer ball as the cortex then you will have quite a good idea of the latter's structure. Obviously, given this folded packaging of the cortex inside the skull, the cortex appears in cross-section as a multilayered structure, **9.4**, but in fact it is best thought of as an approximately spherical sheet with multiple folds.

We can work out how big this soccer ball would have to be in order to have the same surface area as average cortex, which is about 2500 cm². Given that the surface area of a sphere is equal to $4\pi r^2$ (where r is the sphere's radius) this implies that the soccer ball which has been deflated and squeezed into your skull originally had a radius of about 45 cm, **9.5**. This is very much bigger than a normal soccer ball and it is about three times bigger than the radius of the skull, which is about 15 cm in adults. Clearly, crumpling up the cortex avoids us having to carry around a very large head.

The cerebral tissue lying just beneath the cortex is the ***white matter***, so named because its natural color is white. It is composed of billions and bil-

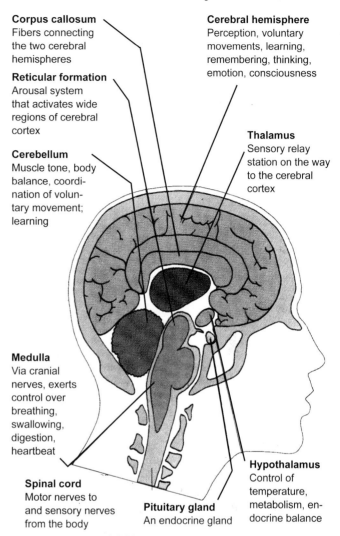

Corpus callosum
Fibers connecting the two cerebral hemispheres

Reticular formation
Arousal system that activates wide regions of cerebral cortex

Cerebellum
Muscle tone, body balance, coordination of voluntary movement; learning

Medulla
Via cranial nerves, exerts control over breathing, swallowing, digestion, heartbeat

Spinal cord
Motor nerves to and sensory nerves from the body

Cerebral hemisphere
Perception, voluntary movements, learning, remembering, thinking, emotion, consciousness

Thalamus
Sensory relay station on the way to the cerebral cortex

Hypothalamus
Control of temperature, metabolism, endocrine balance

Pituitary gland
An endocrine gland

9.3 Diagrammatic cross-section of the brain

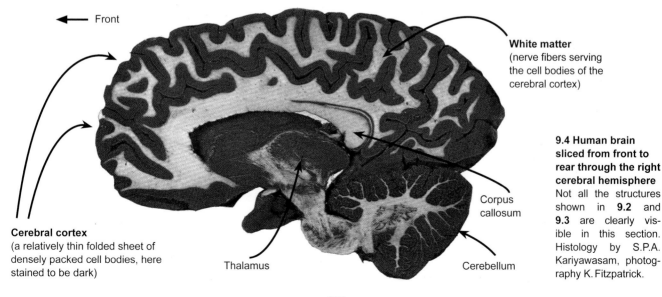

← Front

White matter
(nerve fibers serving the cell bodies of the cerebral cortex)

Cerebral cortex
(a relatively thin folded sheet of densely packed cell bodies, here stained to be dark)

Thalamus

Corpus callosum

Cerebellum

9.4 Human brain sliced from front to rear through the right cerebral hemisphere
Not all the structures shown in **9.2** and **9.3** are clearly visible in this section. Histology by S.P.A. Kariyawasam, photography K. Fitzpatrick.

9.5 The cortex as a crumpled sphere
Most of the surface area of the brain is hidden among its folded convolutions. If we were to remove the brain from the skull and blow it up to make a sphere with the same surface area as the brain then that sphere would have a radius three times bigger than the radius of the brain. This implies that the sphere would occupy a volume 27 times that of the human head.

lions of ***nerve fibers***. Each fiber can be thought of as a tiny cable which carries messages either from place to place within the brain or between the brain and the rest of the body. The cortex has fibers within it too, but its principal constituent is a huge number of ***cell bodies***. Cell bodies and nerve fibers are the two main parts of the type of brain cells called ***neurons***, **9.12**.

Visual Pathways to the Brain: Overview

The major visual pathway carrying the messages from the eyes to the brain is shown in broad outline in **9.1** and in fuller detail in **9.6** in which the eyes are shown inspecting a person. The locations of the various parts of this person "in" the striate cortex are shown with the help of numbers.

The first thing to notice is that the eyes do not receive perfectly equivalent images. The left eye sees rather more of the scene to the left of the central line of sight (regions 1 to 2**)**, and vice versa for the right eye (regions 8 to 9).There are also other subtle differences between the left and right eyes' images, called ***binocular disparities***, in the case of three-dimensional scenes. These are not shown in **9.6**. They are described fully in Ch 18 *Seeing with Two Eyes*, which deals with how differences between the images in the two eyes are used for the perception of depth.

Next, notice that the ***optic nerves***, which contain the nerve fibers (axons) emanating from the ganglion cells of each retina, join at the ***optic chiasm***. Some fibers within each optic nerve cross over at this point and therefore send their messages to the cerebral hemisphere on the side of the brain

opposite to where they originated. Other fibers stay on the same side of the brain throughout.

The net result of this crossing-over, or ***partial decussation***, of fibers is that messages dealing with any given region of the field of view arrive, after passing through structures called the ***lateral geniculate nuclei***, at a common cortical destination, *regardless of which eye they come from*. In other words, left- and right-eye views of any given feature of a scene are analyzed in the same location in the ***striate cortex***.

A very important fact about the way fibers feed into the striate cortex is that they preserve the neighborhood relations existing in the retinal ganglion cells. This results in a ***retinotopic map***, **9.6,** in each cortex. The map is stretched or magnified around the fovea, which reflects the increased density of striate cells dedicated to analyzing the central area of the retinal image. There is nearly a three-fold increase in the amount of cortical machinery dedicated to each foveal ganglion cell's output, relayed through the lateral geniculate nucleus, relative to the output of each ganglion cell in the periphery.

The brain mapping shown in **9.7** summarizes classic work done by Korbinian Brodmann (1868–1918), a neurologist who distinguished different brain regions based on differences in the sizes and types of brain cells seen under the microscope.

Another brain map is shown in **9.8**, based on the pioneering work of Wilder Penfield (1891–1977). Penfield drew up his map by observing the results of gentle electrical stimulation of the surface of the cortex, exposed during brain operations on

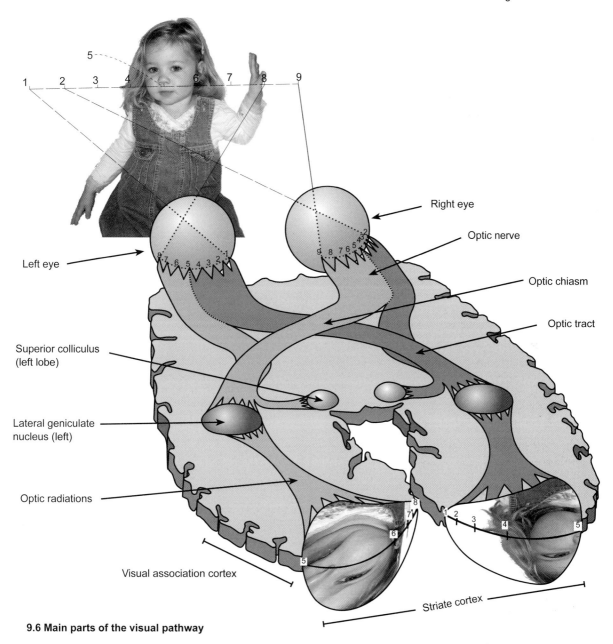

9.6 Main parts of the visual pathway

human patients to relieve epilepsy by destroying brain cells at the focus of the epileptic attack. The surgery was performed under local anesthetic and so the patient was awake to report on conscious sensations created by the stimulation. Electrical "tickling" of the striate cortex made the patient see swirling colored shapes. These shapes were perceived as "out there," somewhere in front of the patient's eyes, and not lodged inside his skull. This corroborates the idea of striate cortex (area V1) being concerned with early edge feature representa-

tions. Stimulation further forward produced much more complex visual sensations.

Indeed, in the inferotemporal cortex, **9.8**, stimulation sometimes caused patients to see whole scenes, complete with details of recognizable objects. So this early evidence supports the idea that the ***inferotemporal lobe*** might be concerned with the integration of the relatively primitive feature analyzes conducted in the early regions, such as V1. More recent work has supported this idea, as described in Ch 10.

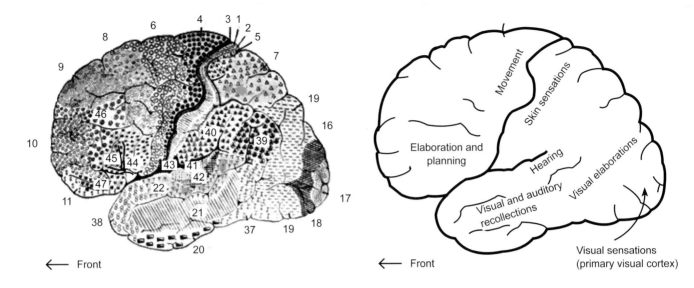

9.7 Brodmann's map of the left cerebral cortex
Each numbered region differs in the size and type of its nerve cells. Area 17 is the striate cortex, areas 18 and 19 are pre-striate cortex.

9.8 Sketch of mapping of cortical functions based on Penfield's brain stimulation experiments in the 1930s
Ch 10 describes modern evidence on the functions associated with different brain regions.

It is clear even from this brief overview that a large proportion of brain tissue is given over to seeing. This should perhaps not surprise us, given the dominant role which this sense plays in our lives. What continues to astonish us is the fact that the brain manages to mediate seeing at all. Most people take its scene representations, the visual world of which we are conscious, very much for granted. But how the brain produces such scene representations is deeply mysterious. Research has provided tantalizing glimpses of what is going on in the brain when we see but our ignorance far outweighs our understanding.

The Layers of the Lateral Geniculate Nuclei

The axons of the retinal ganglion cells leave the retina in a bundle of roughly 1 million fibers that make up each *optic nerve*. They pass through the optic chiasm and terminate on the dendrites and cell bodies of neurons in the left and right *lateral geniculate nuclei*. These sit in the right and left hemispheres and deliver their outputs to the striate cortex, V1. Each lateral geniculate nucleus has four important features.

First, each nucleus consists of 6 bilayers of cells, **9.9**. These are numbered 1 to 6. Each bilayer contains one main layer of cells plus a konio cell sub-layer. Each bilayer carries information from one eye. Thus, in the left LGN, bilayers 2, 3, and 5 receive input from the left eye, whereas bilayers 1, 4, and 6 receive input from the right eye.

Second, all of the layers of one lateral geniculate nucleus receive input from one half of visual space, **9.9a**. For example, even though bilayers 2, 3, and 5 of the left lateral geniculate nucleus receive inputs from the left eye, they only receive information from the left half of the retina in that eye, which corresponds to information from the right half of visual space.

Third, the cells in each bilayer are spatially organized in a retinotopic layout, so that the spatial layout of the retinal ganglion cells is preserved in each bilayer of the lateral geniculate nucleus. Put simply, ganglion cells which are close to each other in the retina have axons which project to points which are close to each other in a single bilayer of the lateral geniculate nucleus. The retinotopic map is magnified around the fovea, which reflects the increased density of ganglion cells dedicated to analyzing this area of the retinal image.

Fourth, different bilayers of the lateral geniculate nucleus are in register with each other, so that the retinotopic map in each bilayer is aligned with that of neighboring bilayers. Bear in mind, how-

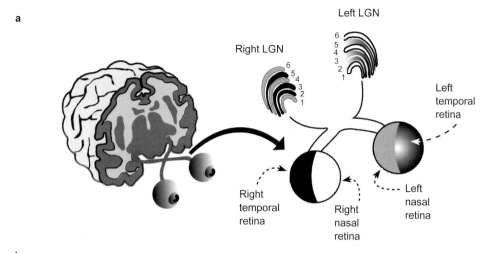

a

Left LGN

Right LGN

Left temporal retina

Right temporal retina

Right nasal retina

Left nasal retina

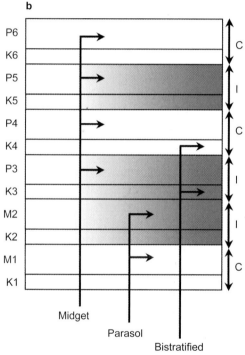

This block diagram is a schematic expansion of the left LGN layers shown in the upper right part of **a**.

9.9 Layers in the lateral geniculate nucleus (LGN)

a Projections from both eyes. Different cell types in both retinas project to specific layers in LGN (see text). The nasal side of retina is on the same side as the nose, opposite to the temporal side of the retina. Redrawn with permission from Bear, Connors, and Paradiso (2001).

b Information from the eye on one side of the head projects to lateral geniculate nucleus layers 2, 3 and 5 of the on the same side of the head (i.e., the ipsilateral LGN), and to layers 1, 4 and 6 of the contralateral LGN. Ipsilateral and contralateral are labeled as I and C. The six konio layers are labeled K1–6, the two magnocellular layers as M1–2, and the four parvocellular layers as P3–6. Reproduced with permission from Mather (2006).

Inputs from Retinal Ganglion Cells to Lateral Geniculate Nuclei

To understand further the structure of the lateral geniculate nuclei requires knowing about the four types of retinal ganglion cell. This is a topic we postponed from Ch 6 to integrate it with the present account.

Midget retinal ganglion cells are the most common type, making up 80% of the total. They have the on/off receptive fields described in Ch 6. They receive most of their input from cones and, like cones, respond with a sustained output but only to high levels of illumination, so they are thought to respond primarily to static form. The midget ganglion cells project (a technical term meaning "send their fibers") to the *parvocellular layers* (Latin: *parvo* = small) of the lateral geniculate nucleus, that is, into bilayers 3–6 in **9.9**, so they are the inputs for roughly 67% of cells in the lateral geniculate nucleus.

ever, that each lateral geniculate nucleus bilayers encodes a different aspect of the retinal image. In effect, each lateral geniculate nucleus contains 12 copies of half of the retinal image (two per bilayer), where each layer encodes a different aspect of the retinal image. Thus, each lateral geniculate nucleus is complex but it is laid out in a very orderly way.

Parasol retinal ganglion cells make up 8–10% of the total. They also have on/off receptive fields. They receive most of their inputs from rods, and like rods, they respond, in both high and low light conditions, quickly and transiently, so they are thought to respond primarily to motion related information. Parasol ganglion cells project to the **magnocellular layers** (Latin: *magno* = large) of the lateral geniculate nucleus, that is, into bilayers 1–2 in **9.9**.

Bistratified retinal ganglion cells make up less than 10% of the ganglion cells in the retina. They respond to short wavelength (blue) light by increasing their firing rate, and to middle wavelength (yellow) light by decreasing it. They project to konio sub-layers 3 and 4 and their response characteristics mirror the blue-on/yellow-off behavior of these ganglion cells. We discuss color vision in Ch 17. The precise source of retinal inputs to other konio layers is not known.

Biplexiform retinal ganglion cells make up less than 10% of all ganglion cells, and are something of a mystery. They are the only type of ganglion cell that connect directly to the receptors (rods, in fact). Each has only an on-center receptive field, and it is thought these cells provide information about ambient light levels.

Feedback Connections to the Lateral Geniculate Nucleus

Surprisingly, only about 10% of the inputs to lateral geniculate nucleus come from the retina! About 30% originates from the place to which the lateral geniculate nucleus delivers its output, the striate cortex, and other inputs arrive from **midbrain** structures such as the **superior colliculus** (see below). Clearly, the lateral geniculate nucleus is not just a relay station for passing messages on from the eyes to V1.

Of these various inputs to the lateral geniculate nucleus, the most puzzling come from the striate cortex. Given that the majority of the output from the lateral geniculate nuclei goes to the striate cortex, it is hard to conceive of the inputs from striate to the lateral geniculate nucleus as anything other than **feedback** connections.

This fits in with evidence showing that striate cells modulate the responsiveness of lateral geniculate nucleus cells. However, precisely what the feedback connections are for is the subject of much research, debate, and confusion. We discuss next one hypothesis that has caused considerable interest in recent years.

Categories First and Fast: Details Later

Simon Thorpe and others have suggested that the brain uses its **feedforward** connections in the retina → lateral geniculate → striate pathway to perform a "quick and dirty" analysis, and then uses its **feedback** connections to refine its interpretation of what is in the retinal image. The feedforward connections transmit information through the visual system in about a tenth of a second and are sufficient to permit a crude categorization of the object or scene present (such as: "It's an animal" or "It's a beach"). Thorpe has demonstrated that people can make such judgements from very briefly flashed images. His proposal, along with others who have pursued similar ideas, is that this fast crude categorization then helps, via the feedback connections, the relatively leisurely analysis of the details in the image.

This certainly makes sense in evolutionary terms because it is better to guess that the blurry streak before you is a snake after a tenth of a second than it is to be certain that it is merely a piece of falling ivy after two tenths of a second.

Van Rullen and Thorpe demonstrated the feasibility of this proposal by creating a computer simulation of image processing early in the visual pathway. This is illustrated in **9.10** and explained further in its legend.

The uncertainty surrounding the role of feedback connections extends beyond the striate cortex. This is because, as a rule of thumb, if brain region A connects to region B then region B usually connects to region A. Indeed, a curious fact about brain connectivity in general is that only a select group of brain areas are not connected with each other!

One recent report has used a brain imaging technique, described in Ch 10, to demonstrate that although lateral geniculate cells are all monocularly driven as far as excitation is concerned, the neighboring layers responding to the left and right eye's inputs can inhibit one another. This may be important in **binocular rivalry**, a phenomenon we discuss in Ch 18.

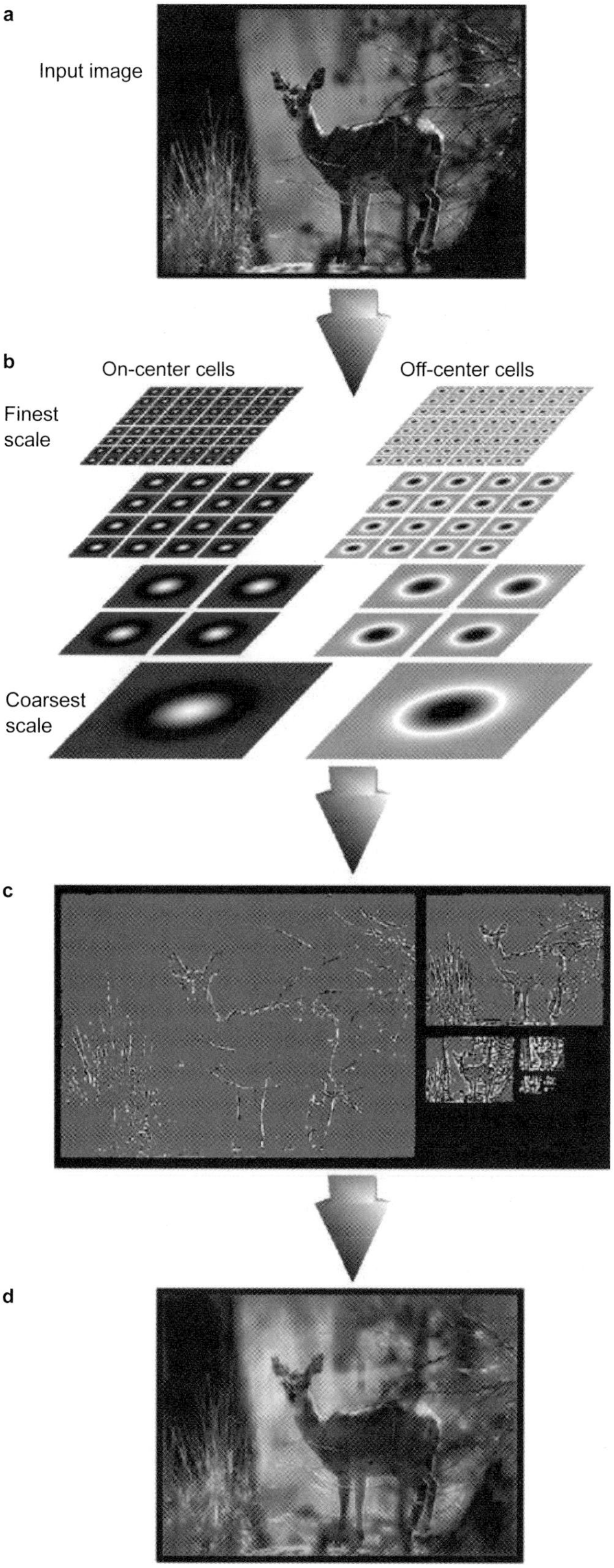

a

Input image

b

On-center cells   Off-center cells

Finest
scale

Coarsest
scale

c

d

All in all, the functions of the lateral geniculate nuclei remain largely mysterious despite decades of research. What computational task is this gloriously precise and intricate structure fulfilling? The answer isn't yet known.

The Superior Colliculus

Before we go on to discuss the way fiber terminations are laid out in the striate cortex, observe that the optic nerves provide visual information to two other structures shown in **9.1** and **9.6**—the left and right halves of the *superior colliculus*. This is a brain structure which lies underneath the cerebrum and it seems to serve a function different from the cortical regions devoted to vision. The weight of evidence at present suggests that the superior colliculus is concerned with what might be termed guiding *visual attention*. For example, if an object suddenly appears in the field of view, mechanisms within the superior colliculus detect its presence, work out its location, and then guide eye movements so that the novel object can be observed directly with the full visual processing power of central (straight ahead) vision. The latter detailed examination seems to be the special role of the visual machinery of the cortex.

The Striate Cortex

Enlarged views of slices of monkey striate cortex are shown in **9.11a,b**. Both sections come from the same region of the cortex, but they are stained differently to emphasize different features. Thousands of small blobs can be seen in **9.11a**. Each

9.10 Model of fast analysis of an image by center surround retinal ganglion cells
The image in **a** is analyzed at eight spatial scales by banks of on-center and off-center cells whose receptive fields are shown schematically **b** (only four scales shown here). The responses of these cells are depicted in **c**, where each of the eight convolution images is associated with one spatial scale. These responses can be recombined to derive an estimate of the original image, as shown in **d**. This simulation shows the system after only 5% of the model center-surround retinal ganglion cells have fired, approximately 50 ms after the image has been presented to the model retina. **d** shows that sufficient information for recognition can be extracted within a very short time interval by a realistic model of early visual processing in mammalian visual systems. Courtesy Van Rullen and Thorpe (2001).

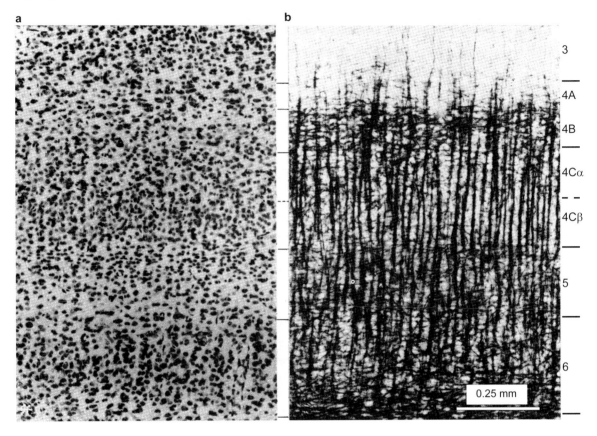

9.11 Monkey striate cortex
Microscopic enlargements of sections stained to emphasize cell bodies in **a** and nerve fibers in **b**. The numbers at the right refer to various layers that can be distinguished microscopically. Courtesy LeVay, Hubel, and Wiesel (1975)**.**

one is an individual cell body of a neuron—the magnification is much less than in **9.12**, which shows just one cell.

If we were to fly a plane through a giant model of the visual cortex then we would soon become aware to its three main organizing principles: ***horizontal layers***, ***vertical columns***, and ***retinotopy***.

In the top layer, we would encounter the dendritic trees of tall ***pyramidal cells***, **9.11b**. Further down, we would encounter the tangle of invading axons from the lateral geniculate nucleus, and, at the bottom, the axons of the pyramidal cells as they exit the cortical layer to deliver information to other brain regions.

As we explore more widely, we would observe that the neurons are arranged in vertical columns. Each column seems to be a modular information processing device dedicated to analyzing one small region of the retinal image for such things as edge orientation, edge width, and color. This small region can be considered as the column's receptive field, which is the combination of all receptive fields of neurons in a given column.

Finally, we would notice that a retinotopic version of the retinal image is laid out across the cortex, in a similar manner to that found in each layer of the lateral geniculate nucleus, with more cells devoted to central than to peripheral vision, **9.9**. The nature of this distortion ensures that if we could keep track of our progress across the retina as we flew across the cortex then we would make little "retinal progress" for each mile of cortical mile traveled when our retinal track was in center of the retina. In contrast, we would make good "retinal progress" for each cortical mile when our image track was at the periphery of the retina.

Having made these important discoveries about the global organizing principles of the cortex, we would be forced to abandon our plane, and to explore further using less leisurely methods.

Layers in the Striate Cortex

The striate cortex is organized into 6 layers. These are relatively thick bands of tissue which have subtly different anatomical properties. The number and letter codes shown on the right in **9.11b** are the technical labels for the various layers.

We will not describe the details here, although it is interesting to note that different layers of the lateral geniculate nuclei terminate in different sub-layers of cortical layer 4. The magnocellular cells terminate mainly in layer 4Cα (4C alpha) and the upper part of layer 9. The parvocellular layers terminate mainly in layer 4Cβ (4C beta) and the lower part of layer 6, and the konio cells terminate in layers 4A, 2, and 3.

Layer 4B is special in that it does not receive incoming fibers, but instead contains a dense network of horizontally-running fibers. This network is visible with the naked eye in brain sections, where it appears as a white stripe. This stripe, called the ***Stripe of Gennari*** after its discoverer, is the structure that gives the striate cortex its name (Latin *strea* = "a fine streak").

Columns in the Striate Cortex

Very different impressions of the anatomy of the striate cortex are given in **9.11a,b**. The section in **9.11b** shows stained to emphasize fibers rather than cell bodies. In fact, the dark stripes running from top to bottom are bundles of several fibers. The predominance of fibers running vertically emphasizes the columnar organisation of the striate cortex.

Look back and consider **9.11a** once again. Previously you probably saw it as undifferentiated, perhaps rather like coarse sandpaper. But now look more carefully and you will be able to distinguish columns of cells, somewhat like strings of onions. It can be helpful in finding them to hold the book almost flat in front of you so that you are looking "up" the columns, as it were, a trick which helps make them discernible. The bundles of fibers shown in **9.11b** deal with individual columns of cells, taking messages up and down the column from cell to cell. Fibers running horizontally also exist, connecting cells in different columns, but these fibers are less striking in **9.11b** than the vertical ones.

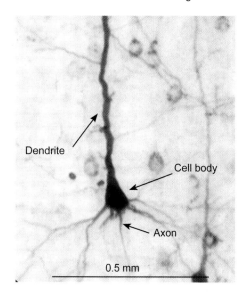

9.12 Microscopic enlargement of a slice of rat brain stained to show a large neuron called a pyramidal cell The long thick fiber is a dendrite that collects messages from other cells. The axon is the output fiber. Courtesy P. Redgrave.

Incidentally, your perception of **9.11a** is now probably quite different from what it was initially. You have "learned" how to see it, helped partly by the information given in the text and partly by your continued re-examination of it. This is a fine example of the enduring consequences from a perceptual interpretation to add to those described in Ch 1 (e.g., the Dalmatian dog example).

Recording from Single Cells in Striate Cortex

The first important property of striate cells discovered by ***single cell recording*** (Ch 3) is that each cell is concerned just with a limited patch of retina, its ***receptive field***. The receptive fields of cells in adjacent columns correspond to adjacent regions on the retina. This is entirely as expected, given all that has been said so far about the retinotopic mapping in the striate cortex—the way neurons from each area of the retina connect up to a corresponding area in the cortex in an orderly way that preserves neighborhood relationships, by and large.

In keeping with this neuroanatomy are results obtained when the microelectrode is driven "vertically" into the cortex perpendicular to its surface, so that the microelectrode passes from one cell to another within a given column. If recordings are taken from each cell encountered in turn then it is

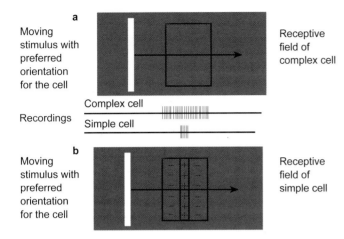

9.13 Complex and simple cells compared
a Complex cell: responds vigorously wherever the line falls on the field;
b Simple cell: responds only when the position of the stimulus fits the excitatory and inhibitory regions of the field.

found that they all have their receptive fields in the same general region of the retina. The whole column of cells is therefore "looking" at the same part of the retinal image (although there is some scatter in the locations of receptive fields). For what is each cell looking?

We described the properties of ***simple cells*** in Ch 3 when we introduced the concept of the receptive field. The hallmark of the simple cell is that its receptive field has separate and adjacent excitatory and inhibitory sub-regions, the effects of which add up comparatively straightforwardly. This means that the simple cell is very demanding

about the position in which a stimulus falls on its receptive field. But other types of cell without this property are to be found. Indeed, it is sometimes said that the variety of receptive field types is so large that if you want to find an edge-driven cell with a particular receptive field then you have a pretty good chance of finding one! Despite this precautionary caveat, we will now describe some classic cell types identified by Hubel and Wiesel in their Nobel Prize-winning research.

The distinguishing feature of a ***complex cell*** is that its receptive field cannot be mapped into segregated excitatory and inhibitory regions. Complex cells are, however, similar to simple cells in that their optimal stimulus must still be either a line, a slit or an edge, and it must still suit the orientation tuning of the complex cell in question (as for simple cells, complex cells are also found with different orientation preferences). But it matters little where the optimal stimulus falls in the receptive field of a complex cell, as **9.13a** makes clear. In that figure, a moving line is shown sweeping across the receptive fields of both a complex and a simple cell, to illustrate the difference between them. The moving line creates a very vigorous response in the complex cell as soon as it appears on the field, and this response is maintained until it leaves the field, having moved right over it. The simple cell, on the other hand, responds only when the moving line happens to "fit" the excitatory and inhibitory regions, and so its response is crucially dependent on the exact stimulus position, **9.13b**.

9.14 Orientation tuning of a complex cell

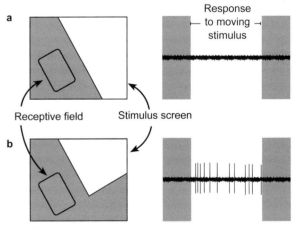

9.15 Hypercomplex cell
a This cell does not respond when a long white edge moves across its receptive field but it does respond when the edge is shortened to form a corner, **b**.

Note that excitatory and inhibitory regions are not shown in **9.13** for the complex cell because its field cannot be plotted out in that way. This is why the discoverers of these cells, Hubel and Wiesel, chose the label *complex* to reflect what they saw as the relative complexity of the analysis performed by these neurons.

The fact that complex cells are orientationally tuned, just like simple cells, is illustrated in **9.14**, which shows how a complex cell tuned to horizontal fails to respond if the stimulus is vertical.

Another kind of cell discovered by Hubel and Wiesel has an even more complex receptive field, which is why they called them ***hypercomplex cells***. They are similar to complex cells in having receptive fields which cannot be mapped into excitatory and inhibitory sub-regions, and similar also in preferring moving stimuli. But hypercomplex cells are distinguished by their added selectivity for stimulus length. That is, they respond poorly if the line stimulus, be it slit, line, or edge, is made too long at one end or both. Thus, the best stimulus for them is either a bar of defined length, or a corner, as in **9.15**. This is why they are often called "end stopped" cells. Again, we have little

idea at present about what hypercomplex cells are for or indeed whether they do amount to a clearly distinct category at all.

The figures in **9.13–15** are schematic illustrations intended to convey the main receptive properties of various cells found in the striate by Hubel and Wiesel. To complement these figures, **9.16** shows a figure published in the *Journal of Neuroscience* in 1998 by Judith Hirsch and colleagues of some neurophysiological data recorded from a simple and a complex cell.

One idea that has been much debated is that the receptive fields of complex cells can be predicted by supposing that they receive their inputs from a set of simple cells whose receptive field locations are suitably located, as shown in **9.17**. However, controversy now surrounds this idea, sparked by evidence that suggests that at least some complex cells do not receive inputs from simple cells. Also, recent work has suggested that it is much more difficult to draw hard and fast boundaries between the simple and complex cell types than originally thought. As with so much about the neurophysiology of vision, the questions far outweigh the answers at present.

a Simple Cell
The square marked out with dashed lines shows a region on the retina in which a flashed bright square (on a dark surround) caused a burst of impulses in the recording trace. The flash duration is shown by the short horizontal bar beneath the trace. A dark square (on a light ground) flashed in this region did not trigger any impulses, indeed it caused inhibition shown by the dip in the trace, a dip called a *hyperpolarization*. If a bright stimulus was flashed in the dark bar-shaped region to the upper left of the high-lighted on-region then the cell was also inhibited (responses not shown here). Thus, this is a simple cell because its receptive field can be mapped into excitatory and inhibitory regions.

b Complex cell
The receptive field for this particular complex cell was constructed from responses to dark stimuli because bright squares had limited action. The responses evoked by dark squares falling in the peak of the field (dotted square) were small, brief *depolarizations* (bottom right) that led to impulses. Thus, this is classed as a complex cell because its receptive field cannot be mapped into *both* excitatory and inhibitory regions.

Note in both **a** and **b** the impulses that occur long after the end of the flashed stimulus, which may be background noise—the *maintained discharge*.

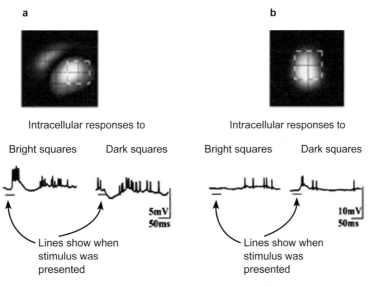

9.16 Responses of simple and complex cells in layer 6 of striate cortex
Adapted from Hirsch, Gallagher, Alonso, and Martinez (1998).

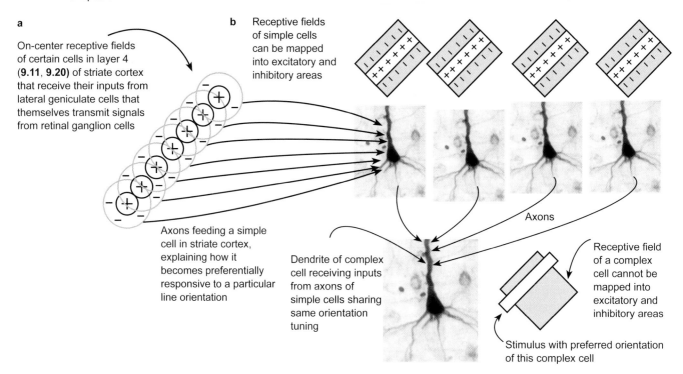

a On-center receptive fields of certain cells in layer 4 (**9.11**, **9.20**) of striate cortex that receive their inputs from lateral geniculate cells that themselves transmit signals from retinal ganglion cells

b Receptive fields of simple cells can be mapped into excitatory and inhibitory areas

Axons feeding a simple cell in striate cortex, explaining how it becomes preferentially responsive to a particular line orientation

Dendrite of complex cell receiving inputs from axons of simple cells sharing same orientation tuning

Axons

Receptive field of a complex cell cannot be mapped into excitatory and inhibitory areas

Stimulus with preferred orientation of this complex cell

9.17 Hypothetical wiring diagram for hierarchy of cells from lateral geniculate nucleus to cortex
a Shows how the response properties of simple cells can be explained by them receiving their inputs from lateral geniculate cells whose receptive fields are positioned on the retina in a line.
b The responses of complex cells might derive from them receiving their inputs from simple cells whose receptive fields fall on adjacent retinal regions.

Orientationally Tuned Columns

The single unit recording results shown schematically in **9.18a** were obtained by Hubel and Wiesel from a single unit microelectrode penetration which traversed quite a large region of striate cortex. Each dot on the microelectrode track represents a cell from which recordings were taken. The optimal stimulus for each cell was found in turn, beginning with one found closest to the point on the cortical surface where the microelectrode entered the brain.

In **9.18a** just the orientation tuning of each cell is illustrated (and not whether it was an "edge," "slit," or "line" detector) by the lines drawn through each dot.

It can be seen that, in the particular penetration illustrated in **9.18a**, the microelectrode at first stayed perpendicular to the surface of the cortex. The key thing to note is that the cells in this part of the track all share the same orientation preference, in this case a preference for vertical. So what this

track shows is a key feature of the organization of the striate cortex: *orientationally tuned columns*.

After the electrode in **9.18a** tracks further into the brain, it leaves the first column it entered and traverses a fold of cortex non-perpendicularly. In this region of the track, the microelectrode found cells with many different orientation preferences. Remember, each dot on the track represents the orientation of the preferred stimulus of a single cell. As can be seen, the orientation tuning of successive cells changed in a fairly regular progression as the electrode traversed over neighboring columns. Some parts of the non-perpendicular track record show successive cells with the same orientation preference: this is because for some short distance the electrode happened to stay within a column. Note that the electrode path is itself always straight, but its angle to the cortical surface changes as it penetrates the brain because the cortex is folded, **9.4**.

To summarize, when the electrode stays within a single column, cells are found to have the same ori-

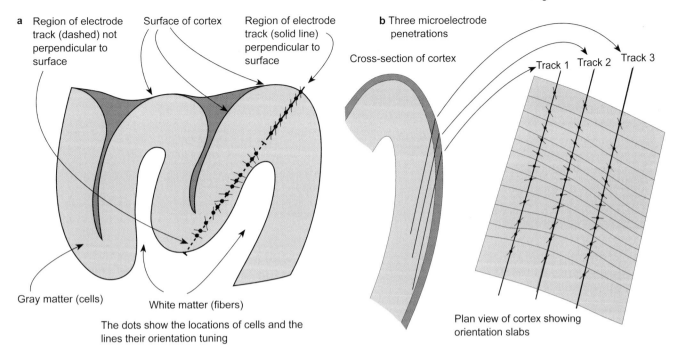

a Region of electrode track (dashed) not perpendicular to surface

Surface of cortex

Region of electrode track (solid line) perpendicular to surface

Gray matter (cells)

White matter (fibers)

The dots show the locations of cells and the lines their orientation tuning

b Three microelectrode penetrations

Cross-section of cortex

Track 1 Track 2 Track 3

Plan view of cortex showing orientation slabs

9.18 Results from microelectrode preparations revealing orientation tuned columns in striate cortex
Adapted with permission from Hubel (1981).

entation preference. But when the electrode starts to pass from one column to the next, cells vary systematically in their orientation preference. This is emphasized in **9.18b**, which provides a summary from three tracks displaced slightly laterally across the cortex.

The only exception to the rule that all cells in a column share the same orientation preference occurs in layer 4, in which the cells are not orientationally tuned at all. Some parts of layer 4, it will be remembered, receive input fibers from the lateral geniculate nucleus, where the orientational analysis has not yet got under way. Note that lateral geniculate cells, like those of the retina, are not orientation tuned: orientation selectivity is a property exclusively of cortical cells in the mammalian visual cortex.

Seeing Vertically Tuned Orientation Columns

One particularly remarkable experiment performed by Hubel and Wiesel, working with a colleague called Michael Stryker, involved an anesthetized macaque monkey whose open eyes were exposed to a pattern of vertical stripes continuously for a period of 45 minutes. The stripes were of irregular width, they filled the whole field of view of each

eye, and they were moved about to make sure that all the vertical line detectors of the monkey's striate cortex would be very active throughout the period of observation.

The monkey was injected just prior to this period with a special chemical which has the property of being much more readily absorbed by active brain cells than by inactive ones. Consequently, during the 45-minute exposure period, the active vertical line detectors excited by the vertical stripes would be absorbing this chemical from the monkey's bloodstream in much greater quantities than other striate cells tuned to non-vertical orientations, which would not be active during exposure to the stimulus.

Following the stimulus presentation, the monkey was instantly killed, and a microscopic examination made of where in the striate cortex the special chemical had collected. Exactly as expected, given the neurophysiological data already described, columns of brain tissue containing the chemical could be identified in slices of the striate cortex. Thus, in **9.19** the labeled dark bands show the regions of high chemical uptake, and they are arranged in columns perpendicular to the cortical surface. These bands therefore signify columns

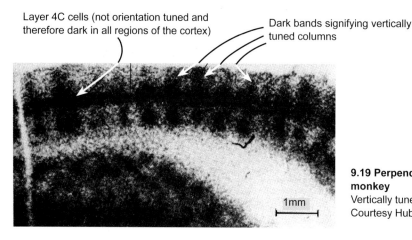

Layer 4C cells (not orientation tuned and therefore dark in all regions of the cortex)

Dark bands signifying vertically tuned columns

1mm

9.19 Perpendicular section through striate cortex of a monkey
Vertically tuned columns are clearly visible, see text.
Courtesy Hubel, Wiesel and Stryker (1974).

of vertically tuned cells which were constantly active during the period of exposure to vertical stripes. The bands are quite broad, and certainly broader than the width of a single column, because each striate cell responds not only to its preferred oriented line stimulus but also to orientations on either side of this optimum (refer back to Ch 3, for a reminder on this point). Consequently, cells with optimal orientations to one or other side of vertical would have been activated to some extent by the vertical stripes and would therefore show up in **9.19** as part of the dark band of active cells.

The bright bands between the dark ones signify columns tuned to orientations around horizontal (say 40–60 degrees), columns which would not have been activated to any appreciable extent by the vertical stripes and so would not have absorbed much of the special chemical.

The dark band cutting through the dark columns and running parallel to the cortical surface in **9.19** shows active cells in layer 4C. These cells are not orientationally tuned and so all of them would have been triggered to some degree by the vertical stripes. Hence, all the cells in this layer appear dark.

Figure **9.19** showed a perpendicular section of the experimental monkey's striate cortex. Also of interest is the view obtained by the non-perpendicular section shown in **9.20**, obtained by slicing almost parallel to the surface of the cortex. Here the active columns do not figure as dark bands as they did in **9.19**, but rather as dark blobs. This is because the columns are now seen in cross-section. In fact, each dark blob represents vertically tuned columns (plus orientation neighbors, of course,

also activated to some degree by the vertical stripes, as just pointed out in relation to **9.19**). Equally, light blobs show inactive columns tuned to within 40–60 degrees or so of horizontal and therefore not excited by the vertical stripes. The whole section presented by **9.20** is thus strong evidence in favor of a distribution of orientation columns of all types over the cortical surface.

Driving an electrode across the cortical surface provides a cross-section of the orientation preferences of all the cells encountered. However, every

The dark band is layer 4C (not orientation tuned)

6

Each dark blob is a cross-section through a vertically tuned column

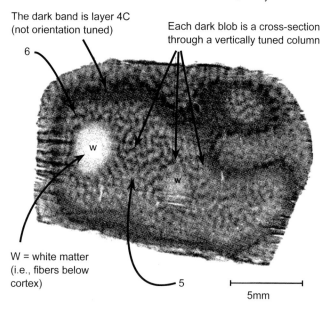

W = white matter (i.e., fibers below cortex)

5

5mm

9.20 Section through the same region of cortex as in 9.19 but now cut approximately parallel to the cortical surface
The section cuts through several layers of cortex, labeled according to number (4C, 5, 6). Courtesy Hubel, Wiesel, and Stryker (1974).

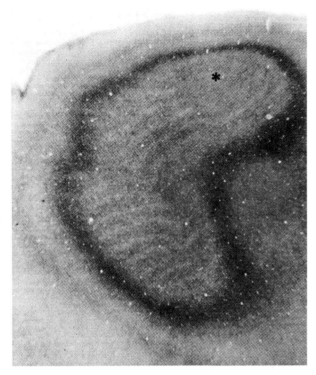

9.21 Section through striate cortex at an angle to its surface
Stained to show bands of left- and right-eye dominant cells. The asterisk marks a region where the section grazes layer 5, where the bands are not visible. The dark staining ring is the Stripe of Gennari (layer 4B) and within this is a region of layer 4C in which the bands can be seen. Reproduced with permission from LeVay, Hubel, and Wiesel (1975).

so often the orientation preference of cells encountered by an electrode "jumps" discontinuously. We have a lot to say about such points, called *singularities*, in the next chapter, Ch 10. We will also describe there where cells are to be found that a tuned to edge features other than orientation, such as color and edge scale (the latter is technically called *spatial frequency*, Chs 4 and 5).

Ocularity

The kind of technique for labeling active tissue explained above has been used to equal effect for discovering the layout of cells that show preferences for inputs from one or other eye. Indeed, Sokoloff, who developed this procedure for labeling active brain tissue, first used it for this very purpose. In his experiment the anesthetized monkey views a richly patterned stimulus which would stimulate almost all line detectors in the cortex,

but the monkey views it with only one eye. The microscopic appearance of the striate cortex upon subsequent examination was found to be rather like that shown in **9.19**, but with the alternating dark and light bars signifying alternating columns of cells, each preferentially tuned to one eye.

It was discovered, however, by Hubel and Wiesel and a colleague, Simon LeVay, that the monocular dominance stripes can be seen directly under the microscope using ordinary staining techniques, without need of the enhancement provided by Sokoloff's procedure for labeling active brain tissue.

It turns out that fibers parallel to the surface of the cortex in layer 4C are denser within the monocular stripes than between adjacent stripes. This means that a suitably stained section of the striate cortex which cuts through layer 4C parallel to the cortical surface reveals alternating dark bands ("fiber-dense": monocular stripes) and light bands ("fiber-sparse": boundaries between monocular stripes). These bands can be seen in the section shown in **9.21**. The general arrangement is made clear in the schematic diagram **9.22**, which shows a reconstruction, from many sections like **9.21**, of

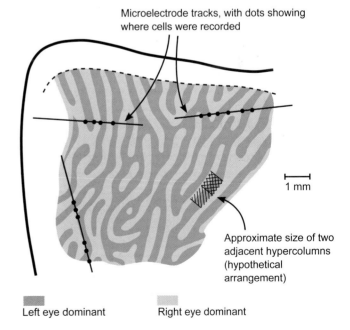

9.22 Schematic diagram of the left- and right-eye dominant bands of cells
Derived from various sections of the type shown in **9.21**.

how the monocular bands appear over a fairly wide region of cortex. It is important to realize in **9.22** that the pale bands are depicted by the boundaries demarcating the left/right monocular stripes (each monocular type indicated by shading).

One particularly fascinating aspect of **9.22** concerns the microelectrode tracks. Hubel and Wiesel collected single cell recordings along these tracks before conducting the microscopic examinations which were used to build up the reconstructed pattern, and then superimposed the tracks upon the reconstruction afterwards. The tracks have dots marked on them and these dots show the points at which cells changed from being preferentially driven by one eye to being preferentially driven by the other. The fascinating finding is that the dots lie exactly on the pale bands, i.e., on the boundaries between the monocular bands.

This matching of neurophysiological and neuroanatomical data is a tremendous achievement. When evidence from two such different approaches blends as neatly as this, the confidence to be had in the overall picture presented by both is greatly increased.

In fact, some cells in the striate cortex are found to be binocularly driven. That is, they respond actively to preferred stimuli in either eye. Others are preferentially driven from just one eye. Therefore the division of the hypercolumn into two strictly defined halves, one left- and one right-eyed, is an over-simplification.

Nevertheless, since the classic work done by Hubel and Wiesel, the monocular dominance (i.e., predominant sensitivity to stimuli in one particular eye) of certain regions of the hypercolumn has been amply corroborated, using a recently developed brain imaging techniques such as **optical imaging**. These new studies have also added many new fascinating details. We will describe those in the next chapter on brain maps.

Hypercolumns

The evidence presented so far indicates that each small region of striate cortex contains a full set of columns of all orientations from each eye. These clusters of columns were dubbed **hypercolumns** by Hubel and Wiesel. The basic idea is that each hypercolumn contains a mass of cells of different types which together process that hypercolumn's

patch of the retinal image. The different types include cells responsive to different scales, color, orientation, and ocularity (Ch 10).

The hypercolumn idea is captured in the so-called **ice cube model** of the striate cortex illustrated in **9.23**. In this model, each hypercolumn is conceived of as an image-processing module crunching data emanating from a small region of retina that has arrived from one lateral geniculate nucleus. The striate region of the right cerebral cortex is shown divided up into squares. Each square represents just one hypercolumn with an area of about 0.5 mm² on the cortical surface. Each hypercolumn is about 3–4 mm deep (remember that the cortex is about 3–4 mm thick) and each contains tens of thousands of cells, perhaps up to a quarter of a million.

We will call the patch of the retinal image that each hypercolumn deals with its **hyperfield**. These hyperfields overlap to some degree, but essentially each hypercolumn is concerned with just one region of the input image. The hypercolumns Thus, all chatter away simultaneously about what features they are "seeing" in their own hyperfield. It is the job of later processes to sort out from this feature description what objects are present in the scene (Chs 7 and 8).

The question then is: what representations do the hypercolumns build from the information they receive? The current view is: edge feature representations. But bear in mind that many fibers also leave the striate cortex to influence the lateral geniculate nuclei, so really the hypercolumn should be thought of as a data processing engine working in close collaboration with the lateral geniculate nuclei.

One such "conversation" between adjacent hypercolumns concerns linking edge features that go together because they come from the same visual objects (Ch 7). That is, this could be one place in the brain in which the figural grouping principles described in Ch 7 are implemented, perhaps by horizontal fibers running perpendicularly to the columns.

Although the hypercolumns are distributed evenly over the cortical surface, hypercolumns concerned with the central retina differ from those dealing with the periphery in having smaller hyperfields. This is also brought out in **9.23** by the

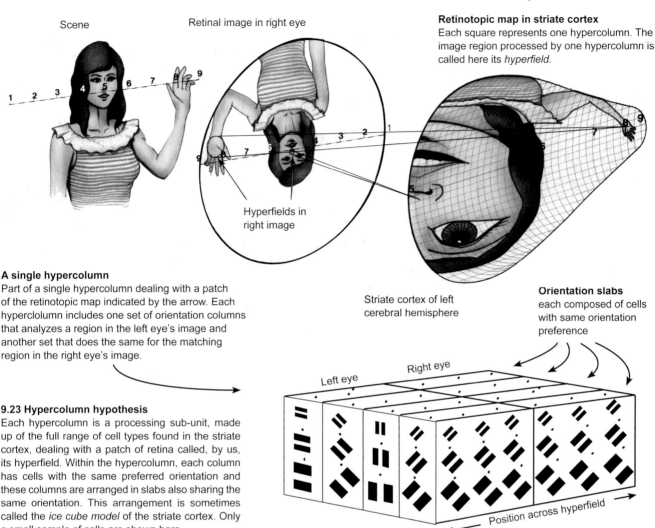

Scene

Retinal image in right eye

Retinotopic map in striate cortex
Each square represents one hypercolumn. The image region processed by one hypercolumn is called here its *hyperfield*.

Hyperfields in right image

A single hypercolumn
Part of a single hypercolumn dealing with a patch of the retinotopic map indicated by the arrow. Each hyperclolumn includes one set of orientation columns that analyzes a region in the left eye's image and another set that does the same for the matching region in the right eye's image.

Striate cortex of left cerebral hemisphere

Orientation slabs
each composed of cells with same orientation preference

9.23 Hypercolumn hypothesis
Each hypercolumn is a processing sub-unit, made up of the full range of cell types found in the striate cortex, dealing with a patch of retina called, by us, its hyperfield. Within the hypercolumn, each column has cells with the same preferred orientation and these columns are arranged in slabs also sharing the same orientation. This arrangement is sometimes called the *ice cube model* of the striate cortex. Only a small sample of cells are shown here.

Left eye

Right eye

Position across hyperfield

spatial distortion in the retinotopic map. Thus, a small region close to the nose (around location number 5 in the retinal image) represents the hyperfield of a central hypercolumn. A peripheral hypercolumn deals with a much larger region of the image—about 6 peripheral hyperfields cover the whole hand. Obviously, if the hand needs to be inspected closely then gaze direction would have to be changed to use the high-resolution of the central hypercolumns.

This is a very important point worth reiterating. The approximate area of cortex in each hypercolumn remains roughly constant over the cortical surface. However, the peripheral hypercolumns have to deal with a much larger area of retina and hence can be concerned only with a relatively crude feature analysis of this larger area. Central hypercolumns on the other hand, with their small hyperfields, can engage in a fine-grain analysis. This accounts for the difference between the finely detailed vision of objects inspected directly (central vision) and the crude appreciation of objects seen off-center (peripheral vision).

Of course, because central hypercolumns have smaller fields, there have to be more of them to cover a given area of the retinal surface. This gives the cortical map its curious spatial distortion: large nose, smaller ears, and tiny hands in **9.23**.

a The viewed scene

c Hypercolumn

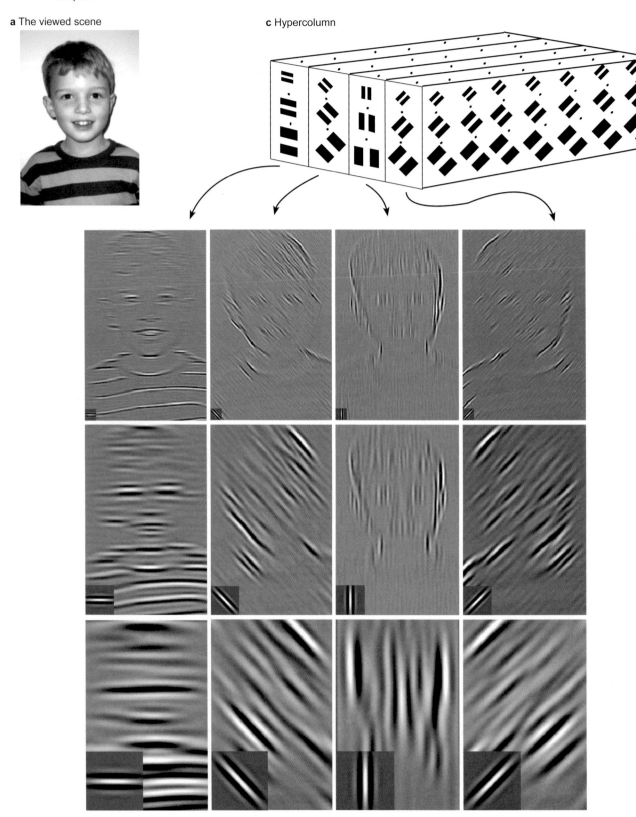

b Convolution images at different scales and orientations

A Cautionary Note

It should be evident from all that has been said that although a very great deal is known about the striate cortex, we still have at best only hazy ideas of what exactly it is doing and how it works. An encouraging sign of progress is that the classic findings of Hubel and Wiesel on hypercolumns fit in remarkably well with certain recent advances in understanding the principles of how feature information can be extracted from visual images. Consequently, the hypercolumn story is included in this book as the current "best bet" about how feature descriptions are obtained in our own visual systems. It could well be that further research over the next few years will substantially alter present ideas about hypercolumns. Indeed, such revision is more likely than not, because our ignorance of visual mechanisms far outweighs our knowledge.

But revision of theories in the light of new data is the everyday business of science. We think it is preferable to give an account of where the latest research seems to be heading rather than to stay too cautiously within the confines of well-established knowledge. In this way readers can share in some of the excitement of current research into vision, an excitement bred of a fascinating blend of psychological, physiological, and computational discoveries.

At the same time, the non-specialist must recognize what the full-time scientist takes for granted: the provisional nature of scientific theories (but see Ch 23, p. 549). There is a constant need to seek out evidence relating to them, and they must be relinquished when the data show that they are in fact unsustainable, despite their achievements to date in helping us understand some aspect of nature. Present evidence favors hypercolumns as image processing modules extracting feature representations; future evidence might not. The hypercolumn idea has been chosen as a worthwhile topic for this book because whatever its ultimate longevity or otherwise, it provides a framework for discussing the fundamental problems to be faced in building up a feature description of a visual scene.

Hypercolumns at Work on a Whole Image

Although we don't know how hypercolumns work, we can at least get a feel for some of the rich sources of data that their various cell types create by showing some computer-generated images of convolutions of various receptive field types with some images.

Consider the images of **9.24** which show convolution images representing the outputs of simple cell receptive fields when convolved with an input image. The inset in each convolution image shows the kind of receptive field used in each case. That is, the receptive field types shown in the insets have taken measurements everywhere over the input image, and an excitation-minus-inhibition count has been taken for every location. The size of each count is expressed in the convolution images on a scale from white (large, positive), through gray (zero), to black (large, negative). Thus, each point in each convolution image represents the response of a single simple cell, which is centered over the corresponding region of the retinal image. (Consult Chs 4–6, if you need a reminder on "convolution images.")

9.24 Hypercolumns at work on a whole image

a Retinal image

b Each point in the image is processed by simple cells with receptive fields with different sizes and orientations. Here, we have used three receptive field sizes, and four orientations. Each point in each convolution image has a gray level which reflects the response of a simple cell with a receptive field centered on that point. The particular receptive field used to generate each convolution image is shown in the corner of that image (see text for an account of how gray level of pixels in each convolution image corresponds to simple cell response). Note that different receptive field types "pick out' different types of image features. For example, the horizontal stripes of the boy's shirt are emphasized by horizontally oriented receptive fields. Each point in the retinal image is analyzed by a "complete set" of receptive field types (a total of 12 types in this simple model). The corresponding set of simple cells is located in a small region of striate cortex, and constitutes a hypothetical hypercolumn. Thus, each hypercolumn provides a battery of cells that appear to be delivering oriented measurements of image intensity changes (Chs 4–6) from which can be built a reasonably complete description of the visual "features" in one small region of the retinal images—the hypercolumn's hyperfield.

c The particular rigidly ordered arrangement of simple cells shown here is inspired by the *ice cube model* of striate cortex. Receptive fields of increasing size have been drawn in different layers for display purposes only; in reality, receptive field size varies across the cortex, rather than within each hypercolumn, Ch 10.

In **9.24**, simple cell response is coded in terms of pixel gray level, where mid-gray represents a zero response, and white represents a large response from a cell with the receptive field shown in the corner of each figure. Note that we show only simple cells with a central excitatory region and inhibitory flanking regions, and not vice versa. Of course, simple cells with a central inhibitory region and excitatory flanking regions exist in the striate cortex (Ch 3), and the responses of these cells are represented in **9.24** by pixel intensities from gray to black. This needs explaining a bit further, which we do next.

In the case of the receptive field insets, black shows where intensities in the input image inhibit the cell in question. In the case of the convolution images, black shows where cells with these receptive fields are more inhibited than excited. In a computer edge detecting system using such fields, these areas can be coded with negative numbers. But neurons cannot have negative (sub-zero) firing rates. Recollect from Ch 5 that the way the brain seems to cope with this problem is having cells of two complementary types. Where one has excitation the other has inhibition. In **9.24**, we show just one type—those with an excitatory central stripe (shown as white) with inhibitory flanks (shown as black). The complementary type would have the reverse: inhibitory central stripe (black) with excitatory flanks (white). Although we do not show fields of this type in **9.24**, the black regions in the convolution images are where such cells would be *active*. Thus, the blacks and the whites in the convolution images do in fact code the outcome of using both types of cells.

How do the convolution images of **9.24** relate to the hypercolumns? The answer is that each hypercolumn would have inside it a pattern of activity corresponding to just one area of each convolution image, the area covered by its hyperfield. For example, the hypercolumn shown in **9.23** has a hyperfield centered close to the nose in the input image. This is the region of the input "looked at" by this particular hypercolumn's cells.

The hypercolumn shown in **9.24** is in fact a much abbreviated one, for simplicity. Only one half of it is shown (remember each hypercolumn has a left- and right-eye section), just four different orientations are illustrated, only "line detecting"

cells are present, and just three widths (scales) of these. But this is a sufficiently rich hypercolumn for our present purposes. Bear in mind that each row of cells in this hypercolumn would be showing an activity profile which is just a small slice of the overall activity profiles shown in each convolution image.

It is easy enough to understand, at least intuitively, why the horizontal stripes in the boy's shirt in the input image in **9.24a** show up strongly in the convolution images for horizontally oriented cells. But it is not so easy to grasp why these cells should be strongly active in other areas where horizontally oriented edges are not so evident. At least, it is not so easy to grasp if one comes to these simple cells with a "line detector" conception of their function. But if one recognizes simple cells to be measurement devices of oriented changes in image intensity gradients then the surprise is diminished. After all, the initial measurements must be sufficiently rich to allow all sorts of features to be extracted, both large and small, fuzzy and sharp. In the example of **9.24**, the the two eyes together form a horizontal feature that is sufficient to stimulate strongly the largest horizontally oriented receptive fields, hence the black bar in the convolution image for those cells. Clearly, these convolution images are not to be regarded as representations of the edge features we see in the input image. Rather, they show measurements of image intensity gradients from which feature representations have to be built, as discussed in Ch 5.

Bear in mind also that the convolution images of **9.24** give us an idea of the activity profiles set up all over the striate cortex within the mass of hypercolumns. Remember that adjacent hypercolumns have overlapping hyperfields (i.e., that the slices shown on the convolution images will overlap with other slices, so that there is a degree of duplication within the striate cortex). At least, this is how we currently think hyperfields are arranged, but the question of degree of overlap is a research issue at present, about which one cannot yet be definite. In any event, **9.24** shows one hypothesis about the the kind of data obtained by the hypercolumns, data which are then interpreted to arrive at a feature description for each region of the input.

Hypercolumns Summarized

By now you should have an idea of the hypercolumn as a processing sub-unit, everywhere replicated over the striate surface, so that each part of the input image can be analyzed. Each hypercolumn has within it a set of simple cells (as well as others, such as the complex and hypercomplex types), and these cells provide a large number of measurements of the image intensity gradients on the hypercolumn's hyperfield. From these measurements, various feature descriptions are built up, and by the time these are complete, the type, orientation, scale (fuzziness), and contrast of the feature in the hyperfield is known.

The interpretation of these measurements is made much easier by the neat layout of simple cells in the hypercolumn, which means that zero crossings (or perhaps peaks and troughs, Chs 5,6) of activity can readily be identified. Also, the orderly layout brings together cells dealing with similar orientations which may help in local figural grouping of local edge representations, such as those proposed in the raw primal sketch idea, Chs 4 and 5. These considerations suggest that processes taking place within each hypercolumn culminate with the striate cortex as a whole sending to other brain sites representations about what edge-based *local* features are present.

But at some point edge representations need to be built up that extend beyond the confines of the individual hyperfields. Some grouping operations may be implemented in the interactions between data being processed in adjacent hypercolumns. Certainly there must exist somewhere in the brain feature representations of the kind of perceptual entities proposed in the full primal sketch, Chs 4 and 5, such as long lines or edges, straddling many hyperfields. These larger entities must in turn be grouped into still larger ones, until the materials are to hand for recognizing objects, Chs 7 and 8. After all, we are aware of much more than tiny local details of the kind dealt with by hypercolumns considered individually.

Thus, there must be further stages of processing which simultaneously scan all the evidence represented in all the convolution images of the kind shown in **9.24**. These processes presumably "read off" from the mass of data a full representation of the whole image—a representation that finally emerges into our consciousness as awareness of the scene being viewed. These longer range grouping operations may not be implemented in the striate cortex, V1, but in the regions to which it outputs project, the prestriate areas and beyond, Ch 10.

The neural code in which feature representations are written is not yet known. Finding it is a contemporary challenge for vision neuroscience.

Some Considerations for the Future

This chapter has given a bare outline of the visual machinery of just some parts of the brain. It must be remembered that there exist in the striate cortex and superior colliculus many variations on the basic cell types described, and that in all probability other cell types wait to be discovered. And it must constantly be kept in mind the presently mysterious roles of the feedback fibers that project from the striate cortex to the lateral geniculate nucleus, as well as the roles of all the other brain sites concerned with vision.

As if all these uncertainties were not enough, we also note that, although the discovery of the columnar structure of the cortex has passed its 50th birthday, having been first reported in connection with the motor cortex by Vernon Mountcastle in 1955, we still have no deep understanding of the functions of the column. The hypercolumn idea is a hypothesis that attempts to give some leverage on this problem but, when all is said and done, it is still a sketchy proposal.

Indeed, Jonathan Horton and Daniel Adams entitled their recent critical review article: *The cortical column: a structure without a function* (see *Further Reading* overleaf).

In the next chapter, we explore how columnar structure in the cortex (as distinct from columnar function) is related to the ***packing problem***. This is the problem of fitting into the striate cortex the many different cell types dealing with a given patch of retina.

Further Reading

Bear MF, Connors BW and Paradiso MA (2001). *Neuroscience: Exploring the Brain.* 2nd edition, Lippincott Williams and Wilkins, Baltimore.

Hirsch JA, Gallagher CA, Alonso J-M and Martinez L (1998) Ascending projections of simple and complex cells in Layer 6 of the cat striate cortex. *Journal of Neuroscience* **18**(19) 8086–8094.

Horton JC and Adams DL (2005) The cortical column: a structure without a function. *Philosophical Transactions of the Royal Society of London (B)* **360** 837–862. *Comment* This is an advanced but well written critical assessment of how much and how little we understand about what cortical columns do.

Hubel DH (1981) *Evolution of Ideas on the Primary Visual Cortex, 1955–1978: A Biased Historical Account* Nobel Prize Lecture.

Hubel DH (1995) *Eye, Brain, and Vision* (Scientific American Library, No 22) W. H. Freeman; 2nd edition. *Comment* This superb book by one of the pioneers of visual neuroscience is no longer in print but can be read online at http://hubel.med.harvard.edu/index.html.

Hubel DH, Wiesel TN and Stryker MP (1974) Orientation columns in macaque monkey visual cortex demonstrated by the 2-oxyglucose technique. *Nature* **269** 328–330.

LeVay S, Hubel DH and Wiesel TN (1975) The pattern of ocular dominance columns in revealed by reduced silver stain. *Journal of Comparative Neurology* **159** 559–576.

Martin AR, Wallace BG, Fuchs PA and Nicholls JG (2001) *From Neuron to Brain: A Cellular and Molecular Approach to the Function of the Nervous System* (Fourth Edition) Sinauer Associates ISBN 0878934391. *Comment* A comprehensive and clear introduction to neuronal function.

Mather G (2006) *Foundations of Perception.* Psychology Press.

Olshausen BA and Field DJ (2005) How close are we to understanding V1? *Neural Computation.* **17** 1665–1699. *Comment* This is an advanced paper that assesses just how much we *don't* know about visual processing mechanisms in the striate cortex.

Van Rullen R and Thorpe SJ (2001) Rate coding vs temporal order coding: what the retinal ganglion cells tell the visual cortex. *Neural Computation* **13**(6) 1255–1283.

Van Rullen R and Thorpe SJ (2002) Surfing a spike wave down the ventral stream. *Vision Research* **42** 2593–2615.

Blurred and shrunken version of a part of 5.22 on p. 204

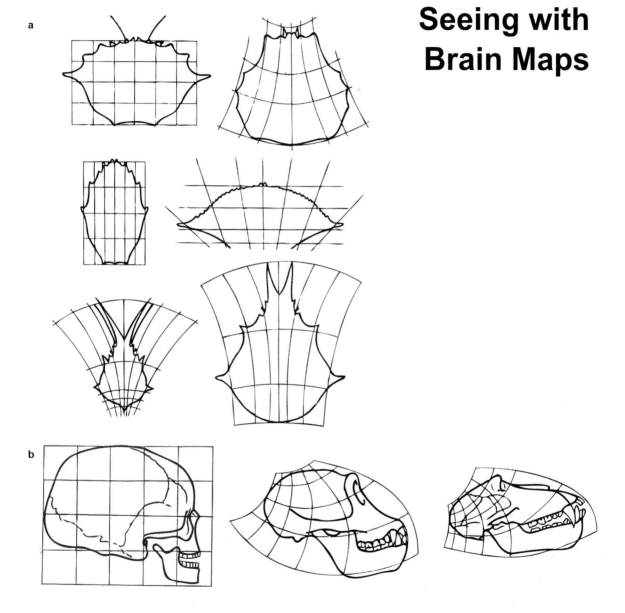

Seeing with Brain Maps

10.1 Examples of mappings in nature: crab carapaces and skulls
a Six crab carapaces, each of which can be obtained as a continuous transformation of the other five.
b Skull of human, chimpanzee, and baboon, each of which can be obtained as a continuous transformation of the other two. Redrawn from Thompson (1917) with permission of Cambridge University Press.

We have already seen that the retinal image is represented as a ***retinotopic map*** both in the lateral geniculate nucleus and the striate cortex. In this chapter we will examine how the brain attempts to represent information extracted from the retinal image into a collection of brain maps. We will see how this results in a tight squeeze, called the ***packing problem***, and how the brain attempts to solve this problem by a division of labor between different cortical areas.

Nature's Maps

Nature is full of maps. Look at the crab shells (carapaces) in **10.1**. Despite their quite different appearance, each shell can be transformed into the others with smooth (continuous) mappings. These shells are maps of each other, and the mechanism that provides the transformation (mapping) from one shell into another is evolution. These shells are typical of examples of D'Arcy Thompson's work on animal forms from 1917, work which pre-dated Watson and Crick's discovery of DNA by some 30 years. Thompson's work suggests that evolution works by making continuous geometric transformations to existing body plans. Evolution does not create a new appendage if it can get away with

Why read this chapter?

The brain is full of maps. In some cases, it is obvious what each map represents. For example, the signals from the retina project to cells in layer 4 of the *primary visual* or *striate* cortex arranged as a *retinotopic* map of the retinal image. Those signals are then analyzed for various properties, such as edge orientation, motion, and color. As the striate cortex is essentially a sheet of neurons, there is simply not enough room to represent every orientation and every color for every point on the retina in a simple retinotopic layout; this is called the *packing problem*. In attempting to solve this problem, something has to give. The result is that primary visual cortex contains *singularities* of various types. These can be clearly seen in maps obtained using optical imaging techniques. We compare the singularities found in the striate cortex to singularities found in other maps, such as fingerprints, and we present two relevant theorems. We then consider other visual areas such as *inferotemporal cortex*, where it is much less obvious what is represented, and what type of map is present. However, it is evident that the retinotopic map found in the primary visual cortex is much less dominant in these areas, suggesting that these maps may be feature maps in which adjacent cortical patches represent similar features. Indeed, there is some evidence that these maps represent similar views of faces, or even actors frequently seen in the same context, such as the cast of a TV series.

modifying an existing one. That is why all vertebrates have no more than four limbs, even though some animals would clearly benefit from a few more (but note that evolution is adept at *reducing* limbs to the point of extinction if required, as in snakes).

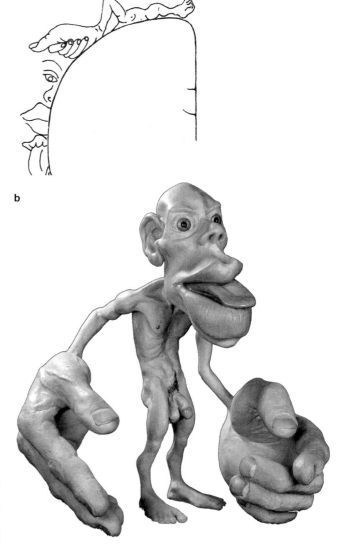

a

b

10.2 Sensory maps

a Cross-section of the sensory cortex showing the mapping of touch receptors in the skin. For location of the sensory cortex in the brain, see Ch 1, **1.5**. Courtesy Wikipedia.
b Sensory man. Each part of the human body has been magnified according to the amount of neuronal machinery dedicated to sensing touch there. Reproduced with permission from Natural History Museum, London.

It is not just crab carapaces that seem to be continuous transformations of each other. Thompson also showed that mammalian skulls are also continuous transformations, as shown in **10.1b**. Indeed, not a single new bone has been added to the 28 bones of the skull for over 400 million years ago. Accordingly, the skulls of all mammals (including our own) can be obtained by the continuous mapping from a skull that first appeared in an air-breathing fish in the pre-Devonian seas (i.e., more than 415 million years ago). Thus, whilst the genetics of shape seem to permit quantitative or continuous changes in size and extent, it seems most reluctant to allow qualitative or discontinuous changes (e.g., in bone number).

Sensory Maps

Given that nature is full of maps, it should perhaps come as no surprise to find that the brain is full of maps. Some of these are literally as plain as the nose on your face. For example, a map which represents inputs from the skin exists on the sensory cortex, as shown in **10.2**. Within this map, body parts which are highly sensitive, such as the lips and hands, are represented by large areas, whereas less sensitive areas, such as the legs and back, are represented by small areas. If we were to make a model of a man with body parts in proportion to the areas in this map then he would look like "sensory man" shown **10.2b**.

The differential magnification of various regions of the sensory map is a key feature of brain maps. Making fine discriminations with the hands is obviously important to us, so hand skin is thus packed with a high density of receptors each of which requires its own neuronal machinery, hence the relatively large region of brain required for the hand. If the level of sensory discrimination found in the hands were found all over the body then the brain would have to be several times the size it is now. This would clearly be very wasteful given the high metabolic demands of brain tissue, as well as demanding an impractical head size.

The Striate Map

The striate cortical map of the retinal image in **10.3** has four important properties. First, the face shown on the cortical surface is magnified, but otherwise perfect. None of it is missing, and every

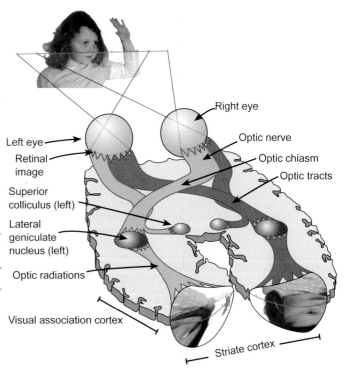

10.3 Schematic projection of a scene into the two retinas and from them into the brain
For more details see **1.6** and **1.7** (pp. 5–6).

part of the face is correctly positioned in relation to every other part. This results from the ***continuous mapping*** of retinal points to corresponding cortical points within each cerebral hemisphere. Second, the cortical representation is upside-down (as indeed are the retinal images too, but this is not shown in **10.3**). Third, the cortical representation of the scene is cut right down the middle, and each cerebral hemisphere processes one half (actually, there is slight overlap of the central region of the scene as a central strip is represented in both hemispheres). Fourth, and perhaps most oddly, the cut in the cortical representation places the regions of the scene on either side of the cut farthest apart in the brain!

All in all, the cortical mapping of incoming visual fibers is curious, but orderly. The orderliness of the mapping is reminiscent of the "inner screen" proposed in Ch 1. But the oddities of the mapping might give any die-hard "inner screen" theorist pause for thought. As we said in Ch 1, the first "screen" we meet in the brain is a strange one indeed.

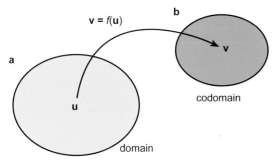

10.4 Mapping
A mapping is defined by a function *f*, which maps each point **u** in one space (the ***domain* a**) to a corresponding point **v** in another space (the ***codomain* b**). If we represent the position of a point in **a** by the variable **u**, and if **u** gets mapped to a corresponding point **v** in **b** then we can define the ***mapping*** or ***function*** **v** = *f*(**u**). Note that both **u** and **v** are typically vector quantities, and each refers to a specific location (e.g., **u** = (*x, y*) maps to a point **v**=(*x′, y′*).

What is a Map?

It has been said that "the map is not the country." This truism is a wise reminder that any map is a *representation* of the spatial distribution of some property in the world (e.g., annual rainfall). More pertinent for our purposes, it is a reminder that the map cannot be the thing itself; a map can only be an abstraction of the mapped-thing, and inevitably omits certain properties of the thing itself (e.g., the wetness of rain!). Even if the map had "the scale of a mile to the mile" (as in Lewis Carroll's *Sylvie and*

Bruno, 1889), it would still represent only certain properties of the (literally) underlying country.

A somewhat rarefied definition is that a map is an abstraction of a particular space which preserves only specific types of information. For example, a road map is an abstraction of a country which preserves information about how roads are laid out; the map does not preserve information about height above sea level or annual rainfall. Most importantly, a road map preserves information regarding distance, so that short roads are represented by short map roads, and long roads are represented by long map roads. We would be shocked to find that a relatively short map road was a relatively very long road in reality. However, most maps in the brain are of this distorted type, as we observed with "sensory man" and the striate map.

More formally, a map is a mathematical ***function*** which transforms points from one space called the ***domain*** (e.g., as the retina) to another space called the ***codomain*** (e.g., the striate cortex), **10.4**. For example, we know that there is a massive expansion in the representation of the fovea in the striate cortex, whereas peripheral retinal areas are squeezed into a small region of the striate cortex. But because one space (the retinal domain) gets squeezed and stretched into another space (the cortical codomain) there are no breaks in the map. That is, every pair of nearby points in the retina corresponds to two nearby points on the striate cortex. If these "neighborhood relations" are preserved this implies that the mapping from retina

10.5 Continuous mapping of images
If every pair of points that are nearby in **a** corresponds to a pair of nearby points in **b** then the mapping from the ***domain* a** to the ***codomain* b** is ***continuous*** (see arrows). If we represent the position of a point in **a** by the variable **u**, and if **u** gets mapped to a corresponding point **v** in **b** then we can define the ***mapping*** or ***function*** **v** = *f*(**u**). The image **b** was obtained by a smooth deformation of **a** (**b** is just a squashed version of **a**). This implies that that the mapping from **a** to **b** is continuous. In this case, the mapping of points back from **b** to **a** is also continuous; this is not always true, see **10.6**.

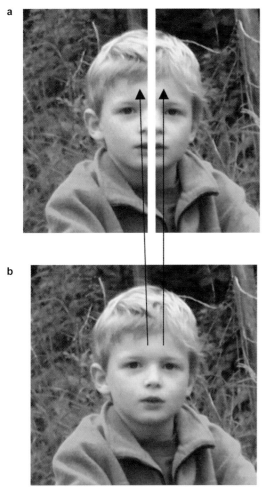

a

b

10.6 Discontinuous mapping
If even just one pair of nearby points in **a** does not correspond to nearby points in **b** then the mapping from the domain **a** to the codomain **b** is **discontinuous** (see arrows). Here, **b** could not be obtained by any continuous deformation of **a**. In this case, the two images in the codomain **b** correspond to the "images" in the left and right halves of the primary visual cortex (see **10.5**).

to the striate cortex is **continuous**, **10.5** (but see below).

By now it should be apparent that a map specifies three entities: two spaces (e.g., **a** and **b**), and the *direction* of the mapping (either from **a** to **b**, or from *b* to *a*). The direction is important because it is often the case the mapping from **a** to **b** is continuous (as in **10.5**), but the mapping from **b** to **a** is discontinuous (as in **10.6**).

Indeed, the split illustrated in **10.6** is reminiscent of the way the retinal image is split vertically between the left and right half of the striate cortex, so that nearby points either side of the

vertical mid-line on the retina map to different hemispheres, **10.3**. Consider, for example, a pair of points which straddle the exact vertical mid-line of one retina, say, the left retina. Now, where do the left and right retinal points map to in the striate cortex? Thanks to the *optic chiasm*, **10.3**, the left point maps to a point in the left striate cortex, whereas the right point maps to a point in the right striate cortex. Here is a clear breakdown in the continuous mapping of retinal neighborhoods. Indeed, if any pair of nearby points on the retina does not correspond to two nearby points in the striate cortex then the mapping of points from retina to the striate cortex is said to be *dis*continuous.

So, strictly speaking, the mapping from retina to the striate cortex is discontinuous, **10.6**, and it is only the mapping from each *half* of the retina to the striate cortex that is continuous. The particular mapping from retina to cortex gives rise to a *retinotopic map* of the retinal image on the cortical surface, **10.7**. Even though we now know this map is discontinuous, we will use the term "retinotopic map" to refer to the continuous mapping from each vertical half-retina to the striate cortex.

Within this retinotopic map, a greater area of cortex is devoted to central vision than to peripheral vision: hence the relatively swollen face and the diminutive arm and hand in **10.3** and **1.7**. This curious mapping doesn't of course mean that we actually see people in this distorted way—obviously we don't. It is just a graphic way of showing that a disproportionately large area in the brain is assigned to the inspection of what lies straight ahead. This dedication of most cortical tissue to the analysis of the central region of the scene reflects the fact that we are much better at seeing details in the region of the scene than in regions that project towards the edge of our field of view.

If we now consider the *inverse mapping* from cortex to retina then it is not hard to see that this is also discontinuous. While most pairs of nearby points in striate cortex map to nearby points on one retina, this is not true of all pairs of points. If two nearby striate points are located in different *ocularity stripes* (Ch 9; also **10.20**) then, by definition, they correspond to different eyes (retinas). Thus, the mapping from the striate cortex to retinas is also discontinuous.

a

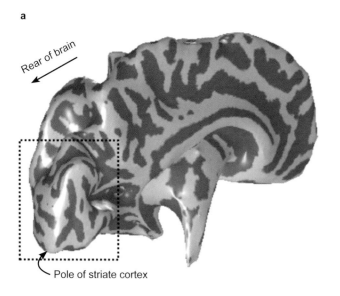

Rear of brain

Pole of striate cortex

b

Vertical meridian in upper right visual field

12.5⁰
10.0⁰
7.5⁰
5.0⁰
2.5⁰

Horizontal in visual field

c

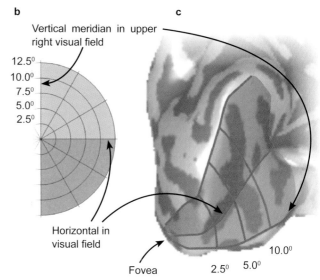

Fovea

10.0⁰

2.5⁰   5.0⁰

10.7 Mapping visual field to cortex
a View of inner (medial) surface of the left hemisphere, with the rectangle framing the visual areas at the back of the brain.
b Right visual hemi-field. The right half of the visual field (which projects to the left side of the retinas of both eyes) is mapped to the striate cortex, **c**, of the left hemisphere. The fovea of both eyes corresponds to the center of the visual field, where the radial lines converge to a point.
c Magnified version of the rectangle in **a**. The lower (pink) half of the visual hemi-field corresponds to the upper part of the calcarine sulcus, and the upper (green) half of the visual hemi-field corresponds to the lower part of calcarine sulcus. The visual field corresponding to the fovea projects to the pole of the striate cortex. The center of the visual field in **b** gets mapped to a disproportionately large area in the striate cortex. This mapping effectively magnifies the region of the visual field which projects to the fovea, and "shrinks" regions which project to periphery of the retina. It is helpful to the understanding of this figure to compare it to the schematic projection in **10.3**. Courtesy Tony Morland.

A Map is a One-Way Street

Notice that we initially considered the striate cortex to be a map of the retina, but we then considered the retina as a map of the striate cortex. Thus, mappings can go in either direction, and it is important to know which way is which. This is why the space we usually call the map (e.g., striate cortex) and the space (e.g., retina) represented in the map have special names. The map (e.g., striate cortex) is called the ***codomain*** and the space represented in the map (e.g., retina) is called the ***domain***, **10.4**.

Most importantly, the correspondence between points in the domain and codomain is defined by a ***mapping function***, or ***mapping*** or ***function***. A mathematical short-hand way to remember this is that the points in the domain **u** (e.g., retina) map to corresponding points in the codomain **v** (e.g., striate cortex), so that $\mathbf{u} \rightarrow \mathbf{v}$. The precise manner in which each point **u** maps to a point **v** is defined by a mapping function f, such that $\mathbf{v} = f(\mathbf{u})$, **10.4**.

Continuous Brain Maps

Why does the brain bother to construct retinotopic maps of retinal space? Could it do its job equally well without having to retain the retinotopic organization of information in the retinal image?

The short answer is that we simply don't know for sure, although there are some plausible ideas. For example, one hypothesis is that maps economize on the length of nerve fibers between closely neighboring areas that have to "talk" to one another. Whatever the truth of that, three observations suggest that maps represent a valuable generic computational device for the brain.

First, the brain has many types of map, for audition, touch, and motor output. Second, the brain has many copies of each map. Third, these maps are, insofar as is possible, continuous. Together, these suggest that the spatial organization of information in the form of maps is extremely important for the brain.

Having said that, within the visual system

retinotopy seems to be a luxury that only the early visual areas (such as the lateral geniculate nucleus and the striate cortex) can afford. It seems likely that the reason for this is not that the brain loses interest in retinotopy in other visual areas, but that it gains interest in representing "higher order" properties (such as face identity) or non-spatial properties (such as color). In attempting to represent non-spatial properties in a map, it seems the representation of retinal space must be sacrificed.

Indeed, most visual maps do not preserve very well the retinotopy observed in the striate cortex. As we shall see, this does not necessarily imply that such maps are discontinuous. It suggests that some other property is represented, and this has the side effect of reducing the integrity of the retinotopic map in these brain areas.

The Orientation Map

As described in the last chapter, Hubel and Wiesel showed not only that most cells in the striate cortex have a preferred orientation, but also that cells which are near to each other have *similar* preferred orientations, **10.8**. As an electrode is moved across the striate surface, the preferred orientation shifts gradually, most of the time. Curiously, and it is an important clue to the fact that there is a *packing problem,* every so often a major "jump" in orientation preference occurs.

Since Hubel and Wiesel's pioneering research using microelectrodes, a new technique called *optical imaging* has allowed beautiful high resolution

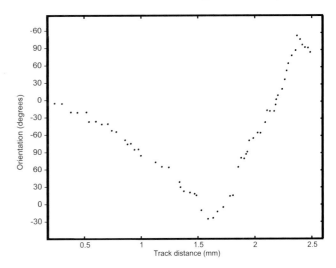

10.8 Nearby cells in the striate cortex tend to have similar orientation preferences
As an electrode is moved across the striate cortex, the preferred orientation of cells changes continuously most of the time. Redrawn with permission of the Nobel Foundation from Hubel (1981).

images of orientation columns to be obtained in awake animals, **10.9**. The details of this method are beyond the scope of this book, but the results essentially confirm and expand on the observations made by earlier researchers.

Two different ways of drawing an orientation map are shown in **10.10** and **10.11**. In **10.10,** the preferred orientation of the cell at each location in the cortex is indicated by a line with an orientation that matches the preferred orientation of cells in that region. However, a more arresting method

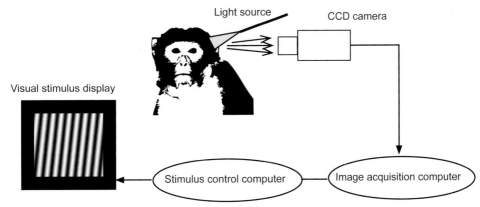

10.9 Optical imaging technique
The amount of light reflected by brain tissue is related to local neuronal activity levels. By presenting lines at different orientations, the activation of different regions within the striate cortex can be measured. After several orientations have been presented it is possible to construct the kind of orientation maps shown in **10.10** and **10.11**. Adapted with permission from Wang, Tanifuji and Tanaka (1998).

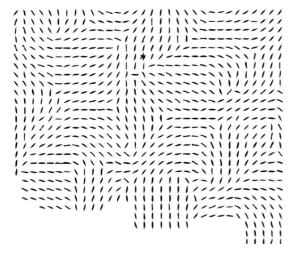

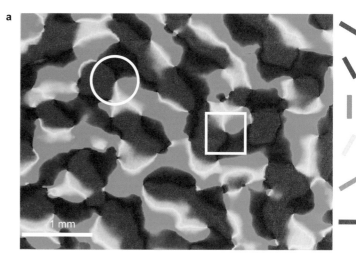

a

10.10 Orientation map in visual cortex (V2)
Surface of part of the V2 viewed from above. Here, the orientation of lines drawn on the cortical surface indicates the orientation to which that cortical region is most sensitive. Courtesy Swindale, Matsubara and Cynader (1987).

of drawing orientation is shown in **10.11**. Here, each preferred orientation is represented by a single color. Using color to represent orientation makes the spatial layout of orientation very obvious.

The main characteristic of the orientation map is that the representation of orientation varies

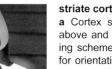

b A real pinwheel. Courtesy Ken Bloomquist.

10.11 Orientation map in the striate cortex
a Cortex surface viewed from above and using the color coding scheme shown on the right for orientation. The singularities form the centers of patterns that have come to be known as *pinwheels*. The circles enclose two singularities with opposite signs (circle = positive sign, square = negative sign). With permission from Blasdel (1992).

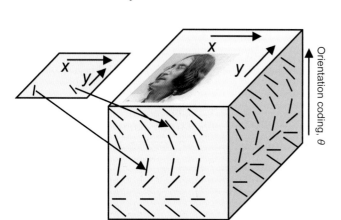

10.12 Three-dimensional parameter space
A line on a plane is defined by three parameters, two for its position (*x, y*) and one for its orientation (*θ*, pronounced *theta*). In order to represent every orientation at every position we require a three-dimensional parameter space in which position is defined by two position parameters (*x, y*) and orientation is represented by a third (depth) parameter. Any line on the plane with position (*x, y*) and orientation *θ* corresponds to a single point with coordinates (*x, y, θ*) in the 3D parameter space. This is *NOT* the way the striate cortex represents orientation.

smoothly from place to place across the cortical surface, but there are occasional "jumps" or **discontinuities,** where the representation of orientation changes abruptly. In the context of brain maps, discontinuities are also called **singularities,** and we will use both terms interchangeably here. Note that the map is continuous with respect to retinal position (except for the split between hemispheres described earlier). However, the map is discontinuous with respect to orientation which is why **pinwheel** patterns, **10.11**, emerge in the orientation map. These discontinuities in orientation occur as a side effect of trying to solve the packing problem.

The Packing Problem

In order to make sense of the visual world we need to be able to see every orientation irrespective of where it is located on the retina. This means that we need to be able to represent every possible orientation for every retinal location. In theory, this could be achieved by representing different orientations at different "vertical" depths within

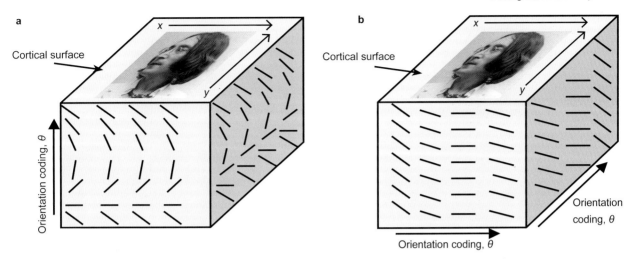

10.13 Two possible ways of mapping x, y **position in the retinal image along with orientation,** θ
The scheme shown in **b** is the one found in the striate cortex.

each cortical column (i.e., at each cortical location). This would perfectly solve our problem, as shown in **10.12**.

Yet, as we have seen, each column in the striate cortex represents a *single* orientation, that is, all the cells in any given column that are sensitive to orientation have the same orientation preference, as in **10.13b**. So the brain does not use the physical dimension of cortical depth to represent orientation, in the way shown in **10.12** and **10.13a**. This means that the cortex must try to pack all orientations into each cortical location representing a patch of retina (without using cortical depth to represent orientation), which is why this is called the ***packing problem***.

In other words, the cortex must *cover* every value of orientation at every cortical location. It is therefore said to maximize its ***coverage*** of orientation (it also tries to maximize the coverage of several other properties, as discussed below).

As we have already seen, the brain not only attempts to maximize coverage, it also seems to try to keep cells with similar preferred orientations near to each other, which gives rise to the orientation map described above. And it also tries to keep cells with nearby receptive fields close to each other, which gives rise to the familiar retinotopic map. In essence, the brain attempts to maximize the continuity of its various maps.

The packing problem can now be succinctly summarized as: the problem of representing a number of parameters (e.g., retinal location, line orientation) in a 2D cortical map (e.g., V1) whilst maximizing the continuity and coverage implicit in that map.

The reason that the packing problem is important is because it applies to all visual areas and to the representation of many different physical properties. In order to see just how big a problem the packing problem is, let's consider it in terms of perception.

The Packing Problem and Perception

If the cortex did not have to represent orientation then it could simply assign nearby regions of the cortex to represent nearby regions of the retinal image (as in fact it does, as best it can, but we are here trying to understand the discontinuities, the points of failure, if you like). In this way, each location on the cortex would be a representation of a specific location on the retina. Each retinal location is defined in terms of two position parameters x and y, and each retinal point maps to a corresponding point with horizontal position x' and vertical position y' on the cortex.

Because the retina can be laid out as a flat sheet, it is said to be two-dimensional, which is why just two parameters (x, y) are required to define the locations of different points on the retina. (These two parameters could be written as a vector quantity $\mathbf{u} = (x, y)$, but we will use (x, y) here, to remind ourselves that we are dealing with two parameters.)

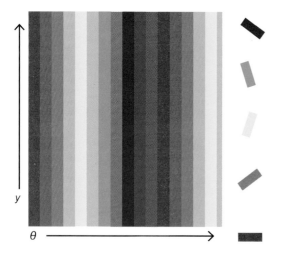

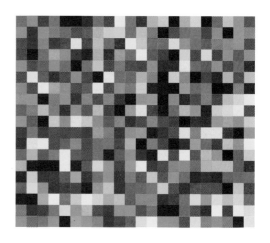

10.14 Mapping by ignoring one spatial parameter
Color code for orientation is shown by the bars at the right. Whereas any orientation can occur at any vertical position y or any horizontal position x on the retina, the brain could simply ignore, say, the retinal x coordinate. In this case, all orientations could be represented at each y-position without any discontinuities in orientation. However, an animal with such a brain would know which orientation θ had been presented to the retina, and it would know where along the retinal y-axis it had been presented, but it would have no information regarding the position along the x-axis of the oriented line. This would clearly be unsatisfactory.

10.15 Mapping with random assignments
Same color code as in **10.14**. Another unsatisfactory solution to the packing problem. If each point on the cortex represents a randomly chosen value of x, y and θ then it would (on average) represent all possible values of x, y and θ. It would therefore maximize coverage, but would do so by sacrificing continuity (see text).

Both **10.14** and **10.15** redrawn with permission from Chklovskii and Koulakov (2004).

Similarly, the cortex is effectively a two-dimensional sheet. Even though the cortex has many folds it too can be laid out as a thin sheet. We already know that the brain does not use the depth of this sheet to represent different orientations: each small region on the cortical surface is used to represent only one orientation, **10.13b**.

Thus as with the retina, locations on the cortical surface also require two parameters (x' and y'). In other words, even though the cortex is folded in three dimensions and even though it has a finite third dimension of depth, we have access to only two of those dimensions to represent orientation.

Parameters

Thus far we have referred to horizontal and vertical position as the parameters x and y. However, orientation is just another parameter. Representing two position parameters in a two-dimensional cortical sheet is relatively easy. Representing three parameters (x and y plus orientation, θ) in a two-dimensional sheet is difficult, to say the least. In essence, the brain is attempting is to represent a

total of three parameters (x, y, θ) in a two-dimensional sheet.

Now, three parameters will go into a two-dimensional space, but not without a fight. Forcing three parameters into two-dimensional space (the cortex) is such an awful squeeze that it will not go without introducing discontinuities ("jumps") in the cortical representation of at least one of the parameters. Because the cortex is especially keen to maintain a smooth map of retinal space, it allows these jumps or discontinuities to occur in the representation of orientation (rather than in the representation of x or y). In fact, it can be proved mathematically that it is not possible to represent (all values of) more than two parameters in a two-dimensional space without introducing discontinuities. This may be easier to see if we represent the three parameters (x, y, θ) in a single three-dimensional ***parameter space***.

Parameter Spaces

In a parameter space, every parameter has its own ***dimension***, so we need three dimensions to represent three parameters. A parameter space is not the same thing as physical space, although it can clearly be represented as a cube in physical space if there are three parameters. Nor is parameter space a loose analogy; it is a precise representation of a number of different parameters in which each parameter varies continuously along exactly one dimension.

The three-dimensional parameter space that includes x, y, and θ has a single point for every possible combination of (x, y, θ). Thus, every possible value of (x, y, θ) is represented by a unique point in this three dimensional parameter space, **10.12**.

We can now state succinctly the packing problem: it consists of attempting to pack all three dimensions of this parameter space into the two dimensions available in the striate cortex. There are many logically possible solutions to the packing problem, some of which are explored here.

First, we could simply ignore one parameter, such as x. In the map shown in **10.14**, y is represented along the vertical axis and orientation (θ) varies (continuously and cyclically) along the horizontal axis, but x is simply not represented. In this case, activity of one cell in the map would tell us the orientation and vertical position of a line on the retina, but we could not tell where it was along the horizontal (x) axis. This defines a continuous map but as it excludes one spatial coordinate (the x-axis) it hardly represents a satisfactory solution to our task of representing orientation at every retinal location.

Second, we could use each point on the cortex to represent a randomly chosen combination of (x, y, θ), **10.15**. This would ensure that all combinations were represented, so it would maximize coverage. However, nearby columns would represent disparate combinations of parameter values because they would correspond to random points in the three-dimensional parameter space (x, y, θ). Consequently, there would be no continuity with respect to any parameter in the cortical map and this would complicate considerably the scheme for computing feature orientation described in Ch 3.

Third, we could treat two parameters with a high priority and the third with low priority. In other words, we could maintain the continuity of two parameters, and do the best we can with the third parameter. This is the solution implemented

by the brain, as shown in **10.10** and **10.11**. The cortex maintains continuity of the position parameters x and y and then does the best it can with respect to the orientation parameter. The high priority parameters constitute the ***primary indices*** of the map, whereas the low priority parameters constitute an example of the ***secondary indices*** of the map. Thus, the striate map is said to have a primary index that is purely spatial because it consists of the x and y parameters, and a secondary index that is orientation, θ.

Later, we will see how the brain may choose to relegate both spatial parameters to the status of secondary indices. However, at least at this early stage of the visual system, the brain prefers to maintain spatial continuity of events on the retina whilst sacrificing continuity in other parameters, such as orientation.

Characterizing Singularities

We know that the packing problem results in the emergence of discontinuities (singularities) with respect to orientation. However, it turns out that the brain is not the only system to be confronted with a packing problem.

Look at the fingerprints in **10.16**. As in the cortex, there is a continuous change in orientation, except for a distinctive singularity where orientation changes abruptly. Because the fingerprint "represents" orientation as a physical mark on the skin, the map of orientation resembles **10.10**, where orientation is drawn as small lines.

Fingerprint Theorem and Penrose's Rule of Thumb

In a fascinating piece of detective work, Elsdale and Wasoff proved a theorem which specifies the minimum number of singularities in patterns such as fingerprints. The story begins with a pathologist named Penrose who noticed that the number of singularities is equal to the number of fingers, minus one (and the finger number varies according to various genetic abnormalities). In pathology, the hand print is characterized in terms of the number and type of singularities, so it is useful to know that there should be no less than some minimum number in order to tell if any have been missed. So this was a useful empirical rule of thumb (no pun intended), which became known as Penrose's rule.

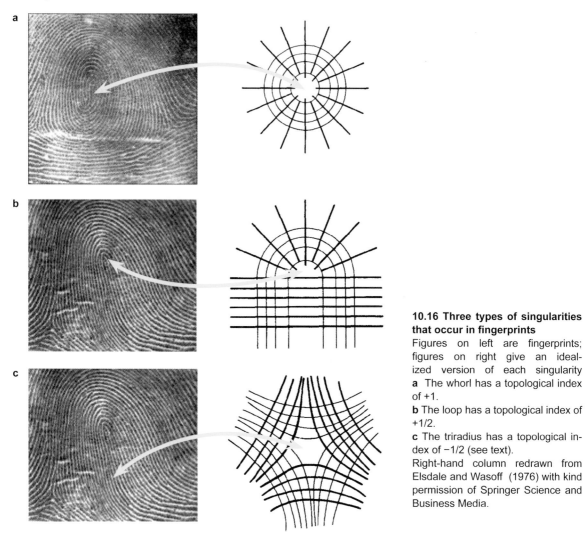

10.16 Three types of singularities that occur in fingerprints
Figures on left are fingerprints; figures on right give an idealized version of each singularity
a The whorl has a topological index of +1.
b The loop has a topological index of +1/2.
c The triradius has a topological index of −1/2 (see text).
Right-hand column redrawn from Elsdale and Wasoff (1976) with kind permission of Springer Science and Business Media.

Elsdale and Wasoff noticed that the number of fingers on a hand acted as a boundary condition on the number and type of singularities that could occur. Specifically, the fingerprint orientation is constrained to lie parallel to the finger tips and the wrist, and perpendicular to the other boundaries of the hand. We do not know why this should be so, but it is so.

Singularities are of different types. In fact, singularities come in four flavors, which are denoted +1, +1/2, −1/2, and −1, for reasons that will be explained shortly. Three of these types are found in fingerprints, as shown in **10.16**.

More precisely, Penrose's rule states that, after adding up the numbers associated with all singu-larities on a hand, the number of excess −1/2 singularities is given by the number of fingers minus one. The term "excess" is used because a +1/2 and a −1/2 singularity count as zero (0 = (+1/2)+(−1/2)). This means that there should be at least four −1/2 singularities on a normal hand, and any number above four should be matched by an equal number of +1/2 singularities.

Elsdale and Wasoff tested their theory empirically using elongated cells called **_fibroblasts._** which, like the lines on a fingerprint, try to line up with each other, and display the same variety of four singularities, **10.17**. The advantage of using fibroblasts is that the number of "fingers" can be altered at will. This is achieved by surrounding fibroblasts

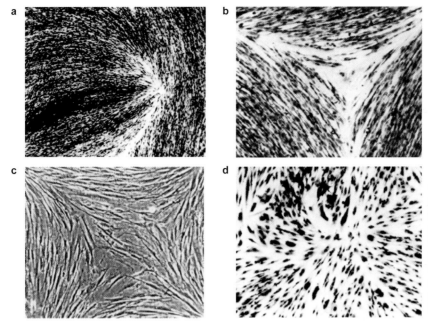

10.17 Fibroblasts
All four possible singularities are found when elongated fibroblast cells organize themselves. The topological index of the singularity in each figure is: **a** +1/2, **b** −1/2, **c** −1, **d** +1. From Elsdale and Wasoff (1976) with kind permission of Springer Science and Business Media.

with a boundary that has a series of gaps and borders. As fibroblasts line up with borders, these act like fingertips, whereas fibroblasts lie perpendicular to gaps which act like finger edges.

Penrose's rule was proved in a mathematical theorem by Elsdale and Wasoff in 1976, and we will refer to it as the ***fingerprint theorem***.

Types of Singularity

The assignment of numbers such as +1/2 to a singularity tells us how orientation varies as we move around its center. This number is called the ***topological index*** of a singularity. First we describe how the sign (+ or −) of the topological index is derived, then how its value (1/2 or 1) is derived.

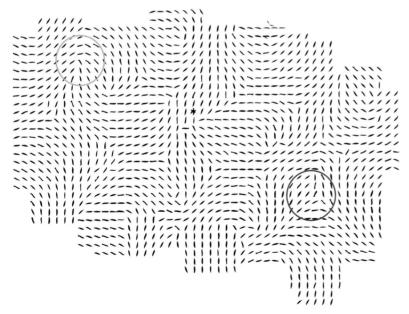

10.18 Two different singularities in the orientation map
The upper (blue) circle is centered on a singularity with a topological index of +1/2, whereas the lower (red) circle is centered on a singularity with a topological index of −1/2. Redrawn with permission Swindale, Matsubara and Cynader (1987).

241

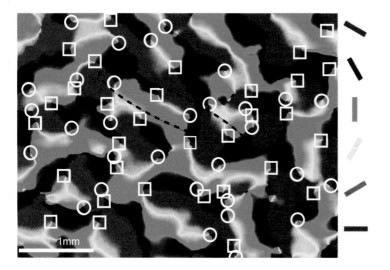

Draw a small circle around a singularity, as shown in **10.18**. Now move around that circle in a clockwise direction. If the underlying representation of orientation also changes in a clockwise direction then the singularity is positive (+). If the underlying representation of orientation changes in an anti-clockwise direction then the singularity is negative (–). This process may be made easier to visualize if you keep a pencil aligned with the underlying representation of orientation, to see how the orientation of the pencil changes with position on the circle.

In order to find the value of a singularity, we need to keep track of how much the pencil's orientation changes as the pencil is moved around the complete circle. In the case of singularities found in striate cortex the pencil rotates through no more than 180°.

For example, a clockwise journey around the upper (blue) singularity in **10.18** yields a clockwise rotation of the pencil of 180°. In contrast, a clockwise journey around the lower (red) singularity in **10.18** yields an anti-clockwise rotation of the pencil of 180°. The reason that a pencil rotation of 180° defines the singularity as ±1/2 is that 180° equals *half* of the full 360° of rotation.

In summary, if a circle is drawn around a singularity, and a pencil, which is maintained aligned with the underlying representation of orientation, is moved around that circle in a clockwise direction then a clockwise pencil rotation of 180° corresponds to a singularity with a topological index of +1/2, whereas an anti-clockwise pencil rotation

of 180° corresponds to a singularity with a topological index of –1/2.

A full rotation of 360° is said to have an index of +1. An example of this the real *pinwheel* in **10.11**. Even though no singularities with a topological index of –1 or +1 have been found in the cortex, it is possible to find these singularities in fingerprints, **10.16a**, and in the hair patterns on top of most people's heads.

The term pinwheel is commonly used to describe the orientation singularities found in the striate cortex. This is, strictly speaking, a misnomer because, unlike real pinwheels, cortical orientation singularities do not have a topological index of +1.

Note that both **10.16a** and **10.17c** (and, of course, the real pinwheel in **10.11**) have a topological index of +1, even though the line elements in **10.16a** are perpendicular to those in **10.17c**. This is an intentional feature of the topological index, which measures only the amount of orientation change in one rotation around a singularity.

Singularities and the Sign Principle

Unfortunately, nothing resembling the fingerprint theorem has been proved for the striate cortex. However, a similar story to that surrounding the fingerprint theorem does exist with respect to the striate cortex.

In examining the relation between nearby singularities, Tal and Schwartz pointed out the following two things in 1997 (both of which had been noted by other scientists such as Swindale). First, for any pair of neighboring singularities it is usually pos-

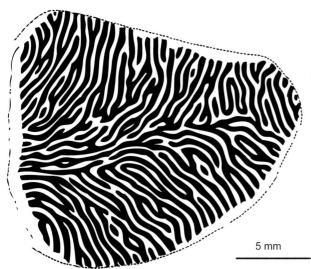

Reconstruction of layer 4C ocular dominance columns over area 17 (V1) of the right occipital lobe made from a series of reduced silver sections. Alternate columns receive input principally from left and right eyes. Note the relative constancy of column widths suggesting left and right eyes project to equal areas of V1. Redrawn courtesy of the Nobel Foundation from Hubel (1981).

5 mm

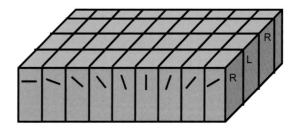

10.21 Ice cube model of the striate cortex
Redrawn courtesy of the Nobel Foundation from Hubel (1981).

sible to draw a smooth curve between them such that every column along that curve has roughly the same orientation preference. Any regions in which orientation preference remains constant is called an iso-orientation domain (*iso* means "same").

Second, two singularities joined by such a curve usually have topological indexes with different signs.

Once again, it turns out that this empirical observation corresponds to a theorem, known as the ***sign principle***. The sign principle states that nearby singularities must have topological indices

with opposite signs, as shown in **10.19**. Tal and Schwartz showed that the sign principle could be applied to orientation maps.

The sign principle originates, not in the world of human fingerprints, but in the world of the physics (random optical wave fields) where it was proved in 1994.

Packing Ocularity

A more recent image of ocularity stripes confirms the findings of Hubel and Wiesel shown in **10.20**. Ocularity represents another parameter to be

R   L   R   L   R   L   R   L   R

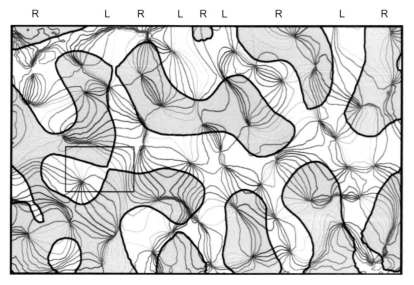

10.22 Relation between maps of orientation and ocularity
Black curves show boundaries of ocularity stripes, whereas each thin colored curve shows location of cells with the same preferred orientation, and curves converge at orientation singularities. Notice how each iso-orientation curve tends to cross an ocularity boundary at about 90 degrees, and that singularities are near the centers of ocularity stripes. R = right eye dominant; L = left eye dominant. Adapted with permission from Hübener et al. (1997).

squeezed into the striate cortex. In fact, most cells (except those in layer 4) respond to inputs from both eyes. However, the extent to which cells in each column respond to one or both eyes varies more or less continuously from column to column. Thus, some cells respond to inputs from only one eye, and are said to be *monocular*. However, most cells respond to inputs from both eyes and are said to be *binocular*. If we use a strict criterion to classify columns as either left or right, depending on which eye they respond to, then the layout of these columns on the cortex can be observed, **10.20**. The resultant pattern of *ocularity stripes* suggests that the brain attempts to ensure that every part of the visual field is processed by a pair of left-right stripes.

If squeezing the two parameters of location and orientation caused a packing problem in the striate cortex then adding ocularity naturally makes the problem more severe. One hypothetical solution is known as the *ice cube model* is shown in **10.21**. Some economy does seem to be achieved by ensuring that each iso-orientation domain in the orientation map tends to cover a pair of left-right ocularity columns. This ensures that each orientation is represented for each eye in a small patch of cortex. This model was first proposed by Hubel and Wiesel in 1971. More recent findings suggest a much more fluid arrangement in the striate cortex, as shown in **10.22**.

A Caveat: Old World monkeys (like us) have ocular dominance columns but most New World monkeys do not. In spite of this structural difference, the visual abilities of these animals are similar. In contrast, orientation columns have been found in all species so far examined.

Spatial Frequency Maps

We need to be able to perceive a line at any orientation in both eyes, in any position on each retina, but we also need to be able to perceive lines of different widths (recollect the scale problem discussed in Ch 5). For this reason, neurons in the striate cortex do not only have preference for certain retinal positions, line orientations and for information from one eye—they also prefer oriented edges with a specific width or *spatial frequency* (Ch 4). Like orientation, the representation of spatial frequency appears to vary continuously, except at spatial frequency singularities, **10.23**. Given that all the above parameters need to be represented for every retinal location, the packing problem appears to be quite overwhelming.

Color Maps

Fortunately, color does not present the same type of packing problem as other parameters, such as orientation. Color seems to be represented exclusively in columns at the center of orientation singularities, **10.24a**. Given the number of parameters that are represented in the striate cortex, this can only be a good thing. However, as the cells at the center of singularities cannot (by definition) have a preferred orientation, this means that color

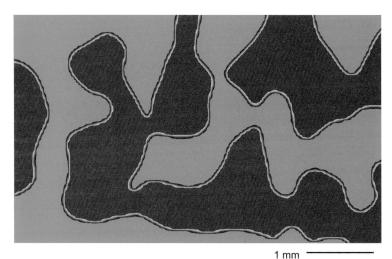

10.23 Spatial frequency map
Green regions respond to low spatial frequency (0.2 cycles/degree), whereas red regions respond to high spatial frequency (0.6 cycles/degree). Adapted with permission from Hübener, Shoham, Grinvald, Bonhoeffer (1997).

1 mm ———————

a
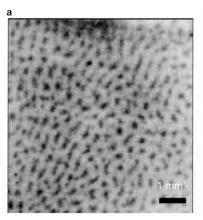

b
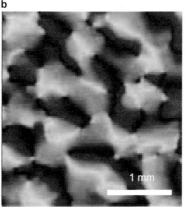

c

10.24 Three different maps of the striate cortex
a Cytochrome oxidase enzyme "blobs" indicate regions of high metabolic activity and of color sensitive cells.
b Orientation map. **c** Ocularity map. Reproduced with permission from Swindale (1998).

is not represented along with any specific orientation. Thus, it seems as if your perception of the orientation of a line defined only in terms of its color could not be mediated by the striate cortex.

Polymaps in Hyperspace

Three different maps of the same area of cortex are displayed in **10.24** including orientation, ocularity, and color (where color is represented at the center of orientation singularities).

If these three maps are overlaid, as they are in the striate cortex of course, they form a ***polymap***. From this polymap, we gain a vivid impression of the scale of the packing problem faced by the cortex, as shown in **10.25**. It seems as if the striate

cortex should be like Dr Who's Tardis, which has much more space inside that is apparent from its size! Sadly, the striate cortex is no Tardis, and the packing problem leads to a very crowded striate cortex indeed.

The striate cortex represents at least four parameters, two for location (x, y), orientation θ, and spatial frequency. Ideally, the striate cortex would exist in a universe in which space had four dimensions, instead of the miserly three dimensions of this particular universe. This four-dimensional hyperspace would allow complete continuity and coverage of four parameters.

However, a packing problem would be encountered even in a four-dimensional universe, if, for

10.25 Polymap
Relationship between spatial frequency and orientation maps in cat area 17. Each colored curve defines the location of cells that have the same preferred orientation. Each grey blob defines regions where cells have similar preferred low spatial frequencies, and lighter regions are where cells have similar preferred high spatial frequencies. Note that the iso-orientation curves and the borders of the spatial frequency domains tend to cross at right angles, and that each pinwheel-center tends to be at the center of either a low or high spatial frequency domain. Reproduced with permission from Hübener, Shoham, Grinvald, and Bonhoeffer (1997).

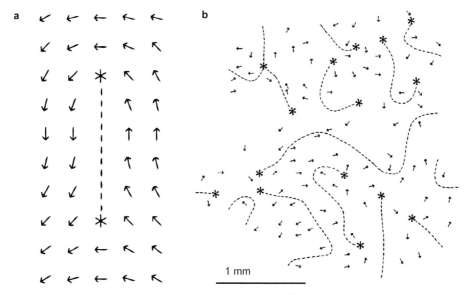

10.26 Relation between orientation and direction maps in area 18 (V2)
a Hypothetical combined map of orientation and direction. Preferred orientation of cells varies continuously around each *point* orientation singularity (marked with a *). Pairs of orientation singularities are joined by a line, which is a direction singularity. Preferred direction of cells varies continuously around each *line* direction singularity (marked as dashed line). But whereas preferred orientation varies continuously across the line singularity, preferred direction flips 180° across the same line singularity. **b** Pairs of orientation singularities in cat visual cortex are joined by curves, and each curve is a direction singularity (line direction reverses across each direction singularity). Redrawn with permission from Swindale, Matsubara, and Cynader (1987).

example, the striate cortex attempted to represent motion as well. In that case, a five-dimensional universe would be required. Envisioning even more physical dimensions is forlorn because as soon as another is added the brain will inevitably require access to a new parameter (e.g., texture, depth), which would require an additional dimension. So, there is no escape by seeking more physical dimensions. The brain needs to find some trick to solve, or at least reduce, the packing problem.

As mentioned earlier, some argue that the cortex is not really trying to solve the packing problem, but that the maps result from trying to minimize the amount of neuronal "wire-length" used. However, a key finding by Swindale et al. in 2000 suggests that the cortex attempts to maximize coverage. Specifically, they found that if any one of the maps of orientation, spatial frequency, or ocularity is altered by even a small amount then this reduces the amount of coverage. With a little thought it can be seen that this result strongly suggests that the cortex must have maximized coverage before any of its maps had been altered. This, in turn, suggests that the intact cortex maximizes coverage when left to its own devices.

Relation between Orientation and Direction Maps

One way to reduce the packing problem is to ensure that related parameters are packed in an efficient manner. This is precisely what happens with orientation and direction, **10.26.** Just as similar line orientations in a small region on the retina are represented by nearby cells in visual cortex, so similar line *directions* are represented by nearby cells.

Whereas each singularity in the orientation map is a point, each singularity in the direction map is a line. Thus, the direction map is continuous everywhere except at such line singularities, where the representation of direction suddenly jumps by 180°. However, the representation of orientation does not alter much across such line singularities, which ensures that the orientation map remains continuous even though it is overlaid on a discontinuous direction map. This can be achieved because every orientation defines two possible directions.

For example, this line | has a vertical *orientation*, but is consistent with two possible *directions*

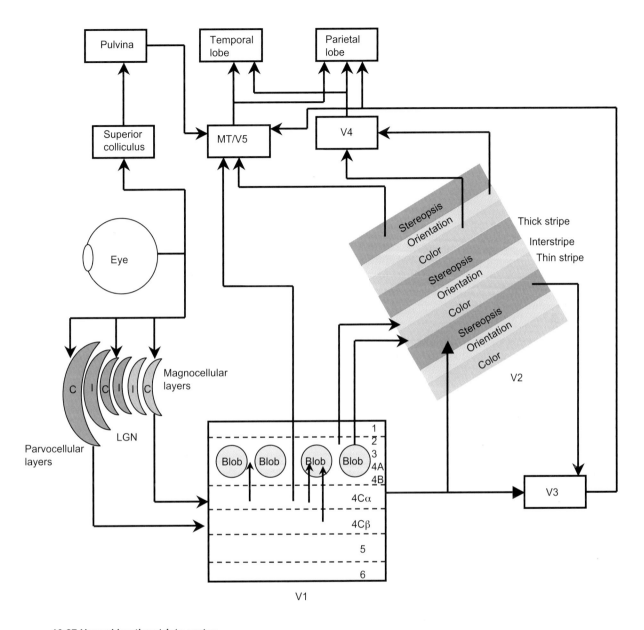

10.27 Unpacking the striate cortex
Information from different ganglion cells in each eye projects to specific layers of the LGN (C = contralateral, I = ipsilateral, see **9.9**). These then project to V1, which projects to discrete stripes of V2, which innervate V3–V5, areas specialized for processing different types of information.

(up and down). In other words, there are 180° of orientation, but 360° of direction. Thus, by using nearby points on the cortex to represent mutually consistent orientation and direction, the brain makes efficient use of cells which code for orientation and direction.

Unpacking the Striate Cortex

The brain alleviates the packing problem within the striate cortex by off-loading information to extrastriate visual areas. Each of these areas seems to be specialized for processing particular types of information. Thus, color information gets delivered to an area called V4, motion to V5, and object identity to inferotemporal cortex. This would be fine, were it not for the fact that these extrastriate visual areas represent more than two parameters or they extract new information (parameters) from their inputs, which creates a whole new packing problem in each extrastriate area.

Maps Beyond the Striate Cortex: Secondary Visual Cortex

The *secondary visual cortex*, known as V2, envelopes the striate cortex (the latter is also known as *primary visual cortex* or V1). In area V2, we witness the start of a division of labor which seems to become more marked as we progress through the visual system.

Information in V2 is organized into parallel stripes which run perpendicular to the V1/V2 border. Staining for the cytochrome oxidase enzyme (the same enzyme that revealed blobs in V1) reveals dark *thick* and *thin* stripes, separated by stripes which are *pale* because they contain relatively little cytochrome oxidase; the pale stripes are also called *interstripes*. These three types of stripes received projections from specific parts of V1, see **10.27**.

The thin stripes receive their inputs from the color blobs of V1. As might be predicted from the response characteristics of cells in the blobs of V1, cells in thin stripes are sensitive to color or brightness but not orientation.

The thick stripes receive their inputs from layer 4B of V1, which is part of the magno pathway. Consequently, cells in the thick stripes respond to specific orientations, but not to color. A key property of these cells is that they are **binocular**. They are driven by inputs from both eyes, and respond best when a line projects to a slightly different position on each retina. This difference or *disparity* in the retinal positions of lines suggests that these cells represent **binocular disparity** (Ch 18).

The pale interstripes (of V2) receive their inputs from the complex and hypercomplex cells in V1. As these cells receive their inputs from the parvocellular layers of the LGN, they respond best to oriented lines.

Despite this division of labor within V2, adjacent stripes respond to same region of retina, ensuring that the retinotopic map found in V1 is approximately preserved in V2.

Area V4

The thin stripes and interstripes of V2, both of which belong to the parvocellular stream, project to the "color area" V4. The quotes around color are used because cells in V4 do not respond only to color; they also respond to simple shapes and objects.

In terms of the parameter spaces defined in Ch 11, if V4 responded only to color then it would represent a one-dimensional subspace. However, as V4 also responds to simple objects it has to find a way to squeeze retinal location, color, and shape

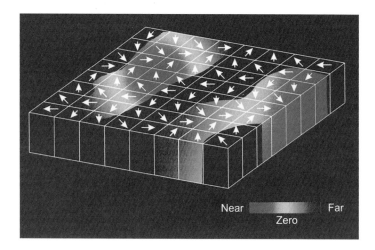

10.28 Area MT (V5) codes for motion and stereo disparity
Schematic diagram in which arrows show the preferred direction and color shows the stereo disparity of cells. The terms "near" and "far" refer to the amount of stereo disparity (see Ch 18 on *Seeing with Two Eyes*). Reproduced with permission from DeAngelis and Newsome (1999).

parameters into a two-dimensional space. As we shall see, V4 seems not to have an accurate retinotopic map, probably because its primary concern is color. Thus, in V4, we may be witnessing the decline of retinal location as the primary index, and the rise of feature-based primary indices, such as color and shape.

Area V5

The thick stripes of V2, which belong to the magnocellular stream, project to the "motion" area V5. Again, the quotes are used to indicate that cells in V5 not only respond to motion; they also respond to other parameters, especially stereo disparity, as shown in **10.28**. Like V4, this area may provide a nice example of how the spatial primary index apparent in V1 and V2 is being replaced by the non-spatial parameters of motion and stereo disparity.

Losing Retinotopy

There is little doubt that the striate cortex provides a reasonably accurate map of the retinal image. However, other visual areas seem to supplant this retinotopy with some other property.

Evidence for the loss of retinotopy can be obtained from observing how a straight line electrode track across a given cortical region maps to a corresponding track on the retina. In a simple and elegant experiment, Zeki did just this, **10.29**. He drove a microelectrode across and just beneath the surface of striate cortex, and then probed the retina to find out where each electrode location mapped to its retinal location. Of course, given a perfect retinotopic map, a straight line penetration across striate cortex should define a straight line on the retina. Essentially, this is what he found.

Zeki then repeated this process with straight line penetrations across visual regions V2, V4, and

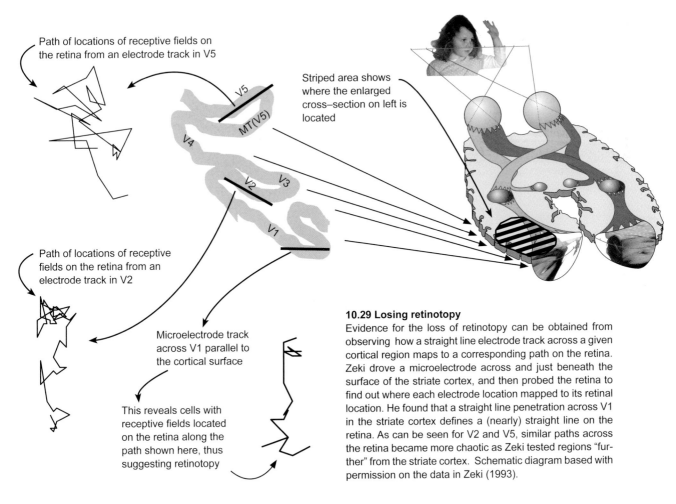

Path of locations of receptive fields on the retina from an electrode track in V5

Striped area shows where the enlarged cross–section on left is located

Path of locations of receptive fields on the retina from an electrode track in V2

Microelectrode track across V1 parallel to the cortical surface

This reveals cells with receptive fields located on the retina along the path shown here, thus suggesting retinotopy

10.29 Losing retinotopy
Evidence for the loss of retinotopy can be obtained from observing how a straight line electrode track across a given cortical region maps to a corresponding path on the retina. Zeki drove a microelectrode across and just beneath the surface of the striate cortex, and then probed the retina to find out where each electrode location mapped to its retinal location. He found that a straight line penetration across V1 in the striate cortex defines a (nearly) straight line on the retina. As can be seen for V2 and V5, similar paths across the retina became more chaotic as Zeki tested regions "further" from the striate cortex. Schematic diagram based with permission on the data in Zeki (1993).

V5. As can be seen from **10.29**, the corresponding paths across the retina became more chaotic as Zeki tested regions "further" from striate cortex.

One explanation for this loss of retinotopy is that the primary and secondary (and tertiary, etc.) indexes gradually swap places as we move away from the striate cortex. That is, the striate cortex strives hard to retain information regarding where on the retina events occur, as evidenced by its spatial primary index.

In contrast, an area such as V5, is probably more concerned with the motion of events on the retina, and is less concerned with exactly where such events occur. In other words, as we move away from the striate cortex, a new agenda gradually supplants the simple low-level description offered by the striate cortex.

While this represents a speculative account, we think it has some merit. The deep question is: if non-striate areas do not have retinal location as their primary index, what is their primary index? We explore this question in the remaining sections of this chapter.

Inferotemporal Cortex

The stunning response characteristics of cells in inferotemporal (IT) cortex speak for themselves, **10.30**. Here each point in a small region of IT seems to represent a different view of a face. But perhaps the most surprising finding is that nearby cells represent similar *views* of the face. In other words, it is as if the primary index of IT is not retinal location at all, it is a highly abstract parameter, namely viewpoint.

Is this cell the legendary **_grandmother cell_**? Well, strictly speaking, such a cell would respond to *any* view of one person. The view-sensitive cells just described were found in monkeys.

More recently, a candidate for the grandmother cell seems to have been discovered in a human. While undergoing surgery for epilepsy, the output of cells in IT was monitored while the patient was shown images of various faces. One cell seemed to respond only to pictures of Jennifer Aniston, who featured in the TV series *Friends* (and who is not a grandmother at the time of writing!). Interestingly,

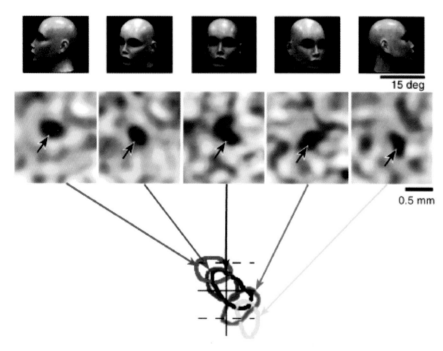

10.30 Face-view cells in inferotemporal cortex
The view shown at the top evokes activity in cortex (middle row) at a location shown by the dark spot. The cortical locations of these spots have been overlaid at the bottom to shown how similar views evoke activity in adjacent cortical regions. This suggests that nearby cortical locations represent similar face views. Reproduced with permission from Wang, Tanaka and Tanifuji (1996).

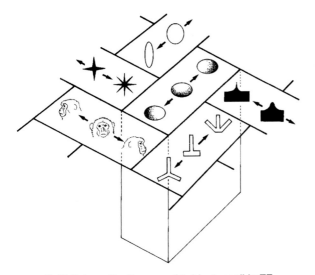

10.31 Schematic diagram of "object map" in TE
Adjacent columns do not always encode similar objects but different layers of a single column encode similar objects. Note that similarity between the layers is greater than similarity between columns. (The anterior region of inferotemporal cortex is called TE.) Reproduced with permission from Tanaka (2003).

even person identity. It is worth noting that this approach has received support from models of learning in cortex.

A variety of other high order features seem to be represented in IT, including hands, faces, walking humans, and objects, **10.31**. Moreover, different types of objects are represented in different regions of IT. However, with each small region there does seem to be a coherent map of related

nearby cells did not seem to respond to different views of Aniston, but they did respond to other characters in the same show. This suggests that nearby cells here represent many parameters that occur nearby in time (where the word *parameter* here implies identity, a very high order parameter indeed).

The reason this is interesting is because the view-selective cells described above may also be organized on the cortex according to temporal context. Nearby cells represent similar views. But notice that views represented by cells that are next to each other on the cortex are usually experienced as being next to each other in time. In other words, the temporal proximity of views experienced in everyday life may be reflected in the physical proximity of cells in the cortex that represent those views.

One objection to this is that the temporal sequence of different views of a turning head occurs over seconds, whereas the appearance of Aniston followed by her co-stars occurs over minutes. However, it remains an intriguing possibility that the high order "feature maps" observed in IT are supported by the temporal proximity of visual inputs rather than by their similarity in terms of some high order parameter, such as viewpoint or

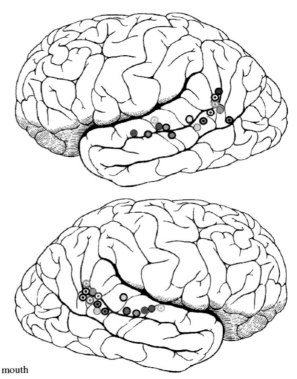

mouth
● Calvert *et al.* lip reading (STG)
◐ Calvert *et al.* lip reading (AG)
○ Puce *et al.* mouth movement
● Puce & Allison mouth movement

eyes
◉ Puce *et al.* eye gaze
◉ Wicker *et al.* eye gaze
⊙ Hoffman & Haxby eye gaze

body
⊙ Howard *et al.* body movement
⊙ Bonda *et al.* body movement
⊙ Senior *et al.* body movement
⊙ Kourtzi & Kanwisher body movement
◉ Grossman *et al.* body movement

hand
○ Neville *et al.* ASL
● Bonda *et al.* hand action
○ Grezes *et al.* hand action
● Grezes *et al.* hand movement
◐ Grafton *et al.* hand grasp
● Rizzolatti *et al.* hand grasp

10.32 Studies reporting preferred stimuli for cells in IT cortex in the two hemispheres
Each colored disc represents a report of a different kind of feature. Reproduced with permission from Puce and Perrett (2003).

objects, **10.32**. As with other extrastriate areas, the retinotopy of the maps in IT seems to be quite eroded, but not absent. However, given that it is no easy matter to discover what it is that each cell represents, making precise feature maps of large regions of IT is very much "work in progress" at the moment.

We began this chapter by stating that the brain is full of maps. Sadly, we do not know what most of them represent, nor what they are for. But the hunt is on in many laboratories world-wide, and a Nobel Prize surely awaits anyone who answers this fundamental challenge of current brain science.

Maps and Marr's Framework

Within Marr's framework for studying seeing, the packing problem is a problem at the hardware level. This does not make it an unimportant problem. Far from it: if we want to understand the brain we need to examine it at all three levels of analysis: computational theory of seeing tasks, algorithms for implementing those theories, and hardware for running the algorithms.

It is definitely not the case that the hardware level simply consists of explaining how particular components work, such as transistors in a computer or nerve cells in a brain. There are problems to be solved at the hardware level which are as challenging as those at the computational and algorithm levels used to analyze seeing tasks.

In the case of the brain, hardware problems arise in terms of how to implement seeing algorithms in ways that cope with the constraints imposed by nerve cells. These constraints include the fact that nerve cells are extremely slow devices (certainly compared with today's computing components). This necessitates massive parallelism, that is, the brain has to process visual information from different retinal regions and concerning different scene attributes at the same time.

Massive parallelism brings with it the problem of managing a massive wiring problem, because computations in one brain region need to be completed efficiently both within that region and information needs to be transferred between regions.

Both of these requirements demand schemes for arranging brain cells such that messages can be sent efficiently along nerve fibers between them.

Another constraint arises from using brain cells specialized for particular forms of processing visual information (e.g, the coding of edge location, orientation, motion, color and so on). Present knowledge of brain cells for seeing indicates that they do not change their tuning properties from moment to moment (although they can show learning effects over time). This is quite unlike the way that a computer's memory elements (called registers) can be assigned very different roles in a computation as the computation proceeds. Having cells specialized for particular functions may be one of the main reasons why the brain has maps that bring cells of particular types to work together in specialized brain regions.

The numerous maps found in the brain constitute an effective hardware solution to these (and doubtless other as yet unrecognized) problems. But using maps raises some serious problems in how to maximize coverage and continuity. Insight into these problems requires analysis at a high level of abstraction, which means using mathematics. This is illustrated by the account given above explaining why singularities arise in brain maps as a side effect of solving the packing problem. This account shows that the hardware level can get just as abstract and intricate as the computational theory of a seeing task.

In fact, we can apply Marr's logical distinction between the theory and algorithmic levels to facilitate the analysis of hardware problems. Thus, for the packing problem, the hardware problem solved by brain maps consists in how to organize cells so as to maximize coverage and continuity. At an algorithmic level, many (essentially similar) algorithms have been proposed which could achieve the emergence of maps that satisfy these goals. Crucially, many of these algorithms look as if they could plausibly be implemented by neurons, because computer simulations show that they require only local interactions between neuron-like elements and local learning rules. This is important because we are not force to invoke a "blind watchmaker" for global order to be achieved in the organisation of maps.

So here we have logical distinctions made by Marr's framework usefully applied to studying brain maps. This works because of the recursive nature of problems encountered in brain science (we first noted this in Ch 2). It appears that as we delve ever deeper into the problem of understanding the brain, we can always profit from the logical distinction emphasized by Marr between theories and algorithms for implementing those theories.

All this shows that we should beware the temptation to think of hardware problems as "second class" relatively trivial implementation problems. If we want to understand the brain then we will have to understand why various brain structures for seeing have evolved.

This requires both a deep understanding of the computations the brain performs for solving seeing tasks and the brain structures (maps) that carry out those computations.

Further Reading

Ballard DH (1987) Cortical connections and parallel processing: Structure and function. In *Vision, Brain and Cooperative Computation* 563–621 Arbib MA and Hanson AR (Eds) 581–584.

Blasdel GG and Salama G (1986) Voltage sensitive dyes reveal a modular organization in monkey striate cortex. *Nature* **321** 579–585.

Blasdel GG (1992) Differential imaging of ocular dominance and orientation selectivity in monkey striate cortex. *Journal of Neuroscience* **12** 3115–3138.

Chklovskii DB and Koulakov AA (2004) Maps in the brain: What can we learn from them? *Annual Review of Neuroscience* **27** 369–392.

DeAngelis GC and Newsome WT (1999) Organization of disparity-selective neurons in macaque area MT. *Journal of Neuroscience* **19**(4) 1398–1415.

Elsdale T and Wasoff F (1976) Fibroblast culture and dermatoglyphics: The topology of two planar patterns. *Development Genes and Evolution* **180**(2) 121–147. *Comment* How boundary conditions determine the number and type of singularities.

Hubel D (1981) Evolution of ideas on the primary visual cortex 1955–1978: a biased historical account. Nobel Lecture 8 December 1981. http://nobelprize.org/nobel_prizes/medicine/laureates/1981/hubel-lecture.html

Hübener M, Shoham D, Grinvald A, and Bonhoeffer T (1997) Spatial relationships among three columnar systems in cat area 17. *Journal of Neuroscience* **17** 9270–9284.

LeVay S, Hubel DH, and Wiesel TN (1975) The pattern of ocular dominance columns in macaque visual cortex revealed by a reduced silver stain. *Journal of Comparative Neurology* **159** 559–576.

Livingstone MS and Hubel DH (1987) Psychophysical evidence for separate channels for the perception of form, color, movement, and depth. *Journal of Neuroscience* **7**(11) 3416–3468.

Logothetis NK, Pauls J, and Poggio T (1995) Shape representation in the inferior temporal cortex of monkeys. *Current Biology* **5** 552–563. *Comment* Early study on preferred shapes of cells in IT cortex.

Merzenich MM and Kaas JH (1982) Reorganisation of mammalian somatosensory cortex following nerve injury. *Trends in Neurosciences* **5** 434–436.

Oram MW and Perrett DI (1994) Responses of anterior superior temporal polysensory (STPa) neurones to "biological motion" stimuli. *Journal of Cognitive Neuroscience* **6** 99–116.

Puce A and Perrett D (2003) Electrophysiology and brain imaging of biological motion. *Philosophical Transactions of the Royal Society of London* (B) **358** 435–444.

Purves D, Riddle DR and LaMantia AS (1992) Iterated patterns of brain circuitry (or how the cortex gets its spots). *Trends in Neuroscience.* **15** (10) 362–368. *Comment* Describes differences in maps found in different species of primate. A paper that is under-cited given its implications.

Swindale NV, Matsubara JA and Cynader MS (1987) Surface organization of orientation and direction selectivity in cat area 18. *Journal of Neuroscience* **7** (5) 1414–1427. *Comment* Describes relation between orientation and direction maps.

Swindale NV (1996) The development of topography in the visual cortex: a review of models. *Network: Computation in Neural Systems* **7** 161–247.

Swindale NV (1998) Cortical organization: Modules polymaps and mosaics. *Current Biology* **8**

270–273. *Comment* Commentary on the nature of maps in striate cortex.

Swindale NV, Shoham D, Grinvald A, Bonhoeffer T, and Hübener M (2000) Visual cortex maps are optimized for uniform coverage. *Nature Neuroscience* **3** 750–752.

Tal D and Schwartz EL (1997) Topological singularities in cortical orientation maps: the sign theorem correctly predicts orientation column patterns in primate striate cortex. *Network: Computation in Neural Systems* **8** 229–238.

Tanaka K (1997) Mechanisms of visual object recognition: monkey and human studies. *Current Opinion in Neurobiology* **7** 523–529.

Tanaka K (2003) Columns for complex visual object features in the inferotemporal cortex: clustering of cells with similar but slightly different stimulus selectivities. *Cerebral Cortex* **13** 90–99.

Thompson DW (1917) *On Growth and Form.* Dover reprint of 1942 2nd ed (1st ed. 1917).

Tootell RBH, Switkes E, Silverman MS, and Hamilton SL (1988) Functional anatomy of macaque striate cortex II Retinotopic organisation. *Journal of Neuroscience* **8** 1531–1568. *Comment* A classic early paper on mapping striate cortex.

Wang G, Tanaka K and Tanifuji M (1996) Optical imaging of functional organization in the monkey inferotemporal cortex. *Science* **272** (5268) 1665–1658. *Comment* Results of attempting to find maps in IT cortex.

Wang G, Tanifuji M and Tanaka K (1998) Functional architecture in monkey inferotemporal cortex revealed by in vivo optical imaging. *Neuroscience Research* **32** 33–46.

Zeki S (1993) *A Vision of the Brain.* Wiley, John & Sons, Incorporated. *Comment* A beautiful book, highly recommended.

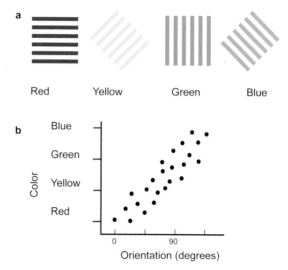

4.30 Generalized contingent aftereffect
This is a colored version of the figure that is introduced and discussed on p. 104.
a Instead of associating two colors with two orientations, a spectrum of colors is associated with the full range of orientations (only four displayed here).
b A graph of orientation versus color for this set of stimuli shows a large correlation between orientation and color.

11

Seeing and Complexity Theory

11.1 Image properties at a particular image location
Every point in the image of this flower has several properties, such as brightness, color, and edge orientation. These properties are known as parameters. A full description of this image therefore consists of a complete set of parameter values for every point in the image.

We saw in Ch 3 how the outputs of certain visually responsive cells in the Striate cortex, called simple cells, tell us relatively little about the retinal image when each cell is considered in isolation. This led us to the conclusion that the output of many such cells must be taken into account in order to work out what feature in the image produced a given set of observed neuronal firing rates. A logical extension to such a scheme is feeding the outputs of these neurons to a set of "high-level" neurons, such that each neuron in this high-level set responds only to a particular category of retinal image. This can be extended *ad infinitum* (or until we run out of neurons), and, in principle, eventually leads to the existence of neurons that are so specialized that they respond only, say, to your grandmother.

More importantly, such a cell responds to your grandmother irrespective of visual irrelevancies, such as the size of her image (perhaps she is far away, or has been on a diet), color (e.g., she has just seen a ghost), and orientation (e.g., she is a trapeze artist, or perhaps lying down or bending over). This hypothetical grandmother cell has given its name to the whole idea cells being specialized for representing specific objects irrespective of the diverse nature of their retinal images.

Why read this chapter?

By now we hope we have persuaded you that seeing is a *hard* problem for the brain. In this chapter, we make use of *complexity theory* to provide a formal account of *precisely* how hard it is. If we consider each small image patch as a sub-image then the appearance of each sub-image is defined in terms of a number of different properties or *parameters*, such as color and orientation. In essence, we show that the number of different sub-images that can occur in a single patch increases *exponentially* with the number of parameters considered, where each sub-image is defined by a unique combination of parameter values. Thus, seeing is an exponential problem, and most exponential problems are *intractable*, in principle. Despite this, almost all different sub-images can, of course, be differentiated by the human visual system. We offer several reasons why, despite its fundamentally exponential nature, seeing is indeed possible. We also explore why the outputs of single neurons tuned to a specific parameter (e.g., orientation) cannot tell the brain very much about the value of that parameter, and how this can be resolved by combining the outputs of more than one neuron.

The grandmother cell idea is usually associated with a particular type of hierarchical model of visual processing in which increasingly specialized cells are constructed by combining the outputs of cells lower down in the visual hierarchy. In this chapter, we explore the viability of such a model from a particular perspective: **complexity theory**. In essence, we consider the simplest question that can be asked of such a model: are there enough cells in the human brain to recognize every object, irrespective of properties such as size, color, and orientation? If you want to skip the rest of this chapter, and cut a long (but hopefully interesting) story short, then the final paragraph in this chapter gives the answer. If you want to find out why that is the answer then read on.

There are two approaches to exploring the problem of vision, **theory driven** and **data driven**. The data driven approach examines how the brain is put together, how each neuron responds to retinal images, how visual perceptual phenomena offer clues about mechanisms, and then tries to work out how the whole system functions. In contrast, the theory driven approach examines the problem of vision from an abstract perspective. The theory driven approach emphasizes the nature of the *problems* to be solved, and which constraints might be employed to solve those problems. Using this approach, it is possible work out how *any* system, man-made, alien, or biological, could take advantage of these constraints to solve the problem of

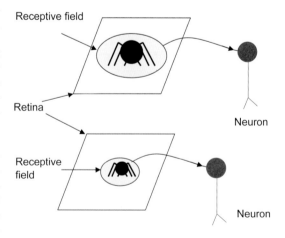

11.2 Template-neurons
Each neuron responds to a spider of a specific size, as defined by the neuron's receptive field on the retina.

vision. That said, one of these systems is of course of particular interest to us, the human brain.

In each chapter in this book, we will usually begin by exploring a given problem of vision with the theory driven approach Specifically, we will adopt a *computational* approach The advantage of this theory driven computational approach is that we can concentrate on the problem of vision itself, without being impeded by having to refer constantly to structure of the brain or to myriad perceptual phenomena. This is not to say that we will not take account of the brain. Ultimately, it is the brain we want to understand. However, our plan is to provide a theoretical analysis of how vision might be solved *in principle*, and then to relate this theoretical analysis to the known neurophysiology of the brain and to perceptual phenomena. Thus, this theory driven computational approach aims to determine not only how vision could be implemented *in principle*, but also how vision can be implemented in the brain *in practice*.

The Concept of Complexity

The essence of the problem posed by vision is shown in **11.1**. If we consider each small image patch as a *sub-image* then the appearance of each sub-image is defined in terms of a number of independent properties or *parameters*, such as color, orientation, and texture. In essence, we show that the number of different sub-images that can occur in a single patch increases exponentially with the number of parameters considered, where each sub-image is defined by a unique combination of parameter values. Every point on the flower defines a sub-image and is associated with *different* values of these parameters. Note that this results in many overlapping sub-images, each of which has its own parameter values.

If we assign a number to represent the value of each parameter for every sub-image in the image then we would have a very large description indeed. Yet this is effectively what the brain does when it correctly perceives what is present in the scene. For example, in order to record the color and orientation of every sub-image in this image, we would need to assign two numbers (one for color and one for orientation) to every sub-image in the image. In perceiving the object that created this image, the brain makes properties of every

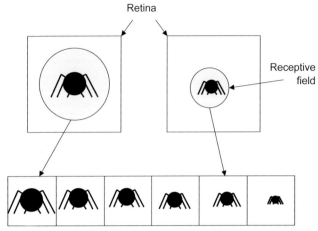

11.3 One-dimensional (1D) parameter space
This shows how 1D value units (template-neurons) encode a parameter with non-overlapping values (size, here).

sub-image *explicit* using a representation which assigns a value to every parameter in every sub-image in the image. How does the brain represent all this information without running out of neurons? This is the main question explored in this chapter.

The answer takes the form of a *counting argument,* which consists of counting the number of possible sub-images to be recognized, and compares this to the number of neurons in the brain. Such arguments have a long history in mathematics, and can be used, for example, to show that there are quantities (i.e., most real numbers) which are literally *uncomputable* because if we write one computer program to compute each real number then we would run out of programs long before we ran out of real numbers.

In tackling this question we need to explore the concept of *complexity* from a scientific perspective. This will reveal that although vision feels effortless to us—because we open our eyes and there is the visual world in all its glory—vision is in fact a highly complex problem.

Fortunately, from a purely mathematical perspective, vision is complex in a way that we can quantify in a rigorous manner. In order to explore precisely *why* vision is hard, we need to back up a little, and to re-introduce the idea of vision as a form of pure template matching. For consistency, we will stick with the example of vision in spiders used in Ch 3. However, we will disregard the am-

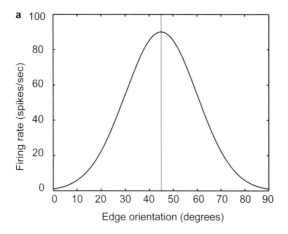

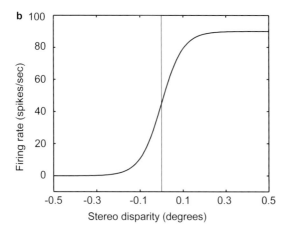

11.4 Response properties of two types of units
a Value unit that responds to orientation.
b Variable unit that responds to stereo disparity.

biguity problem discussed there associated with the use of templates, and we will handicap our spider by preventing its retinas from moving around. These restrictions will make little difference to our final conclusion, but will help in simplifying the arguments developed here.

Template-neurons, Variable Units, and Value Units

Imagine a spider that needs to recognize other spiders using a spider-template. In principle, a spider-template could be constructed in a similar manner to the "bar-detector" receptive field de-

scribed in Ch 3. Thus, a neuron could be connected to the retinal receptor mosaic in such a way that the neuron's receptive field looks like a typical spider shape. We will use the term ***template-neuron*** for these hypothetical neurons, and we will assume that each neuron can detect a spider image defined by its receptive field template, as shown in **11.2**.

It is important to note that this template-neuron, and much of the material introduced here, is in the form of a thought, or *gedanken,* experiment. There is some evidence that template-neurons, or neurons that look like they could be template-neurons, exist. However, for the present, their existence is unimportant. What we are trying to do is to establish whether or not they would be any use *if they did exist.* If we can establish that they would be of use in principle then we can ask whether the neurons in the brain that look like they might be template-neurons are indeed template-neurons. Now let us return to our spider example.

The immediate problem to be faced is that the retinal image of a distant spider is smaller than that of a nearby spider, so we need a different spider template for each possible size. However, size varies continuously, and we would therefore need an infinite set of templates if we had one template-neuron for each size.

We can reduce this to a manageable number by ***discretizing*** the continuous range of sizes to, say, 6 different sizes, as shown in **11.3**. Notice that each template-neuron is actually responsible for a small *range* of sizes, and that the ranges of different templates do not overlap; this implies a local representation using narrow tuning curves, as described later in this chapter. Now, if a spider comes into view then exactly one of the spider template-neurons will be activated. Each neuron with a spider-shaped receptive field responds to the image of a spider at a specific size, and ignores other properties, such as color.

Any neuron that responds to only one property (e.g., size) is said to have a ***univariate*** response function (where uni- means one), and because it responds to only a small *range* of values it is called a ***value unit***, **11.4a**. In contrast, ***variable units*** respond to *all* values of a *single* parameter, as illustrated in **11.4b**.

Parameter Spaces and Multivariate Value Units

Now suppose the spider is on hilly terrain, so that other insects can appear at almost any orientation. This can easily be accommodated by introducing templates at different orientations; 10 different orientation ranges should suffice. But how many template-neurons are required for recognition at each of 10 sizes *and* at each of 10 orientations?

Well, for every possible size, 10 oriented templates (at that size) are required: one template for each orientation. As there are 10 sizes, and 10 different oriented templates are required at each size, this makes a total of 10 (sizes) × 10 (orientations) = 100 templates. If each template is implemented as a template-neuron then the resultant 10 × 10 array of template-neurons **spans** the **two dimensional space** of all possible combinations of 10 size and 10 orientation values. As we have seen, the properties size and orientation are more properly called **parameters**, which is why the two dimensional space of all possible sizes and orientations in **11.5** is called a **parameter space**.

In order to ensure that a spider can be recognized at *any* size and orientation, it is necessary to have a template for *every* possible combination of size range and orientation range. As each of the 100 template corresponds to a square in the two dimensional parameter space, a 10 × 10 grid of templates effectively "tiles" every point in this parameter space. Each template is responsible for a small square region of parameter space.

Here, rather than responding to only one parameter, each template-neuron responds to (a small range of) two parameters. Each template-neuron is a value unit because it responds if the parameter values associated with the retinal image fall within a small range of values. However, these template-neurons are not univariate because they respond to more than one parameter (e.g., size *and* orientation), and are said to have **multivariate** response functions (where multi- implies many).

Note that the template-neurons do not have to be arranged in a 2D grid as shown in **11.5**. The important thing is that there is one template-neuron for each square on the grid because this ensures that each combination of size and orientation can be detected by one neuron.

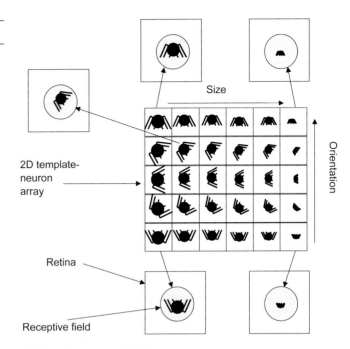

11.5 Two-dimensional (2D) parameter space
This shows a 2D array of template-neurons. Each template-neuron encodes a specific retinal image (e.g., a spider at one size and orientation). Each large square represents an example retinal image with a circular template-neuron receptive field overlaid on the retina. In this example for the sake of drawing clarity, there are 6 sizes and 5 orientations, rather than the 10 parameter values used in the text when calculating combinations.

How Many Template-neurons?

Unfortunately, spiders do not vary only in orientation and size *on the retina*: they also vary in terms of other parameters, such as their physical size in the world, shape, color, and in the way they move. So there are many more parameters than we could have taken into account thus far. Let's choose color as the next parameter to utilize, as jumping spiders have color vision, and, for consistency, we will assume spiders come in one of 10 possible colors. Now, if recognition of a spider at any one of 10 retinal sizes, orientations, and *colors* is required then how many templates are required? Well, for a specific size and orientation, 10 color templates are required, one for each color at that size and orientation. We already know that a 10 × 10 two-dimensional grid of templates is required to recognize the 100 different combinations of size and orientation. We can thus imagine the 10 color templates required for each size and orientation

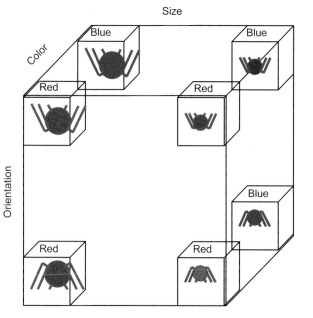

11.6 Three-dimensional (3D) parameter space
Spider templates required to recognize a spider in any combination of three sizes, orientations, and colors. Here the number of parameters is $k = 3$ (size, orientation and color) and the number of values each parameter can adopt is also $N = 3$ for display purposes ($N = 10$ in text). This 3D parameter space of templates therefore contains $M = N^k = N^3 = 3 \times 3 \times 3 = 27$ templates.

sent the 10 values each parameter can adopt so that in our example $N = 10$ (e.g., there are 10 differently sized templates). If the number of combinations of all possible values of each parameter is represented by the symbol M then we have already shown that $M = N \times N \times N = 1000$. This can be written in terms of powers of N, because $N \times N \times N$ is just N cubed (N raised to the power 3) or N^3. Notice that the value of the exponent (3) is equal to the number of parameters (k). This allows us to write: $M = N^k$. This is *very* bad news because it means that the number M of templates required increases ***exponentially*** as the number k of parameters increases, **11.7**.

Understanding the equation $M = N^k$ is crucial for understanding why vision is so very hard; not just for modern computers trying to recognize, say, faces, not just for spiders trying to find a mate, but for *any* system trying to recognize *anything*. The problem presented by this equation cannot be solved by building bigger or faster computers. The exponential nature of vision represents a fundamental problem, for which a complete solution cannot in principle be found using conventional

extending out from each cell on the grid, creating a *three*-dimensional grid or parameter space, **11.6**.

This ***three-dimensional parameter space***, divided into 10 intervals along each dimension, defines a total of 10 (colors) × 10 (sizes) × 10 (orientations) = 1000 small *cubes*, each of which corresponds to one template, which in turn corresponds to one template-neuron.

In summary, if each template is the receptive field of one neuron then we would need a total of 1000 neurons to recognize a spider at any possible size, orientation, and color, as shown in **11.6**. By now you will have noticed a disturbing pattern in the way the number of required template-neurons increases each time we add a new parameter, such as color. In essence, every time we *add* a new parameter, the number of required template-neurons is *multiplied* by a factor of 10! This is very bad news.

We can state this more succinctly if we introduce a little algebra. The symbol k is used to represent the 3 parameters of size, orientation, and color, so that $k = 3$. The symbol N is used to repre-

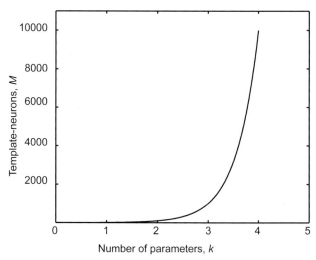

11.7 Exponential growth
If the image of an object is defined in terms of k parameters, and each parameter can adopt N values then the number M of possible images increases as an ***exponential*** function N^k of k ($N = 2$ in graph). Reducing N has little impact on this; it is k that dominates how the number of possible images increases. A faster computer would not help very much because every time we add a new parameter, we multiply the processing time (number of combinations) by a factor of N (e.g., $N = 10$).

computational systems such as brains and computers (but see later for partial solutions).

As we have already noted, each time a new parameter is added we *add* 1 to the value of k, but this has the effect of *multiplying* our old value of M by a factor of N. We can see how the number of templates increases as k increases by plotting a graph of M versus k, where $M = N^k$ as in **11.7**. Note that the value of M increases at an alarming (exponential, to be precise) rate as k increases. We can get an idea of just how alarming this is by evaluating M for plausible values of k and N.

Spider Brain Runs Out of Neurons

If we keep the value of $N = 10$, so that there are 10 possible values for each parameter (e.g., color) and we assume that each object to be recognized is defined in terms of $k = 3$ parameters then the number M of combinations of parameter values is $M = 10^3 = 1000$, as described above. Now if $k = 4$ then $M = 10^4 = 10,000$, and if $k = 5$ then $M = 10^5 = 100,000$. We are now probably at the limit set by the number of neurons in a spider's brain.

In other words, if a spider attempted to recognize an object which was defined in terms of 5 different physical properties or parameters, with a resolution of 10 values per parameter then it would require 100,000 neurons (assuming the spider used one neuron per template).

Note that 5 is not a large number of properties to recognize, and it is therefore not a large value for k. If the object to be recognized could appear at any location on the retina then this adds two new parameters, location along the horizontal axis of retina and location along the vertical axis of retina, making a total of (1) size, (2) orientation, (3) color, (4) location along horizontal axis of retina, and (5) location along vertical axis of retina.

If the object is to be recognized even when it is rotated in depth then this adds another two parameters, making $k = 7$, which implies that $M = 10^7$ or 10 million templates are required. At this point, our spider has long since run out of neurons, and this is assuming a lowly $N = 10$ values per parameter, which is almost certainly an underestimate of what is needed.

Given this theoretical insight, it is now possible to offer a plausible reason why spiders have retinas that can move horizontally, vertically, and rotate. It could be that these movements serve to align the spider's template-neurons with the input image. This would reduce the number of parameters that must be represented by individual neurons by three (two for position, and one for orientation). In our spider, with 10 values for each of its $k = 7$ parameters this reduces the number of required neurons from 10^7 (10 million) to 10^5 (one hundred thousand), a *saving* of 9,900,000 (about 10 million) neurons!

Human Brain Runs Out of Neurons

The conclusion so far is that template matching is unlikely to be a viable strategy for an animal with only 100,000 neurons, although it may be just about viable provided neurons can be saved by introducing special purpose mechanisms such as mobile retinas (and provided the intrinsic ambiguity of neuronal outputs can be overcome, see Ch 3). But what about us, with our legendary large brains? Could template-neuron matching work for us? Sadly, the exponential nature of vision ensures that even we simply would not have enough neurons to go around.

Let's take up the argument well after the point where the spider ran out of neurons. For $k = 7$ parameters, we would require 10 million template-neurons if we allocated one neuron per template. If we are able to recognize objects at any one of 10 speeds then this adds one to k, making $k = 8$. If we can recognize an object at any one of 10 movement directions "around the clock" and 10 movement directions in depth (i.e., "around a clock" viewed edge-on) then this adds two parameters, making $k = 10$. Where are we up to? Well, with $k = 10$ parameters, $M = N^k = 10^{10}$, which is 10,000,000,000, or ten billion template-neurons (a billion = a thousand million). This implies that we need 10^{10} multivariate value units (neurons) *to recognize a single object!*

To continue with this neuron-counting theme, if we can recognize one thousand objects (and there is evidence that we can recognize many thousands of faces, for example) then each object would require its own set of ten billion templates, yielding a grand total of $M = N^k = 10^{10} \times 1000$ objects = 10,000,000,000,000,000 or ten trillion

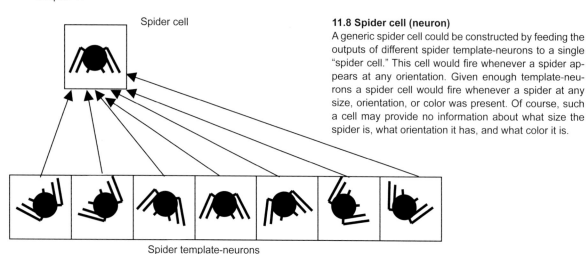

Spider cell

Spider template-neurons

11.8 Spider cell (neuron)
A generic spider cell could be constructed by feeding the outputs of different spider template-neurons to a single "spider cell." This cell would fire whenever a spider appears at any orientation. Given enough template-neurons a spider cell would fire whenever a spider at any size, orientation, or color was present. Of course, such a cell may provide no information about what size the spider is, what orientation it has, and what color it is.

template-neurons (a trillion = a million millions). And this is for a mere 10 values of each parameter for each object. But, you might say, if we allocate one template per neuron then surely the massive human brain would have enough neurons. Sadly, the number of neurons in the human neocortex (the bit that counts for this computation) is a mere 20 billion (20×10^9), according to current best estimates, nowhere near enough to supply the ten trillion required by template matching. The conclusion is inescapable: even the human brain does not have enough neurons to go around, and is defeated (when viewed from the perspective of numbers of template-neurons needed) by the exponential nature of vision.

Grandmother Neuron Saves the Day?

Let's set aside the stupendous size of these numbers for a moment and consider the fact that we could, in principle, construct a neuron which responds to only one object, a spider for example. This can be done by feeding the output of each template-neuron into a single hypothetical "spider-neuron," see **11.8**. This spider-neuron's connections are set in such a way that if *any* of the 10^7 template-neurons fires then so does the spider-neuron (i.e., this spider-neuron is said to perform the **logical OR function** on its inputs). This spider-neuron would fire irrespective of the size, orientation, color, or any other parameter value associated with the retinal image, provided that image contained a spider.

Such a neuron is said to be **invariant** with respect to all the parameters encoded by the

template-neurons. Crucially, this spider-neuron fires *if and only if* a spider is present in the retinal image. However, because the spider-neuron is invariant with respect to most parameters, it is also completely **insensitive** to their values. In other words, the firing of the spider-neuron signals the presence of a spider in the retinal image, but it carries no information regarding the size, color, or location of that spider in the image. This information is implicit in the particular template-neurons that caused the spider neuron to fire, but the spider-neuron does not have access to this information. Generally speaking, we would like a perceptual system which was simultaneously **invariant** with respect to perceptually irrelevant parameters such as color, while retaining **sensitivity** to perceptually salient parameters (such as location). Unfortunately, the notion of perceptual salience is not fixed even from moment to moment, which suggests it is necessary to have access to the values of all parameters associated with a specific object (i.e., to all template-neuron outputs).

Taking this argument into the context of human vision, it would be useful to have a neuron which fired if and only if, say, your grandmother were present. For this reason, these hypothetical neurons are often referred to as grandmother neurons or **grandmother cells**, **11.9** (although that figure actually shows *daughter* cells, which are usually happier than grandmother cells).

One objection to the grandmother cell idea is that it requires at least one neuron per object. However, when set against the ten billion neurons required to detect each object over its entire range of parameter values, it seems niggardly to object to an idea because it requires one extra neuron per object!

In practice, such a scheme ought to include more than one grandmother neuron per grandmother. If you only had one grandmother neuron then the death of this neuron could result in a condition which we will call your *grandmother agnosia,* because it precludes recognition of a single unique object, your grandmother (the term *agnosia* is used to describe an inability to recognize objects).

As with the spider-neuron, the grandmother neuron raises many problems. The firing of your grandmother neuron would inform you that your grandmother is present. However, it is not only necessary to know your grandmother is present, you would also like to know where she is in the room, whether she has lost her spectacles again, and what her orientation is (for example, a *horizontal* grandmother is not a good sign).

Local and Distributed Representations

The grandmother cell lies at one extreme on a continuum of possible schemes for representing information. If each object is represented by a single cell then the representation of that object can be localized in that cell, which is why the grandmother cell is called a *local representation*. Recollect that this concept was introduced in Ch 3. In contrast, a *distributed representation* consists of the collection of states of a population of neurons, which usually have overlapping tuning curves (see *coarse coding* and **3.17** in Ch 3: we return to this issue at the end of this chapter).

Vision and Complexity Theory

This section can be skipped unless the reader is particularly interested in further details of *complexity theory* as it applies to vision.

Complexity theory describes how quickly a small or *tractable* problem can turn into a hard or *intractable* problem. Problems, and equations, come in three basic flavors, *linear*, *polynomial*, and *exponential*. The reason that problems and equations have the same flavors is because the size of a problem (like vision) increases or *scales* at a rate defined by whether its equation is linear, polynomial, or exponential, as shown in **11.10** and previously in **11.7**.

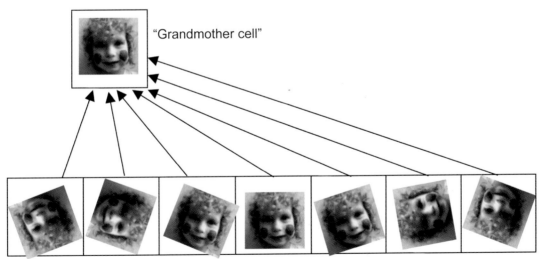

Face template-neurons

11.9 "Grandmother cell"
Extending the logic of **11.8** to faces, a generic grandmother cell (here a daughter cell in fact!) could be constructed by feeding the outputs of different face template-neurons to a single "grandmother cell." Of course, such a cell may provide no information about what size your grandmother is, what orientation she is in, and what color she is (these last two are especially important considerations when it comes to grandmothers).

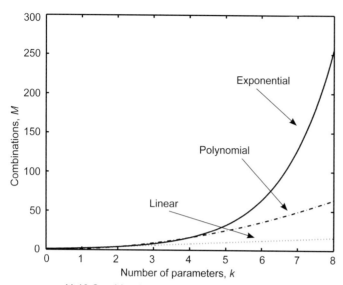

11.10 Combinations grow with the number of parameters
As the number k of encoded parameters (e.g., size, orientation) increases, so the number M of template-neurons required to recognize images increases exponentially ($M = N^k$). For comparison, the number of required neurons required if this relationship were polynomial ($M = k^N$) or linear ($M = kN$) are also shown.

At a practical level we are concerned with the size of our problem, but we are also interested in how this size *scales* with the number k of parameters and with the number N of value ranges for each parameter. Knowing how a problem scales tells us not only if it can be solved for given values of k and N; it also tells us how quickly the problem becomes intractable as these values increase.

For a problem that scales as $M = N^k$, the value of the **base** N has relatively little impact on its size M. In contrast, the value of the **exponent** k is crucial. As k increases, M heads skyward at an alarming rate, almost irrespective of the value of the base N, as shown in **11.10**. In fact, the value of N was set to $N = 2$ for display purpose in **11.10**; otherwise, the exponential curve would be so high that the other curves would be indistinguishable from the horizontal axis.

As we have just seen, the size of our vision problem increases exponentially as the number of parameters increases. Thus, if we double the number of parameters from $k = 3$ to $k = 6$ then the problem size (the number of template-neurons required) increases from $M = N^3$ (i.e., 1000) to $M = N^6$ (i.e., one million), if $N = 10$, as in **11.7**.

In contrast, if the size of our vision problem scaled *linearly* with respect to the number k of parameters then doubling the number k would simply double the number M of required templates. Examples of linear equations are $M = 2k$, or $M = Nk$, where N is assumed to be constant. Supposing $N = 10$, and using $M = Nk$, doubling the value of k from 3 to 6 implies a doubling in the problem size from $M = 30$ to $M = 60$.

Finally, if our problem scaled as a *polynomial* function of k then we might have (for example) $M = Nk^2$. In this example, the problem size scales in proportion to the square of the number k of parameters. (The key difference between a polynomial and an exponential function is that k is raised to some fixed power (e.g., k^2) in a polynomial, whereas k is the power to which some other constant is raised (e.g., 2^k or N^k) in an exponential function). Given a value of $N = 10$, and using $M = Nk^2$, if k increased from 3 to 6 then the problem size M would increase from 90 to 360.

Basically, if our problem scales linearly with k then life is sweet. If it scales as a polynomial function of k then life is tolerable. But if our problem scales exponentially then the number of required template-neurons would increase as an exponential function of k, and life and vision would be, as it is, *hard*.

It might seem odd to have a section of complexity theory in a book on vision. However, before we considered complexity theory, we might have guessed in some vague way that vision is hard, but we weren't sure just how hard. Now we know: *very* hard; specifically, exponentially hard.

This is not an indication that we should pack up and go home, but it is an indication that we should not be surprised to find the problem of vision cannot be solved with mere brute force computing power, in the form of bigger computers, parallel computers, or faster computers, or equivalently bigger brains, parallel brains, or faster neurons. Instead, and in the absence of almost 3.5 billion (a billion = a thousand millions) years of evolution, the problem of vision requires some hard thinking, a rigorous scientific exploration of the only existence proof that vision is possible (i.e., animal brains), and a theoretical framework sufficiently powerful to meet the challenge thrown down by the results of complexity theory.

Naturally, there are several objections to this analysis. The most compelling objection is that, however impossible it may seem, we do *see*. This and other less obvious objections will be considered next.

Objectionable Objections

We call the following objections *objectionable* because, although they seem quite plausible, they actually make no difference at all.

Objection 1: The brain is a parallel device

The brain, unlike most computers, processes many bits of information simultaneously, or *in parallel.* But this is irrelevant. Given that we have assigned one template per neuron, there are not enough neurons to represent every possible combination of parameter values. This is true whether the brain considers all possible templates at the same time (i.e., in parallel), or not.

Objection 2: The brain is too slow

Surely, one might say, if we could speed up the processing rate of neurons, or build a super-fast computer, then the exponential nature of vision would no longer be a problem. One of the great fallacies of our age is that all problems can be solved with more speed. If a computer could compare the retinal image to each template at a rate of one million per second then it would take 2.8 hours to search the N^{10} templates associated with 10 parameters, and millions of years to compare the N^{16} templates associated with 16 parameters (assuming $N = 10$ in both cases)! As shown in **11.11,** a faster computer would be of no help.

How about if we had a parallel computer like the brain, but we made it super-fast? Well, if we increased the brain's processing speed by a factor of 1 million then it would still require the same number of neuron templates, so that's no help.

Valid Objections

We now consider a set of objections which, unlike those above, cannot be cast aside. However, it should be noted that even though some of these objections do reduce the size of the problem, none of them reduces its complexity, and so it remains a problem with exponential complexity.

Objection 3: The brain uses univariate neurons

We have assumed that each neuron is multivariate. In essence, this means that each neuron simultaneously represents a small range of values for each of k different parameters, so that each neuron responds to a specific set of parameter values. This implies that each neuron represents parameter values which correspond to a small *volume* of parameter space.

As we have seen already, if we want to represent every size and orientation using multivariate value units at a resolution of 10 values per parameter then we need $10 \times 10 = 100$ template-neurons, and this set of 100 templates effectively tiles the 2D parameter space.

However, if each neuron is *univariate*, so that it responds to only a small range of just one parameter, and disregards the values of all other parameters, then only a small number of neurons would be required. For example, the size parameter could be represented by 10 univariate neurons, where each neuron responds to a small range of sizes. Similarly the orientation parameter could be represented by another set of 10 univariate neurons.

k	Linear $N \times k$	Polynomial k^N	Exponential N^k
10	1.1×10^{-10} days	2.8 hours	2.8 hours
20	2.3×10^{-10} days	118.5 days	3.0×10^{6} (3 million) years
30	3.5×10^{-10} days	18.7 years	3.2×10^{16} years
40	4.6×10^{-10} days	332.6 years	3.17×10^{26} years

11.11 The problem of recognizing an object described in terms of k parameters
If our "template" is defined in terms of k parameters, and if we could search through 1 **million combinations of parameter values per second** (e.g., 1 million possible images per second) then the time required to search through all combinations (for $N = 10$, and $k = 10$ to 40) is given in the table [$10^{-10} = 0.0000000001$; $10^{6} = 1,000,000$].

1D orientation
subspace

2D parameter
space

Size

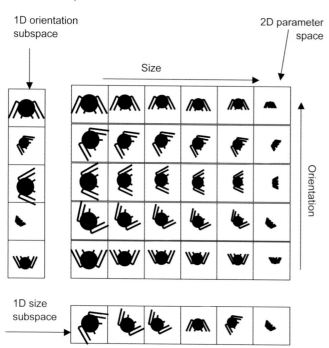

Orientation

1D size
subspace

subspace (see **11.12**). This sounds very good news. So what's the catch?

Here's the catch. If the image contains *two* lines, a thick red line and a thin blue line, then we would not know if we were looking at thick red line and a thin blue line or thin red line and a thick blue line.

In order to see why this is so, let's consider hypothetical animal-specific univariate template-neurons. If the *two* objects in the retinal image are a *red spider* and a *blue bird*, **11.13**, then the color neurons corresponding to the templates "blue" and "red" would be activated, and neurons corresponding to the templates "spider" and "bird" would be activated. Unfortunately, the resultant set of neuronal activations carries absolutely no information about which object has which color. If we only made use of such univariate neurons then we wouldn't know if we were looking at a red bird and a blue spider, or a red spider and blue bird.

11.12 One-dimensional (1D) subspaces

If we allocate one template-neuron for each possible size and orientation then we need 6(sizes)x5(orientations) = 30 neurons to tile this 2D parameter space (or 10 × 10 = 100 if we had 10 values per parameter of size and orientation). However, if we had 6 neurons, each of which responded to size only, irrespective of orientation, then we would need only 6 neurons to encode size. This set of neurons is shown as a horizontal 1D array. Similarly, if we had 5 neurons, each of which responded to orientation only, irrespective of size, then we would need only 5 neurons to encode orientation, shown as a 1D vertical array. This makes a total of 5 + 6 = 11 neurons to encode size and orientation (or 10 + 10 = 20 neurons if we had 10 values per parameter). The 1D array of neurons that encode orientation only is called a **parameter subspace**, as is the array of neurons that encodes size only. However, this reduction in the number of required neurons leads to the binding problem (see **11.13**).

We would therefore need a total of 10 + 10 = 20 template-neurons to represent any combination of size and orientation, as in **11.12**.

Note that each set of 10 univariate neurons corresponds to a *one-dimensional **subspace*** of the full two-dimensional (or *k*-dimensional) parameter space. If we add another parameter such as color then a total of 10 + 10 + 10 = 30 univariate template-neurons is required. Thus, instead of representing *k* = 3 parameters with 1000 template-neurons which together tile a 3D parameter space, we need only a total of 30 neurons where each subset of 10 neurons tiles a one-dimensional parameter

11.13 Binding problem

If the brain used only univariate neurons, so that each neuron encoded only one parameter (such as color), then this figure would activate the following neurons. The blue bird would activate a "blue neuron" and a "bird neuron," whereas the red spider would activate a "red neuron" and a "spider neuron." Given that these four neurons are active, how does the brain know if the retinal image contains a red spider and a blue bird, or a red bird and blue spider? This is an example of the binding problem. Photograph courtesy Charlie Towers.

In essence this single-parameter representation provides no mechanism for *binding* together parameter values as they appear in the retinal image (e.g., red and spider). Hence, this type of representation gives rise to the ***binding problem***: if the brain had only univariate neurons then how would it know which parameter values are bound together in the retinal image? One intriguing solution to the binding problem involves neural synchronization, as discussed in Singer and Gray (1995) and in Singer et al. (1997).

Objection 4: The brain does not need to encode all possible parameters

As we have seen in the case of the spider, reducing the number *k* of parameters by even a small amount can have a dramatic effect on the number of hypothetical template-neurons required. What is true of the spider is also true of any other vision system. Like spiders, and contrary to a common misconception, we cannot easily recognize a given object when it is presented on any position on the retinal image. In other words, we do not have ***position invariant object recognition***.

This is not to deny that we can recognize objects in our peripheral vision (i.e., in peripheral parts of the retina), but object recognition is much worse using peripheral vision. When we want to "home in" on a specific object we use the central region of the eye, the ***fovea***, which is specialized for seeing fine details. We move our eyes to bring the object of interest on to the fovea, a process called ***foveation***, **11.14**. This precludes the need to recognize any object in any position in the retinal image.

The ability to foveate effectively saves us two parameters (horizontal and vertical retinal position), or at least it saves having to code each parameter at a high resolution. Thus, we might

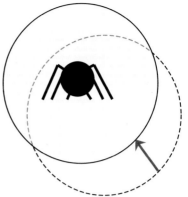

11.14 Foveation
A suitable eye movement can bring the spider image to the center of the retina.

have template-neurons for each object at each size orientation, but most of these would be concentrated around the fovea. In this way, the horizontal and vertical position parameters would effectively be "engineered away" in large part, because the ability to foveate means that we do not require template-neurons for corresponding to the two parameters of retinal position. This consideration underlies our earlier discussion of why the spider has evolved the means to move its retinas (under its fixed lenses, Ch 3).

Objection 5: The brain does not need to encode all possible parameter values

Even if a specific parameter must be represented by the brain then it may be that certain parameter values can be represented with low resolution. This would mean that very few neurons would be used to encode values of that parameter. In the limit, a low resolution implies that certain parameter values are not represented at all. Parameters may be represented with low resolution for two different reasons.

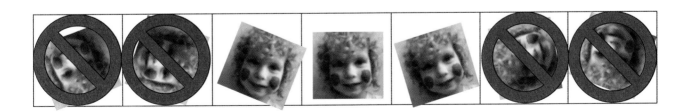

11.15 Humans do not need to encode all face orientations

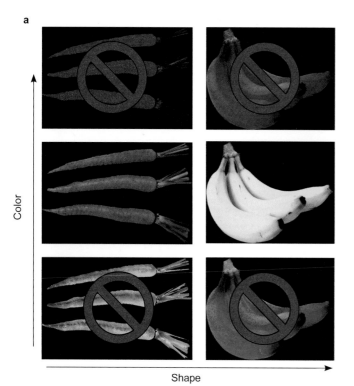

11.16 Cutting down on encoding combinations
a Encoding only yellow bananas and orange carrots.

b Two-headed child?

The first reason can be exemplified using faces, which are almost always the right way up, although they may also appear in profile. Psychological evidence suggests that we are especially poor at recognizing inverted faces, although we can easily recognize faces in profile. Representing faces over a small range of orientations provides a saving of one parameter, which can have a substantial effect on the number of hypothetical template-neurons required, **11.15**.

The second reason reflects the observation that certain parameters are surprisingly stable. For example, certain retinal images are highly stable with respect to changes in object orientation, so that quite large changes in view point yield only minor changes in the retinal image.

Such **generic viewpoints** may form the basis of an efficient strategy for recognizing objects, because neurons which represent such viewpoints could be activated over a relatively wide range of viewing angles. The existence of stable generic views would ensure that certain ranges of the orientation parameters would require relatively low

resolution with no corresponding loss in ability to recognize objects at different viewpoints.

Objection 6: The brain does not need to encode all combinations of all parameter values

We call this the *red banana objection*. Given the co-occurrence of the color yellow and the curved shape of a banana, there seems little sense in allocating many neurons to the representation of red bananas; they are simply too rare, from a purely statistical point of view, **11.16**. And if we have few neurons for bananas other than ones responsive to yellow then we are likely to be slow to recognize a red banana.

This suggests that there could be entire regions of parameter space which are "off-limits" from a practical perspective. Red bananas, blue spaghetti, flying pigs, winged worms, six-foot babies, green humans, and two-headed children (**11.16b**) would exist only in highly restricted regions of parameter space.

Even coming up with examples is difficult precisely because such objects are outside of our everyday experience. Indeed, from a strictly statistical point of view, such objects should be literally unimaginable!

In fact, most of the objects that could exist do not. In other words, most of possible parameter space is entirely unpopulated by objects that exist in the physical world. (Perhaps our dreams permit some regions of this uncharted parameter space to be explored, and perhaps that is why dreams can appear so bizarre.)

In turn, this suggests that we do not need to represent most regions of parameter space; we only need to represent those regions associated with objects we experience in our everyday lives. This produces substantial savings in terms of the amount of neuronal resource allocated to representing objects.

Objection 7: The brain uses population coding

We have assumed that neurons have multivariate response functions, and that each neuron is responsible for an *exclusive* range of parameter values. If there are two parameters then this implies that each neuron is responsible for one small region of parameter space. As this is the region of parameter values to which a neuron is tuned, we refer to it as a *parameter field*. (A parameter field really represents the cross-section of a two-dimensional tuning surface, as we will see later in this chapter.)

For example, a neuron which responds to red-to-orange lines between 40° and 45° in the image on the retina has a receptive field define in the conventional way. However, the same input also defines a small region in parameter space. In this case, the two-dimensional parameter space consists of color and orientation, and the neuron responds to parameter values within a region which includes red-orange along the color axis and 40–45° along the orientation axis. Recollect that we call the receptive field in parameter space the *parameter field*.

If each neuron is responsible for a parameter field which does not overlap with the parameter fields of other neurons then this is called *fine coding*, **11.17b**. For example, if there are two parameters then each of the template-neurons described here has a parameter field that does not overlap with any other parameter field. If there are three parameters then each of the template-neurons described here has a parameter field that is a cube which does not overlap with any other cubic parameter field. This can be extended to include four

or more parameters, in which case each parameter field is called a *hypercube*. Thus, fine coding consists of contiguous, non-overlapping parameter fields.

A more realistic scenario is that different neurons can respond to the same value of a given parameter, but their outputs reflect their different *preferred* parameter values. For example, a neuron might have a *preferred orientation* at 45°, but it also responds to orientations at 43°. Crucially, such

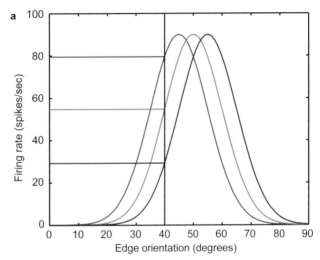

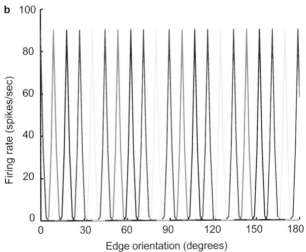

11.17 Coding resolution
a Coarse coding Each value unit has a broad tuning curve which overlaps with the tuning curves of its neighbors. An edge orientation at 40° activates many units (like the three shown here).
b Fine coding Each value unit has a narrow tuning curve which does not overlap with the tuning curves of its neighbors, so a single edge orientation activates only one unit.

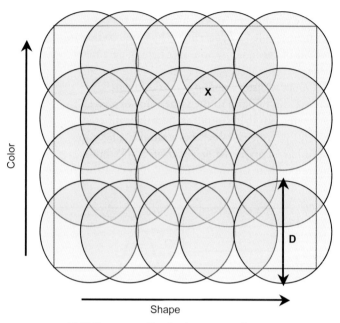

Color

Shape

D

X

11.18 Coarse coding for two parameters
The overlapping parameter fields in this parameter space ensure that each combination of size and color activates more than one neuron. For example, the size and color defined by the point marked x activates four neurons. This is called coarse coding. The parameter D defines the receptive field size.

neurons have *overlapping* parameter fields. These are usually circular, although this has little impact on the argument. Once we have overlapping parameter fields then we enter the world of **coarse coding**, **11.17**, **11.18**.

Using coarse coding, the firing rate of any one neuron does not, of itself, tell us what is in the image. We have to observe the different firing rates of an entire population of neurons in order to interpret what might be in the image. If this sounds familiar (Ch 3), it is because coarse coding includes *distributed representations* and *population coding*—concepts we introduced in Ch 3. To put it another way, population coding and distributed representations are specific examples of coarse coding, **11.17**, **11.18**.

When considering two parameters, each parameter field may be circular, and these circular receptive fields overlap in parameter space. This means that many neurons are activated by a single retinal image of an object. The activation level of each neuron indicates the degree to which the retinal image matches the parameter values at the center of that neuron's parameter field, **11.18**.

What has any of this to do with the exponential nature of vision? In principle nothing, but in practice a great deal. Basically, coarse coding does alleviate the problem of vision, but only a bit. Using fine coding with small non-overlapping parameter fields, we know that the number of neurons required for k parameters is $M = N^k$, if each parameter is divided into N intervals, and if we allocate one template-neuron to each interval. Using coarse coding, as we *increase* the size of each parameter field, so the number of required template-neurons *decreases*.

At first sight, this looks like a major victory for coarse coding against *the curse of dimensionality* (as it is sometimes called). However, it can be shown that, for parameter fields with diameter D, the number M of required neurons is $M = D \times C^k$, where $C = N/D$, and N is now the number of template-neurons per parameter (as a reminder, the value of M obtained with fine coding was $M = N^k$).

Aside from the multiplier D (which we will ignore for now), coarse coding has effectively reduced the base from N to $C = N/D$, *but it has left the exponent k unaltered*. Recall from our initial discussion of exponential problems that it is the value of exponent (k) that matters, not the value of the base (N or $C = N/D$).

As for the multiplier D in $M = D \times C^k$, this is no help at all because it *increases* the number of required neurons by a factor of D. Again, this is actually of little consequence because it does not affect the value of the exponent k.

Whichever way we choose to cut it, we are forced to conclude that, while coarse coding might reduce the overall size of the problem, it has at best only a marginal impact on the fundamentally exponential nature of vision.

Objection 8: Neurons as memory registers
Suppose that the brain uses neurons like a computer uses its memory registers. A computer register is a set of slots for storing information. Each slot can contain a zero or a one, that is a binary digit or *bit*. Clearly, if there is one slot then two possible values can be stored (i.e., 0 and 1). If there are two slots then four possible values can be stored ([00], [01], [10] or [11]). More generally, if there are k slots in the memory register then 2^k possible values can be stored, where these k values are known as a *binary string* or *binary number*.

What has this to do with vision? Well, if each distinct binary number could somehow be associated with a different image, and if each neuron could adopt two possible states (0 and 1), then k neurons could represent up to $M = 2^k$ images. A more realistic assumption is that each neuron can adopt up to $N = 10$ distinct states, which would permit a set of k neurons to represent $M = N^k$ images. Notice how we have sneakily used the same notation as above, but here N is no longer the number of values of each parameter (e.g., speed); it is the number of distinct neuronal states. This is convenient, because the N different states of each neuron could be assumed to represent the N different values of a single parameter. Similarly, k is no longer the number of parameters, it is the number of neurons; but we could assume that each neuron represents a single parameter. Now we have k parameters each of which is represented by a single neuron, and N values for each parameter, where each parameter value corresponds to a distinct state. This seems to suggest that we do *not* need an exponential number of neurons to represent the exponential number of possible images, after all! In fact, it suggests that we need exactly k neurons to represent k parameters.

There are at least two problems with this approach. First, it assumes that neurons act as variable units; that is, each neuron can encode all values of a single parameter. In practice, on the whole, neurons seem to behave like value units, which each neuron encodes a small range of specific values of a small number of parameters.

Second, notice that this is a reduced version of the subspaces described above, in which we allocated one neuron (value unit) for each small range of parameter *values,* rather than one neuron (variable unit) per parameter, as here. This similarity should give a clue to the second flaw in this scheme. Recall that the subspace scheme gives rise to the binding problem. For example, if we have one neuron for speed and one for orientation then a line moving across the retina can be encoded by the joint activity of these two neurons. But if there are two lines then, never mind the binding problem, we also have a representational bottleneck. How could two such neurons represent more than one line?

The simple answer is that they cannot, because each neuron cannot adopt more than one state

at one time. However, one solution would be to reduplicate this pair of neurons, and to allocate one neuron-pair for each line on the retina. Again, the correspondence between the direction and orientation of each line would require some form of binding to ensure that the correspondence between orientation and direction of each line was correctly maintained. And if we wanted to encode three parameters per line (e.g., orientation, direction, and color) then we would need three neurons per line.

More generally, encoding k parameters per line would require k neurons, and each line would require its own set of k neurons with some binding mechanism to keep track of which neurons are associated with the same line. Rather than considering individual lines, we can consider whole objects. Now if each object is defined by k parameters, and if each parameter is encoded as one of N possible values then we require k neurons per object. If we can recognize one million objects then this requires a total of k million neurons (plus a means of binding neurons in the same set together). In summary, this scheme entails three assumptions: first, neurons act as variable units (which seems to go against the prevailing evidence); second, if an object is encoded in terms of k parameters then it has its own dedicated set of k neurons; and, third, some form of binding mechanism exists to ensure that all members of the same set remain bound together. These may not represent insuperable problems, but their solution is far from obvious.

Finally, a computer scientist might object that this chapter is just a simplistic account of the number of possible images that can be recorded by a digital image. For example, if each pixel can adopt $x = 10$ grey levels, and there are $z = 1000$ pixels in an image then the number of different possible images is simply enormous $y = x^z = 10^{1000}$, which is much larger than the number of images we have been considering up to this point. So our counting argument does *not* require that the brain must recognize *all possible images*. This is because most images that are possible are basically visual gibberish (e.g., random grey-levels), and are never encountered in practice. Instead, the counting argument considered here tests the idea that the brain could allocate one cell to each of the possible images that could be generated by a small number (k) of different physical parameters, each of which can adopt one of a small number of values ($N = 10$ here). In this case, the number of possible images to be recognized is "merely" $M = N^k$. And just in case you missed it, the answer is: no, there aren't enough neurons to go around.

Objections: Summary

To sum up, none of above objections overcomes the fundamental fact that vision is exponential. However, some objections remove the need to represent large chunks, and in some cases, almost all, of parameter space.

From a theoretical perspective, the fact that vision remains exponential despite all these objections might suggest that we really should hang up our hats and go home defeated. However, lessons from other research fields where similar problems are encountered suggest that any objection which removes huge chunks of parameter space may make the problem tractable, *in practice.*

By analogy, the brain may apply all sorts of ***heuristics*** to the problem of vision, where a heuristic is defined as a rule of thumb—a generally applicable bit of good sense but not an absolute certainty to work. Heuristics have been found to help solve the problem of vision and yield good, if not optimal, solutions. In the context of vision, these heuristics are often formulated in terms of constraints or priors, and these will form the basis of another chapter, Ch 13 *Seeing As Inference.*

Ambiguity in Simple Cell Responses

In Ch 3 we examined the ambiguity in simple cell responses using the example of how a change in a simple cell's activity could be caused either by a change in orientation or a change in contrast of the stimulus landing on the cell's receptive field. We now look at this issue again, setting it within the context of the concepts developed in this chapter.

Consider the orientation ***tuning curve*** of the simple cell shown in **11.19**. If an edge oriented at either 53° or at 127° (i.e., 37° either side of 90°) is presented to this cell's receptive field then the cell's output for *both* orientations would be 12 spikes/second.

If the cell's output depended only on edge orientation, and not on contrast, then this form of ambiguity could be resolved by considering the output of just two cells, as shown in **11.20**. In this case, the output from a pair of cells acts a signature

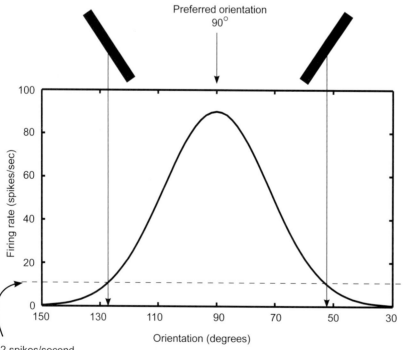

11.19 Ambiguity problem in simple cell orientation tuning curve
The graph shows the responses of a simple cell as a function of stimulus orientation. The orientation that produces the highest response rate for this cell is 90°. The black bars show oblique stimuli (about 37° either side of 90°). They produce a much lower response rate than that for the preferred orientation of 90°. However, the main point illustrated here is that although the bars are of different orientations they generate the *same* response of 12 spikes/second. Hence the cell's response is ambiguous: which orientation caused the response?

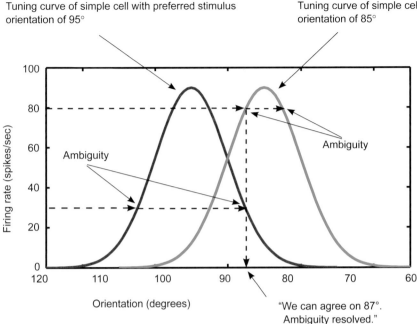

Tuning curve of simple cell with preferred stimulus orientation of 95°

Tuning curve of simple cell with preferred stimulus orientation of 85°

Ambiguity

Ambiguity

"We can agree on 87°. Ambiguity resolved."

Firing rate (spikes/sec)

Orientation (degrees)

11.20 Resolving simple cell ambiguity
The ambiguity in the output of a single cell can be resolved by considering the output of a pair of simple cells with different preferred orientations. The two tuning curves shown here belong to cells with preferred orientations of 85° and 95°. The responses of the two cells cell are 80 and 30 spikes/second. Even though each cell's response does not specify a unique orientation on the associated tuning curves, taken together they uniquely specify an orientation of 87°.

which is uniquely associated with a single orientation. This kind of ambiguity is not unique to cells in the visual system, but is a general property of all sensory cells which have univariate tuning curves (i.e., cells which respond to just one parameter, as in **11.19**, **11.20**). This turns out to be an example of a general rule is: if k is the number of parameters to which a cell is tuned, then resolving the response ambiguity requires $k + 1$ cells.

We will illustrate this with the example of a cell tuned to show a preference for a specific *speed* and a specific *movement direction* of a line moving across the retina. (Ch 14 deals with motion at length.) The tuning curve of such a cell can be drawn as a three-dimensional "hill" or ***tuning surface***, where each location on the "ground plane" corresponds to a specific line speed and direction on the retina, and where height indicates the response of the cell, **11.21a**. So here we have a cell that responds to two ***parameters***, speed and direction. The parameter values which induce the largest response from this cell are its ***preferred*** parameter values. These preferred values define a specific location on the ground plane, a location which corresponds to the peak of the tuning surface.

How ambiguous is the response of a cell like this? Well, a line moving across the retina at a speed of 6 degrees/second and a direction of 5° de-fines the point located at the base of the thick vertical line on the left in **11.21a**. We denote the location of this point by the coordinates $(s, d) = (6, 5)$, where s stands for speed and d stands for direction. The response of the cell is given by the height of the response surface above this location, about 20 spikes/second in this case. However, inspection of **11.21a** reveals that this response of 20 spikes/second could have been produced by many different combinations of speed and direction.

In fact, if we "walk around" the hill at a height of 20 spikes/second then this defines a circle around the response surface. This circle defines a corresponding circle on the ground plane, the circle labeled **1** in **11.21b**. Crucially, every point on this circle defines a *different* combination of speed and direction, and a retinal line with any one of these combinations of parameter values would yield a response of exactly 20 spikes/second. Therefore, the cell's response of 20 spikes/second is consistent with an *infinite* number of pairs of parameter values (i.e., those pairs which lie on the circle on the ground plane in **11.21**; as examples, just four cases are picked out with vertical lines). So the response of such a cell is pretty ambiguous! This situation contrasts with the situation illustrated in **11.19**, **11.20**. In those figures, cells are sensitive to a single parameter (e.g., orientation)

a

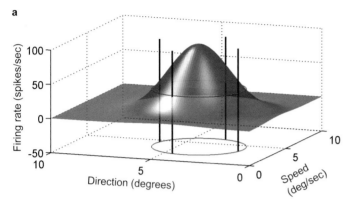

b

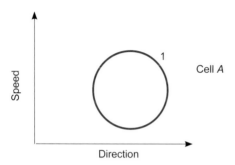

11.21 2D tuning surface of a cell tuned to speed and direction on the retina

a A line moving across the retina at a speed of 6 degrees/second and a direction of 5° defines a single point on the "ground plane." Let's suppose a stimulus induces a response of 20 spikes/second. This defines a specific height on the 2D tuning surface of a single cell, defined by the circle on the surface. This circle has also been redrawn on the ground plane. However, if we know only the cell's response (after all, that is all the brain has to go on) then all we know is that this response could have been caused by a stimulus with any combination of speed and direction that lies on the circle in the ground plane. The vertical lines show how 4 examples of particular speed/direction stimuli link the circle in the circle ground plane to the circle on the tuning surface.

b Plan view of circle on the ground plane. Any combination of speed and direction that lies on the circumference of this circle gives rise to the same neuronal response of 20 spikes/second. The quantities speed and direction are known as stimulus parameters.

a

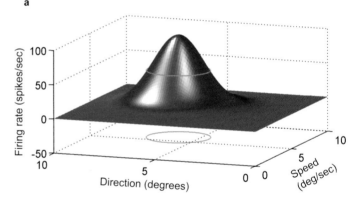

b

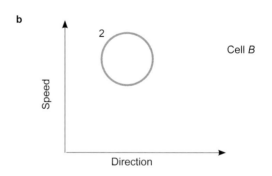

11.22 Another 2D tuning surface of a cell tuned to speed and direction on the retina

a The 2D tuning surface of this cell is tuned to a retinal speed and direction different from that shown in **11.21**.

b Plan view of circle on the ground plane. As for **11.21**, any combination of speed and direction that lies on the circumference of this circle gives rise to the same neuronal response.

and each cell's output is consistent with exactly *two* values of the orientation parameter.

We discovered above that the ambiguity in interpreting a one-parameter ($k = 1$) cell's output was resolved using two cells (i.e., $k + 1 = 1 + 1 = 2$ cells). So for a two-parameter cell ($k = 2$) our formula of $k + 1$ tells us that we need three cells. But let's begin by showing that using only two cells is not enough to resolve the ambiguity.

In **11.22** we show a second cell with a different preferred speed s and direction d to the cell defined in **11.21**. In order to keep track of which cell we are referring to, let's label the first cell as cell A, and the second cell as cell B, and let's refer to the circles defined by the cells' responses as **1** and **2**, respectively. Let's suppose that the line moving across the retina stimulates A a little less than B, so A's response is 20 spikes/second and B's response is 30 spikes/second.

As before, we will consider the effects of a stimulus with values $(s, d) = (6, 5)$. The response of cell B is given by the height of B's tuning surface above the location $(s, d) = (6, 5)$, which is about 30 spikes/second, **11.22a**. Again, from the brain's viewpoint, this response is ambiguous because it could have been generated by a

a

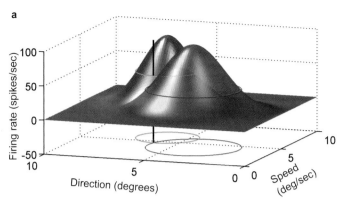

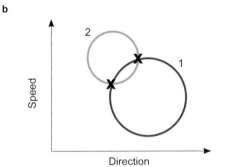

b

11.23 The two receptive fields from 11.21 and 11.22

a A single stimulus evokes a response in each cell, corresponding to the height of the circle drawn on each tuning surface.

b Plan view of circles on the ground plane. These circles intersect at two points on the ground plane (marked **X**). Each intersection point defines exactly one retinal speed/direction pairing that could have evoked the observed response. Although the ambiguity is considerably reduced, it is not completely resolved, and clearly another cell is needed.

line moving across the retina with many different combinations of speed and direction. As for cell *A*, such speeds and directions include the actual retinal stimulus, with parameter values $(s, d) = (6, 5)$, as well as any stimulus with parameter values that lie on the circle on the ground plane in **11.22a**, marked as circle **2** in **11.22b**.

Just as *A*'s response defines the circle **1** in parameter space, so *B*'s response defines a circle **2** in parameter space (smaller, because *B*'s response to $(s, d) = (6, 5)$ happens to correspond to a height "further up its hill").

We now show in **11.23** both cells *A* and *B*. We know, in our example, that the different responses of cells *A* and *B* are generated by the same retinal stimulus with parameter values $(s, d) = (6, 5)$. So the circles **1** and **2** must pass through the location of the true parameter values $(s, d) = (6, 5)$, marked with a vertical line in **11.23a**. This location defines one of the intersection points between circles **1** and **2** in **11.23**. But notice that there is another intersection point, with a location on the ground plane that is also consistent with the two observed cell responses of 20 and 30 spikes/second. This second intersection point implies that there is a *second* pair

of parameter values defining a line moving across the retina with another speed and direction that could have generated the observed cell responses. Thus, having the responses of two cells to a single moving retinal line reduces the ambiguity to two possible speed/direction combinations. But how do we know which intersection to choose? Clearly, we need more cells. So let's now see if our rule of $k + 1$, which here predicts $2 + 1 = 3$ cells, works in this case.

A third cell *C* is shown in **11.24a**. This happens to have a response of 60 spikes/second to our retinal moving line with parameter values $(s, d) = (6, 5)$. The vertical line that ascends from the location $(s, d) = (6, 5)$ pierces the tuning surfaces of cells *A*, *B*, and *C* at heights of 20, 30, and 60 spikes/second, respectively. We know that each of these three observed responses could have been generated by many different combinations of speed and direction, those lying on the circle for each cell.

We also know that each of these three responses was in fact generated by a stimulus with parameter values $(s, d) = (6, 5)$, which is why each of the three circles must, by definition, pass through the location $(s, d) = (6, 5)$ on the ground plane. To be sure, as we saw in the two-cell example above, each *pair* of circles also passes through a second location (marked with a **x** in **11.24b**), which could have generated the responses of that pair of cells. But only one location, marked with a black dot in **11.24b**, has all three circles pass through it, because that location defines the true parameter values, the parameter values $(s, d) = (6, 5)$ of the line actually moving across the retina.

Thus, from the brain's point of view, knowing the response of one cell is massively ambiguous, knowing the response of two cells is still ambigu-

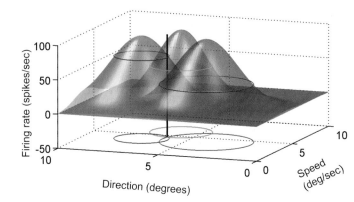

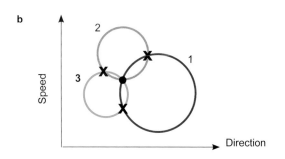

Direction (degrees)

Speed (deg/sec)

Firing rate (spikes/sec)

Speed

Direction

11.24 Using three cells to resolve ambiguity

a A single stimulus evokes a response in each cell, corresponding to the height of the circle drawn on each tuning surface. The vertical line links the stimulus to the tuning surfaces.

b View from above of the circles in the ground plane in a. Each circle represents retinal speed/direction pairings that yield the *same* response from one cell. The points marked **X** define speed/direction pairings consistent with the responses of exactly two cells. Only the point shown with a black dot is consistent with the responses of all three cells.

rameter values define the true situation "out there in the world." In our present example, agreement was discovered by finding the single location on the ground plane where all three circles in **11.24** intersect. (One way of implementing constraint satisfaction in a biologically-plausible algorithm is to use the Hough transform, as explained in Ch 8 *Seeing Objects*).

How does this $k + 1$ rule relate to the main thrust of this chapter, which has explored the combinatorial explosion problem for template theories? Notice that the rule applies to each small region on the retina, so that $k + 1$ cells are necessary to recover the values of k parameters for each region. Moreover, we now know that if cells act as template-neurons then the number of cells required to encode all possible values of k parameters grows *exponentially* with the number of parameters. For example, we could add the parameter of edge orientation, which is known to affect simple cells, so that $k = 3$ (speed, direction, and orientation). If each cell can signal $N = 10$ possible values of each parameter then we would need N^k cells, or $10^3 = 1000$ cells for each small region on the retina. Then, having used 1000 cells to "tile" the 3D parameter space associated with each retinal position, a stimulus would activate some subset of these cells, and we would need to observe the responses of at least four (i.e., $k + 1$) cells in that active subset in order to estimate the values of the three parameters of speed, direction and orientation.

As if this were not bad enough, one crucial caveat ensures that, while $k + 1$ cells are *necessary* to recover k parameter values, they are not usually *sufficient*. This is because the response of each cell contains random noise.

One method for reducing the effects of noise is to *average* responses from many cells. The underlying assumption here is that the noise in any given

ous because they could have been generated by two *possible* combinations of parameter values, but knowing the responses of three cells implies a single combination of parameter values: the true values of the retinal line's speed and direction.

This is a convincing demonstration that the $k + 1$ rule works. In this test case, we have shown that if each cell encodes *two* parameters then exactly *three* cells are necessary to recover unambiguously the values of the two parameters in the retinal image.

The reasoning used in this example is often described in the following way: each activated cell provides a measurement that limits, or technically *constrains*, the possible stimuli that could have caused that response to a subset of all possibilities. A sample of cells with overlapping tuning surfaces provides a ***set of constraints***. Finding stimulus parameters which satisfy the constraints from all cells is a form of processing called ***constraint satisfaction***. Putting this informally, it is as though a populations of cells is being asked: given your current responses, which stimulus parameter values can you all agree on? If agreement is possible, as here in **11.24**, then it is assumed that those stimulus pa-cell's response is independent of that affecting other cells' responses. This means that the noise variations tend to cancel out when an average is taken. This is one reason why it may be a good idea for the brain to consider the responses of entire populations of cells when attempting to recover the parameters of the stimulus that caused those cells to respond, as described in Ch 3. Using averaging to get round the problem of noise is explored Ch 5 in connection with the task of edge detection from noisy images.

A Contradiction?

At first sight, it may appear that we have a contradiction here. First, (1) we have concluded that if each neuron encodes k parameters then we need the output of $k + 1$ neurons to overcome the ambiguity inherent in the output of (anything less than $k + 1$) individual neurons. We have also concluded in this chapter that if each neuron prefers a specific value for each of k parameters and if there is one neuron for every *combination* of N values of each parameter then we would require $M = N^k$ neurons to *tile* the k-dimensional parameter space. However, if we use coarse coding, where the broadness of the tuning curve is denoted by a diameter D, then we concluded that (2) we can reduce the number of required neurons to $M = DC^k$, where $C = N/D$, and N is now the number of neurons per parameter.

However, there is no contradiction here. Statement 1 states that $k + 1$ template-neurons are required to decode *a single instance* of an input image which is defined in terms of k parameters. In contrast, statement 2 says that the number of template-neurons required to ensure that *some* subset of (at least) $k + 1$ template-neurons is activated by *every* input image is $M = DC^k$.

The point is: we know the output of $k + 1$ template-neurons is required to resolve ambiguities but we need to tile the entire parameter space (using $M = DC^k$ template-neurons) because we don't know *which particular subset* of $k + 1$ template-neurons will be active before the image lands on the retina. So statement 1 tells us how many neurons are required to decode *a given input*, whereas statement 2 tells us how many neurons are required to ensure that some subset of (at least) $k + 1$ neurons will be activated by *every* possible input image.

Template-neurons and Grandmother Cells

One obvious question is: do template-neurons or grandmother cells exist in the human brain? This is a controversial issue, but recent research has thrown up some remarkable findings. It has been known for years that cells exist which respond to a hand, a face, or even an entire human walking along, Ch 10 *Seeing with Brain Maps.* Surprisingly, with every year that passes, more cells of this type are discovered, usually in the temporal cortex, with ever greater specificities. For example, in 2006, cells were discovered that fired if Jennifer Aniston (from the TV series *Friends*) was present in the retinal image. However, the precise computational role of these cells is simply not known, although it is naturally the subject of much conjecture. Such cells are clearly candidates for grandmother cells, but whether or not they really are grandmother cells is unknown at present.

Spiders and Edges

If you think the example of recognizing a spider is a little fanciful then consider what would happen if we wanted to recognize a line, or an edge (see Ch 5), or equivalently, one segment of the leg of a spider. In this case the line could vary in terms of about $k = 6$ parameters: orientation, length, width, speed, color, and contrast (in practice, color seems to be encoded by cells in the "blobs" in primary visual cortex, Chs 9, 10. If each parameter can adopt only one of $N = 10$ values then we would need $M = 10^6$ (1 million) template-neurons to represent any line that could possibly occur in *each and every* region of the retina (or least in the fovea).

Now suppose we wanted to recognize two lines joined at an angle, or equivalently, a whole leg of a spider (if we assume that a spider's leg has two visible segments). This adds one parameter, making $k = 7$. Note that we did not consider spiders whose legs could bend in this chapter. If we had, it would have added one extra parameter, per leg; that's 8 extra parameters. By now we hope you start to get an impression how spiders and simple lines, or pairs of lines, or octets of lines (e.g., spiders) are not so different. They all demand exponential numbers of template-neurons because the problem of vision is fundamentally exponential in nature.

Historical Note

John Mollon reported in 2008 his findings on the history of the idea of that each object is represented by a single entity in the nervous system. One of the earliest mentions of the idea that he could find was a proposal by the Swiss naturalist and philosopher Charles Bonnet (1720–1793), who held that there were nerve fibers or "fibrils" that vibrated in sympathy with specific objects or words. Ramon y Cajal (1892) suggested that pyramidal cells of the neocortex were *psychic cells*, and Sherrington (1940) proposed the idea of a ***pontifical cell***.

More recently, a student of Pavlov, Jerzy Konorski (1903–1973), coined the term ***gnostic unit*** in 1967 to capture the idea of a neuron which encodes a single object, and the term "grandmother cell" was coined by Jerry Lettvin around 1961. In a parallel development Barlow proposed the idea of ***cardinal cells*** in 1972.

The general idea of a grandmother cell is the logical and seductive conclusion implied by the increasingly sophisticated receptive fields discovered by Hubel and Wiesel in the early 1960s. These findings suggested a hierarchical organization for visual processing. If this hierarchical organization is pushed to its logical limit then it has seemed inevitable to many that grandmother cells must exist. But only time will tell whether this is true. See *Further Reading* to find fascinating recent commentaries on Barlow's landmark paper, including his own statement on his current views on this topic.

Sparse Coding

The arguments given in this chapter indicate the implausibility of grandmother cells, but there is one consideration in their favor that we have not yet considered: the desirability of ***sparse coding***. In essence, a sparse coding of inputs implies that most neurons are inactive most of the time. There are good computational reasons for the brain to use sparse coding. But Peter Lennie (2003) has argued there are also sound energy-supply reasons for most neurons being inactive most of the time. He points out that the adult human brain uses about 20% of the energy supply of the body at rest. In a child this baseline figure rises to 50%.

About 13% of the brain's supply of energy is used for sending signals down neurons, and the rest is for maintenance. The energy costs of a single spike are such that this 13% can support an average firing rate of only 0.16 spikes/second for each neuron in the neocortex (and the neocortex uses about 44% of the brain's overall energy).

Given that an active neuron has an average firing rate of around 50 spikes/second, this implies that only around 3% of neurons can be active at any one time. More importantly, this type of analysis suggests that the luxury of having a big brain incurs two related energy costs. First, there is the substantial cost of maintaining neurons in their resting state. Second, there is the cost of actually using one of these expensive devices, a cost which is so high that we can afford to use only about 3% of them at any one time (neocortical neuronal activity uses up around 2.5% of the body's total energy in an adult, and 6.5% in a child, i.e., 13% of 20% for an adult, and 13% of 50% for a child).

One unfortunate implication of this is that if you ever manage to engage your brain in a task that involved all your neurons firing at 50 spikes/second then your brain would boil. On the other hand, sparse coding might allow you to recognize your grandmother without cooking your brain.

In summary, the idea of grandmother cells faces considerable computational problems, but it does at least have the merit of being consistent with sparse coding and with the constraints on the amount of energy available to the brain.

Concluding Remarks

We have considered in this chapter how the brain deals with the combinatorial explosion of possible images that the world presents to the retina. In so doing, we are forced to conclude that there is no short cut, no computer or brain that is fast enough, parallel enough, or smart enough to sift through every possible image of every possible object in every possible size, color, and orientation, etc., to recognize objects in less time than it takes for a star to run out of light. Given the issues explored above, it seems remarkable that we can see at all.

Further Reading

Ballard DH (1987) Cortical connections and parallel processing: Structure and function. In Arbib MA and Hanson AR (Eds) *Vision, brain and cooperative computation.* MIT Press, pages 563–621. *Comment* This paper provides a good account of the exponential problems covered in this chapter plus other material not covered here. It also provides a clear and wide-ranging account of how the generalized Hough transform can be used to model neural computations in vision.

Barlow HB (1972) Single units and sensation: A neuron doctrine for perceptual psychology, *Perception* **1** 371–394 *Comment* Classic paper that discusses the relationship between the firing of single neurons in sensory systems and subjectively experienced sensations. Commentaries celebrating this landmark paper are in *Perception* **38** 795–807.

Lennie P (2003) The cost of cortical computation. *Current Biology* **13** 493–497.

Martin KAC (1994) Brief History of the "Feature Detector." *Cerebral Cortex* **4**(1) 1–7. *Comment* A good historical overview of the grandmother cell idea.

Mollon J (2008) The Doctrine of the Gnostic Unit. Seminar at a meeting of the *Yorkshire Vision Network* 18 March 2008.

Quian Quiroga R, Reddy L, Kreiman G, Koch C and Fried I (2005) Invariant visual representation by single neurons in the human brain. *Nature* **435** 1102–1107.

Sherrington C (1940) *Man on his nature.* Cambridge University Press.

Singer W and Gray CM (1995) Visual feature integration and the temporal correlation hypothesis. *Annual Review of Neuroscience* **18** 555-586.

Singer W, Engel AK, Kreiter AK, Munk MHJ, Neuenschwander S, and Roelfsema PR (1997) Neuronal assemblies: necessity, signature and detectability. *Trends in Cognitive Sciences* **1**(7) 252-261.

12

Seeing and Psychophysics

12.1 Hubble telescope photograph
Courtesy NASA.

Imagine you are with a friend watching the stars come out as twilight fades into night, **12.1**. Your friend calls out that he can see a star in a certain location in the sky. You try hard but can't make it out, even though you are sure you are looking in the right place. You conclude that your friend has a much better visual system than your own when it comes to detecting faint light sources.

This kind of situation is not unusual. We are quite often called upon to detect faint stimuli, which relies on an ability to discriminate between the presence or absence of such stimuli. So detection and discrimination are intimately related, although they are often treated as distinct abilities. An example of discrimination is the ability to judge which of two stars is brighter. Psychologists have evolved some clever techniques for making careful measurements of people's ability to detect and/or discriminate stimuli. We explain these psychophysical methods and their theoretical underpinnings in this chapter. Throughout this chapter we use the example of luminance, but these methods can be applied to any perceived quantity, such as contrast, color, sound, and weight.

Psychophysical methods have to be highly sensitive if they are to do justice to the sensitivity of the human visual system. For example, the fact that photoreceptors can signal the presence of a single

Why read this chapter?

What is the dimmest light that can be seen? What is the smallest *difference* in light that can be seen? These related questions are the subject of this chapter, which uses image light intensity (*luminance*) as an example, even though the ideas presented here apply to any stimulus property, such as motion or line orientation. We begin by introducing the idea of a *threshold* luminance, which, in principle, defines the dimmest light that can be seen. When we come to describe how to measure a threshold, we find that it varies from moment to moment, and between different individuals, depending on their willingness to say *yes*, rather than on their intrinsic sensitivity to light. Two solutions to this dilemma are described. First, a *two-alternative forced choice* (2AFC) procedure measures sensitivity to light. Second, *signal detection theory* provides a probabilistic framework which explains sensitivity without the need to invoke the idea of thresholds. We discuss how these methods can be used to redefine the notion of a threshold in statistical terms. Finally, we give a brief historical overview of psychophysics.

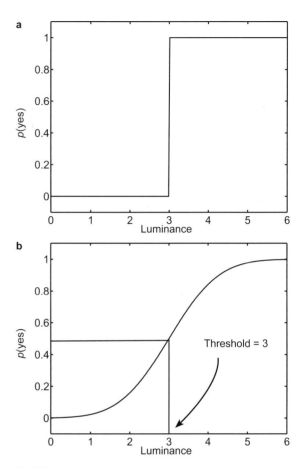

12.2 Transition from not seeing to seeing

a In an ideal world, the probability of seeing a brief flash of light would jump from zero to one as the light increased in luminance.

b In practice, the probability of a *yes* response increases gradually, and defines a smooth S-shaped or sigmoidal ***psychometric function***. The threshold is taken to be the mid-point of this curve: the luminance at which the probability of seeing the light is 0.5. Most quantities on the abscissa (horizontal axis) in plots of this kind are given in log units, so that each increment (e.g., from 3 to 4) along this axis implies a multiplicative change in magnitude. For example, if the log scale being used was to the base 10 then a single step on the abscissa would mean the luminance increased by a factor of ten.

photon was first deduced using a simple psychophysical method by Hecht, Schlaer and Pirenne in 1942 (a photon is a discrete packet or quantum of light). It was confirmed about 40 years later by measuring of the outputs of single photoreceptors. This high degree of visual sensitivity is why we can easily see a white piece of paper in starlight, even though each retinal photoreceptor receives only 1 photon of light every hundred seconds from the paper (with no moon light and no stray light from man-made sources). To put it another way, each set of 100 photoreceptors receives an average of about 1 photon per second in starlight. Whichever way we choose to put it, the eye is a remarkably efficient detector of light. (Note that only the rods, Ch 6, are active in the extremely low light level of starlight.)

The smallest luminance that we can see is called the **absolute threshold**, whereas the smallest perceptible *difference* in luminance is called the **difference threshold**, **difference limen**, or **just noticeable difference** (JND) for the historical reason that it was deemed to be just noticeable. But exactly what do we mean by "see" in terms of absolute thresholds and JNDs? This is a question that caused researchers in the 19th and early 20th centuries to develop what we will refer to as **classical psychophysics**.

Classical Psychophysics: Absolute Thresholds

Suppose you were presented with a series of brief light flashes of increasing luminance, starting from a luminance that lies below your absolute threshold. As the luminance is increased you might expect that there comes a point at which you are suddenly able to see the light flashes, so that the probability of seeing light flashes increases from zero (no flash is seen) to certainty (probability = 1; every flash is seen), **12.2a**. However, in practice this isn't what happens. Instead, the probability of seeing light flashes gradually increases as luminance increases, as depicted by the **psychometric function** in **12.2b**.

This psychometric function implies that there comes a point, as the light flash luminance increases above zero, at which it starts to become visible, but it isn't seen reliably on every presentation. For many presentations of the same (low) luminance, sometimes you see the flash, and sometimes you don't. As the luminance is increased, the probability of seeing the flash increases gradually from zero to one. If you respond *yes* when the light is seen, and *no* when it is not, then your absolute threshold is taken to be the luminance which yields a *yes* response 50% of the time.

A major problem in measuring absolute thresholds plagued psychophysics for many years:

different people have different criteria for saying *yes*. Suppose two people see the same thing when exposed to a very dim light. One of them may be willing to say *yes*, whereas the other may say *no* because s/he requires a higher degree of certainty. (Remember, we are talking here about very low luminances, for which percepts are anything but vivid.) Let's call these two people, which we assume have identical visual systems, Cautious and Risky, respectively. Both Cautious and Risky see the same thing, but Cautious demands a great deal of evidence before being willing to commit. In contrast, Risky is a more flamboyant individual who does not need to be so certain before saying *yes*.

Let's put Cautious and Risky in a laboratory, and show them a set of, say, 10 lights with luminance values ranging from zero, through very dim, to clearly visible. We present each of these lights 100 times in a random order, so there will be 1000 trials in this experiment (i.e., 100 trials per luminance level × 10 luminance levels). As shown in **12.2b**, at very low luminance levels, the light cannot be seen, and so the proportion of *yes* responses is close to zero. As luminance increases, the proportion of *yes* responses increases, until at very high luminance values the proportion of *yes* responses is close to unity. If the proportion of *yes* responses for each if the 10 luminance levels is plotted against luminance then the data would look rather like the psychometric function in **12.2b**.

However, if we compare the corresponding curves for Cautious and Risky, **12.3**, we find that both psychometric functions are identical in shape, but that Cautious' curve is to the right of Risky's curve. This graph implies that a light which induces a *yes* response of say 20% of the time from Risky might induce a *yes* response 5% of the time Cautious. Thus, all of Cautious' responses "lag behind" Risky's responses in terms of luminance. Recall that both Risky and Cautious see the same thing, but Risky is just more willing to say "*Yes – that almost invisible dim percept was caused by a light flash rather than being an artefact of my visual system.*"

To sum up: the absolute threshold is the luminance which evokes a *yes* response on 50% of trials, but Risky's *estimated* threshold is much lower than Cautious', *even though they both see the same thing*. The difference between the estimated thresholds

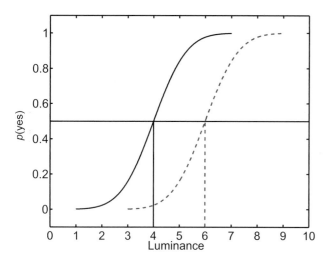

12.3 Individual differences in responding
The proportion of trials on which two observers, Cautious (red dashed curve) and Risky (black solid curve) are willing to respond *yes* to the question: "Did you see a light?"

of Cautious and Risky arises only because Risky is predisposed to say *yes* more often than Cautious. This is clearly nonsensical because we are interested in measuring the sensitivities of Cautious and Risky's visual systems, not their personalities. Fortunately, it can be fixed using **signal detection theory (SDT)** and a procedure known as **two-alternative forced choice (2AFC)**, both of which are described later.

Just Noticeable Differences

The difference threshold or JND is usually defined as the change in luminance required to increase the proportion of *yes* responses from 50% to 75%. Consider two observers called Sensitive and Insensitive, whose psychometric functions are shown in **12.4a**: Sensitive (solid black curve) and Insensitive (dashed red curve). A small increment in luminance has a dramatic effect on the proportion of *yes* responses from Sensitive, but the same luminance change has a relatively small effect on the proportion of *yes* responses from Insensitive. Therefore, Sensitive requires a smaller increase in luminance than Insensitive to raise the proportion of *yes* responses from 50% to 75%. It follows that Sensitive must have a smaller JND than Insensitive, even though both observers may have the same absolute threshold, as in **12.4b**. Thus, the absolute

threshold and the JND are independent quantities: knowing one tells us nothing about the value of the other one, in principle, at least. In practice, a low absolute threshold is usually associated with a small JND, but it is possible for an observer to have a low absolute threshold and a large JND, or vice versa.

To illustrate the independence between absolute thresholds and JNDs, notice that the curves in **12.3** have the same slopes. This means that Cautious and Risky have the same JND despite their different absolute thresholds. The point is that the JND is defined as the difference in luminance

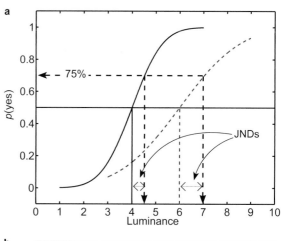

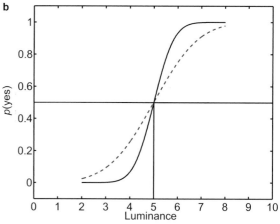

12.4 Individual differences in psychometric function
a Two observers with different JNDs *and* different absolute thresholds.
b Two observers with different JNDs and the same absolute threshold.
Note: 12.3 shows two observers with the same JNDs, but different absolute thresholds.

required to increase the proportion of *yes* responses from 50% to 75%, a change which is unaffected by the *location* of the curve. Thus, the data from Cautious and Risky give biased estimates of their absolute thresholds, but accurate estimates of their JNDs.

Caveat: The definition for the JND given above is: the change in luminance required to increase the proportion of *yes* responses from 50% to 75%. However, the value of 75% is quite arbitrary, and is used for historical reasons; it could just as easily have been 80% or 60%. The main thing for meaningful comparisons is to have a *consistent* measure of the steepness of the psychometric function. In fact, for reasons that will become clear, we will use 76% as the upper limit for the JND later in this chapter.

Signal Detection Theory: The Problem of Noise

The single most important impediment to perceptions around thresholds is *noise*. Indeed, as we will see, the reason why psychometric functions do not look like **12.2a** is due to the effects of noise. This is not the noise of cat's howling, tires screeching, or computer fans humming, but the random fluctuations that plague every receptor in the eye, and every neuron in the brain. The unwanted effects of noise can be reduced, but they cannot be eliminated. It is therefore imperative to have a theoretical framework which allows us (and the brain) to deal rationally with this unavoidable obstacle to perception. Such a framework is **signal detection theory**. We will explain its details in due course but for the moment we give a non-technical overview of the main ideas.

Think about what the brain has to contend with in trying to decide whether or not a light is present. The eye is full of receptors with fluctuating membrane potentials (receptors do not have firing rates), which feed into bipolar cells with fluctuating membrane potentials, and these in turn feed retinal ganglion cells with fluctuating firing rates, whose axons connect to brain neurons with their own fluctuating firing rates. All these fluctuations are usually small, but then so is the change induced by a very dim stimulus. More importantly, these fluctuations happen whether or not a stimulus is present. They are examples of noise. What chance does the brain stand of detecting a very dim light

in the face of this noise? Surprisingly, the answer is, "a good chance." However, *certainty* is ruled out because of the effects of noise.

The amount of noise varies randomly from moment to moment. To illustrate this, let's concentrate first on the receptors. They have a resting potential of about -35 mV (milliVolts), and their response to light is to *hyperpolarize* to about -55 mV. However, to keep things arithmetically simple (and to help later in generalizing from this example) we will define their mean resting potential as 0 mV (just imagine that +35 mV is added to each measurement of receptor output). It will help to have a symbol for a receptor's membrane potential and we will use r for this purpose (think of r standing for *response*). The variable r is known as a **random variable** because it can take on a different value every time it is observed due to the effects of noise. We will use u to denote the mean value of r.

The key point is that variations in r occur due solely to noise, as shown in **12.5a**. If we construct a histogram of r values (as in **12.5b**) then we see that the mean value of r is indeed equal to zero. For simplicity, we assume that this distribution of r values is **gaussian** or **normal** with mean $u = 0$ mV, **12.5c**. In order to help understand how this histogram is constructed, note that the *horizontal* dashed (blue) line in **12.5a** which marks $r = 10$ mV, gets transformed into a *vertical* dashed (blue) line in the histogram of r values in **12.5b,c**. As this distribution consists entirely of noise, it is known as the **noise distribution**.

So much for the noise distribution of responses when no stimulus is present. What happens when a light flash is shown? The flash response is added to the noise to create another distribution, called the **signal distribution**. Obviously, if the luminance of the flash is too low to create any effect in the visual system then the signal distribution is identical to the noise distribution. However, as the luminance of the flash increases, it begins to create a response, and so the signal distribution becomes shifted, **12.6a,b**.

Let's now apply these basic ideas about noise by considering the output of a single receptor in the eye to a low intensity light. The brain is being asked to solve a difficult statistical problem: given that a certain receptor output r was observed during the last trial (which may last a second or two),

a

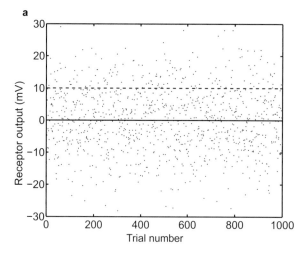

b

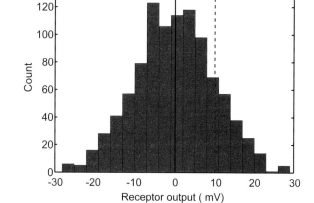

12.5 Noise in receptors
a If we measured the output *r* of a single photoreceptor over 1000 trials then we would obtain the values plotted here, because *r* varies randomly from trial to trial. Note that this receptor is assumed to be in total darkness here. The probability that *r* is greater than some **criterion** value *c* (set to 10 mV here) is given by the proportion of dots above the blue dashed line, and is written as $p(r > c)$.
b Histogram of *r* values measured over 1000 trials shows that the mean receptor output is $u = 10$ mV and that the variation around this mean value has a **standard deviation** of 10 mV. The probability $p(r > c)$ is given by the proportion of histogram area to the right of the criterion *c* indicated here by the blue dashed vertical line.
c The histogram in **b** is a good approximation to a **gaussian** or **normal distribution** of *r* values, as indicated by the solid (black) curve. Notice that this distribution has been "normalized to unit area." This means that the values plotted on the ordinate (vertical axis) have been adjusted to that the area under the curves adds up to unity. The resulting distribution is called a **probability density function** or *pdf*.

c

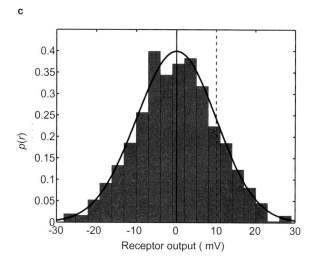

was a light present or not? Remember that *r* varies randomly from second to second whether or not a stimulus is present, due to noise. So some trials are associated with a receptor output caused by the dim light, and other trials have a larger receptor output even if no light is present, again, due to noise. Of course, on average, the receptor output tends to be larger when the dim light is on than when it is off. However, on a small proportion of trials, the presence of noise reverses this situation, and the receptor output associated with no light is *larger* than it is with the dim light on.

The upshot is that the observer has to set a **criterion** receptor output, which we denote as *c*, for

saying *yes*. As this receptor output corresponds to a specific luminance, the observer effectively chooses a criterion luminance, **12.7**. Sometimes this criterion yields a correct *yes*. On other trials, the fluctuating activity levels leads to an *in*correct *yes*.

These ideas allow us to see why the psychometric function has its characteristic *S*-shape. Whether or not these responses are correct, the proportion of trials on which the criterion is exceeded, and therefore the proportion of *yes* responses, gradually increases as the luminance increases, **12.8**.

Having given an outline of SDT, we are now ready to examine it in more detail.

a

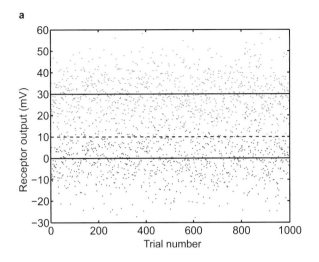

b

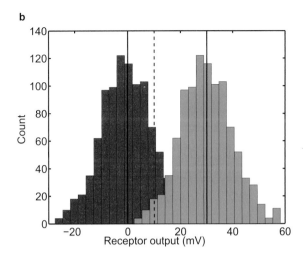

12.6 Signal added to noise creates the *signal distribution*
a Measured values of receptor output *r* with the light off (lower red dots) and on (upper green dots), with the mean of each set of *r* values given by a horizontal black line.
b Histograms of the two sets of dots shown in **a**.

Signal Detection Theory: The Nuts and Bolts

In essence, SDT augments classical psychophysical methods with the addition of ***catch trials***. If an absolute threshold is being measured then each catch trial contains a stimulus with zero luminance. If a JND is being measured then each trial contains two stimuli, and the observer has to state if they are different (with a *yes* response) or not (*no* response), and a catch trial has zero difference between two stimuli. The effect of this subtle change is dramatic, and leads to an objective measure of absolute thresholds, JNDs, and a related measure of sensitivity, known as *d′* (pronounced d-prime).

The question the observer has to answer on each trial is the same as before: did I see a light (absolute thresholds), or a difference between two lights (JNDs)? If an observer responds *yes* on a catch trial then this implies that internal noise has exceeded the observer's criterion for deciding whether or not a light is present (absolute threshold), or whether two lights have different luminances (JND). Thus, the use of catch trials effectively permits the amount of noise to be estimated.

Histograms and pdfs

Understanding SDT requires knowing some technical details about distributions. Readers familiar with standard deviations, and the relationship between histograms and ***probability density functions*** (pdfs) may want to skip this section.

We can see from **12.5a,b** that values of the receptor output *r* around the mean (zero in our example) are most common. If we wanted to know precisely how common they are, we can use **12.5b**. For example, each column or ***bin*** in **12.5b** has a width of 3 mV, so that values of *r* between zero and 3 mV contribute to the height of the first column to the right of zero. This bin has a height of 119, indicating that there were a total of 119 values or *r* which fell between zero and 3 mV. As we measured a total of 1000 values of *r*, it follows that 119/1000 or 0.119 (i.e., 11.9%) of all recorded values of *r* fell between zero and 3 mV.

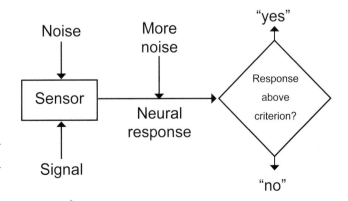

12.7 Overview of a perceptual system faced with noise

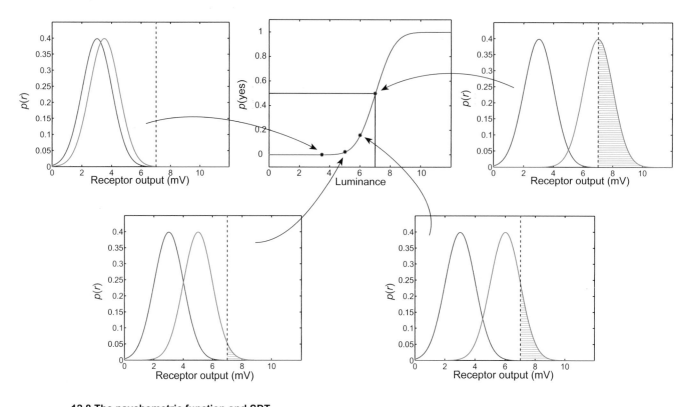

12.8 The psychometric function and SDT
As the receptor output increases the signal distribution (green) moves to the right, increasing the probability of a *yes* response for a *fixed* criterion of *c* = 7 mV.

Thus, the probability that *r* is between zero and 3 mV is 0.119. Now suppose we wanted to know the probability that *r* is greater than, say, 10 mV, given that the light is off. At this stage, it will abbreviate matters if we define some simple notation. The probability that *r* is greater than 10 mV is written as $p(r > 10|off)$: the vertical bar stands for "given that." Thus, $p(r > 10|off)$ is read as "the probability that $r > 10$ given that the light is off." This is a ***conditional probability*** because the probability that $r > 10$ is *conditional* on the state of the light.

From **12.5a**, it is clear that $p(r > 10|off)$ is related to the number of dots above the $r = 10$ mV dashed line. In fact, $p(r > 10|off)$ is given by the *proportion* of dots above the dashed line. Now, each of these dots contributes to one of the bins above the vertical blue line in **12.5b**. It follows that $p(r > 10|off)$ is given by the summed heights of these bins to the right of 0.0 mV, expressed as a proportion of the summed heights of all bins (which must be 1000 because we measured 1000 *r* values). The summed heights of these bins comes to 150, so it follows

that there are 150 *r* values that exceed 10 mV, and therefore that $p(r > 10|off) = 150/1000 = 0.15$. This procedure can be applied to any value or range of *r* values. For example, if we wanted to know the probability that *r* is greater than 0 mV then we would add up all the bin heights to the right of 0 mV. On average, we would find that this accounts for half of the measured *r* values, so that $p(r > 0|off) = 0.5$.

At this stage we can simply note that each bin has a finite width (of 3 mV), so that $p(r > 10|off)$ is actually the *area* of the bins for which $r > 10$, expressed as a proportion of the *total area* of all bins. This will become important very soon.

If we overlay the curve which corresponds to a gaussian curve then we see that this is a good fit to our histogram, as in **12.5c**. In fact, the histogram in **12.5c** has the same shape as that in **12.5b**, but it has been set to have an area of 1.0 (technically, it is said to have been "normalized to have unit area," as explained in the figure legend). Each column in **12.5b** has an area given by its height multiplied by its width (3 mV in this case). If we add up all the column heights (of the 1000 bins) and multiply by the column widths (3 mV) then we obtain $3000 = 1000 \times 3$, which is the total area of the histogram. If we divide the area of each column by 3000 then the total area of the new histogram is one (because $3000/3000 = 1$). If you look closely at **12.5c** then you will see that this has been done already. Instead of a maximum bin height of 125 in **12.5b**, the maximum height in **12.5c** is around 0.4; and, as we have noted, instead of an area of 3000 in **12.5b**, the area of **12.5c** is exactly one.

This transformation from an ordinary histogram to a histogram with unit area is useful because it allows us to compare the histogram to standard curves, such as the gaussian curve overlaid in **12.5c**. As we reduce the bin width, and as we increase the number of measured values of r, this histogram becomes an increasingly good approximation to the gaussian curve in **12.5c**. In the limit, as the number of samples of r tends to infinity, and as the bin size tends to zero, the histogram would be an exact replica of the gaussian curve, and in this limit the histogram is called a ***probability density function***, or ***pdf***.

Just as the histogram allowed us to work out the probability of r being within any given range of values, so does the pdf, but without having to count column heights. For example, the probability that $r < 10$ mV is given by the area under the pdf curve to the left of the dashed blue line in **12.5c**. If we start at the left hand end of the curve and work out areas to the left of increasing values of r, we end up with a curve shaped like the psychometric function. Because this curve gives the cumulative total of areas to the left of any given point it is known as a ***cumulative density function*** or ***cdf***. This area is can be obtained from a standard table of values relevant to the gaussian distribution found in most textbooks on statistics.

One aspect that we have not yet discussed is the *amount* of random variability in values of r. This is revealed by the width of the histogram of r values. A standard measure of variability is the ***standard deviation***, which is denoted by the Greek letter σ (sigma). Given a set of n (where $n = 1000$ here) values of r, if u is the mean then the standard deviation is

$$\sigma = \sqrt{(1/n \sum_i (r_i - u)^2)},$$

where the symbol $\sum$ stands for summation. In words, if we take the difference between each measured value of r and the mean u, and then square all these differences, and then add them all up, and then take their mean (by dividing by n), and finally take the square root of this mean, *then* we obtain the standard deviation.

The equation for a gaussian distribution is

$$p(r) = k \exp(-(u - r)^2/(2\sigma^2)),$$

where $k = 1/[\sigma \sqrt{(2\pi)}]$ ensures that the area under the gaussian curve sums to unity. But if we define a new variable z for which the mean is zero and the standard deviation is one

$$z = (u - r)/\sigma$$

then we can express the gaussian in its standard form as

$$p(z) = k_z \exp(-z^2/2),$$

where

$$k_z = 1/[\sqrt{(2\pi)}].$$

The standard form of the gaussian distribution has a mean of zero, a standard deviation of $\sigma = 1$, and an area of unity; this is the form used in most statistical tables. Any data set of n values of r can be transformed into this normalized form by subtracting its mean u from all n values of r, and dividing each r value by the standard deviation σ of the n values. The resultant data have a mean of zero and a standard deviation of unity. This was done in order to transform our raw values of r in **12.5a,b** to the normalized values shown in **12.5c**. Thus, when a normalized gaussian curve is overlaid on **12.5c**, the fit is pretty good. Conversely, we can go the other way and scale a normalized gaussian curve to get a rough fit to our raw data. This is achieved by adding the data mean u to the normalized gaussian mean of zero, and by multiplying the standard deviation of the normalized gaussian (which is unity) by the standard deviation of our data, as in **12.5b**.

In order to give an impression of what happens to a gaussian curve as we vary the standard deviation, two gaussians with standard deviations of $\sigma = 5$ mV and $\sigma = 10$ mV are shown in **12.9**. Note that the heights of these two curves are different because they both have unit area. Forcing

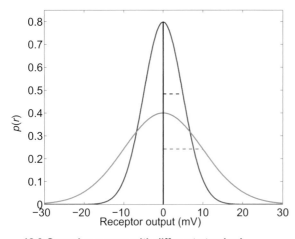

12.9 Gaussian curves with different standard deviations
The narrow gaussian has a standard deviation of $\sigma = 5$ mV, and the wide gaussian has $\sigma = 10$ mV, as indicated by the horizontal dashed line attached to each curve. The curves have different heights because they have different standard deviations, but the same area (unity), which means we can treat each distribution as a probability density function (pdf). The abscissa defines values of r, and the ordinate indicates the probability density $p(r)$ for each value of r.

a distribution or a histogram to have unit area is useful if we wish to interpret areas as probabilities, and it also allows us to treat both gaussian curves as pdfs with different standard deviations.

Before moving on, a few facts about pdfs are worth noting.

First, the total area under a pdf is unity (one). This corresponds to the fact that if we add up the probabilities of all of the observed values of r then this must come to 1.0.

Second, as with the histogram example above, any *area* under the curve defines a probability. Because the area of a bin equals its width times its height, it follows that the *height* alone of the curve cannot be a probability. The height of the curve is called a ***probability density***, and must be multiplied by a bin width to obtain a probability (corresponding to an area under the pdf).

Third, for a gaussian distribution with mean u and standard deviation σ, the area under the curve between u and $u + \sigma$ occupies 34% of the total area. This implies that the probability that r is between u and $u + \sigma$ is 0.34. As half of the area under the curve lies to the left of u, this implies that the probability that r is *less* than $u + \sigma$ is $0.84 = (0.50 + 0.34)$. More generally, the probabil-

ity that a given value r is less than or equal to some reference value x is the area under the curve to the *left* of x

$$p(r \le x) = \phi((x - u)/\sigma),$$

where the function ϕ (the Greek letter, phi) returns the area under the curve to the left of x for a gaussian with mean u and standard deviation σ. Note that the quantity $z = (x - u)/\sigma$ expresses the difference $(x - u)$ in units of σ, is known as a ***z-score***, and was defined on the previous page.

The function ϕ is a called a cumulative density function, because it returns the cumulative total area under the gaussian pdf to the left of z. For example, $p(r \le 10) = \phi(z = 1) = 0.84$.

Signal and Noise Distributions

Returning to our light example, consider what happens if the light is on. This situation is shown by the upper (green) set of dots in **12.6a**, and the corresponding histogram of r values on the right hand side of **12.6**. Let's assume that turning the light on increases the mean output to $u = 30$ mV. For simplicity, we will assume that the standard deviation remains constant at $\sigma = 10$ mV. We refer to this "light on" distribution of r values as the ***signal distribution***.

In order to distinguish between the signal and noise distributions, we refer the their means as u_s and u_n, respectively, and to their standard deviations as σ_s and σ_n, respectively. However, as we assume that $\sigma_s = \sigma_n$, the standard deviation will usually be referred to without a subscript.

Now, we know that if the light is off then the mean output is $u_n = 0$ mV, but at any given moment the observed value of the output r fluctuates around 0 mV. Let's assume that we observe a value of $r = 10$ mV. Does this imply that the light is on or off? Before we answer this, consider the values that could be observed if the light is off in comparison to values that could be observed if the light is on, as shown in **12.6a**. A value of $r = 10$ mV is one standard deviation above the mean value $u_n = 0$ mV associated with the light being off (because $\sigma = 10$ mV here), but it is two standard deviations *below* the mean value of $u_s = 30$ mV associated with the light being on. So, even though an observed value of $r = 10$ mV is unlikely if the light is off, such a value is even *more* unlikely if the light is on. Given that the observer is required to respond *yes* or *no* for each trial, these considerations mean that an ***ideal observer*** should respond *no*. But as we shall see, most observers are not ideal, or at least not ideal in the sense of minimizing the proportion of incorrect responses.

The Criterion

As explained above, given an observed value of the receptor output r, deciding whether or not this output means the light is on amounts to choosing a criterion, which we denote as c. If r is greater than c (i.e., $r > c$) then the observer decides that the light is on, and responds with a *yes*. Conversely, if r is less than c (i.e., $r < c$) then the observer decides that the light is off, and responds with a *no*.

As a reminder, over a large of number trials, say 1000, we present a light or no light to an observer. On each trial, the observer has to indicate whether or not a light was seen. As the light is either on or off, and as the observer can respond either *yes* or *no*, there are four possible outcomes to each trial:

1) light on, observer responds *yes*, a ***hit***, H
2) light on, observer responds *no*, a ***miss***, M
3) light off, observer responds *yes*, a ***false-alarm***, FA
4) light off, observer responds *no*, a ***correct rejection***, CR.

Response	Stimulus present	Catch Trial Stimulus not present
"Yes"	Hit	False alarm
"No"	Miss	Correct rejection

If we measure the observer's responses over a large number of trials then we can obtain estimates of each of these quantities. Each quantity corresponds to a region of one of the histograms in **12.6b**, which has been redrawn in terms of gaussian pdfs in **12.10**. Here, the light-off or noise pdf lies to the left of the light-on or signal pdf. Following the line of reasoning outlined in the previous section:

H: the hit rate equals the large (red) area of the signal pdf to the right of the vertical blue criterion,

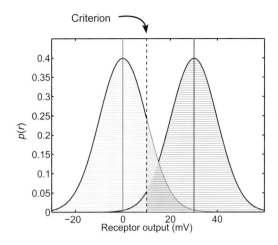

12.10 Estimating *d'*

The distance *d'* (d-prime) between the peaks of the noise (left) and signal (right) distributions can be estimated from a knowledge of two quantities: the *hit rate* **H**, and the *false alarm rate* **FA**. The dashed (blue) line is the criterion *c*, and the observer responds *yes* only if the receptor output *r* is greater than *c* (i.e., if *r* > *c*). The hit rate H is equal to the area of the (red) region of the signal pdf to the right of the criterion, and FA is equal to the area of the (yellow) region of the noise pdf to the right of the criterion.

M: the miss rate equals the small (green) area of the signal pdf to the left of the criterion,

FA: the false alarm rate equals the (yellow) area of the signal pdf to the right of criterion,

CR: the correct rejection rate equals the large (light blue) area of the noise pdf to the left of the criterion.

If we choose $c = 10$ mV then an observed value of $r = 20$ mV would allow us to respond *yes*, because $r > c$. If we adopted a criterion of $c = 10$ mV, how often would we be correct in responding *yes* given that the light is on? This is given by the conditional probability $p(yes|on)$, and is equal to the proportion of the signal pdf shaded red in **12.11**.

Note that the hit rate H = $p(yes|on)$ can be made as large as we like simply by *decreasing* the value of the criterion c, which moves the vertical dashed line leftward in 12.10. For example, if c is set to −20 mV then almost all observed values of r are above c, so we respond *yes* for almost any observed value of r, 12.11a. It's as if we adopt an extremely *laissez faire* or risky approach, and treat almost anything as a sign that the light is on. This implies

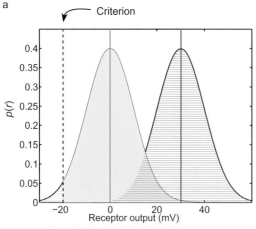

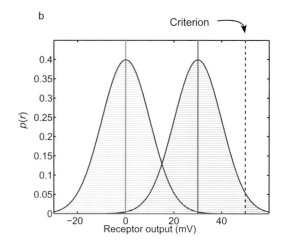

12.11 Effect of criterion
The criterion c is given by the position of the blue dashed line.
a A low criterion of c = –20 mV yields a large hit rate H (the red area of signal pdf to the right of c), but also yields a large false alarm rate FA (the yellow area of the noise pdf to the right of c).
b A high criterion criterion of c = 50 mV yields a low FA (the area of the noise pdf to the right of c, which is so small it is not visible here), but also a low H (the red area of the signal pdf to the right of c).

that if the light is on then we respond *yes*, so that our hit rate becomes close to 100%, for example, $p(yes|on) = 0.910$. This may seem like good news but it is accompanied by some bad news.

Setting c = –20 mV guarantees a high hit rate, but it also ensures that we almost always respond *yes* even when the light is off, resulting in a high false alarm rate FA = $p(yes|off)$. For example, if r = –10 mV then it is not likely that the light is on. But we would respond *yes* if r = –10 mV and if our criterion (c = –20 mV) is set to yield a high hit rate. In other words, the probability $p(yes|off)$ of responding *yes* given that the light is off is close to unity (given by the yellow area in **12.11a**). Thus, setting the criterion c to a very low value increases the hit rate, the yellow region in **12.11a**, but it also increases the false alarm rate.

If we now reverse this strategy and set c to be very high (say, c = 50 mV) then life does not get much better, **12.11b**. In this case, it is as if we adopt an extremely cautious approach, and will not interpret even large values of r as indicating the light is on. Consequently, we rarely respond *yes* when the light is on, so the false alarm rate FA is almost zero (which is good). However, a high criterion means that we rarely respond *yes* even when the light is on, yielding a hit rate H close to zero (which is bad). This situation is shown

graphically in **12.11a,b**. Crucially, there is no value for c which guarantees that our decisions are always correct. However, there is a value of c which guarantees that we are right as often as possible. If the light is on during half the trials then this value is exactly mid-way between the noise and signal distributions.

In summary, a low criterion (*laissez faire* or very risky) yields a large hit rate but a large false alarm rate, whereas a high (very cautious) criterion yields a low hit rate but a low false alarm rate. Ideally, we would like to have a large hit rate and a low false alarm rate. Given that neither a very high nor a very low value for the criterion seem satisfactory, it follows that there must be a value somewhere between these extremes which yields a sensible compromise, which turns out to be the midpoint between the means of the signal and noise distributions. This is explored further in a later section. For now, we explore how a fixed criterion yields the curved *S*-shaped psychometric function in **12.2b** and **12.15d**.

Sensitivity and *d´*

As the luminance is increased from zero, so the proportion of correct *yes* responses (hits) gradually increases, and asymptotes close to a value of unity. Crucially, the *rate* at which the proportion of *yes*

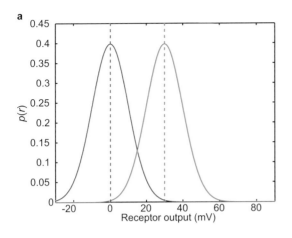

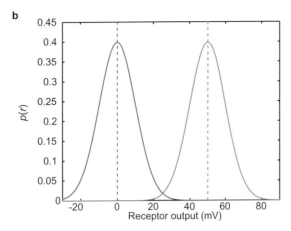

12.12 Individual differences in sensitivity
The left hand red noise distribution of r values is the same for two observers with **a** low and **b** high sensitivities. When presented with the same luminance, the low sensitivity observer's green right hand signal distribution has a mean of $u_s = 30$ mV, whereas the corresponding mean for the observer **b** with high sensitivity is $u_s = 50$ mV. The distance between the signal and noise distributions is measured in units of the standard deviation of the noise distribution, and is called d', which indicates the observer's ability to detect stimuli. If the signal and noise distributions have a common standard deviation of 10 mV then the low sensitivity observer in **a** has $d' = (30 \text{ mV} - 0 \text{ mV})/10 \text{ mV} = 3$, whereas the high sensitivity observer in **b** has $d' = (50 \text{ mV} - 0 \text{ mV})/10 \text{ mV} = 5$.

responses increases is a measure of the sensitivity of a observer, but because sensitivity has its own technical meaning (defined below) we will use the term *responsiveness* for now. In order to understand why the rate at which the proportion of *yes* responses increases is related to responsiveness, a change in perspective is required. Until now we have been considering how the distance between the noise and signal distributions increases as luminance increases for a single receptor within the eye of a single observer. We now consider how the distance between the noise and signal distributions varies between different observers given a *fixed* luminance value.

Let's assume we have two observers, S_{hi} and S_{lo}, whose receptors have high and low degrees of responsiveness, respectively. If we present both observers with a light that has the same luminance then this means that S_{hi} is more likely than S_{lo} to detect the light (assuming that they use the same criterion level for responding *yes*).

When S_{lo} is presented with the light, the mean of S_{lo}'s distribution of receptor values increases from a baseline value of $u_n = 0$ mV to a new value of $u_s = 30$ mV, **12.12a**. In contrast, when S_{hi} is presented with the light, the mean of S_{lo}'s distribution

of receptor values increases from a baseline value of $u_n = 0$ mV to a new value of $u_s = 50$ mV, **12.12b**. Clearly, a change of 50 mV should be more detectable than a change of 30 mV, if the distributions of both receptors have the same standard deviation (as is assumed here). The difference between the mean of the noise (light off) distribution and the mean of the signal (light on) distribution (expressed as a fraction of the standard deviation of the noise distribution) has a special symbol, d', which, as noted above, is pronounced d-prime. The quantity d' is a measure of how responsive each observer is to a given *change* in luminance.

The definition of d' is as follows. Given a noise distribution with mean u_n and signal distribution with mean u_s, where both distributions share a common standard deviation, σ

$$d' = (u_s - u_n)/\sigma.$$

Thus, d' is a measure of the distance between the means of the noise and signal distributions, expressed in units of standard deviations (of those distributions). For example, the low sensitivity observer in **12.12a** has

$$d' = (30 \text{ mV} - 0 \text{ mV})/10 \text{ mV} = 3.$$

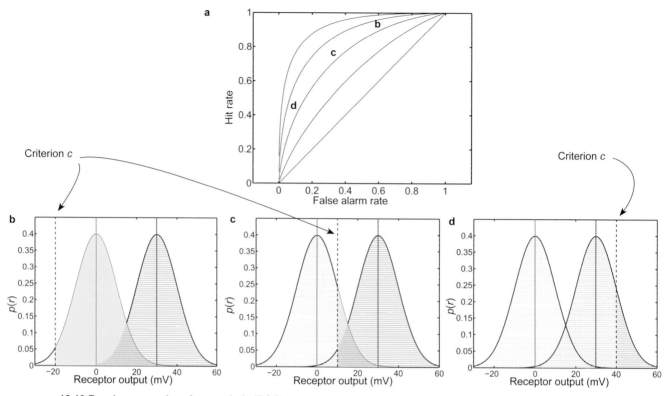

12.13 Receiver operating characteristic (ROC)
a ROC curves for different values of d'. Each curve is obtained by sweeping out criterion values c from high to low (**b-d**) and measuring the hit rate (H) and false alarm rate (FA) at each value of c. From left to right in **a**, the bowed ROC curves correspond to values of $d' = 2$, 1.5, 1.0, 0.5 and 0, where zero defines the diagonal line. If $d' = 0$ then the signal and noise pdfs are the same, and the diagonal line is obtained, which implies that the hit rate H and the false alarm rate FA are equal. The letters on the middle ROC curve in **a** correspond to the H and FA rates deriving from the different positions of the criterion c in the graphs labelled **b-d**. Notice that the criterion, but not d', varies between graphs **b-d**.

Notice that d' is measured in terms of standard deviations, so that one can think of a d' of, say, three, as meaning three standard deviations. However, because d' is a ratio of values (that are, in our example, expressed in units of mV), d' is technically said to be a dimensionless quantity.

If a small increase in luminance induces a large increase in d' (as for S_{hi}) then this change should also induce a large increase in the probability of a *yes* response. Therefore, responsive observers display a rapid increase in *yes* responses as luminance increases, and this is seen graphically in the steep (black) psychometric function of such an observer, **12.4a,b**. Conversely, a less responsive observer would have a small increase in d' (as for S_{lo}) with a correspondingly small increase in the probability of a *yes* response, with the result that the psychometric function would be shallow, as shown by the dashed (red) psychometric function in **12.4a,b**.

How to Increase d'

One potentially confusing fact should be made clear. There are two ways to increase d': find a more sensitive observer (as above), or keep the same observer and increase the change in luminance. The point is that d' is a useful measure of observer sensitivity only for a *fixed* luminance level. Within a single observer, as the change in luminance increases, so too does d', as shown in **12.8**.

Measuring d'

This is all very fine, but how do we actually measure the value of d'?

The quantity d' is defined with respect to a given ***reference luminance*** level I_1, which we have implicitly assumed to be zero up to this point. If $I_1 = 0$ then the noise distribution really is just noise. More generally, we would like to measure

d' for non-zero reference luminances. And if we set $I_1 > 0$ then the distribution associated with the reference luminance is still (confusingly) called the noise distribution. Given a **comparison luminance** I_2 (which is larger than I_1) associated with the signal distribution, we want to find the difference between the means d' of the noise and signal distributions. This means that each trial consists of two stimuli, one with a reference luminance and the other with a comparison luminance. Responses to these non-catch trials are associated with the signal distribution.

Crucially, on catch trials, both luminances are set to be the same as the reference luminance, and responses to catch trials are associated with the noise distribution. On every trial, the observer's task is to answer *yes* if the stimuli appear to have different luminances, and *no* if they do not. The observer does not know about the distinction between reference and comparison stimuli—this is purely for the experimenter's benefit. In practice, the comparison and reference luminances vary from trial to trial, and data for a specific luminance difference is extracted later so that the d' for that luminance difference can be estimated.

As we have already noted, an observer's responses are affected by his/her criterion, so we need to be able to disentangle the effects of the criterion and d'. Using fairly mild assumptions regarding the shape and standard deviations of the noise and signal distributions, it turns out that d' can readily be estimated from the quantities listed in 1-4, on page 291. In fact, we need only the hit rate H and the false alarm rate FA.

Specifically, the distance d' between the centers of the noise and signal distributions can be related to H and FA if we split d' into two parts:

1) the distance $(c - u_n)$ from the noise mean $u_n = 0$ mV to the criterion $c = 20$ mV, plus

2) the distance $(u_s - c)$ from the criterion $c = 20$ mV to the signal mean $u_s = 30$ mV.

Without going into details, these distances can be computed from the FA and H rates, as

$$d' = \phi^{-1}(H) - \phi^{-1}(FA),$$

where ϕ^{-1} (the inverse of the cumulative density function ϕ) maps probability (area under a curve)

to distance along the abscissa for gaussian pdfs, and can be found in standard statistical tables. Thus, together, the hit and false alarm rates suffice to provide d' from the sum of the (signed) distances

$$u_s - u_n = [(c - u_n) + (u_s - c)],$$

irrespective of the criterion adopted by a given observer.

The reason we only need the hit rate and false alarm rate is because, given that H + M = 1, it follows that if we know H then we also know M = 1 − H. Similarly, FA + CR = 1, so if we know FA then we also know CR = 1 − FA. So we could use either H paired with FA, or M paired with CR, in order to estimate d'. By convention, H and FA are used.

Measuring the Bias or Criterion

The H and FA rates can also be used to estimate the criterion c, which is also called the **bias** in this context. The bias is given by

$$c = [\phi^{-1}(H) + \phi^{-1}(FA)] / (-2).$$

A bias of $c = 0$ indicates no response bias, and implies that the criterion is exactly half way between the means of the noise and signal distributions. If c is positive then it lies to the right of this half-way point, and therefore indicates a bias for responding *no*. If c is negative it lies to the left of the half-way point, and therefore indicates a bias for responding *yes*. Finally, note that the bias c and the sensitivity d' of a given observer are independent quantities.

Receiver Operating Characteristics

Almost all accounts of SDT in the literature also describe a graph which defines the **receiver operating characteristics** (ROC). The ROC graph acts as an intermediary representation between signal/noise pdfs and the psychometric function. It is not, in our opinion, necessary for understanding the main principles SDT. However, it is included here for the sake of completeness.

Each ROC curve, **12.13a**, shows what happens, for a given d', to hit rates and false alarm rates as the criterion level c is increased from a low (risky) to a high (cautious) level. The shape of this graph is bowed, with the sharpness of the bow dependent on the value of d', or equivalently, for a given luminance of the signal. Thus, each ROC curve in **12.13a** is obtained by sweeping through all possible values of c, and for each value of c, plotting pair-wise values of H and FA. For any given value of d', we know that decreasing c has the effect of increasing H, but also of increasing FA, and the precise nature of the relationship between H and FA is given by an ROC curve. For example, the top-most ROC curve in **12.13a** is obtained by starting with a large value of c (say, 30 mV), which gives a low H and a low FA. As c is decreased, H rises rapidly, but then FA also begins to rise. Finally, at values of $c = 30$ mV or above, both H and FA are close to unity.

Two-Alternative Forced Choice (2AFC) Methods

The trouble with absolute thresholds described at the outset of this chapter (Risky vs. Cautious responding) was recognized in the very early days of psychophysics in the 19th century. A modern solution, SDT, has already been described.

However, an earlier approach tackled the problem by removing the observer's freedom to choose when he/she could see a stimulus. Instead, the observer is presented with a pair of stimuli, and is asked to indicate which one is brighter (**12.14**; the pair can be presented simultaneously or one after the other, and other attributes than brightness can be used, but the principle remains the same). Now the observer has to decide, not whether the stimulus can be seen, nor whether the difference in luminance between two stimuli can be seen (as in classical psychophysics and SDT), but *which* of two stimuli is brighter.

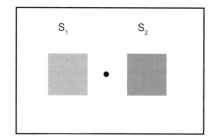

12.14 Forced choice stimulus experiment
The observer fixates the central dot when a warning tone is sounded. Shortly after, S_1 and S_2 appear briefly on the screen and the observer's task is to decide which is brighter.

It might appear that saying whether or not the difference between two stimuli can be seen (a *yes/no* task), and saying which of two stimuli is brighter (a 2AFC task), are similar tasks. However, a *no* response in the *yes/no* task allows the observer to indicate that both stimuli appear equally bright. In contrast, an "equally bright" response is not possible in the 2AFC task, which requires that the observer chooses the stimulus that is perceived as brighter, even if they appear to be equally bright. One easy way to remember this is that classical psychophysical methods and SDT usually require *yes/no* responses, whereas 2AFC requires the observer to *choose* a stimulus.

Finding the JND with 2AFC

In order to fully understand how the difference threshold or JND is estimated using the 2AFC procedure, let's consider a specific example.

As with SDT, on each trial, the observer is presented with a pair of stimuli, S_1 and S_2. Stimulus S_1 has a reference luminance I_1, and stimulus S_2 has a comparison luminance I_2. As before, these labels are for the experimenter's benefit, and the observer is unaware of their existence; as far as the observer is concerned he/she just has to choose the brighter of two stimuli, **12.14**. The observer is presented with many pairs of stimuli, using a range of different reference and comparison luminances, in random order. In order for us to estimate the JND for a single reference luminance I_1 we extract only those trials which contain I_1, and consider the observer's responses as the comparison luminance I_2 is varied.

We label the receptor outputs associated with the luminances I_1 and I_2 as r_1 and r_2, respectively. We also label the distribution of r_1 values associated with I_1 as the noise distribution p_1, with a mean of u_1, a standard deviation of

$$\sigma_1 = 10 \text{ mV},$$

and a variance of

$$v_1 = \sigma_1^2.$$

Similarly, we label the r_2 values associated with I_2 as the signal distribution p_2, with a mean of u_2, a standard deviation of

$$\sigma_2 = 10 \text{ mV},$$

and a variance of

$$v_2 = \sigma_2^2.$$

These signal and noise distributions are the same as those used in SDT, and we have replaced the mean symbols u_n and u_s with u_1 and u_2, respectively, because we now consider two luminances with labels I_1 and I_2.

In a 2AFC experiment, the observer is assumed to obey the following optimal rule: *choose the stimulus that is associated with the largest response.* Notice that we use the phrase "associated with" rather than "caused by" because the effects of noise mean that we cannot state with certainty that an observed response was caused by a stimulus rather than noise. Also, bear in mind that the observer does not know which response belongs to I_1 and which belongs to I_2, the observer just compares the receptor responses he/she gets by looking at both stimuli, and chooses the stimulus that is associ-

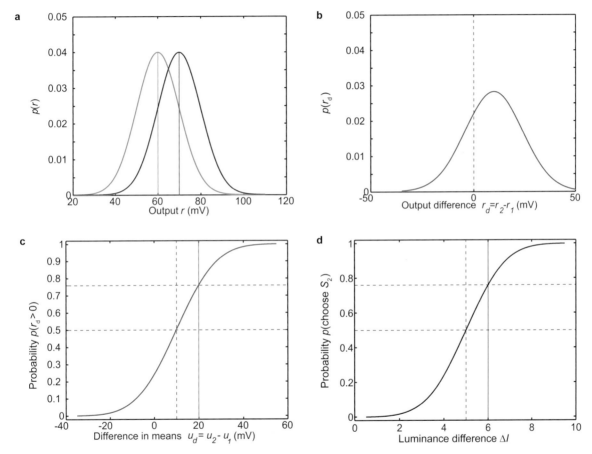

12.15 Using 2AFC to estimate a difference threshold

a Distributions of outputs for two different luminances I_1 and I_2 chosen so that $I_2 - I_1$ = JND. The distribution means are u_1 = 60 mV and u_2 = 70 mV, and both distributions have a standard deviation of σ = 10 mV, so the means are separated by one standard deviation, $u_d = u_2 - u_1$ = 10 mV.

b Distribution of the difference $r_d = r_2 - r_1$ derived from the distributions in **a**. This difference distribution has a mean of $(u_2 - u_1)$ = 10 mV, and a standard deviation 14 mV = $\sigma\sqrt{2}$. This distribution shows that the most likely value of r_d is u_d. The probability that $r_2 > r_1$ is the same as the probability that $(r_2 - r_1) > 0$, which is given by the area under the curve to the right of $r_d = 0$. If an observer chooses I_2 when $r_d > 0$ then this area corresponds to the probability of choosing S_2, p(choose S_2).

c This area is plotted as a function of the difference u_d in means, which increases as the difference $\Delta I = I_2 - I_1$ in luminances increases.

d Using this decision rule, the probability p(choose S_2) increases as the difference ΔI increases. Note that the values on the abcissa of **c** and **d** are different, and that the plotted curves are essentially scaled versions of each other. This implies that $u_d = k\Delta I$, where k is a constant of proportionality.

ated with the largest response. This implies that if the receptor response r_2 to I_2 is greater than the response r_1 to I_1, (i.e., if $r_2 > r_1$) then the observer chooses I_2, otherwise he/she chooses I_1.

This rule can be rewritten as: choose stimulus S_2 (which has luminance I_2) if $r_2 > r_1$, or equivalently, if $(r_2 - r_1) > 0$ mV. Of course, the observer's decision is correct only if $I_2 > I_1$, so this rule is not guaranteed to succeed on every trial, but it is guaranteed to maximize the proportion of correct responses. Thus, using this rule, the probability of choosing S_2 is the probability that $(r_2 - r_1) > 0$ mV. Let's evaluate this probability, **12.15**.

We begin by treating the *difference* $(r_2 - r_1)$ as a new random variable, which we define as

$$r_d = (r_2 - r_1),$$

where r_d has a mean of

$$u_d = (u_2 - u_1),$$

and a variance v_d. (Variance is defined as the standard deviation squared, so $v = \sigma^2$ and $\sigma = \sqrt{v}$.) Both r_1 and r_2 are random variables with gaussian distributions, and the distribution of the difference r_d is therefore also gaussian

$$p(r_d | u_2 - u_1) = k_d \exp[-((u_2 - r_2) - (u_1 - r_1))^2/(2v_d)],$$

where the constant $k_d = (1/\sqrt{(2\pi v_d)}$ ensures that the distribution has unit area, and its variance can be shown to be $v_d = (v_1 + v_2)$. The conditional probability term $p(r_d | u_2 - u_1)$ makes explicit the dependence of r_d on the difference in means $(u_2 - u_1)$, and is interpreted as: the probability density of observing the value r_d given the difference in means $u_d = (u_2 - u_1)$.

The equation above can be written more succinctly as

$$p(r_d | u_d) = k_d \exp[-(u_d - r_d)^2/(2v_d)].$$

If the variances v_1 and v_2 are equal then their standard deviations σ_1 and σ_2 are also equal. We can label these using two symbols without subscripts as

$$v = v_1 = v_2,$$

and

$$\sigma = \sigma_1 = \sigma_2,$$

where $\sigma = \sqrt{v}$. Given that

$$v_d = (v_1 + v_2),$$

this implies that

$$v_d = 2v = 2\sigma^2.$$

By definition,

$$\sigma_d = \sqrt{v_d},$$

which implies that

$$\sigma_d = \sqrt{(2v)}$$
$$= \sigma\sqrt{2}.$$

Rearranging this yields

$$\sigma = \sigma_d/\sqrt{2}.$$

Notice that the distribution $p(r_d)$ describes the probability density of a *difference* $r_d = (r_2 - r_1)$ in receptor output values, and that the standard deviation σ_d of this distribution is larger than the standard deviations σ of the pdfs of r_2 and r_1 by a factor of $\sqrt{2}$.

As noted above, the probability of choosing S_2 is $p(r_d > 0 | u_d)$ (i.e., the probability that r_d is greater than zero). This can be evaluated using the integral of the distribution $p(r_d | u_d)$, which yields

$$p(r_d > 0 | u_d) = \phi[u_d/(\sigma\sqrt{2})],$$

where ϕ is the cumulative density function of a standard gaussian distribution (i.e., with zero mean and unit standard deviation). This function returns the area under the distribution to the left of u_d with mean zero and standard deviation $\sigma\sqrt{2}$; an area numerically equal to the area under the gaussian distribution to the right of zero with mean u_d and also with a standard deviation of $\sigma\sqrt{2}$.

We now have an equation which tells us how the proportion $p(r_d > 0)$ of S_2 responses increases as the difference between distribution means increases, which is driven by the difference between the comparison and reference luminances. More importantly, we can use this equation to find the JND, the change in luminance required to shift the mean receptor output value by one standard deviation σ, a change which also increases the proportion of S_2 responses from 50% to 76%, as described next.

If we increase I_2 from below I_1 to above I_1 then we obtain the typical sigmoid function value describing the probability that the observer chooses S_2. Once I_2 becomes greater than I_1 then choosing S_2 is the correct response, which we assume is described by the probability $p(r_d > 0)$ defined above. If $r_d = 0$ then the means of the receptor outputs are $u_1 = u_2$ and so $p(r_d > 0) = 0.5$. If we could increase the luminance I_2 until

$$u_2 = u_1 + \sigma$$

then this would correspond to increasing u_2 by one standard deviation. One standard deviation σ in the noise distribution corresponds to a change $\sigma_d = \sigma\sqrt{2}$ in the distribution of differences r_d. The corresponding change in $p(r_d > 0)$ is given by the area under the distribution of r_d values between $r_d = 0$ and $r_d = \sigma\sqrt{2}$, which evaluates to $\phi(1/\sqrt{2}) = 0.212$. In other words, if the difference

$$\Delta I = I_2 - I_1$$

is changed so that $p(r_d > 0)$ increases from 50% to 76% then this corresponds to a change in u_2 from

$$u_2 = u_1$$

to

$$u_2 = u_1 + \sigma.$$

Given that the JND is the change in luminance required to increase u_2 by one standard deviation, this implies that the JND is that luminance change sufficient to change p (choose S_2) from 50% to 76%. (*Note:* Be careful not to confuse this case with the fact, described earlier, that the area between the mean and one standard deviation away from the mean of a gaussian is 34%; that is, 84% of a gaussian lies to the left of one standard deviation above its mean.)

It is worth summarizing which quantities are invisible to the experimenter, which ones can be measured, and which ones can be estimated from those measurements.

Basically, the responses of the receptor (r_1 and r_2), the distributions of these responses (p_1 and p_2), their means (u_1 and u_2), variances (v_1 and v_2), and standard deviations (σ_1 and σ_2), are all invisible to the experimenter. The quantities known to the experimenter are the observer responses to the luminances (I_1 and I_2) of the reference and comparison stimuli. By varying these over a range of values, the experimenter can measure the probability $p(r_d > 0)$ that the observer chooses S_2 at each luminance difference $\Delta I = (I_2 - I_1)$. If this prob-

ability is measured for many values of the comparison luminance I_2 then the results can be used to plot a graph of p(choose S_2) versus ΔI. With the machinery described above, the experimenter can use this function to estimate the JND as the standard deviation of the distribution of responses to the reference stimulus. This is done by "reading off" the luminance change ΔI required to increase p(choose S_2) from 50% to 76%.

Why Variances Add Up

We begin by showing that, if the distributions of r_1 and r_2 values have variances v_1 and v_2 (respectively), then

$$v_d = v_1 + v_2,$$

where v_d is the variance of the distribution of differences r_d. We can assume that the mean of each distribution is zero, without affecting our result.

For brevity, we define

$$h = 1/n,$$

so the variances of v_1, v_2 and v_d are defined as

$$v_1 = h\Sigma_i \, (r_1(i) - r_{m1})^2$$

$$v_2 = h\Sigma_i \, (r_2(i) - r_{m2})^2$$

$$v_d = h\Sigma_i \, [\, (r_2(i) - r_{m2}) - (r_1(i) - r_{m1}) \,]^2.$$

Expanding the right-hand side of the final equation yields

$$v_d = h\Sigma_i \, [(r_2(i) - r_{m2})^2$$
$$+ (r_1(i) - r_{m1})^2$$
$$- 2(r_1(i) - r_{m1}) \, (r_2(i) - r_{m2})].$$

If we insert some new summation signs then we can rewrite this as

$$v_d = h\Sigma_i \, (r_2(i) - r_{m2})^2$$
$$+ h\Sigma_i \, (r_1(i) - r_{m1})^2$$
$$- 2h\Sigma_i \, (r_1(i) - r_{m1}) \, (r_2(i) - r_{m2}).$$

We can recognize the first two terms on the right-hand side as v_1 and v_2. The third term is, on average, equal to zero, because both r_1 and r_2 are *independent* random variables, which implies that the sum of their products is zero, so that

$$v_d = v_1 + v_2.$$

Table Summarizing the Differences Between Classical Psychophysics, SDT and 2ADFC

	Classical Psychophysics: Absolute Threshold
Trial contains	A single stimulus.
Question	Did you see a light?
Response	*yes* or *no*.
Yields	Absolute threshold, defined as the luminance that yields a *yes* response on 50% of trials.
Comment	Biased by observer's willingness to respond *yes*, which affects position of psychometric function.

	Classical Psychophysics: JND
Trial contains	A pair of stimuli, one with a reference luminance and one with a comparison luminance.
Question	Did you see any difference between the stimuli?
Response	*yes* or *no*.
Yields	JND, defined as the change in comparison luminance required to increase proportion of *yes* responses from 50% to some fixed percentage (usually 75% or 76%).
Comment	As this does not depend on the position of the psychometric function, it is unaffected by observer bias.

	Signal Detection Theory (SDT)
Trial contains	A *single stimulus*, which is either a non-zero luminance stimulus, or, for catch trials, a zero intensity stimulus. Or Two *stimuli*, a reference luminance, and a comparison luminance, which are the same on catch trials.
Question	Did you see a light? Or Which stimulus is brighter?
Response	*yes* or *no*.
Yields	d', the distance between the observer's noise and signal distributions for a given reference-comparison luminance difference.
Comment	d' is measured with respect to a reference luminance, which can be zero. For non-zero reference luminances, d' is obtained by using a pair of stimuli, where the luminance of these stimuli is the same on catch trials. Observer responds *yes* (difference seen) or *no* (difference not seen) on each trial. A luminance difference that yields a value of $d' = 1$ corresponds to the JND for the reference luminance under consideration.

	Two-Alternative Forced Choice (2AFC)
Trial contains	Two stimuli.
Question	Which stimulus is brighter?
Response	Choose brighter stimulus.
Yields	Absolute threshold, JND.
Comment	The two stimuli are labeled as reference and comparison stimuli. If reference luminance is zero then the comparison luminance which yields a correct response on 76% of trials is the absolute threshold (and the JND for a reference luminance of zero). If reference stimulus is non-zero then the change in comparison luminance which increases the proportion of correct responses from 50% to 76% is the JND for that reference luminance

Given that the standard deviation of the distribution of differences is $\sigma_d = \sqrt{v_d}$ this implies

$$\sigma_d = \sqrt{(v_1 + v_2)}.$$

Now, if $v_1 = v_2$ then we can use the symbol v to stand for both v_1 and v_2 then

$$\sigma_d = \sqrt{(v + v)}$$

$$= \sqrt{(2v)} = \sigma \sqrt{2}.$$

Thus, if the distributions of r_1 and r_2 both have a standard deviation σ then the distribution of differences $r_d = (r_1 - r_2)$ has a standard deviation

$$\sigma_d = \sigma \sqrt{2}.$$

Using 2AFC for Absolute Threshold

If we set the luminance of the reference stimulus S_1 to be zero (i.e., $I_1 = 0$) then it is not physically possible for I_2 to be less than I_1 (i.e., I_2 cannot in this case be negative), and the observer cannot be correct on more than 50% of trials unless I_2 exceeds I_1. Therefore, a graph of correct responses versus the luminance difference $I_2 - I_1$ has a horizontal line at 50% until $I_2 > I_1$, after which the curve begins to rise above 50%. The absolute threshold I_{abs} is given by the value of I_2 which causes the observer to choose S_2 (i.e., to be correct) on 76% of trials. Note that this absolute threshold is the same as the JND for a reference luminance of $I_1 = 0$. That is, a change to the luminance I_1 that is "just noticeable" is I_{abs}. The 2AFC procedure has been found to provide consistently lower absolute thresholds than the methods used in classical psychophysics.

Before moving on, one further subtlety should be noted. The *x*-axis of the distributions of r_1 and r_2 are in units of receptor membrane voltage, whereas the *x*-axis of the psychometric function is in units of luminance. This change is justified if we assume that mean receptor output is a ***monotonic*** function of luminance: an increase in luminance always induces an increase in mean receptor output.

Relation between 2AFC and SDT

The methods of 2AFC and SDT are related inasmuch as 2AFC can be considered as an SDT experiment in which the observer places the criterion mid-way between the means u_1 and u_2 of the noise and signal distributions, respectively. In both cases,

the probability of a correct response is given by

$$p(u_s > u_n) = \phi(z),$$

where

$$z = (u_s - u_n)/\sigma.$$

In the case of 2AFC the JND is given by the increase in luminance required to raise p(*correct choice of brighter stimulus*) from 50% to 76%. In the case of SDT, the JND is given by the increase in luminance required to raise p(*correct yes*) from 50% to 76%, which corresponds to $d' = 1$.

Psychophysical Methods for Measuring Thresholds

There are basically five methods for measuring thresholds: the method of constant stimuli, the method of limits, and the method of adjustment, staircase, and two-alternative forced choice (2AFC). These will be described only briefly here. Only the 2AFC method is untroubled by the problem of observer bias.

Method of constant stimuli: This consists of using a different magnitude on each trial, chosen randomly from a previously arranged set. The threshold is taken as that stimulus value which elicits a yes response on 50% of trials. For difference thresholds, the question put to the observer on each trial requires a yes/no response, such as, "Is the stimulus on the left greater than the one on the right?". *Problem:* much experimental time can be wasted showing stimuli far from the observer's threshold range.

Method of limits: The magnitude of the stimulus is increased until it is observed, and a note of this value of recorded. Then the stimulus magnitude is decreased until it cannot be detected, and a note of this value of recorded. The threshold is taken as the mean of the two recorded values. For difference thresholds, it is the difference between two stimuli that are increased/decreased until the difference becomes detectable/undetectable. A procedure of this general type is used when you are asked by an optician to look at the standard letter chart for assessing vision. This has letters of varying sizes from very large at the top to very small at the bottom. The task is to identify the letters beginning with the large (well above threshold) letters. When you start making errors as the letters get

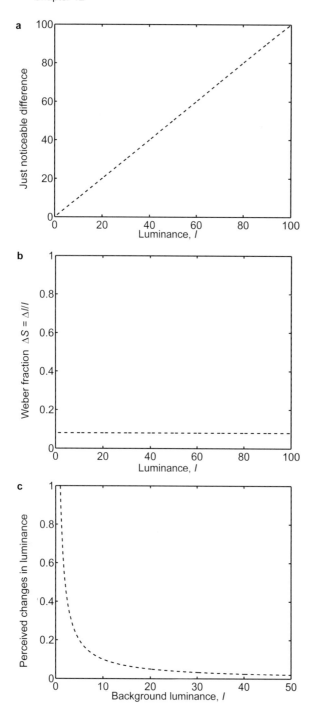

smaller then the optician has a measure of your threshold for ability to see small stimuli clearly (see Ch 4, **4.14**). Note that the optician moves only from large to small differences, and not also vice versa, and so this is not a proper method of limits procedure. *Problem*: much experimental time can be wasted showing stimuli far from the observer's threshold range.

Staircase method: The problem of wasted time in the method of limits can be fixed using a *staircase* in which the magnitude is changed in large steps until the observer changes decision (e.g., from "I saw it" to "I didn't see it"). Then smaller steps are employed for moving the magnitude to and fro above the change of decision point, with step sizes guided by the observer's responses. This keeps the stimuli in the threshold range. *Problem*: The simple staircase just described can mean that the observer never sees a clear-cut stimulus. This problem can be fixed by occasionally presenting stimuli well away from threshold.

Method of adjustment: The observer is asked to adjust the stimulus magnitude, starting from a very low value, until the stimulus can be detected, and a note of this value of recorded. This is repeated, starting from a very high value, and the value at which the stimulus cannot be detected is recorded. The threshold is taken as the mean of the two recorded values. For difference thresholds, the observer is asked to adjust the magnitude of a

Table: Weber fractions

Quantity	Weber Fraction
Electric shock	0.013
Heaviness	0.020
Line Length	0.029
Vibration (60 Hz)	0.036
Loudness	0.048
Brightness	0.079
Taste (salt)	0.083

12.16 Weber's law
a The change in stimulus magnitude ΔI required to yield a just noticeable difference (JND) in perceived magnitude ΔS is proportional to stimulus magnitude I.
b A graph of JND versus stimulus magnitude I defines a horizontal line with height equal to the Weber fraction for the quantity under consideration (luminance here).
c The law of 'diminishing returns' implied by Weber's law. If stimulus magnitude is increased by a fixed amount then the perceived change in magnitude decreases. For example, lighting a candle in a dark room has a dramatic effect, but doing so in daylight has little effect, so there is less 'bang per buck' as the background luminance increases.

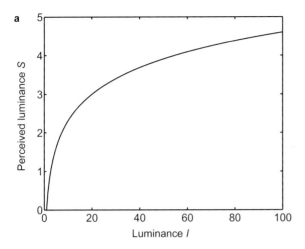

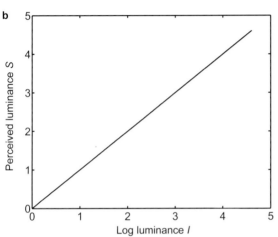

12.17 Fechner's Law
a Fechner's law, $S = k \log I/I_0$, states that the perceived magnitude S of a stimulus with physical magnitude I is proportional to the log of I/I_0, where I_0 is the absolute threshold (set to $I_0 = 1$ here), and k is a constant of proportionality. If $I = 10^S$ then small increments in S induce large increments in I.
b A graph of log luminance versus perceived luminance gives a straight line with slope equal to k. The graph of log I versus S yields a straight line because log $I = S$ (obtained by taking logs of both of $I = 10^S$).

variable stimulus until it appears different from a *reference stimulus* magnitude. *Problem*: The advantage of this procedure is that it is quick but it is regarded by some as particularly vulnerable to observer bias.

Two-alternative forced choice (2AFC): This method was described above. We emphasize that it is the *only* one of the four methods which is uncontaminated by the observer's predisposition (bias) for responding *yes*.

Weber's Law and Weber Fractions

We now provide a brief history of some milestones in psychophysics, beginning with the work of Ernst Heinrich Weber (1795–1878).

If we gave you a box containing 100g of sugar, how much sugar would we have to add in order for you to notice the difference in weight? It turns out that the answer is about 2g. However, if the box contained 200g of sugar then we would have to add 4g for you to notice a difference. And if the box contained 300g of sugar then we would have to add 6g for you to notice a difference.

By now you should have spotted the pattern, a pattern first noted by Weber, who found that the JND in weight increased in proportion to the initial weight, **12.16**. This proportionality implies that the fraction formed by (JND in weight)/(initial weight) is a constant. The constant fraction is equal to about 0.02 in this case because all of the fractions just listed come to

$$2/100 = 4/200$$
$$= 6/300$$
$$= 0.02.$$

In honour of his observations, such fractions are called **Weber fractions**, and they are different for different types of stimuli. (See Table on Weber fractions.)

As stated above, the common definition of a just noticeable difference (JND) is that change ΔI in a stimulus with intensity I which is *perceived* on 75% of trials. In other words, an increment of one JND in stimulus intensity means that the *perceived magnitude* of a stimulus increases by a fixed amount.

So we can think of the ratio $\Delta I/I$ as an increment in perceived stimulus magnitude, which is defined as

$$\Delta S = \Delta I/I.$$

Dividing both sides by ΔI yields

$$\Delta S/\Delta I = 1/I.$$

Consider what this equation implies: as I increases, the ratio $\Delta S/\Delta I$ decreases. So, the perceptual impact ΔS of a fixed change in magnitude ΔI, decreases as I increases. In other words, the amount of "bang per buck" decreases rapidly as I increases. This makes perfect sense if you consider the large perceptual impact of lighting a candle in a dark room, which has a small background luminance I. The candle adds a small amount ΔI of luminance to a small background luminance I, so the ratio $\Delta S = \Delta I/I$ is large. In contrast, lighting a candle in daylight adds the same amount of light ΔI, but this is added to a large background luminance I. In this case, the perceptual impact is small, and is commensurate with the value of small value of the ratio $\Delta S = \Delta I/I$.

Fechner's Law

Putting all this mathematically, if we solve for S in terms of I (by integrating both sides of $\Delta S/\Delta I = 1/I$ with respect to I) then we obtain a law named after Gustav Fechner (1801–1887). This law states that the perceived magnitude S of a given stimulus is proportional to the log of its physical magnitude I

$$S = k \log I/I_0,$$

where k is a constant of proportionality, and I_0 is the smallest value of I that can be detected (i.e., the absolute threshold). Note that the value of k is unique to the particular quantity measured (e.g., brightness and loudness have different values of k).

[See the next section if you need a reminder about logarithms.]

Fechner's law implies that each time the physical magnitude (weight, for example) is doubled this *adds* a constant amount ΔS to the perceived magnitude S. As an example, next, we show that

$$\Delta S = k \log 2.$$

If one physical magnitude I_1 is twice as large as another magnitude I_2 (i.e., $I_1 = 2I_2$) then the difference in perceived magnitude is

$$
\begin{aligned}
\Delta S &= S_1 - S_2 \\
&= k \log I_1/I_0 - k \log I_2/I_0 \\
&= k \log I_1/I_2 \\
&= k \log 2 = \Delta S.
\end{aligned}
$$

Logarithms

If you need a reminder of logarithms, here it is. For simplicity, let's assume that $S = \log I$, so we have implicitly set values in Fechner's law to $k = 1$ and $I_0 = 1$, **12.17a**. Here we make use of logarithms to the base 10. In essence, if $S = \log I$ then S is the power to which 10 must be raised to get I, or

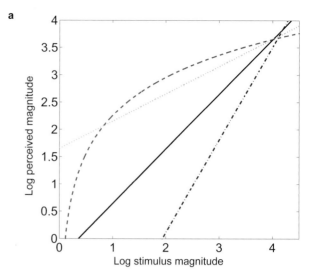

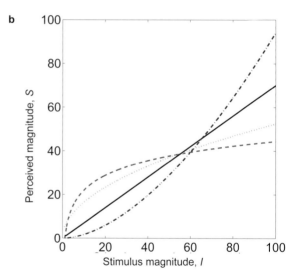

12.18 Stevens's law
Plots for length (black solid curve, $n = 1$), brightness (green dotted curve, $n = 0.5$), and electric shock (blue dash-dot curve, $n = 1.7$). For comparison, Fechner's law for brightness is shown as the red dashed curves.
a The relation $S = mI^n$.
b Graph of log stimulus magnitude versus log perceived magnitude. In each case, the slope of the line is an estimate of the exponent n, and the ordinate intercept corresponds to the constant m.

equivalently, $I = 10^S$. This can be confirmed if we take the log of both sides of this equation, which yields $\log I = S$ (because the log of 10^S is S).

What happens to S if we double the magnitude I? If the magnitude of the stimulus has an initial value of $I_1 = 4$ so that

$$I_2 = 2I_1 = 8,$$

then we have

$$\log I_2 = \log 2I_1$$
$$= \log 2 + \log I_1.$$

Subtracting $\log I_1$ from both sides yields

$$\log I_2 - \log I_1 = \log 2$$
$$= 0.3,$$

where, by definition, $S_1 = \log I_1$ and $S_2 = \log I_2$ so that $S_2 - S_1 = \log 2$. Thus, *multiplying* I by a factor of 2 *adds* an amount $\log 2$ ($= 0.3$) to S. If we substitute $I_1 = 4$ and $I_2 = 8$ then $S_1 = \log 4$, $S_2 = \log 8$, so that

$$S_2 - S_1 = \log 8 - \log 4$$
$$= \log 8/4$$
$$= \log 2.$$

More importantly, doubling the value of I *adds* an amount 0.3 ($0.3 = \log 2$) to S_1, irrespective of the initial value of S_1.

Taking a different perspective, what do we have to do to the magnitude I in order to add one to the perceived magnitude S? If we add one to S_1 so that

$$S_2 = S_1 + 1,$$

then

$$\log I_2 = 1 + \log I_1.$$

Subtracting $\log I_1$ from both sides yields

$$\log I_2 - \log I_1 = 1.$$

This can be rearranged to yield

$$\log I_2/I_1 = 1.$$

If we now raise both sides to the power of the logarithm's base (10, here) then we have

$$I_2/I_1 = 10^1 = 10.$$

Therefore, adding one to S can be achieved by making the ratio $I_2/I_1 = 10$; that is, by making I_2 ten times larger than I_1. This can be expressed as "adding one to S adds one log unit to I".

Stevens's Power Law

Stanley Smith Stevens (1906–1973) showed that Fechner's law could not account for many types of stimuli, and reformulated the relation between sensation magnitude S and physical intensity I in terms of the ***power law***

$$S = mI^n.$$

This states that the perceived sensation S is the physical intensity I raised to the *power n*, where the value of the exponent n depends on the specific stimulus being measured, and m is a constant which depends on the particular units of measurement used (e.g., feet versus metres).

This law has since become known as ***Stevens's law***. Some quantities, such as brightness, are reasonably consistent with both Fechner's and Stevens's laws, as shown in **12.18**. This is because the exponent of $n = 0.5$ for brightness in Stevens's law implies that the perceived magnitude increases with the square root of physical magnitude, and describes a function which is reasonably well approximated by a Fechner's logarithmic function. However, other quantities cannot be accommodated by Fechner's law, but are consistent with Stevens's law. For example, taste ($n = 1.3$) and electric shock ($n = 3.5$) cannot be described by Fechner's law, **12.18**.

Linking Weber's, Fechner's, and Stevens's Laws

A simple interpretation of Stevens' law is that the *relative* change in perceived magnitude $\Delta S/S$ is proportional to the *relative* change in magnitude $\Delta I/I$, so that

$$\Delta S/S = n\Delta I/I,$$

where n is a constant which depends on the physical quantity being measured. Note that this looks quite similar to Weber's law ($\Delta S/S = constant$), but its implications are quite different. This is because Weber's law states that the relative change in perceived magnitude is *constant* for a given physical quantity. Recall that we obtained Fechner's law $S = k \log I$ from Weber's law by integration.

A similar operation on the relation

$$\Delta S/S = n\Delta I/I,$$

yields

$$\log S = \log I^n + \log m,$$

(where $\log m$ is a constant of integration). By taking antilogs, we obtain Stevens' law $S = mI^n$.

The value of the exponent in Stevens's law $S = mI^n$ can be estimated by plotting the log S against log I, which yields

$$\log S = n \log I + \log m.$$

This has the same form of the equation of a straight line, traditionally written as

$$y = Mx + C,$$

where M is the slope of the line, and C is the value of the intercept on the ordinate (y) axis. Thus, Stevens's law predicts that a graph of log S versus log I has a slope $M = n$, and an intercept on the ordinate axis of $C = \log m$.

Conclusions

This chapter has introduced some core ideas about seeing. In particular, it has explained the fundamental problem of noise in sensory systems. Noise needs to be taken into account both when building models of the visual system and when developing methods for measuring fine discriminations.

Further Reading

Bialek W (2002) Thinking about the brain. In *Physics of biomolecules and cells: Les houches session LXXV* H Flyvbjerg, F Julicher, P Ormos and F David, F (Eds) (EDP Sciences, Les Ulis; Springer-Verlag, Berlin). *Comment* A general paper which is quite technical, by an insightful scientist, who discusses thresholds as one of the problems the brain has to solve. Online at http://www.princeton.edu/~wbialek/publications_wbialek.html.

Green DM and Swets JA (1966) *Signal detection theory and psychophysics*. New York: Wiley. *Comment* The classic book on SDT, fairly technical, but also very readable.

Hecht S, Schlaer S and Pirenne MH (1942) Energy, quanta and vision. *Journal of the Optical Society of America* **38** 196–208.

Heeger D (2007) Signal Detection Theory http://www.cns.nyu.edu/~david/handouts/sdt/sdt.html *Comment* A very clear account of SDT, with notes that can be downloaded.

Regan D (2000) *Human perception of objects*. Sinauer Associates, MA. *Comment* A splendid wide-ranging book on vision with a discursive section on psychophysics on pp. 8–30. Also see the many appendices for good tutorials and insights into modeling perception.

13

Seeing as Inference

13.1 Hill or crater?
Try turning the book upside down. Courtesy U.S. Geological Survey.

Seeing presents an extremely difficult task for the human brain. Indeed, if it were not for the fact that we can see, we might reasonably think it is impossible. This is because, to put it bluntly, the image on the human retina is flat but it is derived from a world which is not: from a world that contains depth. Recovering depth information from a flat retinal image would be impossible if the brain did not take advantage of certain fragments of knowledge it has about the structure of the world in order to make its best guess about what is really out there.

These fragments of knowledge take the form of **constraints**, **priors**, or **heuristics**; it matters little what they are called. What matters is that they are fragments of knowledge that cannot be obtained from any given retinal image. They are the extra knowledge that the brain must use in order to make sense of the jangle of images that besiege the retina. Even with the help of these constraints, knowing which object or scene structure is present is literally a matter of guess-work, albeit well informed guesswork, about the world out there. The brain, at its best, attempts to act like an optimal perceptual

Why read this chapter?

Most retinal images do not contain enough information to be interpreted unambiguously, suggesting that vision should be in principle impossible. In practice, it is possible because the processes of interpretation utilise knowledge or *constraints* in the form of *prior assumptions* about what any image might contain. These priors are an essential ingredient in the *Bayesian framework* for solving the problems of seeing. Priors make it possible for the human brain to see, even when there is very little to be seen. This ability to see beyond the information in an image gives rise to many illusions. It allows us to see objects where we should see only phantoms, and this represents the most useful illusion. Using the example of object recognition, the various components of the Bayesian approach to seeing are explored and their implications for vision are discussed. The chapter contains some quite technical details but they are presented in a way designed to allow the general reader or beginning student to grasp the key ideas while skimming over the most technical material. The core ideas underlying the Bayesian approach are important for understanding what it means to claim that seeing depends on inference.

inference engine in arriving at an interpretation of retinal images. Technically, this is described as the brain trying to operate as an **ideal observer.**

Let's begin with some simple examples. Look at the image in **13.1**. It looks like a convex bump, a shallow hill, but there is something a little odd about it. Now turn the book upside down. The hill is actually a crater, which is why it looked a little odd as a hill. The reason you perceive the image as a hill is because your perceptual system insists on assuming that light always comes from above, which is generally a very reasonable assumption. In the language of vision science, the assumption that light comes from above constrains the interpretation of the image. But if you make use of this constraint then you are forced to perceive the upside down image of a crater as a hill. This is a good example of how your prior experience of the physical world (specifically that light comes from above) results in a constraint which overrides the correct interpretation of an image.

Conflicting Constraints

Sometimes the light-from-above constraint can itself be overridden. The images in **13.2** depict different views of a face mask as it rotates through 180 degrees. Thus, **13.2a** is a convex face, whereas **13.2d** is a concave (hollow) face. Remarkably, the face appears to be convex in both cases. As Richard Gregory has pointed out, your perceptual system insists on assuming that both faces are convex, which is a reasonable assumption because faces are almost always convex in natural scenes. However, if we make use of this constraint for the hollow face in **13.2d** then we are forced to interpret the light as coming from below, even though it is actually coming from above.

Thus, the face-convexity constraint overrides the light-from-above constraint. This is a case of the human perceptual system being confronted with a dilemma in the form of two conflicting constraints. On the one hand, it can correctly interpret **13.2d** as a concave face with light from above. On the other, it can incorrectly interpret **13.2d** as a convex face with light from below. As we can see, the human perceptual system solves this problem of constraint satisfaction by choosing the latter as the more probable interpretation. This is presumably because the human perceptual system

a   b   c   d

13.2 Face mask seen from different positions
The light source is coming from above in all figures. **a-c** These views create perceptions of a normal convex face. **d** This view from the rear of the mask does not produce the perception of concave face, as it 'should' do. Instead, a convex face is seen. In order to achieve this incorrect perception, the brain is forced to conclude that the light source is coming from below. Courtesy Hiroshi Yoshida.

has decided it is more likely that light comes from below than it is that a face could be concave.

While it is fairly easy to see how constraints are used to find the most probable interpretation in the above examples, it is less obvious that a similar process of 'constrained guessing' is involved in every act of seeing. In fact, almost all problems in vision can be described of 'constrained guessing' or *perceptual inference*.

Look at the World Wildlife Fund (WWF®) panda logos in **13.3**. The line around the panda's head was deleted around 1986 and yet this has

very little impact on our ability to perceive the boundary of the head. This is because we infer this boundary by essentially re-inserting our own version of a boundary that we know must be there, a boundary that is clearly more apparent than real. We met this phenomenon earlier when considering the task of edge detection (see Ch 5, Kanizsa's triangle).

Reality, Virtual Reality, and Demons

Why is seeing almost impossible? Because the image on your retina, which was probably

1961   1978   1986   2000   2006

13.3 Logo for the World Wildlife Fund
Over many years the WWF®—the global conservation organization—has trimmed away the line from around the panda's head. Remarkably, this has had very little impact on our ability to perceive the boundary of the panda's head. © 1986, WWF—World Wide Fund for Nature (also known as World Wildlife Fund). ® Registered Trademark owner.

a

b

13.4 Any given perception can be produced by an infinity of different scenes
The image of the A (left) is a photograph taken from above of the carefully arranged objects on the right.

generated by the three-dimensional scene you think you see before you, could have been generated by an infinite number of different three dimensional scenes. This is not like the movie *The Matrix*, where every three-dimensional scene perceived by every character in the movie was created by a vast computer. This is more scary.

As every student of philosophy knows, the skeptics argued that there may be no three-dimensional scene before you. But if there were, how could you tell which one it was? We do not want to get all philosophical (this is a book on vision after all), so from a purely practical point of view, how could you tell?

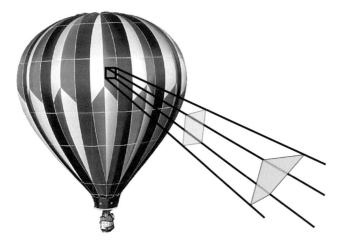

13.5 Infinite ways of creating an image
The image of this balloon could have been produced by a balloon, or by an infinite number of other three-dimensional surfaces. As an example, a square area in the image of the balloon could have been produced by many different surfaces in three dimensions.

Look at **13.4a**. What do you see? The object looks like the letter A. It is in fact a carefully arranged set of three pieces of plasticine, but from a specific angle it looks like the letter A. We met a classic example of this sort of phenomenon in Ch 1 in the form of Ames's chair.

Well, you might say, that's just one untypical trick; it just shows someone could spend hours making three sausages look like a letter A, and then only from one viewpoint. Let's therefore consider a more complete deception.

Take an image, any image, a balloon, for example, **13.5**. Each small element or pixel in that image has a grey-level that is determined by the particular piece of the world from which it is derived. If the grey-level of one pixel corresponds to a region of the balloon surface then that pixel has a specific grey-level that is consistent with the color of the ballooon.

We are now going to do something devious. We are going to alter the world in such a way as to leave the retinal image unchanged while fragmenting the balloon surface. For each pixel, find the piece of world that projects to it, so that each piece of the world effectively has its own pixel, **13.5**.

Now replace that piece of the world with something else, anything else, with the following proviso. The new piece of world must generate the same grey-level in its pixel as did the old piece of world, and the new piece of world must be at a different depth from every other new piece of world. In other words, whereas the old world was made up of smooth surfaces with adjacent pieces that were at roughly similar depths, the new world consists of fragmented surfaces with pieces literally all over the place (in terms of depth). The image is the same, but the world that generated it just changed dramatically. What is more, every new piece of world could be replaced with an *infinite* variety of other pieces without altering the retinal image.

But surely, we hear you say, if the observer were to walk into the world we have just created then he would see that this new world was "visual gibberish" (as also for the letter A in **13.4**). Not necessarily. We can push this argument to its logical conclusion by creating a fast-working "vision demon" that reconfigures the world every time you move your head. Every reconfigured world would be visual gibberish, but the demon works in such a way that the image on your retina always appears to depict an internally consistent three-dimensional scene. Farfetched, perhaps, but not impossible. Given that it is not impossible, how could you tell whether or not the world before you is generated by a zippy demon?

That's the point, how could you tell? The short answer is that you cannot, at least, not with certainty. The best the brain can do is to assume that the physical world is stable, and that it has constraints that affect how it behaves. If the brain can get to know these constraints then it can use them in order to make sense of the images on your retina. In making sense of retinal images, the brain is really generating a "best guess" or **inference** about what generated the retinal image, based on the information in that image combined with previous experience of interpreting retinal images. This experience could be acquired within an organism's lifetime or over eons via the processes of natural selection and evolution.

Object Recognition Is an Ill-Posed Problem

There is a technical term for the fact that the brain cannot tell which surfaces gave rise to an image without extra knowledge. It is that the vision problem is **ill-posed**. In other words, it is not possible *in principle* to discover the three-dimensional locations of points in the world which gave rise to the image without some prior knowledge about the laws of physics govern how the world is put together.

For example, most surfaces are smooth, which means that nearby points on a surface have similar depths in the world. This smoothness is a general consequence of the coherence of matter, which in turn is a consequence of gravity. If we lived in outer space where small particles of matter floated freely around us then there would be no surfaces, just particles, and each particle would be at a different depth.

The fact that we can assume surfaces are smooth turns out to be an enormous help in making sense of retinal images. Of course, there are other constraints, but spatial smoothness seems to be particularly important in overcoming the ill-posed nature of object recognition. Without this knowledge it would be almost impossible to make sense of images. In other words, we depend on this knowledge which cannot be obtained from the image itself, in order to interpret retinal images. Recognizing that vision as an ill-posed problem makes it an obvious candidate for analysis within the **Bayesian framework**.

Bayesian Framework

There are basically two types of information: information we *have*, and the information we *want*. In essence, the Bayesian framework allows the information we have to be combined with extra knowledge in order to obtain the information we want.

Within vision, the information we have is the retinal image (more precisely, it is the *likelihood* of observing that image; we will define this technical term later). But in fact, the information we want concerns the three-dimensional scene from which that image was derived. The extra knowledge is the collection of constraints about how the physical world is put together.

We will describe how to make use of the Bayesian framework to combine the information we have (the retinal image) with extra knowledge (constraints gleaned from previous experience) to obtain the information we want (the three-dimensional structure of surfaces in the world).

Ambiguous Images and Conditional Probability

What do you see in **13.6**? Perhaps some sort of hat? If you have read a wise and charming book

13.6 What is it?
What you see here depends on your prior experience. Specifically, whether or not you have read *The Little Prince*. See **13.7** on p. 314.

13.8 The Little Prince
The principal character in a book by Antoine de Saint Exupéry of this title.

called *The Little Prince*, **13.8**, then it is quite likely that you see not a hat but a quite different object, **13.7** (on p. 314). If you have not read that book, then you would deem the interpretation in **13.7** to be highly improbable.

So suppose we ask you: what is the probability of there being an elephant in **13.6**? Depending on your prior experience of such shapes, you might reply about 0.1 (if you had not read *The Little Prince*) or about 0.9 (if you had read *The Little Prince*). Now suppose we were to ask you a subtly different question: what is the probability of this shape given that there is an elephant in the picture? In this case, your answer would not depend much on your prior experience and might be "about 0.9."

What we have just done is to extract from you two different ***conditional probabilities***. Compare these two probabilities carefully. Note that if you have read *The Little Prince* before, then your prior experience effectively biases you to perceive the shape as an elephant. This form of bias is actually useful information, as will become evident below.

As *The Little Prince* says, "*Grown-ups never understand anything by themselves, and it is tiresome for children to be always explaining things to them.*" If you have children then you know that this is true.

Bayesian Object Recognition

Given an image of an object on the retina, how does the brain know what shape that surface has in three-dimensions? In **13.9** three objects are shown which produce the same image. To us, this image would be interpreted as a cube, but that is just because we are experienced at interpreting such images as cubes. But, as we have just explained, the image could have been generated by any one of the three objects shown.

Armed with this information, you would be able to answer the question: "How likely is it that this image would be observed, given the each of the three candidate objects?" You would be forced to reason that the same image would result from all three objects, and therefore that the image is equally likely for all three objects. You would, of course, be correct. The likelihood of observing this image for all three objects is the same. This means that these three likelihoods (one for each object) alone do not provide enough information to decide which object is in the world. At this stage, this is the only information we have, but it is not the information we want.

But, surely, that's just silly you might say, we just asked the wrong question. Rather than asking "What is the likelihood of the image given each object?" we should ask "What is the probability of each object given the image?" (note the subtle change from "likelihood" to "probability" here, see below). If we had the answer to this question then we could simply choose that object which most probably generated the image. This is the information we want.

The reason for asking the wrong question was twofold. First, it is the only question to which we have a ready, and at first sight, useless, answer (an answer which represents the information we have). Second, the answer to the right question (the information we want) cannot be obtained without knowing the answer to the wrong question. In order to see why, we need to introduce the notion of conditional probability.

Conditional Probability

We will use the symbol s to stand for the object, and s can be any one of three objects s_1, s_2, and s_3, **13.9**. We will use the symbol I to stand for the image, which in principle could be an image of anything, but we will assume it is a specific image which we call I. (More formally, the variable s can adopt any one of three values $s = s_1$, $s = s_2$, or $s = s_3$, and the variable I can adopt only one value here). If the world contains only one of the three objects (s_1, s_2, s_3) then which object produced I?

We already know that each of the three objects yields the same image. If we use the symbol p to denote probability then the **_conditional probability_** of observing the image I given the object s_1 can be written as $p(I|s_1)$. This is just a shorthand way of writing the statement "the probability of observing the image I given the object s_1," and the vertical bar symbol | stands for "given." The corresponding conditional probabilities for s_2 and s_2 are written as $p(I|s_2)$ and $p(I|s_3)$.

Thus, the answer to the wrong question "what is the likelihood of I given each object s_1, s_2 and s_3?" is given by the values of the three conditional probabilities $p(I|s_1)$, $p(I|s_2)$, and $p(I|s_3)$.

What about the answer to the right question, which represents the information we want. Recall that the right question is, "what is the probability of each object given the image?" In this case, the answer is given by three different probabilities, $p(s_1|I)$, $p(s_2|I)$, and $p(s_3|I)$.

Notice that the information we have (e.g., $p(I|s_1)$) and the information we want (e.g., $p(s_1|I)$) look very similar when written in this form. Their similarity is due to the fact that they are both conditional probabilities. For example, the quantity $p(I|s_1)$ is a conditional probability because the probability of I is subject to the condition that a particular object s_1 is present, and the quantity $p(s_1|I)$ is a conditional probability because the

probability that the object s_1 is present is subject to the condition that the particular image I is observed.

A conditional probability differs from our usual notion of probability, although it can be argued that all probabilities are really conditional probabilities. For example, when tossing a fair coin C_1, the probability of observing a head H is $p(H) = 0.5$ (i.e., there is a 50/50 chance of observing a head). But supposing we had used some other coin C_2 which is biased so that it produces a head 90% of the time. For this coin $p(H) = 0.1$. Clearly, the value of $p(H)$ depends on which coin is used, and we need a notation which captures this dependency. For this reason the statement "the probability of observing a head

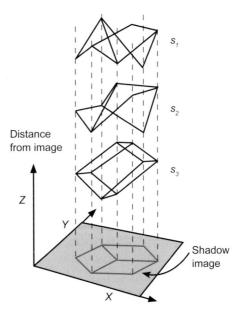

13.9 Three wire frame objects forming the same image
The three wire frame objects labeled s_1, s_2, and s_3 cast the same shadow image in the image plane XY (shown gray). This illustrates that the shadow image could have been generated by an infinite number of different objects, only three of which are shown. The conditional probability (the **_likelihood_**) of the observed image I is the same (about unity) for all three objects, so that $p(I|s_1) = p(I|s_2) = p(I|s_3)$. However, the conditional probability of each object given the observed image (the **_posterior probability_**, $p(s_1|I)$, $p(s_2|I)$, $p(s_3|I)$) depends on our previous experience of how common each object is in the world, as indicated by each object's **_prior probability_**, $p(s_1)$, $p(s_2)$, and $p(s_3)$. Using Bayes' rule, we can combine the likelihood with the prior of each object to make an informed guess as to which object is most likely to have generated the observed image. This "guess" is usually defined by the object with largest posterior probability. With permission from Ernst and Bülthoff (2004).

13.7 Elephant swallowed by a snake
This shows that the shape shown in **13.6** is not a hat.

H given that we have used the coin C_1 is 0.5" is written as $p(H|C_1) = 0.5$. Similarly, the statement "the probability of observing a head *H* given that we have used the coin C_2 is 0.9" is written as $p(H|C_2) = 0.1$.

To return to our vision example, we know that one type of conditional probability (e.g., $p(I|s_1)$) refers to information we have, and the other type (e.g., $p(s_1|I)$) refers to information we want. In order to distinguish between these two different types of conditional probabilities, they have special names. The information we have (e.g., $p(I|s_1)$) is called the ***likelihood***, and the information we want (e.g., $p(s_1|I)$) is called the ***posterior probability***.

Now, the probability of observing the image *I* given each of the three objects can be written as the three likelihoods $p(I|s_1)$, $p(I|s_2)$, and $p(I|s_3)$. In principle, each of the three likelihoods should be equal to unity, because we are certain to observe the same image *I* if any of the three objects is present in the world. However, in practice, we will assume these likelihoods are just less than unity, so that it is almost certain that we would have the image *I* for each of the three objects. This is

because the eye is basically a device for measuring luminance, and any measurement device contains intrinsic statistical variability or noise which ensures that its output is not completely reliable. For example, the lens of the eye is far from perfect and therefore introduces many local distortions to the retinal image. This is why $p(I|s_1)$ must be less than unity (1.0). For simplicity, we assume that the three likelihoods are all equal to 0.8, which we can write as $p(I|s_1) = p(I|s_2) = p(I|s_3) = 0.8$.

The fact that all three likelihoods are equal means that, if we had to use the likelihoods to decide which object is present, then we could not do so. Thus, the image *I* cannot alone be used to decide which of the three objects gave rise to that image.

Seeing through the Eyes of Experience

As experienced perceivers, we would insist that the image *I* was almost certainly produced by only one object, s_3. This is because we have prior experience of what sorts of images are produced by different objects.

This prior experience biases our interpretation of the image, so that we tend to interpret the image *I* as the object s_3 that is most consistent with our prior experience of such images. Specifically, previous experience allows us to assign a value to the probability that each object is encountered in the world. For example, common objects like convex human faces have a high probability (close to unity), whereas concave human faces have a low probability (close to zero). Because such a probability is based on prior experience, and does not depend on the current image, it is called a

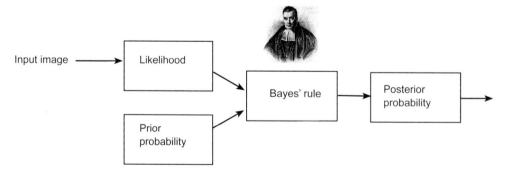

13.10 Bayes' rule
This rule provides a principled method for obtaining the information we want (the ***posterior probability***) by combining the information we have (the ***likelihood***) with information gained from previous experience (the ***prior probability***).

prior probability. Thus, on the basis of prior experience, objects s_1 and s_2 may be assigned (for example) the low prior probabilities of $p(s_1) = 0.17$ and $p(s_2) = 0.33$, whereas object s_3 may be assigned a high prior probability of $p(s_3) = 0.50$.

We now have three key ingredients of the Bayesian framework: the *likelihood*, the *posterior probability*, and the *prior probability* (these last two terms are often abbreviated to the 'posterior' and 'prior').

Bayesian Inference

Now comes the crucial step: *Bayesian inference*. In essence, this involves using the known prior probability of each object to weight the known likelihood in order to obtain the unknown posterior probability, as shown in **13.10**.

Put another way, we can use our prior experience of how common objects are in the world in order to convert the information we have (the likelihood of the image) to the information we want (the posterior probability of observing each object).

Notice that we actually want a posterior probability for each object, and this is obtained from the likelihood and prior probability (both of which we know) for each object. In algebraic form, this can be written as

$$p(s_1|I) = p(I|s_1)p(s_1)$$

$$p(s_2|I) = p(I|s_2)p(s_2)$$

$$p(s_3|I) = p(I|s_3)p(s_3).$$

One way to remember this is by using the names of each term:

$$posterior = likelihood \times prior.$$

Why Do We Want Posterior Probabilities?

We want posterior probabilities because each posterior probability tells us the probability of a different object given the image I. If we know this probability for each object then we can simply choose that object which is most likely to have generated the image.

For example, we might end up with the posterior probabilities

$$p(s_1|I) = 0.13$$

$$p(s_2|I) = 0.27$$

$$p(s_3|I) = 0.40.$$

Given this set of values, we would choose s_3 as the object most likely to have yielded image I. As the object must be one of the three objects, the sum of posterior probabilities should be one, but it is not as explained in the next section.

How could we arrive at these values? Well, we already know that the likelihoods $p(I|s_1)$, $p(I|s_2)$, and $p(I|s_3)$ are all equal to 0.8, and our previous experience tells us that each of the objects has a prior probability of $p(s_1) = 0.17$, $p(s_2) = 0.33$, and $p(s_3) = 0.50$. Using Bayes' rule we can combine the information we have (the likelihood) with information based on prior experience (the prior), to obtain the information we want (the posterior) about each candidate object:

$$p(s_1|I) = p(I|s_1)p(s_1) = 0.8 \times 0.17 = 0.13$$

$$p(s_2|I) = p(I|s_2)p(s_2) = 0.8 \times 0.33 = 0.27$$

$$p(s_3|I) = p(I|s_3)p(s_3) = 0.8 \times 0.50 = 0.40. \quad (13.1)$$

As stated above, given these three posterior probabilities we would choose object s_3 because $p(s_3|I)$ is greater than $p(s_2|I)$ and greater than $p(s_1|I)$. It is therefore more probable that the image I was generated by object s_3 than either of the objects s_1 or s_2. The largest value of these probabilities is known as the *maximum a posteriori* (*MAP*) probability, and the object s_3 associated with this value is the *MAP estimate* of the object presented to the retina. This, then, is Bayesian inference.

We have come a long way. However, in order to make the above account accessible, we omitted a few small but important details.

Coming Clean About Bayes' Rule

At this point we need to come clean about the exact form of Bayes' rule. For the example of an object s_1 and an image I, the exact form of Bayes' rule is given by

$$p(s_1|I) = \frac{p(I|s_1)p(s_1)}{p(I)}.$$

The term $p(I)$ is the probability of observing

the image I, and is rather confusingly called the **evidence**. Strictly speaking, we do not know the value of $p(I)$, but we do know that whatever value it has, it is a constant, because it refers to the probability of an image which we already have. We will call this constant $k = p(I)$. For this reason we should really write

$$p(s_1|I) = \frac{p(I|s_1)p(s_1)}{k}.$$

However, as I is the image we have already observed, we may as well assign the probability of observing it as unity (one). Because we assign the value $p(I) = 1$ when it is usually not equal to one, our posterior probabilities do not sum to one, as they would do if we knew the correct value of $p(I)$. So, for our purposes, we can assume that

$$p(I) = k = 1,$$

which allows us to write Bayes' rule as

$$p(s_1|I) = p(I|s_1)p(s_1).$$

This omitted constant k means that you may also see Bayes' rule written as

$$p(s_1|I) \propto p(I|s_1)p(s_1),$$

where the symbol $\propto$ stands for "is proportional to." As we are only interested in comparing posterior probabilities associated with different objects for a given image, we do not care what value the constant k has. In other words, we are really only interested in the relative sizes of the posterior probabilities. The object which is associated with the largest value of the posterior (e.g., $p(s_3|I)$) remains the largest irrespective of the value of the constant k.

While we are in a confessional mood, we should also reassure the observant reader that the problem of object recognition as posed above could be solved simply by changing the orientation of each object. Alternatively, we could simply walk around each object, which would quickly remove any doubt as to which of the three objects is being observed.

The point being made is that any data are intrinsically ambiguous in some respect, and that prior knowledge must be brought into play in order to correctly interpret such data. To be

sure, walking around an object does make its 3D shape apparent. However, more generally, given some sensory data it is not always possible to perform the equivalent of "walking around the object" to gain more information. One such case is examined in some detail in Ch 15, where the problem of estimating the speed of a moving line is considered. In that example, the data arriving on the retina is all there is, and there is no equivalent to "walking around the object." In such cases, and most cases are like this, Bayesian methods can be employed to combine the sensory data with prior experience to arrive at a statistically optimal estimate of what is really "out there."

Bayes' Rule and Maximum Likelihood Estimation

In order to stress the role of prior experience in interpreting images, we used an example in which the likelihoods all had the same value. However, a more realistic example would use different values for these likelihoods. Surely, in this case, the likelihoods could be used to choose an object from the candidate objects s_1, s_2, and s_3? Let's see what happens if we do this.

Suppose we had an image I, and three candidate objects, except that now the likelihood of the image given each object is

$$p(I|s_1) = 0.75$$

$$p(I|s_2) = 0.29$$

$$p(I|s_3) = 0.64. \qquad (13.2)$$

If we used these three likelihood values to choose an object then we would choose s_1 because this has the largest value. Note that we have chosen s_1 on the basis of the wrong question. That is, we have chosen s_1 because, out of the three objects, it is s_1 that is associated with the largest probability of the image given each object (i.e., the largest likelihood). In other words, s_1 is that object which makes the observed image most likely when compared with the likelihoods associated with the other two objects.

In the absence of any other information, we would be forced to use the relative values of the likelihoods associated with different objects to choose one of those objects. In practice, this is

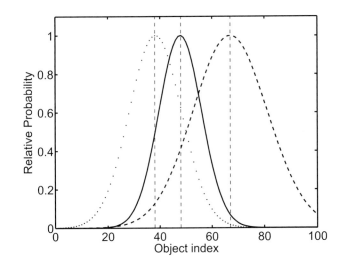

13.11 Graphs of the prior (right), likelihood (left), and posterior (middle) probability density functions (pdfs) for 100 objects
(If you are unfamiliar with the idea of a pdf then just think of it as a histogram). The frequency with which each object is encountered is indicated by the right-hand (dashed) curve. This implies that the object with index number 67 is most commonly encountered, as this index corresponds to a peak in the prior probability function (right-hand curve). When presented with a specific image, the likelihood of each object is given by the height of the left-hand (dotted) curve, the likelihood function. As the peak of this curve corresponds to object index 38, this implies that the *maximum likelihood estimate* (*MLE*) of the presented object is object number 38. The posterior probability of each object is given by multiplying corresponding points on the prior (right curve) and likelihood (left curve) functions. This yields the posterior probability (middle curve) function. The peak of the posterior function corresponds to object number 48, so that this object is the *maximum a posteriori* (*MAP*) estimate of the presented object. Notice that the MLE and MAP estimates are not the same, and that the width of the posterior distribution is less than that of the likelihood function, implying that the MAP estimate is more reliable than the MLE. (For display purposes, each curve has been re-scaled to ensure that its maximum value is unity.)

often the case, and choosing that object which maximizes the likelihood of the image is called *maximum likelihood estimation* (*MLE*), and this object is called the *maximum likelihood estimate*.

Using Bayes' Rule with Unequal Likelihoods

If we do have additional information (and in this case, we do) then we can use this to answer the right question, "what is the probability of each object given the image?"

If the likelihoods are unequal (as here) and if we have a prior for each object then we can work out the posterior probability of each object, as before. Given the prior probabilities used above, and the values for likelihoods given in equations (13.2), the posterior probabilities are:

$$p(s_1|I) = p(I|s_1)p(s_1) = 0.75 \times 0.18 = 0.14$$

$$p(s_2|I) = p(I|s_2)p(s_2) = 0.29 \times 0.32 = 0.09$$

$$p(s_3|I) = p(I|s_3)p(s_3) = 0.64 \times 0.50 = 0.32. \quad (13.3)$$

There are two things to notice here. First, the answer (s_1) we get using the MLE equations (13.2) is different from the answer (s_3) we get with our two examples of Bayes' rule equations (13.1) and equations (13.3). This is because the value of the prior is larger for s_3 than it is for s_1 and s_2 in equations (13.3).

Second, we get the same answer (s_3) using Bayes' rule in equations (13.1) and equations (13.3) whether or not the likelihoods are equal (at least in these examples). If the likelihoods are equal, as in equations (13.1), then the prior determines which object is chosen. Even if the likelihoods are different, as in equations equations (13.3), a strong prior can over-ride the evidence before us, as in equations (13.3) Thus, even though s_3 has a smaller likelihood than s_1, the large value of the prior for s_3 over-rides the effect that the likelihood has on the posterior, so that s_3 is chosen.

Using Bayes' rule, the combined effect of the likelihood and the prior determine which object is chosen. This is as it should be because we are trying to base our decision on a combination of current evidence (the image) and previous experience in order to choose which object is most likely to have generated the image.

Bayes' Rule and Probability Distributions

Instead of choosing between a finite number of possible objects, consider the near infinitude of possible objects generated by morphing between objects. For example, a gradual deformation could transform any object into any other object in **13.9**, and the resultant intermediate objects could give rise to identical images.

Let's consider a more manageable set of, say, $N = 100$ morphed objects, a set that we will define as $\mathbf{s} = (s_1, s_2, \ldots s_N)$. As these objects are generated by a gradual morphing transformation we can number them from 1 to 100 and be reasonably confident that objects with similar number will have similar forms. We can use the index i to denote each object's number, so that s_i refers to the ith object in the set $\mathbf{s}$ of $N = 100$ objects.

We can now assign a prior probability to each object on the basis of how often we have encountered this object in the past. Let's assume that object number 67 has been seen most often, and that objects with higher and lower identity numbers have been seen with a frequency that is given by the right-hand (dashed) curve in **13.11**. As this curve represents how often each object was encountered in the past, it makes sense to assume that it also represents a fair representation of how often each object will be encountered in the future (if we make the reasonable assumption that the past and the future are statistically identical). Accordingly, this curve represents our prior belief about how frequently each object will be encountered, and is called the ***prior probability distribution***, or ***prior probability density function***, or simply a ***prior pdf***.

On a technical note, if N were infinitely large then this distribution would define a ***continuous probability density function*** or pdf, as shown in by the right-hand (dashed) curve in **13.11**. As stated above, this prior pdf reflects our previous experience of objects we have encountered. (We will not dwell on the distinction between the discrete distribution obtained with a finite umber of objects and the continuous distribution obtained with an infinite number of objects here, and will abuse several notational conventions by referring to both as pdfs.)

In this world of N possible objects, an image of an object is associated with a likelihood for each object in $\mathbf{s}$. If we plot a graph of $p(I|s_i)$ against all the objects in $\mathbf{s}$, ordered according to their subscripts $i = 1, 2, \ldots N$, then this defines a curve, which is known as the likelihood function, as shown by the left-hand (dotted) curve in **13.11**.

Recall that the information we have is the likelihood but if we also know the prior these can be combined to obtain the information we want,

the posterior. Therefore, the prior pdf and the likelihood function define a posterior pdf:

$$p(\mathbf{s}|I) = p(I|\mathbf{s})p(\mathbf{s}).$$

Even though this looks superficially like the other examples of Bayes' rule that we have encountered, it is important to bear in mind that we are now dealing with probability distributions (pdfs) and not probabilities as single numbers (scalars). The subtle but important difference is in the use of the bold $\mathbf{s}$, which indicates we are considering the probability of every object in $\mathbf{s}$.

One way to interpret this is to consider the value of $p(\mathbf{s}|I)$ for every object in $\mathbf{s}$. For example, if we consider the 9th object in $\mathbf{s}$ then the posterior probability (a scalar, in this case) is

$$p(s_9|I) = p(I|s_9)p(s_9).$$

Note that we have a similar expression for each object in $\mathbf{s}$. Thus, the ith scalar value $p(s_i|I)$ in the posterior pdf is given by multiplying the ith value $p(I|s_i)$ in the likelihood function by the corresponding value $p(s_i)$ in the prior:

$$p(s_i|I) = p(I|s_i)p(s_i).$$

Now consider what happens when we are presented with an object s_i, or more correctly with the image I of an object on the retina. Of course, what we want to know is which object gave rise to this image. But because all images are noisy and nothing is certain (except death and taxes, of course), we will have to settle for knowing which object most probably gave rise to this image: the object s_i that gives the largest value of the posterior probability $p(s_i|I)$.

In order to find that object we need the value of $p(s_i|I)$ for all objects under consideration (i.e., all objects in $\mathbf{s}$); and in order to find each value of $p(s_i|I)$ we need the values of the likelihood $p(I|s_i)$ and the prior $p(s_i)$ for each object s_i, because Bayes' rule states that $p(s_i|I) = p(I|s_i)p(s_i)$. Once we have the prior and likelihood of each object we can evaluate the posterior pdf as

$$p(\mathbf{s}|I) = p(I|\mathbf{s})p(\mathbf{s}),$$

which defines the left-hand (dotted) curve in **13.11**. The peak, or maximum, of this posterior

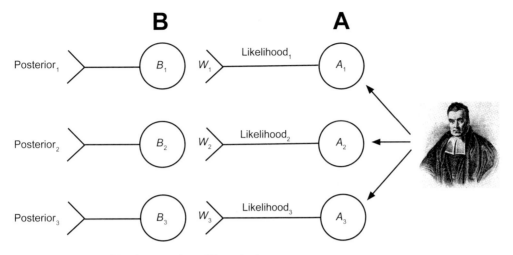

13.12 Imaginary neuronal implementation of Bayes' rule
Given input image I of an object s (where s is the Reverend Bayes in this case), each of N imaginary likelihood neurons $\mathbf{A} = (A_1, A_2, \ldots A_N)$ responds with an output proportional to the match between its preferred object s_i and I (only three such neurons are shown here). Each neuronal output corresponds to the likelihood $p(I|s_i)$. Each neuron in $\mathbf{A}$ is connected to exactly one imaginary posterior neuron in the set $\mathbf{B} = (B_1, B_2, \ldots B_N)$, and the strength of the connection between neuron A_i and B_i is W_i. As the output of A_i is given by the product $A_i = B_i W_i$, we can choose to interpret W_i as the prior probability of A_i's preferred object s_i, which implies that the output of each posterior neuron B_i is a posterior probability $p(s_i|I)$.

pdf curve defines the object that most probably gave rise to the observed image I. The value of this peak is the maximum a posteriori or MAP probability, and the object which corresponds to this peak is known as the MAP estimate of the object which actually generated the image I.

As it often aids understanding to witness a specific implementation of an idea, we offer the following speculative account of how Bayesian inference could be implemented in the brain.

Bayesian Neurons?

Consider a set of $N = 100$ purely imaginary neurons $\mathbf{A} = (A_1, A_2, \ldots A_N)$, each of which is 'tuned' to one object. Specifically, each neuron A_i has a maximal response to the ith objects in the set $\mathbf{s} = (s_1, s_2, \ldots s_N)$ of $N = 100$ objects. We will call these neurons likelihood neurons, for reasons which will become apparent.

Now, when an object's image appears on the retina, each likelihood neuron has an output which reflects the degree of match between the retinal image I and the object represented by that neuron. It is as if each neuron is shouting, "That's my object." But of course, some neurons shout louder than others, because some matches are better than others. If we plot the 'loudness' (the response) of

each neuron then we might obtain the left-hand curve in **13.11**.

If each neuron is acting as a crude template matching device then its response is easily interpreted as a measure of how likely it is that the image I was generated by the object s_i. Accordingly, and with but a small leap of imagination, we can see that the response of neuron A_i is the likelihood $p(I|s_i)$ that the image I was generated by object s_i. If we plot the set of $N = 100$ neuronal responses to the image I then we can see that they define a curve which is actually identical to the likelihood function, the left-hand (dotted) curve in **13.11**. In other words, the outputs of this set of neurons provides the distribution $p(I|\mathbf{s})$.

At this point, we can clarify why we chose to label likelihood as "the information we have." Of course, in general terms, it is no such thing. *Unless*, that is, we have a battery of neurons each of which is tuned to a specific image. In this case, the output of each neuron could plausibly approximate a likelihood $p(I|s_i)$. Given such a battery, the likelihood can indeed be interpreted as "the information we have."

If we were to use the likelihood function to decide which object was present then we would naturally choose the object with the loudest neuron: the neuron with the maximum output value. Because this represents the maximum value of the likelihood function, it corresponds to the maximum likelihood estimate (MLE) of the object. As we have seen before, we are often forced to make a decision based on the likelihood function. However, if we have access to the prior pdf then we can make use of this additional source of information to arrive at a more informed guess at the identity of the object.

In order to gain some insight into what might constitute a prior pdf, let's invent another set of imaginary neurons, $\mathbf{B} = (B_1, B_2, \ldots B_N)$, which we will call *posterior neurons*. Crucially, each posterior neuron B_i receives its input from exactly one likelihood neuron A_i, as shown in **13.12**. Now the output of a posterior neuron B_i is given by the output of the likelihood neuron A_i that connects to it, multiplied by strength W_i of the connection between A_i and B_i.

Crucially, we will assume that this connection strength is proportional to the number of times this neurons object s_i has been seen before. One interpretation of this is that W_i is proportional to the prior probability $p(s_i)$ of encountering the object s_i. Note that a graph of connection strengths, ordered according to neuron number as before, yields the distribution of prior probabilities, the prior pdf, as shown by the right-hand (dashed) curve in **13.11**.

Given that the output of a posterior neuron is the output of a likelihood neuron multiplied by the connection strength that connects them then we have:

posterior output B_i =

likelihood output $A_i \times$ connection strength W_i.

If we assume that all constants are equal to unity then this can be rewritten as

$$p(s_i|I) = p(I|s_i)p(s_i),$$

so that each posterior neuron has an output equal to the posterior probability $p(s_i|I)$. If we consider the entire set of $N = 100$ outputs then we have

$$p(\mathbf{s}|I) = p(I|\mathbf{s})p(\mathbf{s}).$$

Recall that the reason we wanted the posterior for each object in $\mathbf{s}$ was so that we could select that

object which has a maximum value of $p(s_i|I)$. We can find this graphically if we plot the posterior neuronal values ordered according to neuron number. The resultant graph is the posterior pdf; see the middle curve in **13.11**.

The peak in the graph of posterior probabilities identifies the object s_i which is most probably responsible for the observed image I, and the object s_i is the maximum a posteriori probability (MAP) estimate of the object presented to the retina.

Though we have used an imaginary scenario of likelihood and posterior neurons here, it is important to note that the posterior pdf is given by multiplying points in the likelihood function with corresponding points in the prior pdf.

The Bayesian strategy outlined above can be applied to object recognition as here, but it can also be applied to the more general problem of parameter estimation, as we shall see in the context of motion in Ch 15. In such cases, the variable s represents the value of a particular parameter, such as the motion of an edge, the slant of a surface, or the color of an object.

Bayes' Rule in Practice

As a theoretical framework for thinking about vision, it is hard to beat the Bayesian framework. (We discuss in Ch 23 where work using Bayes' theorem fits within Marr's framework.) However, it must be said that the number of practical applications of the Bayesian framework within vision has not been overwhelming. Indeed, most of the applications of the Bayesian framework have been published within the computer vision literature. This is because, when faced with the practical problems inherent in trying to build a seeing machine, the Bayesian framework provides a powerful and elegant tool.

As regards human vision, most applications of the Bayesian framework involve the problem of how to combine information from different sources or perceptual cues. For example, imagine you have a shape-from-texture module in your brain, and a separate shape-from-stereo module. If you look at a slanted surface it is quite likely that these two modules, each of which depends on a different perceptual cue, would provide slightly different estimates of that surfaces slant. How

13.13 Reverend Thomas Bayes (1702–1761)

13.14 Hermann von Helmholtz (1821–1894)

would your brain reconcile these two contradictory pieces of information? In other words, how would your brain perform cue-combination to arrive at a sensible estimate of slant?

The Bayesian framework provides a method for finding the statistically optimal estimate of slant based on the two estimates of slant provided by the shape-from-texture and shape from-stereo modules. This is the estimate that would be made by an optimal inference engine, the ideal observer referred to above. It is possible to compare the slant estimated by an ideal observer with the slant estimated made by humans in order to test whether or not our perceptual systems act according to Bayesian principles; whether or not we are ideal observers. These experiments are described later (Ch 20) because, even though they are based on Bayesian principles, the equations used for cue-combination appear quite different from the Bayes' rule described in this chapter.

Historical Note: Reverend Thomas Bayes

The Reverend Thomas Bayes (1702–1761), **13.13**, did not know, nor did he care as far as we know, about vision. In his life, he published only one mathematical paper on Newton's calculus or fluxions as it was then called ("An Introduction to the Doctrine of Fluxions"). But his main achievement was a paper published in 1764 after his death, "Essay towards solving a problem in the doctrine of chances." This paper was submitted to the Royal Society of London by his friend Richard Price. In his letter to the Royal Society, Price says that the paper describes:

> a method by which we might judge concerning the probability that an event has to happen, in given circumstances, upon supposition that we know nothing concerning it but that, under the same circumstances, it has happened a certain number of times, and failed a certain other number of times.

Given this description, which to our modern eyes appears convoluted to the point of obscurity, it seems remarkable that the editor published the paper. Happily, Bayes' paper was published, and it has since inspired an enormous amount of research, in fields as diverse as quantum mechanics, engineering, computer science, artificial intelligence, biology, and even vision. Given its universal applicability, we should not be surprised to learn that even the search engine Google makes use of the Bayesian framework.

Historical Note: Hermann von Helmholtz

Hermann von Helmholtz (1821–1894), **13.14**, was a genius. He contributed massively to several areas of science and would probably have won three separate Nobel Prizes if they had existed in his time. From his many experiments on perception, Helmholtz deduced that the perceptual system must be engaged in processes that he famously called ***unconscious inference***. For

Helmholtz, the most compelling demonstration that perception requires active work (or, we can say equivalently, computation or inference) came from Wheatstone's stereoscope (Ch 18). That device showed how depth sensations could be generated from two flat images, one fed to each eye.

It is possible to claim, and many have, that Helmholtz's characterization of perception as unconscious inference reveals that his approach sits comfortably with current research exploring the Bayesian approach to perception. Indeed, it is possible to regard Helmholtz as laying the conceptual foundations needed for conceiving that perception may be Bayesian in nature. However, Gerald Westheimer has written a critique arguing against over-simplifying the links between Helmholtz's views and the current Baysian approach to perception—see *Further Reading*.

A distinguished contemporary champion of Helmholtz's characterization of *perception as unconscious inference* is Richard Gregory. His way of developing that approach has been to promote the view of *perception as hypothesis.* He regards the advent of Bayesian approaches as a welcome development that is very much in keeping with his own outlook. See *Further Reading* if you wish to follow up Gregory's recent contributions to this debate.

Proving Bayes' Rule

It is strange that few textbooks give a simple proof of Bayes' rule. We offer one in the Box overleaf.

Further Reading

Ernst MO and Bülthoff HH (2004) Merging the sense into a robust percept. *Trends in Cognitive Sciences* **8**(4) 162–169. *Comment* A good starting point for reading the literature on the application of Bayesian ideas to cue combination—see Ch 20.

Doya K, Pouget A, and Rao RPN (Eds) (2007) *The Bayesian Brain: Probabilistic Approaches to Neural Coding.* MIT Press. *Comment* An enthralling collection of current research papers using Bayesian methods to analyze neuronal and human behavior. Highly technical, but with some good tutorial material, and highly recommended.

Gregory RK (2005) Bayes Window (1) *Perception* **34** 1421–1422. *Comment* The first in a series of lively non-mathematical editorials discusssing links between Bayes' rule and *perception as hypothesis,* a theme of much of Gregory's work. Subsequent items in his series in *Perception* are: **35** 143–144; **35** 289–290; **35** 431–432; **37** 1; **37** 641.

Kersten D and Yuille A (2003) Bayesian models of object perception. *Current Opinion in Neurobiology* **13** 150–158. *Comment* A general introduction to Bayes in the context of perception and natural images with examples of illusions which may result from Bayesian inference in perception. Also contains a comprehensive bibliography with comments on many published papers.

Kersten D, Mamassian P and Yuille A (2004) Object perception as Bayesian inference. *Annual Review of Psychology* **55** 271–304. *Comment* An in-depth overview and analysis of the Bayesian framework in the context of object recognition.

Knill DC and Richards W (Eds) (1996) *Perception as Bayesian inference*. Cambridge University Press.

Körding KP and Wolpert DM (2006) Bayesian decision theory in sensorimotor control. *Trends in Cognitive Sciences* **10**(7) 319–326.

Westheimer G (2008) Was Helmholtz a Bayesian? *Perception* **37** 642–650, *Comment* An interesting essay casting doubts on the oft–made suggestion that Helmholtz's approach to perception can be considered Bayesian in the sense of that term in the current literature on perception.

Proving Bayes' Rule

In the main text we consider three objects s_1, s_2, and s_3, but now we also have two images I_1 and I_2. Due to the effects of noise, each image could have been generated by any one of our three objects. We want to prove Bayes' rule

$$p(s_1|I_1) = p(I_1|s_1)p(s_1)/p(I_1), \qquad (1)$$

which can be obtained if we can first prove that

$$p(s_1|I_1)p(I_1) = p(I_1|s_1)p(s_1), \qquad (2)$$

and then divide both sides by $p(I_1)$ to obtain Bayes' rule.

	I_1	I_2
s_1	3	1
s_2	2	5
s_3	7	4

Examples

	I_1	I_2
s_1	a	d
s_2	b	e
s_3	c	f

Symbols

The examples (left) show the number of times the objects s_1, s_2, and s_3 have been observed with each of two images, I_1 and I_2. On the right are the matching symbols we use when discussing the general case.

The table shows that we begin by assigning letters a-f to the number of times each combination of images (I_1 and I_2) and objects (s_1, s_2, and s_3) occur together in our experience. So, for example, we have observed image I_1 in the presence of object s_1 exactly three times. The corresponding position in the table has been assigned $a = 3$. The total number of observations (image/object pairings) is $T = 22$, so the probability that s_1 and I_1 were observed together is $a/T = 3/22$. This is the probability that two things occurred together, which is why it is called a joint probability, and is written

$$p(I_1,s_1) = a/T$$
$$= 3/22$$
$$= 0.14. \qquad (3)$$

First, let's find the prior for one object, s_1. Object s_1 has been observed a total of $a+d = 3+1 = 4$ times; for $a(3)$ of those times it yielded image I_1, and for $d(1)$ of those times it yielded image I_2. The total number of observations is given by the sum $a+b+c+d+e+f = 22$, a number which we denote as $T = 22$. Thus, out of the 22 observations, $a+d = 3+1 = 4$ of them were of object s_1. So the probability that any one of the 22 observations includes s_1 is $4/22 = 0.18$. This represents the prior probability for s_1

$$p(s_1) = (a+d)/T$$
$$= 4/22 = 0.18. \qquad (4)$$

What about the likelihood, $p(I_1|s_1)$? Well, the probability of observing image I_1 given s_1 is the number of times s_1 and I_1 occur together (a) expressed as a proportion of the number of times s_1 occurs ($a+d$):

$$p(I_1|s_1) = a/(a+d)$$
$$= (3/22) / [(3+1)/22]$$
$$= 3/4$$
$$= 0.75. \qquad (5)$$

So what happens if we multiply the likelihood by the prior?

$$p(I_1|s_1)p(s_1) = [a/(a+d)] \times [(a+d)/T]$$
$$= a/T$$
$$= 3/22$$
$$= 0.14. \qquad (6)$$

Now a/T is the number of times that I_1 and s_1 were observed together expressed as a proportion of the total number T of observations, which we know from equation (3) is the joint probability $p(I_1,s_1)$. Thus,

$$p(I_1,s_1) = p(I_1|s_1)p(s_1). \qquad (7)$$

If we apply the same reasoning to find the prior $p(I_1)$ and the posterior $p(s_1|I_1)$ then we end up with the prior for I_1

$$p(I_1) = (a+b+c)/T$$
$$= (3+2+7)/22$$
$$= 0.55, \qquad (8)$$

and the posterior

$$p(s_1|I_1) = a/(a+b+c)$$
$$= 3/(3+2+7)$$
$$= 0.25. \qquad (9)$$

Multiplying $p(s_1|I_1)$ and $p(I_1)$ together yields

$$p(s_1|I_1)p(I_1) = a/T$$
$$= p(I_1,s_1). \qquad (10)$$

We can summarize this as

$$p(I_1,s_1) = p(s_1|I_1)p(I_1). \qquad (11)$$

Combining equations (7) and (11), both of which equal $p(I_1,s_1)$, yields

$$p(s_1|I_1)p(I_1) = p(I_1|s_1)p(s_1) \qquad (12)$$

Finally, if we divide both sides by $p(I_1)$ then we obtain Bayes' rule

$$p(s_1|I_1) = p(I_1|s_1)p(s_1)/p(I_1)$$
$$= 0.75 \times 0.18 / 0.55$$
$$= 0.25, \qquad (13)$$

which agrees with the answer obtained in equation (9).

14

Seeing Motion, Part I

14.1 Clowns in motion
Image blur caused by the rapid motion of clowns as they jump and run. It is a remarkable fact of human vision that many frames of video image sequences are as blurred as this and yet we see objects with sharp edges in movement.

If you could not perceive motion, what would you see? Clues come from examining people with a stroke which destroys a particular brain region, known as **V5** or **MT**. One such person is known in the scientific literature by her initials, LM. She is a 43-year-old woman who suffered a stroke which damaged posterior parietal and occipital regions on both sides of her brain. The main effect of this stroke was complete loss of the ability to perceive motion. Many tests have been carried out on LM, and they all attest to her inability to perceive motion. But no test result can give a more vivid account of her mental world than the following remarks by the scientists who spent years studying LM:

> She had difficulty, for example, in pouring tea or coffee because the fluid appeared to be frozen like a glacier....In a room where more than two people were walking...she usually left the room because "people were suddenly here or there but I have not see them moving." From Zihl et al. (1983).

This, then, is how the world would appear if you could not perceive motion.

As is often the case when examining the bizarre effects of brain lesions, we are tempted to ask: "How is it possible *not* to see motion?" But to LM, it is quite likely that she would legitimately ask the question: "How is it possible to *see* motion?"

Of the two questions, LM's is more interesting from a scientific point of view. Because we appear to see motion so easily, we wonder how it is possible not to see motion. But the fact that we see motion easily does not imply that seeing motion is easy for the brain, nor that the computational problem of seeing motion is trivial.

So LM's question is the one we should be asking: "How is it possible to see motion?" As with most visual functions, we have some clues, we know some of the computational problems that need to be solved, we know some places in the brain where these problems are solved, but as to the precise mechanisms, we struggle to integrate these into a coherent theoretical framework.

Seeing begins with the retinal image, so that we tend to think of vision in terms of static retinal images. But of course, the eye is not a camera taking snapshots of the world; nor is it a movie camera, because the eye does not take a series of discrete snapshots which are delivered one by one to the brain. Instead, while the eye is stationary, the neuronal machinery within the eye ensures that any changes within the retinal image are transmitted to the brain as a continuous stream of visual information.

We also tend to think of vision as essentially spatial. This stems from two possibly related causes.

First, our language is spatial. We say things like "The man is swimming," rather than "The swim is manning." In other words, the permanent entity (e.g., the man) is the object of our attention, and the transient entity (e.g., swimming) is "attached" to it. This linguistic bias may stem from the way the physical world is structured. Although we will not dwell on it, an influential alternative hypoth-

Why read this chapter?

The problem of how to detect motion, like most problems in vision, appears deceptively easy. However, the true difficulty of motion detection becomes apparent when we try to build a simple motion–detecting device known as a *Reichardt detector*. After describing this in detail, we then consider evidence that a particular kind of Reichardt detector is implemented by neurons in the retina and the lateral geniculate nucleus of the rabbit. This analysis provides a particularly clear example of the usefulness of Marr's three levels: *computational theory*, *algorithm*, and *hardware*. Next, we explore a powerful tool for investigating motion, the *motion aftereffect*, and we discuss neuronal mechanisms that could be responsible for this aftereffect. After giving a more formal definition of motion in terms of *temporal contrast sensitivity* and *temporal frequency*, we return to the problem of how motion direction can be detected given that neurons see the world via a spatially restricted *receptive field*, like a small circular *aperture* placed over the retinal image. The problem of how to detect motion through such an aperture is known as the *aperture problem*. A computational analysis identifies constraints, which lead to the the idea of finding an *intersection of constraints* to solve the aperture problem. We then consider where in the brain the aperture problem may be solved, and conclude that, on balance, a brain region called V5 (also called MT) seems to be critically involved. This is supported by an elegant experiment in which electrical stimulation of V5 was found to bias the perceived direction of motion. This is a crucial experiment because it implies a *causal* link (as opposed to a correlation) between activity in V5 and motion perception.

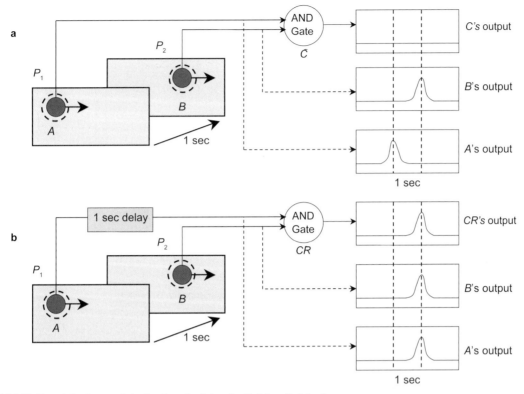

14.2 Motion detector module: basic principle of a Reichardt detector

a A small disc moves in one second (1s) from position P_1 where it stimulates photoreceptor A to position P_2, where it stimulates photoreceptor B. The timing of their outputs reflects this, as shown in the graphs on the right. The detector labeled C fires if and only if it receives outputs from A and B at the same time. It is thus called an "*AND* Gate". In this case C does not fire as its inputs from A and B do not arrive at C at the same time.

b The *AND* gate can be made to detect the motion of the disc if a delay of one second (1s) is introduced into the output of A. This ensures that the outputs of A and B are above baseline at the same time, causing the *AND* gate CR to fire.

esis is that this bias may be a consequence of the structure of our language, an idea known as the ***Whorfian hypothesis***.

Second, it is easy to lay out several objects on a table and to see that they have definite spatial relations to each other; it is less easy to lay out a set of objects in time (e.g., a sequence of objects or events) and to retain the precise temporal relations between them. Thus, our linguistic bias may result from this difference in the accessibility of objects laid out over space as opposed to objects laid out over time. The short answer is that nobody knows whether it is language that determines our perception of the world, or the other way around. The fact remains that in our culture, language and our perception of the world is undeniably dominated by spatial rather than temporal relations.

Motion presents two fundamental problems to the visual system. First, how can motion on the retina be detected? Second, how can motion information be used by the brain to solve tasks as fundamental as standing up, tracking objects, and recognizing objects in the ever-shifting world that flits across our retinas? In this chapter and the next we concentrate on how motion is detected.

Motion as a Correspondence Problem

It has been said that all problems in vision are correspondence problems: object recognition (Ch 8) involves finding a correspondence between an object and the brain's representation of that object; stereopsis (Ch 18) involves finding a correspondence between left and right retinal images, where these images occur at one point in time; finally, detecting motion involves finding a correspondence at successive times, as shown in **14.2**.

Of course, recognizing that detecting motion is a type of correspondence problem is no substitute

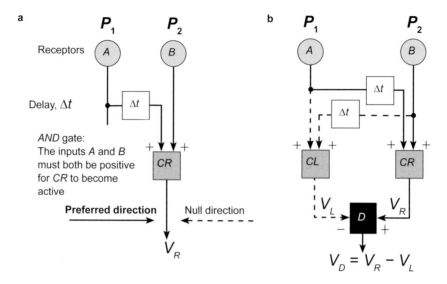

14.3 Reichardt motion detector
This consists of a pair of motion–detecting modules, which detect motion in opposite directions, as in **b**.
a This module has a detector *CR* that detects motion between the retinal points P_1 and P_2 in the rightward (its *preferred*) direction, and has an output V_R.
b A pair of motion–detecting modules, each of which has the same basic structure shown in **a**. One cell *CR* detects motion between the retinal points P_1 and P_2 in the rightward direction, and has an output V_R. The other cell *CL* detects motion between the retinal points P_1 and P_2 in the leftward direction, and has an output V_L. The output of the comparator *D* is the difference $V_D = V_R - V_L$ between these outputs, and is positive for rightward motion and negative for leftward motion.

for finding a method that will detect motion, but it does place the problem of motion detection in a computational context.

Reichardt Motion Detector

The computational problem of motion detection can be formulated as detecting changes in position over time. This can be solved using a simple algorithm which can be implemented in the form of a ***Reichardt detector***, first described by Reichardt in 1961. In order to examine it, we will consider an image feature (such as an edge or a spot of light) which moves 2 millimeters (2 mm) across the retina from point P_1 to point P_2 over a time interval of one second (1s), and we will label the photoreceptors at P_1 and P_2 as A and B, **14.2**. The outputs of A and B are fed into a special cell C that is made to act as an *AND* gate. This means that C fires if and only if both its inputs are present.

Consider first **14.2a**. The inputs from A and B arrive at C at different times. The one from A arrives first and then a second later, the one from B. The *AND* condition is therefore not met and so C does not fire.

But what would happen if we inserted a delay of exactly a second on the output of receptor A to C, **14.2b**? The spot of light would stimulate A, and A's output would start traveling toward CR, which we now call *CR* because, as we will see, it detects motion to the *R*ight).

Due to the delay, A's output will arrive at the CR in a second's time. The spot continues on its way across the retina, and, a second later, receptor B is stimulated and also sends its output but without delay to CR. As A's output was delayed by a second, but B's output was not, both outputs arrive at CR at *the same time*. Thus, the delay ensures that CR receives both inputs simultaneously. The *AND* condition required by CR is therefore met and CR thus fires. This looks like a good candidate for a motion detector.

We should note three properties of this simple motion detector. First, it only detects motion at one speed, say 2 millimeters per second (2 mm/s). If the image feature travels at 1 mm/s (i.e., too slowly) or at 3 mm/s (i.e., too fast) then the inputs from A and B will not arrive at the detector CR at the same time. As CR "looks for" simultane-

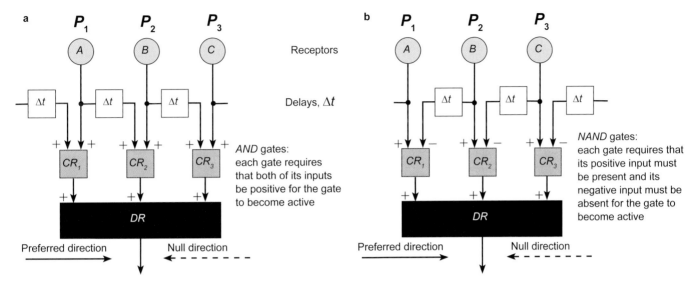

a **P₁**   **P₂**   **P₃**   b **P₁**   **P₂**   **P₃**

14.4 Two algorithms for detecting motion direction to the right
This figure deals with just one of the two motion–detecting modules that comprise the Reichardt detector in **14.3b**.

a *Joint excitation method.* This is an extension of the scheme illustrated in **14.2b** and **14.3a**. The *CR AND* gates become active if and only if their inputs from the receptors arrive simultaneously and are both positive. This happens if a spot of light moves to the right from point P_1 to P_2 then P_3 and thus stimulates the receptors *A*, *B* and *C* successively, because the delays Δt then ensure that inputs to each *AND* gate arrive successively. The opposite, or *null*, direction is if the spot moves from P_3 to P_1 because then the *AND* gates never receive simultaneous inputs from the receptors. The detector *DR* receives excitatory inputs from the *AND* gates, hence its *preferred direction* is to the right. *DR* is acting here as an *OR* gate, which mans it fires if any of its inputs become active.

b *Inhibitory veto method.* Each *NAND* gate becomes active if and only if its inputs from the receptors arrive simultaneously and the delayed input is *not* negative. This happens if the motion is to the right because then the delayed *inhibitory veto* signal to each *NAND* gate arrives too late to stop the excitatory input from exciting it. The *null direction* is in the opposite direction (to the left) because then the delayed signals to the *NAND* gates arrive at the same time as the excitatory ones and these delayed signals then veto the *NAND* gates becoming active. The *NAND* gates send signals to the direction selective motion detector *DR*, whose preferred direction is to the right, as in **a**.
Adapted from Barlow and Levick (1965).

ous inputs from *A* and *B*, *CR* cannot signal that it has detected motion if the speed of motion differs from a specific value (e.g., 2 mm/s).

Second, *CR* only detects motion in one direction (from point P_1 to point P_2). This is called its **preferred direction**. Now, consider what happens if the feature moved at the "right" speed of 2 mm/s and along the "right" path from point P_1 to point P_2, but in the opposite direction (i.e., from P_2 to P_1), technically called the **null direction**. As the feature moved across *B*, *B*'s output would be received by the detector *CR* almost immediately. Let's start a timer at this point in time, so *CR* receives a signal from *B* at time zero. As the spot continued on its way it would reach *A* a second later, but, due to the delay we inserted earlier, *A*'s output would be delayed by another a second before reaching *CR*. Thus, the detector *CR* would receive another input two seconds after the timer

was started. Even though *CR* is constantly comparing the inputs from *A* and *B*, these inputs do not coincide at any single point in time, and so no motion signal is generated by *CR*.

A problem with this scheme is that, as we have noted before, photoreceptors are intrinsically noisy, and for this reason it is all too easy for cell *C* to signal the conjunction of two noise signals arriving at the same moment when there has been no motion. One way to reduce the number of such false motion signals consists of using cells in **opponent pairs** whose preferred directions of motion are opposite, **14.3**.

We label the pair of motion detector cells in **14.3** as *CL* (leftward motion) and *CR* (rightward motion). Both *CR* and *CL* detect motion at the same speed, and along the same path (between point P_1 and point P_2) but in *opposite directions*. Now we can feed the outputs of *CL* and *CR* to a comparator unit, which we will call D. The comparator D keeps a running average of the *difference* between the inputs from CL and CR. Keeping a running average ensures that transient noise-induced increases in the output of A or B get lost in the averaging process.

However, if the input from CR is consistently greater than that from CL then this difference will not be averaged out and D's output increases. In contrast, if the input from CL is consistently greater than that from CR then D's output decreases. Thus, the output of D reliably signals the direction of motion of a feature traveling along the path which connects P_1 and P_2 at a specific speed. Note that, the words *transient* and *consistently* are used in a relative sense here. In the world of motion detection, a transient signal may last less than a millisecond, whereas a consistent input may last a few milliseconds.

This is the Reichardt detector in its simplest form. It consists of two photoreceptors, two opposing direction-selective units (e.g., CL and CR), and a unit D which compares the outputs of the direction-selective cells.

Inhibitory Algorithm for Motion Direction Selectivity

There is another way of using the idea of inserting a delay between two receptors to detect motion. The one just described is exploited in the block diagram in **14.4a** and we call it the ***joint excitation method.*** This is because two simultaneous excitatory inputs to each *AND* gate must be present for each gate to be activated. This may be the way some motion detecting neurons in monkey cortex work, as discussed by Movshon, Thompson and Tolhurst in 1975.

An alternative scheme was proposed in 1965 in a classic paper, still well worth reading today, by Barlow and Levick. It uses the idea of an ***inhibitory veto***, implemented as a "*Not AND*" gate, called a *NAND* gate, **14.4b**. Thus, the motion detector modules CR_1, CR_2 and CR_3 fire if they receive one signal which is excitatory and if a second inhibitory signal is *absent*. Hence, the requirements of the *NAND* gate shown in **14.4b** are, in order from the left: (*A and not B*), (*B and not C*), and so on.

In considering this inhibitory veto method, it can be helpful to think of the excitatory inputs

"getting through" the *NAND* gates, as it were, if there is no inhibition from the delayed signal.

A biologically plausible implementation of this scheme uses inhibitory and excitatory inputs to neurons, hence the use of plus and minus signs in **14.4b**. As explained in the legend to **14.4b**, if the delayed signal from receptor A arrives simultaneously with the signal from receptor B then the two signals cancel, as one is inhibitory and the other excitatory.

Notice in **14.4b** that using the inhibitory veto idea to build a detector selective for rightward motion requires that the delays be fed in the opposite direction to those used in the joint excitation method. This can be seen by comparing **14.4a** and **14.4b**.

There are certain subtleties about the way the inhibitory veto method just described differs from the joint excitation method. For example, each joint excitation module in **14.4a** detects motion in the preferred direction at one speed only. In contrast, each inhibitory veto module in **14.4b** has a positive output for motion at any speed in the preferred direction, but it has a zero output for motion in the null direction at only one speed.

We have explained the inhibitory veto scheme because Barlow and Levick found that it accounted nicely for certain properties of motion direction selective cells in the rabbit's retina. For example, they found that if a stationary spot was flashed briefly between two nearby points on the receptive field of one of these cells then the cell gave a burst of impulses, **14.5b**. This would not be expected by the joint excitation scheme shown in **14.4a** because the motion detecting units are *AND* gates that require *both* their inputs to be positive for them to fire. This conjunction could not happen for a stationary spot. Barlow and Levick also found that when a spot was moved across the receptive field in the null direction the cell's normal ***maintained discharge*** was suppressed. (The latter discharge is the cell's activity level in the dark, i.e., with no stimulus present.) This suppression is consistent with inhibition arriving at the cell arising from movement in the null direction.

Having found strong evidence in favor of an inhibition veto mechanism mediating directional selectivity in the rabbit's retina, Barlow and Levick then considered where neurons might send the

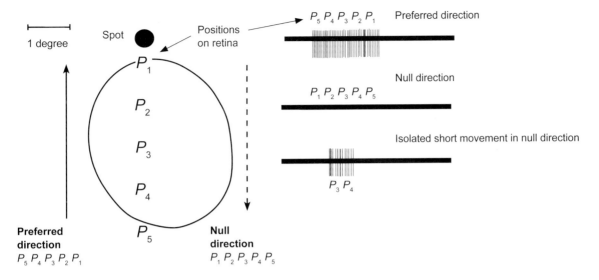

14.5 Direction selective recording from a retinal ganglion cell in the rabbit
The receptive field of the cell is shown by the lozenge–shaped outline on the left. When the spot was moved upward from position P_5 to P_1, the cell responded vigorously, as shown in the burst of action potentials in the upper trace on the right. When the spot moved from P_1 to P_5 there were no action potentials, middle trace. Therefore, movement along the path P_5 to P_1 is said to be in the preferred direction of the cell, whereas the movement P_1 to P_5 is in the null direction. The lower trace shows a burst of impulses for the movement from P_3 to P_4 when it is made in isolation rather than as part of a continuous sweep in the null direction. This burst was interpreted by Barlow and Levick as indicating that lack of inhibition from movement from P_1 to P_3 allowed excitation from the sudden onset of P_3 to escape being vetoed by inhibition. Schematic diagram, not drawn precisely to scale, based on data in Barlow and Levick (1965).

inhibitory signals. Recollect from Ch 6 that *receptors* in the retina send signals to *bipolar cells* and then these bipolar cells send signals to *retinal ganglion cells*, **6.4**, p. 135. This so-called *vertical organization* is enriched by *horizontal*

cells that carry lateral messages at the level of the receptor-bipolar junctions, and by *amacrine cells*, which carry lateral messages at the bipolar-ganglion cell junctions. Barlow and Levick speculated that it is the horizontal cells that carry the inhibitory veto signals mediating motion direction sensitivity.

A great deal of subsequent research, using sophisticated techniques unavailable to Barlow and Levick, has come to a different conclusion: it is amacrine cells of the starburst type that supply the inhibitory veto signals. *Starburst amacrines* get their name from their characteristic pattern of branching dendrites, **14.6**. They provide a remark-

14.6 Dendritic field of starburst amacrine cell
Schematic plan view looking down on the retina, based on Masland (2005). Recordings indicate that an individual segment of the dendritic field, such as the one picked out with the arrows, can operate as a separate processing unit working in isolation from others. Recordings also show that much more inhibitory neurotransmitter is released from the tips of the dendrites when a stimulus moves in the direction shown by the arrows than when a stimulus moves in the opposite direction. The scale bar shows 50 µm (in other words, 50 microns or 50 millionths of a meter). Schematic redrawn with permission from Masland (2005).

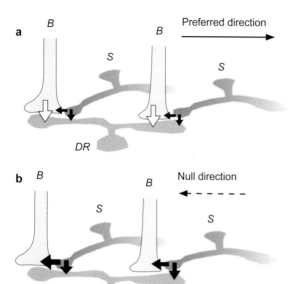

a *Preferred direction.* The unfilled arrows indicate excitation from bipolar cells *B* at their synapses with the direction selective unit *DR* (so labeled because its preferred direction is to the *R*ight). The small black arrows symbolize the small amount of inhibition arriving from the starburst cell *S*. This inhibition is not only weak, it also arrives too late to veto the excitation

b *Null direction.* The larger black arrows indicate strong inhibition arriving at the synapses of bipolar cells *B* from the dendrites of the starburst cells from a movement in the null direction. This vetos the excitatory influences that would otherwise have been passed from the bipolars to the direction selective unit *DR*. Hence, there are no unfilled arrows in this figure. Adapted with permission from Fried, Munch and Werblin (2002).

ably rich set of interconnections in the retina. It has been estimated that if all the starburst fibers in each retina were laid end to end then the total length would come to about 2 km, which suggests that they must be doing an important job. Equally remarkably, it seems that different sections of their dendritic fields can operate as independent processing units. How they mediate lateral inhibitory veto signals is explained in **14.6.**

Fried, Munch and Werblin reported in 2002 inhibitory and excitatory currents in starburst dendrites. The results led them to the scheme shown in **14.7**, which illustrates what they think happens at the ends of the dendrites of a starburst cell. Notice the close similarity between the inhibitory veto algorithm shown in **14.4b** with the physiological results summarized in **14.7**. In the latter, the axons from the bipolars labeled *B* are conveying excitatory signals to a direction selective retinal ganglion cell labeled *DR*. Hence, the bipolars in **14.7** are equivalent to the receptors in **14.4b**. (Recollect that bipolar cells receive their inputs from receptors.) Also, the retinal ganglion cell *DR* in **14.7** is equivalent to the *DR* box in **14.4b**. Finally, the dendrites of the starburst cells labeled *S* in **14.7** are conveying the time-delayed inhibition shown by the boxes labeled Δ*t* in **14.4b.** There is thus a remarkable parallel between the algorithm and certain structures in the rabbit's retina. In short,

the results illustrated in **14.7** strongly suggest that the starburst amacrines are a neurophysiological implementation of the inhibitory veto algorithm in **14.4b.**

Neurophysiological Implementation of Reichardt Detector

What happens to the outputs of motion direction selective retinal ganglion cells shown in **14.7**? Their axons go to the lateral geniculate nucleus (LGN; recollect that this nucleus receives inputs from the retinas via fibers in the optic nerve, p. 209, **9.6**). There is good evidence that in the rabbit certain LGN cells implement the kind of opponent movement direction coupling embodied in the Reichardt detector, **14.3b**.

Thus, in 1969 Levick, Oyster and Takahashi reported findings that are summarized in **14.8a**, which shows two retinal ganglion cells sending signals to a cell in the LGN. The two retinal cells shown in **14.8a** are sensitive to motion in the same retinal region but they have opposite preferred directions. One retinal input to the LGN cell is inhibitory, the other excitatory. Hence, these two inputs are in competition with one another. Only if the excitatory signal is large enough to dominate does the LGN cell fire and send a signal to the rabbit's brain that motion in the preferred direction of the excitatory cell has taken place. This is a neat way to implement with neurons the Reichardt algorithm shown in **14.3b**. Strong evidence for the inhibitory coupling at the LGN cell is shown in **14.8b**.

The LGN neuron depicted in **14.8a** encodes just one direction of motion. Other neurons of this type encode movements in other directions.

Detecting Motion Using Temporal Derivatives

The Reichardt detector is not the only game in town. An alternative method consists of finding the change in luminance over time at a single image location, which is known as the ***temporal derivative***. Finding the temporal derivative is analogous to finding the spatial derivative, and

can be obtained using a temporal version of the receptive fields described in Chs 6 and 7. Marr and Ullman introduced a scheme of this sort in 1981 and its principle is illustrated in **14.9**.

The main difference between the Reichardt detector and temporal derivatives is that for the latter there is no "motion correspondence problem," inasmuch as it is not necessary to ensure that the same feature activates two separate photoreceptors at successive times.

Note that this type of mechanism needs to be associated with the contrast sign of the luminance edge. The detectors in **14.9** work if the luminance change defining the edge is Dark-Side-on-Left/Light-Side-on-Right, as illustrated by the lumi-

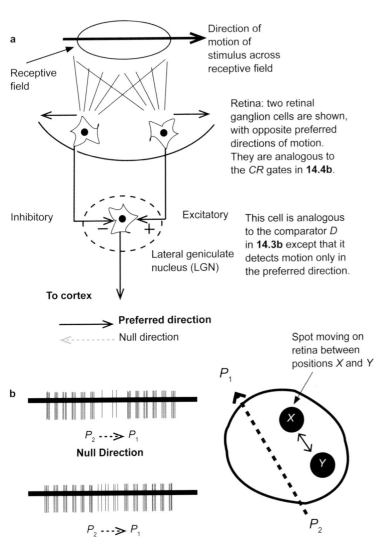

14.8 Motion detection in rabbit

a Implementation of the scheme shown in **14.3b** in rabbit's visual system. See text.

b Recordings from a single cell in rabbit's LGN with a receptive field indicated by the closed curve on the right. A small dot is moved to and fro between retinal positions X and Y which causes the repetitive bursts of action potentials shown in the recordings on the left. Each burst arises when the oscillating dot was moving in the cell's preferred direction. The gaps between the bursts occurred when the dot was moving in the null direction. The line under the recordings shows two periods marked with an arrow in which a second spot was moved across the receptive field in the null direction from position P_2 to P_1. During those periods the bursts that would be expected from the first dot moving in the preferred direction were greatly reduced. This result is consistent with motion in the null direction inhibiting the LGN cell via the pathway shown schematically in **a**.

What happened when retinal ganglion cells were stimulated with the dot moving to and fro between X and Y combined with motion from P_2 to P_1? This movement in the null direction did not inhibit them as strongly as it did the LGN cells. This result supports the idea that in the rabbit the LGN cells are the site of a noise suppression process of the kind shown in the Reichardt detector in **14.3b**.

Redrawn with permission from Levick, Oyster and Takahashi (1969).

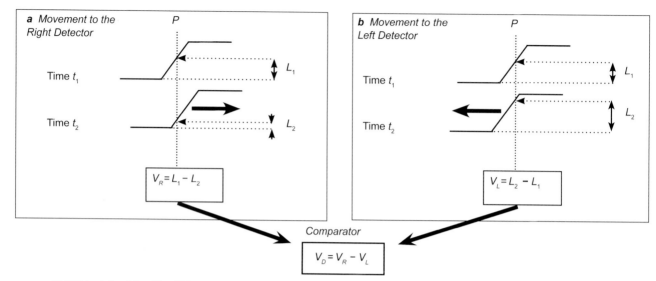

14.9 Principle of the Marr/Ullman temporal derivative scheme for detecting motion

a *Movement to the right detector*. Each ramp shows an image luminance profile of an edge. The position of the edge changes between two points in time, t_1 and t_2. The box below each ramp represents a neuron which compares the luminances at image point P at times t_1 and t_2 by subtracting them. Thus, the output of this neuron is $V_R = L_1 - L_2$. This will produce a positive output if the luminance ramp moves to the right, as shown. Hence, this neuron acts as a *Movement to the Right Detector* for the kind of luminance edge shown.

b *Movement to the left detector*. The same idea can be used to detect movement to the left by a neuron that performs subtraction using the luminances the other way round, as it were. Thus, the output of $V_L = L_2 - L_1$, which produces a positive output if a luminance ramp of this type moves to the left.

Notice that the subtractions specified for these neurons would produce negative values for movement in the direction opposite to the directions shown. However, neurons cannot "fire negatively." They are either active or they are silent. So having this pair of neurons, one for each direction, neatly solves this problem. Technically, it can be said this pair of neurons encode the positive (V_R) and the negative (V_L) parts of the temporal derivative of the image luminance change shown. A decision on which movement direction has occurred is taken by the Comparator which determines which neuron is more active. This task could be implemented by inhibitory links between the two neurons, similar to those shown in **14.8**. This scheme needs to make allowance for the contrast sign of the edge, see p. 333.

nance ramps in both **a** and **b**. If the ramps had the opposite contrast sign (Light-Side-on-Left/Dark-Side-on-Right), then the subtraction $V = V_R - V_L$ would also have to be reversed to detect movement-to-the right and movement-to-the-left respectively. Hence, there is a need to keep track of the contrast sign of the edge.

Mather reported in 1984 psychophysical evidence indicating that human vision uses temporal derivatives as one of its methods for motion detection. He demonstrated that when observers are shown *stationary* stimuli whose luminances are changed over time then they reported seeing motion. Moreover, the direction of this perceived motion was as predicted by the Marr/Ullman model.

Mather's evidence adds to earlier reports by various authors of perceived motion from stationary stimuli. Particularly interesting is Anstis's (1990) finding that a species of motion aftereffect can be

generated from luminance changes in a stationary stimulus. Motion aftereffects usually arise from moving stimuli, as we describe next.

Motion Aftereffect

After watching movement in one direction for a prolonged period, a subsequently viewed stationary scene appears to be moving in the opposite direction. This ***motion aftereffect*** is strong and is often experienced in everyday situations.

It was known to the ancients, with Lucretius providing the first clear account around 56BC. Aristotle had described a motion aftereffect much earlier, around 339BC, but without stating its direction.

The motion aftereffect is commonly known as the waterfall illusion because it was famously described in 1834 by Robert Addams, after viewing the Fall of Foyers in northern Scotland:

Having steadfastly looked for a few seconds at a particular part of the cascade, admiring the confluence and decussation of the currents forming the liquid drapery of waters, and then suddenly directed my eyes left, to observe the vertical face of the sombre age-worn rocks immediately contiguous to the waterfall, I saw the rocky face as if in motion upward, and with an apparent velocity equal to that of the descending water.

You can witness a particularly vivid version of the movement aftereffect for yourself using the spiral shown in **14.10**. Copy this out on to a circular piece of paper and then find a way of spinning it slowly for a few minutes; or you can visit the website cited in the legend of **14.10**. Stare at the centre of the disc while it is rotating. The rotation of the spiral is perceived as a continuous contraction of the spiral rings toward the disc centre, if **14.10** is rotated anti-clockwise. Keep looking for about one minute, to get a really good aftereffect. When this adaptation period is up, stop spinning and see what happens. What you should see is illusory movement in the opposite direction—the spiral seems to be continuously expanding rather than contracting. (If the rotation during adaptation was clockwise, then the aftereffect is seen as a contraction of the spiral.) This motion illusion is very vivid for a few seconds and then it weakens, with quite a long tail-off, perhaps up to 20 seconds or so during which a slow movement drift can just be detected.

After one of these adaptation periods, try looking away from the stationary spiral at any convenient stationary surface. You will find that the illusory movement can still be seen quite readily. Notice also that although the aftereffect gives a very clear illusion of motion, the apparently moving features nevertheless seem to say still. That is, we are still aware of features remaining in their "proper" locations even though they are seen as moving. What we see is logically impossible, a paradox.

This perceptual paradox, like so many others, suggests that the visual system can detect one feature attribute, here motion, quite independently of others. And if one part of the feature-representation system is "suffering an illusion" for some reason, this does not mean that other aspects of the total feature representation are "forced into line," and adjusted to obtain an overall perception which

is "coherent." The visual system is quite happy to live with paradox. Its individual parts often seem to work like separate modules, in some respects at least, indeed almost as separate sensory systems. One is reminded here of remarks made in Ch 1 about impossible objects, and in Ch 2 about a slant-from-texture module as an example of a vision module.

The most commonly accepted explanation of the motion aftereffect is illustrated in **14.11**. The origins of this idea are credited to Sigmund Exner, who speculated in 1894 about the processes that might be involved, well before the invention of microelectrode recording and the discovery of motion detecting neurons. The key idea in Exner's kind of explanation is that different directions of stimulus movement are detected by different neurons, and that pairs of neurons dealing with opposite directions are coupled together. A prolonged period of exposure to one stimulus direction recalibrates and/or fatigues one member of each pair, leading to an imbalance in their activity which produces the motion aftereffect. Let us follow through the details of this explanation. It is important because it illustrates an opponent-process system at work, and such systems are common in perceptual mechanisms. Indeed, the Reichardt detector described in the previous section is an example of one such system.

14.10 Spiral for the movement aftereffect
http://lite.bu.edu/vision-flash10/applets/Motion/Spiral/Spiral.html has a spiral movie that gives a vivid motion aftereffect.

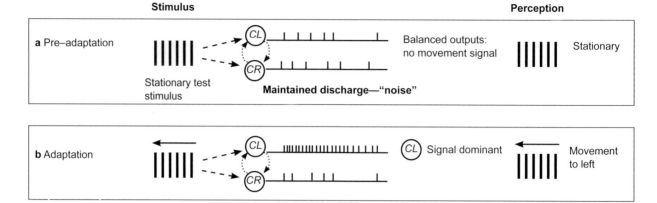

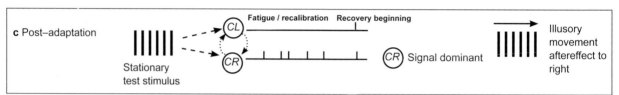

14.11 Explanation of the movement aftereffect
The basic idea is credited to Exner writing in 1894. This figure explains his idea in terms of modern knowledge about neurons sensitive to movement reported by Barlow and Hill in 1963, **14.12**. *CL* and *CR* show coupled pairs of direction–tuned movement detecting units, which are analogous to the motion modules in the Reichardt motion detector in **14.3b**. The details of the explanation are set out below.

First, consider the pair of neurons *CL* and *CR* shown schematically in **14.11a** responding to a stationary grating in the pre-adaptation period. The *CL* neuron detects motion to the left, the *CR* neuron motion to the right. As would be expected for movement-detecting cells, a stationary stimulus such as this hardly excites them at all and they are shown responding with a similar low frequency "maintained" (or "noise") discharge rate. It is assumed that when their outputs are roughly equal in this way, later processes (such as those described in **14.8**) work out that the stimulus is stationary, as indeed it is.

Second, **14.11b** shows the same two cells responding to an adapting stimulus, a grating moving to the left for a prolonged period. The *CL* neuron is shown responding vigorously at first, with its response diminishing somewhat as the motion stimulation continues (this diminution could be due either to the recalibration and/or the fatigue processes discussed in Ch 4). The *CR* neuron keeps firing at its maintained discharge rate, as though it were simply "seeing" a stationary stimulus. As the firing rate of the *CL* neuron is much larger than the maintained discharge of the *CR* neuron, the result is perceived movement to the left, a "veridi-

cal" perception. This kind of opponent-direction coupling is similar to that used in the Reichardt detector, **14.3b**.

Third, **14.11c** shows what happens when the adapting stimulus is removed and replaced with a stationary grating. The *CL* neuron no longer responds because its sensitivity has been lowered (fatigue theory) or adaptively changed (recalibration theory) during adaptation. This leftward-motion neuron does not even show a maintained discharge rate until it recovers. The *CR* neuron, on the other hand, keeps on firing with a maintained discharge rate as it has done throughout. But now this maintained discharge is treated by later processes as a movement signal because it is neither canceled as in **14.11a**, nor exceeded, as in **14.11b**, by activity in the *CL* neuron. As a result, an illusory movement to the right is perceived—the motion aftereffect. This lasts until the *CL* cell recovers (fatigue theory) or its response again adaptively altered (recalibration theory).

A variation on the above explanation is to suppose that the leftward and rightward neurons are coupled together with inhibitory connections, shown in **14.11** as the dotted arrows joining the pairs of cells. If these connections transmit inhibition according to the degree of activity of each cell then the most active cell would "come out on top" as it were, having silenced its "opponent." In this way, the later processes of interpretation might be relieved of having to judge the relative activities of the *CL* and *CR* neurons because only one signal would "come through." The aftereffect might then be the product of the *CL* cell being released from inhibition for a moment or two because of the fatigue/recalibration affecting the *CL* cell during adaptation. We shall not go into the details of this scheme, but we mention it to illustrate the computational power of connections between neurons. A pair of simple inhibitory fibers might go a long way in "interpreting" or "balancing out" or "canceling" opposing movement signals.

Readers familiar with electronics might notice a similarity between the opponent processes we have described and push-pull amplifiers. Indeed, there may be good computational reasons for using neurons as opponent push-pull pairs, as suggested by Smith in his book of 2004. Smith noted the similarity between the opponent pairs of transistors (or vacuum tubes) used in such an amplifier and the opponency of neurons. Engineers use opposing pairs of transistors in an amplifier for several reasons, but mostly because its output is *linearly* related to its input. In other words, a push-pull amplifier just gives a version of the input without adding in any extras (known as distortion products) that were not in the original input. As it is with amplifiers, so it is with brains. The brain does not want neurons with outputs that contain distortion products, and one way to minimize these is to pair up neurons which encode different ranges of the same signal (e.g., left versus right motion), which ensures that the difference between their outputs is an accurate representation of their input. More generally, this push-pull amplifier principle may underlie other perceptual effects (e.g., color aftereffects) thought to be associated with opponent processes (see Ch 17, p. 409).

Finally, note that the pair of neurons shown in **14.11** is only one pair among many. Each pair deals with a different direction along which movement can occur. So far we have supposed that the direction of the motion aftereffect can be adequately explained just in terms just of the individual pairs operating independently. However,

this scheme is in fact too simple. Exner reported in 1887 that if horizontal and vertical gratings were moved, respectively vertically and horizontally, behind a circular aperture then the movement perceived was in a diagonal direction. The motion aftereffect to this movement was in the direction of the opposite diagonal. Results of this kind led Mather in 1980 to propose a distribution model in which motion is computed from activities in the opponent pairs tuned to motion directions "all around the clock," as it were. Mather suggests that the motion aftereffect is the result of a shift in the pattern of activities in this distribution of opponent pairs.

For a rotating spiral stimulus such as **14.10**, cell pairs dealing with all motion directions in each retinal region would be stimulated during the adaptation period. The motion direction computed locally from the distribution of pairs for every region would produce a contracting movement of the spiral as a whole (if that spiral is rotated anti-clockwise). This would then be followed by the illusory expansion, the motion aftereffect.

The individual pairs might be located in the striate cortex in monkeys and in man, which has been shown to have directionally sensitive cells. This is suggested both by neurophysiological findings (see later) and by experiments in which only one eye is exposed to an adapting motion. The aftereffect from this monocular motion can be seen when the eye that received the adapting stimulus is closed, and the other eye is opened. This phenomenon is called inter-ocular transfer of the motion aftereffect and it forces the conclusion that there must be some central brain sites involved (see reviews in Mather, Verstraten and Anstis, 1998). However, this does not rule out retinal sites as well. Indeed, as we have explained at length above, in some species, such as the rabbit, retinal cells with the required properties have been discovered, **14.7**. A further complication is that each cell pair would have a receptive field of limited size and so there needs to be a full set of pairs devoted to analysing the movement occurring in *each region* of the retinal image.

Historical Note on the Motion Aftereffect

The motion aftereffect has a distinguished place in the history of vision research. Up until around

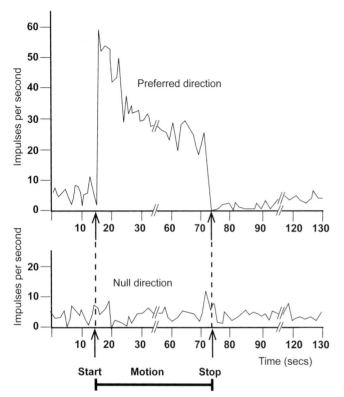

14.12 Adaptation in motion direction selective cells in rabbit's retina

The upper graph shows the firing rate of a retinal ganglion cell when exposed to prolonged motion stimulation in its preferred direction in the period between the Start and Stop labels, during which the firing rate dropped by more than 50%. After the motion had stopped there was a period of about 30 seconds when the firing rate was close to zero. The lower graph shows that when the motion was in the opposite (null) direction the cell showed few if any responses above the maintained discharge level. Schematic diagram based on data in Barlow and Hill (1963). These are the kind of data on which the schematic diagrams in **14.11** are based.

1960 a new paper on it appeared on average every two years, producing a total of about 50 papers, according to Mather. A good number of these were concerned with using the motion aftereffect to assess the effects of drugs, brain damage, or personality. After 1960 or so there was an exponential explosion of interest, with the total reaching over 400 by the year 2000.

This dramatic growth reflects in part the massive increase in scientific publications generally during the last four decades. But there is a particular reason for the recent interest in the motion aftereffect: it has been one of the most valuable aftereffects serving as a psychologist's microelectrode, Ch 4.

Figure **14.12** shows a classic example of how neurophysiological results have been treated as revealing the underlying neural mechanisms mediating the motion aftereffect. Mather et al.'s (1998) book is a good starting point if you want to know more about this story, and indeed the motion aftereffect generally, including its long history as revealed in a fascinating account by Wade and Verstraten in that book.

What Have We Learned So Far?

The scientific hunt to pin down the mechanism of motion direction sensitivity has been vigorously pursued since Barlow and Levick's seminal papers in the early 1960s (Niedeggen and Wist, 1998). The outcome is a fine example of an analysis of a visual problem (motion direction detection) at Marr's three levels.

The computation theory level is concerned with analysing the task of extracting motion signals from an array of receptors. Reichardt's analysis considered motion detection as the task of detecting a change of position over time. In contrast, the temporal derivatives analysis treats this task as detecting a change in luminance at a given position.

The Reichardt detector is often referred to as a type of **correlation model** because it depends on the correlations between the outputs of receptors in different retinal locations. It can be proved that, for motion over small retinal distances, this type of mechanism is optimal; that is, no other mechanisms can detect motion better than a correlation model. In fact, recent theoretical work suggests that the differences between the Reichardt and temporal derivative schemes may be more apparent than real. A formal analysis of motion detection by Bialek in 1997 suggests that the way in which motion is best computed depends on the amount of noise in the image. Specifically, if the image contains much visual noise then a Reichardt detector is optimal, whereas for low noise situations a temporal derivative is optimal. We speculate that a single mechanism exists that may be responsible for both types of computation, and that this mechanism has a "sliding scale" which adjusts itself between the two modes, correlation or temporal derivatives, depending on the amount of noise in the image.

Our account of research on motion direction selectivity above also illustrates the importance of keeping the algorithm level of analysis distinct. We described how a correlation model can be implemented in two different ways. One used *joint excitation* and the other *inhibitory veto*.

Finally, as far as the hardware level is concerned, neurophysiological work on the rabbit's retina and LGN provides clear evidence for inhibitory veto being used, whereas evidence for a joint excitation mechanism has been found in some direction selective cells in the cortex.

In considering different ways of implementing a computational theory, it is not unusual to find that deciding on important details of algorithms raises new computational problems to be solved. We have not delved sufficiently deeply into the joint excitation and the inhibitory veto algorithms to be able to illustrate this point but it is an important one. Marr's framework can and should be applied within the many layers of sub-problems implicit in any biological system as complex as the brain.

Seeing Bullets

A retinal ganglion cell only detects motion over a small range of speeds. For example, if a bullet went past your eyes you would not see it. Similarly, if you look at the minute hand of a clock it appears to be stationary, even though it is in motion at the rate of six degrees of rotation per minute. It is tempting to think that you do not see the bullet because it moves too fast, and that you do not see the minute hand move because it is too slow.

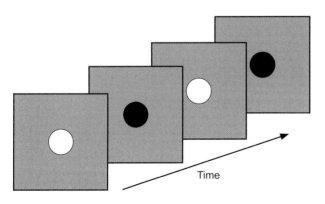

14.13 Measuring visual performance at a single flicker rate
A small disc changes from white to black and then back to white. If this black–white–black cycle takes 0.1 second then the disc has a temporal frequency of 10 Hz.

However, from the bullet's point of view, your visual system is simply too sluggish, and from the minute hand's point of view, your visual system is just too speedy. Of course, all these points of view are equally valid, and serve to emphasize that the visual system sees motion within a small band of speeds. If our visual systems were built differently then we would be able to see bullets in flight (though this might need new kinds of biological components!), or a minute hand moving, or grass, and even trees, growing. All things are relative. Thus, all events that move above or below a certain speed limit cannot be seen. These upper and lower limits define a band of speeds at which we can see motion. Because the visual system only admits, or *passes*, motion within this *band* of speeds, it is described as a ***temporal band-pass filter***.

Measuring Temporal Contrast Sensitivity

This band-pass behavior can be plotted on a graph, with minute-hand motion at one extreme and bullet-like motion at the other, **14.14**. This graph is constructed using a small disc which flickers from black to white at different ***flicker rates*** or ***temporal frequencies***, **14.13**. A single ***cycle*** consists of the disc changing from black to white and back to black again (other starting and finishing position can be chosen but they will have the same cycle time). If the disc passes through one cycle in a tenth of a second (i.e., 0.1 second), it is said to have a ***period*** of a tenth a second, and a temporal frequency of ten cycles per second, or ten Hertz (10 Hz). Thus, the temporal frequency is given by 1/(the cycle period) which here yields

$$1/(0.1 \text{ second}) = 10 \text{ Hz}.$$

The ***sensitivity*** of the visual system at each temporal frequency is measured in terms of the lowest luminance ***contrast*** that can be seen at that temporal frequency. Recall that contrast is the difference between two luminances divided by their sum, see p. 78. The contrast of the disc stimulus can therefore be varied by altering precisely how white and how black the disc is. The value of the lowest contrast that can be seen at one temporal frequency defines a ***threshold*** for that temporal frequency, such that stimuli with contrasts below this threshold cannot be seen. The lower this threshold is at a given temporal frequency, the more sensitiv-

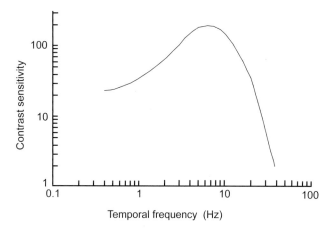

14.14 Temporal contrast sensitivity function
Sensitivity is defined as the inverse of the contrast thresh-old, so that, for example, sensitivity at 10 Hz = 1/(contrast threshold at 10 Hz). As the contrast needed to detect temporal frequency of 10 Hz is relatively low, this means that sensitivity to 10 Hz is relatively high. Compare the sensitivity plot shown here with the contrast threshold plot in **14.16**. Note both axes are logarithmic. After de Lange (1958).

ity the visual system has at that temporal frequency. We can thus plot sensitivity as the inverse of the contrast threshold at each temporal frequency (e.g., sensitivity at 2 Hz = 1/(contrast threshold at 2 Hz)), as shown in **14.14**.

Visual System as a Spatiotemporal Filter

The visual system acts as a temporal band-pass filter, but as we have seen before (Ch 4, p. 98–101) it can also acts as a spatial band-pass filter. This prompts the question: does the temporal sensitivity of the visual system depend on which spatial scale is considered? The answer is *yes*.

Rather than using a flickering disc to measure temporal contrast sensitivity, we can use a *sinusoidal luminance grating*, **14.15**. In this case, one spatial period is the distance between successive troughs or between successive peaks in the luminance grating. Thus, when held at a fixed distance from the eye, each grating has a specific *spatial frequency* (**4.5**, p. 80). If this grating is moved to the left or right at a constant speed then this provides a specific temporal frequency. For example, we might use a stimulus which has a spatial frequency of 8 cycles per degree (cycles/degree) which passes before the eyes at a speed sufficient

to provide a temporal frequency of 6 cycles per second (i.e., 6 Hz). That is, an entire spatial period (e.g., peak-to-peak distance) passes before the eye in 1/6th of a second, so that 6 periods pass before the eye each second (i.e., at a rate of 6 Hz).

In order to fully characterize the spatiotemporal sensitivity of the visual system, it is necessary to measure sensitivity at all combinations of spatial and temporal frequency. As before, sensitivity is defined as the lowest contrast that can be seen at a given spatial frequency. The results can be plotted as a three-dimensional graph in which each point on the horizontal "ground" plane defines a value for spatial and temporal frequency, and height defines the sensitivity of the visual system at that spatial and temporal frequency, as shown in **14.16**.

The Aperture Problem

The ability of the visual system to see stimuli with different spatial and temporal properties depends on cells within the visual system tuned to detect motion. As we have discussed above, each cell has a specific receptive field, which defines the area of the retina that a given cell can "see". You can think of this *receptive field* as a small **aperture** though which the cell "looks" at the world, **14.17**. This creates few problems when considering spatial properties of cells, such as the orientation of lines. However, this limited view creates severe problems when considering temporal properties of stimuli, such as motion.

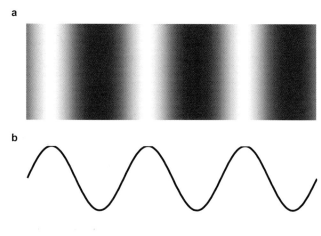

14.15 Sinusoidal grating
a Gray–levels.
b Sinusoidal luminance function showing the way the gray–levels vary across the image.

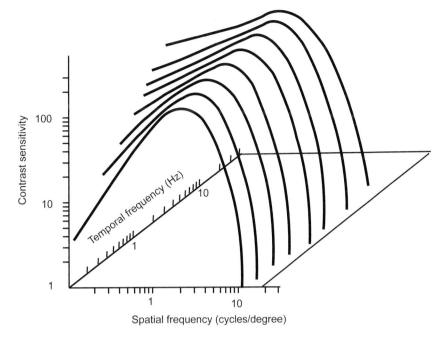

14.16 Spatiotemporal contrast thresholds of the visual system
Each point on the "ground plane" defines a specific spatial and temporal frequency, and the contrast threshold of the visual system at that spatial and temporal frequency is given by the height of the plotted surface (beware: high means most sensitive, because the contrast axis is plotted "upside down," i.e., from high contrasts going *up* to low contrasts). Curves shown here are schematic approximations of experimental data. Note all three axes are logarithmic. Schematically redrawn with permission from Kelly (1979).

A critical *computational problem* for the visual system consists of detecting the direction of motion of a moving luminance edge. Because each cell receives information from a single receptive field which acts like a small aperture, the motion of an object is ambiguous. For example, if the diamond shape in **14.18** moves across the aperture to the right (Direction 1) then the border "seen" crossing the aperture looks like the series of successive snapshots. However, if the diamond shape

moves downward (Direction 2 in **14.18**) then the snapshots as seen through the apertures are exactly the same as when the diamond moved to the right (Direction 1). That is, the edge still appears to move south-easterly. In fact, provided the diamond moves in *any* direction between south and east the line in the aperture appears to move in the same south-easterly direction. Note that the apparent motion direction of a line seen through an aperture is always **perpendicular** (at right angles) to the

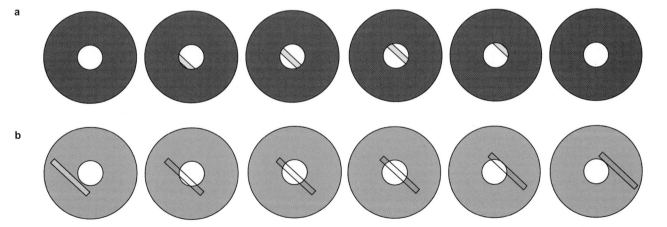

14.17 Viewing a moving object through an aperture
In both **a** and **b** the bar moves horizontally to the right across an aperture. Even though the bar is moving identically in **a** and **b** when seen through the aperture in **a**, it appears to be moving in a "north–easterly" direction, which is the direction perpendicular to the exposed border of the object. Viewing an object through an aperture gives a "neuron's eye view" of the world because each neuron only "sees" events within its own receptive field.

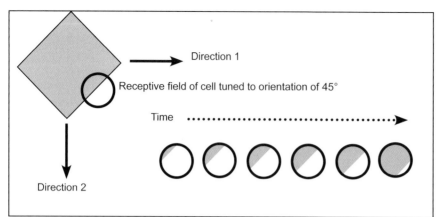

14.18 Aperture problem
If an object moving at a fixed speed passes across a neuron's receptive field centered on one border of that object then the neuron "sees" exactly the same motion whether the object moves to the right (Direction 1) or downward (Direction 2). Therefore, this neuron cannot, in principle, determine the motion of the object.

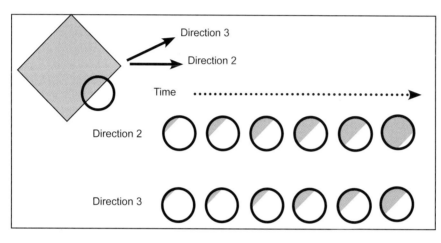

14.19 Aperture problem
contd
If the object moves along Direction 3 at the same speed as in **14.18** then the border appears to move across the aperture in the same direction as in **14.18** but the rate of motion, as measured through the aperture, is reduced. The two global motions make the border "slide" across the aperture yielding a local motion in each case perpendicular to the border, but this motion is slower for Direction 3 than Direction 1.

orientation of that line. Thus, it is not possible to ascertain the motion of an object by observing the motion of a part of that object through a small aperture.

This may strike you as a surprising outcome. To emphasize what is going on **14.19** shows what happens when the diamond moves upward to the right, labeled Direction 3. The diamond moves at exactly the same speed as before, but in a different direction. The border "slides" over the aperture and is again "seen" as an edge moving south easterly, but it appears to move across the aperture more slowly.

How can this motion direction problem be solved? If we have access to two apertures, both of which show different parts of the same moving object, then it is possible to estimate the global motion of that object. The computational problem of estimating the *global motion* direction of an object from the different *local motions* apparent

through two or more apertures is called the *aperture problem*.

Solving the Aperture Problem Using Intersection of Constraints

The global motion of an object is defined by two numbers, its *speed* and its *direction*, which together make up its *velocity*. Because velocity represents two quantities, it can be written in the form of a *vector*, which is basically an ordered list of numbers. For example, if an object moves at a speed of n meters per second in the direction α (the Greek letter alpha) then we can write its velocity as the *motion vector* $\mathbf{v}_0 = (n, \alpha)$. Note that $\mathbf{v}_0$ is written in bold typeface, which is standard notation for vectors.

In order to gain an intuitive understanding of motion, any motion vector can be represented as an arrow in a *velocity diagram*, where the direc-

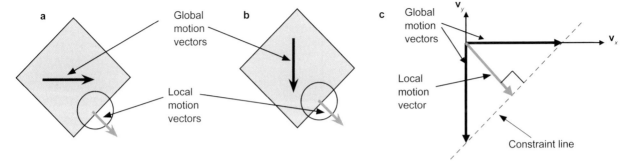

14.20 Local and global motion vectors
a The motion of an object is its *global motion*, as depicted by the horizontal arrow within this rhombus, whereas the motion seen through the circular aperture is called a *local motion*, as depicted by the diagonal arrow from the edge appearing in the circular aperture.
b A different (vertical) global motion from that in **a** can give rise to the same local motion as in a.
c Both local and global motions can be represented as vectors (arrows), which are drawn with a common origin. Each motion is represented by a single vector, where its length depicts speed and its direction depicts the direction of motion. If vectors are plotted like this, then they are said to be plotted in *polar coordinates*. The symbols v_x and v_y refer to horizontal motion and vertical motion, respectively. Thus, the global motions in **a** and **b** are represented by a vertical and horizontal vector respectively. The local motion as seen through the aperture is represented by a diagonal vector. A line drawn perpendicular to a local motion vector is a *constraint line* (dashed). As the constraint line joins two global motion vectors, both of these are therefore consistent with the observed local motion vector. Note that the defining feature of the constraint line is that it is perpendicular to the local motion vector.

tion of the arrow points in the direction of motion, and the length of the arrow represents speed. One of the two objects in **14.20** moves down and the other to the right, so the vectors corresponding to their global motions are represented respectively by an arrow that points down and an arrow that points to the right in **14.20c**. Now, add an arrow for the single local motion vector, which is the same for both downward and rightward global motions of this object. We always find that all three arrowheads lie on a straight line, called a *constraint line*, **14.20c**. Crucially, this constraint line is always perpendicular to the local motion vector. This implies that, if we only have access to the local motion vector (as is typically the case), and we draw a constraint line perpendicular to this vector, then we know that the global motion vector of the object must fall somewhere on this constraint line. Thus, both the rightward global motion vector of **14.21a** and the downward global motion vector of **14.21b** touch the constraint line defined by the (same) local motion vectors in **14.21a** and **14.21b** because this vector is consistent with both global motion vectors.

We do not know where the global motion vector is, but we now know that it touches the constraint line perpendicular to any local motion vector. Thus, a single local motion vector, derived

from a single aperture, *constrains* (but does not specify) the possible values of the global motion vector which we seek.

We can see this more clearly if we adopt a slightly different perspective on the three motion vectors just described. Consider the single object with a rightward global motion shown in **14.21a**. Two apertures are shown, and the local motion apparent in each is perpendicular to the line within each aperture. If we draw one of these local motion vectors as an arrow and we draw a constraint line perpendicular to it then we know that the global motion vector must touch this constraint line.

Thus, one local motion vector is consistent with every global motion vector that touches that line. If we now draw the arrow defined by the other local motion vector and draw a constraint line perpendicular to it, then we know that the global motion vector must touch this constraint line also. Clearly, if the global motion vector must touch both constraint lines, and the constraint lines cross at one point, then this point must be where the arrowhead of the global motion vector is located.

In other words, where the two constraint lines cross corresponds to the global motion of the object. This crossing point is called the *intersection of constraints* (*IOC*). A second example of

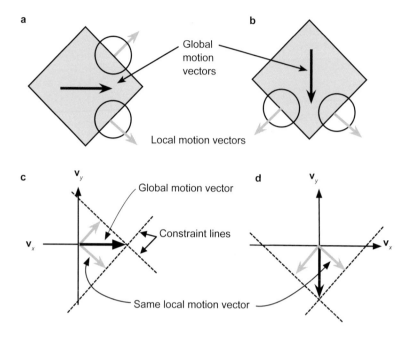

14.21 Intersection of constraint lines (IOC)
Each local motion vector associated with a moving object defines a constraint line, and the intersection of these lines defines the global motion vector for that object. Notice that the polar plots in **c** and **d** both have one (south–easterly) local motion vector in common, but that this turns out to be associated with a rightward global motion vector in **a** and **c** but with a downward global motion vector in **b** and **d**.

this idea is shown in **14.21b** in which the object is moving down. Why does this work? We explain the mathematical reasons in Ch 15.

Where is Motion Detected?

Having considered how the problem of detecting motion can be solved, we next examine which parts of the visual system might solve this problem. We already know that the magnocellular retinal ganglion cells receive inputs from a larger number of photoreceptors than the other ganglion cells. This gives them poor spatial resolution but makes them ideal for detecting motion. In some animals, such as the rabbit, ganglion cells with circular receptive fields perform the first steps in the computation of motion direction. In others, such as monkeys, ganglion cells respond equally well to motion in *any* direction, and for monkeys the motion direction computation starts in the striate cortex.

The elongated receptive fields of the simple and complex cells found in the striate cortex (Ch 8) collate information from many ganglion cells to form receptive fields sensitive to orientation. Some of these striate cells are also tuned to motion direction. These motion selective cells are found in layer 4B of the striate cortex, and some of these project directly to the "motion" area MT, also known as area V5 (MT stands for medial temporal sulcus; it is just lucky that M also stands for motion). Other cells in layer 4B project to the thick stripes of area V2, which then project to area MT, which seems to receive a relatively pure magnocellular input.

Skip from Here to Summary Remarks?

The next few sections contain details that the beginning student may wish to postpone reading.

Space-Time Diagrams and Spatiotemporal Receptive Fields

The precise mechanism used in primate vision to collate information from motion sensitive (but not directionally selective) ganglion cells to form direction-sensitive cells found in the striate cortex is not known. There are in fact many ways in which the ganglion cells could be combined, at least, in principle. Before we describe these, it will be useful to describe the properties of these cells in terms of *space-time diagrams*.

Imagine a ball moving vertically at a constant speed, which is illuminated with a strobe light which flashes at the rate of one flash per second. Using this set-up, we would see a series of snapshots in which the ball appears to jump between consecutive positions. The faster the ball moves, the larger these jumps appear to be. Now if we plot the ball's position on the vertical axis of a graph, where time increases along the horizontal axis then we would obtain a space-time diagram of the ball's motion, **14.22**. Because we take a snapshot every second we effectively *sample* the ball's position once a second, so that points plotted on our graph are always separated horizontally by one second gaps.

To take an extreme example, if the ball is stationary then at time zero then we would plot a point at $t = 0$ on the vertical axis for the fixed position to which we will assign the arbitrary value $x = 0$. We would thus plot our first point at the coordinates $(t, x) = (0, 0)$. At time $t = 1$ another snapshot is obtained, and the ball has not moved, so that $x = 0$ still. But time has moved on, so that $t = 1$. We would thus plot our second point at the coordinates $(t, x) = (1, 0)$. If we continue this process a series of horizontally aligned dots is obtained, **14.22** lowest plot, which indicates that the ball's position is the same in every snapshot.

If the ball moves at 1 meter per second (1 m/s) then the first snapshot at time $t = 0$ would be at $(t, x) = (0, 0)$, as before. But after one second, $t = 1$ and the ball has moved one meter ($x = 1$), so that the point plotted on the graph has coordinates $(t, x) = (1, 1)$. After two seconds the ball has moved 2 meters and the plotted point would have coordinates $(t, x) = (2, 2)$. The resultant series of dots would all lie on a straight line with a slope of 1, which indicates that the ball was moving at 1m/s, **14.22** middle plot. (The slope is defined by the angle of the line plotted and here it is 1 because for every unit moved sideways the graph moves upward by 1 unit.) Finally, if the ball moves at 2 meters per second (2 m/s) then the space-time plot of its position over time would yield a straight line with slope of 2, which corresponds to 2 m/s, **14.22**

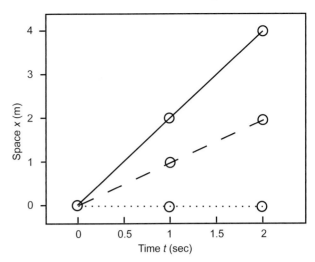

14.22 Space–time graph of the motion of a ball
Each line represents the ball moving at a constant speed: 2 m/s (solid line), 1 m/s (dashed line), 0 m/s (stationary, dotted line).

upper plot. Thus, space-time graphs depict the motion of points over time, such that the slope of the line in each graph increases with the speed of the object.

What has all this to do with receptive fields? Well, just as we can plot the motion of an object over time using a space-time diagram, we can also plot the spatial and temporal properties of a receptive field using space-time diagram.

Spatiotemporal Receptive Fields

As a simple example, the receptive fields of half of a Reichardt detector can be plotted in a space-time graph as follows. Recall that a Reichardt detector detects motion at a fixed speed between two fixed image points. Let's suppose that one Reichardt detector is tuned to detect a bright spot (e.g., a white ball) which moves at a speed of 2 m/s between two image positions, which we denote as x_1 and x_2. We can construct a *spatiotemporal receptive field* for this Reichardt detector as follows (see **14.23**).

We begin with two purely spatially defined receptive fields connected to two cells which act as "spot detectors." One spot detector has its receptive field at position x_1 and the other cell has its receptive field at x_2. Let's call these cells C_1 and C_2, and let's assume that x_1 and x_2 are exactly 1 meter apart. Note that this example assumes a *big* retina, but that does not affect the point being made. If a

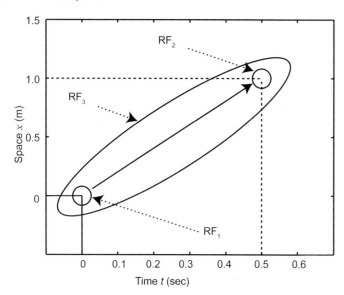

14.23 Spatiotemporal receptive field
The spatiotemporal receptive field RF_3 of cell C_3 comprises two component spatial receptive fields RF1 and RF_2, which belong to cells C_1 and C_2, respectively. The connections between C_1, C_2 and C_3 ensure that if C_2 is active half a second after C_1, then C_3 is activated. Thus, a stimulus moving at 2 m/s over RF1 and then over RF_2 is detected by the spatiotemporal receptive field RF_3 of C_3.

spot of light travels from x_1 to x_2 at 2 m/s then it would activate C_2 half a second after it activated C_1, because a spot moving at 2 m/s takes half a second to travel the one meter between C_1 and C_2. If a third "Reichardt detector" cell C_3 adds the output of C_1 to the output of C_2 but only after C_2's output has been delayed by half a second, then C_3 will be maximally activated by a spot which passes over x_2 half a second after it moves over x_1. In other words, C_3 is activated by a spot which moves from x_1 to x_2 at the speed to 2 m/s. This implies that cell C_3 effectively has a spatiotemporal receptive field which can be plotted in a space-time graph, as shown in **14.23**.

The spatiotemporal receptive field of cell C_3 consists of two component receptive fields, one associated with cell C_1 and one with C_2. The receptive field of C_1 is located at x_1 which we will define as $x_1 = 0$. The receptive field of C_2 is located at x_2 which we will define as $x_2 = 1$ meter, so that x_1 and x_2 are 1 meter apart. We know that these cells must be activated half a second apart in order to detect objects moving across the image at 2 m/s. In terms of our space-time graph, we can thus place C_1's receptive field at the coordinate $(t,x) = (0,0)$ and C_2's receptive field at the coordinate $(t,x) = (0.5,1)$, as shown in **14.23**.

Note that the location of a spatiotemporal receptive field in this space-time graph is determined both by its spatial location, which is a property of C_1's purely spatial receptive field position, and by

the time at which it activated, which is a property of C_3's temporal characteristics. In other words, the spatial position of C_1's receptive field is determined only by C_1, whereas the temporal delay introduced by C_3 on inputs from C_1 is determined by C_3, and it is this delay which defines the position of each component receptive field along the horizontal (time) axis.

We have constructed a spatiotemporal receptive field for a Reichardt detector using two purely spatial receptive fields. However, as mentioned above, there are many ways to combine receptive fields in order to construct spatiotemporal receptive fields. The precise manner in which spatiotemporal receptive fields are implemented in biological vision systems is not known, but it is encouraging that the basic receptive field components required to make spatiotemporal receptive fields are readily available within the early stages of the visual system.

Where Is the Aperture Problem Solved?

The Reichardt detector can detect local motion, but because its receptive field is restricted to a single image region, it cannot determine the global motion of an object. In this respect, the Reichardt detector is analogous to the direction-selective cells in layer 4B of the striate cortex. Like Reichardt detectors, these striate cells respond to local, but not global, direction of motion. Thus, if two receptive fields are stimulated by a diamond shape moving from left to right (as in **14.20a**) then each would signal that the local motion is either to the "north east" or "south east," whereas the global motion is to the right (east). So it would seem that the aperture problem is not solved in the striate cortex.

What about the motion area V5 (MT) discussed in Ch 8? A convincing test of MT's response to global motion is given by the use of *plaid patterns*. A plaid consists of two sets of parallel stripes called *gratings*, where each grating moves in a different direction, **14.24**. If a small aperture is drawn around one set of moving gratings then this gives the clear impression of stripes moving in a direction perpendicular to their orientation. In this case, we perceive motion as if we were a cell in the striate cortex, and we make the same mistake as that cell. Now, if the two gratings are superimposed then the motion appears to be to the right. However, the response of a motion-selective cell in the striate cortex remains unaltered; unlike us, it is still fooled by the motion of lines in its preferred direction.

Of course, if we perceive the plaid direction correctly then there should be some part of the brain that underpins this perception. This visual intel-

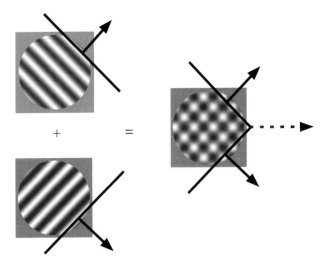

14.24 A plaid pattern
The two sinusoidal gratings (left) can be superimposed to form a plaid pattern (right). Cells in the striate cortex tend to respond to the local motion of each grating even after it has been combined to make a plaid, whereas we perceive this plaid as having a global motion to the right. Like us, cells in area MT (V5) respond to the global motion of the plaid, and not to its constituent local motions.

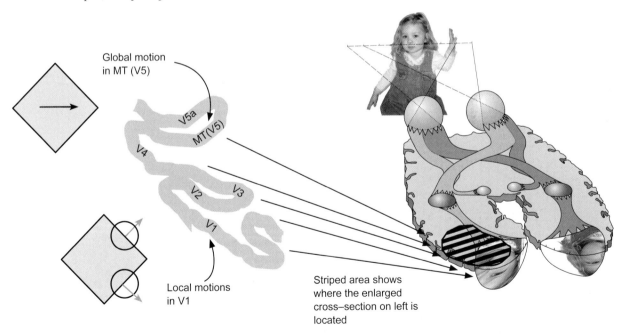

14.25 Solving the aperture problem in the brain
The worm–shape represents the visual cortex in a horizontal slice through the brain. Its approximate location in the brain is shown by the black stripes in the upper right figure. Different regions in the slice are labeled V1 to V5a. Cells in V1 seem to be "fooled" by the local motion vectors: they respond to the local direction of edges on an object rather than to the global motion of that object. In contrast, cells in area MT (V5) respond to the object's global motion but see text for cautionary remarks. Left part of this figure redrawn with permission from Zeki (1993).

ligence is found in the motion area MT, **14.25**. Unlike striate cells, MT cells do not respond to the direction of grating motion within the plaid, but to the global motion implied by the combined gratings that make up the plaid, **14.24**.

However, this neat story, which emerged about a decade ago, has become complicated by recent findings. Pack and Born found in 2001 that even MT cells are fooled by the **vector sum** (see Ch 15, p. 359) within the first 60 ms (6 hundredths of a second) of motion, but thereafter, they settle on the global direction of motion. It is as if the cells within MT use a simple sum of outputs from individual striate cells at first, but then come to a more considered, and more accurate, decision regarding global motion. This speculative account is itself complicated by findings by Guo et al. in 2006 which show that some *striate* cells also respond to *global* motion. It is therefore possible that lateral connections within striate cortex are responsible for the integration of local motion signals within the striate cortex.

But even this complex account is further complicated by studies which find that human perception sometimes agrees with the vector sum, and sometimes with the intersection of constraints (but see Ch 15, p. 369–372, for a Bayesian reconciliation of these contradictory findings).

Comparing Human and MT Neuron Performance

In this, and previous chapters, we have referred to cells which are sensitive to orientation, color, size, or motion. We expect and hope that these cells somehow underpin the perception of those properties, but we have not presented any *evidence* that this is so. In a series of stunning experiments, it was found not only that MT cells of a rhesus macaque monkey (*Macaca mulatta*) are sensitive to motion, but that some MT cells have a sensitivity matched only by the monkey itself. However, the real clincher, the *coup de grace*, was that experimental manipulation of MT cells changed the motion perceived by the monkey.

We should point out that, even though MT cells are involved in solving the aperture problem, the experiments reported here do not involve the aperture problem, but refer to the general problem of detecting motion along a specific orientation.

In order to compare the performance of MT cells to that of monkeys, Britten, Shadlen, Newsome and Movshon showed in 1992 used stimuli consisting of moving dots, **14.26**. The monkey's task was to indicate the direction of motion of the dots, which varied from trial to trial. However, this direction was by no means obvious because not all of the dots in this type of stimulus move in the

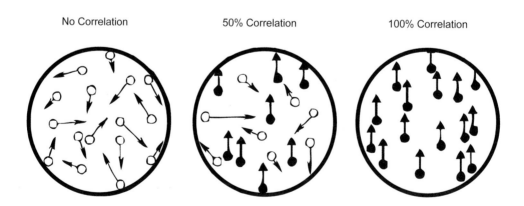

14.26 Britten et al.'s dot stimuli
The amount of correlation within each stimulus is defined as the percentage of dots which move in the same direction. A correlation of zero (left) looks like a de–tuned TV screen, a correlation of 50% (middle; filled dots moving in same direction) seems looks similar to this, but seems to drift in one direction, and a correlation of 100% (right) looks like a moving sheet of dots. Redrawn with permission from Britten et al. (1992).

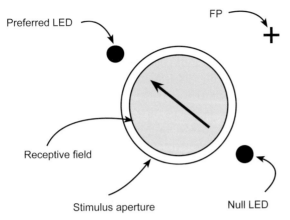

14.27 Britten et al.'s stimulus set-up
Each moving dot stimulus was presented within a small aperture which enclosed the receptive field of one neuron when the monkey looked at a cross at the fixation point (FP). Small lights (LEDs) on opposite sides of this aperture correspond to the null and preferred direction of this neuron. The monkey was trained to look at one of these lights after each stimulus was presented in order to indicate the direction in which motion was perceived. Redrawn with permission from Britten et al. (1992).

same direction. For example, if 90% of the dots move upward and the remaining 10% move in random directions then it is fairly easy to see that the overall or global direction is upward. However, if only 10% of the dots move upward and the remaining 90% move at random then the global direction of motion is much less apparent. Because some proportion of dots move in the same direction on each trial they are *correlated*, and a stimulus that has 10% of its dots moving in the same direction is said to have a *correlation* of 10%. By varying the amount of correlation (the proportion of dots moving in the same direction) from trial to trial it is possible to build up a picture of how sensitive the monkey, and an individual MT *neuron*, is to motion.

Let's start with the monkey. On each trial a *fixation cross* appears somewhere on screen in front of him, labeled as FP in **14.27**. After he looks at the fixation point, moving dots appear on the screen for 2 seconds. These dots are restricted to a small aperture to one side of the fixation point, as explained a below. After 2 seconds the dots disappear and two small lights appear on opposite sides of where the aperture had been. The monkey is trained (using rewards and a huge amount of patience) to indicate in which direction the dots appear to move by looking at one of the two lights. Note that the monkey can choose only one of two opposite directions, defined by the positions of the lights. This is called a *two-alternative forced choice* (**2AFC**) procedure and is designed to minimise the effects of bias in the monkey's response (Ch 12, p. 296).

As we might expect, the reliability of the monkey's response increases as the correlation between dot directions increases, **14.28**. Here, performance is measured as the proportion of trials on which the monkey gets the correct answer. Note that the monkey is correct 50% of the time even with very low correlations because the 2AFC procedure forces the monkey to make a decision on every trial, and even guessing yields 50% correct responses. As the correlation increases, so the monkey's guesses become more accurate, and performance increases accordingly. The open circles in **14.28** show performance at a range of different correlation values. These circles are joined by a curve which

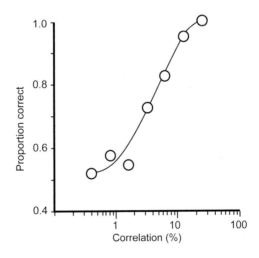

14.28 Psychometric function from a monkey
The accuracy of a monkey's responses increases as the correlation between dot directions increases from 0% to 100%. Note logarithmic axis. Redrawn with permission from Britten et al. (1992).

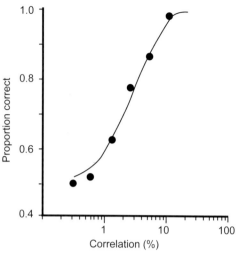

14.29 Neurometric function of a monkey's MT neuron
As in **14.28**, the accuracy of an MT neuron increases as the correlation between dot directions increases from 0% to 100%. Note logarithmic axis. Redrawn with permission from Britten et al. (1992).

has a characteristic "S" shape, an *ogive*. This type of performance profile is found when measuring many different abilities (such as detecting lights at different intensities), and is known as a ***psychometric function*** (**12.2**, p. 282). Note that the abscissa (horizontal axis) in **14.28** (and also **14.29**) is a logarithmic scale, so that equal intervals imply that correlation is multiplied by 10.

The amazing and audacious thing about this experiment is that the psychometric function of *single* neurons was measured, **14.29**. With such novel experiments come novel words, and the term ***neurometric function*** was coined to describe the performance profile of a single neuron. The circular aperture used to present moving dots to the monkey (as described above) was adjusted so that it corresponded to the receptive field of one neuron in area MT. This ensured that the performance of a single MT cell could be isolated.

One problem is that neurons cannot make judgments in the same way that monkeys can. A monkey can indicate which of two directions he thinks the dots move, but a neuron cannot, because the output of a neuron is its firing rate, which varies continuously. However, it is possible to estimate the decision a neuron *would have made* on the basis of its response to dots presented on different trials.

Each MT neuron has a preferred motion direction, such that, if the motion of dots is in the

preferred direction of a neuron then its output (firing rate) is high. If the dots move in the opposite direction, or null direction, then the neuron's output is low. Without going into details, a neuron's response to dots in its preferred and null directions can be used arrive at a discrete decision (e.g., left/right or up/down) for each trial. Once Britten, Shadlen, Newsome and Movshon had a decision for each trial, they could construct a neurometric function for each neuron, just as they constructed a psychometric function for the monkey.

Now, if the monkey relies on the output of cells in MT to make a decision regarding the direction of dots then we would expect the performance of cells in MT to be comparable to the performance of the monkey. This is what is found, as can be seen by comparing **14.28** with **14.29**, which gives a convincing impression of the similarity between the monkey's performance and the performance of cells in MT.

Correlation and Causality

This elegant experiment is certainly impressive, but it still amounts to circumstantial evidence. An MT neuron has motive, opportunity, and no alibi, but that does not mean that it is *responsible* for the monkey's decisions. This is a common problem within the realms of brain science, and indeed in science generally. The activity of a cell or a brain region can be highly correlated with the perception of some stimulus or property (e.g., color), but this *correlation does not imply causality*. Even if the activity of a neuron and the behavior of the monkey were *perfectly* correlated then this would not imply that the neuron *caused* the behavior.

One simple counter-example should persuade you of the truth of this common misconception. Suppose that a neuron were discovered which fired every time a monkey raised its arm. A naïve observer might infer that the neuron caused the arm to be raised. Then a neurophysiologist casually comments that the neuron under consideration is purely sensory in nature; in fact, it is a stretch detector which tells the brain how much tension there is in a specific arm muscle. In this case, the inferred causality was exactly the "wrong way around." It was not the neuron firing that caused the arm to be raised, but the arm raising that caused the neuron to fire.

More generally, there is no reason why the neuron firing and the arm raising should be causally connected at all. For example, we might observe that the length of a railway track increases and contracts throughout the day and that this is perfectly correlated with the rate of evaporation from a nearby lake. Did the expanding track cause the evaporation, or did the evaporation cause the track to expand, or did the sun coming out from behind the clouds cause both to increase and decrease in perfect synchrony? *Correlation does not imply causality.*

The usual way to try to break out of this straightjacket is to intervene in one of the events and see if this affects the other one. That is, if manipulating one event (e.g., MT neuron firing) leads to a change in some other event (e.g., perceptual decision on motion direction) then it is likely that these events are causally connected.

MT Neurons Cause Motion Perception

In order to test if the activity of MT neurons affects the decisions made by a monkey, scientists had to find a way to go beyond measuring correlations between neuron and monkey behavior. In 1992, using the same procedure as described above, Salzman, Murasugi, Britten and Newsome

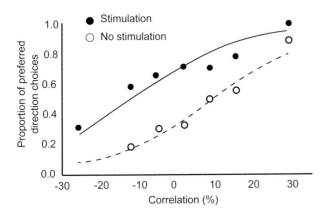

14.30 Effect of microstimulation on monkey psychometric function
This graph shows the change in the proportion of "preferred" choices caused by microstimulation of area MT. Redrawn with permission from Salzman et al. (1992).

found that they could perturb the activity of cells in MT by injecting a small electrical current into MT. Over a number of trials, they found that this *microstimulation* of neurons in MT influenced the monkey's decision.

As with many brain regions, cells in MT are arranged in columns, and adjacent columns have similar preferred directions. Thus, even though the microstimulation affected several columns, these all had similar preferred directions. Salzman, Murasugi, Britten and Newsome identified the preferred direction of a specific column of neurons by testing them with moving dots in their collective receptive field. They found that microstimulation of a specific region within MT biased the monkey's decision *toward* the preferred direction of the column centered in this region, **14.30**. It is as if microstimulation of a column of MT cells acted as a proxy for additional dots moving in the preferred direction of cells in that column.

The amount of influence that each set of neurons had varied from place to place within MT, but the change in behavior could be equated with adding between 7% and 20% to the correlation of the dots displayed. In other words, stimulating MT had the same effect on the monkey's decisions as adding up to 20% to the correlation of the dots.

Note that stimulation of MT neurons affected the monkey's response by introducing a *bias*, rather than a change in the *discriminability* of motion. This can be seen in **14.29**. This shows that the proportion of monkey responses in the preferred direction of neurons increases if those neurons are microstimulated. It's as if microstimulation of MT neurons introduced a predisposition for perceiving dots moving in the preferred direction.

Given these results, it is much harder to argue that cells in MT do not underpin the monkey's decision about the perceived direction of dots. However, we use the vague word "underpin" to indicate that we still much to learn about the details of the neuronal mechanisms that compute motion, even though we are fairly confident that these mechanisms depend critically on neurons in area MT.

Confirming evidence that area MT mediates motion perception comes from the neuropsychological study reported at the beginning of this chapter.

Summary Remarks

We began this chapter by asking: how is it possible to see motion? We can say that motion represents a well defined problem in vision, and one which has witnessed considerable progress, from theoretical and psychophysical, and physiological perspectives. Despite the fact that there is much that remains mysterious about motion perception, it represents a prime candidate for analysis using Marr's computational framework.

At the computational level, the problem of motion is well defined. At the algorithmic level, one method for solving this problem is to detect the same spot at two different locations as it moves across the retina. We considered two ways to implement this algorithm, the joint excitation method (Reichardt detector) and the inhibitory veto method. We also considered a different algorithm, the temporal derivatives method, but did not delve into how this might be implemented.

In the second part of the chapter we explained space-time diagrams and spatiotemporal receptive fields as they have been used as an alternative way of thinking about motion detection mechanisms. We concluded the chapter with an account of some classic neurophysiological research on a brain region specially associated with motion detection, MT (V5). This region appears to be critically involved in solving the aperture problem in monkey cortex.

Further Reading

Anstis S (1990) Motion aftereffects from a motionless stimulus. *Perception* **19** 301–306.

Barlow HB and Levick WR (1965) The mechanism of directionally selective units in rabbit's retina. *Journal of Physiology* **178** 477–504. *Comment* A classic early paper that is still well worth reading.

Born RT and Bradley DC (2005) Structure and function of visual area MT *Annual Rev Neuroscience* **28** 157–810.

Britten KH, Shadlen MN, Newsome WT, and Movshon JA (1992) The analysis of visual motion: a comparison of neuronal and psychophysical performance. *Journal of Neuroscience*

12 4745–4765. *Comment* A classic and ground–breaking paper.

Bruce V, Green PR, and Georgeson MA (2003) *Visual Perception: physiology psychology and ecology.* 4th Edition Psychology Press: Hove U.K. *Comment* An excellent but in many ways quite advanced textbook. A good starting point for pursuing the temporal derivatives approach to motion detection. Also a good source of references for further reading on perception generally.

Fried SI, Munch TA, and Werblin FS (2002) Mechanisms and circuitry underlying directional selectivity in retina. *Nature* **420** 411–414. *Comment* An excellent commentary on this paper is provided by Sterling P (2002) *Nature* **420** 375–376.

Guo K, Roberston R, Nevado A, Pulgarin M, Mahmmood S, and Young MP (2006) Primary visual cortex neurons that contribute to resolve the aperture problem. *Neuroscience* **138** 1397–1406. *Comment* Quite technically demanding but thorough account of motion processing in MT.

Kelly DH (1979) Motion and vision. II. Stabilized spatio-temporal threshold surface. *Journal of the Optical Society of America* **69**:1340–1349.

de Lange H (1958) Research into the dynamic nature of the human fovea-cortex systems with intermittent modulated light. I. Attenuation characteristics with white and colored light. *Journal of the Optical Society of America* **48** 777–784.

Levick WR, Oyster CW, and Takahashi E (1969) Rabbit lateral geniculate nucleus: sharpener of directional information. *Science* **165** 712–714. *Comment* A classic paper.

Marr D and Ullman S (1981) Directional selectivity and its use in early visual processing. *Proceedings of the Royal Society of London, Series B*, **211** 151–180.

Masland RH (2005) The many roles of starburst amacrine cells. *Trends in Neurosciences* **28**(8) 395–396.

Mather G (1980) The movement aftereffect and a distribution-shift model for coding the direction of visual movement. *Perception* **9** 379–392.

Mather G (1984) Luminance change generates apparent movement: implications for models of directional specificity in the human visual system. *Vision Research* **24** 1399–1405. *Comment* This reports Mather's research testing whether the temporal derivatives method of Marr and Ullman (op. cit.) is implemented in human vision. It is also summarizes the early research of Anstis and others in this area.

Mather G, Verstraten F, and Anstis S Eds (1998) *The Motion Aftereffect.* The MIT Press: Cambridge, Mass. *Comment* A splendid collection of papers reviewing research on the motion aftereffect and fine source book for references on this illusion.

Mikami A, Newsome WT, and Wurtz RH (1986) Motion selectivity in macaque visual cortex II. Spatiotemporal range of directional interactions in MT and V1. *Journal of Neurophysiology* **55** 1328–13310.

Niedeggen M and Wist ER (1998) The physiologic substrate of motion aftereffects. In Mather, Verstraten, and Anstis (1998).

Pack CC and Born RT (2001) Temporal dynamics of a neural solution to the aperture problem in visual area MT of macaque brain. *Nature* **409** (6823) 1040–1042.

Salzman CD, Murasugi CM, Britten KH, Newsome WT. (1992). Microstimulation in visual area MT: Effects on direction discrimination performance. *J Neuroscience* **12**(6) 2331-2355. *Comment* A classic paper showing a causal link between perception and neuronal activity.

Smith CE (2004) *Neural Engineering: Computation Representation and Dynamics in Neurobiological Systems.* Cambridge, Mass.: MIT Press.

Wade N and Verstraten FAJ (1998) Introduction and Historical Overview [of the motion aftereffect]. In Mather, Verstraten, and Anstis (1998)— see ref above, pages 1–24. *Comment* This provides entertainingly and at length the brief historical context given here on the motion aftereffect.

Zeki S (1993) *A Vision of the Brain.* Wiley-Blackwell. *Comment* A lovely book with many fine illustrations.

Zihl J, von Cramen D, and Mai N (1983) Selective disturbance of movement vision after bilateral brain damage. *Brain* **106** 313–340. *Comment* This is the source for the quotation on the first page of this chapter about the patient LM.

Zohary E, Shadlen MN, and Newsome WT (1994) Correlated neuronal discharge rate and its implications for psychophysical performance. *Nature* **370** 140–143.

15

Seeing Motion, Part II

15.1 Peregrine falcon with wings in rapid motion, causing extreme image blur

In Ch 14 we introduced the motion aperture problem and its solution using the intersection of constraints. Here, we begin by exploring this in more detail. We then consider from a Bayesian perspective how populations of neurons might compute motion.

Intersection of Constraints: Geometric Account

Let's suppose that we know the global motion of an object, and we draw this as a rightward arrow, as shown in **15.1**. This figure depicts a circle, but for the sake of argument, imagine that it is a many-sided polygon so that each aperture contains a straight line. We define motion to the right as motion at 0°, upward motion as 90°, and so on, "around the clock." If a line on the object at 120° (roughly like this backslash \) is viewed through an aperture then this defines a local motion vector at 30° (because the local motion vector is perpendicular to the line in the aperture, so the local motion vector is has a direction of 120 – 90 = 30°, a bit like this forward slash /). But what of the length of the local motion vector, which is proportional to the speed apparent through the aperture? Let's take two extreme examples to explore the relation between direction, local speed, and global speed.

Example 1 The line in the aperture is almost horizontal, as shown in the aperture marked A in

Why read this chapter?

We continue the analysis of the *aperture problem* by giving both a geometric account and an algebraic account of why the *intersection of constraints* embodies a solution to the aperture problem. We also describe an alternative, and less accurate, method which involves *vector summation*. We then consider how a population of motion-sensitive neurons could be used to estimate the direction of motion of a line moving across the retinal image. This involves a description of *neuronal tuning curves, noise, probability density functions,* and *maximum likelihood estimation*. With these various statistical tools at our disposal, we return to the aperture problem. Results from psychophysical experiments suggest that the aperture problem is sometimes solved using vector summation and sometimes by using an intersection of constraints method. We describe how these apparently contradictory results can be reconciled within the *Bayesian framework* if the human brain acts like an *ideal observer*. This chapter could be skipped on first reading of the book.

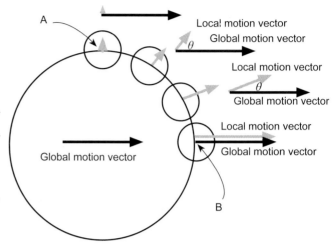

15.2 Relation between local and global motion vectors for a circular object
This figure depicts a circle, but for the sake of argument, imagine that it is a many-sided polygon so that each aperture contains a straight line. As the angle θ (theta) between the local and global motion vectors increases, so the local motion, as seen through an aperture, decreases. For example, the local and global motion vectors at A are almost perpendicular, and the local motion is almost zero. In contrast, the local and global motion vectors at B are parallel, and the local motion is the same as the global motion. The local motion vector shrinks as we move from B to A.

15.2. In this case, the local motion vector is almost perpendicular to the horizontal global motion vector, and there would be almost no motion apparent through the aperture, because the motion of an almost horizontal line sliding along in a rightward direction is extremely difficult to detect. In the limiting case, if the line in the aperture were *exactly* horizontal then no rightward (or leftward) motion could be detected.

Example 2 The line in the aperture is vertical, as shown in the aperture marked B in **15.2**. In this case, the local motion vector is **colinear** with the global motion vector (i.e., the local and global vectors lie along the same line). Moreover, the speed of the line in the aperture would be identical to the speed of the object, because this corresponds to a vertical line moving rightwards on an object that is also moving rightwards.

From these two examples, we can see that, for an object with constant velocity, as the line in the aperture changes orientation, so the apparent speed of that line varies. Specifically, the local speed varies between zero (if the local motion vector is perpendicular to the global motion vector) and the global speed of the object (if the local motion vector is colinear with the global motion vector). In the first case, the angle between the local and global motion vectors is 90°, and in the second case, this angle is 0°. As the apparent speed of a line is proportional to the length of its local motion vector, it follows that the length of the local motion vector *increases* as the angle between the local motion vector and the global motion vector *decreases*, as shown in **15.2.**

The exact relationship between the local and global motion vectors can be derived as follows. As we have seen, the speed of a line L_1 in an aperture depends on its orientation relative to the global motion direction. In fact, the aperture effectively "picks out" one component of the global motion vector. Specifically, the aperture picks out that component which is perpendicular to L_1 itself. As the local motion vector is just one component of the global motion vector, it follows that there is another component which cannot be "seen" from the motion of L_1 when viewed through an aperture. Together these two vector components ***decompose*** the global vector in the sense that the global vector can be reconstituted from the vector sum of these two vector components. This unseen vector is colinear with L_1 because only motion that is colinear with L_1 is invisible from the aperture's point of view. Thus, the aperture effectively decomposes the global motion vector v_0 of L_1 into two perpendicular vectors, which we will call v_1 and v_1'. The vector v_1 is known because it is the local motion vector and is perpendicular to L_1. The unknown motion vector v_1' is colinear with L_1 and is therefore perpendicular to v_1. We are not done yet.

The key to finding the global motion vector v_0 is to notice that v_1 and v_1' form two sides of a right-angled triangle because they are perpendicular components of the global motion vector. It follows that the global motion vector v_0 corresponds to the hypotenuse of this triangle. We still don't know v_1' nor v_0, but we now know that v_0 and v_1' meet at the same point, and that this point must lie on the constraint line perpendicular to v_1. However, given a specific local motion vector v_1, there exist many combinations of v_1' and v_0 which are consistent with the local motion vector v_1, as shown in **15.3** (contrast with **15.4**).

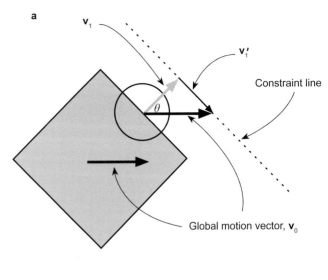

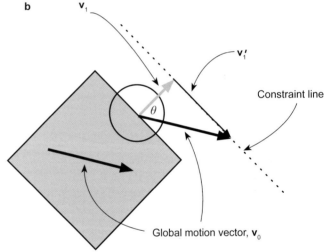

15.3 Global motion vectors consistent with a single observed local motion vector
a and **b** both depict the same object border seen through the same aperture. In **a**, this line's local motion vector v_1 is a component of the object's global motion vector v_0. In **b**, the *same* local motion vector v_1 is a component of a *different* global motion vector. There are thus many global motion vectors consistent with any single observed local motion vector v_1, but all of these putative global motion vectors end at the constraint line (dotted) perpendicular to v_1 in both **a** and **b**.

Intersection of Constraints: Algebraic Account

We know that the global and local vectors constitute two sides of a right-angled triangle, so we can work out precisely how the local and global speeds are related, as shown in **15.4**. If we define the angle

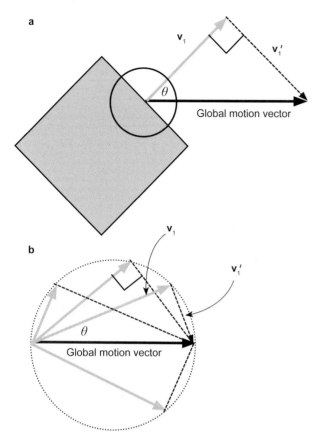

a

b

15.4 Local motion vectors consistent with a single global motion vector

This should be compared to **15.3**, which shows global motion vectors consistent with a single local motion vector. If we knew the global motion vector $\mathbf{v}_0$ of an object then we would find that the local motion vectors $\mathbf{v}_1$ measured through an aperture would depend on the angle θ between $\mathbf{v}_0$ and $\mathbf{v}_1$. Specifically, each putative local motion vector would touch the circle whose diameter is the global motion vector $\mathbf{v}_0$ as shown in **b**. Within this circle, geometry dictates that a line drawn perpendicular to any local motion vector $\mathbf{v}_1$ must connect $\mathbf{v}_1$ to the diameter $\mathbf{v}_0$ and is therefore a constraint line.

between the local and global vectors as θ then we have

$$\cos \theta = \frac{\text{local vector length}}{\text{global vector length}}.$$

Recall that the lengths of the local and global vectors correspond to the speeds of the local and global motions, so this equation can be rewritten as

$$\cos \theta = \frac{\text{local speed}}{\text{global speed}}.$$

This can be rearranged to give the local speed as

$$\text{local speed} = \text{global speed} \times \cos \theta.$$

If we define the local speed as s_1 and global speed as s_0 then we have

$$s_1 = s_0 \cos \theta.$$

Again, this is consistent with our informal proofs above that the local speed is zero if the $\theta = 90°$ (because cos 90 = 0, and global speed × 0 = 0), and that the local speed equals the global speed if $\theta = 0°$ (because cos 0 = 1, and global speed × 1 = global speed).

How does this help recover the global motion? If we only have one aperture, it does not. But if we have two apertures then it does. For the first aperture, we can define θ_1 as the angle between local and global motion directions, so that

$$s_1 = s_0 \cos \theta_1.$$

This can be rearranged to yield

$$\cos \theta_1 = s_1/s_0.$$

Note that we only know the value of s_1 in this equation, which therefore defines how $\cos \theta_1$ and s_0 are related. We can't use this equation to obtain $\cos \theta_1$ or s_0, but we can use it to obtain different values of $\cos \theta_1$ for different putative values of s_0.

Accordingly, if we plot the value of θ_1 for different values of s_0 then we obtain a straight line, as shown in **15.5**. This is done by "stepping through" different values of s_0 and using the equation $\cos \theta_1 = s_1/s_0$ to find corresponding values of θ_1 (taking the inverse cosine function of both sides $\theta_1 = \text{acos} (s_1/s_0)$). This is the same *constraint line* as described in the previous chapter, **14.20**. Note that this constraint line is plotted in ***polar coordinates*** in both **14.20** and **15.5**. This means that, as we step through different putative values of s_0, if we obtain the value of $\theta_1 = 30°$ from a value of, say, $s_0 = 2$ then this is plotted as a point 2 units away from the origin at an angle of 30° to the horizontal axis. If we do this for a range of values of s_0 then we obtain a corresponding set of θ_1 values, which can be plotted in polar coordinates to obtain a straight constraint line. We can repeat this process for the second aperture which defines the equation

$$s_2 = s_0 \cos \theta_2,$$

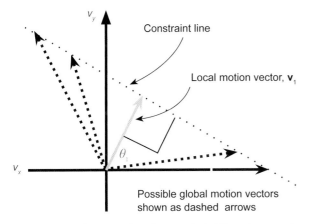

Constraint line

Local motion vector, $\mathbf{v}_1$

θ_1

v_x

v_y

Possible global motion vectors shown as dashed arrows

15.5 Constructing the constraint line given an observed local motion vector
For a given local motion vector $\mathbf{v}_1$ with length s_1, the relation between the angle θ between $\mathbf{v}_1$ and the global motion vector $\mathbf{v}_0$ is given by the equation $\cos\theta = s_1/s_0$, where s_0 is the length of $\mathbf{v}_0$. We can insert different values of s_0 into this equation and obtain corresponding values of θ_1. Each pair of values (θ_1, s_0) so obtained defines a putative (dashed) global motion vector $\mathbf{v}_0$. If we plot these pairs of (θ_1, s_0) on a graph where θ_1 is the angle of a line, and s_0 is its length then they define a constraint line, which has the property that the global motion vector must touch this line. Dashed arrows correspond to putative global motion vectors. Only one value of θ is shown, θ_1.

where s_2 is the local speed and θ_2 is the angle between the local and global motion directions. We can then plot the constraint line associated with the equation $\cos\theta_2 = s_2/s_0$. Where the two constraint lines intersect defines the values of θ_0 and s_0 which are consistent with the local motion in both apertures, **15.6**. Specifically, the global speed s_0 is given by the length of the line which connects this intersection point with the origin, and the global motion direction is given by the direction of this line. For the correct value of s_0 this is given by

$$\cos\theta_0 = s_1/s_0,$$

which is also equal to $\cos\theta_0 = s_2/s_0$.

Thus, given two apertures, which show different parts of the same object, we can estimate the global motion of that object.

Vector Summation

Before describing vector summation, we emphasize that, unlike the intersection of constraints, this method does not usually give the correct estimate of global motion, **15.6c**. We provide this account

because there is evidence that certain cells (described below) seem to compute the vector sum in order to estimate global motion.

As we have already seen, an object's global velocity can always be decomposed into two **velocity components**, and this decomposition can be effectively carried out using two apertures. Now, the two velocity components can be drawn as two sides of a parallelogram. The length of each side corresponds to the speed of one component, and the orientation of each side gives the direction of that component. Given that each side of this parallelogram is a vector, the sum of these vectors is the diagonal of the parallelogram, which is known as the **resultant vector**, see **15.6c**. Using this scheme, the object's estimated speed corresponds to the length of diagonal of the parallelogram, and its direction is given by the direction of this diagonal, as shown in **15.6c**. Because the global motion estimate is obtained as the sum of two local motion vectors, this is known as **vector summation**. (The terms vector sum and **vector average** are both used in the literature. In fact, even though they share a common direction, the *speed* implied by the vector sum is twice that implied by the vector average.)

What Are Cells in MT Trying to Tell the Brain?

It is not enough for cells in area V5 (MT) to respond to motion; they must be able to tell the rest of the brain what they have found. This, of course, is true of all neurons in the brain. Just as every neuron **receives** as well as **transmits** information, every neuron must be able to understand or **decode** the chattering of the neurons that project to it. In the case of MT neurons, the evidence suggests that they attempt to tell the rest of the brain about motion direction and speed.

Consider a set of, say, 60 neurons in MT. In order to find out what MT neurons have to say, we need to be able to decode their outputs. Each neuron has a preferred direction and speed. In order to simplify the account in the next few sections, and unless stated otherwise, we will ignore the preferred speed of neurons and consider their preferred directions only. Specifically, we will assume that each of the 60 neurons has a preferred direction which corresponds to one of the minute ticks on the face of a clock (i.e., the preferred directions of any two neurons are 6° apart, on average). Each

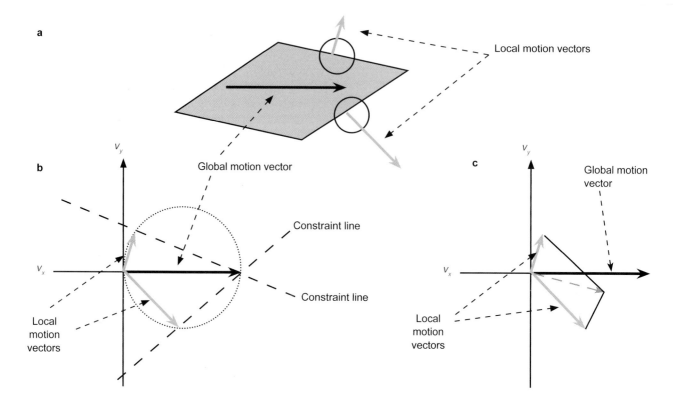

15.6 Comparison of vector summation (VS) and intersection of constraints (IOC) for estimating global motion
a Global and motion vectors for a rhombus with sides that are not perpendicular.
b The IOC method is guaranteed to give the correct estimate of the global motion vector, even if two local motion vectors are not perpendicular, because it is based on the physics of motion.
c Vector summation consists of finding the resultant vector of the two local motion vectors. This is achieved by constructing a parallelogram from the two local motion vectors, as shown. The global motion vector estimated by vector summation is then given by the diagonal of this parallelogram (which is also the resultant vector of the local motion vectors). The true global motion vector is drawn as a horizontal arrow for comparison. Note that the global motion vectors estimated using IOC and VS are the same only if the local motion vectors are perpendicular, as in **15.4**.

neuron has a fairly broad tuning so that a range of different motion directions stimulate it, **15.7**. We assume that these 60 neurons belong to the same hypercolumn, and that they all have receptive fields which are in the same region on the retina. In this way, for any motion that occurs within this region of the retina, several neurons, with similar preferred directions, will be activated. We define the collective responses of this set of 60 neurons as a ***response profile vector***. We use this term because if neurons are ordered according to their preferred directions then their collective response to a single stimulus direction defines a profile. This profile consists of an ordered set of firing rates, which is just an ordered set of numbers, and any ordered set of numbers is a vector, **15.8**.

Neuronal Voting

Given the outputs of these 60 neurons in the form of a response profile vector as in **15.8**, we can now ask questions: (1) what information is available from this response profile vector about the motion on the retina?, (2) how can this information be recovered in principle?, and (3) how can this information be recovered in the brain?

Before we proceed it is important to recognise that there is one major spanner in the neuronal machinery of the brain: ***noise***. Not audio noise, but noise in the form of random fluctuations in cell activities. If it were not for the ubiquitous presence of noise then all we need do is identify that neuron with the highest output and conclude

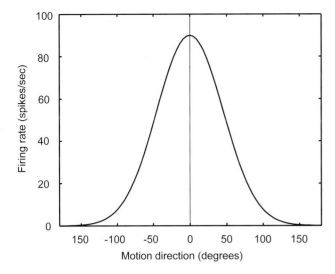

15.7 Tuning curve of a neuron
As the direction of motion of a stimulus varies so the mean response of a neuron increases and then decreases. The gaussian tuning curve of this neuron peaks at 0°, the preferred direction of this neuron. The width of the tuning curve is defined by its standard deviation σ, which is 20° in this case, and the variance is defined as $v = \sigma^2$.

directions of neurons with the largest outputs contribute most to the final estimate of direction. This is often referred to as a ***voting scheme***, but it is not a democratic "*one person, one vote*" scheme; it is more akin to a shouting match in which the final message is dominated by the loudest words (that is, by neurons with the largest output). As neurons typically have a preferred speed *and* direction, the preferred stimulus of a neuron has two components (i.e., speed and direction). The preferred stimulus of a neuron is therefore specified as a vector quantity. Accordingly, when all the shouting is over, the result is a single "average" value for speed and a single "average" value for direction, which together define an "average" pair of speed/direction values known as the ***population vector***.

One major drawback of this scheme occurs if there happen to be a large proportion of neurons with the same preferred direction α (alpha). Now consider what happens if the direction of motion β (beta) on the retina is close to, but not equal to, α. Each of the, say 1000, neurons with preferred

that the retinal motion is in the direction preferred by this particular neuron. But noise ensures that the most active neuron during one presentation of retinal motion will not be the most active neuron during a second presentation of an identical retinal motion. Just as you can hear your friend shout "Noisy, isn't it?" in a noisy street on one occasion but not on another due random variations in the traffic passing by, so any given MT neurons will "see" its preferred direction on one presentation but not on another due to any number of possible sources of neuronal variation (e.g., blood supply, temperature). Noise is the bane of computation, both in your computer and in your brain. It is everywhere, and it cannot be ignored.

The Average of the Response Profile Vector

As we saw in Ch 3 when computing bar orientation from simple cell responses, one way to minimize the effects of noise is to use some form of average. This principle can be applied in a very similar way to our motion response profile vector, as follows. For each neuron, its contribution to the final estimate of motion direction is given by the value of its preferred direction *weighted by the firing rate of that neuron*. In this way, the preferred

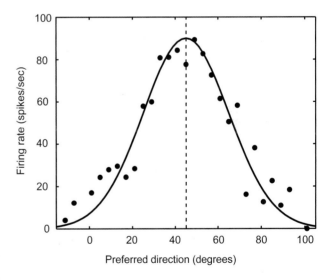

15.8 Response profile vector of neurons with different preferred directions to a stimulus moving at 45°
Each plotted point represents the response (firing rate) of a neuron with a different preferred direction. This preferred direction is given by the horizontal position of each plotted point. For example, given a stimulus at 45° degrees, the response of the neuron with a preferred direction of 80° is about 20 spikes/second. If a smooth curve is drawn between the points (as above) then it defines an idealized response profile vector which has exactly the same shape as the tuning curve of a single neuron (see **15.7**).

direction α will respond, but with a relatively low output. On the other hand, each of the, say 100, neurons with preferred direction β responds with a large output. However, the total output of the 1000 neurons with preferred direction α can swamp the "voices" of the 100 neurons which are tuned to the actual direction of motion β on the retina. This would give the wrong impression that the direction of motion on the retina is α, simply because there are many more neurons tuned to α than to β. Additionally, even if the distribution of preferred directions is uniform (as we have assumed here) then the population vector is less accurate than the ***maximum likelihood estimation*** (MLE).

Estimating Motion Direction: Maximum Likelihood Estimation

Maximum likelihood estimation is a method which forms part of the Bayesian approach to vision. Recall from Ch 13, that Bayes' rule can be considered as a method for combining information from images that we currently *have* with information gained from *prior* experience to obtain the information we *want*. Recall also that maximum likelihood estimation is, in a sense, a poor man's Bayes' rule. For the present, we are poor men because we are forced to use it. However, relative to our "population voting" cousins, we are rich!

The information we have is 60 neuronal output values, which we will denote as $\mathbf{r}$, where $\mathbf{r}$ represents the ***response profile vector*** of 60 output values, so $\mathbf{r} = (r_1, r_2, \ldots, r_{60})$. (Note that $\mathbf{r}$ is in bold typeface, which is the standard convention for variables that are vectors). The information we know from prior experience is that all directions are equally probable (this is why we will use maximum likelihood estimation, as explained below). The information we want is the direction of motion on the retina, which we denote as s^* (for stimulus). Our best estimate of s^* is denoted $\hat{s}$ (pronounced s-hat), and we use s to stand for any value of direction (i.e., s is a variable). In terms of Bayes' rule we have

$$p(s|\mathbf{r}) = p(\mathbf{r}|s) \, p(s) \, / \, p(\mathbf{r}).$$

Here $p(s|\mathbf{r})$ is the ***posterior probability***; it is the probability that the retinal motion direction has a value s given that the 60 neurons have a collective output $\mathbf{r}$. The quantity $p(\mathbf{r}|s)$ is the *likelihood*; it is the probability that the 60 neurons have a collective output $\mathbf{r}$ given that the retinal motion direction is s.

It turns out that we can ignore $p(s)$, the probability that the direction of motion has a value equal to s. Because we assume that all motion directions are equally probable, this implies that if we constructed a histogram of s values then it would be a flat line, which is why the variable s is said to have a ***uniform prior probability distribution***. In other words, the value of $p(s)$ is the same irrespective of the value of s, and we can therefore treat it as a constant and set its value to one (i.e., $p(s) = 1$ for all s). We can also assume that $p(\mathbf{r})$, the probability that the 60 neurons have a collective output equal to $\mathbf{r}$, is a constant equal to one. This is usual in using Bayes' theorem for ***parameter estimation***, as here. In this case, the parameter whose true value we are trying to estimate is s, the direction of a stimulus.

As can be seen just by checking the equation above with both $p(s)$ and $p(\mathbf{r})$ set to unity (1), these fortunate simplifications mean that the posterior probability $p(s|\mathbf{r})$ reduces to the likelihood $p(\mathbf{r}|s)$. That is, $p(s|\mathbf{r}) = p(\mathbf{r}|s) \times 1/1$. In such cases, the estimate $\hat{s}$ of s^* found by the Bayesian formulation of a problem is the same as that found by maximum likelihood estimation.

As we are using maximum likelihood estimation, we seek that value $\hat{s}$ of s which maximizes the probability of obtaining the observed value of $\mathbf{r}$ for a single presentation of a stimulus. In other words, we want to know which value of s makes it most likely that we would obtain the value of $\mathbf{r}$ that was actually observed when the stimulus was presented. If this sounds a little bizarre then we advise a brief re-read of Ch 13 at this point.

Tuning Curves

In order to proceed, we need to back up a little so that we can examine the ***tuning curve*** of a neuron.

The tuning curve of a neuron is defined as the mean response or firing rate of that neuron to repeated presentations of a stimulus, such as a specific motion direction s.

The mean response is defined in terms of the ***tuning function*** $f(s)$, where $f(s)$ typically has a gaussian (bell-like) shape, as shown in **15.7**. The

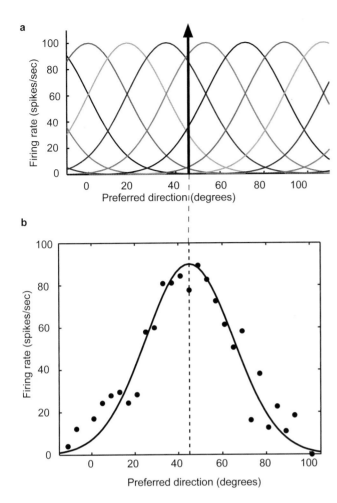

a

b

15.9 Noise tuning curves and population response profile vector
a Each gaussian represents the tuning curve of a single cell.
b If cells are ordered according to their preferred direction then a single stimulus moving at 45 degrees induces a mean response profile vector (dots) identical in shape to the tuning curve of each cell in **a**, and identical in position to the tuning curve of a cell with preferred direction at 45 degrees (which is the most likely stimulus value s in this case). This is true for any symmetric tuning curve.

equation for a gaussian function is

$$f(s) = r_{max} \exp(-(s-u)^2/(2v)),$$

where r_{max} is a constant which specifies the maximum firing rate, and u is the preferred direction shown in **15.5**. This equation states that the output $f(s)$ of the tuning curve is largest if $s = u$, because then $(s-u)^2 = 0$ and $\exp(0) = 1$, so $f(s) = r_{max}$ as shown in **15.7**. The term v is the **variance**, which is equal to the square of the **standard deviation** σ

of the tuning curve, $v = \sigma^2$, and defines the width of the tuning curve.

Tuning Curve Templates: As stated above, if we plot the output of a set of neurons which have been ordered according to their preferred directions, as in **15.9**, then we obtain a response profile vector. This profile is gaussian, just like the tuning curve of a single neuron. In fact, it is *exactly* the same as the tuning curve of a single neuron.

Thus, the gaussian tuning curve defined above for a single neuron is observed once more, as shown by the solid curve in **15.9b**. However, now each point on the curve corresponds to the mean output $f(s = 45)$ of a *different* neuron with a specific preferred direction u in response to a stimulus with direction $s = 45$.

In other words, if the ith neuron's mean output value is given by $f_i(s)$ then the array of mean output values defines a gaussian curve. This is a handy way of visualizing a tuning curve, and consists of plotting responses as a function of preferred direction.

How does this help in finding the true value s^* of the stimulus direction? Recall that, in practice, we do not know where solid curve drawn through the response profile vector is located in **15.9b**. Let's call this curve a ***tuning curve template***, for reasons which will become apparent.

All we have is the response profile vector itself (i.e., the noisy response of each of a set of neurons). But if we had a principled way of finding out where the tuning curve template is located along the array of neurons then we could use the location of its *peak* as an estimate $\hat{s}$ of the true direction of motion s^*.

Note that the ordered array considered here and shown in **15.9** is for explanatory purposes only. Of course, the neurons are not necessarily spatially ordered in the brain, and the method, described below, used to find the location of the tuning curve template does not depend on any such ordering. However, in order to gain insight into this method we need to make use of this imaginary ordered array of neurons.

So, let's use the response profile vector to find out where the tuning curve template is located along the array of neurons. In essence, this can be done by sliding the tuning curve template along the array of neurons in order to find which position (preferred direction s) is a *best fit* to the

a
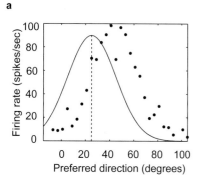

b
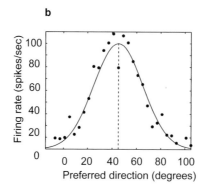

c
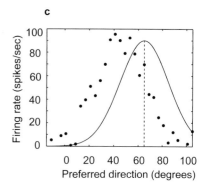

15.10 Template fitting
Given a stimulus moving at 45 degrees, the response profile vector of neurons with different preferred directions can be fitted with a gaussian template that is identical to the tuning curve of each neuron.
a Template centered on 25° = poor fit. **b** Template centered on 45° = good fit. **c** Template centered on 65° = poor fit.

response profile vector **r** of the array of neurons, as shown in **15.10**.

All that remains is to define what we mean by "best fit." This remaining task of defining a "best fit" will occupy several pages, but bear in mind that all the ensuing mathematics is a means to an end, namely, finding a position for our tuning curve template which is a best fit to the responses of a population of neurons. We will then take the position of the center of this "best fit" tuning curve template as our estimate $\hat{s}$ of the true stimulus direction s^*. To return to the topic of tuning curves, the presence of noise ensures that each presentation of the same stimulus value s elicits a different response r. Like the tuning curve, this noise typically has a gaussian or **normal distribution** with zero mean, so that the probability of observing noise with amplitude n is given by

$$p(n) = k_n \exp(-n^2/(2v_n)),$$

where $k_n = 1/\sqrt{(2\pi v_n)}$ is a constant, and v_n is the variance of the noise distribution, **15.11**. This equation states that the noise is mostly close to the mean of zero, with values far from zero occurring with decreasing probability. Because n varies from trial to trial in an unpredictable manner, it is known as a **random variable**. Quantities such as $p(n)$ are actually **probability densities** (Ch 12), but as we usually treat them as if they are probabilities we will not lose sleep over this subtle distinction here.

Note that both the tuning curve and the noise

are defined in terms of gaussian functions. *Do not be fooled by this coincidence*, however common it is in the literature. The tuning curve $f(s)$ defines the **deterministic** part of the response of a neuron to a stimulus; it is not a probability, it is equal to the *mean response* (firing rate) which is determined by the stimulus value s. In contrast, the noise distribution defines the *probability* $p(n)$ that noise of amplitude n is observed for a single presentation of a stimulus with value s. Having distinguished between the *probability* $p(n)$ (of observing noise with value n) and the *value* $f(s)$ (of the tuning curve at s), we can now combine the deterministic part of a neuron's response $f(s)$ given by the tuning function with the noise n to obtain the probability of a response $r = f(s) + n$.

Since the output r of a neuron is the sum of a deterministic term $f(s)$ and a random variable n, r is also a random variable. As

$$r = f(s) + n,$$

it follows that

$$n = r - f(s),$$

so the probability of a response r is

$$p(r|s) = k_n \exp(-(f(s) - r)^2/(2v)).$$

This equation states that the probability of observing a response r given a stimulus value s is $p(r|s)$. Note that this is a **conditional probability** (Ch 13) because the probability of r depends on a *given* value of s. As shown in **15.11**, this probability shrinks as the difference between the response r and the mean value $f(s)$ of r increases.

Notice that $p(r|s)$ is a *scalar* (i.e., a single value) if we consider a single value of s, and note also that this should be written in the form $p(r|s = 10)$ or $p(r|10)$ for example. However, $p(r|s)$ also defines a *distribution* if we consider s over a range of values, such as the values on the horizontal axis of **15.12**. Here we use $p(r|s)$ to refer to scalars and to distributions, and we rely on the context is used to remove any ambiguity. The distributions of the output r and the noise n differ only by their means ($f(s)$ and zero, respectively).

For a given stimulus with direction s^*, the response profile vector

$$\mathbf{r} = (r_1, r_2, \ldots, r_{60})$$

of 60 neuronal outputs has a typical profile **r**, as shown in **15.8**. The probability of observing a specific *response profile* **r** (not single response r) given a stimulus value s, is given by $p(\mathbf{r}|s)$. Crucially, if the noise in each neuron's response is independent of the noise in the other 59 neuronal responses then we can us a "product trick" to evaluate the prob-

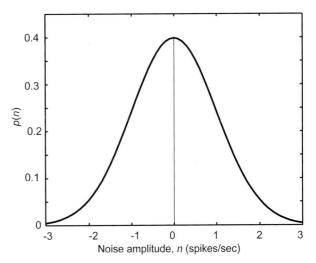

15.11 Gaussian noise

Probability of observing noise at amplitudes between -3 and 3 spikes/second. The most probable value of noise n is zero, and this probability decreases as the value of n moves away from zero. The curve depicted defines a gaussian or normal distribution of values of n. This distribution has a mean value of zero. Its width is defined by the standard deviation σ, which is $\sigma = 1$ spike/second in this case. The variance *var* of the values of n is given by the square of the standard deviation, $v = \sigma^2$. The curve plotted here is known as a probability density function (pdf), because it defines the probability of observing values along the horizontal axis.

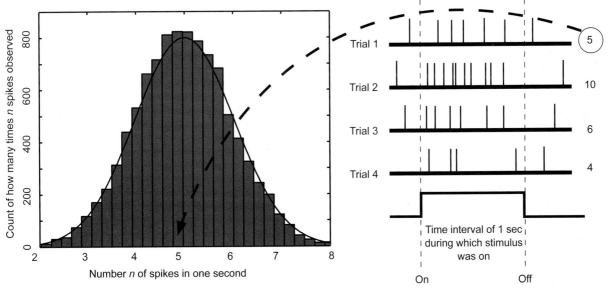

15.12 Response histogram

The presence of noise ensures that repeated presentations of a stimulus with the same direction elicits different responses from a single neuron on each trial. The right-hand panel depicts the output of one neuron to four presentations of the same stimulus. The left-hand panel depicts a histogram of outputs measured over a large number of trials. The solid line represents the probability density function (pdf) of the responses. Note that this histogram has the same width (standard deviation, σ) as the noise distribution (pdf) shown in **15.11**. This is because the response pdf results from adding noise to the deterministic output defined by a neuron's tuning curve.

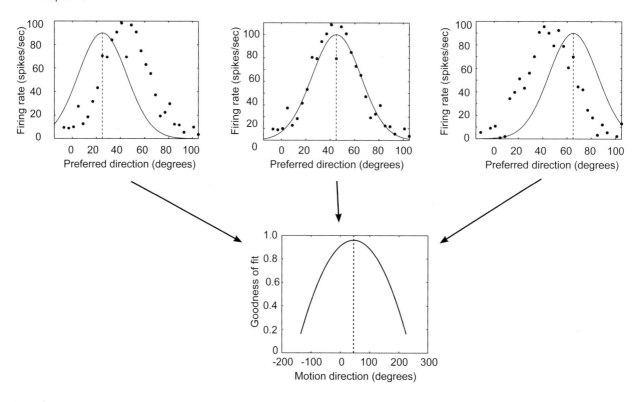

15.13 Estimating direction using a gaussian template
As the template is slid across the response profile vector, the goodness of fit increases and then decreases. The position of the template center when the fit is best represents the estimated value of direction, and is exactly the same as the maximum likelihood estimate of direction.

ability $p(\mathbf{r}|s)$ as

$$p(\mathbf{r}|s) = p(r_1|s)\ p(r_2|s)\ \dots\ p(r_{60}|s).$$

In other words, the probability of observing the response profile $\mathbf{r}$ is given by the products of the probabilities of observing each of the responses r_i in $\mathbf{r}$. This equation can be rewritten more succinctly as

$$p(\mathbf{r}|s) = \prod_i p(r_i|s),$$

where the capital Greek letter $\prod$ (pi) indicates that all terms should be multiplied together (e.g., the probabilities of the responses of different neurons), and the subscript identifies each of the 60 neurons. Given that responses have a gaussian distribution, so that

$$p(r_i|s) = k_i \exp(-(f_i(s) - r_i)^2/(2v_i)).$$

Now $p(\mathbf{r}|s)$ can be rewritten as

$$p(\mathbf{r}|s) = \prod_i k_i \exp(-(f_i(s) - r_i)^2/(2v_i)).$$

We can use this product trick for the same reason that you can almost certainly agree that the probability of obtaining two heads in a row when flipping a coin is 1/4 or 1/2 × 1/2. Your intuitions in this matter are related to a fundamental tenet of probability theory: *if two events are independent (as with coin flipping) then the probability of any two outcomes is just the product of the individual outcome probabilities*. So it is with neuronal outputs. If the noise in their outputs is independent then the probability of any set of (e.g., 60) output values is just the product of the probabilities of each individual outcome.

Now we can find the maximum likelihood estimate (MLE) $\hat{s}$ of the true value s^* by substituting different values of s into the last equation above. Note that we have a response value r_i for each neuron, and we assume sensible values for their variances v_i the responses. So what we need is a value for s which makes $p(\mathbf{r}|s)$ as large as possible. We might expect, for example, that substituting the true value s^* of s into this equation yields a high value for $p(\mathbf{r}|s)$. The effects of noise ensures that the value of s that makes $p(\mathbf{r}|s)$ largest may not be s^*, but it will almost certainly be close to s^*. This is why the value $\hat{s}$ of s that maximizes the likelihood $p(\mathbf{r}|s)$ is known as the maximum likelihood estimate of s^*. Note that the process of finding a value $\hat{s}$ of s that maximizes $p(\mathbf{r}|s)$ is a purely *computational* process (involving evaluation of $p(\mathbf{r}|s)$ for different values of s), and does not therefore require that neurons are spatially ordered according to their preferred directions.

The best fit template tuning curve in **15.13** also provides an answer to the question "which value of s maximizes the probability of obtaining the observed output values $\mathbf{r}$?" In short, the location of the peak of this curve defines an estimate $\hat{s}$ of s^*. The "goodness of fit" of this curve is given by its deviation from the plotted outputs of different neurons, and is exactly equal to $p(\mathbf{r}|s)$, as shown in **15.13**. This is because $p(\mathbf{r}|s)$ is a measure of how neuronal outputs deviate from a perfect gaussian profile expected under zero noise conditions. The position of this template tuning curve is therefore a "best fit" curve for the output values elicited by a particular presentation of a stimulus with value s^*. As $p(\mathbf{r}|s)$ in the equation above is a likelihood, the value of s that maximizes this likelihood is called the maximum likelihood estimate (MLE) of the true value s^*.

MLE and Least-Squares

That's it, we're almost done, aside from two more tricks, this time to avoid an explicit search for the best estimate $\hat{s}$ of s^*.

Here's the first trick. Rather than retaining the exponential function, we take the logarithm of each term. We can do this because any change in s that increases $p(\mathbf{r}|s)$ also increases $\log p(\mathbf{r}|s)$; and therefore the value of s that maximizes $p(\mathbf{r}|s)$ also maximizes $\log p(\mathbf{r}|s)$. Taking logarithms effectively "un-does" the exponential because, for example, if

$$y = \exp(x)$$

then $\log(y) = \log(\exp(x))$, which yields x. Taking

logs also transforms our *product* of 60 terms into a *sum* of 60 terms because, for example,

$$\log(a \times b) = \log(a) + \log(b).$$

Thus taking the logarithm of $p(\mathbf{r}|s)$ defined above yields

$$L(s) = k_{logn} - \sum_i ((f_i(s) - r_i)^2/(2v_i)),$$

where $L(s) = \log p(\mathbf{r}|s)$, and is known as the **log likelihood function**. The symbol $\sum$ (the Greek capital letter sigma) indicates summation of terms indexed by the subscript i. Notice that we have "wrapped up" all the irrelevant constants into a single constant $k_{logn} = -60 \log(k_n)$. We do not care about the value of k_{logn}, as will become apparent. If we ignore the constants then this equation states that, "*the log probability that each neuron has an output value r_i is equal to minus the squared distance between r_i and $f_i(s)$ (the height of the neuron's tuning curve at s), where this distance is "discounted" by the variance v_i in each neuron's output*". This is quite a mouthful, which is why we like to use mathematics, and not English, when doing mathematics. But we confess, it can be helpful to have things "spelled out" like this.

Irrespective of the value of the constant k_{logn}, if we remove the minus sign from above then the value of s that maximizes $L(s)$ also *minimizes*

$$\sum_i (f_i(s) - r_i)^2/v_i.$$

As this is a sum of squared terms, the value of s that minimizes it is called the **least-squares estimate** of s^*.

This has a more intuitive appeal than the maximum likelihood formulation, even though the two are exactly equivalent if, as here, the noise is assumed to be gaussian and with the same variance for each neuron. This least-squares equation states that the best estimate $\hat{s}$ of s^* is given by the value of s which minimizes the (squared) difference between the observed output r_i and the mean output $f_i(s)$ of each neuron when presented with a stimulus with direction s.

Finding the Maximum Likelihood Estimate $\hat{s}$

From a purely practical perspective, there are two ways to find the maximum likelihood estimation $\hat{s}$

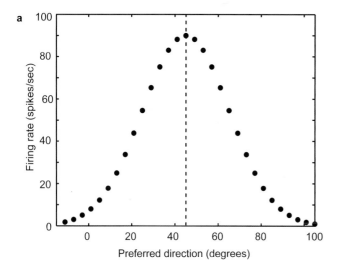

a

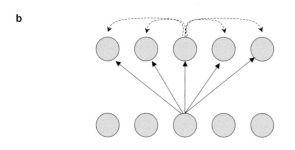

b

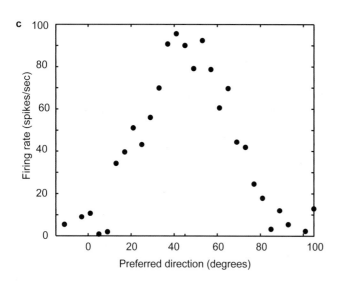

c

of s^*, **exhaustive search** and **gradient ascent**. Usually both of these provide exactly the same result.

Exhaustive search consists of discretizing values of s by dividing s into, say, 60 directions from s_1 to s_{60} such that $s_1 = 0°$, $s_2 = 6°$, ..., $s_{60} = 360°$, as above.

The maximum likelihood estimate $\hat{s}$ of the true value s^* of s is then given by that value of s which yields the largest value of $p(\mathbf{r}|s)$. This requires a repeat of the process described above for each value of s, until we have a value of $p(\mathbf{r}|s)$ for all 60 values of s, $p(\mathbf{r}|s_1)$, $p(\mathbf{r}|s_2)$, ..., $p(\mathbf{r}|s_{60})$. Given this data, we can plot a graph of s against $p(\mathbf{r}|s)$ and then read off the value of s that provides the largest value of $p(\mathbf{r}|s)$; such a graph would be as shown in the lower panel of **15.13**. This value is the maximum likelihood estimation $\hat{s}$ of s.

This has effectively been done in **15.13**, where the likelihood (goodness of fit) is evaluated for many values of s. The maximum likelihood estimate s^* (the best we can do in finding the true value, s) corresponds to the peak of the plotted function in the lower panel of **15.13**.

Gradient ascent consists of iteratively updating our estimate $\hat{s}$ of s^* until $p(\mathbf{r}|s)$ cannot be increased any further. For example, we might begin with $s = 20°$. After evaluating the likelihood $p(\mathbf{r}|20)$, we then test to see if $p(\mathbf{r}|21)$ is greater than or less than $p(\mathbf{r}|20)$. We keep on changing the value of s in small steps until the likelihood $p(\mathbf{r}|s)$ cannot be increased any further. The value of s which maximizes $p(\mathbf{r}|s)$ is the maximum likelihood estimate $\hat{s}$. In terms of **15.13**, this corresponds to iteratively taking small steps along the horizontal axis until the peak value of the plotted function is found.

So far, we have defined the computational

15.14 Model of neuronal implementation of maximum likelihood estimate (MLE) of the direction of a stimulus
a Response profile of model neurons in upper layer after processing by lateral connections. The peak of activity corresponds to the MLE of the stimulus direction.
b Recurrent model network. Every model neuron in lower layer is connected via an excitatory connection to every model neuron in upper layer, which have lateral inhibitory connections to each other. Noisy inputs from model neurons in lower layer get "cleaned up" by upper layer. Only some connections are shown for clarity.
c Response profile of model neurons in lower layer is noisy.

problem, and we have considered two algorithms for finding the maximum likelihood estimate of motion direction. Next we will consider how maximum likelihood estimation might be implemented in the brain.

Neuronal Maximum Likelihood Estimation Using Lateral Connections

Surprisingly, the maximum likelihood estimation method can be implemented in a network consisting of two layers of model neurons, as shown in **15.14**. The upper layer consists of neurons with different direction preferences; these could be the 60 MT cells described above, for example. The lower layer could also consist of 60 cells (only nine are shown in **15.14**). Each neuron in the lower layer projects to every neuron in the upper layer via *feedforward connections* (connections from only one lower neuron are shown). The connections between lower and upper layer neurons determine how much each lower neuron affects each upper neuron. If this were all there was to the network then each neuron in the upper layer would receive a *weighted average* of the outputs in all the neurons in the lower layer.

Now comes the clever bit. Each neuron in the upper layer receives a weighted average of all unit outputs (including itself) in this layer via *lateral connections*. These lateral connections are arranged to give large inputs from nearby neurons, and small inputs from distant neurons. In this way, each upper neuron takes a weighted average of the outputs of all upper neurons.

Once the upper neurons have received their inputs from the lower neurons, the lateral connections ensure that each upper neuron receives a weighted average of the upper neuron outputs, each of which is weighted average of lower neuron outputs. After this, neurons in the upper layer continue to affect each other via the lateral connections, but the neuronal activations quickly "settle" into a stable profile.

To cut a long story short, all this averaging between and within layers has the effect of smoothing out the initial noisy activation profile of the upper neurons. The resultant activity profile in the upper layer is almost identical to the curve in **15.9b**, whose peak corresponds to the maximum likeli-

hood estimate of s^*. This simple mechanism thus provides a plausible means by which neurons could compute the maximum likelihood estimate of motion direction. More recent work along similar lines has shown how full Bayesian estimates can be implemented in the brain.

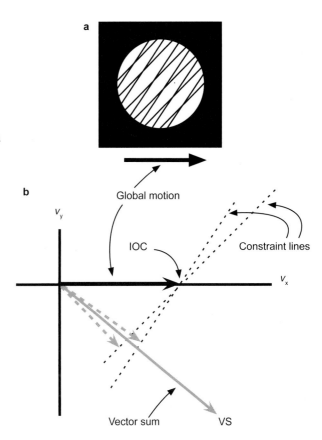

15.15 Global motion estimated from the intersection of constraints and vector summation for a plaid pattern
a Plaid consisting of two sets of lines. The global motion defined by this plaid is to the right, as indicated by the arrow, and the solid arrow in **b**.
b Motion estimated using vector summation (VS) and intersection of constraints (IOC). Each of the two line orientations in **a** yields a local motion vector (dashed arrow) in **b**. The sum (solid gray arrow) of these two vectors points about mid-way between them, and is the VS estimate of global motion. Each local motion vector is, by definition, perpendicular to its constraint line (dashed). The point where these constraint lines meet defines the IOC estimate of global motion (solid black arrow). The IOC always provides the correct estimate. With permission from Weiss, Simoncelli and Adelson (2002).

369

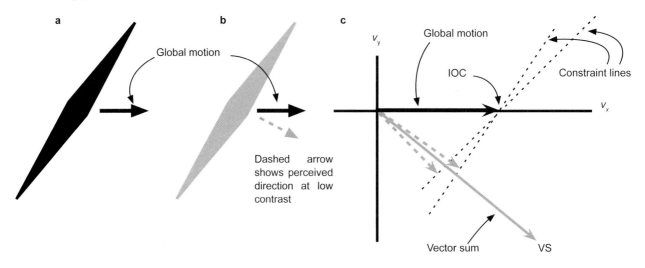

15.16 Effect of contrast on perceived global motion
a At high contrast a horizontally moving parallelogram is perceived as moving horizontally.
b At low contrast, the same parallelogram still moving horizontally is perceived as moving diagonally (dashed arrow).
c Global motion as predicted by intersection of constraints (IOC) and vector sum (VS). The perceived motion of the high contrast stimulus is predicted by intersection of constraints, whereas the perceived motion (dashed arrow in **b**) of the low contrast stimulus is predicted by the vector sum of the local motion vectors (dashed arrows).
Redrawn with permission from Weiss, Simoncelli and Adelson (2002).

Reconciling Intersection of Constraints and Vector Summation

Having described motion in terms of Bayes' rule, we can now examine a Bayesian model of how the aperture problem might be solved. As mentioned above, some studies find that human perception is consistent with intersection of constraints and other studies report that it is consistent with vector summation. In an elegant analysis, Weiss, Simoncelli and Adelson showed in 2002 how these two sets of apparently contradictory findings can be reconciled using a Bayesian analysis.

The perceived direction of a rhombus depends on two factors: its contrast, and how "fat" (square) it is. Note that we can equate the sides of a rhombus with the gratings of a moving plaid pattern, and that each rhombus generates two constraint lines (as two sides share one orientation, there are only two side orientations and therefore two constraint lines).

This can be seen by comparing the plaid of **15.15** with the rhombus of **15.16**. A high contrast "thin" rhombus moving horizontally appears to move horizontally; a result consistent with the intersection of constraint lines **15.16c**. However,

if the same stimulus is presented at low contrast then it appears to move diagonally, a result consistent with the vector sum of local motion vectors associated with the rhombus sides, **15.16c**. (These stimuli can be seen on the Web site: http:www.cs.huji.ac.il/~yweiss/Rhombus).

An ideal observer was derived using a Bayesian analysis. Recall that a Bayesian analysis includes a prior probability, a likelihood, and a posterior probability (the thing we want). For this ideal observer, the prior probability was defined such that speeds close to zero were considered highly probable, and all motion directions were considered equally probable. This can be seen as the identical prior probability *distributions* shown on the left of **15.17a** and **15.17b** (the use of such distributions is the norm in Bayesian analysis). This prior distribution specifies the probability that any given motion direction and speed will be observed.

In this context, the likelihood is the probability of the observed "noisy" neuronal response to motion given values for the line's speed and direction. This likelihood defines a single constraint line for each local motion vector. Note that the likelihood is a probability distribution. Thus, for any point (v_x, v_y) in the middle and right-hand graphs

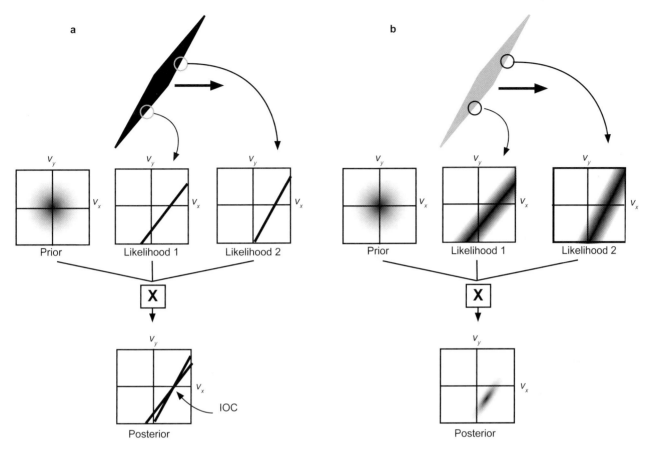

15.17 Bayesian motion estimation

a *High contrast rhombus* The motion of a high contrast rhombus defines two constraint lines. One of these lines corresponds to the likelihood depicted in the panel Likelihood 1. Each point in this panel defines a global motion vector **v**, and the gray-level r of each point corresponds to a likelihood $p(r|\mathbf{v})$, the probability of observing r given **v**. If each point corresponds to a neuron with a preferred motion defined by the location of that point, then the panel gives an impression of the levels of activity in an array of neurons with different preferred directions and speeds. Thus, the dark line in the panel for Likelihood 1 indicates a subset of neurons (points), each of which is sensitive to motion defined by the location of that point. A similar interpretation applies to the panel Likelihood 2. The prior probability of any given motion vector is indicated by the gray-level of each point in panel Prior; notice that all directions are deemed equally probable with this prior. If corresponding points in the three panels (Likelihood 1, Likelihood 2, and Prior) are multiplied (symbolized by the box with X in it) then the lower panel Posterior is obtained. This Posterior panel shows the posterior distribution, and the gray-level of each point gives a posterior probability $p(\mathbf{v}|r)$, whose maximum value corresponds to the intersection of constraints (IOC).

b *Low contrast rhombus* The motion of a low contrast rhombus defines two broad constraint lines. The gray-level r of each point in the panel Likelihood 1 corresponds to a likelihood $p(r|\mathbf{v})$. The decreased contrast of the rhombus introduces uncertainty, resulting in the broad lines of panels Likelihood 1 and Likelihood 2. The Posterior panel shows the posterior distribution $p(\mathbf{v}|r)$, with a maximum value that corresponds approximately to the direction of the vector sum. Thus, like us, the Bayesian ideal observer sees a global motion consistent with the intersection of constraints for high contrast stimuli, but a global motion direction consistent with the direction of the vector sum for low contrast stimuli. Redrawn from Weiss, Simoncelli and Adelson (2002).

of **15.17a** or **15.17b**, the gray-level defines the probability of the observed "noisy" response given the global motion is $\mathbf{v} = (v_x, v_y)$ (where darker gray implies higher probability).

For high contrast data, the likelihood is identical to the well-defined constraint lines we are used to drawing, **15.17a**. However, for low contrast,

and therefore noisy data, the likelihood for each border defines a broad, fuzzy constraint line, as shown in **15.17b**.

Crucially, the intersection of constraint lines for the high contrast stimulus corresponds to the global motion of the rhombus (i.e., the correct answer), **15.17a** (bottom). However, the intersection of the fuzzy constraint lines for the low contrast stimulus defines an egg-shaped region roughly in the direction of the vector sum of the two local motion vectors (i.e., the wrong answer), **15.17b**.

In summary, Weiss et al. showed that human motion detection seems to act like an ideal Bayesian observer, and that the apparently contradictory results obtained in different studies are entirely predictable, indeed, *desirable* from a purely Bayesian perspective.

Summary Remarks

We have considered how a population of neurons, each of which is tuned to a different motion, can be used to estimate the motion that appears on the retina. This has involved a review of tuning curves, and a detailed account of maximum likelihood estimation, which included a review of histograms as a species of pdfs. Finally, the thorny question of how all this elegant Bayesian mathematics might be implemented in the brain was addressed, and it turns out that a good approximation to MLE can be obtained using recurrent networks of model neurons.

In conclusion, motion is a good example of the computational approach and the usefulness of Marr's three levels. We began with the formulation of solution to a well-defined problem. The next step was to consider possible instantiations of the proposed solution at the algorithm level (e.g., tuning curves and firing rates). Finally, we considered how the algorithm might be realized in recurrent neuronal networks—the hardware level.

Further Reading

Born RT and Bradley DC (2005) Structure and Function of Visual Area MT. *Annual Review of Neuroscience* **28** 157–189. *Comment* A comprehensive overview with the stated goal of finding out what MT does.

Deneve S, Latham PE and Pouget A (1999) Reading population codes: a neural implementation of ideal observers. *Nature Neuroscience* **2** (8) 740–745. *Comment* A technical paper which provides an account of how a recurrent network can implement maximum likelihood estimation.

Pouget A, Zhang K, Deneve S, and Latham PE (1998) Statistically efficient estimation using population coding. *Neural Computation* **10** 373–401. *Comment* Formal account of how recurrent nets can implement maximum likelihood estimation.

Weiss Y, Simoncelli EP, and Adelson EH (2002) Motion illusions as optimal percepts. *Nature Neuroscience* **5**(6) 598–604. *Comment* A somewhat brief account of a technical method.

16

Seeing Black, Gray, and White

16.1 Problem of lightness perception
All the letters have the same luminance. However, their lightnesses (the technical term for perceived luminances) depend upon the luminances of the background on which they lie. The reasons why the visual system should work in this way are discussed in this chapter.

Imagine you are looking at a writing desk illuminated by a table lamp and on the desk lies a black-edged blotter holding a white sheet of paper, as in **16.2**. If you came across such a scene in real life, you would have no trouble seeing that both edges of the blotter were black and that the paper was white. But this apparently simple matter is in fact deeply mysterious. The reason is that the black edge of the blotter lying directly under the lamp sends more light to the eye than the far edge of the paper—which yet appears white. This is odd, to say the least. How can a surface appear black when it is sending more light to the eye than one which appears white? If the blackness/whiteness of a surface was seen simply according to the amount of light entering the eye from that surface, then the physically black surface under the lamp should appear whiter than the white surface distant from the lamp!

The scene shown in **16.2** is a perfectly usual one and no tricks are being played. Indeed, all the clever tricks happen inside the viewer's visual system. Somehow this system has taken notice of

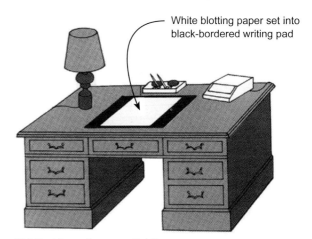

White blotting paper set into black-bordered writing pad

16.2 Problem of uneven lighting
The black border under the lamp reflects more light than the white blotting paper on the far side away from the lamp—and yet the visual system takes account of the different intensity of the light falling on each and gets the right answer. How it does this is the problem of lightness perception.

Why read this chapter?

Surfaces differ in the amount of light they reflect. In this chapter we consider how this important attribute, called *surface reflectance*, can be recovered from images. The term *lightness* is used for the correlate of surface reflectance in human vision. The central problem is that the amount of light arriving at the eye from a surface (the *luminance*) is determined partly by surface reflectance and partly by the amount of light striking the surface (the *illumination*). Thus, if the eye is looking at a point on a surface then the luminance of the corresponding point in the retinal image is determined partially by the ambient illumination and partially by the reflectance of the surface. How, then, does the human brain perceive black, grays and white? The Land-Horn theory is described and reviewed as a model of human lightness perception. It explains some, but not all, of the many lightness illusions. These illusions provide clues about the way human vision works. They demonstrate that lightness computations are intimately linked to processes that decide which image regions belong to which scene surfaces, and also to the 3D structure of surfaces. The chapter ends with a review of models of a famous lightness illusion, Mach bands. The overall conclusion of the chapter may come as a surprise: how we see black, grays and whites is still far from being explained.

the fact that the black edge under the lamp is more highly illuminated than the white paper distant from the lamp, and that once the factor of illumination is taken into account, it is "proper" to see the surfaces as respectively black and white because this is how they really are.

The problem which the visual system faces in this kind of situation is depicted in **16.3a**, which shows in schematic form the **luminance profile** across the blotter. **Luminance** refers to the amount of light coming from a surface. One factor that contributes to luminance is the amount of light falling on the viewed surface—its **illumination**. Another is the proportion of this incident light which is reflected from the surface, its **reflectance**, **16.3b**. The relationship between these three is specified in the following equation:

Luminance (L) = Illumination (E) x Reflectance (S)

The perceptual correlate of reflectance is termed **lightness**, and this varies from black (low reflectance; blacks reflect less than about 3% of incident light), through gray (medium reflectance), to white (high reflectance; whites reflect more than about 90% of incident light).

The visual system is remarkably good at computing lightness irrespective of illumination, and it is this fact to which **16.1** and **16.2** draw attention. Somehow in **16.1** we see the brightly illuminated black letters as black even though the luminance of their surfaces is greater than that of the dimly illuminated white letters. For **16.2** the visual system has to work from an intrinsically ambiguous luminance profile (**16.3a**), and yet from this it manages to extract the two distinct stimulus attributes (**16.3b**) of surface reflectance (perceived as lightness) and illumination. The latter is perceived as **brightness**: we can discriminate between a brightly lit white and a dimly lit white, or a brightly lit black and a dimly lit black. This capacity is illustrated dramatically in **16.1**.

The **transmittance** of whatever lies between a surface and the eyes can also affect the intensity of light entering the eyes, **16.4**. Air absorbs insignificant amounts of light energy in near viewing which is why its transmittance doesn't need to be taken into account in **16.2**. However, air does absorb some wavelengths more than others over great distances, which yields the depth cue of **aerial perspective**.

So far we have considered only surfaces that reflect all wavelengths of light equally

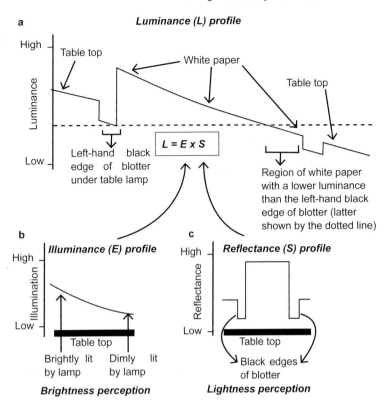

16.3 Luminance is illumination multiplied by reflectance
The luminance (L) profile in **a** is determined by the proportion of the illumination (E) reflected from scene surfaces. This proportion is known as surface reflectance (S) and is illustrated in c. Hence, $L = E \times S$.

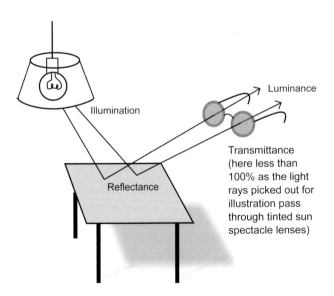

16.4 Factors determining achromatic surface luminance
These factors are the intensity of the illumination, surface reflectance, and the transmittance of any transparent surface between the surface and the eye.

well, so that we see them black, gray or white. These are called **achromatic** surfaces. But of course, most surfaces in our visual world absorb some wavelengths more than others, with the result that we see them as colored. We will consider color **perception** later in this chapter after having examined in detail the perception of black, grays, and white.

Lightness Constancy

The gray level of each pixel in the image of a surface depends on the three factors just described: the reflectance of the surface, its illumination, and the transmittance of the medium between the surface and the image. We will ignore transmittance and concentrate on how the visual system manages the remarkable feat of disentangling the effects of illumination from those of reflectance to arrive at a representation of a surface's reflectance.

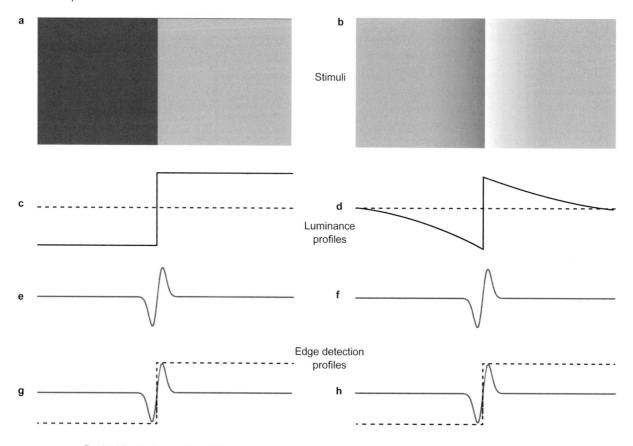

Dashed lines show extrapolations from the central edge (edge detection profiles in **e** and **f** also shown)

16.5 Craik-Cornsweet-O'Brien illusion
a and **b** are two stimuli whose luminance profiles are shown in **c** and **d**. The lightness of the two gray areas toward the left and right edges of **b** appear different, but they are in fact the same, as can be revealed by masking the central boundary with a pencil or a finger. The other sections of this figure illustrate the principle of how the Land-Horn computation accounts for this illusion, see text.

At first sight it would appear impossible to decide whether a given luminance value derives from a surface with high reflectance that is dimly illuminated, or a surface with low reflectance that is highly illuminated. Indeed, there exists an infinite number of pairings of reflectance and illumination that could produce any given luminance. Despite this ambiguity, **_lightness_** (defined as the perceptual correlate of surface reflectance) is generally quite accurate almost irrespective of the illumination. This is called **_lightness constancy_** and is famously due to our ability to "discount the illuminant," a phrase introduced by the great German scientist Hermann von Helmholtz (1821–1894). Clearly, the visual system must rely on assumptions to escape the apparent intractability of the problem.

Computation of Lightness

So how can the visual system extract the reflectance of the various surfaces from the image luminance profile of a scene, given that the luminance profile is contaminated with the unwanted contribution made by the varying illumination over the scene? We begin by describing an early computational theory by way of introducing some key ideas. This theory was suggested by Edwin Land following his psychological studies on the perception of lightness. Berthold Horn implemented it in a successful and elegant computer program which, as David Marr pointed out, has some features which seemed to make it consistent with known mechanisms in the retina. We will describe how this computation works because it illustrates a fascinating interplay of psychology, neurophysiology, neuroanatomy, and computational theorizing. The theory can explain some illusions of human lightness perception. However, it fails to explain various other lightness illusions and so it has been discarded as a general theory. Nevertheless, the theory and its shortcomings are instructive, and so we begin with it.

The Land-Horn theory exploits the constraint that variations in illumination are relatively gradual, whereas variations in reflectance are rather sudden. This can be seen in **16.2b**, where the reflectance transitions from black to white and vice versa are steep, whereas the fall-off in illumination from left to right is relatively gradual. Given this consideration, one way to "filter out" the component due to varying illumination is as follows.

First, *detect edges*. A great deal of attention was given in Ch 5 to the problem of detecting edges. It turned out to be much more complicated than one might have guessed, and required a good deal of intricate seeing machinery. Nevertheless, a theory of edge detection was finally arrived at which delivered edge descriptions with various parameters, including edge scale ranging form shallow/coarse to steep/fine. This parameter of scale is needed for the next stage of the Land-Horn computation.

Second, *ignore all shallow (low contrast) edges*. This means discarding all information about luminance differences between adjacent points in the input image if these differences fail to exceed a certain threshold value. In this way, gradual luminance transitions are eliminated while the sudden ones are preserved.

Third, *build up the required lightness profile by reconstituting between edges*. A simple way to think about this is "joining up" areas between above-threshold edges, giving these areas lightness values determined by the size of the luminance differences forming the edges.

Craik-Cornsweet-O'Brien Illusion

A lightness illusion that the Land-Horn theory can explain, at least in some of its forms, is illustrated in the two stimuli in **16.5a,b**. They look similar: a light gray area adjoining a dark gray area in each case. But they are in fact very different, as their associated luminance profiles reveal, **16.5c,d**. Thus,

the left-hand stimulus is composed of adjoining light and dark gray areas, but the equivalent areas in the right-hand stimulus are everywhere the same shade of gray except for the presence of an "edge" at the boundary between them, **16.5d**. Check this surprising fact for yourself by covering the boundary in **16.5b** with a pencil, which will enable you to see the two areas in their true lightnesses.

This illusion, called the ***Craik-Comsweet-O'Brien illusion*** after its joint but independent discoverers, is one of the most impressive of all visual illusions. It can be interpreted as a particularly revealing instance of how the visual system can be fooled by using a computational stratagem which is usually successful but which is occasionally inappropriate for the prevailing input. The argument here is that, like many other illusions, it offers a valuable clue about how the visual system works in normal cases. And the illusion suggests that the lightness computation carried out by the visual system may indeed be along the lines proposed by Land-Horn, because it can be explained as follows.

In both **16.4a** and **b**, the visual system first detects the edge separating the two gray areas, discarding all other luminance information. This produces the edge-detection profiles shown in **16.5e** and **f**, which can be seen to be very similar. Next, **16.5g,h** show "reconstitution" of the areas on either side of the edge to produce the perceived lightness profiles illustrated in **16.4c,d**. Thus, this theory holds that the reasons **16.4a,b** look alike is that the visual system applies to both a lightness computation designed to eliminate problems due to varying illumination, and this lightness computation is misled into "thinking" that the edge profile in **16.5d** comes from the boundary between two regions of different reflectance. It therefore produces a lightness profile accordingly, but an illusory one in this instance.

If this is the correct account of what is going on then Craik-Cornsweet-O'Brien illusion shows a clever way of tricking the visual system by turning one of its own tricks against it.

Stimuli similar to the Craik-Cornsweet-O'Brien figure are sometimes found in ordinary real-life objects, as Floyd Ratliff pointed out over 50 years ago. Consider, for example, the piece of Ting white porcelain shown in **16.6**. Its glaze is a uniform white, but the incised lotus design appears lighter

16.6 Chinese Ting Yao saucer
Courtesy Percival David Foundation of Chinese Art.

than its background because the incisions have edges shaped rather like those of the Craik-Corn-sweet-O'Brien illusion. They thus produce shadows of the required kind to generate the same illusory lightness effect.

Of course, it must be remembered that we are talking solely about the computation of lightness. Hence, when mention is made of "discarding all illuminance information," this is for the purposes of the lightness computation alone. Illumination too must be noted for the purposes of its percep-tual correlate, brightness perception, and for that, luminance gradients other than those marking edges presumably play a crucial part. Also, gradual transitions in luminance might be independently detected and used for the perception of 3D shapes. Refer back to Ch 1 for examples of how shadows and shading, instances of illumination differences, can be powerful cues to depth.

Finally, the assumption that illumination varies gradually is not always valid, as sometimes shadows have quite sharp edges.

Implementing the Land-Horn Theory

Consider the luminance profile made up of a gradual illumination change superimposed upon a sudden reflectance change shown in **16.7a**. The objective of lightness computation is to extract the reflectance profile from this ambiguous input. A gray level description is shown in **16.7b**, in the usual form of levels of activity in the receptor mo-saic (albeit a smaller mosaic, for reasons of space). Its numbers show that the illumination compo-nent appears in the form of a set of small steps, of 60/57, 57/54, and 24/21. On the other hand, the reflectance step is much greater: 54/24.

The next thing to note is that the gray level description is convolved with on-center units. A central strip of convolution cells from a convolu-tion array is shown in **16.7c**. For this convolution, negatives were allowed, and so some negatives ap-pear in the convolution cells. Try working out for yourself why the numbers are as they are, referring back to Ch 5 for guidance.

In **16.7d**, convolutions with "biological" cent-er-surround units are shown, their distinguishing feature being that no negatives are allowed. Thus, any cell that would have been negative is set to zero. Convolutions with on-center and off-center units of this type are both illustrated, again with the results shown from just one strip across a full convolution array. As can be seen, only positive numbers appear in each strip, but where they ap-pear is different in each case. Thus, the negative numbers of **16.7c** appear as positive numbers in **16.7d** for the off-center units. And the positive numbers of **16.7c** appear as positive numbers in the on-center strip of **16.7**d. This might sound complicated, but the complexity is more apparent than real, as studying **16.7d** will show you.

Having obtained our on-center and off-center convolutions, the next step is to apply a threshold to get rid of the small changes in image intensity produced by the illumination gradient. In **16.7e**, the threshold is set to +2. In practice, the threshold would need to be dependent on the exact viewing circumstances. For example, if the general level of illumination is high then variations in illumination are also likely to be high, and if the threshold was left low these variations would then masquerade as genuine reflectance changes. But how the threshold might be set and reset to suit the prevailing illu-mination conditions is a detail we leave aside here. The result of applying the +2 threshold is shown in **16.7f**: it leaves just the large edge-measurements of +10 as the only ones appearing in each convolu-tion array.

Once the threshold has been applied, the final step is to build up the required lightness profile by extending the activity outward from the above threshold edges. This operation could be done in two new sets of arrays, termed respectively the whiteness and blackness arrays in **16.7g,h**.

The whiteness array shows the results of extend-ing out from the edge recorded by the on-center

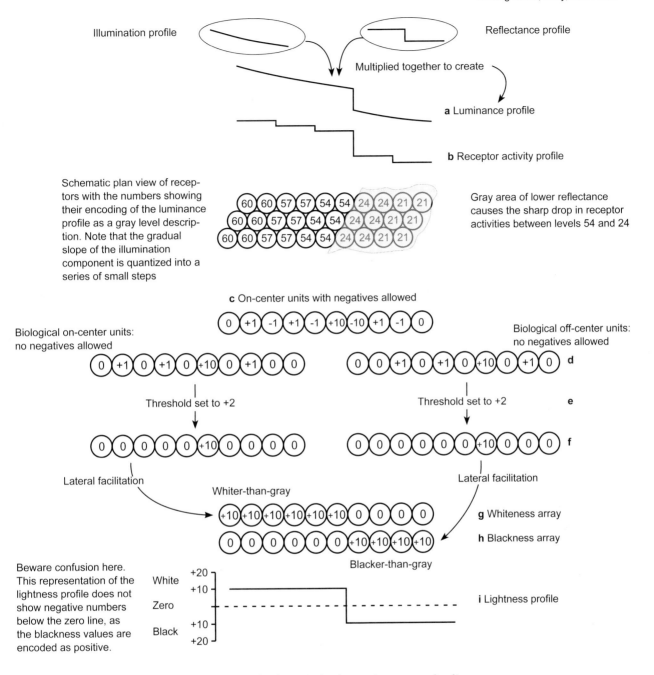

Illumination profile

Reflectance profile

Multiplied together to create

a Luminance profile

b Receptor activity profile

Schematic plan view of receptors with the numbers showing their encoding of the luminance profile as a gray level description. Note that the gradual slope of the illumination component is quantized into a series of small steps

Gray area of lower reflectance causes the sharp drop in receptor activities between levels 54 and 24

c On-center units with negatives allowed

0 +1 -1 +1 -1 +10 -10 +1 -1 0

Biological on-center units: no negatives allowed

Biological off-center units: no negatives allowed

0 +1 0 +1 0 +10 0 +1 0 0

0 0 +1 0 +1 0 +10 0 +1 0 **d**

Threshold set to +2

Threshold set to +2 **e**

0 0 0 0 0 +10 0 0 0 0

0 0 0 0 0 0 +10 0 0 0 **f**

Lateral facilitation

Lateral facilitation

Whiter-than-gray

+10 +10 +10 +10 +10 +10 0 0 0 0

g Whiteness array

0 0 0 0 0 0 +10 +10 +10 +10

h Blackness array

Blacker-than-gray

Beware confusion here. This representation of the lightness profile does not show negative numbers below the zero line, as the blackness values are encoded as positive.

White +20 / +10
Zero
Black +10 / +20

i Lightness profile

16.7 Land-Horn lightness computation implemented using center-surround units

units, and the blackness array does likewise for the off-center units. Somehow activity in the whiteness array is not allowed to spread in the wrong direction, across the white/black border, and vice versa for the blackness array. This essential restriction could be achieved by having each white/black pair of cells dealing with the same part of the gray level description mutually inhibitory, so that whichever

cell is more active wins out and inhibits the other one to a zero level. Thus, any facilitation passed across the edge within either array would never exceed the value of the inhibiting opponent cell, and so it would be "vetoed." Without going into details, we can see that, in principle, whiteness and blackness arrays of the kind displayed in **16.7g,h** can between them represent the lightness profile

a The starting state

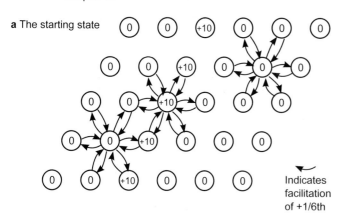

Indicates
facilitation
of +1/6th

b The finishing state

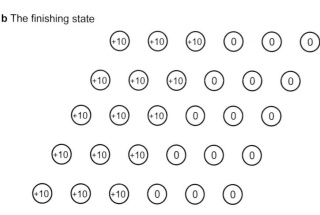

16.8 Deconvolution by lateral facilitation
a Starting state: connections for just a few cells are shown but all cells would be connected up identically.
b Finishing state after many iterations—see text. All connections have been removed here for clarity.

shown in **16.7f**. Note the clever way in which the use of whiteness and blackness cells means that activity is always positive: it is just that positive activity means something different in the two cases.

The remaining task is to explain how to achieve the spread of activity necessary for reconstitution. This process is an example of what is technically called *deconvolution* because it is the inverse of convolution. We won't delve deeply into this, but note that reconstitution could be created within the whiteness and blackness arrays using a process of *lateral facilitation*, **16.8a**, rather than the lateral *inhibition* observed in bipolar cells, Ch 5.

Notice that in **16.8a** the network of connections required for doing this is within the array itself, unlike the center-surround convolution case **16.7**, where connections went from the receptors to the convolution cells. Here, to mediate the process of deconvolution, the fibers linking up cells run between adjacent

cells within each "deconvolution array." Note that in **16.8a**, each deconvolution cell is shown with a number in it—the post-threshold edge information fed into the array by the preceding stage. The array shown in **16.8a** could be either a blackness or a whiteness array: it does not matter for the purposes of exposition, as they both work in a similar fashion and both deal solely with positive numbers.

Each cell in **16.8a** both influences its neighbors and is influenced by them. Thus, "double" connections are shown between cells. Not all connections are given in **16.8a** for reasons of simplicity, but you should imagine all adjacent cells are connected up as indicated for the three chosen examples. The +1/6th fractions shown for each connection indicate the weighting of the lateral facilitation mediated by the connections. That is, each cell excites its neighbors by 1/6th of its own activity level. Moreover, each receives excitation from its neighbors 1/6th of their level. So each cell is "helping out" its neighbors and in turn being "helped out" by them: hence the suitability of the term "lateral facilitation."

An important point to grasp is that the process of mutual facilitation is iterative, so it keeps going repeatedly until a steady state is achieved by the network. For example, consider the cell in the lower left-hand corner of **16.8a**, which is shown with its connections to neighboring cells. It starts off from a level of zero because this is what was fed into it by the preceding thresholding stage. It thus begins by offering no excitation to its neighbors because, of course, $0 \times 1/6\text{th} = 0$. On the other hand, it receives excitation from two neighboring cells, which started from a level of +10 because they are "on" an edge. As $+10 \times 1/6\text{th} = 1.67$, the total facilitation received by this unit is twice this value, i.e., 3.34.

Having now taken on an activity level of 3.34, it can proceed to facilitate its neighbors more distant from the edge, cells which initially received no excitation at all because when the deconvolution started they were entirely surrounded by cells at zero levels of activity. Perhaps you can grasp intuitively that, with this kind of iterative updating of facilitation, a spread of activity sweeps through the network like a bushfire. Soon all the cells would assume the same value on the same side of the border—and the lightness profile would be built up as required, as shown in **16.8b**.

But you might reasonably ask at this point: what stops the facilitation rushing across the edge boundary? Why do not *all* cells end up with +10? One possible answer is that the opposing deconvolution array stops this happening by inhibiting such a spread. Remarks were made above to this effect when discussing **16.7i**, and one must suppose that if, for example, **16.8a** is a whiteness array, then the blackness array provides the vital inhibition, which stops the spread going too far.

The blackness array would also have to provide some inhibition to stop the whiteness array going "mad" and rising via the process of mutual facilitation up above +10 To "super-white" levels of activity, from starting points of a much lower kind. After all, the cells on the edge would initially receive no facilitation from their neighbors, but once these neighbors had been brought above zero they would facilitate the very cells by which they had themselves been activated; and, if these latter were not "held down" by inhibition from the blackness array, then the whole thing would get out of hand.

Thus, although the visual system solves the problem of not being allowed negative levels of activity by inventing a blackness channel to complement a whiteness channel, the price that has to be paid is that the two arrays need to "speak to one another" so that silly results do not ensue. This "conversation" probably takes the form of cross–couplings of some kind between the whiteness and blackness arrays.

Reconstituting between edges by deconvolution creates a point-by-point representation of lightness across a surface. This form of representation is structurally similar to the intensities found in the pixels of a photographic image. But there is a logical hazard to be avoided here: it is dangerous to regard perception as taking the retinal image and recreating images of it in the retina/brain pathway. Perception is not a species of photography, as Ch 1 emphasized. Hence, it reasonable to ask: what has been achieved by modeling deconvolution in the way shown in **16.7f,i** and **16.8**?

One answer is that the activities of units shown in **16.8b** provide an *explicit* representation of lightness. That is, each point in the representation stands for the reflectance of the corresponding point on the surface. It is not an image representation but a type of scene representation.

That said, it is also reasonable to ask: what other way might the brain use for representing surface lightness? An alternative might be that the surface

regions identified in the full primal sketch (Ch 7) could have lightness parameters attached to them. Those parameters would also serve the function of being an explicit representation of lightness. But this is just speculation and it has anyway nothing to say about the neural code for these parameters. The honest truth is: vision science presently has little or no idea what the brain code is for lightness perception.

Despite this limitation in our understanding, it has been worthwhile to describe this neurophysiological implementation of the Land-Horn theory. This is because it is a nice example of how a computational theory can, in principle, be realized in neurons. Not least, it shows that the Land-Horn theory is a biologically plausible candidate for modeling lightness computation processes in human vision.

Scaling Problem

Let's recapitulate what we have done so far. We have examined the task of finding out, from image luminance values, the reflectances of surfaces (i.e., the proportion of incident light that surfaces reflect). One constraint for solving this task is that variations in illumination usually change gradually across a scene, whereas reflectance changes tend to be sudden. The Land-Horn algorithm provides a way of exploiting that constraint. It is time now to examine in more detail exactly what is represented in the outputs of the Land-Horn computation shown in **16.7g,h**. The question we consider is: how might those outputs be used for establishing surface lightness? Recollect that surface lightness is the perceptual correlate of surface reflectance.

In **16.7g,h**, different regions of the stimulus are labeled with different numbers. These numbers are mapped on to a scale in the graph in **16.7i** that goes from black to white. In this example, the numbers for the higher luminance region of the stimulus (on the left side of the central edge) fall in the whiter part of the scale. The numbers for lower luminance regions of the stimulus (on the right-hand side of the central edge) fall in the darker part of the scale. But these numbers do not in themselves tell us how or light or dark these stimulus regions are perceived to be.

For example, consider two surfaces with different reflectance values, such that one of these has a

16.9 Applying different anchors within different region of illumination
The three disks are an identical shade of gray. Gilchrist (2006a,b) suggests that for scenes such as this they appear to have very different lightnesses because they are interpreted in relation to the different highest-luminance anchors in the different regions in which they lie.

reflectance that is ten times that of the other. This difference could generate lightness perceptions such that one surface is perceived as white and the other is perceived as light gray. But it would also be possible in principle to interpret this difference such that the lightness of one surface is perceived as light gray and the other is perceived as dark gray. It has been proposed that one way to resolve this dilemma is to find a single scene surface which acts as a reference or anchor when calculating the lightnesses of all others. Finding this surface is called the *scaling problem*.

Anchoring Rules for Solving the Scaling Problem

Hans Wallach, whose many contributions to the study of lightness perception began in the 1940s (see Wallach, 1976), suggested that the surface with the highest luminance is seen as white. Land also used this maximum luminance rule, as it is sometimes known. An alternative is the average luminance rule which uses the average luminance computed from the entire image to define middle-gray as the anchor. This is sometimes referred to as the gray world assumption. All these rules

are called *anchoring rules* because they propose that the scaling problem is solved by anchoring all lightness values to particular value of luminance.

Alan Gilchrist in 1994 devised a way of testing which anchoring rule might be used by human vision. He asked observers to place their heads inside a large hemisphere with an interior that was middle gray on the left and black on the right. It turned out that people perceived the middle gray as white and the black as middle gray. This is against expectations drawn from the average luminance rule, which predicts that the middle gray should appear light gray and the black should appear dark gray, with no white seen at all. Gilchrist concluded that our perceived gray scale is anchored from the maximum luminance, not from the average luminance.

Natural scenes are, of course, much more complex than Gilchrist's simple painted hemisphere. This led him to argue that the visual system uses a different anchor for each region of illumination within the scene, **16.9**. The problem then becomes: how can the visual system work out what regions of the image are to be aggregated for the purposes of using the appropriate anchor for them?

This is a non-trivial question. One way of tackling it is to use grouping rules of the kind introduced in Ch 7 when dealing with the problem of *Seeing Figure from Ground*. We will provide some examples in the next section, but before leaving the topic of anchoring we note that Paola Bressan has suggested a development of Gilchrist's anchoring model. He proposes that to account for certain lightness illusions it is necessary for the entities within any given grouped framework to be anchored to the maximum luminance *and* to the average luminance of their surround. We won't explain the details of his scheme (see Bressan, 2006). We mention it only to emphasise the complexities of lightness perception. As Gilchrist has remarked, how ironic it is that we use the phrase "black and white" to refer to things that are simple or crystal clear, when the processes of seeing blacks and whites are in fact shrouded in mystery.

To illustrate this further, a problem with the maximum luminance anchoring rule is that illumination sources often generate the highest image luminances and yet they are not seen as white surfaces but as what they are: sources of illumination.

Luminance Ratios and Reflectance Ratios

The Land-Horn theory relies on the constraint that variations in illumination usually change gradually across a scene, whereas reflectance changes tend to be sudden. Having identified edges and eliminated shallow ones, the final stage of the algorithm we described above used deconvolution to build up the required lightness representation by reconstituting between edges. There is, however, a different way, of proceeding that doesn't use deconvolution.

Consider **16.10**, which shows three patches of different reflectances illuminated by a lamp on the left so that the illumination gradually declines across the scene. A cross-section through an image of this scene is shown, thereby displaying a luminance profile across the image, with the luminances at the edges of the patches picked out for special attention. The remarkable fact is that the luminance ratios at the edges of the patches are almost the same as the reflectance ratios of the patches. The reason is that luminance variation due to illumination is so small as to be negligible in the small regions used when calculating the luminance ratios at edges. Hence, the luminances at edges can

be ascribed almost entirely to reflectance changes, as shown in the algebra in **16.10**. This permits a different algorithm for exploiting the constraint that illumination changes are usually gradual compared with reflectance changes, which are generally rather sudden.

The first step is as before: select the steep edges by imposing a threshold, as in **16.7**. The next step is to calculate the ratios of luminance either side of the selected edges, which gives us an estimate of the ratios of the reflectances of the patches in the scene. So far so good, but this still leaves us with the scaling problem.

For example, consider two surfaces with different reflectance values, such that one of these has a reflectance that is ten times that of the other. This reflectance ratio could generate lightness perceptions such that one surface is perceived as white and the other is perceived as light gray. But it would also be possible in principle to interpret this difference such that the lightness of one surface is perceived as light gray and the other is perceived as dark gray. Hence, there is still a need to find an anchor for mapping the reflectance ratios into lightnesses, that is, to perceived reflectances.

Ch 17 gives more details of Land's ***retinex*** theory which exploits the fact that luminance ratios are similar to reflectance ratios at edges.

Other Contrast Illusions

Can the Land-Horn theory explain illusions other than the Craik-Cornsweet-O'Brien illusion? If so, that would strengthen its claims as a model of human lightness processes. Hence, we turn next to investigating whether the theory can explain perhaps the simplest contrast illusion of all, one first shown in Ch 1 in **1.14b** and repeated here in **16.11a**. The small inset gray squares are of equal luminance, as shown in the luminance profile just beneath them. This is difficult to believe because the one on a white ground looks so much darker than the one on a black ground. Why should this illusion, this failure of "correct" lightness perception, come about?

In fact, this illusion, which is an example of a simultaneous contrast illusion, is a necessary consequence of the Land-Horn lightness computation. Its existence therefore adds credibility to the idea that our own visual system works in a similar

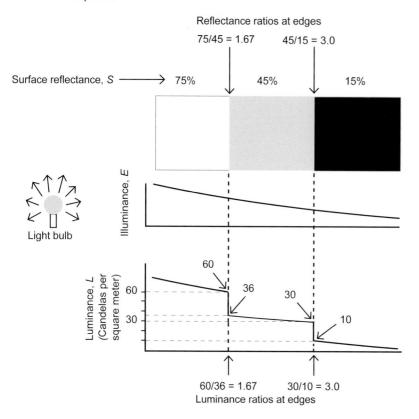

Reflectance ratios at edges

75/45 = 1.67   45/15 = 3.0

Surface reflectance, S ⟶   75%   45%   15%

$L = S \times E$. Hence:

$$\frac{L_1}{L_2} = \frac{S_1 \times E_1}{S_2 \times E_2}$$

As $E_1 \approx E_2$ at edges, the E values cancel and we get

$$\frac{L_1}{L_2} \approx \frac{S_1}{S_2}$$

Light bulb

60   36   30   10

60/36 = 1.67   30/10 = 3.0
Luminance ratios at edges

16.10 Luminance ratios are similar to reflectance ratios at edges
Luminance is the energy emitted or reflected by an extended source of light. It is measured in candelas per square meter of surface, cd/m². For example, a computer monitor screen will typically have a luminance of about 63 cd/m² and paper in moonlight 0.2 cd/m² (Mather, 2008). Luminance from a surface is determined by its *reflectance* (the proportion of light it reflects) and the amount of light falling on it (its *illuminance*). The middle graph shows an illuminance profile arising from a light source on the left so that the illuminance decreases from left to right. If the white, gray, and black patches shown at the top were illuminated with this light source then the luminance profile that would arise is shown in the bottom graph. Note that the luminance ratios at the edges between the patches are almost the same as the reflectance ratios at those edges. See text.

fashion. Once again, we have here an illusion that is a necessary penalty for adopting a strategy that usually delivers a reasonably truthful lightness perception. We will now explain in some detail why this must be so.

First, consider **16.11c**: it shows the result of subjecting **16.11a** to an edge-detecting convolution of the kind thought to be performed in the retina by bipolar cells. Thus, **16.11c** is a convolution image (Ch 5) and is to be interpreted in the following way: think of it as composed of a mass of dots, so small and so closely packed together that each individual dot cannot be distinguished as a separate entity. Each dot represents the output from a pair of bipolar cells, one an on-center unit and one an off-center unit. If both bipolar cells

are firing at zero level, then the dot representing them is set to a mid-gray level. If one cell is firing at a non-zero level, then the dot is either whiter-than-mid-gray, if the on-center unit is more active, or blacker-than-mid-gray, if the off-center unit is more active. The convolution image as a whole is therefore signaling where the edges are, and this can easily be seen both in the convolution image itself and in the profile of activity across the central slice of this image shown just beneath it, **16.11d**. This edge profile is, of course, reminiscent of the edge profiles shown in connection with the Craik-Comsweet-O'Brien illusion **16.4**.

Second, consider **16.11e**: this shows a profile from reconstituting from the edge convolution image of **16.6c**, using the deconvolution procedure

a The inset gray patches have identical luminances but appear of different lightnesses

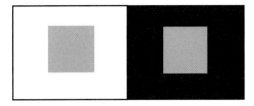

b Luminance profile of **a**

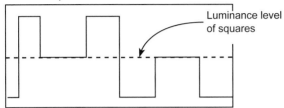

Luminance level of squares

c Edge detector convolution of **a**

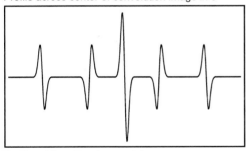

d Profile across center of convolution image in **c**

e Profile arising after deconvolving **c**

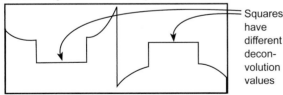

Squares have different deconvolution values

f Squares from **e** shown on a common background

16.11 Explaining a simple contrast illusion using the Land-Horn lightness computation

that produced the results set out in **16.5g,h**. Recollect that we are showing the results of deconvolution simply to illustrate a species of reconstitution, without promoting this as a theory of the way human vision actually does reconstitution for lightness perception.

The important thing to notice in the profile in **16.11e** is the way the deconvolution values arising from the squares get "pulled up" (right square), or "pulled down" (left square). The computer program which did this deconvolution was written by our colleague John Mayhew and its output is shown at a point in the computer's deliberations where full reconstitution was not yet obtained.

Figure **16.11f** shows squares reflecting the different convolution values appearing for the matching regions the profile **16.11e**. These squares have different intensities, thus simulating a contrast illusion of the type we see ourselves when we look at the input stimulus—**16.11a**. We have set the squares of **16.11f** on a shared background; otherwise our visual systems will treat **16.11f** as just another input figure for its own lightness computation and give us another contrast illusion which we might be misled into thinking was a "physical" event rather than a psychological one.

Thus, we see here the Land-Horn lightness computation producing a contrast illusion, and it does so as an inevitable consequence of its stratagem of using an edge-detecting phase followed by a reconstitution phase. This is an impressive parallel with what we experience. It supports the idea that the visual system itself operates in at least a broadly similar fashion in some circumstances.

So far, so good. But we have only scratched the surface of the extensive range of lightness illusions exhibited by human vision. How well does the Land-Horn theory work for other classic lightness illusions? Consider, for example, **16.12**. The two gray triangles both have two sides adjacent to black and one side adjacent to white, so they would be expected to receive the same amount of local contrast effects. They look very different however. It is not easy to see how the algorithm illustrated in **16.11** would produce different lightnesses for the inset triangles. The same general remarks can be made of **16.13**.

What seems to be going on in **16.12–13** is that the contrast effects are localized to the surfaces in which the various patches seem to belong. This generalisation is brought out clearly in Todorovic's lightness illusion, **16.14**. This conclusion indicates that if the Land-Horn is to work as a model of human lightness perception then the computation has to be run as a within-surface procedure. That in turn enforces a requirement that image regions have to be assigned to surfaces before the lightness computation is conducted. Whether that can be done using the kind of figure/ground grouping processes described in Ch 7, and thus without the involvement of high-level object recognition processes, is a current research question.

The idea of localizing contrast effects in surfaces was raised in a classic paper by Ken Nakayama, Shinsuke Shimojo, and Vilayanur Ramachandran published in 1990, that discussed the perception of transparency and its relation to depth, subjective contours, luminance, and neon color spreading.

3D Factors in Lightness Perception

Developing further the point just made, we now describe various demonstrations showing that factors to do 3D surface perception are critically involved with lightness computation.

For example, both **16.15a** and **b** depict the same vertical junction, with the same horizontal luminance profile as in the Craik-Comsweet-O'Brien illusion, **16.5**. In **16.15a**, the luminance profiles of the curved surfaces of both 3D cylinders look the same, as indeed they are. In **16.15b**, the front faces

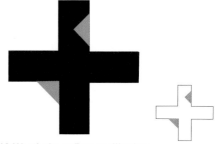

16.12 Wertheimer-Benary illusion
The gray triangles have the same luminance (see small insert) but appear very different.

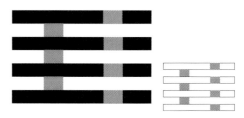

16.13 White's illusion
The gray rectangles have the same luminance (see small insert) but appear very different.

of both cubes appear to be different, but they are the same, and the same as in **16.15a**. Place a pen along the vertical junction between the objects in **a** and **b** to confirm this. What's going on here?

Well, the Craik-Cornsweet-O'Brien junction in **16.15a** is what would be expected at the interface of two adjacent cylinders with the same uniform reflectances. It is therefore "correct" that

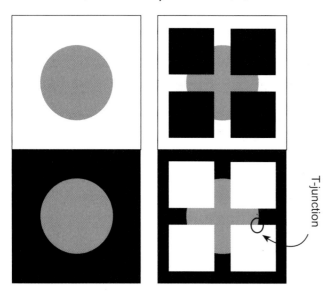

16.14 Contrast effects localized within surfaces
Left Standard simultaneous lightness contrast illusion: the discs have the same gray levels but the lower one is set on a dark surround that causes it to appear lighter. **Right** Todorovic's figure: the standard contrast illusion is preserved, it is perhaps even stronger. This is odd because the borders of the discs are created largely by the white and black squares. Hence, it would be expected that the lightness contrast effect should be reversed from that seen in on the left if lightness was determined by contrast effects arising from longest border contrast relationships. The effect suggests that contrast effects are localized within the surfaces on which the discs are seen to lie—and that is on the surfaces behind the partially masking squares. Todorovic suggests that T-junctions (example in circle in lower right figure) are important cues used by the visual system to decide which objects are masking others, and hence which image regions should be grouped together, including for lightness perception. Redrawn with permission from Todorovic (1997).

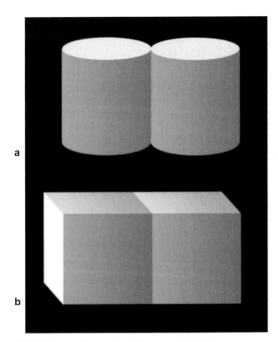

16.15 Three-dimensional factors in lightness perception
The luminance profile across the junctions between the curved (top) and planar (bottom) surfaces both have the shape used in the Craik-Cornsweet-O'Brien illusion, but the perceptions of lightness in each case are very different. Reproduced with permission from Knill and Kersten (1991). http://persci.mit.edu/people/adelson/publications/gazzan.dir/gazzan.htm

we perceive each cylinder to be uniformly gray. In contrast, the Craik-Comsweet-O'Brien junction in **16.15b** is what would be expected at the junction of two planar surfaces with different reflectances.

In other words, the same Craik-Comsweet-O'Brien junction is consistent with two physical scenarios: two curved cylindrical surfaces with the same reflectance, and two planar surfaces with different reflectances.

The most important lesson to be drawn here is that what we see does not depend only on the presence of a junction with the Craik-Comsweet-O'Brien luminance profile. It is also affected by the apparent curvature of the 3D surfaces which meet at this junction. In short, the lightness we perceive on each side of the Craik-Comsweet-O'Brien junction depends on the 3D context of that junction.

This means that any model which does not take account of 3D structure would place the same lightness interpretation on the surface of **16.15a**

and **b**. A similar argument can be applied to **16.16** where an identical image luminance change is associated with either a change in reflectance across a planar surface (lower circled edge) or with a change in surface orientation for a surface with constant reflectance (upper circled edge).

The depth cues in **16.15** and **16.16** derive from the outline shapes of the objects depicted. David Buckley, John Frisby, and Jonathan Freeman have shown that depth relationships signaled by stereoscopic cues can play a similar role. These are cues that arise from the slightly different images that the two eyes receive, because they are located at different positions in the head. We discuss them fully in Ch 18.

Consider next the small 3D tent-like figure on the right in **16.17b**. Imagine looking down on it from above with two eyes so that its 3D structure would be readily discerned using stereoscopic (two-eyed or binocular) depth cues. You would see the small gray flaps sticking up from the sides of the tent.

Now imagine closing one eye to eliminate the stereo effect. You would then see a flat picture

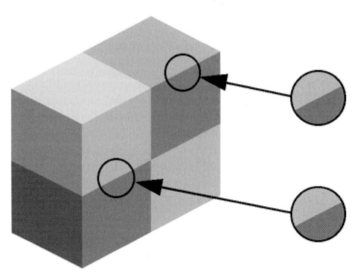

16.16 Lightness illusion created by Ted Adelson
The same change in image luminance (circled) can be induced either by a change in surface orientation of a surface with constant reflectance (top circle) or by a change in reflectance of a surface with constant (i.e., planar) orientation (bottom circle). A model that does not take account of three-dimensional shape or surface orientation cannot, in principle, correctly interpret both types of image luminance changes. With permission from Adelson (2000).

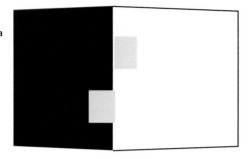
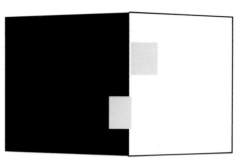
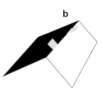

16.17 Is the contrast illusion determined by the depth plane in which the gray rectangles lie?
a Stereo pair: crossed-eyes fusion yields the 3D percept sketched in **b**. Devised by John Mayhew and John Frisby.

similar to one of those shown in **16.17a**. In each of these pictures, the gray flaps are seen as flat patches lying on either a black or a white background. In these images, the standard simple contrast illusion (**16.11**) is evident. That is, the gray patch on the black surround appears to be lighter than the gray flap on the white surround, but that is an illusion.

So that is what happens in each of the images on the left of the tent considered on their own. But what happens if the tent is seen not as a flat picture but as a 3D entity, as in the small insert in **16.17b**? In this case, the gray flaps are seen as continuations of the planes formed by the sides of the tent. Hence, if depth-from-stereo cues are used to decide planes within which lightness computations are to be conducted, the gray flaps should processed in relation to the sides of the tent on the opposite side of the midline of the stimulus. As this opposite field is of "opposite" luminance to the one which surrounds the gray patches on each retina, the binocular lightness illusion should be the reverse of the monocular one—if Gilchrist's hypothesis of lightness computations taking place in depth planes is correct. To our eyes it does not reverse, but stubbornly maintains the same basic nature binocularly as it has monocularly.

(You can see the 3D perception of the tent if you are able to cross your eyes so that the left image in **16.12a** goes into the right eye, and the right image in **16.12a** enters the left eye. This is called "crossed eye" fusion of a *stereogram*. We discuss it in Ch 18. Stereograms are artificial stimuli that mimic the different images received by the two eyes in normal viewing. We show many stereograms in Ch 18 whose depth effects can be seen with the red-green glasses supplied with this book. However, the large black region of **16.17b** and the

need to control precisely the intensities of the gray flaps makes red-green printing unsuitable for this figure.)

Can Gilchrist's hypothesis about 3D depth planes as a grouping rule for lightness computations be saved in the face the tent-like figure in **16.17**? One possible way is to suppose that the stereogram in **16.17a** does not provide sufficiently strong cues for the flaps to be grouped with the plane on the opposite side of the tent. But that seems a little *ad hoc* to us. We have tried building the tent object in **16.17b** but it is exceedingly difficult to control lighting so that the two flaps are equally illuminated.

Kanizsa's Triangle

Kanizsa's triangle, **16.18a**, was introduced in Ch 7. An explanation given was that its illusory light-dark contours might be the result of the visual system's attempt to interpret the stimulus in terms of the "object hypothesis" of "triangle masking discs." This account can now be seen to be a similar to that set out above for **16.18c**. That is, the figural grouping clues provided by the broken lines and the sectored discs indicate a masking object, and then its lightness is computed using the appropriate context. The key contextual elements in this regard are the black sectored discs and they lead to a lightness contrast effect that propagates throughout the region asserted to be the masking object.

There is a terminological and potentially confusing side-issue here as to whether the "masking object hypothesis" postulated regarding Kanizsa's triangle is properly called "high level." The figural grouping processes set out in Ch 7 which would be triggered by Kanizsa's figure run without any use

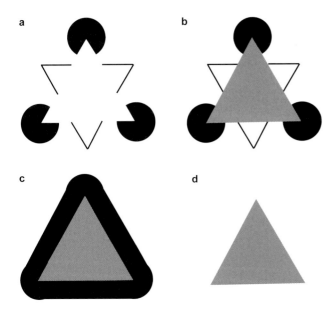

16.18 Kanizsa's triangle as a lightness illusion

16.19 Isoluminant version of Kanizsa's triangle
See p. 418. The lightness aspect of the illusion is much reduced or absent.

16.20 Kanizsa's triangle made with low-contrast dots
The lightness effect attached to the masking triangle is reduced in comparison with **16.18a**, which is as expected if the lightness is a species of contrast effect.

of object-specific knowledge, and without any dependence on identifying what the masking object is (here a triangular shape). Hence, they are often regarded as in the domain of low-level or "early" visual processes.

Notice that it is not necessary to have an "incomplete" masking object in order to get the lightness illusion. Even if a clearly visible masking triangle is substituted, an illusory lightness increment is still apparent, **16.18b**.

It could be argued that **16.18b** is a different illusion from **16.18a**, with only superficial similarities. Nevertheless, **16.18b** shows how Kanizsa's triangle, **16.18a,** can in principle be considered as a lightness contrast illusion with an incomplete border (compare **16.18b,c,d**). It is just that, in the case of **16.18a**, it is an incompletely bordered white area that receives a lightness contrast effect by extrapolation from edge-based contrast effects initiated in the regions within the sectored discs.

It is of interest in this regard that an ***isoluminant*** version of Kanizsa's figure does not create the illusory lightness effect, **16.19**. In this case, the sectored discs are a different color from the ground on which they lie, but they have the same luminance. We find that lightness aspect of Kanizsa's triangle is much weaker or even no longer apparent. This is instructive because it shows that a masking object assertion on its own does not create the illusory contour that is the distinctive attribute of Kanizsa's triangle. Why not? The inference we draw is that those contours are the result of a lightness contrast effects, and not solely the product of a masking objects being hypothesized.

This conclusion is supported by the finding that the lightness effect is lost if the sectored discs are made out of dots, rather than solid black, **16.20**. A masking triangle is still seen as an abstract shape but without an associated lightness effect. This makes sense from the vantage point of regarding the normal Kanizsa's triangle as a species of contrast illusion, because dot textures of the kind used in **16.20** generate poor lightness contrast effects.

Yet it needs to be noted that there is as yet no consensus on a "good" theory for the illusory contours seen in Kanizsa's triangle, and it would be unwise to be dogmatic about its "correct" explanation. Our own preference is to push the lightness computation style of theorizing as hard as possible to see what its limits are before bringing in, as essential, factors to do with high-level object hypotheses. Moreover, treating Kanizsa's triangle as a lightness contrast illusion illustrates the important role of extrapolation effects in the lightness computation (see earlier remarks on deconvolution, **16.8**).

Lightness Computation Using Image Statistics

Zhiyong Yang and Dale Purves proposed in 2004 a different way of trying to explain the lightness illusion in **16.11a**. Their idea is that the visual system exploits the statistics of stimulus features in the natural environment to generate its lightness percepts.

They hypothesize that the visual system has acquired, during evolution and/or just from visual experience, expectations of patterns of luminance that occur in images of natural scenes. When deciding what lightness value to attach to any given "target" image region, the probabilities of different lightness values for that image region are estimated in relation to the contextual luminance surrounding the target image region.

We won't go into details but simply say that they propose an algorithm for selecting a value for the perceived lightness of the target region based on how often that value occurs in the context of the image luminance around the target. They also provide evidence that their approach can account for the illusions in **16.12–13**.

This approach rejects all attempts of the kind described above to devise methods based on explicit constraints, for example the constraints identified by Land and Horn for "discounting the illuminant." It also rejects the idea of finding ways of running lightness computations within regions which have been perceptually grouped together, and is unconcerned with neurophysiological mechanisms to do with edge detection or lateral inhibition (an on-center retinal ganglion cell can be described as receiving "lateral" inhibition from its off-surround.) In short, this approach is claimed to be a radical departure from those hitherto explored in the area of lightness perception.

What is to be made of Yang and Purves's ideas? On the positive side, they provide evidence showing that their approach can account for the illusions in **16.12–13**. On the other hand, they do not run their algorithm over the 3D lightness illusions shown in **16.15–16**. The latter figures strongly suggest 3D factors are crucially important in lightness perception. The Yang-Purves algorithm, in its current form, would not cope with those phenomena because it does not take account of 3D surface orientation. Whether a similar approach based on

the statistics of lightness for known 3D surfaces may work is an open research question.

We have a more general concern about the Yang-Purves statistical approach. The fact that some lightness illusions are produced by the Yang-Purves algorithm indicates that image statistics distributions encode some relevant knowledge. But what exactly is the knowledge that is exploited when using the statistical relationships between image intensities? Could it be that the image statistics capture the same constraints identified by Land-Horn? If so, then using the image statistics in the way proposed by Yang and Purves amounts to an algorithm, not a computational theory in Marr's sense. This is not to diminish the value of their interesting work but it is to ask: *why* does their algorithm work? And could their algorithm be a good technique for "discounting the illuminant" for the purposes of building a surface lightness representation. The answers to these question are not yet clear.

Meanwhile, Purves and his colleagues have no doubts at all that they are working on the right lines. They believe they are launching a new paradigm for vision research generally, which they entitle the ***empirical theory of vision***. Purves and Lotto (2003, p.10) summarize this approach as follows:

> the percepts that are entertained would be expected to correspond neither to the retinal image as such nor a veridical representation of the physical world generating the stimulus. Rather, they would accord with the accumulated experience of what the retinal stimulus in question had typically signified in the history of the species and the individual.

It will be interesting to see how this approach fares over the next few years. For a critical review of Purves and Lotto's empirical theory of vision, which includes an outline of its historical antecedents, see Gilchrist's (2003) review of their book.

Skip to Summary Remarks?

The next few sections delve into some quite technical accounts of perhaps the most famous lightness illusion of all that goes under the title of ***Mach bands***. You may wish to skip to *Summary Remarks* on p. 395 and then on to the next chapter.

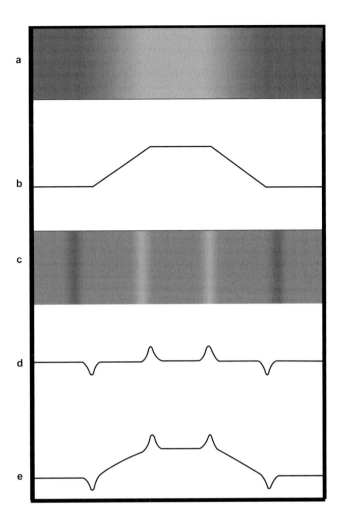

16.21 Mach bands

a Ramped changes in luminance give the illusion that there is a pair of thin dark and light bands.

b The lightness profile of the image in **a** shows that there are no bands on the page.

c Output of a model on-center retinal ganglion cell with a center-surround receptive field. The bands are clearly evident, suggesting that they might arise in this stage of processing.

d Profile of output of a model ganglion cell. The peaks correspond to the light bands and the troughs to the dark bands seen in **a**.

e Result from model retinal ganglion cell receptive field in which, when illuminated by uniform illumination across the entire field, the influence of the excitatory region is not completely cancelled by the influence of the inhibitory region. The perceived lightness profile includes Mach bands, but also includes the ramped changes in lightness apparent in **a**.

Mach Bands

Ernst Mach (1838–1916) was an Austrian sensory physiologist whose investigations of various visual phenomena led him to conclude that there must be processes of *lateral inhibition* operating within the visual system. He arrived at this conclusion in the 1860s, well before microelectrodes were available. (Remember, center-surround units are built on the basis of lateral inhibition: e.g., an off-surround can be said to laterally inhibit an on-center unit.)

The kind of phenomenon that impressed Mach, and to which his name is now firmly attached, is illustrated in **16.21a**. Each strip of gray is in fact of uniform luminance, as the luminance profile shows, but appearances are very different. Thus, the boundaries between strips are apparently bounded by light and dark bands, the illusory

Mach bands, so that the perceived lightness profile is as modelled in **16.21c**. (To create the latter, the overall output increases from left to right because a small imbalance was introduced such that the excitatory center and the inhibitory surround of the receptive field did not quite cancel each other.)

Explaining Mach bands is far from straightforward. This difficulty is analogous to some diseases which seem to have no cure: the less known about the disease, the more varied and bizarre are the treatments, so confidently proposed for its cure. So it is with Mach bands. We describe a number of possible explanations below, with the caveat that all are speculative, in recognition of the fact that Mach bands still pose a tricky research question more than 150 years after their discovery.

Mach Bands and the Land-Horn Computation

First, it might be that Mach bands are an inevitable consequence of the Land-Horn type of lightness computation, described earlier. On the other hand, they may represent a failure of the lightness computation which is not intrinsic to the Land-Horn approach, but instead is the product of a rather poor implementation of this stratagem in the visual system. For example, it may be that the reconstitution network is not well designed to reconstitute from the edge information it receives at the boundaries of the gray strips. That is, one side of the gray strip is associated with only on-center units having high outputs, whereas the other side is associated with only off-center units having high

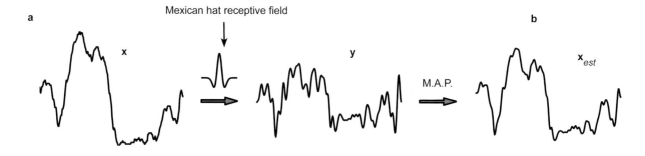

a Mexican hat receptive field

x

y

M.A.P.

$\mathbf{x}_{est}$

16.22 Bayesian image reconstruction
a A luminance profile **x** obtained by taking a cross-section through a gray level image. The receptive field of a ganglion cell is approximated by a center-surround (Mexican hat) filter, which when convolved with the image $\mathbf{x}_{est}$ yields the outputs **y**, which represents the outputs of adjacent ganglion cells.
b If the brain's job is to reconstruct **x** from **y** then this can be achieved using Bayesian inference, for which the most probable image $\mathbf{x}_{est}$ is obtained from the ganglion cell outputs **y**. This image $\mathbf{x}_{est}$ is known as the maximum *a posteriori* probability (M.A.P.) estimate of **x**. Reproduced with permission from Jaramillo and Pearlmutter (2006).

outputs. This may cause problems in reconstituting from the edges, with the blackness and whiteness arrays illustrated in **16.7g,h** getting into a competitive struggle which ends up with Mach bands being seen.

Mach Bands and Bayes' Theorem

Rather than considering Mach bands as an inconvenient artefact produced by the machinery of vision (e.g., center-surround receptive fields), Jaramillo and Pearlmutter argued in 2006 that these bands are a necessary consequence of *optimal inference*. By this it is meant that the brain is constantly trying to guess or *infer* what is out there in the word, but it is hamstrung by the ubiquitous effects of noise. Fortunately, there are principled methods for dealing with noise, which gives rise to the idea of optimal inference (or "best guess'; see Ch 13).

In a Bayesian model, Jaramillo and Pearlmutter provide a compelling demonstration that if the brain tries to reconstruct the luminance profile of an image then its best guess results in the appearance of Mach bands. Here's what they did.

The cross-section of any retinal image provides a luminance profile, which we denote by the vector of luminance values $\mathbf{x} = (x_1, x_2 \ldots x_N)$, where x_2 represents the luminance value at the second photoreceptor, for example, **16.22a**. However, due to the presence of ganglion cells with center-surround receptive fields, the brain receives information equivalent to the outputs **y** of an ordered set of adjacent "Mexican hat" filters (see Chs 4, 5). Thus, **y** is effectively a filtered version of the image profile **x**.

The problem confronting the brain is to find an estimate $\mathbf{x}_{est}$ of the original luminance profile **x** from the filtered version **y** of that profile, **16.22b**. The reconstruction $\mathbf{x}_{est}$ can only be an educated guess due to the presence of noise, but it turns out that one particular educated guess yields Mach bands.

Within the Bayesian model proposed, the posterior distribution (the thing we want) $p(\mathbf{x}|\mathbf{y})$ is the conditional probability of the observed image **x** given the filtered image **y** (which is, after all, what the brain is given by the eye).

The likelihood in this case is $p(\mathbf{y}|\mathbf{x})$, the conditional probability of the filtered image **y** given the original image **x**. It is assumed that the noise in the filtered image **y** has a gaussian or *normal* distribution. This is a common assumption, which presumes that the convolution adds small amounts of noise to the convolved image (Chs 3–5).

Finally, the prior (the thing based on prior experience) is given by a gaussian distribution, which amounts to the assumption that all images are essentially made up of gray levels chosen from a gaussian distribution of values. That is, it is assumed that if we were to plot the gray levels of a

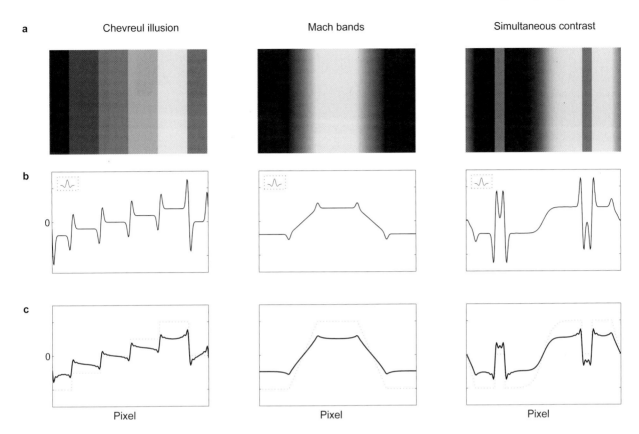

a Chevreul illusion Mach bands Simultaneous contrast

16.23 Three classic lightness illusions
a From left to right, the Chevreul, Mach bands, and simultaneous contrast illusions.
b The result of convolving the cross-section of each image (depicted as a dotted line in **c**) with a Mexican hat filter (see inset), typical of retinal ganglion cells.
c The result of deconvolving each profile in **b** with a deconvolution filter obtained from Bayesian inference yields a reconstructed image luminance profile (solid line) which corresponds to the illusion we see. The original image luminance profile is also shown (dotted line). Reproduced with permission from Jaramillo and Pearlmutter (2006).

large number of images in the form of a histogram then it would have the typical bell-shape of a gaussian distribution.

Given Bayes' rule, we can write

$$p(\mathbf{x}|\mathbf{y}) = p(\mathbf{y}|\mathbf{x})\,p(\mathbf{x})/p(\mathbf{y}),$$

which doesn't look like much of a help at this stage (even if we assume $p(\mathbf{y})=1$, see Ch 13). However, once the assumptions regarding the gaussian form of noise in $\mathbf{y}$ and the gaussian form of $\mathbf{x}$ are included, a relatively standard equation follows (though not so standard that we could repeat it here without several pages of explanation). The upshot of this equation is that it can be solved to find an *inverse filter*. This filter, when applied to $\mathbf{y}$, attempts to reverse the effects of filtering the original image $\mathbf{x}$. The technical term for this type

of operation, ***deconvolution***, was introduced in earlier (p. 380).

The inverse filter is estimated as that particular inverse filter which provides the most probable image $\mathbf{x}_{est}$ given the filtered image $\mathbf{y}$, under the gaussian assumptions stated above. Because the result $\mathbf{x}_{est}$ of deconvolution is associated with the maximum value of *a posterior* probability distribution it is known as the ***maximum a posteriori probability*** or MAP estimate.

As can be seen from three examples in **16.23**, applying this inverse filter to the filtered image $\mathbf{y}$ restores an accurate estimate $\mathbf{x}_{est}$ of the original image $\mathbf{x}$, provided that original image conforms to the assumptions made in estimating the inverse filter.

A key assumption used in estimating this inverse filter is that images are essentially gaussian. However, images such as the luminance steps which yield Mach bands are far from gaussian. We should therefore not be surprised to find that an inverse filter that works well on gaussian images, fails on an image of luminance steps. However, it is the nature of this failure that is of most interest here. Specifically, when the inverse filter, which was derived to deconvolve gaussian images, is applied to luminance steps, it yields Mach bands, **16.23b**.

Note that the *same* type of inverse filter, when applied to the convolved images of three classic contrast illusions, recreates those illusions. In each case, the result is an estimated "perceived" image that is close to the one perceived by the human visual system. In other words, the deconvolved images contain the same illusory gray levels that we experience when we look at these images, **16.23**.

The implication of these results is that the brain may attempt to reconstruct the retinal image from the convolved image it receives via the optic nerve using the simple assumption that the noise in image gray levels is gaussian. When confronted with an image which does not conform to this assumption, the brain gets it "wrong", and experiences a lightness illusion.

In evaluating Jaramillo and Pearlmutter's theory, does it matter that image gray levels are *not* generally distributed in a gaussian fashion? The answer to this question is not obvious. It may be that this assumption is made by the visual system as a "best bet" because it has no better one available. Whether things would turn out differently in terms of predicting lightness illusions if a more empirically accurate distribution was assumed, and an inverse filter could be devised for this distribution, is an open research question.

Mach Bands: 3D Factors

The theories described above account for effects observed in 2D images, but do not address the possibility that Mach bands are an illusion related to 3D effects.

As we discussed above, Yang and Purves have shown that various 2D lightness illusions can be explained in terms of the statistics of 2D gray level images. Purves extends this general line of reasoning to argue that Mach bands may result from the statistics of 3D surfaces, **16.24**. Unlike 2D

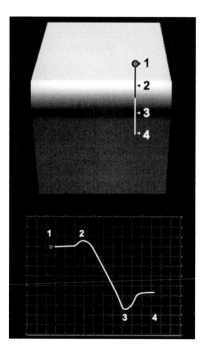

16.24 Non-illusory bands similar to Mach bands

An object under natural lighting gives rise to to peaks and troughs in image luminance that are a physical consequence of the object's shape. It has been proposed that these non-illusory bands underpin the illusory bands seen in Mach bands. Reproduced with permission from Purves and Lotto (2003). ©2003 National Academy of Sciences, USA.

lightness illusions, no formal statistical analysis of 3D effects exists, and these arguments remain speculative.

Mach Bands: A Summary of Theories

The Land-Horn theory is based on the observation that illumination changes are relatively gradual, whereas lightness changes are relatively steep. In a world where these assumptions are valid it will often work well, and provide "illusions" when the assumptions are invalid (but see remarks earlier in this chapter about numerous lightness illusions that are not predicted by the Land-Horn computation).

Similarly, Jaramillo and Pearlmutter's Bayesian model can be considered as a computational-level theory. It predicts our perceptions on the assumptions that the brain is essentially an optimal inference engine and that the distribution of image gray level values and the noise added by retinal processing is gaussian. It also creates illusions when its assumptions are violated. So how can we choose between these two theories?

The gaussian gray levels assumption is more general in scope and to that extent might be deemed to more likely to be the one exploited by the human visual system. Also, it might be argued that, being based on Bayes' theorem, it has a better theoretical rationale.

However, it is interesting to ask: is the Bayesian account a computational theory in Marr's sense of the term? Has it identified key constraints arising from the analysis of a particular task, and shown that exploiting those constraints entails the price of suffering from various visual illusions? The task consists of recovering the luminance profile within a retinal image. The constraints used to solve this problem are that image noise is gaussian, and that images are gaussian. While the first of these assumptions seems reasonable, the second is not consistent with typical image gray levels. Thus, the "gaussian image" assumption is used as a constraint, but this assumption is not consistent with a computational analysis of the problem. The assumption "works," insofar as it yields various lightness illusions. However, this type of model would be more convincing if the illusions could also be demonstrated for models which assume non-gaussian distributions, especially if those non-gaussian distributions coincided with the particular distributions which characterize typical images.

Summary Remarks

Perhaps you have been rather surprised by what has been said in this chapter. After all, how many non-specialists would suspect that there was much problem about anything so mundane as seeing black, grays, and white? And yet some remarkably intricate machinery is necessary to extract even this elementary information from the inherently ambiguous retinal image. Indeed, the processes are so intricate, and so poorly understood, that it has seemed wise to postpone until this late chapter a description and discussion of processes involved.

One of the curious things described in this chapter has been that the visual system detects blackness as a property of surfaces in its own right, just as it might detect blueness, or redness. This is the opposite of most people's intuitions. The "inner screen" theory of Ch 1 reflected the "common sense" view that blackness is registered simply as the absence of activity in nerve cells, but this natural expectation is in fact far from the truth. Blackness is coded by activity in a certain population of nerve cells, with greater activity coding a deeper black.

It is not known as yet where this coding takes place, i.e., the brain site whose activity is the neural correlate of conscious awareness of blackness. The cells in the retina that seem related to blackness perception can only be a first step: no one supposes that the retina completes the whole business of lightness perception. Somehow its outputs have to be tied in with a whole host of other visual processes carried out by the brain. What this chapter has been concerned with is but an early stage of the mechanisms for seeing black, grays, and white. This is a fair reflection of current knowledge: we have little idea about brain mechanisms concerned with these perceptual attributes.

If you find it difficult to believe that blackness is coded by activity in certain cells, rather than simply by inactivity in cells which cover the whole black-gray white range, you might find it helpful to consider the following curious fact. A television screen appears a pale gray when switched off, and yet it has no trouble presenting us with blacks if the image being transmitted requires them. This is odd because the electronics of the television set have no way of dimming the screen, only of brightening it up.

So the blacks we perceive in television images are "created" by our visual system as a product of its lightness computation. Blackness is "discovered" where only gray physically exists. If our brain cells responded simply to the physical luminance of points on the screen, as proposed by the "inner screen" theory of Ch 1, then we should never be able to see blacks on the screen at all—only gray through to white.

It is difficult enough, in our experience, to convince people about the difficult problems of understanding how we see objects. But it is even more difficult to persuade them that seeing lightness is problematic. The example given in the previous paragraph, plus the various contrast illusions shown in this chapter, might help undermine the commonsense view and so lead you to an appreciation of the difficulties. But if you need further persuasion, consider the following famous truth.

When you descend into that dark cellar to collect some coal which is stacked up against a white-painted wall, and then return to a sunlit room with a bucketful of the substance, the intensity of the light entering the eye from the sunlit coal is greater than that which came from the dimly lit white cellar wall. And yet the coal still looks black, and the wall still looks white. This illustrates the problem of lightness perception par excellence. Black remains black even though it here has greater luminance than a poorly illuminated white, a superb piece of visual computation.

Further Reading

Adleson EH (2000) Lightness perception and lightness illusions. In *The New Cognitive Neurosciences: Second Edition,* 339–349, MS Gazzaniga and E Bizzi (Eds). MIT Press.

Bressan P (2006) Inhomogenous surrounds conflicting frameworks and the double-anchoring theory of lightness. *Psychonomic Bulletin and Review* **13** 22–32.

Buckley D, Frisby JP and Freeman J (1994) Lightness perception can be affected by surface curvature from stereopsis. *Perception* **23** 869–881.

Gilchrist AL (1980) When does perceived lightness depend on perceived spatial arrangement? *Perception and Psychophysics* **28** 527–538.

Gilchrist AL (2003) Looking Backward. *Nature Neuroscience* **6** 550. *Comment* A review of Purves and Lotto's book cited below.

Gilchrist AL (2006a) Seeing in black and white. *Scientific American* **17** 42–50. *Comment* A fine short accessible introduction to the field.

Gilchrist AL (2006b) *Seeing Black and White.* Oxford University Press. *Comment* The fruit of a twenty-year heroic effort grappling with the problems of lightness and brightness perception.

Jaramillo S and Pearlmutter BA (2006) Brightness illusions as optimal percepts. *Technical Report* NUIM-CS-TR-2006–02CS Dept. NUI Maynooth2006

Knill DC and Kersten D (1991) Apparent surface curvature affects lightness perception. *Nature* **351** 228–230.

Land EH (1983) Recent advances in retinex theory and some implications for cortical computations: Color vision and the natural image. *Proceedings of the National Academy of Sciences USA* **80** 5163–5169.

Mather G (2008) *Foundations of Perception 2nd Ed.* Psychology Press. *Comment* A good textbook with an excellent tutorial section on the photometric system used for measuring light including luminance.

Nakayama K, Shimojo S, and Ramachandran VS (1990) Transparency: relation to depth, subjective contours, luminance and neon color spreading. *Perception* **19** 497–513. *Comment* Commentaries celebrating this classic paper are in *Perception* **38** 859–877.

Purves D and Lotto RB (2003) *Why We See What We Do: An Empirical Theory of Vision.* Sinauer Associates: USA.

Ruderman DL, Cronin TW and Chiao C (1998) Statistics of cone responses to natural images: implications for visual coding. *Journal of the Optical Society of America* **15** 2036–2045.

Todorovic D (1997) Lightness and junctions. *Perception* **26**(4) 379–394. *Comment* Sets out the case for T-junctions being used to segment regions for lightness computations.

Wallach H (1976) *On Perception.* New York: Quadrangle/The New York Times Book Co.

Zhiyong Y and Purves D (2004) The statistical structure of natural light patterns determines perceived light intensity. *Proceedings of the National Academy of Sciences USA* **101** 8745–8750.

17

Seeing Color

17.1 Color aftereffect
Note first the appearance of the standard color photograph, lower left, and its black and white version, lower right. Then fixate the cross in the upper left picture for about one minute, after which transfer your gaze quickly to the cross in the black and white picture, which will then appear colored. This an example of a color aftereffect. The upper left figure, which is serving here as an *adapting stimulus* (see Ch 4), was created by color inversion. See pp. 408-410 for an explanation of color aftereffects in terms of opponent color channels.

Why does a strawberry look red? In order to explore this deceptively simple question, we will need to describe some fundamental properties of light, but first we should ask: why do we care what color a strawberry is?

An evolutionary fitness argument is that the color of a strawberry helps us find it amongst the leaves that surround it, **17.2**. Color can also tell us how ripe the strawberry is. Similarly, the color of a piece of moldy meat might warn us it would be a bad idea to eat it. Additionally, there is evidence that the photopigments in the receptors of fruit-eating monkeys are beautifully tuned to the task of picking out fruit amongst leaves in their particular ecological niche (see *Further Reading*). Finding fruit is just one example of how color can confer evolutionary benefit by facilitating discriminations.

17.2 Why do we have color perception?
The strawberries are much easier to pick out in the colored photograph than in the black and white one.

Why read this chapter?

Seeing color is difficult because the colors we see do not exist in the world. Instead, color is the result of a constructive, inferential process. We begin by defining the three *spectra* involved in color vision: the *illuminance*, *reflectance*, and *luminance* spectral functions. It is the reflectance function that corresponds to the "color" of a surface, but this is a product of the other two spectra, neither of which is easily accessible to the visual system. This is the reason why color vision is such a hard computational problem. For a given colored surface, the spectrum of wavelengths entering the eye (the luminance spectrum) changes dramatically under different illuminant spectra (e.g., tungsten bulb versus sunlight), but our perception of color does not. This stability under a wide range of illumination conditions is called *color constancy*. We describe Land's *retinex theory* of color perception in some detail. We then describe the three types of color cone in the human eye, and explain how their limited sampling of the luminance spectrum gives rise to *metamerism*, the finding that different mixtures of *primary* colored lights appear identical. We describe two different theories of color perception (*trichromacy* and *color opponency*), which have been reconciled in the form of the dual process theory. Finally, we describe the properties of *color opponent* ganglion cells in the retina and cortex, but note that a full understanding their role in mediating color perception will depend on advances at the computational theory level.

Properties of Light

Visible light is a part of the ***electromagnetic spectrum*** of radiation, **17.3**. At one end of this spectrum, radiation has extremely short wavelengths of the sort emitted by nuclear bombs. For example, gamma radiation has a wavelength of 0.001 nm, where 1 nm = 1 nanometer = one thousandth millionth of a meter. At the other end of the spectrum, radiation has extremely long wavelengths, with those around 10 meters long being used for radio stations. In between these extremes, the radiation used in microwaves has a wavelength of about 12 cm. As the wavelength shrinks to 700 nm it begins to become visible to the human eye as red light. Between 700 nm and 400 nm each wave-

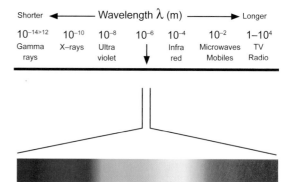

17.3 Electromagnetic spectrum of radiation

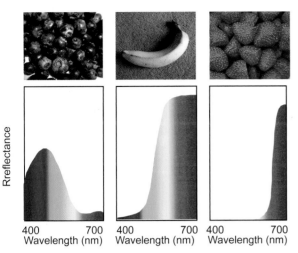

17.4 Reflectance functions
Each panel shows the approximate proportion of each wavelength reflected by the object above it.

Rreflectance

400 700
Wavelength (nm)

400 700
Wavelength (nm)

400 700
Wavelength (nm)

length appears as one of the colors of the rainbow: red, orange, yellow, green, blue, indigo and violet (which can be memorized as the name ROY G BIV). Ultraviolet (UV) light has a wavelength slightly shorter than 400 nm, and is invisible to the human eye. UV light is visible to some animals, such as kestrels which can detect the presence of vole's urine (and presumably voles) because it reflects ultraviolet light.

It is worth noting that light is not colored. Photons are not little colored particles. Color is what we see when light with a particular range of wavelengths enters the eye. We will sometimes slip into the convenient shorthand of speaking as if light is colored, but we must be careful not to mistake perception (color) for physics (wavelength). Now, what about this strawberry?

In essence, a strawberry looks red because (almost) only long wavelength ("red") light is reflected by a strawberry. This implies that a strawberry absorbs all wavelengths except those at the long ("red") end of the spectrum of visible light. Notice that we referred to wavelength*s* and not to a single wavelength. This is because a strawberry reflects a *range* of wavelengths in the red end of the spectrum. Specifically, for each wavelength, the strawberry reflects a proportion of light at that wavelength. We can plot this proportion for different wavelengths in order to see the strawberry's "physical color." The resultant graph defines the *spectral reflectance function* of a strawberry, **17.4,**

which is often shortened to just *reflectance function*, and is denoted by the letter *S*.

For clarity, we will use the term *color* to refer to the *perceived color*. Insofar as it is possible to speak of the physical color of a surface, the reflectance function is the closest thing we have to the physical correlate (defined in physical units) of the perception of the color of a surface. Indeed, if we, as perceivers, somehow had "direct" access to the reflectance function of a surface then color perception would be a trivial matter. But we don't. All that we have is the outputs of three different kinds of cone receptors sensitive to different wavelengths (see Ch 6, p. 136, and later in this chapter). What is worse, the output of each cone depends not only on the reflectance function of a surface (which we don't know), but also on the spectrum of the light source that illuminates that surface (which we don't know either). In other words, the spectrum of light entering the eye from a surface depends on two unknown spectral quantities, the spectrum of the ambient light, and the spectral reflectance function. At this point, we need a more extensive definition of terms:

The luminance of a surface is measured in *candelas per square meter* (cd/m²). For example, white paper in moonlight has a luminance of about 0.2 cd/m² whereas the same paper in sunlight has a luminance of 40,000 cd/m² (see Mather, 2008, for details of physical units for measuring light).

Illuminance: When we refer to the "color" of a light source, we actually mean the amount of light at each wavelength. Thus, each light source, such as sunlight, has a specific amount of energy at each wavelength. Just as the color of a surface is defined in terms of its reflectance function, so the color of a light source is defined in terms of its *illuminance spectrum*: the amount of light energy at each wavelength. Different light sources may have very different illuminance spectra, **17.5.** The illuminance spectrum is denoted here by the letter *E*, and wavelength is denoted by the Greek letter lambda (λ). This notation makes it easy to specify the amount of energy at a single wavelength.

Note that the illuminance spectrum is not a single value, but is essentially an ordered set of illuminance values, each of which specifies the illuminance of a corresponding ordered set of wavelengths λ between 400 nm and 700 nm (in

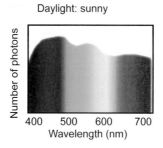

Tungsten light bulb

Daylight: sunny

17.5 Illumination spectra for daylight and a tungsten bulb

The amount of energy at each wavelength λ (lambda) of a light source defines its spectral balance, or, more precisely, its illuminance spectrum. Notice that the spectrum of a tungsten light bulb has a high proportion of energy at the red end of the spectrum in comparison to the daylight spectrum shown. This is why the world can take on a blue–ish hue when going outside after being in a room lit by bulbs. This effect is more pronounced if the transition to daylight is viewed through a camcorder, which does not have the robust color invariance mechanisms of the human visual system. These illuminance spectra are only approximate.

the case of visible light). If the illuminance at wavelength $\lambda_1 = 450$ nm is 2 cd/m², for example, then this is written as $E(\lambda_1) = 2$ cd/m². This should be read as: illuminance E varies with wavelength, where wavelength is denoted by the **variable** λ, so that $E(\lambda)$ is the illuminance at wavelength λ.

Reflectance: As we have already said, this refers to the proportion of light reflected from a surface at each wavelength, **17.4**. This employs a similar notation to that used for the illuminance spectrum. Thus, if the proportion of energy reflected at a specific wavelength $\lambda_1 = 450$ nm is 0.4 (i.e.,

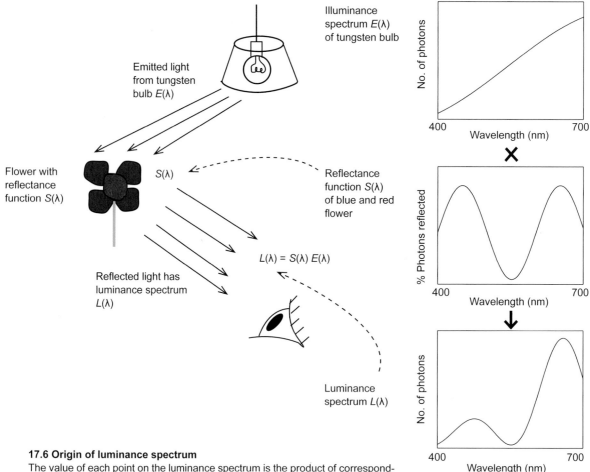

17.6 Origin of luminance spectrum

The value of each point on the luminance spectrum is the product of corresponding points on the illumination spectrum and the reflectance function.

40%) then this is written succinctly as $S(\lambda_1) = 0.4$. The reflectance function is the nearest thing to a physical definition of color, because it defines the proportion of light reflected at every wavelength.

Luminance: If a surface is illuminated with a light source with a particular illuminance spectrum then the light reflected from that surface reflects a certain amount of light energy at each wavelength, which defines its luminance spectrum. For example, if the amount of energy reflected at a specific wavelength $\lambda_1 = 450$ nm is 2 cd/m^2 then this is written succinctly as $L(\lambda_1) = 2$ cd/m^2. Once again, note that the luminance spectrum $L(\lambda)$ is not a single value, but is essentially an ordered set of luminance values, each of which specifies the amount of energy reflected at a single wavelength. Crucially, the luminance spectrum of a surface with a given reflectance function varies, depending on the ambient illuminance spectrum. See **17.6**.

It is conventional to adopt the simplifying assumption that the amount of light $L(\lambda)$ reflected from a surface patch and the amount of light received by the corresponding retinal image patch are the same. Consequently, we assume in general that the medium through which light travels from a surface to the eye does not alter its spectral content. This assumption is invalid when we view surfaces far away, as air absorbs some wavelengths more than others. This is why distant objects can appear to have a slight blueness compared those nearby, a phenomenon called *aerial perspective* that is sometimes exploited by painters. We will ignore this complication, and assume that the spectral luminance spectrum which can be measured in each small region of the retinal image is the same as the luminance spectrum $L(\lambda)$ of the corresponding surface patch.

In summary, the amount of energy reflected $L(\lambda)$ from a small surface patch at each wavelength λ is given by the proportion $S(\lambda)$ of energy $E(\lambda)$ in the light source that is reflected from the surface at that wavelength. As $L(\lambda)$ is the proportion $S(\lambda)$ of $E(\lambda)$, this implies that $L(\lambda)$ is given by the energy $E(\lambda)$ in the light source multiplied by the reflectance $S(\lambda)$ at the wavelength λ, **17.6**:

$$L(\lambda) = E(\lambda)\, S(\lambda).$$

So there we have it, three spectral quantities. One of these we know $L(\lambda)$, but don't want except as data to be interpreted. Two of these we don't

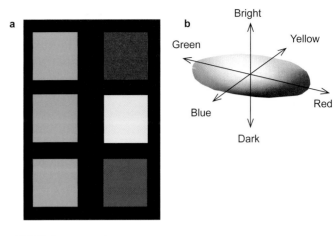

17.7 Color parameters
The three dimensions associated with color are hue, saturation, and brightness.
a Examples of differences in hue (upper pair), saturation (middle pair), and brightness (lower pair).
b 3D graph representing the three color dimensions. Hue varies around the vertical axis. Saturation increases with horizontal distance from the vertical axis. Brightness increases with height on the vertical axis. (Beware: artists often use different terms for these color attributes; see Edwards, 2005).

know, $E(\lambda)$ and $S(\lambda)$, and what we really want to know is $S(\lambda)$. So there we don't have it, after all.

Of course, if we could estimate the illuminance spectrum $E(\lambda)$ then we might be able to combine it with $L(\lambda)$ to find $S(\lambda)$. If we know $E(\lambda)$ then we can rearrange

$$L(\lambda) = E(\lambda)\, S(\lambda),$$

to obtain

$$S(\lambda) = L(\lambda)/E(\lambda).$$

Unfortunately, this assumes that we know the illuminance spectrum $E(\lambda)$, which we don't.

The fact that estimating $S(\lambda)$ entails making allowance for $E(\lambda)$ is often referred to as "*discounting the illuminant*." Thus, if the illuminance spectrum $E(\lambda)$ is known then its effects on the luminance spectrum $L(\lambda)$ measured in the image can be taken into account (discounted), when using this as a starting point to estimate the reflectance function $S(\lambda)$.

Having defined color in terms of physical quantities, we can now give a brief summary of how these map on to the psychological dimensions of color.

Hue, Saturation, and Brightness

Our perceptions of color can be defined in terms of three attributes or parameters, **hue**, **saturation**, and **brightness**, **17.7**.

Hue corresponds to our everyday notion of color, such as red, and corresponds to the reflectance function of a surface.

Saturation corresponds to how "pure" the color is. For example, a poppy and a London bus would be described as pure red, and most people would perceive a desaturated red as being pink. Whilst the mapping is not exact, saturation corresponds to how concentrated the reflectance function is around one wavelength.

Brightness is the perceived intensity of color. For example, a poppy on a sunny day is brighter than it is on a cloudy day. Brightness corresponds to the overall intensity of wavelengths of the luminance spectrum , and therefore also depends on the illuminance spectrum.

Color Constancy

Our ability to discount the illuminant gives rise to the phenomenon of **color constancy**. This is a good name because it refers to the fact that the color (hue) of a surface remains roughly constant despite changes in the illuminance spectrum of the light source (e.g., a light bulb versus sunlight), **17.5**.

Color constancy suggests that the visual system is good at recovering the reflectance function of a surface, despite changes in the illuminance spectrum. In fact, due to limitations in nature of our cones, it is impossible to recover the full reflectance function of a surface. But it is possible to recover something which is a good approximation to the reflectance function despite changes in the illuminance spectrum, and it is this ability that we refer to as color constancy.

In essence, the problem that the brain attempts to solve is this: what fraction of the light entering the eye at each wavelength within the luminance spectrum is due to the surface reflectance function, and what fraction is due to the illuminance spectrum at that wavelength?

For example, suppose there is a high proportion of long wavelength ("red") light in the luminance spectrum. Is this because the surface is red? Or is it because the light source is a tungsten bulb, which

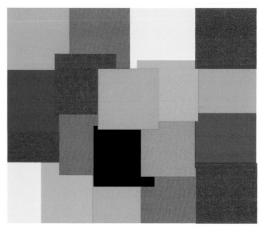

17.8 Mondrian-type pattern
This Is the kind of colored block pattern used by Land.

emits a large proportion of light at long wavelengths, **17.5**?

It is worth noting that color constancy is not perfect; there are limitations on how much variation in the color of the illuminant we are able to discount, as those who visit exotically lit night clubs know well.

Color constancy is analogous to lightness constancy, which, as we saw in Ch 16, refers to the fact that the lightness of an achromatic surface remains pretty constant despite changes in the *amount* of illuminant energy across a scene.

Color Constancy Depends on Context

Pieter Mondrian (1872–1944) was a Dutch artist who specialized in creating paintings containing blocks of different colors, like **17.8**. This was probably fortunate for Edwin Land (1909–1991), who was a vision scientist. Land demonstrated that our ability to see color in a Mondrian figure is constant almost irrespective of the color of light used to illuminate that figure. For example, if you look at one colored patch of a Mondrian when it is illuminated with red light then its color would not appear to change much when the Mondrian is illuminated with green light. This is an example of color constancy.

However, Land discovered that if a patch is shown in isolation from the rest of the figure then a change in illumination color does alter the perceived color of that chosen patch. Similarly, if red, green, and blue projectors are adjusted so that two different patches of a Mondrian reflect identical amounts of red, green, and blue light then the patches appear to have different colors. Thus, the perceived color of a patch depends on its *context*. That is, somehow the visual system takes into account the *relative* intensities of light of different wavelengths reflected from different surfaces.

In fact, Land's observations in this regard were anticipated in 1789 by the French mathematician Gasparde Monge, as John Mollon discovered in a fascinating piece of historical research:

> Monge had arranged a red paper to hang on the wall of a west-facing building that stood opposite the meeting chamber. He invited his fellow academicians to observe the red paper through a red glass. A red paper, as Monge well understood, predominantly reflects red rays. A red glass passes red rays and attenuates others. So we might expect the red paper through the red filter to look a saturated red, hinting perhaps at the bloodshed that was soon to touch even the members of the Academy. But it did not. It looked desaturated, even white. … Similarly counter-intuitive was the appearance of a white object through the red glass. Such an object, Monge reminded his listeners, reflected rays of all colors, while a red glass would pass only the red rays. So there should be two changes in the appearance of the object when the red glass is introduced: it should be reduced in brightness and its color should change from white to red. In fact, the white object seen through a red glass continues to look whitish … (Mollon, 2006).

The reference to bloodshed is because Monge was speaking at the time of the French Revolution.

So it seems that color constancy depends on the presence of many surface patches with different reflectance functions which the brain somehow uses as context in creating perceptions of color. What is needed is a theory which can show *how* context can be used to discount the illuminant. Land's **retinex theory** of color vision does just this, and we describe it later in this chapter.

Seeing Colors in Shadows

We illustrated in Ch 16, p. 394, that changes in image luminance can be caused by a change in the orientation of a surface because this alters the

17.9 Edited images of a Rubik's cube
a The pink patch labeled A and the red patch labeled B share the same reflectance function on the printed page, as can be seen when they are isolated in upper row. This was achieved by copying the color of B to A using a graphics application. **b** Identical image to that in **a**, but using gray–levels. Here, A and B have the same luminance on the page, but appear not to share the same lightness. Thus, the difference in color perceived in **a** may be due to an underlying difference in lightness.

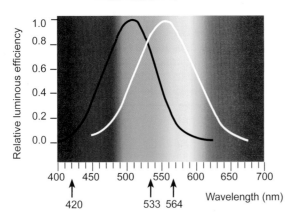

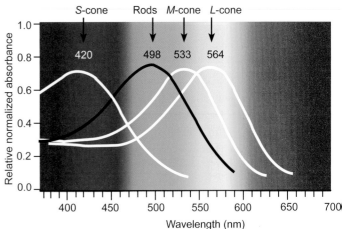

17.10 Human spectral luminous efficiency functions
These functions show relative efficiency in detecting different wavelengths of light. The arrows mark the peak sensitivities of the cones shown in **17.11**. The black curve is for scotopic vision (i.e., when dark adapted, so this is for "night time" vision). The white curve is for photopic vision (i.e., when the eyes are light adapted, so this is for "day time" vision or when artificial lighting is bright). The difference between their peaks is called the **Purkinje shift** and is discussed in the text. The functions are approximations based on *Commission Internationale d'Eclairage* (*CIE*) definitions, and are here superimposed on the colors of the visible spectrum.

17.11 Relative spectral absorbance functions of the three cone types and of the rods
Each curve shows the approximate absorbance of light of different wavelengths relative to the peak sensitivity of each receptor type (hence the vertical axis label uses the term normalized). The numbers above the graphs give the peak sensitivities. Data based on Bowmaker and Dartnall (1980).

amount of light falling on, and hence reflected by, that surface. This led to a discussion, in the context of considering lightness constancy, of the appearance of achromatic surfaces in shadow. It is of interest to ask: what happens to the appearance of a colored surface in shadow?

In **17.9a**, the image patches marked *A* and *B* are printed identically on the page, so they have the same luminance spectrum s. Despite this, they appear dramatically different. How would you describe the difference in appearance using the color dimensions of hue, saturation and brightness set out in **17.7**? A change in hue would not be expected as placing a patch in shadow usually reduces only the amount of illuminant light without altering its spectrum. For us, this consideration is borne out in that the two patches *A* and *B* are best described as a change in saturation (pink vs. red) and also brightness (the "pink" patch appears so bright that it might almost be interpreted as a light source). Some evidence for a brightness factor is given by **17.9b**, which shows the same figure in black and white. There it is apparent that the two squares have different lightness values, and

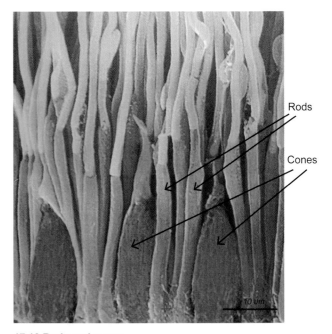

17.12 Rods and cones
Scanning photomicrograph of retina showing the distinctively different shapes of rods and cones. Photomicrograph Ralph C. Eagle.

this may underpin the change in appearance seen in **17.9a**. We modeled **17.9a** on a figure giving a similar effect that was created by Purves and Lotto in 2003.

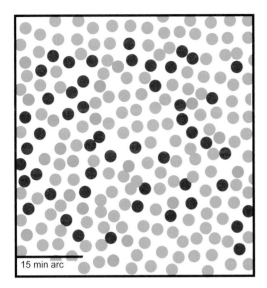
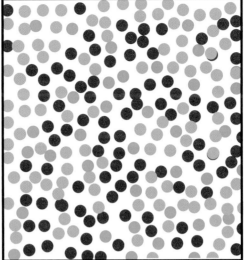
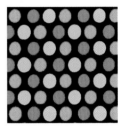

15 min arc

Below
Magnified view of a small patch of a television screen to show the precise and regular layout of spots of red, green and blue phosphors. Courtesy P. Halasz.

17.13 Cone mosaics

These are schematic diagrams of plan views of retinal regions located about one degree from the centers of the foveas of two individuals. The cones are artificially colored as red, green, and blue to show respectively the *L*-, *M*-, and *S*-cones. The cones are packed closely together in reality but are shown here separated for clarity regarding two points. First, the proportion of *S*-cones is relatively small, being roughly 4–6.5% of all receptors. Curiously, no *S*-cones are found in the central fovea. Secondly, and equally curious, is the great variability in the ratio of *L*- to *M*-cones, a ratio extending from under 0.5 to over 16 for people with normal color vision. This is illustrated by the two diagrams shown here: the left-hand one has many more *M*-cones (green dots) than the one on the right, and rather fewer *L*-cones (red dots). This is so despite, we emphasize, both individuals having normal color vision. These diagrams are loosely based on the data of Roorda and Williams (1999). The irregular jittered arrangement of cone receptors contrasts sharply with the precise layout of red, green, and blue phosphors on television screens—see small inset on the right.

Three Color-Sensitive Cones

The retina has two types of receptors, ***rods*** and ***cones*** (as noted briefly in Ch 6). The three types of cone found in the human retina support color vision, which makes humans ***trichromats***. (***Dichromats*** have two cones and thus have abnormal color vision; see Nathans, 1987.) The three types are called *S*-cones, *M*-cones, and *L*-cones, where the capital letters refer to their peak sensitivities in the short, medium, and long wavelength regions (respectively) of the visible spectrum, **17.10–11**. The output of each type of cone provides measurements of incoming light intensity over a broad range of wavelengths (each one covers almost the entire visible spectrum in fact) but with peak sensitivities at different wavelengths: short (420 nm), medium (533 nm), and long (584 nm). Figure **17.12** shows why rods and cones have their names, and **17.13** shows the distribution of the three cone types in a retinal region about 1° from the center of the fovea.

These *S*-, *M*-, and *L*-cones are sometimes, but confusingly, called respectively blue, green, and red cones. The danger in using this terminology is that it confuses cone outputs with color perceptions. As cone outputs are measured in physical units (millivolts) and color perceptions are not (no one is certain what units perceptions should be measured in!), this is an example of a ***category error***. It is thus important to bear in mind that the cones provide just the starting point for color vision, and that their outputs are not simple correlates of our perception of color.

Having broadly tuned cones is an example of ***coarse coding***, a principle that is exploited widely in the visual system and that we describe at length in Ch 3, p. 69, and also Ch 11, p. 270.

The difference shown in **17.10** between the peaks in the spectral luminosity efficiency functions for scotopic and photopic vision causes a change in the appearance of colored surfaces at different illumination levels. For example, reds look much darker than greens at night because the rods, which are the receptors that mediate scotopic vision, are insensitive to long wavelengths and are much more responsive to green light. The reverse happens during daylight: reds then tend to look brighter than greens. The change in peak wavelength sensitivity resulting from adaptation to low light levels is called the ***Purkinje shift***.

The insensitivity of the rods to long wavelengths is exploited in the way some safety critical environments are illuminated. For example, sailors working on the bridge of a ship sailing at night need to maintain their state of dark adaptation in order to have the best possible chance of picking up weak light signals from other ships. At the same time, they need to be able to read their various instruments and their charts. The answer is to illuminate the bridge with red light and to use red lights in instrument displays, as even quite bright red light will not disturb the state of adaptation of the rods due to their insensitivity to long ("red") wavelengths. Failure to observe these precautions has led to some tragic accidents at sea.

Why Do We Have Three Cones?

Having three cone types with broadly tuned and overlapping wavelength sensitivities provides measurements of the spectral luminance spectrum at each location in the retinal image. These measurements are sufficiently good for the brain to build up a useful representation of the spectral reflectance functions ("colors") of most scene surfaces viewed in most circumstances. Yet you may wonder why evolution has not produced very many cone types at each retinal location, with each type sharply tuned to a different narrow band of wavelengths.

In fact, some animals do have more than three cone types. Indeed, some women have four cone types, and so they are technically ***tetrachromats.*** These women have two different types of long wavelength cones, due to the fact that each of their two X-chromosomes code for a slightly different photopigment. Many birds have more than two cone types, with the pigeon having four, and the chicken having five. The increased color perception implied by such a large number of cones may be a reflection of their need to discriminate food in cluttered backgrounds, which would effectively

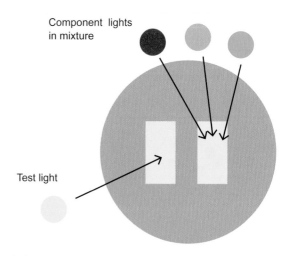

17.14 Color metamerism
The two stimuli in the center are created in different ways. The one of the left is made using a light source of a single wavelength. The one on the right is made from a mixture of three wavelengths. The observer's task is adjust the intensities of the three light sources on the right until the two stimuli look the same.

acts as a camouflage without very good color perception. Presumably having many receptor types comes at the price of reduced visual acuity, because these extra cones have to be fitted into each image patch dedicated to each surface patch, if the color of that image patch is to be sampled by every cone type. Hence, color discrimination ability (many cone types) may reflect an evolutionary trade-off with visual acuity (few cones types).

The current record for the largest number of photopigments goes to the mantis shrimp, which has fifteen in total. Packing in so many cone types led Snowden, Thompson and Troscianko (2006) to suggest that perhaps this shrimp recognises significant entities in its visual world in terms of color rather than shape. At the other extreme, certain nocturnal mammals, such as owl monkeys, have only one type of cone photoreceptor. Color vision cannot be mediated by a single cone type, and so perhaps the rods in these monkeys mediate black and white spatial vision at night and their single cone type does the same during the day.

Most mammals have two color receptors (they are dichromats). For reasons no-one understands, among the New World monkeys all males but only half of the females are (blue-yellow) dichromats.

A remarkable discovery by Shapley (2009) is that replacing a missing gene in adult color-blind monkeys restores normal color vision, despite these monkeys lacking the gene from birth to adulthood. This is surprising because visual deprivation in early life often leads to lifelong consequences.

Color Metamers

Long before the existence of color cone receptors was established, scientists argued about how color was represented in the brain. There were basically two camps, those who argued (in modern terminology) that color relied on three color channels, the **trichromacy theory,** and those who argued that it relied on four color channels, the **opponent process theory**. In fact, as is often the case in psychology, both camps were right and their differences were finally resolved in the form of a **dual process theory**.

The three-channel theorists based their arguments on the finding that the color of any single light can be duplicated exactly with a combination of three different colored lights, **17.14**. These two, apparently identical, light sources are said to be **metamers** because they actually consist of different wavelengths (one comprising a single wavelength and the other of three). In fact, the same type of effect would exist for any finite number of cone types. For example, if we had five cone types (like chickens) then we would need to adjust five colored light projectors so as to mimic any color.

It is important to understand this **metameric matching task**. Consider a light that contains one wavelength which is perceived as yellow, **17.14**. This light generates responses in the short, medium, and long wavelength cone types of r_S, r_M, and r_L (respectively). Now, self-evidently, a mixture of *any* other lights which also induce these same cone responses r_S, r_M, and r_L should also be perceived as yellow. In other words, there are many combinations of primary light intensities that give rise to the same cone responses r_S, r_M, and r_L, and which therefore give rise to an identical perception of color. This fact is the basis of metamerism.

[In thinking about this, distinguish the additive form of color mixing of lights, **17.14**, **17.15a**, from the subtractive mixing that occurs when pigments are mixed, **17.15b**. Adding two different wavelengths together creates a very different stimulus from adding two different pigments together, as explained in the legend to **17.15**.]

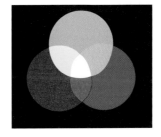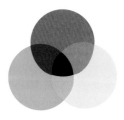

Additive   Subtractive

17.15 Additive and subtractive color mixing
Additive mixing is what happens when light from different sources is combined. Mixing differently colored paints is an example of subtractive mixing. Color pigments absorb wavelengths differentially: a blue pigment appears blue because it absorbs short wavelengths less than other wavelengths. Hence, color perceptions from color pigment mixes depend on what wavelengths are "subtracted" (absorbed) by the addition of each pigment. Painters have to learn the rules of pigment mixing to achieve the colors they require (Edwards, 2004).

If you find it difficult to believe that any color can be constructed from three colored light sources then have a close look at a color television image (try using a magnifying lens close to the screen). At every location in the TV image there are three tiny dots. One emits blue light, another emits red light, and the third glows green. What you receive while watching color television (and you can see any color there) consists at every image location of a combination of three tiny red, green, and blue elements. Your visual system uses these outputs to produce color perceptions.

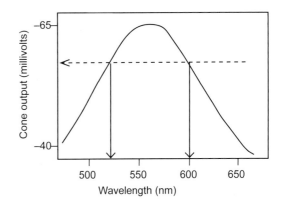

17.16 Ambiguity problem in cone outputs
Schematic graph illustrating that the *L*-cone output indicated by the dotted line could be achieved by either of the two wavelengths indicated by the vertical arrows.

The reason that metamers exist is because each cone has a **broad tuning width**, **17.16**. This implies that the output of any single cone cannot specify the wavelength of light that is stimulating it, simply because *many* different wavelengths can activate it, **17.16**. In other words, if a photon of a particular wavelength is absorbed by a cone, the cone response is identical regardless of the wavelength of the absorbed photon This is called the **principle of univariance**.

Consider, for example, the tuning curve of the *L*-cone shown in **17.10**. If we shine (red) light at a single wavelength of 600 nm on this cone then it has an output which we denote $r(600)$. But we can obtain exactly the same output using another wavelength within the cone's tuning curve. That is, if we keep the intensity of light the same but shift its wavelength then we can find another point along the cone's tuning curve (at about 525 nm) where the cone's output $r(525)$ matches its output $r(600)$ at 600 nm, so that $r(600) = r(525)$, **17.16**. (This ambiguity is similar to the problem encountered when we tried to use simulated "feature detection" neurons to encode retinal images, Ch 3). So the brain cannot tell, simply by knowing the cone's output, if the light is at 600 nm or 540 nm, because, when both lights have the same intenity they give the same receptor output.

Conversely, if we are permitted to adjust the intensity, but not the wavelength, of the light source then we can drive the output of the cone to match any desired value by using a sufficiently bright or dim light (provided the wavelength of this light falls within the tuning curve of the cone). For example, a light at 600 nm which evokes a cone response of $r(600)$, and a light at, say 400 nm evokes a smaller response $r(400)$. But if we increase the intensity of the light at 400 nm then it is possible to drive up the cone's response until it reaches $r(600)$. This is effectively what happens in metameric matching tasks, and it provides a clue as to how metameric matching is possible.

The logic behind the metameric behavior of a single cone described above can be extended to any finite number of cone types, although finding metamers becomes harder if different cone tuning curves overlap, as they typically do in biological eyes. It is worth noting that the metameric matching task works most readily if the patch

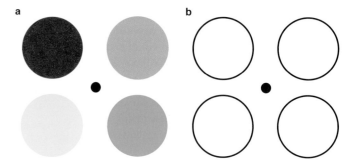

17.17 Color aftereffect
Fixate the central black dot on the left for about 30 seconds. Then transfer your gaze to the black dot on the right and notice the perceived color of the discs.

illuminated by the pair of light sources is viewed in isolation (that is, unlike the situation in one of Land's Mondrians, **17.8**).

To sum up on color metamerism, the output of the three cone types can be represented by three numbers. The fact that metamerism exists implies that these three cone outputs can be obtained by many different combinations of three light sources. This, in turn, implies that these three cone outputs do not uniquely specify the reflectance function of a given surface. In other words, many different reflectance functions can, in principle, give rise to the same three cone output values. This implies that we see surfaces with different reflectance functions as being identical in color. We provide a formal (mathematical) description of metamers later.

Trichromacy Theory of Color

The three-channel or **trichromacy** theory of color perception was, it is generally agreed, first proposed by George Palmer in 1777, who believed that light itself consisted of three different types of colored light, red, yellow, and blue, and that each type was detected by one of three corresponding retinal "particles" (e.g., photoreceptors). Palmer's hypothesis that light consisted of three different types of colored light was odd given that Isaac Newton had conclusively demonstrated that light consists of a continuum of colors in his experiments during 1665–1666. Perhaps Palmer's view about the nature of light arose because he was a merchant of stained glass and not a full-time scientist (though he was clearly a diligent part-time scientist).

The trichromacy theory was supported by Thomas Young's experiments on color mixing, **17.14**, in 1802, and by similar experiments by James Maxwell in 1856. Helmholtz, also working in the 19th century, concluded that the eye contains three types of nerve fiber which measure red, green, and blue light. His estimates of the tuning curves of each of these three nerve fibers are strikingly similar to modern measurements of the tuning curves of the cone system.

Opponent Color Theory

Hering proposed his opponent color theory in 1872. Hering had noted several anomalies which seemed to be incompatible with a simple trichromacy theory of color. These anomalies were that it does not seem possible to experience a "red-ish green" or a "blue-ish yellow" color. Accordingly, he proposed that colors were represented in two opponent pairs, red-green and blue-yellow, which is why we referred to this above as a four channel theory (red, green, blue, and yellow). However, Hering later added a third non-opponent pair: black-white. The core idea underlying Hering's theory, *opponency*, has stood the test of time, so much so that his theory could have been used to predict the characteristics of some of the color sensitive cells in the visual system (see below).

Compelling support for the existence of color opponent channels comes from color aftereffect illusions. Look at the dot at the center of **17.17a** for about 30 seconds, and then look at the central dot **17.17b**. You should see that the hitherto white circles in **17.17b** are tinged with colors post-adaptation. Thus, the red adaptation causes "its" white circle in **17.17b** to appear greenish, and vice versa for the green circle. Equally, adaptation to the blue circle causes a yellowish tinge, and vice versa. Crucially, the aftereffect consists of colors that are opposite members of an opponent pair (i.e., red-green, blue-yellow). A particularly dramatic color aftereffect is shown in **17.1**.

The red-green channel may seem plausible in that its opponents can be linked to the L- and M-cones. Also, the output of the S-cones can be considered as the input to the blue side of the blue-yellow channel. But we have now run out of cone types, so where does the yellow component of the blue-yellow channel come from? Figure **17.18** shows how it is possible to construct a blue-yellow channel which consists of the output

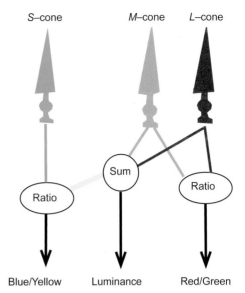

17.18 Three cone types feed three channels
Two opponent channels are illustrated, Blue/Yellow and Red/Green, and one non-opponent luminance channel.

of S-cones, and a yellow channel which consists of the combined outputs of the L-and M-cones. The luminance channel shown is non-opponent, and is derived from outputs of L- and M-cones.

Additional psychophysical support for these opponent channels is the finding that it is possible to "null off" the perception of red from a red light by mixing it with a green light, and vice versa. However, when red and green lights are mixed in suitable proportions to eliminate the perception of either red or green, the result is not, as you might expect, the perception of gray, but yellow. This fact fits in neatly with the scheme shown in **17.18**. This is because even when the outputs of the L- and M-cones have been nulled off within the red-green channel their combined activities are still "in force" to create a signal that leads to the perception of yellow mediated through the blue-yellow channel.

It turns out that these opponent channels are optimal for transmitting as much information as possible, as shown by Ruderman, Cronin and Chiao (1998).

Finally, physiological support for color opponency can be found in the form of color sensitive retinal ganglion cells (see below).

The nature of "opponency" deserves discussion. One possibility is that the outputs of the red and

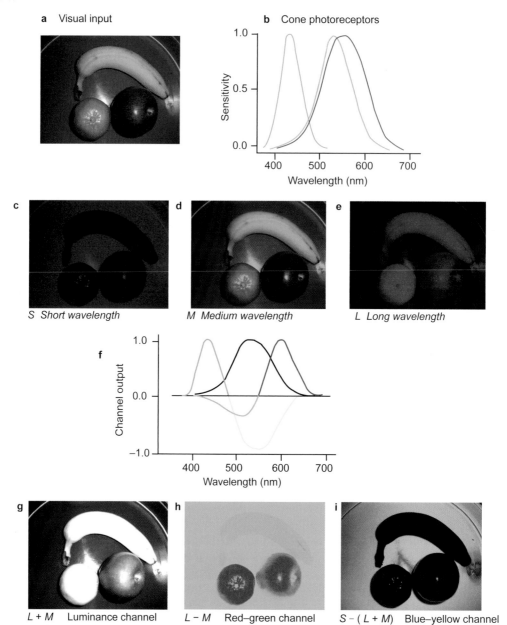

a Visual input

b Cone photoreceptors

c S Short wavelength

d M Medium wavelength

e L Long wavelength

f

g L + M Luminance channel

h L − M Red–green channel

i S − (L + M) Blue–yellow channel

17.19 Dual process theory

a Fruits and vegetables of different colors.

b Normalised absorption spectra of the three cone types (channels) in the eye.

c–e The outputs r_L, r_M, and r_S of the three cone types (channels) L, M, and S represented as three colored images.

f Normalised response profiles of the three types of opponent retinal ganglion and LGN cells (channels).

The blue-yellow and red-green curves show responses of corresponding color-opponent cells.

The black curve shows the response of the non-opponent luminance channel (formed by adding the responses of the red and green channels).

The outputs of these three channels for the fruit bowl image **a** are shown in **g–i**.

g Output of the luminance channel (obtained by summing the outputs of L and M cone types).

h Output of the L-M ("red-green") channel, obtained by subtracting the output of the "green" M channel from the "red" L channel.

i Output of the S-(L+M) ("blue-yellow") channel, obtained by subtracting the output of the "yellow" (L + M) channel from the "blue" S channel, where the "yellow" channel is the sum of the "red" and "green" channels.

Adapted schematically from Gegenfurtner (2003).

green channels are subtracted from each other. However, the red-green channel behaves as if its output is given by the ratio L_L/L_M, where L_L and L_M denote the luminance input to the red and green channels, respectively. A similar case can be made for the input to the blue-yellow channel.

The ratio concept of color opponency has the desirable effect of removing the impact of the overall level of illumination. For example, the ratio of input luminance to the red and green channels is the same irrespective of the overall illuminance because illuminance affects the red and green luminances equally. In contrast, the difference between the luminance values in the red and green channels increases with the overall level of illumination.

A subtractive way of thinking about color opponency can be "rescued" by subtracting the channel output of r_M and r_L (for example) because

$$r_L - r_M = \log L_L - \log L_M = \log L_L/L_M .$$

This mathematical "sleight of hand" is justified on the grounds that the output of a cone increases in proportion to the logarithm of the light energy absorbed by that cone. However, for the sake of simplicity, we assume that cones are *linear* so that their outputs increase in proportion to light energy, and therefore that channel outputs are expressed as ratios as in **17.19**, and in our account of retinex theory given below.

The kind of scheme shown in **17.19** is called a *dual process theory* as it embodies the idea that color is processed in two stages, a trichromatic stage followed by a color opponent stage. This way of thinking about color vision was proposed in 1957 by Hurvich and Jameson in order to reconcile the apparently contradictory trichromacy and opponent theories.

Color Opponent Neurons

Neurophysiological investigation of the visual system has vindicated all three theories: trichromacy, opponent channels, and dual processes. It is notable that all three theories were based on psychophysical findings, and were proposed before relevant modern neurophysiological research.

With regard to the opponent theory, retinal *bistratified ganglion cells* display blue-yellow opponency. The response to yellow light is obtained by combining outputs of cones sensitive to red and

green light. To date, cells of this type have been shown to be excited by blue and inhibited by yellow. This is true for all parts of the receptive field, and therefore such cells display only *spectral opponency* but not *spatial opponency*. This blue-yellow channel seems to be associated with the *konio cells* of the lateral geniculate nucleus, Ch 9.

Retinal *midget ganglion cells* have an output which is the difference between the outputs of cones sensitive to long and medium wavelengths, but only in the center of the retina. Mather has suggested that this restriction to the central retina may be a statistical side effect of the small receptive field sizes found there. These cells have a center-surround receptive field, and display spatial opponency. For example, a red light may excite the central region and inhibit the suround of a cell's receptive field, whereas another cell may show this same behavior for green light. These cells project to the *parvocellular* layers of the lateral geniculate nucleus (see Ch 9, p. 211). A red-green color opponency is also shown by *parasol ganglion cells*.

(Note that the midget and bistratified ganglion cells do not contribute to the *magnocellular* pathway, which begins in the retina with the parasol ganglion cells whose inputs are derived primarily from rods.)

We cannot help but note with disappointment, even dismay, that the number and diversity of various degrees of spatial and spectral opponencies found in the retina/LGN/cortex is bounded only, it seems, by the enumeration of all possible types of combination imaginable. In the absence of a coherent computational theory, simply describing all the types that seem to exist in the visual system is as close as one can get to mere stamp collecting as is possible without actually having a stamp album. While certain species of stamp collecting have their place in science, they are only useful if they contribute to a computational theory of stamps. Accordingly, as one alternative, we would urge the reader to follow the information-theoretic analysis of Ruderman et al. (1998) as a profitable line of research.

The information carried by these ganglion cells gives rise to the color-sensitive cells, **17.20,** found in the *cytochrome oxidase blobs* of the striate cortex (although recent studies have cast doubt on the blobs being specially concerned with color process-

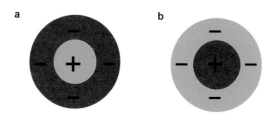

17.20 Receptive fields of two typical spatially opponent midget retinal ganglion cells
a Green light is excitatory in the central region and red light is inhibitory in the surround.
b Red light is excitatory in the central region and green light is inhibitory in the surround.
If the outputs of a pair of these two types are combined then this gives rise to the double–opponent cells found in the color blobs of primary visual cortex. For such a cell, for example, green light is excitatory in the central region and inhibitory in the surround, whereas red light is inhibitory in the central region and excitatory in the surround.

ing). These blobs are almost always located at the center of ocularity columns (which suggests they are monocular), and at the center of pinwheels (which suggests they respond equally well to all line orientations; Ch 9).

Information from the color blobs in the striate cortex is "collated" in the area V2, an area that envelopes the striate cortex (V1). From there, information goes to an area that seems to be specialized for color, known as **V4**.

Brain Areas Specialized for Color: V4, V8

The region of the brain known as V4 has been proposed as the color area, initially by Samir Zeki. Unlike cells in "lower" brain areas (e.g., retina, LGN, striate), cells in V4 exhibit color constancy. That is, when exposed to a patch in a Mondrian, the response of some V4 cells do not vary if the spectral balance of the illuminant is changed.

In general, V4 cell responses seem to coincide with the colors perceived by humans rather than the wavelengths of light to which lower brain areas respond. However, like many areas once thought to be exclusively dedicated to a single property, many cells in V4 also respond to stereo, direction and orientation. Even so, about 50% of cells in V4 are sensitive to color, which is a higher proportion than in other specialized areas (e.g., V1, V2, V3, and V5(MT)). But this is a controversial area of research, and a nearby region V8 has also been proposed as the color area.

The findings that underpin these claims may reflect the functions of the areas, or it may reflect the fact that different researchers use different criteria and experimental protocols for obtaining evidence consistent with a specific region being associated with color processing. Our own viewpoint is that it matters less where a computation occurs than the nature of the computation being carried out.

Skip to Summary Remarks?

The next few sections deal with some rather more technically demanding material. You may wish to proceed to *Summary Remarks* on p. 416.

Retinex Theory and Color Constancy

We said earlier that the fundamental problem of color vision consists of estimating the reflectance function $S(\lambda)$ of a surface (its physical "color") from the observed luminance spectrum $L(\lambda)$. To recapitulate, the luminance $L(\lambda)$ of a surface at each wavelength λ is given by the energy $E(\lambda)$ in the light source multiplied by the reflectance $S(\lambda)$ at that wavelength (i.e., proportion of light reflected at that wavelength by the surface), **17.6**:

$$L(\lambda) = E(\lambda)\, S(\lambda).$$

The luminance spectrum is a spectrum associated with each point in the retinal image, and is effectively *sampled* by the three color cone types. Even though this sampling of the luminance spectrum causes problems, the main obstacle to finding $S(\lambda)$ is that the observed luminance spectrum $L(\lambda)$ is the product of an *unknown* reflectance function $S(\lambda)$ with an *unknown* illuminance spectrum $E(\lambda)$.

In 1977, Land published an account of his **retinex theory**, which shows how it is possible to estimate the reflectance function $S(\lambda)$ of a surface, using a set of plausible and reasonably robust assumptions. The name *retinex* derives from Land's assumption that the theory involves computations within the retina and cortex (**retin**a-cor**tex** = **retinex**).

The solution proposed by Land is to treat the outputs of all, let's say for example, the long wavelength receptors as a separate image or channel, and to solve the lightness problem within this long wavelength channel ("red"). The same is done for the medium wavelength ("green") and short wavelength ("blue") channels associated with re-

Long wavelengths ("red")

Medium wavelengths ("green")

Short wavelengths ("blue")

17.21 Land's three lightness channels
The photograph of the flowers (upper left) is shown on the right in three achromatic versions produced from the red, green, and blue components of a standard digital camera, so that light of just one narrow range of wavelengths is selected in each case. Showing them in black and white emphasizes that each channel encodes intensities for a given range of wavelengths, and not color as such. However, the colored copies of them below the main photograph help reveal what each channel is encoding.

ceptors sensitive to medium and short wavelengths respectively, **17.21**. The overall effect of all this is to *"discount the illuminant"* independently in each of the three color channels. If we assume that each of the three color channels is represented by the wavelength to which it is most sensitive then the essence of Land's retinex method can be described as follows.

Remember, we seek the reflectance function $S(\lambda)$ associated with each point in the image of a scene. For brevity, we will assume that we are dealing with a single image region unless stated otherwise from here on. For simplicity we will treat the three cone types c_S, c_M, and c_L as separate channels, each of which is sensitive to a single wavelength, which we denote as λ_S, λ_M, and λ_L, respectively, so that we can estimate $S(\lambda)$ at only three wavelengths. Therefore, we will have to settle for an approximation of $S(\lambda)$ that consists of the value of $S(\lambda)$ at three wavelengths. It turns out that this approximation is *usually* quite good; that is, the reflectance function $S(\lambda)$ is well approximated by the three sampled values $S(\lambda_S)$, $S(\lambda_M)$, and $S(\lambda_L)$.

As an additional reminder, the eye effectively measures luminance spectrum $L(\lambda)$, the amount of light reflected at each wavelength λ. Let's consider the case for which $\lambda = \lambda_L$. In principle, we could use this to obtain $S(\lambda_L)$ if we knew the illuminant energy $E(\lambda_L)$ at λ_L. Specifically, given that

$$L(\lambda_L) = E(\lambda_L)\ S(\lambda_L),$$

we could obtain $S(\lambda_L)$ by rearranging this equation to obtain

$$S(\lambda_L) = L(\lambda_L)/E(\lambda_L),$$

if we knew $E(\lambda_L)$, which we (still) don't.

Even though we do not know the illuminance spectrum $E(\lambda_L)$ we can find a quantity can act as a proxy for it. If a scene contains just one surface that reflects 100% of the light at the wavelength λ_L then it follows that the amount of light reflected from this surface is proportional to the amount of light source energy $E(\lambda_L)$ at λ_L. This surface effectively provides an **anchor** for this channel. Such a surface will stimulate one c_L cone, whose output $r_L(\lambda_L)$ is larger than that of any other c_L cone, and we use the symbol $R_L(\lambda_L)$ to denote the response of this particular cone.

If we express the response of every cone of type c_L as a proportion of $R_L(\lambda_L)$ then we have

$$r_L(\lambda_L)/R_L(\lambda_L).$$

Now, given our simplifying assumption that cones are linear, this ratio is the proportion of light reflected at the wavelength λ_L and is therefore, by definition, equal to the reflectance function at λ_L

$$S(\lambda_L) = r_L(\lambda_L)/R_L(\lambda_L).$$

We now have an estimate of the "lightness" within the long wavelength channel for a particular point on the surface, an estimate which depends on the assumption that $R_L(\lambda_L)$ is produced by a surface patch that reflects 100% of the light at the wavelength λ_L. In other words, we have an estimate of the proportion $S(\lambda_L)$ of "red" light reflected by this surface point. The same line of reasoning can be used to estimate $S(\lambda_M)$ and $S(\lambda_S)$ as

$$S(\lambda_M) = r_M(\lambda_M)/R_M(\lambda_M)$$

$$S(\lambda_S) = r_S(\lambda_S)/R_S(\lambda_S),$$

to obtain estimates of the proportions $S(\lambda_M)$ and $S(\lambda_S)$ of "green" and "blue" light reflected by this surface point.

So now we have values $S(\lambda_S)$, $S(\lambda_M)$, and $S(\lambda_L)$ for the reflectance function at three wavelengths

λ_S, λ_M, and λ_L. If the reflectance function is smooth (and it usually is) then these three samples are a good approximation to the true shape of the reflectance function. Job done? Not quite.

We need one more assumption. In practice, the response r_L of a cone such as c_L depends on its response to wavelengths other than λ_L. But if that response is determined primarily by the amount of energy at λ_L then we can safely disregard the influence of other wavelengths on r_L. In other words, we assume that the observed response of a cone is determined by the intensity of the wavelength to which it is most sensitive, so that (for example) $r_L = r_L(\lambda_L)$ and similarly $R_L = R_L(\lambda_L)$.

If we consider a single point in the scene which yields three cone outputs (r_S, r_M, r_L) then the surface's reflectance function $(S(\lambda_S), S(\lambda_M), S(\lambda_L))$ at the wavelengths λ_S, λ_M, and λ_L can be approximated by the triplet

$$(S(\lambda_S), S(\lambda_M), S(\lambda_L)) \approx (r_S/R_S, r_M/R_M, r_L/R_L).$$

Thus, from image measurements within three color channels, we can recover a reasonable approximation to the color, or more accurately, the reflectance function $S(\lambda)$, of any point on a surface.

As stated above, a key part of the retinex theory, as described above, is finding a surface patch which reflects 100% of the light at λ_L, for example. This is done in order to estimate the maximum value of the reflectance function $S(\lambda_L)$ at λ_S (because a surface cannot reflect more light than it receives from a light source). In other words, the maximum observed value of $S(\lambda_L)$ acts as an anchor value for one channel. We discussed the problem of finding such anchor values in Ch 16, when discussing the lightness computation of gray-level surfaces. The point we are making here is that it crops up also in the surface color computation, and that Land's solution is to use the *maximum luminance rule* within each wavelength channel. In terms of a single channel, this rule translates into a kind of **maximum channel output rule**. This amounts to an assumption that the surface patch which reflects most light at λ_L (for example) is actually reflecting 100% of the light at λ_L. It is likely that a break down in color constancy occurs if this assumption is violated.

Note that every point in the image can have a different reflectance function $S(\lambda)$. More realistically, each point within a local patch of image can be regarded as having the same reflectance function. This implies that the method outlined above must be applied to every "small patch" of image in order to estimate the color, or, more specifically, it must be applied to the spectral reflectance function of that image patch. Deciding appropriate "small patches" within which to run the computation is a non-trivial task.

Finally, it is noteworthy that Land and others have shown that predictions from his retinex theory account for a range of experimental data. However, sometimes the predictions of retinex-based models deviate from the data. See Land (1977, 1983) if you want to follow up this story.

Color Metamers: A Formal Account

As discussed in earlier, any colored light can be matched by the combined output of three lights with different wavelengths. Suppose we are given a **test light** to match, which excites the short, medium and long wavelength cones by an amount r_S, r_M, and r_L, respectively. We are also given three **primary lights** which look roughly blue, green, and red, corresponding to short, medium, and long wavelength lights. For simplicity, we assume that each light has a single wavelength, which we denote as λ_S, λ_M, and λ_L, respectively. We are asked to adjust the intensity of all three primary lights until the end result matches the test light.

Here, we explain how it is possible to find such a combination to yield a **metameric** match. In order to refresh your memory of the basic strategy behind this formal account, you may wish to re-read the section *Color Metamers* earlier in this chapter.

First, consider how the three primary lights yield an output r_L from the long wavelength cone c_L if all three primary lights have the *same* intensity. The response of this cone c_L to the "blue" primary light with wavelength λ_S is $r_L(\lambda_S)$, the response of that cone to the "green" primary light with wavelength λ_M is $r_L(\lambda_M)$, and its response to the "red" primary light with wavelength λ_L is $r_L(\lambda_L)$. Thus, the observed response r_L induced in c_L by the three primary lights is

$$r_L = r_L(\lambda_S) + r_L(\lambda_M) + r_L(\lambda_L).$$

In the color mixing task, we are allowed to adjust the intensity of each primary light: suppose we choose intensities of a_L, a_M, and a_S for these lights, so that they do not have the same intensities as in the example above. As the response of a cone, say c_L, is proportional to the intensity of light falling on it at each wavelength, the response r_L of that cone is

$$r_L = a_S r_S(\lambda_S) + a_M r_L(\lambda_M) + a_L r_L(\lambda_L).$$

Now suppose we wish to excite c_L so that its output is some new value, r_L'. If we set the blue and green primary lights to some intensity, which we denote as a_S' and a_M', then we can always find a value for the intensity a_L of the red primary which ensures that $r_L = r_L'$. We can work out the value of this intensity by rearranging the equation above:

$$a_L = [r_L - a_S r_S(\lambda_S) - a_M r_M(\lambda_M)]/r_L(\lambda_L).$$

Of course, this also holds true if we had chosen to fix any other two primary light intensities. So already we can see that the desired output of a single receptor can be obtained by an infinite number of different combinations of primary light intensities. How does this generalize to triplets of cones?

Suppose a test light excites the short medium and long wavelength cones and yields outputs r_S, r_M, and r_L, respectively. We now know that *each* of the quantities r_S, r_M, and r_L can be obtained separately by adjusting the primary light intensities, but we have not yet established that *all* of them can be obtained simultaneously using a single triplet (a_S, a_M, a_L) of primary light intensities. Specifically, we know that

$$r_L = a_L r_L(\lambda_S) + a_M r_L(\lambda_M) + a_L r_L(\lambda_L),$$

and that a (possibly different) triplet of light intensities (a_S, a_M, a_L) yields a response from the cone c_M

$$r_M = a_L r_M(\lambda_S) + a_M r_M(\lambda_M) + a_L r_M(\lambda_L),$$

and that a (possibly different) triplet (a_S, a_M, a_L) yields a response from the cone c_S

$$r_S = a_L r_S(\lambda_S) + a_M r_S(\lambda_M) + a_L r_S(\lambda_L).$$

In broad terms, we now have three linear simultaneous equations and three unknowns, the primary intensity triplet (a_S, a_M, a_L). This implies that the primary light intensity triplet can be obtained by solving this set of simultaneous equations (excluding unusual circumstances, which we ignore here; for mathematicians, solving a set of equations like this is like falling off a log.)

The primary light intensities (a_S, a_M, a_L) are called the ***tristimulus coordinates*** of the test light spectrum, and the three corresponding cone responses (r_S, r_M, r_L) are called the ***cone coordinates***. This may be easier to see if we re-express the problem in terms of vector-matrix notation.

We denote the response of the short wavelength cone c_S to the wavelength of each of the three primary lights as the vector

$$\mathbf{r}_S = (r_S(\lambda_S), r_S(\lambda_M), r_S(\lambda_L)).$$

This vector is the response of one cone type c_S to the triplet of primary light intensities, if all primary light intensities are the same. The intensities of the primary lights are defined by the vector

$$\mathbf{a} = (a_S, a_M, a_L).$$

The responses of the three cones can now be written succinctly as

$$r_S = \mathbf{a}.\mathbf{r}_S$$

$$r_M = \mathbf{a}.\mathbf{r}_M$$

$$r_L = \mathbf{a}.\mathbf{r}_L,$$

where the dot denotes the ***inner product***.

Given the two vectors,

$$\mathbf{r}_S = (r_S(\lambda_S), r_S(\lambda_M), r_S(\lambda_L))$$

$$\mathbf{a} = (a_S, a_M, a_L),$$

their inner product is defined by multiplying corresponding elements of $\mathbf{a}$ and $\mathbf{r}_S$, and then summing the results of these products:

$$r_S = \mathbf{a}.\mathbf{r}_S$$

$$= a_S r_S(\lambda_S) + a_M r_S(\lambda_M) + a_L r_S(\lambda_L).$$

If we now define the cone responses as the vector $\mathbf{r} = (r_S, r_M, r_L)$ then we have the column matrix (shown, as conventionally, with the large brackets)

$$r = \begin{pmatrix} \mathbf{a}.\mathbf{r}_S \\ \mathbf{a}.\mathbf{r}_M \\ \mathbf{a}.\mathbf{r}_L \end{pmatrix}.$$

For the mathematically inclined, we can now define a 3x3 matrix M as

$$M = \begin{pmatrix} \mathbf{r}_S \\ \mathbf{r}_M \\ \mathbf{r}_L \end{pmatrix} = \begin{pmatrix} r_S(\lambda_S) & r_S(\lambda_M) & r_S(\lambda_L) \\ r_M(\lambda_S) & r_M(\lambda_M) & r_M(\lambda_L) \\ r_L(\lambda_S) & r_L(\lambda_M) & r_L(\lambda_L) \end{pmatrix}$$

so that we can obtain the three cone responses $\mathbf{r}$ as the vector-matrix product

$$\mathbf{r} = M\mathbf{a}^T,$$

where the superscript T is the ***transpose operator***, which simply turns a row vector into a column vector. Now we can obtain the primary light intensities as

$$\mathbf{a} = M^{-1}\mathbf{r},$$

where M^{-1} is the 3x3 inverse of the matrix M. Note that the elements of M are determined by a combination of the spectral sensitivities of each cone type, and the spectrum of the three primary lights. Both of these are fixed within a given experiment, and so M is a matrix of nine constants. Therefore, the only way to obtain the required cone outputs is to adjust the elements of $\mathbf{a}$ (the primary light intensities). In fact, it can be shown that such values exist for the elements of $\mathbf{a}$ provided the rows of M are ***linearly independent***, which means that no row of M can be obtained as a ***linear combination*** of the other rows. This amounts to the condition that one primary light cannot be metameric with respect a mixture of the other two.

Summary Remarks

The existence of metamers implies that two different spectra can give rise to the same triplet of cone outputs. But a single cone triplet does not necessarily result in a single color being perceived, as demonstrated by the use of Mondrian paintings. Even worse, two different cone triplets can give rise to the *same* color being perceived, depending on the context of the perceived color. The existence of this perceptual spaghetti is what makes color a hard problem for the brain solve, and a hard problem for us to understand how the brain supports color perception.

Perhaps you have been just as surprised by this chapter on seeing color as by Ch 16 on seeing black, grays, and white. After all, how many non-specialists would suspect that so little is understood about such a basic attribute of our perceptual worlds as how we see color? To be sure, a vast amount is known about receptors and the properties of color sensitive cells in the retina and the cortex. What is lacking is a theoretical understanding of why these various cells have the properties that they do. In short, in this area of seeing, knowledge at the hardware (implementation) level in Marr's conceptual framework far outstrips understanding at the computational level. It is worth quoting here the rueful statement recently written in 2007 by two highly respected color vision scientists, Solomon and Lennie:

> … problems … stem principally from our not having a clear idea of the properties to be expected of neurons that are responsible for color perception.

Color perception is in need of a grand theory to organize the huge amount of knowledge about various color effects in terms of a few well defined principles. In short, we need a good computational theory of color perception. Land's work on the retinex theory has proved an influential starting point but it is unable to explain all aspects of color perception. As Hurlbert (2007) points out:

> Mondrians are unlike real colored objects and color adjustment tasks are unlike the tasks of everyday life. Mondrian stimuli do not possess specular highlights, surface irregularities, shading, shadows, mutual reflections, or other reflection features due to three-dimensional shape, all of which, in theory, reveal information about the illumination color.

The lack of a grand over-arching theory of color vision condemns us at present, to a considerable extent, to be recorders of unexplained phenomena.

Further Reading

Bowmaker JK and Dartnall HJA (1980) Visual pigments of rods and cones in a human retina. *Journal of Physiology* **298** 501–511.

Brainard DH (2001) Color vision theory. In *International Encyclopedia of the Social and Behavioral Sciences* Smelser NJ and Baltes PB (Eds.) Elsevier, Oxford, UK, Vol 4 2256–2263. Available from http://color.psych.upenn.edu/brainard/pubs.html. *Comment*: A very thorough computational account of the various problems of color vision, including a detailed vector-matrix formulation of color theory.

Brainard DH and Wandell BA (1986) Analysis of the retinex theory of color vision. *Journal of the Optical Society of America* 1651–1661.

Edwards B (2004) *Color: a course in mastering the art of mixing colors*. Tarcher/Penguin. *Comment* The best account of color use in painting that we have come across. Explains how artists commonly use the terms hue, intensity and value for terms set out in **17.7** of hue, saturation and brightness.

Foster DH (2003) Does color constancy exist? *Trends in Cognitive Sciences* **7(10)** 439–443. *Comment* Questions existence of color constancy.

Gegenfurtner KR (2003) Cortical mechanisms of color vision. *Nature Reviews Neuroscience* **4** 563–572. *Comment* This paper has a similar figure to **17.19**.

Gilchrist AL (2003) Looking backward. *Nature Neuroscience* **6** 550 *Comment* A review of Purves and Lotto's book cited below.

Hurlbert A (2007) Color constancy. *Current Biology* **17**(21) R906-R907.

Land EH (1977) The retinex theory of color vision. *Scientific American* Vol. CCXXXVII 108–128.

Land EH (1983) Recent advances in retinex theory and some implications for cortical computations: color vision and the natural image. *Proceedings of the National Academy of Sciences USA* **80** 5163–5169.

Lay DC (2005) *Linear algebra and its applications*. Addison Wesley. *Comment* A very intuitive introduction to vector-matrix algebra.

MacLeod DIA (2003) New dimensions in color perception. *Trends in Cognitive Sciences* **7**(3) 97–910.

Mather G (2008) *Foundations of Perception, 2nd Ed*. Psychology Press. *Comment* A good textbook with an excellent tutorial section on the photometric system used for measuring light, including luminance.

Mollon J (2006) Monge. The Verriest Lecture. *Visual Neuroscience* **23** 297–309.

Nathans J (1989) The genes for color vision. *Scientific American* February, pp 28–35. *Comment* A good starting point for those interested in color blindness.

Purves D and Lotto RB (2003) *Why We See What We Do: An Empirical Theory of Vision* Sinauer Associates: USA.

Roorda A, Williams DR (1999) The arrangement of the three cone classes in the living human eye. *Nature* **397** 520–522.

Ruderman DL, Cronin TW and Chiao C (1998) Statistics of cone responses to natural images: implications for visual coding. *J Opt Soc Amer* **15** 2036–2045.

Shapley R (2009) Vision: Gene therapy in color. *Nature* **461**, 737–739 *Comment* Replacing a missing gene in adult color-blind monkeys restores normal color vision. A remarkable and surprising result, given that these monkeys lacked the gene from birth to adulthood.

Solomon SG and Lennie P (2007) The machinery of color vision. *Nature Reviews* **8** 276–286 *Comment* Excellent overview of the neurophysiology of color vision.

Zeki SA (1993) *Vision of the Brain* Wiley-Blackwell. *Comment* A lovely book that includes an account of Zeki's work on the neuroscience of color systems.

4.29 McCollough effect (from Ch 4)

First, look at the test stimulus **4.29b**, which is simply a pattern of black and white bars, some vertical and some horizontal. Be sure to confirm that this stimulus appears quite colorless to you.

Next, adapt your visual system using **4.29a**. Look for 10 seconds at the red–vertical pattern and then for 10 seconds at the green–horizontal one, then go back to the red–vertical for a further 10 seconds, and so on. Alternate between the two adapting stimuli for a period of at least 3 minutes. If you can bear it and want to obtain a good effect, do this for 10 minutes. When the adaptation period is up, transfer your gaze back to **4.29b** and note what you see.

After adaptation the black and white test stimulus appears colored; this is the **McCollough effect**. However, these color illusions are not color after-images of the usual sort, Ch 4. If you look carefully you will see that the vertical bars in the test stimulus appear faintly greenish (the vertical adaptation bars are red) and the horizontal test bars appear faintly reddish (the horizontal adaptation bars are green).

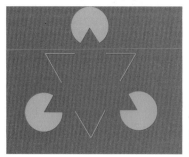

16.19 Isoluminant version of Kanizsa's triangle (from Ch 16)

The lightness aspect of the illusion is much reduced or absent. See Ch 16, pp. 389–391, for a discussion of this figure. Note that printing limitations may not create a perfect isoluminant figure.

18

Seeing with Two Eyes, Part I

18.1 Egyptian ship model
View with the red/green filters supplied with this book. Place the green filter in front of the right eye. If you have normal vision you will see a strong three-dimensional effect, with the prow of the boat sticking out from the page. This picture is an ***anaglyph,*** and the way such figures work is explained in **18.3**. From http://commons.wikimedia.org/wiki/File:Egyptian_ship_model_1.jpg under the *GNU Free Documentation License*. You can see many more anaglyhs at that site, and at others that you can find by searching the Web. See also Julesz's classic book (1971; republished in 2006 by The MIT Press) for many stunning anaglyphs. This book's Web site (http://mitpress.mit.edu/seeing) has all the anaglyphs in this chapter: viewing them on-screen using the filters supplied with this book gives more vivid depth effects than in the printed versions, due to better separation into the left and right eyes of the two superimpose images (see **18.3** for an explanation of this *cross-talk problem.*)

Have you ever wondered why we have two eyes? If you think the answer is obvious, try closing or covering one eye and then walking around. Did the world look very different? Did you bump into anything? We doubt it. The reason is that the visual system is richly endowed with a variety of mechanisms able to exploit **monocular depth cues** for seeing the spatial relationships of the objects in the scene before us. We do not always need two-eyed or **binocular vision** for seeing depth. Painters know well how to use monocular depth cues to create vivid depth effects, **18.2**.

This simple experiment shows that we can get by without binocular vision, so that some of the celebrated cases of monocular achievements become rather more understandable. For example, the Nawab of Pataudi played first class cricket with great skill after losing the sight of one eye. Wiley Post, who made the first solo flight around the world, was similarly one-eyed, thereby demonstrating that adequate depth cues are available monocularly for taking off and landing aeroplanes. Such high-class performances demonstrate that whatever the benefits bestowed by binocular vision, they are not so crucial as to make monocular vision a severe handicap.

This conclusion is supported by the fact that about 4% of people are essentially monocular, in that each eye may work well enough alone but the two eyes do not cooperate as they should. Nevertheless, people with monocular vision get by remarkably well, so much so that many of them have no idea that they suffer a binocular visual deficit until it is detected by routine clinical screening. These considerations suggest that the proper answer

18.2 Vivid depth from monocular depth cues
This is a pavement drawing by Julian Beever. He is an English artist famous for his art drawn on the pavements worldwide exploiting a wide range of monocular depth cues, which include: *linear perspective* (parallel receding lines converge, e.g. the lines forming the sides of the building descending "into" the pavement); *shading*; and *interposition* (near objects mask those behind them). The power of the monocular cues in this drawing makes it difficult to believe that it is drawn on a flat pavement.

Why read this chapter?
We begin by discussing why people have two eyes. We then describe various methods for presenting *stereograms*, which are stimuli that mimic the small differences, called *binocular disparities*, between the images in the left and right eyes. Finding matches between corresponding elements in the two images is called the *stereo correspondence problem*. It is illustrated using random-dot stereograms and the Marr/Poggio model for solving it is described. We use a sequence of stereograms to highlight various aspects of human *stereopsis*. Finally, the elements of *stereo geometry* are set out and we explain how *horizontal disparities* can be used to recover 3D scene structure.

to the question "Why do we have two eyes?" is that having two is good insurance against losing one.

But if we choose the viewing conditions more carefully, one-eyed viewing can give us a clue about another possible reason for having evolved binocular vision. Try covering one eye again, but this time keep your head still and look first with two eyes, and then with just one, at a vase of flowers, a tree, a bush, or any object which has lots of parts arranged at different distances (or "depths"). With one eye it is difficult to discriminate the relative depths of the leaves, petals, etc., but with two eyes it is easy if you have normal binocular vision. The appearance of depth when both eyes are open is called *stereopsis* (from the Greek for "solid sight').

Repeat the one-eyed viewing experiment several times to be sure that you see the no-depth/depth transition as you open the closed eye. Be sure also to look for several seconds with one eye only before looking again with two. This precaution is necessary because the three-dimensional (3D) effect built up with binocular vision, prior to closing one eye, takes that long to disappear.

Also, be very careful not to move your head at any stage. Objects at different depths produce retinal images that move at different rates across the retina when the head is moved. This is another **depth cue** called **head movement parallax**. The brain is quite up to the task of using this cue to generate depth perceptions which can be as vivid as binocular stereopsis. This is why in the opening paragraph you were invited to walk around the room with one eye covered, rather than looking at it from a single stationary position. Walking ensures that the depth cue of head-movement parallax is available to the visual system. Hence, walking helps show how well we can cope when forced to rely on just one eye.

Although walking helped ensure a "depthful" visual world in the initial monocular viewing experiment, the world does not appear totally flat with stationary monocular inspection. To be sure, certain sorts of objects, such as vases of flowers, lose that special kind of "depthfulness" called stereopsis under these conditions, but the world as a whole is still seen in 3D. Some of the monocular depth cues which the brain uses for this kind of distance perception (sometimes called *monocular stereopsis*) are illustrated in **18.2**. The pavement art shown there demonstrates how powerful **pictorial depth cues** can be when used skillfully.

The dramatic depth effects in **18.2** are so vivid that the flatness of the pavement goes unnoticed. However, more usually when viewing paintings of a 3D scene, we see depth within them and yet we also see that it is portrayed on a flat surface. This paradoxical outcome can be eliminated for "normal" paintings by viewing them monocularly through a tube. This removes binocular cues supporting perception of the flat surface of the painting. This trick works especially well if the painting

contains strong shadows. It is well worth trying in a picture gallery. We discuss the combination of stereopsis with other cues to depth in Ch 20.

Anaglyphs

View the anaglyph shown in **18.1** wearing the red/green filters supplied with this book. If you have normal vision then should see a strong 3D effect. This is why the term *anaglyph* is appropriate: it comes from the Greek for a carving in relief.

If you find the 3D effect does not come at first, be persistent. Be sure to hold the filters with the green filter in front of the right eye. Some people find that the depth effect appears only after a few minutes but still find it vivid when they succeed. Once you think you have experienced the effect, try closing or covering one eye. The illusion of depth will then fade away, which demonstrates that it depends on binocular vision and not on any monocular depth cues present in the picture.

The principle underlying anaglyphs is explained in **18.3**. An anaglyph contains two component pictures called the **stereo halves** of a **stereogram**, also known as a **stereo pair**. The red/green printing, coupled with the red/green spectacles, is a way of presenting one stereo half to one eye and the other to the other eye. The term **binocular fusion** is used to refer to the brain combining the information in a stereo pair to create a single percept.

The depth effect seen in the anaglyph has nothing whatever to do with color vision. The colored filters are simply an optical method for separating the two over-printed stereo halves so that just one projects into each eye. Another way of achieving the same result is to print the two halves on film in such a way that they can be separated out using filters that let through only light that is polarized in one direction. This is the optical method usually used for 3D films shown in cinemas.

The reason we see depth in the anaglyph in **18.1** is that its two component images are not identical. This simulates the fact that the two eyes do not receive exactly the same views of a 3D scenes viewed close at hand. This is illustrated in **18.4**, which shows an observer looking down on a solid object—a pyramid with the top sliced off. The particular point he is looking at is called the **fixation point**. This point will change as eye movements take his gaze around a scene. Sketches of his left

Left and right halves of a stereogram, also called a stereo pair.

Anaglyph

The two halves of the stereo pair above are printed together to create the anaglyph. See **18.1** for a larger version and view it with the red/green filters supplied with this book.

Only the red colored image gets through the red filter, and so into the left eye.

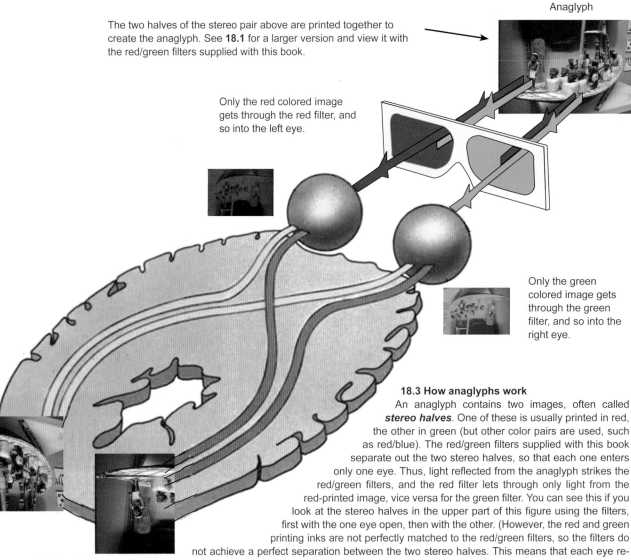

Only the green colored image gets through the green filter, and so into the right eye.

18.3 How anaglyphs work

An anaglyph contains two images, often called *stereo halves*. One of these is usually printed in red, the other in green (but other color pairs are used, such as red/blue). The red/green filters supplied with this book separate out the two stereo halves, so that each one enters only one eye. Thus, light reflected from the anaglyph strikes the red/green filters, and the red filter lets through only light from the red-printed image, vice versa for the green filter. You can see this if you look at the stereo halves in the upper part of this figure using the filters, first with the one eye open, then with the other. (However, the red and green printing inks are not perfectly matched to the red/green filters, so the filters do not achieve a perfect separation between the two stereo halves. This means that each eye receives one image of quite strong contrast together with a second much fainter unwanted one. This *cross-talk problem* is greatest for the green filter for many anaglyhs in this chapter.) There are small differences between the two stereo halves and these differences, called *binocular disparities*, are used by the brain to generate 3D stereoscopic vision.

His right eye's image   His left eye's image

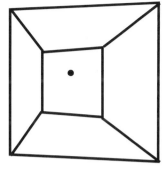
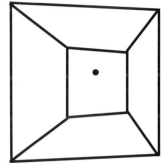

and right views are shown above the observer. You can see that they are not identical. The differences between them are the geometrical consequence of the two eyes being located in different positions in the head so they get slightly differing views. The differences are called ***binocular disparities*** and we will describe their characteristics in detail shortly.

You can easily observe binocular disparity for yourself by looking at objects that are at different distances but not too far away. First look with just one eye and note the positions of a couple of objects in that eye's view. Next, look with just the other eye at the same two objects. If you choose two suitable objects at different distances then you should be able to discern differences between the left and right eyes' images.

An anaglyph simulating the views of the observer in **18.4** is shown in **18.5**. Begin by looking at this without the red/green filters and notice the different positions of the red/green vertical edges of the base of the pyramid. This particular binocular disparity has the result that when you view this anaglyph with the filters (with green filter in front *right* eye, which is the default to be used for all anaglyphs in this chapter) the pyramid seems to be rising above the page. This demonstrates that binocular disparities are used by the visual system to generate 3D perceptions.

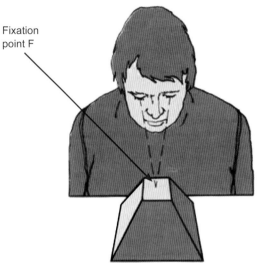

Fixation point F

Defining Binocular Disparity

Binocular disparities are the geometrical consequence of the two eyes being in different positions in the head, so that each eye gets a different view, **18.4**. We need to introduce the basics of ***stereo geometry*** to define the binocular disparity.

Consider **18.6** which shows a cross-section of the two eyes viewing a pair of rods. The optics of each eye is modeled as a pin-hole camera (Ch 2, p. 31), and rays of light are shown arising from the ***fixation point*** on the front edge of the right hand rod. These rays from the fixation point project to the centers of the left and right retinas.

Light rays are also shown arising from a point on the front edge of the left hand rod. These latter rays do not project to matching locations in the two eyes. Matching left and right retinal locations are technically known as ***corresponding points***. If a point in the scene does not project to corresponding points then it is said to have a (non-zero)

18.4 Binocular disparities
The two eyes receive slightly different views of 3D scenes viewed from near. The trapezoidal shapes are perspective projection effects (exaggerated here) arising from some parts of the pyramid being closer to one side of each eye than the other side of that eye (see Ch 19).

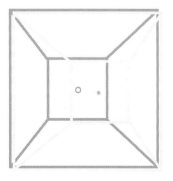

Notice that in this anaglyph, and others in later figures, the red-printed image is seen as dark lines in the eye viewing through the *green* filter. This is because the red print reflects little "green light" and so little gets through the red filter from this component. The same thing happens but the other way round for the green-printed image.

18.5 Anaglyphs of the 3D truncated pyramid in 18.4
This anaglyph has been made low contrast as the cross-talk problem is then less, which helps some people see the depth effect. View with the green filter in front of the right eye. This produces the perception of a truncated pyramid rising up from the page. If you put the green filter in front of the *left* eye then the truncated pyramid appears to descend into the page, for reasons explained on pages 430–431.

binocular disparity. The size of the disparity is often measured by the difference in the left and right retinal locations. A second (but equivalent) way of defining binocular disparity is in terms of the differences in the directions of the light rays illustrated in **18.6**. We will explain both definitions in more detail in this chapter.

Historical Note on Stereoscopes

Thinking about the mysteries of binocular vision has a long history, extending far back into antiquity. Moreover, diagrams similar to **18.6** showing the basic geometry of binocular disparity were set out many centuries before Charles Wheatstone (1802–1875) described in 1838 results from his famous *stereoscope*, **18.7**. (The word comes from the Greek for "solid viewing'), This is remarkable: why did it take so long for someone to ask whether two dissimilar pictures could be binocularly combined to yields the perception of depth? It is a mystery, but Nicholas Wade (1983) has provided a fascinating review of the history.

In fact, many readers probably did not need this book to tell them that binocular vision has something to do with the perception of depth because they learned this in childhood by playing with one or other form of stereoscope. These are optical instruments, using mirrors, prisms or lenses, that enable a different picture to be presented to each eye and so serve the same purpose as anaglyphs.

One type of stereoscope, popular in Victorian times, is shown in **18.8**. The two photographs in the device are of the same scene taken from slightly horizontally different camera positions chosen to imitate the different viewing positions of the two eyes. Thus, the pictures might have been taken with a single camera which was moved sideways between shots by about 6 cm (mimicking the distance between the eyes). Alternatively, a special stereoscopic camera equipped with two lenses about 6 cm apart, and capable of taking two shots simultaneously, might have been used.

Either way, binocular fusion of the pair of photographs in the stereoscope is made possible by a prism-cum-lens serving each eye, **18.8**. This ingenious arrangement, described by David Brewster in 1849, uses the prismatic function of the prism-cum-lens to alter the directions of light rays coming from the two photographs so that the

By convention, we refer here and elsewhere to the fovea as if it lies at the center of the retina, although it actually lies about 5° temporal to the eye's optic axis, i.e., from the true optical center. (Temporal means "toward the temples," i.e., in a direction away from the nose, **18.49**.)

Fixation point on front edge of nearer rod

Left eye

Right eye

Fovea

Fovea

Disparate images of far rod

18.6 Stereo geometry underlying binocular disparities
The eyes are fixating a point on the nearer rod, which therefore projects an image to the fovea (but note remark in upper left label in this figure). The image of the more distant rod falls on different or binocularly disparate locations in the two eyes—further to one side of the fovea in the left eye.

eyes can converge comfortably when fusing them binocularly.

The lens aspect of the prism-cum-lens ensures that the viewer can focus his eyes by using a setting of his own eyeball optics (technically called his state of *accommodation*) that is comfortable for the *vergence angle* between his eyes that is required for fusion. This latter angle is the one between the two dotted lines in the plan diagram in **18.8**. It varies according to whether we are fixating objects near to us (relatively large vergence angle, large accommodation effort required for the normal person), or far from us (relatively small vergence angle, small accommodation effort).

The net result in the Brewster stereoscope, as in other types, is that the observer sees a single binocularly fused scene, with the various objects in the

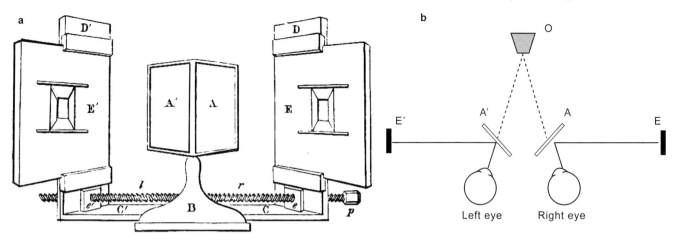

18.7 Wheatstone stereoscope
a The panels E´ and E hold disparate drawings that are reflected by the mirrors A´ and A, one into each eye as the observer faces the device.
b Plan diagram which makes it clear that the fused object O (here the pyramid with its apex sliced off) is seen lying in depth directly in front of the observer's eyes and behind the mirrors A´ and A. The solid lines represent light rays from the disparate drawings E´ and E.

scene appearing at vividly different depths. Hence, the scene perceived in the stereoscope is similar to the one that would have been seen in the original scene from which the photographs came.

A Brewster-type arrangement is the one used in modern stereoscopes available in any good

toyshop. It is well worth having a look at one. The stereopsis effect produced by a good stereoscope can be quite startling, with the vividness of the stereopsis exceeding that obtained with anaglyphs.

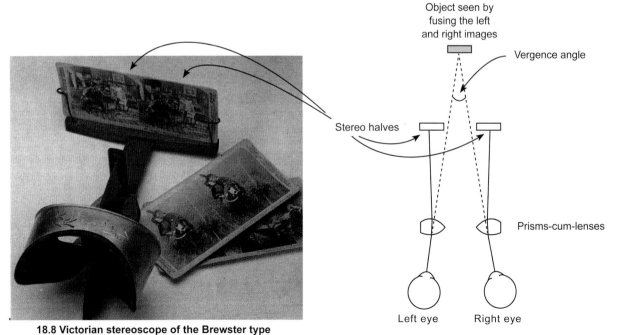

18.8 Victorian stereoscope of the Brewster type
The diagram on the right illustrates the optical principles of the prism-cum-lens used in a stereoscope of this type.

a

Left half forms
image in Right eye

Right half forms image
in Left eye

b Crossed-eyes
free fusion

Fused
binocular
images

Relaxed
vergence

Left eye   Right eye

c Parallel-eyes
free fusion

d

Normal vergence
angle for the plane
containing the
stereogram

18.9 Free fusing using the crossed-eyes method
a A helpful trick in learning to cross your eyes to obtain
binocular fusion of a stereo pair—see text on this page.
b Geometry: note that the left half of the stereogram enters
the right eye with the crossed-eyes method, *vice versa* for
the right half. This means that the depth effect is reversed
unless the right eye's image in placed on the left.
c Relaxation of convergence to obtain stereo fusion.
d Hand-drawn stereo pair for crossed-eyes fusion.

Free Fusing

The simplest and cheapest way of getting the two
halves of a stereo pair to the two eyes separately is
to place the two halves side-by-side and cross the
eyes, **18.9**. Not many people, however, can voluntarily cross their eyes in the controlled way required
for this technique to be successful—at least, not
without considerable practice. But the practice is
well rewarded because it enables you to expand
enormously the range of doodling which you can
engage in during dull moments in talks, seminars,
lectures, etc. Once you can cross your eyes suitably,
you can explore an endless variety of home-made
stereo effects by drawing your own stereo pairs.
Precision in drawing is not required. Even the
hand-drawn squiggles shown in **18.9d** will happily
fuse binocularly, despite the many mismatches of
their various parts, so that the line appears to have a
3D hump shape.

One way to learn how to cross your eyes in the
manner required is illustrated in **18.9a**. Hold a
pencil or some other marker about halfway between
you and the two halves of the stereo pair of the

truncated pyramid shown in **18.4**, upper. Fixate
on the top of the pencil and you are now crossing
your eyes to about the right extent. Keep the tip of
the pencil below the stereo halves, so that it does
not obscure them, and try gradually to pay attention not to the pencil tip but to the stereogram.
With effort and a bit of luck, the stereo halves will
eventually "snap" into fusion. Once this has happened you can take the pencil away and stare at the
glorious stereo effect quite unaided.

At first, you will find that your ability to hold
fusion and focusing will wax and wane, and your
eyes will get tired: so don't overdo it. But eventually, with persistence, you should be able to obtain
stereo fusion without need even of the pencil. Note
that the pyramid in **18.4** will recede into the page,
not protrude above it, with crossed-eyes fusion
because this causes the left stereo half to enter the
right eye, and vice versa for the right stereo half.

A word of warning, however: learning to cross
your eyes can lead to problems for some people, so
if you find it uncomfortable after a few tries then
you are best advised to give it a miss. Indeed, at

least one distinguished vision scientist refuses to learn the technique for fear that it will change forever how his binocular system works, thereby making him an atypical person for studies of stereopsis.

Some readers might prefer to relax the angle of vergence of their eyes to obtain the stereo effect, **18.9c**, rather than over-converge as happens when the eyes are crossed. This is a more unusual ability, however, and may indicate that your eyes have a tendency to swing outward, called a latent squint (technically an *exophoria*; the opposite tendency is an *esophoria*.) If the stereo halves are separated by more than about 6 cm (the approximate separation of the eyes), then it will be necessary to have the eyes point outward from parallel if fusion is to be obtained, if using the under-convergence method. Some ophthalmologists believe that it is impossible for a person with wholly normal eyes to enlarge the angle of his/her vergence voluntarily beyond parallel, i.e., that the most that can be achieved is a relaxation of convergence to the parallel position. In view of this, it seems best that, if you wish to learn the free fusion trick, you practice over-convergence rather than relaxation of convergence. You will not then be limited to fusing stereo halves separated by no more than about 6 cm.

It might help you as you practice to note that one of the major problems to be overcome whether you try to over-converge or under-converge is that you have to uncouple your usual *synkinesis* between vergence angle and accommodation. That is, you have to get used to focusing your eyes for a distance different from the one normally matched with the required vergence angle. So when practicing, do expect to see a rather fuzzy image during the early stages. Later on you will find it will become clear and sharp.

If you manage to achieve under-convergence, you should expect to see the opposite depth effect from someone using over-convergence. This is because under-convergence will result in the left stereo half going to the left eye and the right stereo half going to the right eye, as happens in the case of the anaglyph (when the filters are worn in their usual position—*green* filter in front of *right* eye). On the other hand, if you practice over-convergence, expect to see a reversed depth effect, because the right eye receives the left stereo half, and vice versa (**18.9b,c**). You can try getting reversed stereo

effects with the red/green filters by using them for the various anaglyphs in this book with the red filter in front of the right eye.

Learning free fusion will allow you to see the startling stereo effects in *autostereograms*. In the simplest instances, these are composed of many repeating pattern elements, **18.10**. The basic principle in autostereograms is known as the *wallpaper illusion*: have you ever noticed that a wallpaper with regular repeating elements sometimes can appear nearer to you (or farther away) than it really is? This is because your stereo mechanisms are fusing up the "wrong" pattern elements.

We present here a wide range of different stereograms so that you can appreciate the kinds of stereo effects that are possible and thus some of the phenomena with which any fully satisfactory theory of stereopsis must cope. In discussing these stereograms, the term "point" will be used loosely to refer to whatever type of stimulus element it might be which is combined for the purposes of obtaining local matches. Some stereograms are not referred to in the text but have self-contained captions.

Lenticular Stereograms

You may have experienced displays or table mats that create a vivid depth effect, without any need for special glasses or unusual eye movements. The principle on which they are based is that the right and left images of the stereo pair are printed as a series of vertical strips, such that alternate strips come from the two images. Then, overlaid on the combined image is a plastic sheet with a corrugated surface, designed with optical properties that "bend" the light reflected from each strip so that each eyes sees only the strips from one image. The upshot is that one eye receives one stereo half and the other eye gets the other stereo half. The result is a 3D percept deriving from whatever disparities are embedded in the two stereo halves.

Random-dot Stereograms

The autostereogram in **18.10** is an example of a random-dot *stereogram*. These highlight the *stereo correspondence problem:* how can corresponding points in the left and right images be matched?

The potential scale of this problem is illustrated in the computer generated random-dot stereogram

in **18.11**. This kind of stereogram was magnificently exploited as a stereo vision research tool by Bela Julesz, a Hungarian refugee radar engineer turned visual psychologist. Like any other stereogram, one of Julesz's stereograms consists of two halves to be presented separately to the two eyes, using one or other stereoscopic viewing technique.

Inspect **18.11** with the red/green spectacles, again making sure that the green filter is in front of your right eye (the two halves of this stereogram are printed in black and white below the anaglyph version). If you have normal binocular vision, in the center of the anaglyph you will see a square floating a few centimeters above its surround, as shown in the sketch in **18.11b**. The surround *S* will also appear to be lifted off the page a little, due simply to the two stereo halves being printed on the page in a slightly offset fashion, so giving a horizontal disparity to the whole figure.

The experience of depth is, for some observers, rather more difficult to achieve with a random-dot stereogram of this type than it is with stereograms of natural scenes (e.g., **18.1**). But if you have difficulty in getting the effect, do not give up trying too soon.

The immediate problem posed by a random-dot stereogram such as **18.11** is: how does the visual system work out which dot in the left image is to be matched with which one in the right image? Julesz called this the problem of **global stereopsis** because the visual brain somehow selects the correct overall (or global) solution from the myriad possible *local* matches. We will consider this ambiguity problem in detail but for the moment we describe how random-dot stereograms are constructed.

18.10 Autostereogram (left)
Either crossing your eyes or under-converging allows you to see a vivid stereo effect. Note that although colored red, this figure has nothing in common with an anaglyph—hence, viewing with the red/green glasses is inappropriate. Courtesy Jacques Ninio whose book (1994) has many fine examples.

Making a Simple Random-dot Stereogram

Despite the fact that the two halves of the stereogram in **18.11** look identical, they in fact contain the depth cue of binocular disparity which mimics the horizontal disparities that would be produced by a genuinely protruding square surface. It is as if a speckled square was held in front of a speckled background in such a way that the boundaries of the square merged so well into the texture of the background that this boundary could not be seen with either eye alone. Each half stereogram has a square within it which is shifted in relation to the corresponding area in the other half, but bear in mind that these squares are hidden to monocular view. This will become clearer by describing how the stereogram was made.

The first step in the manufacture of a random-dot stereogram is to create a piece of "random visual texture." In the stereo pair of **18.11** a computer was used to "draw" a random checkerboard of small black and gray squares (or "dots"— hence the name "random-dot stereogram"). But other components could have been used to equally good effect, such as round dots or small lines, as illustrated in anaglyphs later in this chapter. The important thing in a random-dot stereogram is not the shape of the component elements but the fact that they are distributed in a random way, which camouflages in each eye's individual view the area which finally appears in depth. That is, the randomness of the texture ensures that the area where there is disparity cannot be picked out when each half-stereogram is inspected on its own.

However, when an anaglyph is inspected without the red/green spectacles, either with one eye or both, the disparate area can sometimes be detected. But this is a by-product of the printing of one picture on top of the other, and as soon as the anaglyph is viewed through the spectacles, this by-product disappears, and the only remaining clue to the shape of the area which is intended to be seen in depth is the disparity between the two

a

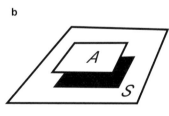

b

c

18.11 Random-dot stereogram

a Anaglyph: if viewed with the filters provided with the book, and if you place the green filter in front of the right eye, then you should see the 3D effect sketched in **c**, assuming you have normal stereo vision. Sometimes the depth effect for random-dot stereograms takes time to build up, so you may need to be patient.

b Sketch showing depth effect from **a**: a central square area of dots *A* appears lifted above its surround *S* (again, if **a** is viewed with the green filter in front of right eye).

c Two halves of **a** printed in black and white and side by side instead of on top of one another, as in **a**. They are arranged so that crossed-eyes fusion produces the same percept as in the anaglyph in **a** and as shown in **b**.

half-stereograms. Ideally, though, there is no trace of the shape to be seen in depth even when the anaglyph is viewed without the red/green spectacles.

Once a piece of random visual texture of some kind has been created, the next step is to make two copies. These copies become the two halves of the stereo pair. In one or both, a certain patch of random texture is shifted slightly horizontally, usually in a different direction in each half. This process is illustrated in **18.12** for a random-dot stereogram which has a square-shaped disparate area, as in **18.11** (but remember that the disparate area can be almost any shape required, as we will see). The disparate area in **18.12** is labeled *A*, its surround *S*. Shifting area *A* results in a certain portion of the texture being "covered up" and thus lost to view. The shifts also result in "holes" (*X* and *Y*) being created in each half-stereogram but these are filled in with new random texture so that they do not exist in the finished product. So it is impossible to see by inspecting each member of the stereo pair on its own which area has been shifted. The key property of a random-dot stereogram is thus ensured.

Note that the dots in *X* and *Y* areas in **18.12** have no "correct" matches in the other image. They are therefore said to be *uncorrelated*. Yet when the stereogram is fused they find a home, as it were, as part of the plane of the surround *S*. This makes perceptual sense because stereo images of natural scenes with opaque surfaces at different depths have many uncorrelated patches of texture—because they appear in one eye's view only. And it makes geometric sense for them to be placed perceptually behind the opaque surfaces whose textures do appear in each image. This is because if they were to be placed in front then they would appear in both images, which is physically impossible for uncorrelated regions in natural binocular viewing.

Some people do not find it easy to visualize the steps in the creation of a random-dot stereogram and the following alternative description is offered. First, obtain two identical copies of a piece of random visual texture: these will be used as the surround. Second, obtain two identical copies of another similar but smaller piece of random texture: these will be used as the disparate areas. Third, lay the small pieces of random texture on top of the first, larger pieces—one small piece on top of each large piece—to obtain the two members of the stereo pair. Make sure that the small pieces are precisely aligned on the larger pieces so that they cannot be distinguished as a different entity when each stereo half is viewed alone. Also, place the small pieces in horizontally shifted positions in the two halves, to give the disparity cue. You now have a random-dot stereogram.

It is time for you to try viewing **18.11** with the filters reversed, so that the green filter is in front of your *left* eye. If you hold the filters in this way and look at **18.11** afresh, you should now see in the center of the anaglyph a "window" through which you can see a square about a centimeter behind the frame of the window. The surround too seems to be a little further away from you than the page on which it is printed.

This reversal in depth occurs because when you reverse the filters you exchange the views which the two eyes receive, so that the disparity now mimics

Two identical copies of a random texture in which a region *A* is selected. *S* labels the texture region surrounding *A*.

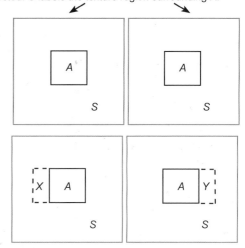

18.12 Making a simple random-dot stereogram
The position of region *A* is shifted in the two copies and the gaps thereby created (*X* and *Y*) are filled in with new random texture of the same type. The result is a pair of stimuli that comprise the left and right halves of the stereogram.

that which would be produced in the two eyes by a genuine window with a square seen through it. The same effect could have been achieved if, in the manufacture of the stereogram, the disparate squares were shifted not toward the mid-line but instead toward the outer edges, as shown in **18.13**. This is equivalent to exchanging the two eyes' views. The reversal in depth can also be achieved by turning an anaglyph upside-down while keeping the red/green filters in the same position.

When we see something lying further away than the point we happen to be fixating (e.g., the receding square seen as a surface behind a window in **18.11**, if we fixate the surround with the glasses reversed), then the further away entity is said to have ***divergent disparity***. This is because we would have to diverge our eyes to transfer our fixation from the surround to this further-away object. Equally, a ***convergent disparity*** cue is one which generates the perception of something nearer to us than the point we happen to be fixating (e.g., the protruding square seen in **18.11** if we fixate the surround with the filters used normally). Whether a convergent or divergent disparity cue is present with respect to the surround of a random-dot stereogram depends simply on the shift imposed on the disparate region, **18.13**. We will explain later the geometric definition of divergent and convergent disparities. For the moment, bear in mind that these terms are used with respect to the point being fixated, and in **18.13** it is assumed the observer is fixating the surround.

The reason shifts in opposing directions give opposite effects is simply geometrical. Receding surfaces cause one sort of shift, protruding the other, in retinal images of genuine 3D scenes. The brain takes advantage of this fact and generates depth perceptions accordingly. In a stereogram with a protruding square, as in **18.11** with green filter over right eye, the square has a convergent disparity with respect to the surround as just explained. However, if after fusion one chooses to fixate the square, then the surround has a divergent disparity with respect to the square. Beware possible confusion. In real life, our vergence angle changes all the time as we look at different fixation points, so the disparity signals landing on our retinas are also constantly changing, even for a stationary scene.

An important point to grasp about random-dot stereograms, and the one which makes them so attractive for studying stereopsis, is that they present the disparity cue to depth in a "pure" form, with monocular cues to depth absent or minimal. For example, **18.11** has no shading or perspective cues to depth. On the other hand, the fact that the texture elements everywhere have the same size is a texture cue to the disparate zone being in the plane of the surround. So this random-dot stereogram presents, like most others, a species of ***cue conflict*** (stereo vs. texture; details in Ch 19), even though the conflict in **18.11** is relatively minor.

The methods just described for creating random-dot stereograms are easy to do with computer applications offering graphics tools, such as those available in Photoshop or Power Point. So if you have these tools available it easy to make your own random-dot stereograms, **18.14**.

Stereopsis Does Not Need Object Recognition

An early conclusion which Julesz drew from his research into random-dot stereograms was that stereopsis can be computed by the visual system without need of a prior stage of object recognition operating on each eye's image. We do not see any "object" within either stereo half of **18.11** viewed alone, but this does not prevent us fusing the two halves to achieve vivid stereopsis. Indeed, it is only

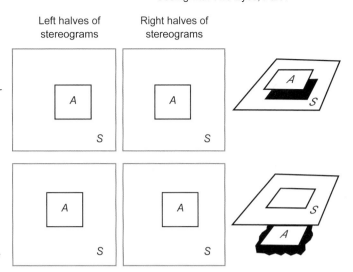

Left halves of stereograms   Right halves of stereograms

18.13 Convergent and divergent disparities
The disparate region *A* is shifted in different directions in the upper and lower stereograms, to create convergent and divergent disparities respectively. In the former, binocular fusion results in *A* being seen floating above its surround *S*. In the latter, *A* is seen as though through a window in the surround *S*.

after stereopsis has been computed that we can say "Ah! There is a square" and thereby succeed in the business of object recognition. These considerations lead to the overall scheme shown in **18.15a** for the sequence of processing operations that take place when a random-dot stereogram is fused.

This conclusion was surprising before Julesz's work because a commonly accepted scheme of processing was that shown in **18.15b**. Thus, many thought that a full scene description was computed for each eye's view separately, including the stage of object recognition, and that it was only then that binocular combination occurred, leading to stereopsis. Random-dot stereograms proved this theory wrong, at least as a full account of stereopsis, because they showed that binocular combination need not happen after object recognition. In the terms of Ch 8 of this book, we can express this by saying that building up high-level object descriptions of the left and right views and then matching these with stored object descriptions is not a necessary prerequisite for stereopsis.

Clinical Tests of Stereopsis

Because one sees nothing but a random texture before stereopsis occurs, random-dot stereograms have found a practical application in eye clinics.

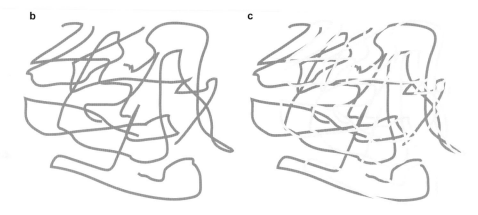

18.14 Making a stereogram using computer graphics tool such as Photoshop or PowerPoint
a Create squiggles or other patterns, using a color similar to the green filter (cyan will do). Counterintuitively, this green pattern is seen through the *red* filter. The squiggles appear dark through the red filter as green does not get through it.
b Make a red copy of **a** and stretch it horizontally using the graphics tool. This pattern appears as dark elements through the *green* filter.
c Superimpose **a** and **b**—and you have a stereogram of a slanted surface. Try creating other depth effects with various other graphical manipulations of one or both images.

They are useful screening stimuli because a patient who is being tested for the presence/absence of binocular vision cannot "cheat" with them, either intentionally or otherwise (as long as the random-dot stereograms are properly made and presented). If the "hidden object" is detected then the unambiguous clinical decision is *Stereopsis Present*.

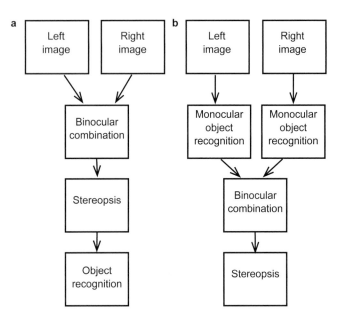

18.15 Two different theories of stereopsis
These flow charts differ in the hypothesized sequence of processing operations.

Various clinical tests of stereopsis exploit the fact that in random-dot stereograms no object is seen prior to the perception of depth. One such test exploits this with stereopsis from a real object, **18.16**, rather than a stereogram. Imagine printing the pattern shown in **18.16a** on one side of a transparent piece of plastic. Call this the *target*. Next, print similar elements around the target but on the *other* side. Someone with normal binocular vision will readily be able to see the target floating either above or below its surround (depending on which side of the plate is viewed). This is because stereoscopic vision allows the observer to see the thickness of the plate. Closing one eye makes the target indistinguishable from its surround, because no monocular depth cues are present. This is also why the target cannot be detected in **18.16b**. The test plates can thus be used to discover if a patient has stereopsis: if he/she has it then he/she can easily pick out the quadrant with the target.

This test is particularly advantageous for testing young children because they do not have to wear red/green or other special spectacles, as with other standard clinical tests of stereopsis. Moreover, children can indicate where the target is by pointing, as in **18.16c**, so no language is necessary—they do not need to describe where the target is. This also makes the test valuable in a cross-cultural context, **18.16d**.

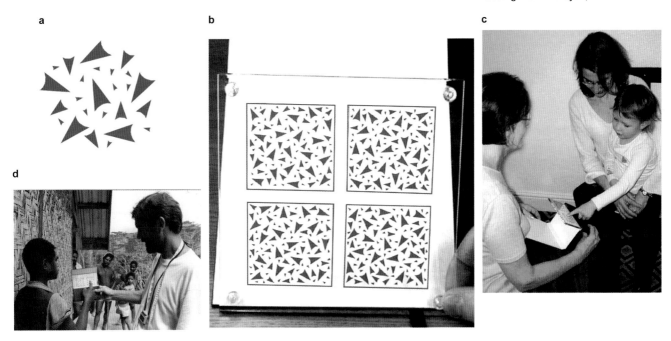

18.16 Frisby stereotest
The various parts of this figure are explained opposite. Visit www.frisbystereotest.co.uk for more details.

The Problem of Global Stereopsis

If binocular combination can be achieved before object recognition, at what level of processing is it done? The obvious answer is to say that it occurs at the "low" level of feature representations. That is, individual small points, dots, lines, or edges are discovered in each view, and then similar features in the two eyes' views are matched up. The argument here is that any given scene feature will project into the left and right images as features of similar shape, size, color, and contrast. Finding matches in this way is referred to as relying on the ***compatibility constraint*** to solve the problem of ***global stereopsis***—finding correct matches from the plethora of ambiguous *local matches* that can be formed from the left and right retinal images.

That is fine as far as it goes—but what if all the image features have the same figural properties, as in **18.17a** in which all the dots have identical color and shape in each eye's view? Because of this similarity, any given dot in one image could in theory be matched with any one of a large number of dots in the other image. This ambiguity is brought out in **18.17b** which shows all possible matches between the left and right images of a sample of four dots. The candidate matches for this small set, selected for illustration of the principles, are shown as the intersections (*nodes*) in a network comprising lines extending from each dot in each image. Try following the line from any one dot and you will see it intersects with all the lines from the dots in the other image, which is why the possible matches are represented by the intersections. As each of the four dots in one image can make a match with any one of the four in the other image, the total number of possible matches is $4 \times 4 = 16$, of which only four are correct matches.

So how can the brain decide which dot in the left stereo half is to be matched with any given dot in the right stereo half? Indeed, if it turned out that stereo matching of **18.17** was impossible it would make more intuitive sense than the result which actually occurs—which is that the brain fuses the stereogram with reasonable ease and produces the correct result of a central rectangle protruding in depth.

Or perhaps, if not impossible to fuse, we might instead have expected **18.17** to yield no more than a random "fog" of dots at many different distances from us, each one resulting from the more or less arbitrary local matches of dots in the left and right images. And yet the brain solves the ambiguity problem and comes up with the perception of

a

b

● Correct matches
● False matches

18.17 Stereo correspondence problem
a Two halves of a random-dot stereogram of a flat surface.
b Four dots selected to illustrate the ambiguity to be resolved
in deciding which left/right dot pairings are to be chosen as
matches. All possible matches between the four example
dots, with each potential *local* match shown at the intersection of "lines of sight" extending from each dot from each
image. Julesz called finding the set of correct matches from
many possible local matches *solving the problem of global
stereopsis.*

depth intended when the stereogram was made.
This is an amazing achievement by the brain. The
advent of random-dot stereograms made the stereo
correspondence problem a major area of binocular
vision research, particularly computational modeling, because the problem to be solved was so
clearly stated.

It is important to realize that the global stereopsis
problem is not confined to unusual and artificial
stimuli such as computer-generated random-dot
stereograms. Consider, for example, natural scenes
such as a leafy tree, or a vase of flowers, or even just
a carpet receding from you into the distance. How
does the brain decide which leaf, or petal, or tuft of
carpet, in one image to match with the appropriate
one amongst all the myriad possibilities in the other
image? Answering this question has been studied
intensively over the past three decades and computational theories for solving it have been proposed,
our next topic.

Computational Solution to the Stereo Correspondence Problem

Another illustration of the stereo correspondence
problem is shown in **18.18**. The left and right
eyes are shown viewing the left and right halves
of a random-dot stereogram, presented to them
through prisms in a Brewster stereoscope (**18.8**).
Imagine the stereo halves in **18.18** as small horizontal slices taken from a full random-dot stereogram. For simplicity, just three representative dots
have been taken out for special attention in each
stereo half. Rays of light (solid lines) are drawn
from these to the left and right retinas. Somehow
the brain matches these dots appropriately: the
correct matches are shown as blue dots in
the visual world built up by binocular fusion.

But notice that the correct matches, whereby
the three dots all lie at the same distance from the
observer, are not the only possible ones. For example, there is nothing in principle to stop, say, the
central dot in the left image matching, not with the
central dot in the right image as it "should"
do, but instead with either the left or right dot in
the right image. These particular false matches, or
ghost matches as they are sometimes called, plus
all other possible ones, are depicted in **18.18** with
red dots. Thus, the problem of global stereopsis can be described as the job of ensuring that
only the correct (blue disc) matches are selected
from amongst all nine possible matches in **18.18**,
six of which are false (red discs).

Of course, **18.18** grossly underestimates the
global matching problem because, for simplicity, it
shows only three dots in each stereo half, and the
possible matches they create. Consider the number
of matches possible in richly-textured random-dot
stereograms of the kind shown in **18.17** and similar figures.

So we have a clearly stated problem to be solved:
the ***stereo correspondence problem***. The next step,
when using the computational approach that is the
theme of this book, is to seek some well-founded
constraints for solving the problem. In this case, the
constraints must be capable of resolving the matching ambiguities.

We have already mentioned one useful constraint, the ***compatibility constraint***, which is:
only allow as candidate local matches, from which

to choose the correct ones, those which could have come from the same entity in the scene. In other words, the initial matches need to have similar figural properties, such as contrast (e.g., allow black-to-black matches but not black-to-white matches as there is no scene entity that could appear black in one image and white in the other). This is fine as far as it goes, but it is not a sufficient constraint to resolve all the ambiguities.

Another useful constraint for limiting the number of potential matches exploits stereo geometry. The idea here is that candidate matches should be limited to those that are "geometrically possible." It turns out to be a fact of stereo geometry that possible matches for a point in one image fall on a line in the other image called the *epipolar line*. So this line is the place to seek possible matches, **18.19**. Small sections of epipolar lines for a surface covered in an array of dots are shown in **18.20**.

Epipolar geometry is a good constraint in tackling the correspondence problem because it makes the search space for correct matches one-dimensional (a line) rather than two-dimensional (an area). In other words, the *epipolar constraint* says: there is no point in searching all over the image for the match of any given point, as the match must lie on a line, the epipolar line.

Finding out where the epipolar lines are requires knowledge of the *viewing geometry parameters*, i.e., knowing the parameters that define where the eyes are looking in the scene. This throws up a chicken-and-egg type problem. We need to find these parameters in order to use the epipolar constraint to help solve the matching problem. But to find these parameters requires that the matching problem has first been solved. Ch 19 describes a way to break this circle. For the moment, we simply say that one way to find required parameters is using a few very distinctive features for which there is no matching ambiguity.

If the distance *d* to the fixation point F is small then epipolar lines will be slightly tilted, as in **18.20** for *d* = 25 cm. But for *d* = 100 cm the epipolar lines are horizontal. (The definition of *d* is the distance from the cyclopean eye, which is the midpoint between the optical centers of the two eyes, to F, wherever F might be located within the viewed scene: see pages 457–459.)

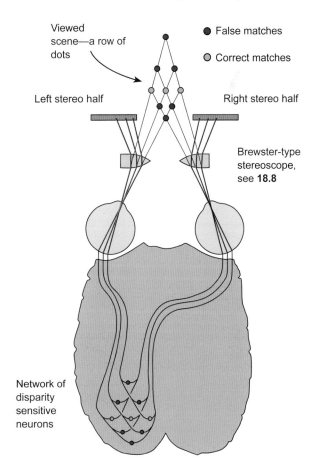

18.18 Stereo correspondence problem
Correct and false matches represented in both "scene space" and a network of neurons tuned to different disparities.

Notice in **18.20** that in some cases there is a difference in the vertical positions of the left image's dots and the right image's epipolar lines. These are examples of *vertical disparities* which we consider later. Notice how the vertical disparities between the dots and their epipolar lines is reduced as distance *d* to the fixation point is increased. This geometric fact is exploited in a theory about how vertical disparities can be used to discover *d*, as explained in Ch 19.

So we now have two useful constraints, feature compatibility and epipolar geometry, but they are still not enough. They have limited the size of the stereo ambiguity problem but not solved it. David Marr and Tomaso Poggio argued in 1976 that there are two other good constraints available, both deriving from the nature of our visual world.

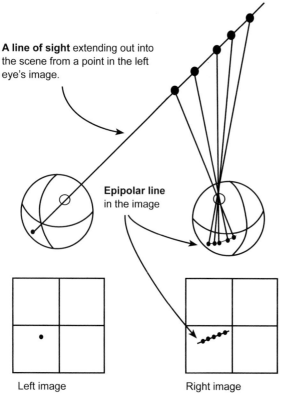

A line of sight extending out into the scene from a point in the left eye's image.

Epipolar line in the image

Left image   Right image

These images are planar approximations to the retinal images falling on the globe of the eyeball.

18.19 Epipolar geometry

A **line of sight** is shown extending out from a dot in the left image through the optical center of the left lens into the scene. Any point in the scene on this line could in principle have given rise to this left image dot—a sample of five are shown. If lines of sight are traced from these dots into the right eye through its optical center then they fall on a line called the **epipolar line**. Thus, the epipolar line defines the *geometrically possible* search zone for seeking matches, in the right image, for the left image dot. The dots along the epipolar lines arise from dots at different distances from the observer. Epipolar lines are nearly horizontal, unless distance d the the fixation point is very small. This shows that disparities in the vertical direction are not determined by the different distances of the points whose images fall on the epipolar line (see also **18.20**). The importance of this fact is explored in Ch 19.

The first is the ***surface smoothness constraint***. We do not normally see a world composed of isolated points (a swarm of gnats and a snow storm are rare counter-examples). Instead, our visual world is composed of larger scale objects, such as people, trees, etc. In other words, visual entities in our world tend to be "smooth," where this term means that variations in their depths tend to be small compared with distance d to the fixation point. Of

course, sudden switches or discontinuities in depth often occur at the edges of objects, but they are not nearly so common as smooth depth changes. The upshot of this constraint is that matches should be preferred that have neighbors with similar depths.

The second of Marr and Poggio's insights was that each retinal image point should be allowed, finally, to be matched up with only one point in the other image. They called this the ***uniqueness constraint***. The idea here is that each retinal image point must physically derive from a scene entity that can be in only one place in 3D space at any one time. Hence, it makes no physical sense to allow any image point to take part in two matches.

A challenge to the idea that human vision relies on the uniqueness constraint is ***Panum's limiting case***, a stereogram in which two lines appear in one image but only one line in the other image, **18.21a**. When viewed binocularly one line is seen in front of the other, suggesting that the single line has served for two matches, **18.21b**. However, if the single line is made wavy, **18.21c** then only one of the two lines is seen as humped in 3D when this stereo pair is fused binocularly. The double-matching explanation shown in **18.21b** predicts that *both* lines seen post-fusion should appear humped, as in the control stimulus **18.21d**. The failure to see two humped lines in **18.21c** suggests that the double-matching explanation of Panum's limiting case is wrong, and that perhaps human vision does use the uniqueness constraint after all (Frisby, 2001).

A Stereo Correspondence Algorithm

We now have a set of four constraints: compatibility, epipolar geometry, surface smoothness, and unique matches. The next step is to put them to work in a biologically plausible matching algorithm.

A neural network algorithm for stereo matching was first tried by Dev in 1975 but we present here a version proposed by Marr and Poggio in 1976 because it was more clearly founded on valid constraints. This algorithm had several stages.

First, the compatibility and epipolar constraints are used to establish all possible local matches, both correct and false. These are represented in a network of computer elements rather like the idealized neurons in the simplified ambiguity

a $d = 25$ cm   **b** $d = 50$ cm   **c** $d = 100$ cm

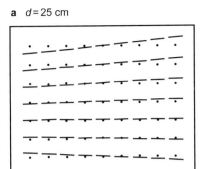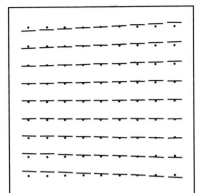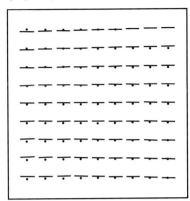

18.20 Epipolar lines for an array of dots on a fronto-parallel surface
The dots show points projected into the left image. Their matches in the right image are to be found somewhere along the line, called an *epipolar line*, closest to each dot in this figure. **a**, **b**, and **c** differ in the distance to the fixation point *d*—notice how the positions of the epipolar lines vary with *d*.

diagrams in **18.18**. One neuron is made active or "alive" for each and every possible local match

Note that in **18.18** each input cable from each retina serves three cells, those that would deal with matches along the same ***line of sight***, **18.19**, from the eye in question. Follow the various cables in **18.18** through to check this fact yourself. Note that the network of neurons is really just a simple replica of the visual world depicted out in front of the eyes, cast in terms of neurons. The general idea is: each active neuron stands in a one-to-one relationship to a particular possible local match.

Second, having established possible local matches, the next step is to "kill off" false matches (i.e., make the elements representing them in the network inactive). This can be done by the combined action of inhibitory and excitatory connections between elements in the network. If all goes well, this process leaves only the correct matches active. Hence, those neurons active in the final state of the network represent a solution to the problem of global stereopsis. It is a solution because the active neurons make *explicit* which feature in one image matches which feature in the other image.

The depth-processing network shown in **18.18** readily lends itself to a neurophysiological interpretation, which is why it is shown as a network of neurons drawn within an outline of a brain. Note

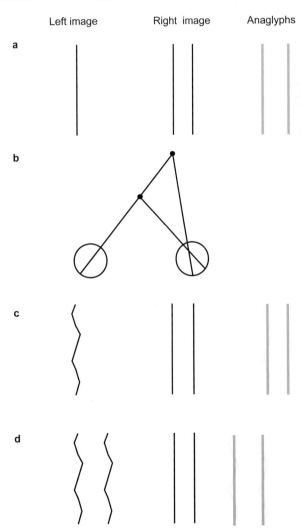

Left image   Right image   Anaglyphs

18.21 Panum's limiting case
The figures in the left and central columns are stereo halves suitable for crossed-eyes fusion. Use the red/green filters to view the anaglyphs on the right.

that signals from the right eye are shown crossing over to the left hemisphere at the optic chiasm.

If the row of three dots had fallen on the other side of the fovea from the one shown, then it would have been necessary to display the depth network in the right hemisphere rather than in the left one. Separate networks exist in each hemisphere for dealing with different regions of the field of view, **1.6**.

The problem now is: what combination of excitatory and inhibitory messages between cells can reliably pick out the correct matches and kill off the false ones? This is the moment to put Marr and Poggio's constraints to work.

Marr and Poggio drew from their *surface smoothness constraint* the algorithmic processing rule: cells in the network representing points at similar depths (and hence similar disparities) should exchange *excitatory* support. The idea is here that each active cell can be thought of as saying to its neighbors:

> Hey guys, I am active, because I have similar features on my left and right eye inputs. And as I am active, it is likely you should be too: therefore I will give you a helping hand and pass you some excitation.

Marr and Poggio derived from their *uniqueness constraint* the processing rule: any given network cell should try to kill off any other cell that wants to use one or other of the features that it has used to form its match.

This can be done by each active cell passing inhibition to all other cells along the connections that represent the lines of sight extending from its own matched points. This is valid because these lines of sight define the matches that any given point can make along an epipolar line, **18.19**.

To see these two constraints at work it is necessary to show a rather larger network of depth-processing cells, **18.22**. This network is wholly similar in structure to the smaller one shown in **18.18**, but with 8 × 8 = 64 cells, rather than just 3 × 3 = 9. Cables are shown coming from the left and right eyes to feed the network. These cables carry either information about a black point in a certain retinal location or information about a white point. This is because the eyes are viewing a black/white random-dot stereogram. Thus, each eye's input can be thought of as a slice from the kind of random checkerboard in **18.11**.

The first thing to appreciate is that those cells that become active initially are those that receive a suitable input from both eyes, i.e., white left/white right inputs, or black left/ black right. Thus, each cell cares not about whether its inputs are black or white, only about whether its two inputs are similar. This requirement ensures, of course, that not all cells become active in the network. It is based on the compatibility constraint defined earlier. Those that do become active are shown either as blue discs, indicating that they are correct local matches, or as outline discs with a red dot in them, to show that they are false local matches.

Of course, the network doesn't "know" at this stage which potential matches are correct and which false—both types are treated identically. Try following an input cable from one eye through the network, and note that either a blue disc or a dotted disc occurs only when this cable meets a cell whose input cable from the other eye is carrying similar information.

Look next at the checkerboard patterns coming from the two eyes. Note that each one has an area of the disparate square. This is a horizontal slice of the square in the random-dot stereogram that appears to stand out in depth because of the disparity of its position in the two eyes. Check for yourself that it is disparate in its location in the checkerboard slices shown in **18.22** for each eye. Small patches of surround area are shown with zero disparity (i.e., their positions match exactly in the two eyes). And finally, note the X and Y areas—the uncorrelated bits of the pattern in each eye that have no matches in the other eye because of the shift imposed on the disparate squares (refer back to **18.12** for an explanation of these areas).

Having explained how the network is set up, and having drawn attention to the disparate area, it is time to explain how it is that the blue disc matches come to be selected from all those initially registered. As we explained above, the two key features of the Marr/Poggio scheme for achieving this are to set up: (1) inhibitory connections between cells lying within the same line-of-sight columns, which implement the uniqueness constraint operating along matches found on epipolar lines, **18.19**; and (2) excitatory connections between cells representing the same disparity, which implement the surface smoothness constraint. The

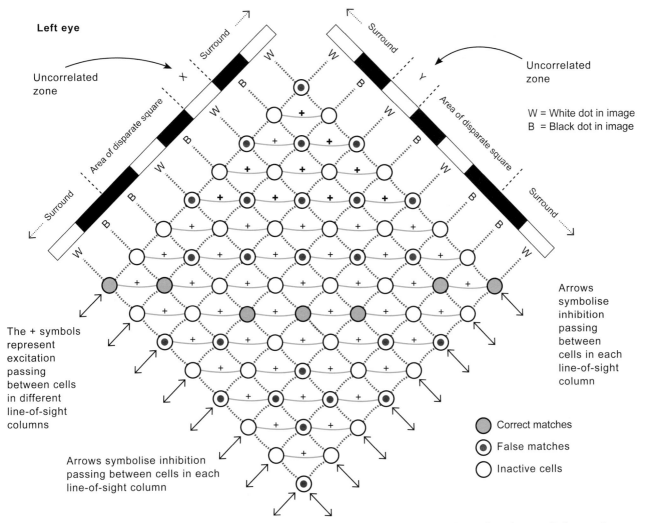

Left eye

Uncorrelated zone

Surround

Area of disparate square

Surround

W = White dot in image
B = Black dot in image

Uncorrelated zone

Surround

Area of disparate square

Surround

The + symbols represent excitation passing between cells in different line-of-sight columns

Arrows symbolise inhibition passing between cells in each line-of-sight column

Arrows symbolise inhibition passing between cells in each line-of-sight column

Correct matches

False matches

Inactive cells

18.22 Resolving the stereo correspondence problem using a network of cells exchanging excitation and inhibition
For illustrative purposes, just a small diamond-shaped vertical slice of the complete network is shown dealing with a very small disparate area. This latter feature makes the correct fusions appear less extensive than some areas of ghost fusions, a result unrepresentative of larger areas.

arrows just outside the network in **18.22** represent inhibitory influences passing up and down the line-of-sight columns, and the + labels on the lateral connections between cells indicate the passage of excitation between adjacent cells coding the same disparity.

Let's recap on what is going on here in the network to implement the constraints. All cells coding disparities along a line-of-sight are set to "fight" to see which one is the "strongest" (most active) and therefore justifies selection against the competition provided by all the others. The fight is carried out by exchange of inhibition, so implementing

the uniqueness constraint. But how can some cells turn out to be stronger than others?

The answer is that some active cells get a boost in activity from the excitation passed to them from "friends" in the same disparity plane lying in neighboring line-of-sight columns. This excitation implements the surface smoothness constraint and, for *correct* matches, the *excitation exceeds the inhibition*.

In **18.22**, facilitation is shown coming from only two neighbors because the network is shown as a two-dimensional slice, not as a complete 3D entity. In fact, exchange of excitation arrives from *all* the disparity neighbors encircling each cell. The excitatory network connections for an individual cell are in fact more numerous than this because excitation is shared between neighboring slices of the network. The 3D sketch of the network in **18.23** illustrates this point. It shows a circular region around each one of three examples of cells encoding potential matches. The small inset gives more details on what is going on for each cell.

Each layer in this structure represents one layer of disparity. The central layer is the one for zero disparity (i.e., the depth at which we are fixating), the upper three layers those for convergent (near) disparities, and the lower three layers those for divergent (far) disparities, **18.18**. For simplicity, excitatory and inhibitory connections are shown for three cells only, but in fact all cells in all depth planes have similar sorts of linkages. However, bear in mind that the excitatory linkages are from nearby ('local') neighbors within a given same-disparity.

Note also that **18.23** shows simply the network structure prior to input of a stereo pair for computation, and the inhibitory connections are shown with lines passing through all the layers of the network. Any active cell on these lines sends inhibition to all cells on them.

Having explained the design of the neural network, it is now time to see it at work on a genuine random-dot stereogram. In **18.24**, the 3D model of a seven-plane disparity network is shown with all connections between cells omitted for clarity. The tiny black dots in each depth plane represent active cells, each dot thus depicting a cell which has become active because its left and right eye inputs match, either as black/black or as white/white. Be careful to note that each black dot shows an active cell regardless of the nature of its match: do not get confused by thinking that the black dots represent only black/black matches.

The initial state of the network before any inhibition or excitation has passed between cells is shown in **18.24a**. It is the first step in the global stereopsis computation—the identification of all possible local matches, be they correct or false. And note just how many false ones there are. Black dots appear almost everywhere, showing just how extensive the global problem is. But note also that the regions of completely dense black, representing the surround of the random-dot stereogram and

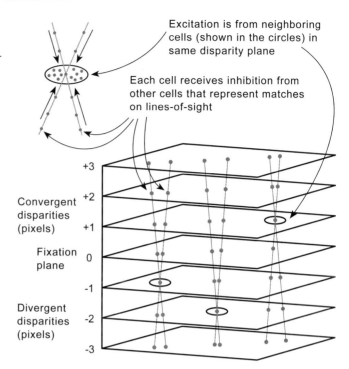

18.23 Solving the stereo correspondence problem

its disparate square area, are also discernible in the zero disparity layer and the second divergent-disparity layers respectively. The task of the computation is to leave these regions intact, in the final state of the network, while killing off the myriad false matches.

The results of the first round of excitation and inhibition are shown in **18.24b**. To get to this state, each cell had to do the following sum: it had a certain number of units of activity according to whether or not it was active in the first place; it then subtracted from this number the total of inhibitory units sent by its competitors down the lines-of-sight which passed "through" it; it added on the units of excitation from its depth-plane neighbors.

It worked out this total and then "asked" whether or not it was above a pre-set threshold value. If it was, it became "active" for the next round; if it failed to exceed the threshold, then the cell was "inactive" for the next round.

The results of the next round of excitation and inhibition are shown in **18.24c**. You may be wondering why more than one round is necessary. In a serial computer simulation of such a network, it is necessary to approach the end state in a series of steps, called *iterations*, each one doing just a bit of the overall computation. The result of each iteration represents a "snapshot," if you like, of the network as it passes through a particular stage on the way to its final state.

Each iteration does exactly the same arithmetic as any other, but its input is different: it works out the sum of inhibition and excitation using the state of the cells produced by the preceding iteration. If this network were to be implemented in a biological system built of nerve cells, the network would settle fairly quickly to its final stable state, with all the connections constantly and simultaneously passing inhibition or excitation according to the influences they are receiving at any one time.

Note that in **18.24** a lot of "killing off" takes place immediately, and then the successful solution begins to grow back from the rather tattered remnants of the first battle. But it grows well enough, and at the end, **18.24g**, the desired state of surround and disparate square are picked out in differ-

ent disparity planes. The battle is won. Matches have been established.

This network is said to exhibit the property of ***cooperativity***. That is, it reaches a state of *global* organization via *local* but highly interactive processes that together cooperate to produce the required solution.

Of course, the local processes have no knowledge about what is going on elsewhere in the network, so this is a "blind" form of cooperation: each node simply does its own thing according to the rules, and the global state of organization "pops out" at the end of it all, as an inevitable but highly desirable consequence. Julesz was the first to propose that cooperativity in this sense was a property of the mechanism of stereopsis. If he is right, then the type of network just described gains credibility as a model of the actual neural computation of stereopsis performed by our visual system.

Marr and Poggio's way of implementing the surface smoothness constraint has been criticised

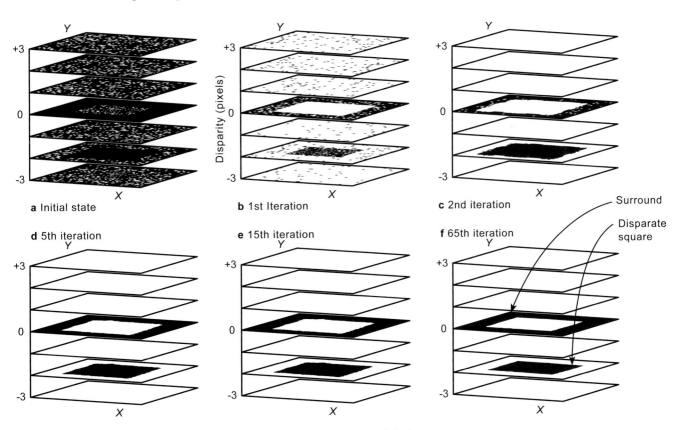

18.24 Evolution of a Marr/Poggio network dealing with a random-dot stereogram
Each rectangle represents a set of cells coding a particular disparity, in the range +3 to -3 pixels (see axis labels in **18.23**), at locations (*X*, *Y*) in the random-dot stereogram.

for being too restrictive, by demanding smoothness defined in terms of same-disparity neighbors. Quite often, "smooth" scene surfaces fail this criterion as they are slanted from the viewpoint of the observer. This means that points on them have different disparity neighbors.

A subsequent stereo matching algorithm, called *PMF* after its inventors Stephen Pollard, John Mayhew and John Frisby who devised it in 1985, avoided this limitation. It did so by allowing excitatory facilitation to be passed between neighboring active cells in *nearby*, rather than just the same, disparity planes in the network. This form of excitation was allowed if the cells exchanging excitation encoded matches from surfaces that were "not too steeply" slanted. The details of how *PMF* was inspired by certain studies of human vision that led to its definition of "steepness" can be found in Ch 19. So here we have an example of the same basic constraint, surface smoothness, being put to work in an stereo matching algorithm but in a different way.

Finding the correct matches is fine as far it goes, but it isn't the final goal of the computation. The next step is to use the disparities of these matched points to compute the *X, Y,* and *Z* scene coordinates of the matched points, as described in the first half of this chapter.

Neurophysiological Mechanisms for Stereopsis

If the Marr/Poggio type of computation is the one used by the visual system for obtaining global stereopsis, then the first requirement is neurons which can detect the local matches. Possible candidates for such cells have been found by neurophysiologists. The single cell recording technique, Ch 3, has revealed cortical neurons in the cat which become active only if their optimal stimulus is positioned very carefully in the two eyes so that it possesses the degree of disparity required for the particular cell in question.

Some results of this kind, obtained by Colin Blakemore in 1970 from the cat, are particularly interesting in the present context because he found what he called direction columns. That is, if his microelectrode stayed perpendicular to the surface of the cortex, so that all the cells he recorded from were in just one column (refer to Chs 8 and 9 for a reminder on cortical columns), then he sometimes found that cells varied in their required disparity along a line-of-sight from one eye, **18.25**.

This raised the possibility that Blakemore had discovered the neurophysiological machinery for providing the line-of-sight inhibition required by the Marr/Poggio computation, because it would clearly be a straightforward matter for inhibitory influences to pass up and down these columns along known fiber tracts within the columns. Lateral connections between columns might then be the mediators of the lateral excitation provided by other cells in the same depth plane. If so, this is an encouraging and exciting link between a computational analysis on the one hand, and a set of neurophysiological and neuroanatomical findings on the other. However, since Blakemore's early work, a great deal of subsequent research has painted a rather different picture—see Ch 19.

What Exactly Is a "Point" for Matching?

There are difficulties with the Marr/Poggio scheme, both computationally and neurophysiologically. These center around the deceptively simple question: what exactly are the "features" that are used

18.25 Disparity sensitive neurons
The dots represent cells whose preferred stimulus required a disparity in position in the two eyes. This would make them most sensitive to scene entities appearing in the depth locations shown. Reproduced with permission from Blakemore (1970). Visit http://www.viperlib.york.ac.uk/ for historic movie footage of early single-cell recording in the cat visual system by Hubel and Wiesel.

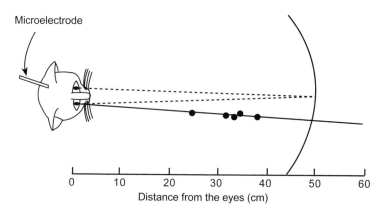

Microelectrode

Distance from the eyes (cm)

to create local matches? Edge points of the kind described in Ch 5 are obvious candidates. But the known neurophysiological disparity units seem much like the various cortical cells described in Ch 9 in their general level of processing sophistication. And if we were right to say there that such cells are but measurement precursors to building up a proper feature description, and that they are not part of the representation of the feature description itself, then the known disparity cells seem poor candidates for mediating local feature matches directly.

Note that for convenience while explaining the Marr/ Poggio computational solution to the problem of global stereopsis, we slid gently and surreptitiously into treating a point-for-fusion as a small black or white zone in either the left or right retinal image. Such zones might conceivably have as their neurophysiological representation cells that signal blackness or whiteness (i.e., neural elements coding one small zone of the image after the lightness computation).

But this view of what constitutes a point-for-fusion is quite at odds with our present neurophysiological knowledge about disparity cells. These are orientation-selective, which suggests that when fibers stemming from retinal ganglion cells are used to build up the activity of disparity cells, the property of orientation selectivity is built in at the selfsame moment. Mayhew and Frisby proposed this approach in their theory of the ***binocular raw primal sketch***, in which disparity processing is intimately intermingled with the computation of edge descriptions (see Ch 19).

The question therefore about what stimulus attributes in the left and right images form the basis for local matches is a tricky one (review in Howard and Rogers, 2002), and a thorough discussion of it is beyond the scope of this book.

Random-Dot Stereograms of Complex Surfaces

The global stereopsis problem and its inherent difficulties are perhaps best brought out not by stereograms containing just a square or a rectangle in depth but by stereograms containing a much more elaborate "hidden object" with elements at many depths, not just two as in **18.11**.

The random-dot stereogram in **18.26** is a case in point. Have a look at it with the red/green filters

as usual and see if you can make out the intricate 3D shape which it contains. But please be patient. Many people take quite a long time to see stereopsis in a complex stereogram of this kind, and may even need several attempts each lasting a few minutes. But as before, effort and patience are well rewarded as the final depth effect is truly remarkable, indeed beautiful. Once the global matching problem has been solved by your brain, you will see a spiral rising up from the page.

To help you fuse **18.26** a similar stereogram is shown in **18.27** but with the outline of the staircase drawn in to help you see it in depth. The latter stereogram is not a true random-dot stereogram because it breaks the rule that the shape to be seen in depth must not be visible in either the left or the right image by itself. However, the monocularly-discriminable contour helps unpracticed observers see the 3D spiral (Frisby and Clatworthy, 1975).

An interesting thing about both **18.26** and **18.27** is that you only realize gradually, as you look at them, that the spiral is made up of stairs. At first, the surface of the spiral looks smooth— like a helter-skelter. But bit by bit the shallow stairs reveal themselves, as the brain continues to refine the its stereo processing. Some people never manage to see the steps because their binocular vision is not sufficiently acute to pick up the very tiny differences in disparity between the dots in one step and the dots in the next.

Try reversing the red/green filters while looking at **18.26** so that the red filter covers the right eye. You should now see a receding spiral staircase, cork-screwing down through the page. The sight is marvelous and some people find it easier to see than the protruding staircase. So if you had to give up on **18.26** with the filters in their normal position (green filter, right eye), do have another try with the filters reversed.

In stereograms of complicated surfaces such as the spiral staircase, different parts of the picture appear to be at a whole range of different distances from you. This contrasts with **18.11**, in which only three distances were used—the page, the surround, and the square or rectangle. The wider selection of distances is easily achieved by shifting different parts of the random texture by different amounts—the bigger the shift, the greater the illu-

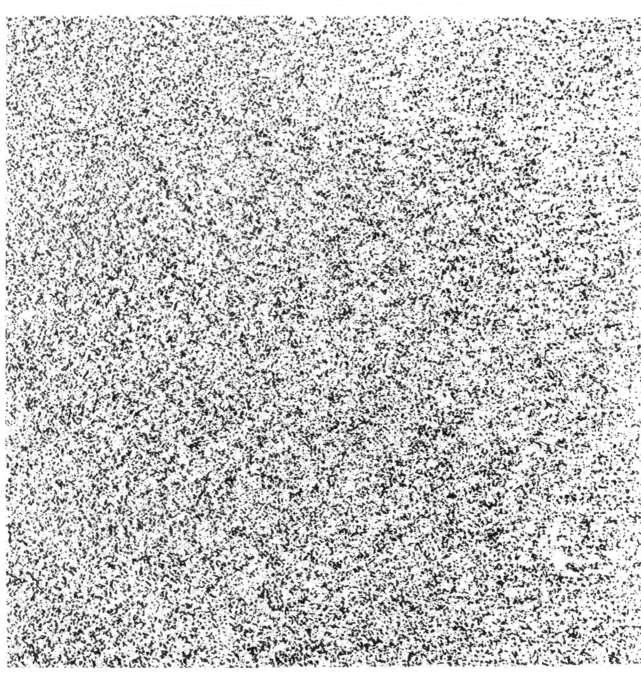

18.26 Random-dot stereogram of a spiral
This is just one of many stunning stereograms that Julesz published in his classic book in 1971, now made newly available in a reprint in 2006. With permission from the MIT Press.

sion of depth that results. This relationship between the size of the shift and the amount of the perceived depth is clearly illustrated by the anaglyph in **18.28**. The upper outline square has been shifted in each

eye's view by much more than the lower square. The resulting larger horizontal disparities cause the upper square to protrude much more.

Learning to See Random-Dot Stereograms

The tactic shown in **18.27** of drawing a monocularly prominent line around the area that is designed to be seen in depth is especially helpful if the disparity between the left and right images is very large. For example, most people find it very difficult to see a shape in depth in **18.29**. However, as shown by Anne Saye and John Frisby (1975), if a square outline is added, as in **18.30**, then the depth effect is much easier to see.

This outline may help by giving the brain appropriate cues for the control of eye movements. One stratagem the brain seems to use to help solve the problem of global stereopsis is to avoid fusing left and right elements with disparities larger than a certain limiting size. This limit is known as **Panum's fusional area**. If a disparity is presented that exceeds this limit then human vision fails to achieve binocular fusion—unless the observer is allowed to alter their vergence angle. That is, features which fall in grossly different depth planes, and so create large disparities, cannot be fused unless suitable eye movements are made to fixate the two features successively.

The idea here is that left/right feature pairs within the limit are fused and thereafter held "locked" together, while a new eye movement enables the fusion of other feature pairs with disparities that exceed the limit. Thus, Panum's fusional limit applies for a given fixation. And once made, a fusion can survive a shift of fixation (although

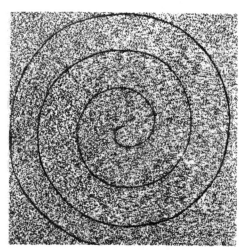

18.27 Stereogram in 18.26 but with contours drawn around the edges of the spiral

there are limits on this also). Evidence supporting the idea that prominent monocular contours help produce stereopsis by facilitating the required eye movements is that the contours help only when the disparities are large. When eye movements are unnecessary for fusion because the disparities are small, then the monocular contours convey no benefits in speeding up fusion.

A word of caution is in order here. As a convenient shorthand we have talked about the binocular fusion of left and right image points while discussing matching. However, matching of points may not always lead to fusion. Binocular fusion is defined as the percept of a single entity from the left and right images. Its converse is **diplopia**—double vision. It is possible to have depth sensations, which Ogle (1954) called **qualitative stereopsis**, from stereograms even though the depth is attached to diplopic entities.

Returning to the issue of why we need to learn how to see random-dot stereograms, it may be that these stimuli in general provide poor cues for vergence eye movements. Perhaps as you get better at fusing random-dot stereograms you are unconsciously learning to "liberate" your vergence mechanisms somewhat, so that you can converge and diverge without the need for first seeing something in each monocular image to converge or diverge upon. The monocular contours of **18.30**, however, do provide some large-scale monocularly visible contours to converge/ diverge to, and this is why they may help naïve observers achieve stereopsis.

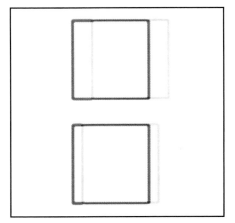

18.28 Two sizes of disparity
The larger disparity for the upper square creates more perceived depth than the smaller disparity of the lower square.

18.29 Large disparity random-dot stereogram
Most people find this stereogram difficult to fuse.

Whatever the merit of this eye movement hypothesis about learning to see random-dot stereograms, the debate is a reminder of the salutary warning first given in Ch 1 about ever-present eye movements. It is convenient when analyzing visual processing to consider first simple cases in which the observer holds fixation on a given point. We have done this throughout this chapter. But of course, in normal vision of normal scenes the

18.30 Stereogram in 18.29 with disparate square outlined
Adding the monocularly-visible contours help fusion.

observer gets a constantly changing flux of retinal images due to body, head, and eye movements. The "real" inputs which our visual system uses are image sequences—vision is a dynamic business. This should not be forgotten.

Monocular contours might play a helpful role in other ways in addition to facilitating eye movements. For instance, they might provide high-level shape information which could in principle guide the solution of the global stereopsis problem. But it is unlikely that much benefit is in fact bestowed along these lines, for two reasons.

First, as already noted, monocular contours can still provide help even if their shape bears no particular relationship to that of the disparate zone, **18.31**.

18.31 Help from a monocularly visible contour
A monocularly-visible contour in the target helps achieve fusion in the large disparity stereogram of **18.29** even if it does not mark out the shape of the disparate target.

Secondly, an experiment by John Frisby and Jeremy Clatworthy gave different groups of subjects different sorts of cues about the spiral stereogram of **18.26**. They found no evidence of benefit from high-level cues such as telling subjects what they "ought" to see, or even showing them a 3D model of the staircase before presenting the stereogram itself. So, at least in the case of stereopsis from random-dot stereograms, high-level cues do not provide much if any benefit. Stereopsis appears to be resolutely a phenomenon of low-level early visual processing.

18.32 Illusory depth contour

18.33 "Pulling up" effect

Illusory Contours in Depth

In **18.32** an ordinary random-dot stereogram is divided in two by a white strip. Look at the anaglyph using the red/green filters and you will see that the white strip cuts a central protruding square in half. But if you look more carefully, you will notice that the square has not really been cut in two at all, at least not as far as its apparent depth is concerned. Rather, the part of the white strip which cuts across the square seems to be in the same depth plane as the square itself. It is as though this central region of the strip gets "sucked up" with the square, despite the fact that it contains no texture, and so offers no disparity cues to justify its allocation ito the same depth as the square.

A possible explanation of this curious depth effect, whose boundaries are marked out by illusory contours defining the depth boundary, is in terms of lateral excitation of the kind already described in connection with the Marr/Poggio computation of global stereopsis. Perhaps cells in the depth-processing network feed out lateral facilitation to disparity units dealing with the region of the central strip and "bring them alive." This is possible because the region of the central strip is essentially ambiguous as far as disparity processing is concerned—it would give rise to white/white matches in all possible depth planes, or else no initial matches at all. Thus, it makes sense that

those cells receiving help from their depth neighbors win out, to give protruding depth to the strip when it passes through the central square, and no depth when it passes through the surround (the latter region "pulling down" the strip, just as the square-in-depth "pulls it up"). The same "pulling up" effect can be seen by looking back at **18.24**. Notice that the dots in the central square-in-depth appear to "take with them" the white ground on which they lie. This could be why in **18.33** we see the dots in the square-in-depth "pull up" the white ground on which they lie.

The effect in **18.33** is an example of ***3D surface interpolation***, in which disparity cues carried by the dots are used to create a surface embracing both those dots and the surrounding blank area around them. A technique that has been used to study this phenomenon is illustrated in **18.34**, in which a region of texture in one half of a stereo pair of images is blanked-off. Tom Collett has used this method but we show examples devised by David Buckley.

Figure **18.34a** depicts a 3D horizontal ridge formed from two slanted planes. In **18.34b** the horizontal blanked-off area in one half of the stereogram covers the join between the planes. The result is that the texture in the corresponding region in the other half-stereogram is "pulled up" by interpolation from the disparity cues in the

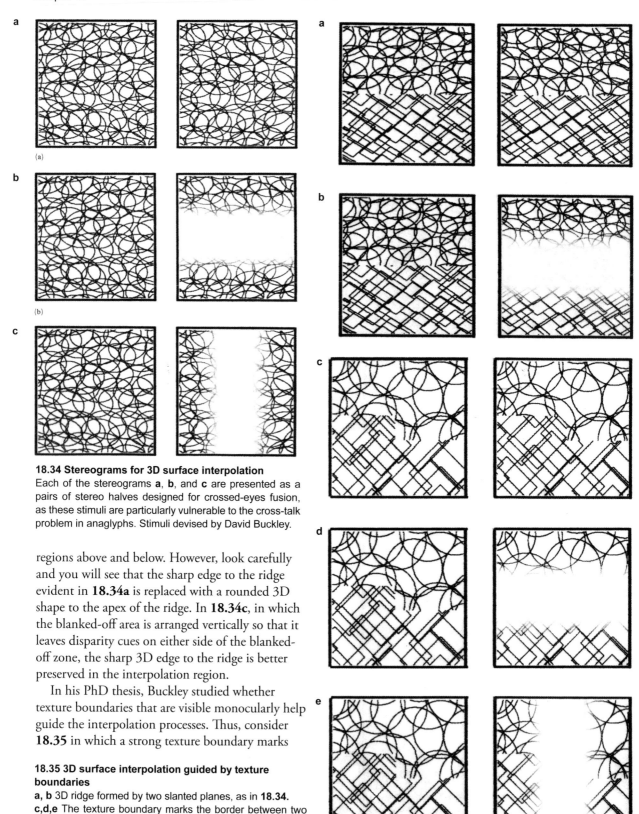

18.34 Stereograms for 3D surface interpolation
Each of the stereograms **a**, **b**, and **c** are presented as a pairs of stereo halves designed for crossed-eyes fusion, as these stimuli are particularly vulnerable to the cross-talk problem in anaglyphs. Stimuli devised by David Buckley.

regions above and below. However, look carefully and you will see that the sharp edge to the ridge evident in **18.34a** is replaced with a rounded 3D shape to the apex of the ridge. In **18.34c**, in which the blanked-off area is arranged vertically so that it leaves disparity cues on either side of the blanked-off zone, the sharp 3D edge to the ridge is better preserved in the interpolation region.

In his PhD thesis, Buckley studied whether texture boundaries that are visible monocularly help guide the interpolation processes. Thus, consider **18.35** in which a strong texture boundary marks

18.35 3D surface interpolation guided by texture boundaries
a, b 3D ridge formed by two slanted planes, as in **18.34**.
c,d,e The texture boundary marks the border between two flat planes set at different depths.

the apex of the 3D ridge. Do you see a sharp apex in **18.35b**, unlike the rounded apex seen in the matching stereo pair **18.34b**? That is what we see. This suggests that texture boundaries can play a role in shaping 3D surface interpolation from disparity cues (Buckley et al., 1989).

It is of interest that texture boundary cues can also guide the interpolation processes if the texture boundary is not just a simple straight line but one that depicts a corner shape, **18.35c,d,e**.

Stereopsis Survives Large Contrast Differences

It is possible to fuse quite readily left and right stereo halves which differ greatly in contrast, **18.36**. Thus, the rule for matching left and right points

18.36 Stereopsis survives the large contrast difference between the two stereo halves

a Reversed contrast: simple line stereogram

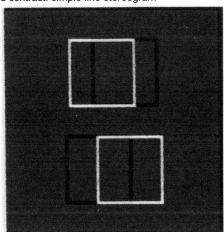

b Luminance profiles across stereo halves of **a**

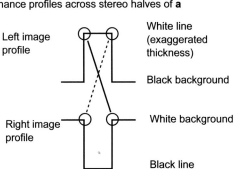

c Reversed contrast: random-dot stereogram

d Luminance profiles across stereo halves of **c**

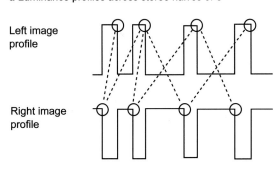

18.37 Stereopsis and reversed contrast
a Stereopsis is possible in a simple stereogram with reversed contrast.
b Paul Whittle has suggested that the brain is fusing white edges (or black edges) when faced figures like **a**.
c Stereopsis is impossible if one stereo half of a random-dot stereogram is a contrast reversal (black-for-white and white-for-black) of the left half.
d Perhaps the reason stereopsis fails for a random-dot stereogram is that reversal of contrast in this case disrupts rather more seriously the locations of such edges in each field, producing too great a variety of different and inconsistent matches of the edges of randomly determined blocks of white or black cells.

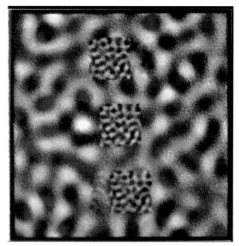

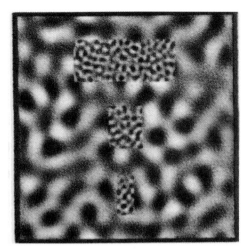

18.38 Stereopsis with rivalrous texture
View with the filters as usual but try closing each eye alternatively to see what each stereo half contains.

18.40 Paradoxical stereopsis effect
The inset speckled figures occur in only one eye's image (except for *cross-talk* between stereo halves in anaglyphs: see legend to **18.3**). Frisby and Mayhew (1978).

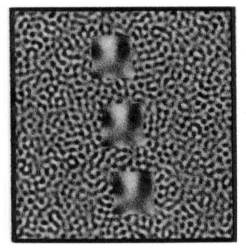

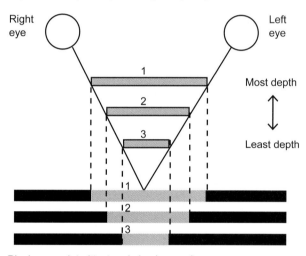

18.39 Rivalrous texture stereogram
Stereopsis is difficult at best and perhaps impossible.

Black—correlated texture in background
Gray— uncorrelated texture

18.41 Explanation of paradoxical stereopsis
The effect might be linked to the uncorrelated regions being treated as objects that are "squeezed up" above the background because an uncorrelated region is the normal consequence of an object lying in front of a background. This explanation predicts that the amount of depth seen should vary with the width of the uncorrelated area, and this is what is seen in 18.40.

is not one which insists that, say, a "white" point of a given brightness in the left half can be matched only with a white point of similar brightness in the right half. Rather, it seems that as long as a white point in one field can find a whiter-than-mid-gray point in the other field, then it will fuse with it. (However, contrast differences between stereo half images can worsen stereoacuity; Halpern and Blake, 1988). What it refuses to do is fuse with a blacker-than-mid-gray, as shown in **18.37c** in which contrast has been reversed. Each element of this stereogram is of opposite black/white "color" in the two halves and stereopsis is impossible.

It has been known for a very long time that reversal of contrast in a simple line-stereogram can be tolerated quite well. This is illustrated in **18.37a**, whose legend discusses Paul Whittle's explanation in 1963—matches are between the *opposite* sides of lines. This explanation thus proposes that the matches are always between contrast changes of the same sign, white-to-black or black-to-white.

Stereograms with Rivalous Textures

If two textures do not fuse happily together when presented stereoscopically, then the resulting perception is said to show **binocular rivalry**. That is, first one eye's view and then the other succeeds in becoming dominant, the two seeming to be in a state of rivalry for "possession" of visual awareness. Surprisingly, it is possible to obtain stereopsis in the face of such rivalry.

Consider **18.38** for example. The details of the textures forming each small inset square are rivalrous because they do not match up systematically in the view of the left and right eyes, and yet the squares as a whole have a disparity shift which successfully yields stereopsis. We see a staircase in depth of the three squares, with the lowest square protruding furthest above the page. Somehow, these squares are binocularly matched overall despite being rivalrous in their details. Presumably this depth effect is mediated because the edges of the squares are used for forming left/right matches.

Interestingly, if the squares are marked out with textures of different spatial frequencies, then rivalry is much more pronounced and stereopsis is much more difficult if indeed not impossible **18.39**.

A surprising effect of this kind is shown in **18.40**. Here the background is matched in its texture in the left and right eyes (and so is non-rivalrous when the stereogram is binocularly combined), but the monocularly discriminable shapes in the left field have no corresponding shapes in the right field whatsoever—check this by closing each eye in turn. And yet stereopsis is obtainable.

Moreover, the degree of perceived depth in **18.40** is a function of the width of the shapes, the narrowest seeming to protrude least (or recede— the depth effect can spontaneously reverse), and the widest seeming to protrude farthest. This is weird. How can depth appear when there is nothing in one eye for the shape-in-depth to fuse with?

A possible explanation of paradoxical stereopsis is illustrated in **18.41**, based upon the idea that the brain treats the rivalrous textures as cases of the uncorrelated regions that occur in stereo images from many scenes (cf. the X and Y areas in **18.12**).

Whether paradoxical stereopsis has anything in common with Panum's limiting case is an open research question.

Stereopsis from Blurred Images

Stereopsis can readily be obtained if both stereo halves are blurred, **18.42**, although this does

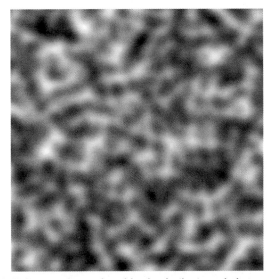

18.42 Stereopsis survives blurring both stereo halves
A blurred version of a random-dot stereogram showing a square in depth, cf. **18.11**. Technically, the low spatial frequency information has been kept (see Chs 4, 5) and the high spatial frequency detail (sharp edges) filtered out.

18.43 Stereopsis can survive blurring of one stereo half
This stimulus simulates an optical prescription called *monovision*, in which one eye has a lens suitable for distance and the other has one for reading (described on p. 452).

18.44 Stereopsis carried by blurred blobs survives noise comprised of edges added to one stereo half
Noise is used here as the technical term for irrelevant content added to an image. The outcome shows that low spatial frequency information can be processed separately from high spatial frequency details.

18.45 Stereopsis carried by blurred blobs is impaired when similar blurred noise is added to one stereo half
Technically, this can be described as disruption of stereopsis when noise of similar spatial frequency and contrast to that carrying the disparity cues is added to one stereo half.

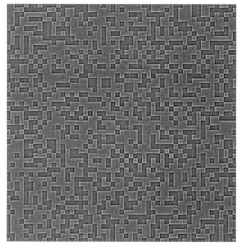

18.46 Stereopsis is possible if only edges are present
Each stereo half of a normal random-dot stereogram has been filtered to select just the high spatial frequencies.

18.47 Stereopsis carried by edges plus burred noise
Stereopsis survives (but probably only for experienced viewers of these sorts of stimuli) when high spatial frequencies carry the disparity cues and low spatial frequency noise of low contrast is added to one stereo half.

reduce ability to discriminate fine differences in depth. More surprisingly, stereopsis also survives blurring of just one stereo half, **18.43**. The latter stimulus partially mimics the visual inputs of people who wear pairs of contact lenses that make one eye have a sharp image for reading and the other a sharp image for distance. Thus, the latter eye receives a blurred image when reading. An optical prescription of this kind is called ***monovision*** and its users appear to manage remarkably well**.**

Much work has been done on disparity channels tuned to particular spatial frequencies (re-read Ch 4 for a reminder on channels and spatial frequencies). Bela Julesz and Joan Miller (1975) explored stereopsis from various kinds of low and high spatial frequency filtered stereograms. Their experimental tactic was to introduce uncorrelated texture (noise) into one stereo half and enquire whether it disrupted stereopsis. Their idea was that stereopsis should survive if the noise had a spatial

c **Superimposing the right and left images so that F$_L$ and F$_R$ are aligned on the point F$_{LR}$**

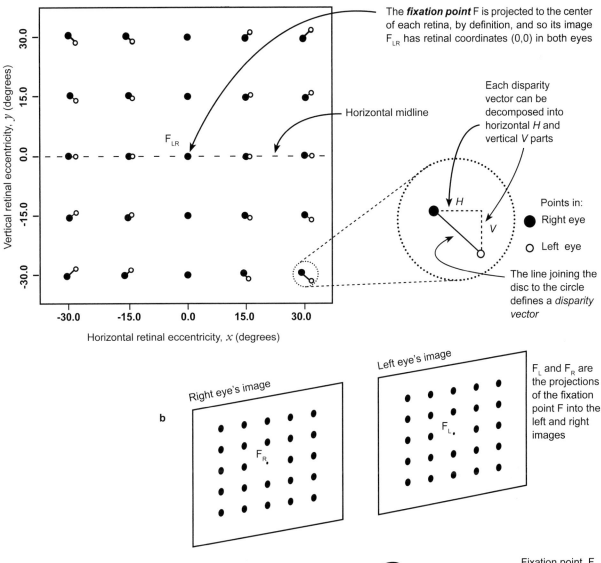

The *fixation point* F is projected to the center of each retina, by definition, and so its image F$_{LR}$ has retinal coordinates (0,0) in both eyes

Horizontal midline

Each disparity vector can be decomposed into horizontal *H* and vertical *V* parts

Points in:

● Right eye

○ Left eye

The line joining the disc to the circle defines a *disparity vector*

F$_L$ and F$_R$ are the projections of the fixation point F into the left and right images

Right eye's image

Left eye's image

b

18.48 Binocular disparities

a The scene: an observer gazing at a fixation point F on a fronto-parallel plane with dots on it.

b Left and right retinal images arising from **a**.

c Retinal images superimposed. The axes of the graph show *retinal eccentricity*, i.e., how far in the vertical and horizontal directions each point is from the center of the retina, **18.49**. The lines, called *disparity vectors*, join corresponding points in the left and right images. Their sizes have been enlarged to make them visible: binocular disparities are usually so small that they are often measured in minutes (60 minutes in a degree) or even seconds (60 seconds in a minute). Hence, this figure reveals the qualitative character of the *disparity vector field* for a fronto-parallel plane viewed as in **a**, not its correct quantitative details.

a

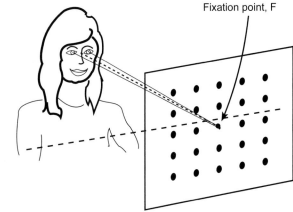

Fixation point, F

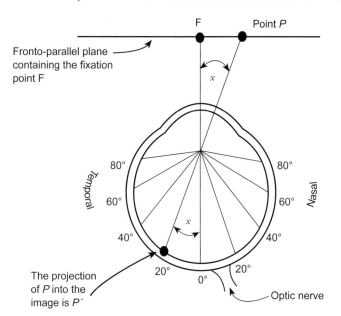

18.49 Measuring retinal eccentricity with an angle
Horizontal section through the left eye. The scene point *P* projects to a retinal point *P'* whose horizontal position, or *eccentricity*, with respect to the center of the retina (labeled 0˚) can be measured using the angle *x*.

frequency outside the tuning range of the channels picking up the texture carrying the disparity signals then. This is exactly what happens in **18.44**.

In contrast, in **18.45** the noise spatial frequency overlaps with the texture carrying the disparities and this destroys stereopsis. The converse cases are shown in **18.46** and **18.47**. Julesz and Miller concluded that spatial frequency channels tuned to disparities do exist but other work has suggested that they are perhaps not as independent as they proposed (see Ch 19 and Mayhew and Frisby's concept of the **binocular raw primal sketch**; another issue discussed in Ch 19 is whether low and high spatial frequency channels deal with coarse and fine disparities respectively).

Defining Binocular Disparities

Consider **18.48a**, in which an observer is fixating a small point F on a flat surface on which there is an array of larger dots. This surface is arranged to be parallel to the vertical plane containing the centers of the two eyes. Planes of this kind have a special name—they are called *fronto-parallel planes.*

The different viewpoints of the two eyes lead to their images being different, **18.48b**. Superimpos-

ing one image on top of the other brings out these differences (disparities) clearly, **18.48c**. (Imagine doing this by peeling the left and right images off their retinas and putting one on top of the other.) To keep track of what is going on, the right eye's dots are marked with black discs, the left eye's dots with outline circles when they fall in different places in the two images.

Each disc/circle pair in **18.48c** shows the locations of the *corresponding points* in the two images deriving from a particular dot on the surface in the scene. The binocular disparity of a scene point is usually defined as the difference in the retinal positions of the corresponding points associated with that scene point.(We say "usually" because there are other ways of thinking about binocular disparities: see earlier remarks on pp. 423–424 where we discussed the question: What exactly is a point for matching?)

The size of binocular disparities can be measured in terms of millimeters but it is conventional to use angles. We need to explain this next.

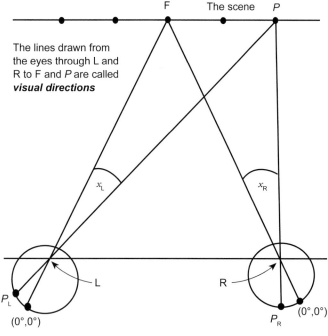

18.50 Geometry of binocular disparities
The scene is the middle row of dots from **18.48a** shown in plan view. Visual directions are shown for only two dots, the fixation point F and the point P. The direction of P is measured as the angles x_L and x_R for the left and right eyes respectively. The binocular disparity of point P is defined as the difference between x_L and x_R and is an example of a point disparity.

We need a simple way of designating retinal position. Examine **18.49**, which shows a horizontal section through the center of an eye of a person fixating the scene point F. A scene point P is shown, which happens to be in the fronto-parallel plane containing F but the basic ideas apply to all scene points. It turns out that a convenient measure of *horizontal* position on the retina is the angle x. That is, the angle x in **18.49** defines the horizontal **retinal eccentricity** of the retinal point P' to which the scene point P projects.

These concepts are illustrated in **18.50** for two eyes fixating a point F. The point P projects to points P_R and P_L on the right and left retinae, respectively. The associated retinal eccentricities are the angles x_L and x_R.

The term *eccentricity* is apt because it measures how far away (how eccentric) the points P_R and P_L are from the centers of the retinae. Hence, in **18.50** x_R tells us how far P_R is horizontally from the vertical midline of the right retina.

Vertical retinal position can be measured similarly, using another angle y, which says how far P_R is vertically from the horizontal retinal midline in the right retina.

Let's revisit **18.48c** now to see how the x and y angles are used to measure the locations of retinal points for the purpose of measuring disparities. This figure depicts the superimposed left and right retinal images on a graph with axes x and y. The origin of this graphical representation is marked by the label F_{LR} because it is where the fixation point F projects in the superimposed images. Hence, we can say that F projects to center of each retina, with retinal eccentricity of $x = 0°$, $y = 0°$ in each retina. That is, the **retinal coordinates** of F_{LR} are (0,0) in each eye. Remember, x and y are angles, and for the special case of the projection of F these are zero.

You may be worried by the fact that the retina is a concave surface, whereas the retinal images in **18.48c** are flat. Modeling retinal images as flat surfaces is acceptable because it is possible mathematically to link points on a spherical image to matching points on a planar image. Also, we will largely be concerned with scene points that project within 20–30° or so of the centers of the retinae: for this central region the retinae are close to being flat.

Notice that just a black disc appears in **18.48c** for the projection F_{LR} of the fixation point F, which makes sense as the point F_{LR} has same coordinates (0,0) in both eyes. Using just a black disc for F_{LR} in **18.48c** brings out the fact that the left image of F is exactly superimposed on the right image of F. This is always the case wherever the eyes are fixating in the scene, because by definition the image of F falls at the center of each retina.

As the projection of F always has the same coordinates in both left and right images its disparity is always zero. (This is not true of people with squints characterized by constant ocular misalignment between the two eyes.)

Now consider the black discs and outline circles on the horizontal midline in **18.48c**. There is quite a large difference in the horizontal positions of these corresponding elements in the left and right images at the edges of the array. That is, they have an easily discernible horizontal disparity.

Notice that the left and right images of the dots in **18.48a** that fall on the vertical retinal midline in **18.48c** are, like F_{LR}, shown with single black discs. This indicates that these points also have identical retinal locations. So the points in the scene giving rise to these images on the vertical midlines of the retinas also have zero binocular disparity.

Next, consider the disc/circle pairs in the corners of **18.48c**. These have the largest differences in position and hence the scene points that gave rise to them have the largest binocular disparities. Also, each outline circle has *both* a vertical and horizontal difference in position.

Finally, notice that the disparities between the left and right images of each scene point projecting near to F are smaller than those of the corner dots. However, one can just make out that these disparities are not quite zero (except for those on the vertical midline—see above). This link between the sizes of binocular disparities and retinal eccentricity is described technically by saying disparities are *scaled* by retinal eccentricity: in general, the more eccentric a point, the larger will be its disparity. We examine this geometric fact in detail in Ch 19.

The disc/circle pair in the lower right corner of **18.48c** is picked out and enlarged, and an arrow is shown joining the two dots. This enlargement shows that the disparity can be "decomposed" into two components, labeled H and V.

The **horizontal disparity** H of a scene point shows how far apart its corresponding image points

Each array is called a *disparity vector field.*

Each dot+line is one disparity vector. The dot represents the projection into the left image of a point on the slanted surface. The matching point in the right image is found at the other end of the line.

Slanted scene surface (thick line) is the same in all three cases.

The head turn is a rotation around the cyclopean eye (small solid dot), which therefore remains in same place.

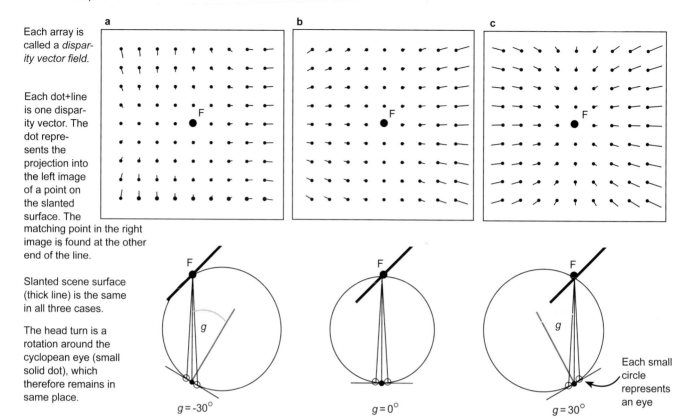

Each small circle represents an eye

18.51 Varying gaze angle *g* to the fixation point *F* on a surface by a head turn
The head turn has been made around a point mid-way between the two eyes, called the *cyclopean eye*. Fixation is held during the head turn on the point F on the slanted scene surface, thus creating the three different gaze angles and the dramatically different disparity vector fields in **a**, **b** and **c**, despite the scene remaining unchanged. The disparity vectors are shown by the lines in each field that connect the left (large dots) and right (other ends of the lines) images of points on the surface. These figures do not portray the texture gradient that would be present if the array of dots were arranged in a regularly spaced array on the scene surface. Reproduced with permission from Gärding, Porrill, Mayhew and Frisby (1985).

are in the horizontal retinal direction. For example, if the horizontal coordinate of a point in the left eye's image is x_L and the horizontal coordinate of the corresponding point in the right eye's image is x_R, then its horizontal disparity H is defined as $H = x_L - x_R$.

The ***vertical disparity*** V of a scene point shows how far apart the corresponding points are in the vertical retinal direction. The vertical coordinates of a point in the left and right images are referred to respectively as y_L and y_R. Hence, $V = y_L - y_R$.

As the retinal eccentricity coordinates (x, y) are measured in angles, both H and V are also angles. For example, suppose a scene point projects a left image point with coordinates $(x_L, y_L) = (11°, 6°)$ and its corresponding point in the right image has coordinates $(x_R, y_R) = (10°, 5.5°)$. In this case, the

disparities would be $H = (11° - 10°) = 1°$ and $V = (6° - 5.5°) = 0.5°$.

The H and V components of a disparity are useful in different ways. The horizontal disparity H contains information about the 3D structure of the scene. The usefulness of H is brought out by the following observation: have you noticed when hanging out washing that the depth of an empty smooth washing line can be difficult to discern? The line provides no horizontal disparities because it has no texture to allow H to be measured for its various parts. Hence, the absence of horizontal disparities H along the line means that its depth cannot ascertained with two-eyed vision. And as monocular cues to the depth of the line are also weak or non-existent, we have problems in seeing just how far away an empty washing line is.

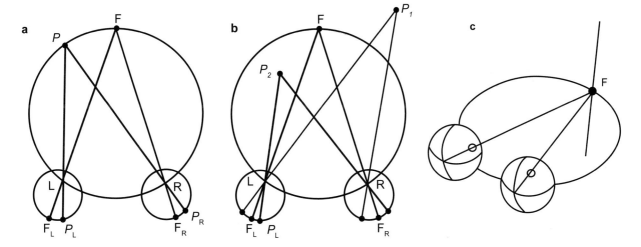

18.52 Vieth-Müller circle

a Plan view of the eyes fixating a point F in the horizontal plane. Any point *P* lying on the Vieth-Müller circle, which passes through the fixation point F and the optical centers of the eyes L and R, has zero disparity (*H* = 0° and *V* = 0°) because its projected images in the two eyes have the same retinal coordinates.

b Same plan view as in **a**. Point *P₁* lies outside the Vieth-Müller circle and is said to have a **divergent disparity**. Point *P₂* lies inside this circle and has a **convergent disparity**. The terms **uncrossed** and **crossed** are also sometimes used for divergent and convergent disparities respectively.

c 3D sketch showing the Vieth-Müller circle passing through the eyes. Scene points on a line tilted slightly backwards and passing through F also project to image points with zero disparities. The tilt arises from the eyes being slightly rotated around the *Z* axis.

However, the size of H is also affected by the rotational states of the eyes in their sockets, that is, where the eyes are fixating. For example, in **18.51a** the observer has turned his/her head to the right while keeping fixation on the same dot on the slanted surface.

This rotation has been made around a point mid-way between the optical centers of the two eyes, called the ***cyclopean eye*** (Cyclops is the one-eyed monster of Greek mythology). This means the eyes have rotated to different positions in their sockets from the straight ahead fixation shown in **18.51b**, or the head turn to the left in **18.51c**. The images projected into the two eyes are different in each case, and this means the patterns of disparities are also different. So eye rotations need to be taken into account in extracting the useful depth information about the scene from horizontal disparities.

The sizes of vertical disparities V are also affected by where in the scene the observer chooses to fixate, that is, by eye rotations. Values of V can be used to find out eye rotation positions (Ch 19) because they are not much affected by scene depth

structure. That is, more or less the same V values would arise in **18.48c** if the array of scene dots in **18.48a** did not lie in the fronto-parallel plane containing the fixation point F but were instead scattered about at different distances from the observer.

Origins of Binocular Disparities

So much for defining disparities in terms of differences in the retinal eccentricities of corresponding points. It is time to explain why they arise. We said at the outset of this chapter that they derive ultimately from the different vantage points of the two eyes but we need to consider this in more detail.

Consider again **18.50**, which shows a plan view illustrating the formation of the left and right images of the middle row of scene dots in **18.48a**. The optics of each eye is modeled as a pin-hole camera (p. 31), with the pin-holes set at the optical centers, L and R, of the lens system in each eye.

The images of the fixation point F fall on the centers of left and right retinas, indicated by their retinal coordinates (0,0). The images of the scene point P are shown at the retinal points P_L and P_R. The different positions of the eyes entail different *visual directions* from each optical center to the scene point *P*.

Visual directions, just like retinal eccentricities, are measured in terms of angles from the centers of the retinas. Hence, when we defined horizontal disparity in terms of different retinal eccentricities, $H = (x_L - x_R)$, we could as well have described this as a definition in terms of different visual directions. This equivalence is an advantage of using angles to measure retinal eccentricities, rather than, say, millimeters on the retinas.

The middle row of scene dots that project to the horizontal midline in **18.48c** have zero vertical disparities *V*. It is worth checking this by inspecting **18.48c** and noting that the corresponding retinal points for those scene dots lie on a horizontal line. Notice in addition in **18.48c** that *V* disparities are also zero for the points lying on a line passing vertically through the centers of the retinas (this is the line for which $x = 0$).

It is interesting to ask here: which points in the scene, in addition to *F*, project to corresponding retinal points for which *H* and *V* are *both* zero? Another way of asking this is: which points in the scene (in addition to the fixation point *F*) project to left and right image points that have the *same* retinal coordinates?

For the simple case shown in **18.52a**, in which the eyes are pointing straight ahead, the answer is: points lying on a circle drawn in the horizontal plane in **18.52** (see also **18.51**) passing through the fixation point F and the optical centers, L and R, of the two eyes. (If the complex optics of the eyes are modeled with a pin-hole camera, Ch 2, then the optical centers are defined as the points marking where the pin-holes lie.) This circle is called the **Vieth-Müller circle** to commemorate two mathematicians who described it in the 19th century.

Points lying on an almost vertical line passing through F also have *H* and *V* disparities equal to zero, **18.52c**, for a fixation point in the horizontal plane. This is not quite vertical because the eyes are slightly rotated around the *Z* axis.

Suppose the observer is not fixating a point straight ahead, as in **18.48a**. It is a geometrical

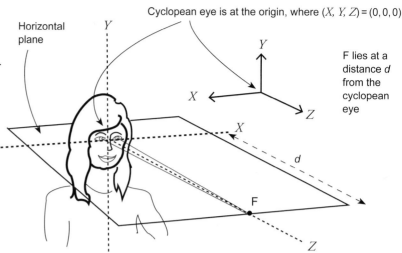

18.53 Headcentric 3D coordinate framework

fact that wherever the fixation point F lies on the Vieth-Müller circle, all the other points on this circle still have zero disparity. Check this for yourself by measuring the angles of the visual directions to any points on the Vieth-Müller circle in **18.52**.

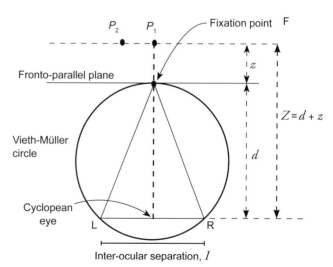

18.54 Plan view illustrating the depth interval z
It is important to note that the depth interval *z* is referred to with lower case to distinguish it from the upper case *Z* used for the headcentric depth axis extending straight ahead into the scene from the cyclopean eye. Hence, for both of the points P_1 and P_2 in this example, $Z = d + z$.

Representing 3D Space

The linking theme of this book is the computational approach to seeing. The first step in devising a computational theory of vision is to identify a clear goal. In the present case, the overall goal is to use the disparities between stereo images to build a representation of the 3D structure of the scene. This leads to the question: what form should that depth representation take?

A helpful starting point for thinking about this is to consider a **3D coordinate frame**. An example is shown in **18.53**, in which three axes set at right angles to one another are labeled X, Y, and Z. Their **origin**, the point where $(X, Y, Z) = (0, 0, 0)$, is shown located midway between the eyes at a point called the cyclopean eye.

The coordinate frame in **18.53** is called **head-centric** because its origin is a point in the head. Our goal can now be stated clearly: it is to represent the 3D location of each scene point with a set of three numbers (X, Y, Z), measured from this origin.

For example, in **18.53** the observer is shown fixating a point F in the **horizontal plane,** which is defined as the plane containing the Y and Z axes. The representation for the fixation point F in this coordinate frame is $(0, 0, d)$ because $X = 0$, $Y = 0$ and $Z = d$, where d is defined as the distance from the origin to the fixation point F, **18.53** and **18.54**. Note that d is the distance to F wherever F might be in the viewed scene—d happens to lie in the horizontal plane in **18.53** and **18.54** but this will not be the case if F lies above or below the horizontal plane.

Notice that moving the eyes to change the point of fixation does not alter (X, Y, Z) for any scene point as long as the head isn't moved at the same time. However, eye movements do change the left and right images, and hence the disparities between them, a fact we commented on earlier.

Using a headcentric frame means that if head position is changed then the X, Y, and Z values recorded for all scene points will also change. A way of avoiding this is to use a coordinate frame whose origin is a point in the scene. The technical term for this type of scene representation is **allocentric** or **geocentric**. There is evidence that some cells in the midbrains of rats encode spatial locations irrespective of the animal's location, suggesting they

mediate an allocentric representation. Humans must have brain structures that serve a similar function because the world appears stable as we move around in it. Even so, this consideration does not rule out the headcentric coordinate scheme in **18.53** serving as the initial representation of 3D space in human vision.

Using (X, Y, Z) to represent the depth of every scene point is an example of a point-by-point or **pointillist** representation. Again, this may be a useful starting point, but it cannot be the whole story as far as human vision is concerned. We see complex 3D relationships between scene points, such as sets of points lying in planes, or forming bumps and hollows, corners, and so on. These spatial relationships are not made *explicit* in a pointillist representation. Human vision must surely have representations for bumps, hollows, etc as they are manifestly parts of our visual world, but we won't go into that issue further.

3D Scene Structure from Horizontal Disparity

We have now defined what is meant by horizontal disparity, H. Our goal is to use H to represent scene depth structure by finding out X, Y, and Z for all scene points. So the next step is to enquire: how to get from H to X, Y, and Z? As usual, what we need is a computational theory of the task, based on a detailed and careful analysis of what the task entails. In this chapter we consider a restricted and simplified viewing situation to give the basic ideas. A fuller story is given in Ch 19.

The simplifications we adopt here are threefold. First, we confine our attention only to scene points projecting to the horizontal retinal axis through the retinal centers of each eye. These points are said to lie in the horizontal plane, **18.53**. As noted above, vertical disparity V is zero for corresponding points falling on the horizontal retinal axes.

Second, we consider only one fixation position, that shown in **18.53** in which the fixation point F lies on a line extending from the cyclopean eye along (or close to) the Z axis.

Third, we recover only the depth of scene points, such as P_1 and P_2 in **18.54**, that lie on or close to this straight ahead line, the Z axis. These simplifications mean that for such points we already know that $X = 0$ and $Y = 0$. Hence, we are left solely with the task of finding Z.

Notice that for the scene point P_1 shown in **18.54** Z is composed of two parts. The first part is *d*, the distance to the fixation point F. The second part is *z* which is defined as the **depth interval** of the scene point P_1 from the fronto-parallel plane containing F, **18.54**. That is, $Z = d + z$.

We will assume that we know the distance *d* to the fixation point F. You could be forgiven for thinking this is cheating. However, the value of *d* is determined by the observer's fixation position: if the observer in **18.54** changed fixation from F to P_1 then *d* would be increased by the difference between F and P_1. Hence, it is reasonable to consider *d* as a **parameter** measuring one aspect of **eye positions**. And indeed, in Ch19 we show how the distance *d* to the fixation point can be obtained from the pattern of vertical disparities *V*—which fits in with what we said above about *V* containing information about eye positions. So we are just postponing describing a theory of how to recover *d* from disparities, not avoiding the issue.

If we assume we know *d* then our quarry becomes the quantity *z* because we are after *Z*, and $Z = d + z$, **18.54**. Our starting point is to assume that *H* for P_1 has been measured as an angle using the definition $H = x_L - x_R$. We will have more to say about what is involved in that measurement process later.

We now need to find a decent computational theory of the task of finding *z* for the situation shown in **18.54**. It turns out that the theory again involves a geometrical analysis. This analysis is straightforward (see Ch 19 for references) but we won't go into the details. We leap immediately to the outcome, which is the following equation:

$$z = \frac{H \times d^2}{I \times 57.3}$$

where *I* is the interpupillary distance, defined as the distance between the optical centers of the two eyes. We have already defined *d* as the distance from the cyclopean eye (which lies at the origin of the headcentric coordinate frame (X, Y, Z) to the fixation point, and we have said that the horizontal disparity *H* has been measured in degrees of visual angle for the point P_1 in **18.54**. It is because we are measuring *H* in degrees that the constant 57.3 appears in the numerator. The reason is that the geometrical analysis takes advantage of using

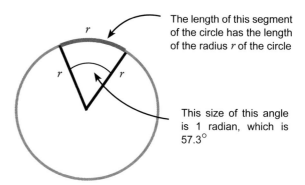

18.55 Measuring angles in radians
Imagine a line that has the length of the radius of a circle being laid out along the circumference of that circle. A radian is defined as the angle subtended at the center of the circle by this line.

an angular unit called the **radian**. There are 57.3 degrees in a radian, so dividing by 57.3 converts *H* from degrees to radians, **18.55**.

We are now ready to give an example of this equation at work. Suppose, in a viewing situation of the kind shown in **18.54**, the observer has an interpupillary distance $I = 6$ cm and is fixating a point 100 cm straight ahead (i.e., $d = 100$ cm). Suppose also that $H = 0.1°$ for P_1, then we have:

$$z = \frac{0.1 \times 100^2}{6 \times 57.3} = 2.9 \text{ cm.}$$

And so, $Z = d + z = 100 + 2.9 = 102.9$ cm. We are done.

You may not be used to thinking about and dealing with equations, so we will draw out what the equation for *z* says in words.

The first point is that if any item on the top line (the numerator) gets larger then the depth interval *z* will get bigger. So the larger the value of the horizontal disparity *H* the larger *z* will be. The same applies to the distance *d* to the fixation point. That is, for any given *H*, the farther away the fixation point the larger will be the depth interval *z*.

Conversely, if *z* is held fixed in a scene and the observer moves away from the fixation point so that *d* becomes larger, then *H* must reduce in size. It is easier to see this if the equation for *z* is rearranged thus:

$$H = \frac{I \times z \times 57.3}{d^2}.$$

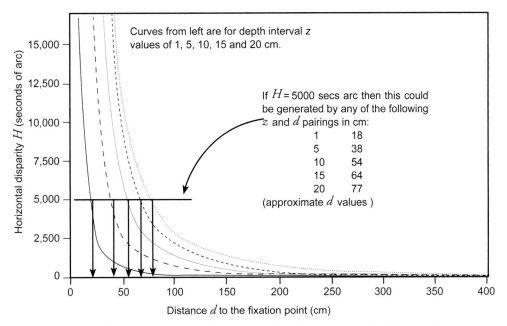

Curves from left are for depth interval z values of 1, 5, 10, 15 and 20 cm.

If H = 5000 secs arc then this could be generated by any of the following z and d pairings in cm:

z	d
1	18
5	38
10	54
15	64
20	77

(approximate d values)

18.56 The ambiguity of H disparity: trading off distance to fixation point d against the depth interval z
This graph shows that it is necessary to know d in order to work out z from H. Courtesy David Buckley.

It is now clear that if d becomes infinitely large then H becomes zero. This situation is approximated by looking at the moon. The left and right retinal images are identical for all practical purposes (i.e. they have no disparities) despite the moon having many deep craters.

However, it is not true, as is sometimes said in textbooks, that horizontal disparities become zero for viewing distances over about 6 meters. You can check this for yourself by looking at a landscape and alternating one-eyed and two-eyed viewing by masking one eye. If a depth interval z in the landscape is sufficiently big then it will still create detectable H disparities and hence binocular depth perception.

It will be apparent, from what has been said about the way z, d, and H are related, that any given value of H measured from the retinal images is ambiguous. That is, it could have arisen from any one of an infinite number of pairings of z and d, **18.56**. (In considering this figure, note that there are 60 seconds in a minute, and 60 minutes in a degree. So the H value of 5000 seconds of arc picked our for illustration could have equally well be described as 5000/(60x60) = 1.39°.)

Notice also the role played by I, the interpupillary distance in the equations for z and for H.

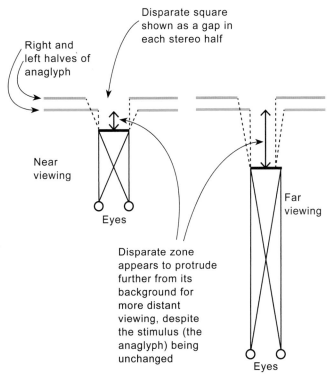

Disparate square shown as a gap in each stereo half

Right and left halves of anaglyph

Near viewing

Eyes

Far viewing

Disparate zone appears to protrude further from its background for more distant viewing, despite the stimulus (the anaglyph) being unchanged

Eyes

18.57 Perceived depth in a stereogram
This figure shows a geometric explanation of why perceived depth in a stereogram varies with viewing distance.

461

Having this term in the equation allows for the fact that the size of H from any non-zero depth interval z is partly determined by how far apart the eyes are. Hence, if two people with different I view the same 3D scene then the person whose eyes are set wider apart will have a larger horizontal disparity H for the same depth interval z. The geometrical analysis allows for this factor and the outcome is having I in the denominator of the equation where it compensates for the variation between observers in the distance apart of the eyes. There is some evidence that people with eyes wide apart have a (small) advantage over people with close-set eyes in discriminating fine depth intervals.

The equation for the depth interval z given above yields only approximate z values but they are accurate enough for most purposes. This is true even if the point P_1 is not immediately behind F but up to ±20° or so from F, e.g., point P_2 in **18.54**.

Viewing Distance Alters Perceived Depth in a Stereogram

Prop the book open on some suitable surface and view one or other of the anaglyphs of this chapter that yield strong stereopsis, such as **18.1**, from near (e.g., 10 cm) and then from far (go back as far as you can while still being able to see the depth effect). If you do this for **18.1** you will find that the prow of the ship sticks out farther from the page the further away you go.

This is an interesting effect to observe, but not perhaps quite as mysterious as one might at first think. This is because an increase in depth is exactly what would be expected simply from the optics of the situation, **18.57**. Since the stimulus itself does not change, of course, as the viewing position alters, **18.57** shows that the location in depth which the disparate square "must" occupy is different for different viewing distances.

This effect can also be interpreted algebraically using the formula for the depth interval z. The size of the horizontal disparity H depicted in the stereogram is scaled by 1/(viewing distance), 1/d. This is an example of the usual change of image size caused by perspective projection: objects appear smaller as they are seen from farther away. However, in the numerator is d^2, which works in the opposite way, to increase the size of z as distance increases. The net effect is that z is scaled by d, which means that

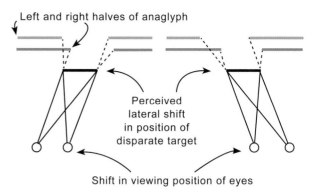

18.58 Shifting viewpoint while viewing a stereogram
Refer to **18.57** for labels explaining the various parts of this figure.

as d increases so too will the depth interval z seen in stereogram.

In a real 3D scene, as distinct from a stereogram, the formula for H shown opposite tells us that horizontal disparity H from a given depth interval z reduces with distance, being scaled by $1/d^2$, not 1/d. This is why when moving toward and from real objects perceived depth remains constant, as long as the system has available an accurate measure of d of course.

The moral of this story is: stereograms are not to be regarded as the same as real objects. They mimic depth effects from real scenes using methods that do not capture all aspects of natural viewing.

To underline the moral just drawn, it is interesting to try the effects of moving your head to and fro sideways while viewing stereograms. Choosing **18.5**, for example, you will find that the truncated pyramid moves with your head movement. This seems odd at first: why does it not stay still, as a normal 3D object would seem to do in this situation? Understanding why it moves is again helped by considering the changed optics of the situation, this time those created by a sideways change in head position. Thus, **18.58** shows that the disparate object "has" to move as it does because this is the only perception consistent with the retinal images.

Finally, it is worth experiencing the effect of running a finger across the surface of a page on which an anaglyph is printed, and also of bending up the corner of the page so that its surface is curved rather than flat. These manipulations can cause interesting distortions of the various depth effects, again for optical reasons due to altered disparity signals sent to the eyes.

Why Two Eyes?

Once again, we have tried in this chapter to display the fruitfulness of linking computational, neurophysiological, and psychological approaches to a perceptual problem. We deepen the story in Ch 19, by describing a theory about how vertical disparities can be used to recover the viewing geometry parameters describing the rotations of the eyes in their sockets. Knowing these parameters is important because they determine the sizes of disparities as well as scene structure, **18.51**. We will see that the parameters in question are distance to fixation d, gaze angle g, and elevation angle e.

A fitting way to end this chapter is to return to the question with which it opened: why two eyes? Now that we have shown how H disparities can be used to recover (X, Y, Z) vectors, the immediate and obvious answer is: we use the depth cue of binocular disparity, provided by virtue of the fact that the two eyes look at the world from different positions, to tell us about the depths of objects in the scene before us.

But this eminently plausible and sensible answer to our question may not in fact capture all the benefit which stereopsis conveys. To begin with, we might note again as we did at the outset that the kind of situation in which disparity provides the only adequate depth cue is the stationary inspection of such things as a bunch of flowers or a tree. In very many situations we can get by with other depth cues. Indeed, we might note that these other cues are sometimes given extra weight by the visual system, in that they may override stereopsis if they are placed in conflict with it (Ch 20). If the prime benefit of having two eyes is seeing depth, it is clearly not the case that this depth information is so stressed by the visual system that it wins out over other depth cues, come what may.

This consideration prompts a somewhat different answer to our question, "Why two eyes?" Could it be, perhaps, that the depth effect is a secondary advantage, and that the prime one is giving the visual system a superb way of grouping together features for the purposes of building up a scene representation at the level of object recognition?

It was explained in Ch 7 that grouping of features is a vital step in scene description, and various grouping rules were described. But these rules failed

18.59 Stereopsis and camouflage breaking
After binocular fusion, but not before, a central square of low spatial frequency blobs is seen behind a protruding network of high spatial frequency edges. This is similar to seeing fish swimming behind lacy pond weed, or looking at objects through a window with net curtains.

in certain circumstances: refer back, for example, to the problem posed by the overlapping leaves in **7.22**. With stereopsis available, this ambiguity could perhaps have been settled by grouping together those features with a similar depth. That is, features belonging to one leaf could have been separated from those belonging to the other leaf, without any need of high-level conceptually driven processing to disambiguate them.

From this viewpoint, perhaps the initial evolutionary advantage of having two eyes was as a solution to the problem of decoding camouflage. Did two-eyed vision really came into its own when it provided a means of grouping together stripe features belonging to the tiger (or other predator, or desirable but hidden prey) and separating them from stripe features produced by the branches, twigs, and leaves of the tree in which he was hiding, ready to pounce? A similar argument can be made for deciding which branch to jump to in a tree while hunting for fruit.

This speculation is certainly in keeping with the outcomes from random-dot stereograms. They show just how superb stereopsis is as a camouflage-breaking system, **18.59**. After binocular fusion the "hidden" object is revealed. Interestingly, stereo photographs are taken from aircraft in order to break down the camouflage of military installations on the ground, a trick that takes advantage of stereopsis for breaking camouflage. So perhaps, with the special kind of depth perception that is stereopsis in its armory, the visual system is much better at solving the figure/ground problem (Ch 7), and thereby better able to carry out its task of seeing *what* is *where*.

Further Reading

Blakemore C (1970) The representation of three-dimensional visual space in the cat's striate cortex. *Journal of Physiology* **209** 155–178. *Comment* Classic early paper on disparity-sensitive neurons.

Buckley D, Frisby JP and Mayhew JEW (1989) Integration of stereo and texture cues in the formation of discontinuities during three-dimensional surface interpolation. *Perception* **18** 563–588.

DeAngelis GC, Cumming BC and Newsome WT (2000) A new role for cortical area MT: the perception of stereoscopic depth. In Gazzaniga MS and Bizzi E (Eds) *The New Cognitive Neurosciences.* MIT Press, pages 305–314. *Comment* Advanced review of research on disparity-sensitive neurons.

Frisby JP (2001) Limited understanding of Panum's limiting case. *Perception* **30** 1151–1152.

Frisby JP and Clatworthy JL (1975) Learning to see complex stereograms. *Perception* 4 173–178.

Frisby JP and Mayhew JEW (1978) The relationship between apparent depth and disparity in rivalrous-texture stereograms. *Perception* 7 661–678.

Gärding J, Porrill J, Mayhew JEW and Frisby JP (1985) Stereopsis, vertical disparity and relief transformations. *Vision Research* **35** 703–722. *Comment* Best read after reading the next chapter.

Halpern DL and Blake RR (1988) How contrast affects stereoacuity. *Perception* **17** 483–495.

Howard I and Rogers BJ (2002) *Seeing in Depth Volume 2 Depth Perception* Published by Porteus Canada. *Comment* Provides an heroically wide ranging literature survey. If you are seeking references to follow up a topic in binocular vision this is a very good place to start looking for them.

Julesz B (1971) *Foundations of Cyclopean Perception.* University of Chicago Press. Republished in 2006 by the MIT Press, Cambridge: Mass. *Comment* A classic book that summarizes much of Julesz's work.

It has many marvelous random-dot stereograms, including the spiral shown here in **18.26**. For an obituary reviewing Julesz's diverse contributions see Frisby JP (2004) Bela Julesz 1928–2003: A personal tribute. *Perception* **33** 633-637.

Julesz B and Miller J (1975) Independent spatial frequency tuned channels in binocular fusion and rivalry. *Perception* **4** 125–143.

Marr D and Poggio T (1976) A cooperative computation of stereo disparity. *Science* **194** 283–287. *Comment* A classic paper describing the stereo correspondence algorithm set out in this chapter.

Ninio J (1994) *Stereoimagie.* Editions du Seuil. *Comment* Contains many marvellous autostereograms, including the one in **18.10.** In French.

Ogle KN (1954) *Research into Binocular Vision.* Saunders: Philadelphia. *Comment* Describes Ogle's classic work on stereoscopic vision.

Saye A and Frisby JP (1975) Role of monocularly conspicuous features in facilitating stereopsis from random-dot stereograms. *Perception* **4** 159–171.

Wade NJ (1983) *Brewster and Wheatstone on vision.* MIT Press: Cambridge, MA. *Comment* If you are interested in the early history of stereoscopes, this is an excellent starting point. It includes an account of the rivalry between these two giants of 19th century vision science.

Whittle P (1963) *Binocular rivalry.* PhD dissertation University of Cambridge UK. *Comment* For a recent paper testing Whittle's ideas about how opposite contrast stereograms might be matched, see: Howe PDL, Watanabe T (2003) Measuring the depth induced by an opposite–luminance (but not anticorrelated) stereogram. *Perception* **32** 415–421.

19

Seeing with Two Eyes, Part II

19.1 Jagged surface of the type used for testing stereo correspondence algorithms relying on the surface smoothness constraint
This is a 3D plot of a rough surface lacking surface smoothness of the kind assumed by Marr and Poggio's stereo algorithm. The PMF stereo algorithm, described in this chapter, copes quite well with this type of surface by defining surface smoothness in terms of a ***disparity gradient limit***, which is far less restrictive than the definition of smoothness used by Marr/Poggio.

The two eyes have slightly different views of the world due to their different positions in the head. This means that the retinal images in the left and right eyes are generally slightly different when we look at a 3D scene. These differences are called ***binocular disparities***, and we examined aspects of them in Ch 18. In this Chapter we delve deeper into factors that determine the sizes of binocular disparities, and how they can be measured and used. This analysis includes a computational theory showing how a representation of 3D scene structure can be obtained from information in disparities. This theory takes into account the complicating factor that the sizes of disparities are partly determined not only by 3D scene structure but also by where in the scene the observer happens to be fixating.

Having explained the theory, we then show how a biologically plausible algorithm of the Hough type (Ch 8) can implement the theory. Next, we review evidence suggesting that the theory is exploited by human stereo vision. Finally, we return to the stereo correspondence problem to extend the treatment given in Ch 18.

Recapitulating Basic Stereo Geometry

We begin by refreshing your familiarity with some fundamental concepts introduced in Ch 18.

Why read this chapter?

We develop Ch 18's introduction to stereo vision by delving more deeply into the geometry of binocular disparities. This entails explaining *binocular disparity vector fields* and how disparity vectors can be decomposed into their horizontal and vertical components. Mayhew and Longuet–Higgins' computational theory for using *vertical disparity* to recover eye rotation parameters, which specify where the eyes are fixating, is set out in detail and implemented in a Hough algorithm. Evidence that human vision uses vertical disparities in the way predicted by the theory is reviewed. This review begins with an analysis of the *induced effect*, a curious illusion in which a surface appears rotated around its vertical axis if one image from a pair of stereo images is stretched in the vertical direction. We then explain how the *PMF algorithm* utilizes the *disparity gradient limit* that characterizes human stereo vision to solve the stereo correspondence problem. Finally, we discuss recent findings on the neurophysiology of stereo vision.

The binocular disparity of any scene point P is defined in this book as the difference in the positions of the projections of P in the left and right images. In **19.2a** the two eyes are shown fixating the center of a square. The upper right corner of this square is labeled P. Due to the optics of the eye, P projects to the lower left corners of the trapezoids shown in **19.2b**. An arrow is shown connecting the corners in the images: this is an example of a ***binocular disparity vector***. If all corresponding points in the left and right images are linked in this way, then we have a ***disparity vector field***.

Each disparity vector can be usefully decomposed into its horizontal H and vertical V parts, **19.2b**. If the retinal coordinates of the projection of P into left and right eyes' images are respectively (x_L, y_L) and (x_R, y_R), **19.3a**, then the horizontal and vertical disparities of P are respectively defined as

$$H = x_L - x_R \tag{1}$$

$$V = y_L - y_R. \tag{2}$$

The retinal coordinates (x_L, y_L) and (x_R, y_R) are measured as angles, **19.3b**: check back to Ch 18 for details.

The usefulness of the decomposition of disparity vectors into H and V relies on the ***principle of separability***. This states that the H and V components convey different kinds of information, as we will see.

The sizes of H and V are partly determined by the prevailing ***viewing geometry***. This term refers to three parameters that specify where the eyes are fixating: ***distance*** d from the cyclopean eye to ***the fixation point*** F, the ***gaze angle*** g, and the ***elevation angle*** e, **19.4**. These parameters are defined with respect to the ***cyclopean eye***, which is a point in the head midway between the optical centers of the eyes, **19.2a**, so they are said to be ***headcentric***.

The viewing geometry parameter d conveys information equivalent to the ***vergence angle*** between the eyes. This is because d can be calculated from the vergence angle using I, the ***interpupillary distance***. The gaze angle g is the parameter that describes where the eyes are pointed in the left–right direction. In **19.2a,b** the eyes

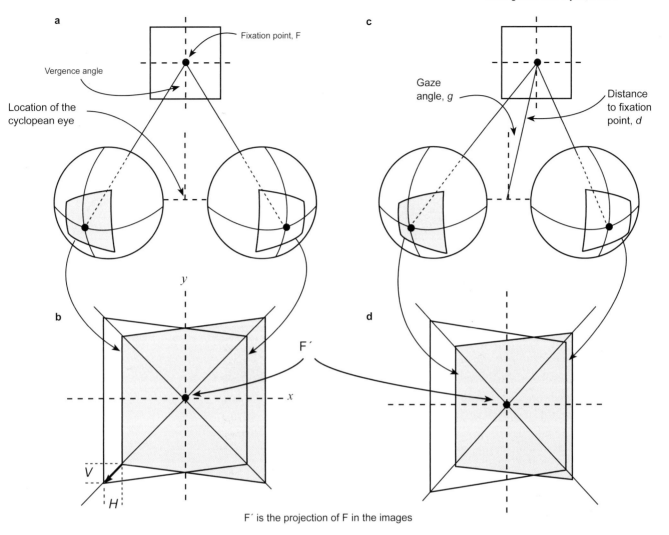

a

Fixation point, F

Vergence angle

Location of the cyclopean eye

c

Gaze angle, *g*

Distance to fixation point, *d*

y

b

F′

x

V

H

d

F′ is the projection of F in the images

19.2 Stereo geometry
All aspects are schematic, i.e., not drawn to scale.
a The eyes are shown as globes with fixation on the center of a square set in a fronto–parallel plane. The angle between the lines of sight to the fixation point is called the ***vergence angle.*** The square is imaged in each eye as a trapezoid. This kind of shape arises because one side of the square is slightly nearer to one eye than to the other and so it casts a larger image. However, the size of this effect is greatly exaggerated for the purposes of illustration.
b The left and right images from **a** are shown superimposed on a flat plane. To assist clarity, the left image is shown shaded. The arrow at the bottom left shows the disparity vector for one corner of the square in the scene. The *H* and *V* components of this disparity vector are indicated.
c The square has moved to the right but the observer has kept fixation on its center by holding the head fixed and using a non–zero gaze angle *g*. This altered viewing geometry changes the disparities from those in **b**, as shown by the superimposed stereo images in **d**.

are looking straight ahead and so *g* = 0: this is called ***symmetric fixation***. In **19.2c,d** the fixation point F has shifted to the right: this is *asymmetric fixation,* for which *g*≠0. Changes in *g* cause changes in the left and right images and hence changes in the disparities between them. This can be seen in the differences between the images in **19.2b,d**.

Elevation angle *e* is the viewing geometry parameter that describes where the eyes are pointed in the up/down direction, **19.4**. It is the angle between the *Z* axis and the plane containing the optical centers of the eyes and F. In human vision, eye elevation above or below the horizontal plane is accompanied by a small ***cyclorotation*** of the eyes

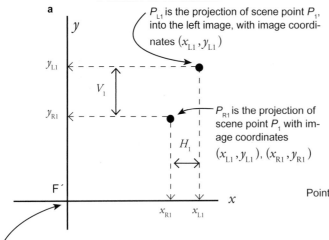

P_{L1} is the projection of scene point P_1, into the left image, with image coordinates (x_{L1}, y_{L1})

P_{R1} is the projection of scene point P_1 with image coordinates (x_{L1}, y_{L1}), (x_{R1}, y_{R1})

F′ is the projection of the fixation point F in the scene into the retinal images. F′ is the origin of each retinal coordinate frame.

19.3 Definition of horizontal and vertical disparities
a (Left) The left and right images are shown superimposed. The dots show the left and right image positions of the projections of scene point P_1 for which $H_1 = x_{L1} - x_{R1}$ and $V_1 = y_{L1} - y_{R1}$.
b (Below) Vertical retinal eccentricity y measured as an angle.

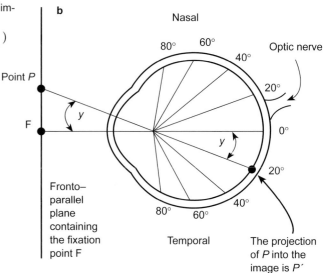

around their optic axes (**2.4b**, p. 31). We will not consider either elevation angle or cyclorotation in detail.

The way of representing eye positions shown in **19.4** is called the ***gun turret model***. This name arises from the analogy that moving the eyes to fixate F is like pointing the barrel of a gun by raising it up/down and swinging it left/right. In this gun turret model, g is defined as the direction of gaze in the plane containing the optical centers of the eyes and F. This plane moves up and down with elevation angle.

The fact that the sizes of H and V are partly determined by the prevailing viewing geometry is illustrated in in **19.5**.

Horizontal disparities

Armed with this knowledge of stereo geometry, we are now able to consider what determines the sizes of disparities for each scene point P. John Mayhew showed in 1982 that for horizontal disparities measured in radians

$$H = x_L - x_R$$

$$\approx H_z + H_g + H_{ecc} \tag{3}$$

$$= Iz/d^2 + Igx/d + Ix^2/d. \tag{4}$$

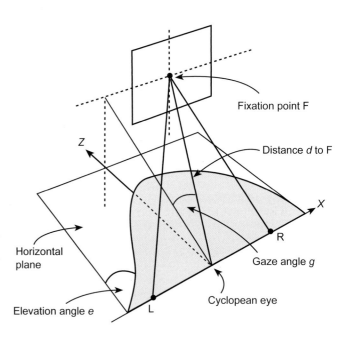

19.4 Gun turret model for representing stereo viewing geometry parameters d, g and e.
The shaded area is part of the plane containing the fixation point F and the optical centers of the left and right eyes, L and R. This plane contains the gaze angle g and the distance d to F.

Equation (3) says that H is made up of three parts, or *terms*. The equation for each term is given in equation (4). This equation appears formidable at first sight but in fact it is quite easy to understand. Each term has a distinguishing feature and we will explain each one in detail.

The key property of equation (3), demonstrated by Mayhew's geometric analysis of disparities, is the **principle of additivity**. This states that horizontal disparity H (and also vertical disparity V, as we will see later) can be treated, to a useful approximation, as the sum of *independent* terms. This means that variations in the key distinctive variables in one term do not cause changes in any of the other terms. Mathematicians love equations with this property as they are very easy to work with.

In thinking about the last point, you may be wondering how it can be true because interpupillary distance I and distance to the fixation point d appear in all three terms in equation (4). But I and d are *global* variables, which means that they help determine the horizontal disparities of *all* scene points. That is, they are not the distinctive variables defining each term in equation (4). We will see how this works out shortly.

Meanwhile, notice that each term in equation (4) is said to be *scaled* by I because each one is multiplied by it. This means that each term, and hence also H which is the sum of those terms, will be larger for people whose eyes are set relatively far apart. This may give them an advantage in making fine depth discriminations on the basis

of horizontal disparity (Frisby, Davis and Edgar, 2003).

Notice also that each term in equation (4) is scaled by $1/d$. This means that they all shrink as the viewing distance d to the fixation point F is increased, as happens if fixation is shifted to a point far away in the scene. If you are looking at the moon then your left and right retinal images of it are identical, for all practical purposes.

To avoid getting lost in equations, we need now to recapitulate on what we are trying to do, that is, we need to restate the *goal of the computation*. It is to find out the 3D structure of the scene using H and V. This entails deciding on a way to represent 3D scene structure. One way is to work out the scene coordinates X, Y and Z for all the

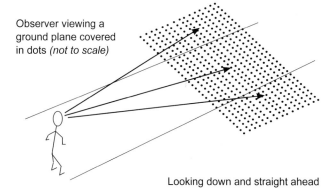

Observer viewing a ground plane covered in dots *(not to scale)*

19.5 Effect of different eye positions on disparity vector fields arising from a ground plane

The visual angle subtended by these images is about 30°. Thus, the retinal eccentricities of the points lying toward the periphery are not large, and yet the disparities are often more than a degree in magnitude.

Note that each disparity vector has vertical and horizontal disparity components. The vertical components are, at some locations, larger than the horizontal ones. Also, the pattern of disparities is markedly affected by changing the direction of gaze despite the scene (a textured ground plane) remaining the same.

These differences illustrate the critical requirement to take into account eye positions in using disparities for recovering 3D scene structure. The eye position parameters needed for doing this are distance to the point of fixation d, gaze angle g and elevation angle e.

Each line represents a disparity vector for one point in the scene. The dot shows the location in the left image of where the scene point projects. The other end of each line is where the scene point projects in the right image. Courtesy Mayhew, Zheng and Cornell (1993).

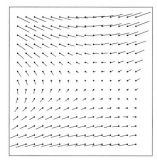

Looking down and straight ahead

Looking down and to the left

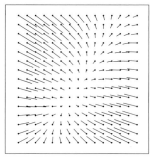

Looking down and to the right

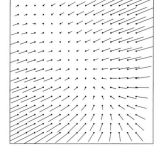

points in the scene, **19.6**. The X, Y, Z coordinate frame has its origin $(0,0,0)$ at the cyclopean eye. The Z axis extends out from the origin into the scene perpendicularly from the X axis, which is the line joining the optical centers of the eyes. The Y axis is perpendicular to the X and Z axes: it extends vertically through the cyclopean eye. Notice that the upper case letters X, Y and Z are used for the scene coordinate frame axes to distinguish them from the retinal coordinate frame axes x and y.

The X, Y, Z representation is a very limited one, being restricted to a point–by–point description of where each scene point is located. It is useful but is best thought of as a possible starting point for developing representations for the richly articulated depth structures that we can so readily see, such as corners or sloping surfaces.

The X, Y, Z representation is an example of a *metric* representation, so called because measurements of lengths (and angles) can be worked out from the X, Y and Z coordinates of scene points. There are other ways to represent scene structure. One important type is called **affine**. This represents only qualitative aspects of the 3D structure, such as recording where there are bumps and hollows without saying how high a bump is, or how deep is a trough. Thus, affine representations do not allow lengths and angles to be computed directly from them using trigonometry.

Horizontal Disparity Equation Again

The first term in Mayhew's equation (4) is the one we met in Ch 18:

$$H_z \approx Iz / d^2. \qquad (5)$$

Its distinctive feature is that it contains information about the depth interval z of P with respect to F, **19.7**. If the observer is looking straight ahead, so that $g = 0$, then z is defined as the depth difference of P from the **fronto-parallel plane** containing F, **19.7a**. If the observer has

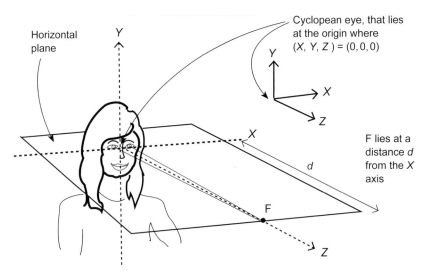

19.6 Headcentric 3D coordinate framework
The rectangle lies in the horizontal plane, for which $Y = 0$.

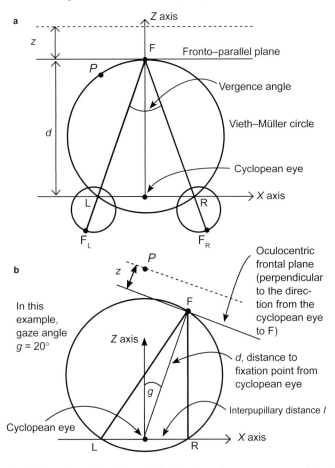

19.7 Plan views showing the depth interval z for a scene point P
a Symmetric (g=0) and **b** Asymmetric fixation. It is important to note that the depth interval z is set in lower case to distinguish it from upper case Z used for the headcentric depth axis extending into the scene from the cyclopean eye.

asymmetric fixation ($g \neq 0$) then z is defined as the depth difference of P from the ***oculocentric frontal plane*** containing F, **19.7b**. In both cases, if $z = 0$ then P is in the scene plane perpendicular to the line extending to F from the imaginary cyclopean eye, whose length is the distance d. We use the convention that z is positive if P lies further away than the plane containing F. Conversely, z is negative if P is in front of F.

The first term H_z was well known before Mayhew's work. He called it H_z to draw attention to its distinctive property—it is the term with z in it. Rearranging H_z gives the equation (see p. 460)

$$z \approx H_z d^2 / I. \tag{6}$$

Notice that because H_z is in the numerator of equation (6), as z grows bigger so too will H_z if d and I remain unchanged. That is, the more depth between the scene point P and the plane containing the fixation point F the larger H_z will be. Of course, if H_z grows larger then, as long as the other terms remain unchanged, H will also grow larger because H_z is one of the three terms that govern the size of H in equation (4).

The quantities z and d are crucial for getting what we want, the X, Y and Z coordinates of where point P is located in scene. We will see shortly that d can be obtained from vertical disparities.

However, the problem is that the horizontal disparities measured from the images for all scene points are not just affected by z and d but also by where the eyes are looking, that is, by the viewing geometry, **19.5**. Equation (3) shows why this is because equation (3) contains two other terms apart from H_z. These are H_{ecc} and H_g. We will now explain these terms in detail. (If you are used to thinking about what equations mean then you can skip from here to the next section.)

Mayhew called the second term in (3) H_g because, although it has x eccentricity in it, its distinguishing feature is the viewing parameter g:

$$H_g \approx Igx / d. \tag{7}$$

Because g appears in the numerator, H_g grows larger as g grows larger. But notice that this effect is scaled by x, where $x = (x_L + x_R) / 2$. This definition of x means that H_g is scaled by the *average* horizontal retinal eccentricity of x_L and x_R. So if $x = 0$ (that is, the average x eccentricities of the

image points lie somewhere on the vertical retinal axis through the center of the retina F) then there is no contribution to H made by g.

The third term in equation (3) is H_{ecc}

$$H_{ecc} \approx Ix^2 / d. \tag{8}$$

Mayhew called it the eccentricity term because it contains neither z or g; its distinctive feature is that it contains x^2. Thus, it contains important information about the retinal eccentricity of the image points to which the scene point P projects, where again $x = (x_L + x_R) / 2$. Thus, think of H_{ecc} as a measure of how far the images of scene point P are from the center of the retina F´ in the horizontal retinal direction. Recollect that the fixation point F projects to F´, **19.2b**.

The H_{ecc} term explains why the corner points in **19.2** have larger horizontal disparities H than points on the edges of the square. This is because the corner points are farther away from F´ than the edge points, so they have larger x and larger x^2, and hence a larger H_{ecc} term. Note that H_{ecc} will be zero if P lies somewhere on the vertical retinal axis through F. This is because in such cases $x = 0$ and any number multiplied by zero becomes zero.

But why does the x scaling in the H_{ecc} term use x^2 rather than just x? Recollect that locations of scene points for which horizontal disparity is zero fall on the Vieth–Müller circle, **18.52a**. The fronto–parallel plane containing the fixation point F grazes this circle. This makes sense because F has zero horizontal disparity by definition.

Now, imagine moving a scene point away from F in **18.54** while keeping it in the fronto–parallel plane. The distance of this scene point from the Vieth–Müller circle will grow as it is moved out from F because it will fall farther and farther away in depth from the Vieth–Müller circle. This depth increase means that the horizontal disparity of this scene point will also grow as it moves farther away from the Vieth–Müller circle. The shape of the growth in its horizontal disparity will be circular, because it reflects the shape of the Vieth–Müller circle. The equation for a circle has an x^2 term and so this is why the H_{ecc} term also has x^2.

Mayhew's analysis of stereo geometry was restricted to fixations of scene points lying in the horizontal plane for which $Y = 0$, **19.6**. However, John Porrill, a physicist colleague, showed how fixations above or below the horizontal plane can be taken into account by adding to equation (3) a fourth term, whose distinctive feature is that it contains the viewing parameter *e*, elevation angle, **19.4**. We will not concern ourselves further with making allowances for elevation angle, nor with the cyclorotation that typically accompanies it in human vision, but see Porrill et al. (1987) if you want to know more about how this can be done.

We hope you now agree that the initially daunting equation (4) for *H* is in fact not quite as opaque as you might have first thought. This is largely because of the principle of additivity, which means that the scene structure quantity *z* and the viewing geometry parameter *g* are nicely tucked up in different terms. Thus, variations in these parameters alter only their associated terms. There are no complicating interactions between *z* and *g*.

Recollect that our goal is to obtain from *H* the quantity *z*, **19.7**. So we need to rearrange equation (4) as follows

$$z \approx Hd^2/I - x^2d - gxd. \qquad (9)$$

Mayhew's equations rely on the *small angle approximation*. This is the assumption that the tangent and the sine of a small angle *θ* (theta) in radians is roughly equal to the angle *θ* in radians. The concept of a radian was defined in **18.55**. There are about 57.3° in 1 radian.) All angles in Mayhew's equations are in radians. He showed that using the small angle approximation simplifies stereo geometry at the price of errors of around only 2% for *g* if *x* is kept within the range ±30°. Hence, the small angle approximation turns out not to be a crippling handicap for modeling human vision.

Computer vision systems do not need to use the small angle approximation because they can exploit arbitrarily high levels of arithmetical accuracy. However, just as for biological vision, their accuracy will be limited by the resolution with which *H* and *V* can be measured.

Mayhew made two other assumptions in deriving his equations. First, he assumed that the depth interval *z* is much smaller than viewing distance *d*. In symbols, *z* << *d*. This does not impose a severe restriction of the generality of his equations. Second, he also assumed that interocular separation *I* is much smaller than *d*;

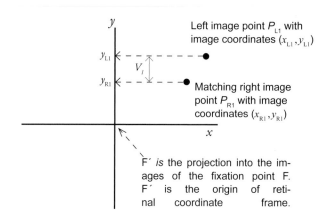

F′ is the projection into the images of the fixation point F. F′ is the origin of retinal coordinate frame.

19.8 Definition of vertical disparity definition
$V_1 = y_{L1} - y_{R1}$. The dots show left and right images of a point superimposed within a shared retinal coordinate frame.

I<<*d*. This too is not a severe restriction because the distance between the eyes is around 6 cm and most fixations will be to scene points much farther away than 6 cm.

Mayhew's Vertical Disparity Equations

Mayhew showed that the principle of additivity also applies to the vertical disparity *V* for a scene point, **19.8**, as follows

$$V = y_L - y_R$$

$$\approx V_g + V_{ecc} \qquad (10)$$

$$= Igy/d + Ixy/d. \qquad (11)$$

The definitions of *x* and *y* are as before

$$x = (x_L + x_R)/2 \quad \text{and} \quad y = (y_L + y_R)/2.$$

So, just as for *H*, the vertical disparity *V* has terms for the gaze angle *g* and retinal eccentricity *x*, and again both terms are scaled by the global variables of *I* and 1/*d*. However, the equations for these terms differ from those in the horizontal disparity equation (4) for *H*.

The gaze angle term V_g is scaled by *y*. What this means is that the effect of *g* on V_g grows larger as the images of the scene point *P* move farther in the vertical retinal direction from the projection of the fixation point F′.

The eccentricity term V_{ecc} grows larger with the product *xy*. This means that *V* increases as the left and right images of scene point P move further from F' in either or both of the horizontal and vertical directions. Notice that V_{ecc} will be zero if either $x = 0$ or $y = 0$.

Porrill showed that Mayhew's analysis of vertical disparities can be extended to cope with a non–zero elevation angle by adding a third term. But, as for horizontal disparities, for simplicity we will deal only with fixations in the horizontal plane, for which elevation angle is zero.

The striking thing about equation (11) for V is that nowhere does z appear in it. This means that V is not affected by the depth of P with respect to F. It is because V is insensitive to z that measurements of V can be used to recover d and g.

Once these viewing parameters are known then they can be substituted into equation (9) to get z.

Notice that equation (9) reduces to equation (6) $z \approx Hd^2/I$ if the observer is fixating a point lying straight ahead ($g = 0$), and the scene point for which z is being calculated is either in front of or behind the fixation point so that $x = 0$ (see p. 460).

Mayhew's equation for V relies on the same assumptions used for his equation for horizontal disparity H. The small angle approximation implies that it is only reasonable to assume that V is unaffected by the scene structure quantity z if the gaze angle g and eccentricity x are within the range ±30°. If extremely high resolution is used for measuring V then a small effect of z on V becomes evident for angles within this range. This small effect is ignored in Mayhew's equations. It is irrelevant for modeling biological vision.

Before describing a practical algorithm for extracting d and g from vertical disparities, we next summarize what we have said so far, and place it in an historical context.

Historical Note

One of the giants of 19th century science, Herman von Helmholtz (1821–1894), was aware that vertical disparities contain information about viewing distance d. Strangely, however, he did not give any details in his hugely influential book *Physiological Optics* Vol 3. His comments were neglected, perhaps because it was widely assumed, incorrectly for near distances, that vertical disparities are too small to be detected by the human visual system.

Mayhew's equations constitute a computational theory of stereo vision. His starting point was a mathematical proof that it is possible in principle to recover the X, Y and Z coordinates of all scene points if both the horizontal and vertical retinal coordinates are used from stereo projections. This proof was devised by a polymath, Christopher Longuet–Higgins, who was at various times physical chemist, mathematician, musical theorist, pioneer in artificial intelligence, and vision scientist.

Longuet–Higgins published, in the same issue of the journal in which Mayhew published his equations, a detailed mathematical analysis of stereo geometry that took into account vertical disparities. His analysis did not arrive at Mayhew's simple equations but they published a joint paper on their work in *Nature*.

This collaboration has been recognized by the underlying theory often being referred to in the literature as the Mayhew and Longuet–Higgins theory of stereo vision. The core idea in the theory, due to Mayhew, is to measure the vertical disparities for a sample of scene points and to use this information to find the viewing geometry parameters d and g.

Knowing these parameters makes it possible to work out the depth interval z for each scene point considered individually using each point's horizontal disparity H and equation (9). Knowing z, d, and g then permit the calculation of X, Y, and Z coordinates for each scene point, and thereby attain the goal: a representation of 3D scene structure.

The same line of reasoning can be used if fixations are not in the horizontal plane by incorporating Porrill's extensions to Mayhew's equations to make allowance for a non–zero elevation angle e to the fixation point.

Using vertical disparities to find d, g, and e is a viable strategy because vertical disparities are not greatly affected by 3D scene structure. In other words, whatever z values scene points might have, their vertical disparities remain much the same.

This is true as long as fixations are within ±30° of the Z scene coordinate axis, and the distance to fixation d and the scene interval z are large compared to the interpupillary distance I. Hence, vertical disparities provide a "separable" source of data about the d, g, and e parameters defining the prevailing fixation position.

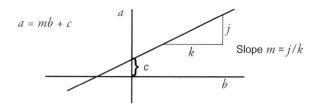

$a = mb + c$

19.9 Formula and graph for a straight line
In the formula $a = mb + c$, m is the slope of the line and c is the intercept with the vertical axis.

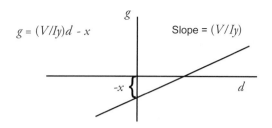

$g = (V/Iy)d - x$ Slope = (V/Iy)

19.10 Relationship between *g* and *d* is a straight line
(V/Iy) is the slope of the line and the value of the intercept is -*x*.

Vertical disparity V_1 measured at the retinal location x_1, y_1

	V_1	x_1	y_1
Radians	0.03	0.50	0.50
Degrees	1.86	28.65	28.65

Table of gaze angles *g* calculated for a range of possible distances to the fixation point *d*, using the V_1 data above. These *d, g* pairs are plotted in the graph below left (solid symbols).

d	*g* (radians)	*g* (degrees)
20	- 0.3	- 17.19
30	- 0.2	- 11.46
40	- 0.1	- 5.73
50	0.0	0.00
60	0.1	5.73
70	0.2	11.46
80	0.3	17.19
90	0.4	22.92
100	0.5	28.65

Vertical disparity V_2 measured at the retinal location x_2, y_2

	V_2	x_2	y_2
Radians	0.01	- 0.50	- 0.50
Degrees	0.08	- 28.65	- 28.65

Table of *d, g* pairs calculated using the V_2 data above. These *d, g* pairs are plotted in the graph below left (open symbols).

d	*g* (radians)	*g* (degrees)
20	0.41	23.74
30	0.37	21.28
40	0.33	18.83
50	0.29	16.37
60	0.24	13.92
70	0.20	11.46
80	0.16	9.00
90	0.11	6.55
100	0.07	4.09

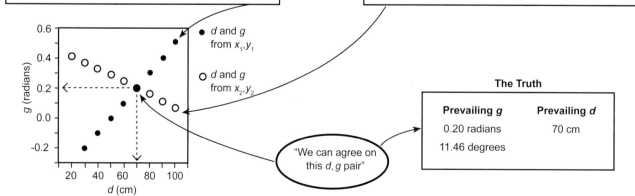

19.11 Hough accumulator recording votes for *d, g* pairs calculated for two image points
The accumulator is shown as the graph in which the intersection point between the two lines provides the true *d, g* pair.

There are three reasons why we have examined Mayhew and Longuet–Higgins' theory in detail. The first is that it has stimulated much psychophysical research aimed at finding out whether human vision implements their theory. The second is that it has been a good way to introduce basic aspects of stereo geometry. And thirdly, it can be implemented in an algorithm of the Hough type that can be run using processing components similar to neurons. It is thus said to be *biologically plausible*.

A Hough Algorithm for Finding *d, g, e*

Sue Peek, a postgraduate student of Mayhew, worked with him on devising a practical algorithm for extracting the viewing parameters *d* and *g* from measurements of vertical disparity, *V*. The algorithm can readily be extended to extract elevation angle *e* also, but to keep things simple we deal only with recovering *d* and *g*. Think of this version as restricting the algorithm to fixations in the horizontal plane *X, Z* in **19.6**, i.e., fixations with *e* = 0.

The starting point is the equation for *V* introduced earlier:

$$V \approx Igy/d \, + \, Ixy/d. \qquad (11)$$

Rearranging this to solve for *g* we get

$$g \approx (V/Iy)d \, - \, x. \qquad (12)$$

This version of Mayhew's equation for *V* has the structure of the equation of a straight line between two variables *a* and *b*, which is

$$a \; = mb + c,$$

where *m* is the slope (or gradient) of the line and *c* is the intercept with the vertical axis, **19.9**.

So in (11), *g* and *d* are the variables equivalent to *a,b* in the general formula for a straight line. Hence, (*V/Iy*) is the slope of the line and the intercept is -*x*, **19.10**.

Let's step back now and ask: what do we know, and what are we trying to find out from equation (12)?

The *knowns* in equation (12) are *V* (measured from the stereo images), *I* (interocular separation, which is a constant for each individual, assumed

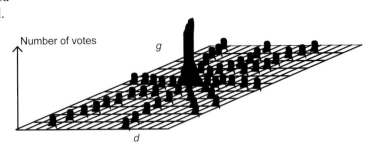

19.12 Hough accumulator for *d, g* as a 3D histogram
The votes for each cell in this schematic plot are shown as the black columns, with the height of the columns showing the number of votes. The votes come from vertical disparity *V* measurements from several image points. It can be seen that the votes fall on lines, as in **19.11**. There is a clear peak for one cell, which encodes the *d, g* pair consistent with all the *V* data. Due to noise in the measurements of *V*, cells surrounding the peak get more votes than cells far away from it, but the peak is still clearly discernible.

to be learned and adjusted gradually as we grow up and our heads grow), and the image coordinates of the point whose *V* has been measured, (*x, y*).

The *unknowns* are *d* (distance to the fixation point) and *g* (gaze angle), which are the quantities we are after.

The immediate problem we have is that the *V* measured for the projected point of any given scene point *P* is ambiguous with respect to *d,g*. This is because both these entities appear in equation (12). This means that any given *V* could have arisen from, say, a large *d* paired with a small *g*, or a small *d* paired with a large *g*. In fact, there are infinitely many pairs of *d,g* values that could have produced any given *V*.

The trick for getting out of this bind is to use more than just one measurement of *V*. Mayhew's key observation here is that *d,g* are *global* viewing parameters, by which is meant that they must have the *same* value for all image points. Hence, the way forward is to find a *d,g* pair that is *consistent* for all measurements of *V*. Only two *V* measurements are needed in principle, to break the ambiguity. (Readers familiar with algebra will recognize this as solving two simultaneous equations.) However, it is best to use many more than two measurements of *V* to help cancel out measurement noise, and thereby improve accuracy in estimating d, g. We are now ready to spell out the various steps of the algorithm in detail.

Step 1 Find a feature in the left image

Record its left image image coordinates, (x_{L1}, y_{L1}), **19.8**.

Step 2 Find the matching feature in the right image

Record its right image image coordinates, (x_{R1}, y_{R1}). This step requires that the stereo correspondence problem (see Ch 18 and later in this chapter) has been solved.

Step 3 Calculate V

$$V_1 = y_{L1} - y_{R1.}$$

Step 4 Calculate values of g for a range of d values for a particular image point with V_1

Do this by substituting d values using (12):

$$g \approx (V_1 / Iy_1)d - x_1.$$

where $y_1 = (y_{L1} + y_{R1})/2$, and $x_1 = (x_{R1} + x_{R1})/2$. Enter the calculated values of g for the various d values into a Hough accumulator table, **19.11**. Think of each entry as a "vote" in favor of a particular d, g pair from V_1.

Step 5 Return to Step 1 and repeat for another feature in the left image

This means calculating V_n using new image data: x_{Ln}, y_{Ln}, x_{Rn} and y_{Rn}. Keep cycling around *Steps 1–5* for all image points, or until some chosen limit is reached for n, the number of image points to be used.

Step 6 Find the peak in the Hough accumulator

If all has gone well, there will be one cell with more votes than all the others, **19.11, 19.12**. This cell will code the correct d, g pair. Goal attained.

Step 7 Use d, g to interpret H and find z

The d, g values can be used to interpret all the H values measured from the images. There will be one H measurement for each scene point P. So the task now is to substitute the d, g and H values into equation (9) to calculate the z value of each P. (Recollect that z is the perpendicular distance of P from the plane containing the fixation point F that is itself perpendicular to the line from the cyclopean eye to F ; details in **19.7**). Knowing z and d allows the X, Y, Z scene coordinates for P to be calculated using trigonometry. We will not go into the details of the latter step.

Why does the Hough algorithm find the correct d, g? The reason why the correct cell gets most votes is that the d, g entries into the accumulator will fall on lines that intersect in only one cell—the cell encoding the correct pair. Just two lines are shown intersecting in **19.11** but in practice there will be very many, **19.12** The votes for incorrect d, g pairs get smeared out over the accumulator, whereas those for the correct d, g pair pile up in just one cell.

It is easy to see from the graph in **19.11** why this algorithm is referred to as finding an *Intersection of Constraints*. This type of algorithm was used for object recognition (Ch 8) and for resolving the aperture problem in motion detection (Ch 14). So the algorithm turns out to be of value in many different computational contexts. It thus, illustrates nicely the value of Marr's insistence that three levels of analysis of complex information processing systems should be kept distinct. In the present case, we have:

Computational Theory

This is Mayhew and Longuet–Higgins' analysis of stereo geometry revealing how vertical disparities V convey information about the viewing parameters d, g, uncomplicated by scene structure.

Algorithm

This is the Hough accumulator, or equivalently the Intersection of Constraints method.

Hardware

We have not yet discussed evidence on the brainware involved in stereo vision. However, one can see intuitively that the cells in a Hough accumulator could be implemented as neurons receiving excitatory inputs from cells sensitive to vertical disparities. Then the cell that receives most excitation will be most active, and hence it will be the one encoding the correct d, g pair. This easy mapping between the Hough accumulator and neurons is why the Hough is said to be a biologically plausible algorithm.

Coping Robustly with Noisy Measurements of *V*

Measuring a vertical disparity *V* is inevitably prone to some degree of error due to various sources of *noise* (see Ch 5 for a general discussion of noise in sensory systems). Hence, we need to ask: how well does the Hough scheme just described perform when asked to deal with noisy measurements of *V*?

The answer from Peek's work was quite well. She tested the scheme using natural images, making some provision for noise by allowing each measurements of *V* to generate votes for cells adjacent to its *d, g* line traversing the Hough accumulator (**19.12**). Even so, subsequent work revealed that it tends to underestimate the parameters distance to fixation *d* and gaze angle *g*. This pointed to the need for a better way of dealing with noise. John Porrill provided a theory for doing that using a Bayesian analysis of the task.

The starting point of Porrill's contribution was to show that there was a better way of rearranging equation (11), repeated here

$$V \approx Ixy / d + Igy / d \qquad (11)$$

which we rearranged to obtain this equation

$$g \approx (V / Iy)d - x. \qquad (12)$$

Now, after adding noise *n,* equation (11) becomes

$$V \approx Ixy / d + Igy / d + n, \qquad (13)$$

which if rearranged to obtain an equation in *g* and *d* becomes

$$g \approx (Vd - nd) / Iy - x. \qquad (14)$$

This is a horrid equation, as the noise *n* is tangled up with the *d* parameter.

Porrill's answer to this was to rearrange equation (14) to obtain:

$$I / d = (-1 / x)(Ig / d) + V / xy + n / xy. \qquad (15)$$

This ensures the noise *n* does not appear in the terms for either *d* or *g*. Introducing new symbols, α (alpha) and β (beta), for the two parameters (*Ig / d*) and *I / d*, equation (15) becomes:

$$\beta = (-1 / x)\alpha + V / xy + n / xy. \qquad (16)$$

This is an eminently suitable linear equation to use for finding the α and β parameters in a Hough

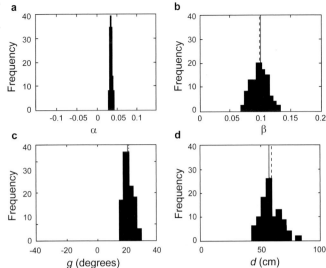

19.13 Results from a simulation using a Hough algorithm to obtain α and β and from them *d* and *g*

The data come from 200 trials. For each trial, a new pair of stereo images was created of a fronto–parallel surface patch, 20 cm in diameter and viewed at a distance of 57 cm with a gaze angle of 20 degrees. That is, the patch has (*d, g*)=(57,20). The patch was textured with 512 randomly positioned dots. Noise was injected by including for each of these 512 dots a further 10 dots portraying incorrect matches distributed uniformly over the possible vertical matching range for each of the 512 correct matches. Thus, the correct *d, g* pairing was signaled by 512 dots mixed in with 512 × 10 = 5120 noise dots. Thus the signal/noise ratio was only 10%. The graphs show that despite this huge amount of noise, accurate estimates of *d, g* were obtained on average. However, the *g* data show less spread over the 200 trials than the *d* data. Reprinted with permission from Porrill, Frisby, Adams, and Buckley (1999).

algorithm. This is because equation (16) satisfies the standard formula for a straight line, *a = mb + c*, with *a* = β and *b* = α. We will have more to say about the α and β parameters later, as their use has also appeared in a different context.

The final step is to derive *d* and *g* from α and β using straightforward algebra. Porrill showed that using α and β in a Hough algorithm produced highly satisfactory results, even in the face of massive noise **19.13**.

The excellent results in **19.13** arise because the correct matches all vote for the same *d, g* pairing, whereas votes from the noise matches are smeared over the Hough accumulator. We will return to Porrill's analysis later, as he had much more to say about the extraction of *d, g* from vertical disparities in the face of noise.

This accuracy in the face of huge noise suggests that the viewing parameters d, g can be obtained from vertical disparities *even if the stereo correspondence problem has not been solved.*

Recollect that this problem is establishing correct matches between entities in the left and right images. It was discussed in Ch 18 and we will return to it later in this chapter. For the moment, we note that if d, g can be found prior to solving the correspondence problem then it allows the epipolar constraint to be exploited in finding correct matches. This is because knowing the epipolar line for each image point requires knowledge of the viewing geometry, that is, the parameters d and g.

Does Human Vision Use Vertical Disparities?

Prior to Mayhew's work it was commonly thought that V disparities were more or less ignored in human stereo vision. Indeed, many vision textbooks made no mention of them, such was the emphasis on horizontal disparities. Often those who did know about them reckoned that they were just too small to be measurable by the human visual system. This was odd because calculations show that vertical disparities can be very large for points projected not very far into the periphery of the retina, **19.5**.

But even if measurable, it was generally thought that vertical disparities were irrelevant to recovering scene depth structure because their sizes are not much affected by the different distances of scene points from the observer. This is due to the eyes having roughly the same vertical positions in the head. It is because the eyes are separated horizontally that horizontal disparities are related to 3D scene structure.

Yet even allowing for the fact that they are not much affected by scene structure, it was still odd to regard vertical disparities as irrelevant to human vision. This is because it has long been known that artificially manipulating just the vertical dimension of one image, while leaving the horizontal dimension the same, causes a curious depth effect. We turn next to examining this phenomenon in detail, and we follow that by discussing other lines of psychophysical evidence about the use made by human vision of vertical disparities.

Ogle's Induced Effect

Mayhew and Longuet–Higgins's theory got off to a good start as a computational model of human vision because, as they pointed out, their theory accounts neatly for a classic phenomenon of human stereo vision. This is the ***induced effect*** studied by Kenneth Ogle in the 1930s.

Ogle found that magnifying one eye's image in the vertical direction, while leaving its horizontal size unchanged, causes a change in perceived depth. If, for example, an observer is looking at a picture of a face set in a fronto-parallel plane, as in **19.14**, and the right image is magnified vertically, then the picture appears slanted away from the eye with the lens.

Ogle used a special sort of lens, called a ***meridional size lens,*** for magnifying an image in just one direction. In modern research, image manipulations of this kind are usually done with computer graphics, as in **19.14**. The induced effect takes some time to emerge for many people, so you may need to be patient when trying to see it.

The depth effect caused by magnification of one image in the horizontal direction is much easier to perceive. Ogle created this effect simply by rotating the meridional size lens by 90 degrees, so that one image is stretched horizontally, **19.15**. Ogle called this the ***geometric effect*** because it is predicted straightforwardly from stereo geometry. Thus, the horizontal disparities scaled by stretching one image closely resemble those obtained when viewing a real surface slanted left/near to right/far.

To help you see why this is so, hold up a book at a slant and observe each image by closing each eye in turn. If you make the slant steep enough you should be able to see that one eye gets a wider image than the other eye. This effect is illustrated in **19.15** by the face in the right image being much wider than that in the left because the picture of the face in the scene is slanted.

So the geometric effect is easy to explain: it simulates the way slanted objects normally project into the left and right images.

The induced effect is quite another matter. Why should a vertical stretch of one image cause a depth effect? There are no natural viewing circumstances that could cause a vertical image stretch without an accompanying horizontal stretch.

Left and right images of a stereogram of a face drawing, with the right eye's image magnified vertically but not horizontally, as required for obtaining the **induced effect** (magnification exaggerated for illustrative purposes). The diagram below illustrates the perceived slant that arises from vertical magnification.

Left and right images of a stereogram of a face drawing as in **19.14** but with the right eye's image magnified horizontally and not vertically (magnification exaggerated for illustrative purposes). This produces a slant, called the **geometric effect**, that can readily be explained by the geometry illustrated in the diagram below.

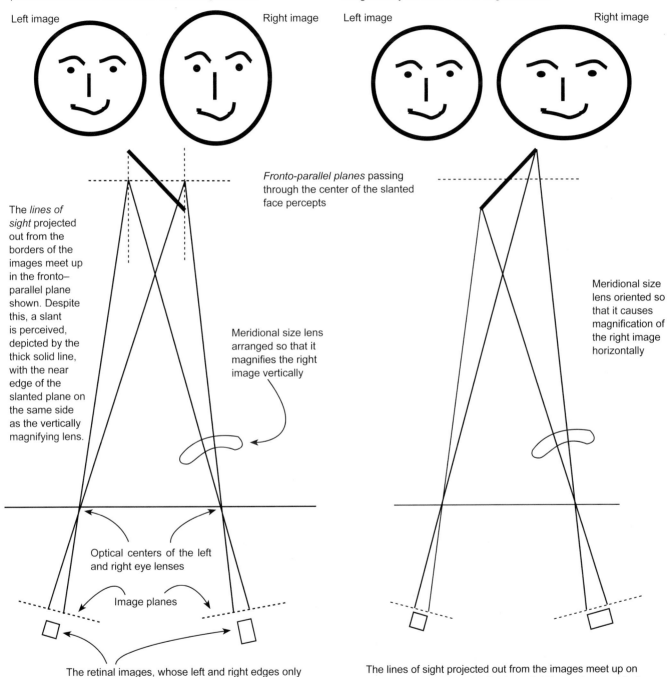

Left image   Right image

Left image   Right image

Fronto-parallel planes passing through the center of the slanted face percepts

The *lines of sight* projected out from the borders of the images meet up in the fronto–parallel plane shown. Despite this, a slant is perceived, depicted by the thick solid line, with the near edge of the slanted plane on the same side as the vertically magnifying lens.

Meridional size lens arranged so that it magnifies the right image vertically

Meridional size lens oriented so that it causes magnification of the right image horizontally

Optical centers of the left and right eye lenses

Image planes

The retinal images, whose left and right edges only are picked out by the lines of sight passing through the optical centers to meet the retinal image planes (dotted lines).

The lines of sight projected out from the images meet up on a slanted plane. This explains in straightforward geometrical terms why horizontal magnification of one image (here the right one) leads to the perception of a slant, with the near edge of the slanted plane on the opposite side to the magnifying lens.

19.14 Induced effect Not drawn to scale.

19.15 Geometric effect Not drawn to scale.

a

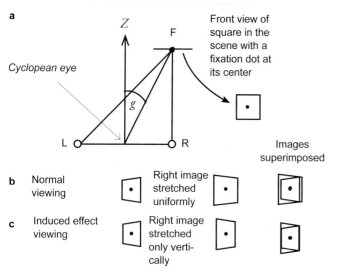

Front view of square in the scene with a fixation dot at its center

Images superimposed

b Normal viewing — Right image stretched uniformly

c Induced effect viewing — Right image stretched only vertically

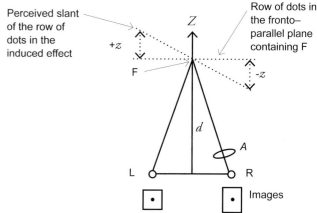

Images

19.16 Explaining the induced effect as a side effect of adopting a non-zero gaze angle *g*

a Plan view of an observer fixating a dot F at the center of a square by adopting the gaze angle *g* with head position kept fixed. This places the right eye R a little closer to the square than the left eye L. Hence, the size of the image of the square projected into the right eye's image is a little larger overall than the size of the left eye's image.

b Relative sizes of the retinal images arising from the viewing geometry in **a**. The larger size of the right image is a uniform expansion, i.e., the vertical and horizontal dimensions are stretched equally. Note that the square in the scene projects as a trapezium in each image, as in **19.2**.

c Relative sizes of retinal images used for the induced effect, in which one image, here the right, is magnified in just the vertical direction. This simulates what happens to the vertical dimension for the non–zero *g* shown. This vertical stretch is a source of evidence about *g*, with details defined in Mayhew's equation for vertical disparity *V*.

19.17 Illusory slant in the induced effect

The observer is shown fixating a dot F set at the center of a row of dots lying in the fronto–parallel plane. The observer is fixating straight ahead, so gaze angle *g* is physically zero. However, the meridional size lens A in front of the right eye R creates vertical disparities that, according to Mayhew's theory, result in a non–zero *g* being calculated. If this *g* is used to interpret the horizontal disparities arising from every dot of the surface, using equation (9) for *z*, the result is that the row appears slanted, Thus, explaining Ogle's induced effect. Note that the *z* value calculated for each dot in the row is scaled by retinal eccentricity, i.e., *z* increases with retinal eccentricity.

Ogle's explanation was that the induced effect resulted from a scaling process aimed at overcoming potentially troublesome size differences between the two images caused by differences in the optics of each eye. His idea was that the visual system measured the vertical sizes of the left and right images and adjusted their representations to make them equal. This adjustment was applied in *both* the vertical and horizontal directions. Comparing the vertical extents of the images to establish overall sizes would work well in normal circumstances. Using them rather than horizontal size avoids complications from mistaking horizontal disparities arising from 3D scene structure as an overall image size difference.

Ogle argued that a side effect of using vertical extents in this way would be the induced effect. This is because the vertically stretched image triggers image size scaling and thereby gets rid of the vertical differences. However, this scaling also alters the horizontal dimension. That is fine when getting rid of overall image size differences caused by optical factors but in the artificial circumstances shown in **19.14** there are no unwanted horizontal size differences to be scaled away.

Hence, the scaling alters the horizontal sizes inappropriately, it thereby injects horizontal disparities, and it is these that produce the slant, suggested Ogle. It is because he thought that vertical stretching induces *H* changes in the way images are represented that he called the slant the *induced* effect.

Mayhew's explanation for the induced effect is quite different. He pointed out that it is the inevitable consequence for any system estimating *d* and *g* using his equation for vertical disparity *V*, when faced with one image being vertically stretched. The reason is that a vertical magnification of one image simulates, in the vertical dimension, what happens when the observer adopts a non–zero gaze angle, **19.16**.

Hence, if vertical disparities are used to calculate gaze angle *g*, then this will lead to the recovery of a false *g* for the images used in the induced effect. When this false *g* is fed into equation (9) to obtain *z*, an illusory slant of the whole field of view is the outcome. To understand this, bear in mind that equation (9) is used separately for every scene point *P*, so that each *P* has its own *z* value calculated and *z* scaled by retinal eccentricity, **19.17**.

Looking at this in more detail, notice that in equation (9)

$$z \approx Hd^2/I - x^2d - gxd,$$

the term containing *g* is scaled by *x*, the retinal eccentricity of the image of the point *P* for which *z* is being calculated.

Hence, the effect of a non–zero *g* on *z* becomes larger as *x* becomes larger. The result is that the incorrect *z* values calculated in the induced effect grow larger with retinal eccentricity. This is why a whole field slant is seen.

You may be wondering that if equation (12)

$$g \approx (V/Iy)d - x,$$

is used for working out *g*, why doesn't the illusory slant in the induced effect also appear when one image is expanded *symmetrically* in the vertical and horizontal directions, as in normal viewing with non–zero *g*, **19.16**? The reason is that a uniform expansion also alters the horizontal disparities, and this change neatly cancels out the effect of the non–zero *g*. In the induced effect, the vertical magnification leaves the horizontal disparities unchanged. So if a non–zero *g* is fed into equation (9) while *H* remains the same, something has to "give" for the two sides of this equation to balance. The something is *z* and the illusory slant is the result.

In thinking about the last point and how equation (12) works, careful attention has to be paid to the signs of the various terms. For example, in this book *z* is defined as positive or negative depending respectively on whether the *P* in question is at a depth greater or less than the distance *d* to F. Recollect also that we defined the horizontal disparity of any scene point *P* as $H = x_L - x_R$. Hence, *H* can be either negative or positive depending on the sizes of the horizontal retinal coordinates (x_L, x_R) which themselves can be either negative or positive, **19.3**. So care is needed here (even talented mathematicians make sign errors from time to time). If you are keen on seeing the details of how Mayhew's equations lead to the induced effect then you could try creating a spreadsheet to do the necessary calculations.

If the human visual system exploits equation (12) to calculate *g* from *V*, does this lead to observers of the induced effect feeling that their eyes are looking to one or other side of straight ahead? The answer seems to be no. It therefore seems that the *g* obtained from *V* is used only for the purposes of interpreting *H*, not also for perceived eye positions.

However, it is interesting to ask: is non–visual information about eye positions *also* used, along with *V*, for interpreting measurements of horizontal disparities? This information could come from stretch receptors in the eye muscles and their tendons (proprioception or *inflow signals*) or from the brain knowing the positions in which it had placed the eyes, via commands sent to the eye muscles (*outflow signals*).

If these sources are used for estimating *g* in addition to equation (12) then the visual system is faced with a cue conflict when dealing with induced effect stimuli. Should it use the non–zero *g* given by equation (12) or use the zero *g* given by eye position mechanisms? Or perhaps some mixture? It could be that one reason the induced effect takes some time to emerge for many observers is because of the time needed to resolve this cue conflict.

Marty Banks and Ben Backus (1998) have (among others) reported evidence that non–visual information about eye positions is used in combination with the visual information embedded in vertical disparities. Their study utilized the induced effect. They found that when the two sources of information are so arranged that they

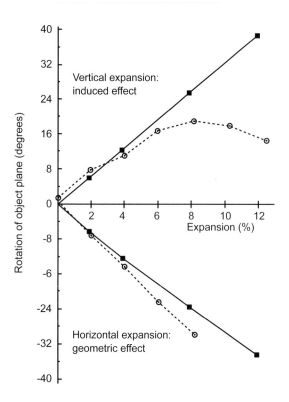

19.18 Geometric and induced effects compared
A sample of Ogle's data from a human observer, with inter–
ocular separation *l* = 6.5 cm, is shown with the dashed lines.
The viewing distance of the stimulus array was 40 cm. The
solid lines show the results of Mayhew's algorithm, also for
l = 6.5 cm and *d* = 40 cm. The geometric effect (lower half
of the graph) of the human shows a linear increase with
horizontal magnification of one image that is close to that of
the algorithm. The induced effect (upper half of the graph) of
the human is initially close to the results from the algorithm
but the human's induced effect no longer increases in size
for vertical magnifications of one image greater than about
6%. This "roll–off" result for the induced effect is discussed
on p. 481-482.

yield consistent data on gaze angle *g* then the
induced effect no longer shows a roll–off for
vertical magnifications of one image larger than
about 4–6%, **19.18**.

Mayhew had argued that this roll–off makes
sense in terms of his theory because in natural
viewing it is impossible to adopt gaze angles that
create vertical magnifications of one image larger
than about 6%. However, Banks and Backus argue
that their data, by showing large induced effects
from large vertical magnifications, strongly support
the cue conflict interpretation. In any event,

their data suggest that non–visual information
is indeed used, along with vertical disparities,
for the purpose of finding out gaze angle when
interpreting horizontal disparities.

Simulating Changes to *d* by Scaling *V*

We have said that the induced effect is evidence
that human vision uses vertical disparity
measurements to estimate gaze angle *g* (page
481). But we showed that the Hough transform
algorithm yields not only *g* but also distance to
fixation *d*. Is there any evidence that human vision
recovers *d* using vertical disparity measurements?
Several studies have investigated this question,
using the strategy of presenting stimuli that
simulate changes in vertical disparity that naturally
arise when *d* is changed.

Consider **19.19**, which compares changes in
vertical disparity caused by changes in *g* with
changes caused by *d*. Thus, **19.19b** shows a vertical
magnification of the right image, as used to create
the induced effect; and **19.19c** shows vertical
disparity scaling simulating a change of *d*. In the

a No image manipulation

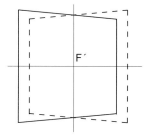

b Right image mag-
nification

c Vertical disparities scaled

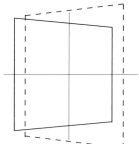
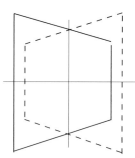

19.19 Two ways of altering vertical disparities
Left images solid lines, and right images dotted.

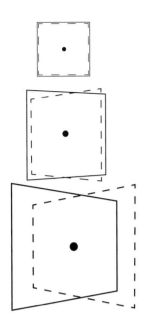

d = very large, e.g., 100 m
Left and right images
identical (though dotted
image shown smaller
to make it visible in this
figure).

d = middle range, e.g., 1–2 m
Left and right images are
different and have both
small vertical and horizontal
disparities.

d = small, e.g., 20–40 cm
Left and right images are
very different and have
relatively large disparities.

19.20 Effects of varying fixation distance d on retinal images an outline square in the frontoparallel plane. The gaze angle g is zero. Notice that the images are increased in size as d is reduced. Not drawn to scale.

latter, the scaling has been shared between the two images, by shifting the locations of the vertical coordinates of all points in such a way that vertical disparities are enlarged. This scaling exaggerates the trapezoidal shape of the images.

Consider **19.20** to get a feel for why the scaling in **19.19c** simulates, for the vertical dimension, a change in distance to fixation d. As d is reduced the projected images of the outline square become larger and the vertical and horizontal disparities also increase in size. It is the enlargement of the vertical disparities that is picked up by equation (13), and interpreted as a reduced d.

So the question becomes: if vertical disparities are artificially scaled, using computer graphics to create images of the kind shown qualitatively in **19.19c**, and if human vision uses vertical disparities to estimate d, what consequences would be expected in how the scene appears? Let's consider the simple case of the observer fixating, with zero gaze angle, a surface of dots lying in the fronto–parallel plane. In this case, the g term in equation (12) is zero so we get

$$z \approx Hd^2/I - x^2d. \quad (17)$$

Now consider what will happen if horizontal disparity H is left unchanged and d is reduced artificially by scaling vertical disparities, as in **19.19c**. As the observer is viewing a fronto–parallel plane, the depth interval z starts out as being zero for every scene point (check back to **19.7** if you need a reminder on the definition of z). However, as d is reduced, equation (14) will lead to z being non–zero. In other words, the dots that started out as appearing in the fronto–parallel plane will no longer appear in that plane. Instead, the surface of dots should appear curved in 3D around a vertical axis, **19.21**. Why curved?

The 3D curve arises from the term $-x^2d$ in equation (17). We explained above that this term served as a corrective factor. It nulls off, for the purposes of finding z, the way in which the horizontal disparities of points in the fronto–parallel plane containing F increase with eccentricity. Under normal viewing this works fine, with z from equation (9) coming out as zero, as it should do for those points on the fronto–parallel plane containing F. However, when H is left unchanged and d is artificially changed by scaling vertical disparities, the calculation of z from equation (9) no longer produces $z=0$. And it is because the eccentricity term $-x^2d$ has x squared in it that the result is a curved change in 3D appearance.

Points appear on a convex surface when d is artificially reduced while H is left unaltered

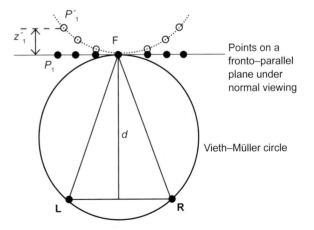

19.21 Effects of scaling vertical disparities
A fronto–parallel plane appears curved in depth when the recovery of d from vertical disparities is changed by scaling vertical disparities while leaving H unaltered.

We should add that whereas **19.21** shows the effect of reducing d by using suitable vertical scaling, it is perfectly possible to get the opposite curvature (i.e., a concave surface rather than the convex one in **19.21**). The way to do this is to begin by showing an observer a fronto–parallel plane at, say, 40 cm, and then scaling the vertical disparities to simulate an enlarged d of, say, 100 cm. This time equation (9) creates negative z values, which places the scene points in question in front of the fixation point, and hence lying in a convex surface.

So much for the theory. Does human vision show the predicted curvature of a fronto–parallel surface when vertical disparities are scaled? The answer is yes. A classic observation made by Helmholtz in the 19th century is that a fronto–parallel field of vertical threads appears bowed around a vertical axis *unless* beads are attached to the threads (off the vertical and horizontal meridia). The role of the beads can be interpreted as providing vertical disparity information required to see the threads in the correct plane.

More direct studies in the 1990s using the kind of image scaling described above have found the predicted effect, for example, those by Brian Rogers, Mark Bradshaw, Andrew Glennerster, and De Bruyn (1998). Interestingly, they found that the size of the bending effect is determined by the size of the field of view and there was a trade–off between d signaled by vertical disparity scaling and the d signaled by manipulating vergence. The first factor emphasizes the importance of retinal eccentricity and helps explain why some earlier attempts to find disparity scaling effect may have been unsuccessful by using stimuli that were too small. The second underlines the fact that studies of this sort inevitably involve complicating cue conflicts between visual and non–visual information used for interpreting horizontal disparities.

Although, for simplicity, we have explained the theory in terms of effects of vertical disparity scaling on the fronto–parallel plane, the scaling is predicted to affect all surfaces. For example, if a vertical ridge is presented then scaling to simulate nearer or farther d makes the ridge appear respectively either shallower or more peaked. This was demonstrated in a study in 1999 by John Frisby, David Buckley, Helen Grant, Jonas Garding, Janet Horsman, Stephen Hippisley-Cox, and John Porrill. Equally, a horizontal ridges adopts a banana shape from d scaling.

That study also lent support to Garding's theory, published in 1995, showing how vertical disparity information can be used within a two–stage process. The first stage, called *disparity correction*, can be thought of as "correcting" for the effects of retinal eccentricity on horizontal disparity. He suggested this stage could be thought of as "unbending" the Vieth–Müller circle without estimating either d or g. The resulting "corrected" horizontal disparity H measurements are sufficient to extract relief properties, such as peaks and troughs in scene surface. For example, a peak in the corrected H measurements relates directly to a peak in the viewed surface regardless of the values of d or g.

The second stage, called *disparity normalization*, takes the processing a stage further to work out the values of d and g. It then uses these values to estimate metric scene surface properties, such as peak amplitudes. The main advantage of a two–stage process for using vertical disparity could be the rapid and easy recovery of relief properties from the first stage. This issue is discussed by Gärding et al. (1985), who also report evidence suggesting that human stereo vision uses vertical disparity information in two stages specified in the theory.

Local vs. Global Manipulations of Vertical Disparity

We pointed out when describing the Hough algorithm for calculating d and g from vertical disparities that d and g are global variables. This means that they affect the disparities of all points in the field of view. Hence, if human vision uses vertical disparities, we would expect that simulating a changed g by altering V values for just a few points, leaving all others unchanged, would not produce the induced effect. And this is exactly the result. The same absence of an effect is found if a changed d is simulated by scaling V for just a few points, rather than for all points. These results contrast sharply with the vivid and immediate effects found for local manipulations of horizontal disparities H. The latter arise because the equation for H contains the H_z term, so local changes to H for a given scene point result in a changed computation of that point's *z* value, and that means a change in perceived depth.

The global character of *d* and *g* means that vertical disparities from all points can be pooled the better to estimate them. A study using the induced effect by John Porrill, John Frisby, Wendy Adams and David Buckley examined the pooling process in human vision by asking what happens when a stimulus contains two intermingled populations of points, each signaling a different *g* value due to each population having vertical magnifications of opposite signs assigned to them. For example, one population had a magnification signaling an induced effect slant of increasing depth toward the right. The other population signaled a slant of increasing depth toward the left. In a series of different conditions, the size of the magnification difference between the two populations was gradually increased. The research question was: would the data from the two populations be pooled over the full range of conflict, despite their increasingly opposing implications?

Porrill's theoretical Bayesian analysis showed that once the difference in magnifications for the two populations exceeds a certain critical value, then either one or the other population should be chosen to indicate *g*. Before this critical *bifurcation* point is reached, data from the two populations should be pooled; after it there should be separation, **19.22**.

This predicted bifurcation can be explained by considering what would be going in a Hough algorithm when faced with increasingly conflicting votes about *g* from two opposing sources of data. Initially, the similarity of the signals from the two populations of votes would cause a single cell to come out on top, as in **19.12**. However, as the conflict grows, so two different cells would each get strong support, unlike the single cell in **19.12**. A choice would then have to be made, perhaps randomly, to select one or other cell. Porrill showed in his analysis that using the Hough algorithm in this way was consistent with a Bayesian model for optimal and robust pooling of noisy data.

This bifurcation behaviour is called *robust* because it is based on rejecting *outliers* as unsuitable for pooling. A outlier is defined as a measurement so far removed from the central

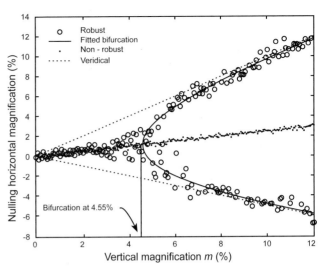

19.22 Simulation showing a bifurcation in pooling conflicting data from two populations of points
The middle line showing non–robust pooling comes from simple averaging of the data on *g* from the two populations. In contrast, robust pooling leads to the data from one or other population being selected. Reproduced with permission from Porrill, Frisby, Adams, and Buckley (1999).

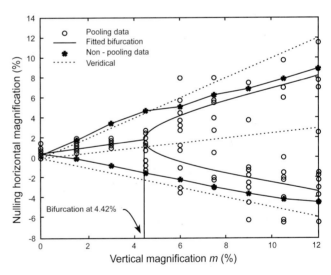

19.23 Combined data from four human observers showing a bifurcation in pooling conflicting data on *g* from two populations of points
Reprinted with permission from Porrill, Frisby, Adams, and Buckley (1999).

tendency of a sample that it makes no sense to pool it in with the others—it almost certainly comes from a different population.

The Porrill et al. study went on to report an experiment investigating whether human vision

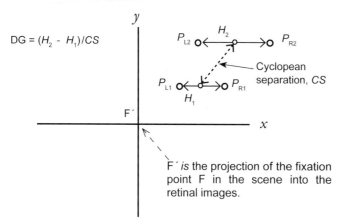

DG = (H_2 - H_1)/CS

Cyclopean separation, CS

F′ is the projection of the fixation point F in the scene into the retinal images.

19.24 Definition of disparity gradient (DG)
Left and right images are superimposed. The image points P_{L1} and P_{R1} are the projections of scene point P_1 and have horizontal disparity H_1. Similarly, points P_{L2} and P_{R2} are the projections of scene point P_2 and they have horizontal disparity H_2. For simplicity, no vertical disparities are shown. The small outline circles mark the retinal midpoints between the left and right images of each scene point. The *cyclopean separation* CS is defined as the distance between these midpoints. The horizontal disparity gradient between two points is defined as the difference between their horizontal disparities divided by the cyclopean separation between them. In symbols, DG = (H_2 - H_1)/CS.

also exhibited bifurcation at a critical point in the vertical magnification difference between two populations of points. Their observers showed an induced effect whose size at first was consistent with pooling but, after the critical point was reached, the induced slant was determined by one or other sub–population, **19.23**. This is evidence that human vision exploits Porrill's theory.

PMF Stereo Correspondence Algorithm

We have already mentioned in this chapter the problem of finding correct matches between entities in the left and right stereo images. Solving this correspondence problem is vital for measuring horizontal disparities, in order that they can be used for building a representation of 3D scene structure. In Ch 18 we described four constraints on which were based the following rules for binocular combination:

> restrict potential matches to those that can be made between *figurally compatible* left/right image entities;

> restrict potential matches to those lying on *epipolar lines*;

> *surface continuity*: allow same–disparity matches to exchange mutual support;

> *uniqueness*: allow each image primitive finally to be matched with only one primitive in the other image.

It turns out that these constraints prove useful but insufficient. Additional constraints were identified and then implemented in a stereo correspondence

algorithm called **PMF** after its originators, Stephen Pollard, John Mayhew, and John Frisby. We select this stereo algorithm for detailed treatment because it has had considerable influence. The next two sections sample heavily from a commentary written by invitation of the journal *Perception* for a special issue celebrating its most cited papers (Pollard, Frisby, and Mayhew, 2009). Many of the figures shown below derive from Pollard's thesis.

Disparity Gradient Constraint

Marr and Poggio's constraint of surface continuity is that differences in the depths of neighboring points in a scene will in general be small compared to fixation distance *d*. They used this observation to justify same–disparity support between neighboring image points in their algorithm. The problem with this support rule is that it is too restrictive. Many smooth scene surfaces that satisfy the surface continuity constraint, such as table tops and floors, would not be favored by Marr and Poggio's algorithm. This is because neighboring points on such surfaces do not generate disparities of the same size. The main distinctive feature of the PMF algorithm is that implements a far less restrictive rule for exchange of support between neighboring points.

PMF was stimulated by a phenomenon of human stereo vision, first noted by Chris Tyler in 1973 but brought to the close attention of Pollard et al. by the work of Peter Burt and Bela Julesz. This phenomenon is that human binocular fusion seems subject to a disparity gradient (DG) limit. DG is defined as the ratio of the horizontal

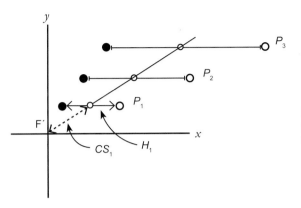

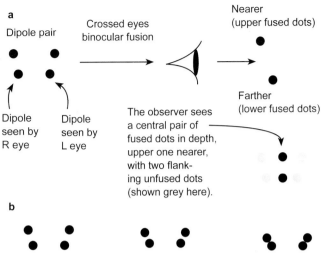

Upper and lower dots are brought closer and closer together (left to right dipole pairs), while leaving their horizontal separations and hence, horizontal disparities unchanged.

19.25 Disparity gradient limit of 1.0

The conventions used in **19.24** are also used here. The horizontal disparity H_1 of scene point P_1 is equal to its cyclopean separation CS_1 from the fixation point F'. Convince yourself of this by measuring them. F' has zero disparity by definition, so the DG of P_1 with respect to F' is 1 because $1=(H_1 - 0)/CS_1$. A DG of 1 is the limit for fusion in human vision. This limit would be exceeded for P_1 with respect to F' if either H_1 was increased in size, or CS_1 reduced. The perceptual result would then be that the left and right images of P_1 would no longer be seen as a single entity. Instead, those two images would both be seen, a condition called *diplopia*. The figure has been drawn so that the scene points P_2 and P_3 also project images with DG=1 with respect to F'. Again, check by making measurements. Notice that these points have much larger disparities than P_1 but they would nevertheless still be seen as fused because their images are further from F', hence, keeping them (just) within the DG limit of 1 with respect to F'. The important point illustrated is that the fusion limit is not fixed by the absolute sizes of disparities but by the ratio between disparity size to cyclopean separation.

19.26 Dipoles used by Burt and Julesz to explore the disparity gradient limit for fusion

a If you can fuse the dipoles (upper left) by the eye crossing technique, then you will see the upper and lower dots at different depths. The depth arises from the differences in the horizontal positions of the dots, yielding horizontal disparities.
b This disparity is kept the same while the upper dots are brought closer and closer to the lower dots. This means that the cyclopean separation between fused dots is reduced, and hence the disparity gradient between dots increases. The point at which fusion breaks down as the dots become closer marks the disparity gradient limit. Burt and Julesz used arrays of dipoles of this kind, with vertical separation between upper and lower dots increased from top to bottom.

disparity difference between a pair of points to their cyclopean separation, **19.24**. The DG limit means if two points exceed it then they cannot be simultaneously fused in human vision. Instead, the fusion of one point breaks down into diplopia (double vision). In human vision the DG limit is about 1, **19.25**. The kind of stimulus used by Burt and Julesz is shown in **19.26**.

The DG limit replaced a concept called **Panum's fusional area**, which held that diplopia emerges if the disparity of a point with respect to the fixation point exceeds an absolute disparity limit. Although it was recognized that the size of Panum's fusional area varies with retinal eccentricity, the idea of a disparity gradient limit makes this relationship precise.

The implication of the DG limit is that there exists a "forbidden cone" around any given point in *cyclopean disparity space*, **19.27**. If any other point enters this cone then its fusion breaks down into diplopia. The forbidden cone in disparity space implies a forbidden cone in scene space, **19.27**.

The upshot is that if the depth transition between a pair of scene points is steep enough then diplopia emerges. However, it turns out that a DG limit of 1 does not impose a severe restriction on the slopes of scene surfaces that can be tolerated for fusion. For example, planar surfaces with maximal slopes of 74 degrees will be tolerated at a viewing distance of 6 interocular units, rising to 84 degrees for 10 interocular units. Hence, at any reasonable viewing distance, only a small proportion of planar slopes

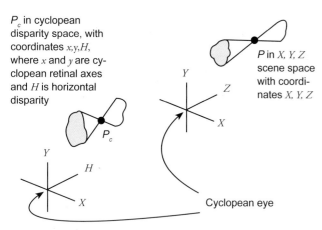

P_c in cyclopean disparity space, with coordinates x, y, H, where x and y are cyclopean retinal axes and H is horizontal disparity

P in X, Y, Z scene space with coordinates X, Y, Z

Cyclopean eye

19.27 Forbidden cones arising from a disparity gradient limit

The scene point P with scene coordinates X, Y, Z projects into the left and right images as P_L and P_R (not shown here). When the images are superimposed (as in **19.2**) the match P_c arising from P_L and P_R can be considered as being in a *cyclopean disparity space* with coordinates x, y, H, where x and y are the retinal axes and H is horizontal disparity. The cone shown in cyclopean disparity space represents the boundaries of the disparity gradient limit with respect to P_c. If any other point enters this cone then it cannot be fused if P_c is fused. Hence, this cone is said to mark a "forbidden" zone for fusion. The disparity cone implies the existence of a forbidden cone in scene space, marking the boundaries around P which must not be transgressed by other scene points if both are to remain fused.

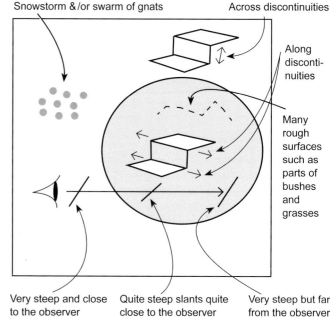

Snowstorm &/or swarm of gnats

Across discontinuities

Along discontinuities

Many rough surfaces such as parts of bushes and grasses

Very steep and close to the observer

Quite steep slants quite close to the observer

Very steep but far from the observer

19.28 Scene structures that do and do not violate the disparity gradient (DG) limit

The rectangle encloses all possible scene structures. Those lying inside the (shaded) oval fall within the DG limit of 1.

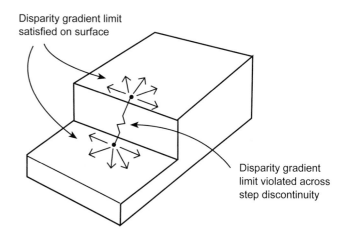

Disparity gradient limit satisfied on surface

Disparity gradient limit violated across step discontinuity

19.29 Disparity gradient along and across a step continuity

will create disparity gradients that violate the limit of 1, **19.28**.

Note that the DG limit is violated across a step discontinuity but not along it, **19.29**. Also, it can be helpful to think about the DG limit as setting bounds on the disparities of points lying on epipolar lines, **19.25**.

Pollard et al., stimulated by Burt and Julesz's psychophysical results, implemented a within–DG limit in their PMF stereo algorithm, by allowing matches to exchange mutual support if the DG between them does not exceed 1. This is a much more "liberal" support rule than the Same–disparity rule of Marr and Poggio, which imposes a DG limit of zero for supporting matches. It is justified by the surface continuity constraint, although one could perhaps better say that it is rooted in a constraint of typical scene surfaces being "not too jagged." For example, **19.31** shows

a stereogram of a quite jagged surface, the vast majority of whose points are correctly matched by PMF.

The reason for PMF's success is that disparity gradients between correct matches are usually less than 1, whereas this is seldom the case for incorrect

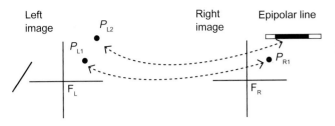

19.30 Disparity gradient limit defines matching range on epipolar line

The lower dashed line indicates that P_{L1} has been provisionally matched with P_{R1}. P_{L2} can give neighbor support to this match if it can itself find a match within a disparity range satisfying the disparity gradient limit. This range is given by the black bar on the epipolar line in the right image which defines where a match for P_{L2} must lie.

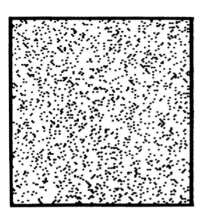
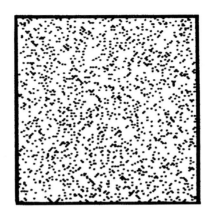
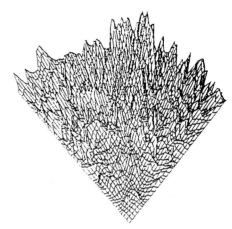

19.31 3D Jagged surface
a Stereo halves of a jagged surface arranged for crossed–eyes fusion (see **18.9**). All correct matches on this surface have been arranged to satisfy PMF's within–DG limit.
b 3D plot of the jagged surface as found by PMF. It contains 93% of the correct matches available in **b**.

matches formed from the same image elements. PMF exploits this by using a local neighborhood support scheme in which within–DG limit matches within a circular support zone contributes to a matching score for each potential match. The support given is scaled inversely by the distance between the matches as the chances of finding a within–DG limit match by chance reduce almost linearly with distance. Matches are then selected iteratively on the basis of the uniqueness constraint. The latter was implemented by choosing the most strongly supported score for both the left and right image primitives.

Why should the limit be set to 1? Enforcing a non–zero DG limit allows a balance to be struck between disambiguating power (smaller DG limit) and allowable surface jaggedness (larger DG limit). In fact, PMF's use of a DG limit can be viewed as a way of parameterising the binocular matching

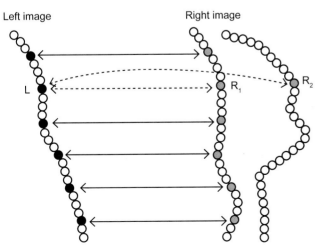

Left image · Right image · L · R_1 · R_2

19.32 Exploiting the constraint of figural continuity
The horizontal lines depict a sample of strongly supported matches between points on a line appearing in the left and right images. However, the point L on this line in the left image has two equally well supported candidate matches (shown with the dashed lines) in the right image, R_1 and R_2. R_2 comes from a second line shown only in the right image. R_1 is the preferred match for L because if selected it preserves the figural continuity of the string of points in the left image. This figural constraint was introduced by Mayhew and Frisby in their *Stereoedge* stereo correspondence algorithm.

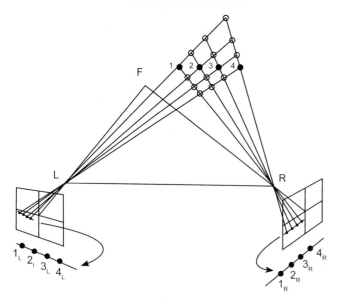

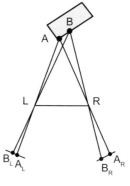

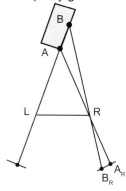

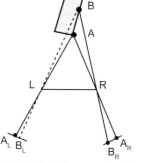

19.33 Ordering constraint
A set of four points is shown on a pair of epipolar lines. The ordering constraint forbids order reversals in the images. For example, the set of matches 1_L1_R, 2_L2_R, 3_L3_R and 4_L4_R would be allowed but not the set 1_L1_R, 2_L3_R, 3_L2_R and 4_L4. The justification for the ordering constraint is given in **19.34**.

19.34 Ordering constraint and disparity gradient
A and B are points lying on an opaque object (shaded). Their images in the left and right image planes are labeled A_L, B_L, A_R and B_R.
a Order in the left and right image planes is preserved if matches have a disparity gradient (DG) less than 2.
b A DG of 2 is the theoretical maximum for matches from opaque surfaces. It is the DG that obtains in Panum's limiting case (**18.21**).
c If DG exceeds 2, and if the points lie on an opaque surface, then one scene point becomes hidden from one eye's vantage point, here the left eye. Hence, for opaque surfaces, re–ordered matches violate a DG limit of 2 (illustrated by showing where the image point B_L would be if it were visible: the order is AB in the left image and BA in the right image). Re–ordering also therefore violates the DG limit of 1 that characterizes human stereo vision.
d Re–ordered image points arise if one scene point is directly in front of another. Such scenes do not produce simultaneously fused points in human vision.

rule of seeking those matches which satisfy surface "smoothness" (mathematically, this is said to be "Lipschitz smoothness"). Indeed, in later versions of PMF it was found useful to change the value of the DG limit from initially "generous" (1.0) to finally more "severe" (say, 0.5) when resolving stubborn ambiguity problems in the final stages of matching.

There were numerous other algorithmic refinements implemented in later versions of PMF. One was to embed explicitly a matching rule of preferring matches that exhibit *figural continuity*, **19.32**. Another was to prefer matches that preserve *ordering*, **19.33**. These constraints are in fact subsumed under the DG limit, **19.34**, but their explicit use improved PMF's algorithmic efficiency.

PMF was cast firmly within Marr's computational approach which articulates the need to understand an information processing problem first and foremost in terms of the identification of constraints appropriate for its solution. Only then can one go on to consider appropriate algorithmic and hardware

implementation details. At the same time, Marr also recognized that experiments can help the identification of good computational constraints. That is, the computational level is logically prior but not *methodologically* so. This was true in the case of PMF, as its design and the identification of the principal computational constraint that underlies it were stimulated by the psychophysically demonstrated DG limit. The latter discovery was made without any concern for developing a computational theory.

It is worth reiterating at this point that the DG limit for human vision was stated (and measured) by Burt and Julesz (1980) in terms of a requirement for binocular fusion, not binocular matching. However, Ogle demonstrated over 50 years ago that human vision can extract some depth information from points that do not appear fused. Thus, human vision is prepared to match points for the purposes of recovering the depth signaled by the disparity between those points even if they are not seen as fused into a single binocular entity after this (inferred) matching has taken place. Hence, in a sense PMF went beyond its remit in terms of support from human vision regarding the DG limit, because PMF utilized the DG limit on perceived fusion as a constraint on the allowable range for exchange of mutual support, which need not have the same limit.

PMF as a Model of Human Stereovision

How does the PMF algorithm stand up as a model of stereo matching in human vision? PMF as presented in the 1985 *Perception* paper was a very simple model. It specified only that unique matches should be sought for those primitives between which the DG limit is satisfied. Contrary to the underlying psychophysics upon which it was based, PMF does not actually prevent fusion between matches for which the DG limit is violated. Rather, it was supposed, very informally, that for human vision the overall percept was that which would result after taking eye movements into account, with each fixation delivering evidence about within–DG matches by way of building up a representation of the 3D structure of the whole scene.

For example, while the PMF computer vision algorithm can simultaneously match image primitives either side of a horizontal depth discontinuity lying between two surfaces at very different depths, **19.29**, the human visual system seems to require eye movements to fuse first one side and then the other. This illustrates that PMF was not offered as a detailed model of how the human visual system might exploit the DG limit for the purposes solving the stereo correspondence problem. Rather, the (unstated) assumption underlying the research on PMF was that human vision might realize the same

underlying computational theory (exploit the same basic constraint) in algorithms that also reflect *additional* constraints on how it operates, located at the algorithmic and/or the hardware levels.

Even though PMF does not provide, and was not intended to provide, a complete working model of the human visual system it has nevertheless provided a useful model for the interpretation of some experimental results on human stereo vision.

For example, Daphne Weinshall presented stereograms which, she argued, show that the human visual system does not impose the uniqueness constraint. These were random patterns of horizontally spaced paired dots where the spacing was different for the left and right eyes. When viewed stereoscopically, for some parameter combinations these patterns resulted in the percept of up to four transparent lacy depth planes rather than the two that would perhaps be expected.

However, Pollard and Frisby were able to show that PMF generates the same number of depth planes reported by Weinshall. This is so despite the fact that PMF utilizes the uniqueness constraint. PMF finds more matching support for one depth plane at some locations and support for other depth planes at other locations, Thus, simulating the appearance of multiple overlapping surfaces.

This illustrates, we suggest, the critical need to test arguments of this kind with computational experiments using well specified algorithms. It is a commonplace in artificial intelligence that seemingly solid "logical" arguments reveal their shortcomings when exposed to computer implementation.

Lacy transparent stereograms have been used elsewhere to try to further understand the role of DGs in human binocular vision. In a series of experiments reported by Gephstein and Cooperman, subjects were presented with random–dot stereograms portraying either a horizontal or vertical cylinder behind a transparent occluding plane. Their results show that as the density of random dots in the occluding surface and/or the disparity between the transparent surfaces increases so the ability of subjects to distinguish the orientation of the cylinder falls off. This is consistent with models of human stereopsis based on either excitatory within–DG limit support or beyond–DG limit inhibition (their preferred model) as the chance of matches violating the DG limit by chance increases with both of the design parameters of the experiment.

Since PMF was introduced in 1985 there has been something of a shift away from an engagement at the theoretical level with few (any?) new constraints identified. The dominating concern in computer stereo vision presently appears to be with algorithmic detail (Pollard et al. 2009).

More recently, Suzanne McKee and Preeti Verghese presented lacy stereograms as evidence that fusion is not actually necessary for depth perception. They used random dot pairs to form lacy surfaces. Each dot pair was located with a vertical displacement in each image such that the DG between dot pairs could be varied systematically (cf. Burt and Julesz's array in **19.26**). Their results suggest that DGs much larger than 1 still give rise to the correct percept of lacy transparency, as judged from control lacy stereograms lacking these large DGs.

PMF has not, however, been run on these kinds of stimuli. Nevertheless, we provisionally share their speculation that PMF would find the two lacy planes even when individual dot pairs have large DGs by virtue of the exchange of within–DG limit support between matches in each plane (compare the transparency demonstrations referred to above in connection with the uniqueness constraint).

McKee and Verghese argue that it is not clear whether PMF would simulate certain other details of their results regarding detecting targets lying in between the two depth planes. Such speculations can only be proved one way or the other by actually running a PMF type algorithm on the stimuli in question.

Matching at a Range of Spatial Scales

Another problem in assessing PMF as a model of the human visual system concerns the choice of image primitives.

In its development PMF was applied to a large number of random–dot stereograms and natural images. For the stereograms, matches were between the raw points from which the stereograms were constructed. For natural images, PMF used an edge detector at a single spatial scale to extract

some matching edge features. As such, PMF did not take into account the potentially important role played by spatial filtering in human stereopsis.

Marr and Poggio advanced a second computational theory of stereovision in 1975. This model proposed disparity channels with spatial frequency tuning coupled to disparity tuning to solve the stereo correspondence problem when used to drive vergence eye movements in a coarse–to–fine cascade.

The underlying idea was that low spatial frequency tuned channels produce coarse images, from which a sparse set of edge points are derived from their large blobs, and hence a greatly reduced matching problem. Once matched, the output of the coarse channels then drives vergence movements to bring higher spatial frequency channels into their more limited disparity range in such a way that they too then have little in the way of a matching problem. This line of thinking thus envisages stereo matching being done at various spatial scales (Ch 5), in a coarse–to–fine hierarchy.

Mayhew and Frisby argued that this theory was flawed on two counts.

First, they noted that coarse spatial frequency channels inevitably blur data within images arising from edges lying in different depth planes in a way that is problematic for guiding the required vergence movements for bringing into register the finer disparity channels (consider such channels faced with transparent lacy textures of the kind discussed above).

Second, the thesis work of their then postgraduate student, Peter Mowforth, had shown that vergence eye movements to large disparities could be driven by higher spatial frequency textures than those predicted by the theory.

So the theory was flawed both logically and empirically. If work grappling with these problems in using spatial frequency channels for stereopsis is being done, then we have missed it.

Mayhew and Frisby concluded from this critique, and from their many psychophysical experiments, that binocular matching in human vision is embedded within processes computing what they called a *binocular raw primal sketch* (for a description of Marr's monocular raw primal sketch, see Ch 2, 5, 6, and 7).

Neurophysiology of Stereopsis

Neurons sensitive to binocular inputs are found widely in the visual cortex. We described some of the main finding in Ch 10 *Seeing with Maps*.

Some of the recent physiological work on disparity coding mechanisms has given new impetus to models for binocular matching that utilize a range of spatial scales (see remarks above). Much of this modeling is guided by computer experiments but it appears to lack analysis at Marr's computational theory level.

That is, although interesting experiments have been conducted using simulations of the properties of disparity sensitive neurons, the computational theory constraints embodied in the design of those neurons does not seem to have received much attention.

The point is often made that if one is interested in brain processes then one has to take into account physiological data. We agree. But we also think that we will not understand why these neurons have the properties they do unless we have computational theories of the task being solved in the way advocated by Marr.

Thinking at the computational theory level may help illuminate the continuing theoretical challenge for the field discussed in the previous section. The disparity sensitive neurons modeled by Qian in 1994 have fields that integrate image luminance information over scales much larger than individual dots that human vision has no problem in seeing as separate entities in 3D when they are given appropriate disparities. So how are their outputs used—what is the functional role of these cells?

We look forward to the day, probably a long way off, when the neurophysiology of stereo vision is well understood. For a recent paper that provides a starting point for reading up on this topic, see Bredfeldt, Read, and Cumming (2009).

Summary Remarks

We remain convinced of the huge importance for understanding vision of finding out the structure of the problems that must be solved. Computational vision experiments can help in this regard, and also in discovering good constraints for solving the problem(s) identified. Marr seems to us

as correct now on this fundamental issue as he was when he argued this case in the 1970s. (And equally correct, it is worth adding, are his antecedents who also argued this case: see Ch 23).

On the other hand, the way Marr and Poggio used the surface smoothness constraint for solving the stereo correspondence problem was weak indeed. Their algorithm works poorly for images of natural scenes because the world is not largely composed of surfaces that generate sufficient same–disparity neighbors within the size of excitatory support region needed for resolving the stereo correspondence problem. As for PMF, the key idea of constraining matching support using a DG limit still seems to be part of the current scene in computer stereo vision, albeit embedded in algorithms of a very different character. PMF has been challenged as a model of human stereo matching although it can cope surprisingly well with certain demonstrations (e.g., extracting surfaces in superimposed lacy planes).

You can follow up many of the details raised in the foregoing discussion of PMF by reading a set of commentaries on the original paper and subsequent developments (Pollard et al., 2009).

Further Reading

Banks MS and Backus BT (1998) Extra–retinal and perspective cues cause the small range of the induced effect. *Vision Research* **38** 187–194. *Comment* This paper is an example of the use vertical and horizontal *size ratios* to describe stereo projections. For an explanation of them see Howard and Rogers (2002). Whether size ratios help or hinder the understanding of vertical disparities is a matter of taste. We find they lead to complicated and confusing equations but others disagree. Also, we doubt that they lend themselves to biologically plausible algorithms for recovering the viewing parameters d, g, which may be why, as far as we are aware, they have not been used for that purpose.

Bredfeldt CE, Read JC, and Cumming BG (2009) A quantitative explanation of responses to disparity-defined edges in macaque V2. *Journal of Neurophysiology* **101** 701-713.

Burt P and Julesz B (1980) Modifications of the classical notion of Panum's fusional area.

Perception **9** 671–682. *Comment* Describes their influential work on the disparity gradient limit.

Frisby JP, Davis H, and Edgar R (2003) Does interpupillary distance predict stereoacuity for normal observers? *Perception* **32** ECVP Abstract Supplement.

Frisby JP and Mayhew JEW (1980) The role of spatial frequency tuned channels in vergence control. *Vision Research* **20** 727–732. *Comment* Describes evidence against Marr and Poggio's (1979) theory as a model of human stereo matching. See also Mowforth et al. below.

Frisby JP and Mayhew JEW (1980) The role of spatial frequency tuned channels in vergence control. *Vision Research* **20** 727–732. *Comment* Describes evidence against Marr and Poggio's (1979) theory as a model of human stereo matching. See also Mowforth et al. below.

Frisby JP and Mayhew JEW (1980) Spatial frequency tuned channels: implications for structrue and function from psychophysical and computational; studies of stereopsis. *Proceedings of the Royal Society London Series* B **290** 95–116. *Comment* Reviews early work on the role of spatial frequency tuned channels. It outlines a cross–channel model for stereo matching and the idea of the orientated spatial frequency tuned channel as a non–linear grouping operator.

Gärding J, Porrill J, Mayhew JEW, and Frisby JP (1985) Stereopsis, vertical disparity and relief transformations. *Vision Research* **35** 703–722. *Comment* Describes their *disparity normalization* theory based on Mayhew and Longuet–Higgins' work on vertical disparities, plus related psychophysical studies.

Glennerster A, Rogers BJ, and Bradshaw MF (1998) Cues to viewing distance for stereoscopic depth constancy. *Perception* **27** 1357-1366.

Howard I and Rogers BJ (2002) *Seeing in Depth Volume 2 Depth Perception* Published by Porteus Canada. *Comment* A heroically wide ranging literature survey. If you need references to follow up in any given area of binocular vision this is a very good place to start looking for them.

McKee SP and Verghese P (2002) Stereo transparency and the disparity gradient limit. *Vision Research* **42** 1963–1977.

Marr D (1982) *Vision.* Freeman: San Francisco. *Comment* Marr's highly influential book.

Marr D and Poggio T (1976) A cooperative computation of stereo disparity. *Science* **194** 283–287. *Comment* Describes their classic work on solving the stereo correspondence problem using a network of units with excitatory and inhibitory interconnections.

Marr D and Poggio T (1979) A computational theory of human stereo vision. *Proceedings of the Royal Society London Series* B **204** 301–328. *Comment* Describes their non–cooperative stereo algorithm based on the idea of coarse–to–fine stereo matching.

Mayhew JEW (1982) The interpretation of stereo disparity information: the computation of surface orientation and depth. *Perception* 11 387–403. *Comment* Describes Mayhew's theory for using vertical disparities to compute stereo viewing geometry parameters of *d* and *g*.

Mayhew JEW and Frisby JP (1981) Toward a theory of human stereopsis. *Artificial Intelligence* **17** 349–385. *Comment* Introduced the concept of a *binocular* primal sketch, that arose from both psychophysical and computational studies.

Mayhew JEW and Longuet–Higgins HC (1982) A computational model of binocular depth perception. *Nature* **297** 376–379. *Comment* A highly cited mathematically more developed version of Mayhew (1982).

Mowforth P, Frisby JP, and Mayhew JEW (1981) Vergence eye movements made in response to spatial–frequency–filtered random–dot stereograms *Perception* **10** 299–304.

Mayhew JEW, Zheng Y, and Cornell S (1993) The adaptive control of a four-degrees-of-freedom stereo camera head. In *Natural and Artificial Low-level Seeing Systems.* Eds. Barlow HB, Frisby JP, Horridge A, and Jeeves MA (1993) Oxford Science Publications. *Comment* This book is a collection of papers given at a wide-ranging symposium of the

Royal Society. The technical content is demanding but the collection is well worth reading. The discussion following Mayhew et al.'s paper on a controversy concerning the effect of eye rotations on the disparity vector field will be helpful to anyone interested in that topic.

Ogle KN (1954) *Research into Binocular Vision.* Saunders: Philadelphia. *Comment* Describes Ogle's classic work on the induced effect.

Pollard SB and Frisby JP (1990) Transparency and the uniqueness constraint in human and computer stereo vision. *Nature* **347** 553–556. *Comment* Demonstrates how PMF can cope with seemingly problematic stereo demonstrations relating to the the uniqueness constraint.

Pollard SB, Mayhew JEW, and Frisby JP (1985) PMF: A stereo correspondence algorithm using a disparity gradient limit. *Perception* **14** 449–470. *Comment* For a set of commentaries on this paper, see a special issue of the journal *Perception* (2009 **38** 879-893) devoted to discussions of its most cited papers.

Pollard SB, Mayhew JEW, and Frisby JP (1991) Implementation details of the PMF stereo algorithm. In *3D Model Recognition from Stereoscopic Cue*s Eds Mayhew JEW and Frisby JP MIT Press: Cambridge, MA Pages 33–310. *Comment* Describes much work done on developing PMF.

Pollard SB, Mayhew JEW, and Frisby JP (2009) Author's update on the PMF stereo correspondence algorithm. *Perception* **38** 879–893. *Comment* This has several commentaries on the PMF algorithm in a special issue of this journal on its "landmark" papers.

Pollard SB, Porrill J, Mayhew JEW, and Frisby JP (1985) Disparity gradient, Lipshitz continuity and computing stereo correspondences. *Proceedings of Third International Symposium of Robotics Researc*h Gouviex France MIT Press Pages 19–26. *Comment* Develops the theoretical basis for the PMF algorithm.

Porrill J, Mayhew JEW, and Frisby JP (1987) Cyclotorsion, conformal invariance, and induced effects in stereoscopic vision. *Frontiers of Visual*

Science: Proceedings of the 1985 Symposium. National Academy Press: Washington, D.C. Pages 90–108.

Porrill J, Frisby JP, Adams W, and Buckley D (1999) Robust and optimal use of information in stereo vision. *Nature* **397** (6714): 6763–6766.

Prazdny K (1985) Detection of binocular disparities. *Biological Cybernetics* **52** 93–99. *Comment* Describes a stereo algorithm similar in some ways to PMF.

Qian N (1994) Computing stereo disparity and motion with known binocular cell properties.*Neural Computation* **6** 390–404. *Comment* Develops a model based on neurophysiological findings.

Tsai JJ and Victor JD (2003) Reading a population code: a multi–scale neural model for representing binocular disparity. *Vision Research* **43** 445–466. *Comment* An example of modeling stereo matching using a range of coarse–to–fine scales.

Tyler C W (1973) Stereoscopic vision: cortical limitations and disparity scaling effect *Science* **181** 276–278. *Comment* Introduces the concept of the disparity gradient limit.

CVonline: The Evolving, Distributed, Non-Proprietary, On-Line Compendium of Computer Vision http://homepages.inf.ed.ac.uk/rbf/CVonline/ *Comment* A free resource, masterminded by Bob Fisher at the University of Edinburgh. It includes *TINA,* which provides the PMF algorithm.

20

Seeing by Combining Cues

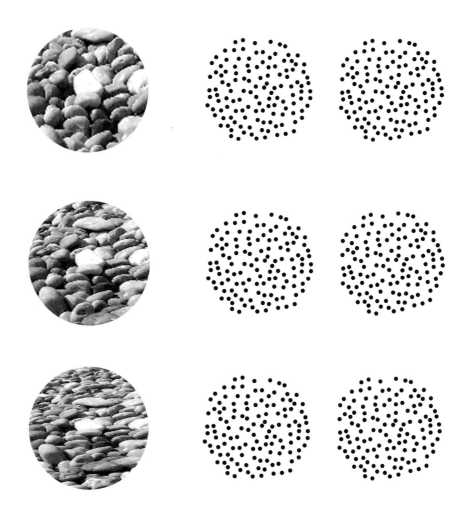

20.1 Images of texture-only and stereo-only shape cues for slanted surface
Column on left: Texture-only cue, with slant increasing from top to bottom.
Two columns on right: Stereograms: if the two members of each stereo pair are combined using crossed eye fusion (Ch 18) they yield perceptions of slant increasing from top to bottom, roughly similar to the slant from the texture-only cues alongside for the pebble scenes. No slant is perceived in each half stereogram, so dot textures of this kind are often described as being effectively stereo-only cues to slant. However, it is more accurate to say that each half-stereogram contains a texture cue to *zero*-slant.

  In general when viewing everyday scenes, information is available from several shape cues about any given surface. This chapter considers how information on shape from different cues is combined.

We have considered in previous chapters how the brain can make use of a single visual cue such as stereo or texture to estimate respectively depth and surface orientation. However, not all visual cues are equal, inasmuch as some cues are more reliable than others. Worse still, the reliability of each cue is not constant, but can vary with surface depth and orientation, as we shall see below. So not only are visual cues unequal, their inequality varies with surface orientation. How might the brain deal with such a computational conundrum?

One emerging framework for investigating how the brain might achieve a rational compromise is based on the work of Thomas Bayes (1702–1761), an English Presbyterian minister and mathematician, whose ideas were introduced in Ch 13.

Perceptual Modules

For the sake of simplicity let's assume that the brain contains different independent functional ***modules*** for using stereo and texture information to estimate surface orientation. The stereo module generates an estimate of surface orientation based only on stereo cues, and ignores cues based on texture. Conversely, the texture module generates an estimate of surface orientation based on texture cues, and ignores cues based on stereo. Thus, given a particular surface, each module computes an

estimate of that surface's orientation, as shown in **20.2**. In a perfect world the modules' outputs would agree perfectly, each confirming the output of the other. In this imperfect world, various slings and arrows ensure that the module outputs usually do not agree.

How should we combine the module outputs to arrive at a good estimate of surface orientation? If we know that one module is more reliable than the other, then one solution is to simply disregard the output of the less reliable module. But this may throw away useful information, and an alternative is to take the mean of the module outputs. If we do not know how reliable each module is, then taking the mean turns out to be the optimal strategy. But what if we do know how reliable each module is?

We won't worry precisely what is meant by reliability for now; the main thing is to understand that the optimal estimate of surface orientation is given by combining the outputs of the two

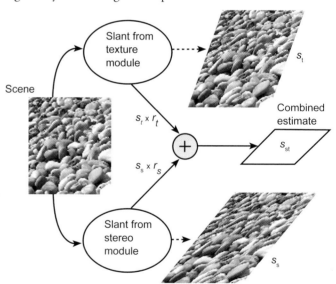

20.2 Cue combination overview
Combining cues from different sources. A stereo module and a texture module (inside your head, for example) observe the same slanted plane, which contains both texture and stereo information. Each module generates its own estimate of surface slant: s_s (slant based on stereo only) and s_t (slant based on texture only). These estimates differ in two ways. First, they usually have different values. Second, they have different reliabilities, r_s and r_t. Bayes' rule implies that the module outputs can be reconciled by weighting each module output by its reliability to yield a final estimate s_{st} of slant, which is based on both stereo and texture information.

Why read this chapter?

In this chapter, we deal with the general question: how should the brain combine two or more unreliable outputs to form a more reliable output? This is explored in the context of texture and stereo disparity. There are many different cues to shape, but some cues are more reliable than others. Using the *Bayesian framework*, we describe how the outputs of two hypothetical brain modules, which compute shape from texture and shape from stereo disparity, can be used to obtain estimates that are more reliable than that of either module considered in isolation. This defines a benchmark for optimal behavior such as would be exhibited by an *ideal observer*. We then present evidence that the human visual system combines texture and stereo information as if it is an ideal observer, and that this ideal observer behavior also applies when information is combined from different vision and touch. Finally, we describe how reliance on different sensory inputs seems to vary between individuals.

a

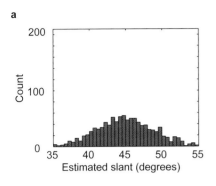

b

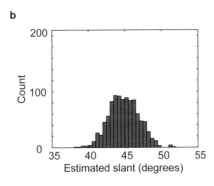

c

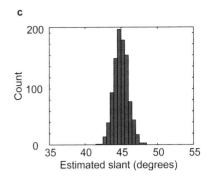

20.3 Increasing reliability from a to c increases the peakiness in a histogram of module output values
As reliability increases, so the spread of values decreases, where 'spread' can be equated with variance. Note that the mean output remains constant at 45°.

modules. In order to understand why this is so, we first need to introduce some notation. To keep things simple, we also assume that the tilt of the surface orientation is known and that we are after the slant (i.e., the amount of slope). We denote the stereo module with M_s, and its output as s_s. Similarly, the texture module is M_t and its output is s_t. Our final estimate of slant is denoted s_{st}.

Now, if M_s has a reliability which is represented by r_s and M_t has a reliability r_t then we could naïvely multiply each module's output by its reliability and then add the results together:

$$s_{st} = r_t s_t + r_s s_s. \tag{1}$$

In essence, this implies that we take more notice of the output of modules that are more reliable. If you want to see where this equation comes from then see Knill and Richards (1966).

Note that s_{st} consists of a bit of s_s and a bit of s_t. This is perfectly sensible provided we ensure that the two bits sum to unity; that is, provided $r_s + r_t = 1$.

For example, if we were to take the average of s_s and s_t then we would implicitly be setting both r_s and r_t to 0.5 (i.e., $r_s = r_t = 0.5$), from which it can be seen that the mean implies $r_s + r_t = 1$.

As another example, if we know that M_s has a reliability of $r_s = 0.7$ and M_t has a reliability of $r_t = 0.3$ then our final estimate of the slant would give a larger weighting to s_s than to s_t

$$s_{st} = 0.7 s_s + 0.3 s_t. \tag{2}$$

Because we are taking different proportions or *weightings* of s_s and s_t, the resultant estimate s_{st} is called a ***weighted mean***.

This is all very fine, but how do we find out how reliable each module is? One simple way would be to present the *same* slanted surface many times, and observe the variability in each module's output, because this variability should give an indication of the reliability of each module.

For example, if we presented the same slant $N = 1000$ times to M_s and then made a histogram of module outputs $[s_s(1), s_s(2), \ldots s_s(1000)]$ then it could well look like **20.3b**. If this module were 100% reliable then we would expect all outputs to have about the same value, and so the histogram would be very "peaky," as in **20.3c**. If the module were pretty unreliable then there would be a wide spread of values, and the histogram would be almost flat like the one in **20.3a**. From this we can see that reliability increases with the peakiness of a histogram, and conversely that *unreliability* increases with the *spread* of a histogram.

Variance and Unreliability

A formal measure of the spread of module output values is ***variance***, which is also a good measure of *un*reliability. At this point we need to define exactly what we mean by variance. In the output of a stereo module:

$$v_s = (1/N)\sum_i (u_s - s_s(i))^2. \tag{3}$$

The symbol $\sum$ (the capital Greek letter, sigma) is used to indicate a sum over all values of s_s, where $s_s(i)$ represents the *i*th output value. The symbol u_s

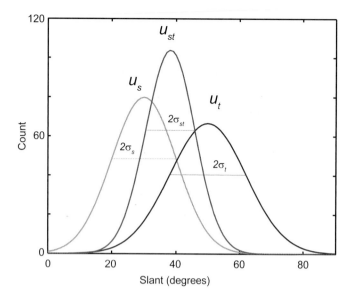

Count

Slant (degrees)

20.4 Output distributions from different modules when presented many times with a surface at a fixed slant

The mean output u_s of a stereo module M_s is the at the center of a distribution of values s_s from this module. This distribution has a width which increases with the variance v_s of these values. Similarly, the mean output u_t of a texture module M_t is the at the center of a distribution of values s_t from this module, which has a variance v_t. If a third module M_{st} (e.g., your brain) uses the outputs s_s and s_t of M_s and M_t to find a weighted mean s_{st} then its output has a mean at u_{st}. Note that the variance v_{st} of M_{st} is less than v_s and less than v_t, so M_{st} has a more reliable output than both M_s and M_t. For example, if the module outputs happen to coincide with their means so that $s_s = 30.0$, $s_t = 50.0$, then $s_{st} = 38.2$. More importantly, the variance in module outputs is $v_s = 100.0$, $v_t = 144.0$, which implies that $v_{st} = 59.0$, a value clearly less that v_s and v_t. Each horizontal line indicates two standard deviations of one distribution, where standard deviation is the square root of variance (e.g., $\sigma_s = \sqrt{v_s}$).

represents the mean of all N output values [$s_s(1)$, $s_s(2)$, ... $s_s(1000)$]. In words, the variance is the difference between each output value $s_s(i)$ and the mean u_s, squared, and summed over all output values; finally, dividing by N gives the mean of these squared differences.

For completeness, note that the square root of v_s is the **standard deviation**, denoted with the lower case Greek letter σ, so $\sigma_s = \sqrt{v_s}$. Even though the definition of variance is a complicated looking equation, bear in mind that it just represents a measure of the spread in a set of module output values.

Similarly, for the texture module we have

$$v_t = (1/N)\sum_i (u_t - s_t(i))^2. \tag{4}$$

So, if we take the outputs [$s_s(1)$, $s_s(2)$, ... $s_s(1000)$] from the stereo module M_s and the outputs [$s_t(1)$, $s_t(2)$, ... $s_t(1000)$] from the texture module M_t then we can use these to find the variance v_s and v_t of each module from equations 3 and 4.

Variance of the Weighted Mean

Crucially, the variance of the weighted mean defined in equation (1) from the two modules is *guaranteed* to be no larger than the variance of M_s and of M_t. If we denote the variance of the weighted mean as v_{st} then it can be shown that

$$v_{st} = 1/(1/v_s + 1/v_t). \tag{5}$$

This implies that we can place more trust in the

weighted mean than we can in the output of either module considered in isolation. This can be seen graphically in **20.4**.

Equation 5 holds true under fairly general conditions. It is obtained by assuming that each individual module's output has a gaussian distribution with variances v_s and v_t.

Difference Thresholds and JNDs

The standard deviation σ in the output of a module is clearly related to how reliably that module's output indicates the value of the module's input. A module with a small standard deviation is more informative than a module with a large standard deviation (you might notice that people are a bit like modules in this regard). For example, consider the module M_s which has a mean output of 30° and a standard deviation $\sigma = 10°$. Recall that the module output is varying all the time due to the intrinsic variability of the module. Given these facts, you might decide the output mean must change by at least one standard deviation (10°) for you take any notice of it; that is, the mean would have to be greater than 30°+10° = 40° or less than 30° – 10° = 20°.

Now, the change in input (e.g., surface slant) required to change the output mean by one standard deviation is regarded as a kind of "perceptual unit." In most modern texts, this unit is called the **difference threshold** or **just noticeable difference** (JND). For example, if a change in slant of 10°

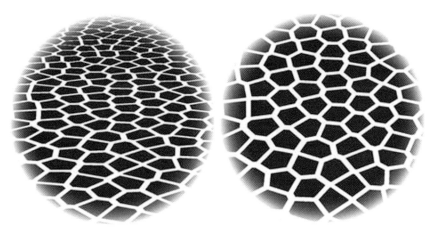

20.5 Example stimuli used in a single trial of a two-alternative forced choice (2AFC) procedure for estimating the texture difference threshold or just noticeable difference (JND)
The observer's task is to indicate which of the two stimuli depicts a plane with the larger slant (the left image here). Unknown to the observer, one of the stimuli is a reference stimulus with a constant slant, and the other is a comparison stimulus with a slant that varies from trial to trial. As the difference in slant increases, the proportion of comparison responses increases as indicated in **20.6**. Reproduced with permission from Knill and Saunders (2003).

alters the mean module output by one standard deviation then we have a JND of 10°. Notice that this expresses difference thresholds relative to the amount of intrinsic variability in the module's output. A module with a large intrinsic variability has a large JND because a large change in slant is required to change its output by one standard deviation.

Variance and JND

If we could measure the variance of a module as described above, then we could proceed. In fact, we cannot measure the variance in the outputs of individual modules directly. In order to do so, we would have to present many stimuli, where each stimulus has a different slant, and get the observer to tell us the slant in degrees. We could then measure the variance in the observer's responses, from which we could obtain the JND. However, this is clearly impractical; observers simply cannot provide such information. So the experimenter has to find a way to extract JNDs in an indirect way.

JNDs and 2AFC

One important method for finding the JND is called the **two-alternative forced choice** (**2AFC**) procedure, described in detail in Ch 12. This consists of presenting a pair of stimuli on each trial, either one after the other or side by side. For our purposes let's assume that the stimuli consist of a pair of slanted planes presented side by side, and

that such a pair is presented on each trial, **20.5**. The observer's task is to indicate which plane is more slanted, so one of the stimuli must be chosen on each trial.

To make this work, we need to define a **reference stimulus** which is a plane with a fixed **reference slant** θ_1 (theta 1), and a variable **comparison stimulus** with a **comparison slant** θ_2 which varies from trial to trial. After the experiment has been run, the effect of changing the comparison slant on the observer's responses can be analyzed. As the comparison slant increases from being less than the reference slant, to being equal to, and then greater than the reference slant, so there is an increasing probability that the observer chooses the comparison stimulus as having the larger slant. A graph of the difference in slant (between the reference and comparison) against the proportion of trials on which the observer chooses the comparison slant has the typical sigmoid shape of a psychometric function, **20.6**. For the sake of brevity, we define responses for which the observer chooses the comparison slant as the **comparison responses**.

We already know that the JND is the change in slant $\Delta\theta = (\theta_2 - \theta_1)$ required to increase the mean module output from some value u_1 to $u_2 = u_1 + \sigma$. If there is such a thing as a texture module in the brain then its output is likely to be measured in terms of impulses per second, rather than degrees. However, recall that the experimenters never get to see the output of a module; they only get to see the

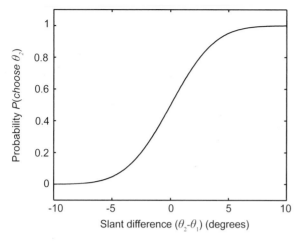

20.6 Probability of choosing comparison stimulus
The probability of choosing the comparison stimulus θ_2 as having the larger slant increases as the difference between the reference and comparison slants increases. If the comparison slant θ_2 is less than the reference slant θ_1 (i.e., negative $\theta_2 - \theta_1$) then the observer tends to perceive the reference slant as larger, and so chooses the comparison slant with low probability. Conversely, if the comparison slant is larger than the reference slant (i.e., positive $\theta_2 - \theta_1$) then the observer tends to perceive the comparison slant as larger, and so chooses the comparison slant with a high probability.

observer's responses. So whether module outputs are in units of degrees or impulses per second is irrelevant, even though we have been pretending each module's output is in units of degrees. It turns out that, whatever the module outputs are, we can find out how much we have to change the slant in order to increase the module's mean by one standard deviation. In a 2AFC experiment, the JND (denoted T) is the change $\Delta\theta$ in slant required to increase the proportion of comparison responses from 50% to 76%.

Estimating Reliability

Using the squared JNDs T_s^2 and T_t^2 as a proxy for the variances $v_s = \sigma_s^2$ and $v_t = \sigma_t^2$, we can find the reliability of each module, as follows.

The inverse of variance is called the **precision**. If we simply weighted the stereo module's output by its precision $w_s = 1/v_s$ and the texture module's output by its precision $w_t = 1/v_t$ then the result would not be a weighted mean because there is no guarantee that the weights w_s and w_t sum to unity. However, we can ensure that the weights do sum to unity if we normalize them as follows:

$$r_s = w_s/(w_s + w_t), \qquad (6)$$

$$r_t = w_t/(w_s + w_t). \qquad (7)$$

All we have done is to re-express each weight as a fraction of the sum of both weights. This ensures that each normalized weight is a proportion, so that $r_s + r_t = 1$.

So now we have a way to evaluate the reliability of each module's output, which means that we can obtain the weighted mean s_{st} defined in equation (1). Thus, Bayes' rule tells us not only that it is indeed sensible to use a weighted mean, it also tells us *precisely* how much weight to attach to the outputs of different modules in estimating this weighted mean.

One important fact is worth repeating at this point: if the outputs of the stereo and texture modules are combined using equation (1), with weight values defined in equations (6) and (7) then the reliability of the weighted mean s_{st} is *guaranteed* to be no less than the reliability of the module output s_s and no less than the reliability of the module output s_t. In other words, the variance v_{st} of s_{st} is no larger than v_s and no larger than v_t. The variance of the weighted mean in equation (1) was given in equation (5), and is repeated here:

$$v_{st} = 1/(1/v_s + 1/v_t). \qquad (8)$$

For example, if the variances of the stereo and texture module are $v_s = 100$ and $v_t = 144$ then $v_{st} = 1/(1/100 + 1/144) = 59$, which is clearly less than 100 and 144.

If we substitute the values $v_s = 100$ and $v_t = 144$ into equations (6) and (7) (where $w_s = 1/v_s$ and $w_t = 1/v_t$) then we get the normalized weights as $r_s = 0.59$ and $r_t = 0.41$. Then, if $s_s = 30.0$ and $s_t = 50.0$, we can use equation (1) to obtain $s_{st} = 38.2$, as shown in **20.4**. It can be shown that the weighted mean s_{st} is optimal in a Bayesian sense.

If we replace the variances with JNDs (squared) then equation (8) becomes

$$T_{st}^2 = 1/(1/T_s^2 + 1/T_t^2). \qquad (9)$$

This equation will prove useful later.

Now, let's explore the use this general method in the context of an actual experiment.

Combining Cues from Stereo and Texture

In a paper written in 2003 Knill and Saunders asked the question: *Do humans optimally integrate stereo and texture information for judgments of surface slant?* In fact, this was the title of their paper. The answer to the question is *"yes,"* but this would have made the paper the shortest and most opaque in history.

The basic idea behind the design of this experiment was to treat the observer as a stereo module on some trials (as in 1, below), as a texture module on other trials (as in 2, below), and as a combined stereo-texture module on others (as in 3, below). In order to do this it was necessary to obtain a measure of each observer's slant JND, for each of a series of four different slants, based on texture cues only, stereo cues only, and combined stereo and texture cues.

On each trial the observer was shown a pair of slanted surfaces on a computer monitor, as shown in **20.1** and **20.7**, and was asked to choose the one that looked more slanted, so this was a 2AFC procedure. There were three different types of surfaces:

(1) In some images, the slant of a surface was conveyed via stereo information only. These images consisted of surfaces with dot textures, so that the shape of individual texture elements did not act as a cue to slant. These images were used to obtain the observer's slant JNDs based only on stereo information, denoted as T_s.

(2) In other images, the slant of a surface was conveyed via texture information only. These images consisted of surfaces with large texture elements, which provided a good cue to surface slant, and were used to obtain the observer's slant JNDs based only on texture information, denoted T_t.

(3) Finally, some images contained both stereo and texture cues. These images were used to obtain the observer's slant JNDs based on both stereo and texture information, denoted as T_{st_used}.

Once we have the JNDs T_s and T_t, we can use equation (9) to make predictions T_{st} of the JNDs used T_{st_used} for the stereo-texture condition, and

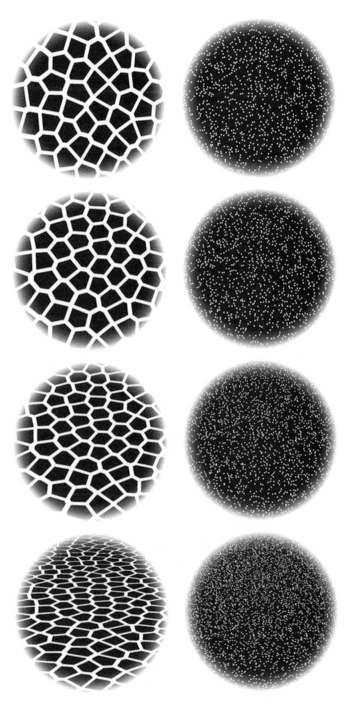

20.7 Images of stereo-only and texture-only cues
Each circular image represents the stimulus presented in a trial. *Left*: Texture-only cue, with slant increasing from top to bottom. *Right*: Stereo-only cue: only one image from each stereo-pair is shown here to give an impression of the type of random-dot stimulus used, but see also **20.1**. From Knill and Saunders (2003).

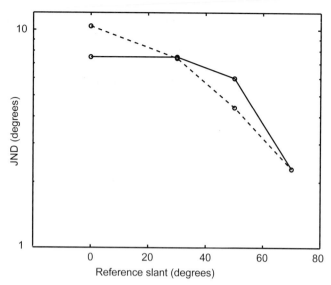

20.8 Comparison of JNDs for combined stereo-texture cues (solid) with JNDs predicted (dashed) from stereo-only and texture-only JNDs
Data replotted from Knill and Saunders (2003).

then compare these predicted values with those obtained in the stereo-texture condition, as shown in **20.8**. Note that this procedure is repeated for all four of the reference slants used, so there are four values for each of the measured JNDs T_s, T_t, T_{st_used} and for the predictions T_{st} of T_{st_used}.

Now, it is a fact that the reliability of information from texture cues increases rapidly with surface slant, whereas information from stereo cues increases more slowly with surface slant. So we would expect the observer's JNDs for texture to decrease with slant; in other words, as the information from texture cues becomes more reliable the observer's JND for this cue should decrease. This is indeed what happens, as shown in **20.8**.

In order to fully understand this graph, let's back up a little to see how a specific point on each of the four curves in **20.8** was obtained. First, consider the stereo-only curve, which describes how the JND of texture changes as slant increases. The point on this curve corresponding to a slant of 70° indicates a JND of about 8°. How was this value obtained? Here's how.

On each of 64 trials, an observer was shown the reference stereo-only slant of 70° plus a comparison stereo-only slant, one after the other, and asked which of two planes was more slanted. The order in which these slants were presented was rand-

omized across trials, and the comparison slant was adjusted automatically across trials in order to get a good estimate of the JND. As the comparison slant increased from say 60° to 80° (i.e., 10° either side of the reference stimulus), so the proportion of comparison responses increased. As described above, the resultant sigmoid curve is the cumulative area under a gaussian, which has a standard deviation equal to $\sqrt{2}$ times the JND T_s; and the JND is related to the change in slant required to increase the proportion of comparison responses from 50% to 76%. The change in slant required to increase the comparison responses from 50% to 76%. was measured as 8°. Therefore, the JND is is 8°.

This was repeated for each of the four reference slants used. Naturally, the stimuli from all four reference slants were inter-mixed, and the observer was unaware of which image was the reference image in each trial. When averaged across all seven observers used in this experiment, the result is an estimate of the JND at each reference slant for stereo-only stimuli, as indicated by the dashed line in **20.8**. This was repeated using texture-only stimuli, and using combined stereo-texture stimuli. The JND for each stimulus type defines a different curve in **20.8**.

Now comes the clever bit. At each reference slant, the JND T_s for stereo-only and the JND T_t for texture-only stimuli were combined to make a prediction T_{st} of the JND T_{st_used} obtained with the combined stereo-texture stimuli. Specifically, the prediction T_{st} of the JND T_{st_used} is defined by equation (9), which has been re-arranged here,

$$T_{st} = \sqrt{[1/(1/T_s^2 + 1/T_t^2)]}. \qquad (10)$$

The key question is: *do humans act like ideal observers when given both texture and stereo cues?* For our purposes, this amounts to the question: is the JND T_{st_used} measured using the stereo-texture images predicted by the JNDs T_{st} predicted from the two single-cue conditions?

The graph of slant versus JND graph in **20.8** shows that the values of all three measured JNDs decrease with slant, indicating that slant is more accurately perceived for large slants. More importantly, the values of T_{st} predicted from T_s and T_t in the single-cue conditions agree well with the observed values of T_{st_used} obtained in the stereo-texture condition (Knill and Saunders show that the

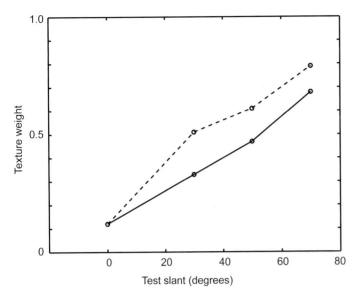

20.9 Predicted (dashed) and observed (solid) reliance on texture cue for different values of slant
Data averaged over observers. Data replotted from Knill and Saunders (2003).

predicted and observed JNDs are not statistically significantly different).

In a second experiment, which we report only briefly here, Knill and Saunders made use of single-cue stimuli (as above) and *cue-conflict* stimuli. The single-cue stimuli were used estimate the weightings given to stereo and texture information as slant increases. A cue-conflict stimulus has stereo information consistent with one slant, but texture information consistent with a different slant. By measuring the perceived slant of such stimuli, it is possible to estimate the relative weighting to given to stereo and texture actually used in images that contain both stereo and texture cues, for each value of slant tested.

Given that the reliability of texture information increases dramatically as the slant increases, this implies that the relative weighting given to texture information should increase accordingly in images containing both stereo and texture cues. More importantly, the weightings given to stereo and texture information as derived from the single-cue stimuli used allowed Knill and Saunders to *predict* the value of the texture weighting for each value of slant used by observers with the cue-conflict stimuli. As can be seen from **20.9**, these predictions were impressively accurate.

Combining Cues from Vision and Touch

The experiment described above was preceded by a similar experiment by Ernst and Banks in 2002 which explored how visual and haptic (touch) information is combined. The design of these two experiments was very similar. Observers viewed the reflection of the visual stimulus, presented on a cathode ray tube (CRT) binocularly in a mirror, **20.10**.

Ernst and Banks used a virtual reality set-up to find out how much reliance observers placed on their visual and haptic senses. On each trial, the observer's task was to indicate which of two

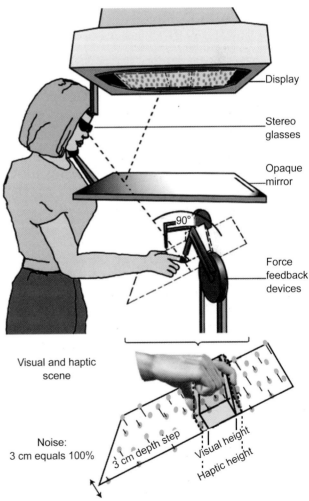

20.10 Experimental set-up for comparing visual and haptic (touch) performance
With permission from Ernst and Banks (2002).

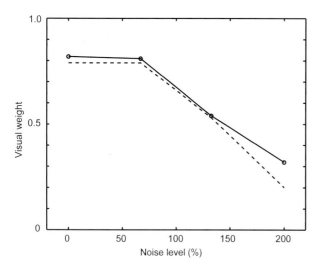

20.11 Predicted and observed weights given to visual information
Both the predicted (dashed) and the observed (solid) weights decrease with increasing visual noise as the reliability of visual information decreases. Replotted from Ernst and Banks (2002).

blocks was wider, using a 2AFC procedure. There were three conditions:

(1) *Stereo-Haptic.* Observers viewed a stereo image of a block, while simultaneously grasping a virtual block between finger and thumb (a force-feedback device simulated feeling real blocks of various sizes as specified by the experimenters for any given trial).

(2) *Haptic-Only.* Observers grasped the virtual block between finger and thumb while not being able to see the block.

(3) *Stereo-Only.* Observers viewed a stereo image of a block, with no touching of the block.

In order to vary systematically the reliability of information, increasing amounts of visual noise were added to the stereo images. This noise was added in such a way that the apparent depth of points in the stereo images contained increasing amounts of "depth jitter" as the amount of noise was increased.

The JNDs used in the stereo-only condition (with noise), and in the haptic-only condition were estimated for four levels of visual noise. These JNDs were combined to find the predicted JND weights for the combined stereo-haptic condition with different amounts of added visual noise. The observed JNDs in the stereo-haptic condition were compared with the predicted JNDs derived from the two single-cue conditions.

Using a similar strategy to that outlined above for Knill and Saunders' second experiment, Ernst and Banks compared the estimated weighting given to visual information with that predicted from cue-conflict stimuli (although historically, Ernst and Banks' experiment preceded that of Knill and Saunders). The agreement between the predicted and estimated weights is striking, as shown in **20.11**.

The question Ernst and Banks asked was: do humans integrate visual and haptic information in a statistically optimal fashion? The answer to their question is given in **20.11**, and in the title of their paper: "Humans integrate visual and haptic information in a statistically optimal fashion."

Individual Differences in Cue Combination

So far we have considered cue combination experiments in which the data were averaged over observers. This approach has proved useful but it neglects the fact that different people systematically give different weights to different cues.

For example, when stereo and texture cues are put in conflict, it has been found that some people give more weight to stereo and less weight to texture and they are therefore called *stereo dominant.* Others give more weight to texture and hence are called *texture dominant.* Why there should be these individual differences is not clear. It is tempting to speculate that any given observer's visual system might be better at dealing with some cues than others, in the sense of giving less noisy outputs. If so, it makes sense, given the theoretical overview provided here, that the cue providing the more reliable data should be given most weight—and hence that it should be dominant in cue conflict situations. An example of an experiment that has investigated individual differences in cue combination is described in **20.12**.

Vertical ridges

a Cues conflict: Stereo 9 Texture 3 Outline 3

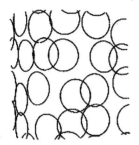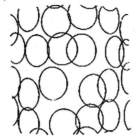

b Cues consistent: Stereo 9 Texture 9 Outline 9

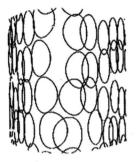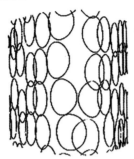

Horizontal ridges

c Cues conflict: Stereo 9 Texture 3 Outline 3

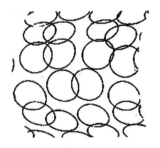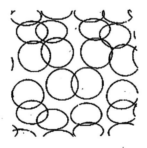

d Cues consistent: Stereo 9 Texture 9 Outline 9

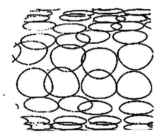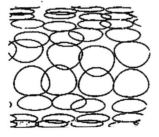

20.12 Exploring stereo and texture cue combination using stereograms

These are examples of stimuli used in a series of experiments run by David Buckley, John Frisby and their colleagues. Each picture **a—d** comprises the two halves of a stereogram. If these are fused in a stereoscope then a 3D curved ridge is seen. (If you can achieve stereo fusion by crossing your eyes, Ch 18, then each picture creates the percept of a curved ridge protruding above the page.)

The stereo pairs in **a** and **b** portray vertical ridges, those in **c** and **b** horizontal ones. All the stereograms contain a stereo cue for a ridge with amplitude 9 cm if viewed from 57 cm. This is shown by the label Stereo 9 above each picture. In addition, a texture shape cue is embodied in each picture and also an outline shape cue.

In **b** and **d** the texture and outline cues are Texture 9 and Outline 9, respectively, so they are drawn to be *consistent* with the S9 stereo cue for a 9 cm amplitude.

In **a** and **c** the texture and outline cues are for the shallower ridge amplitude of 3 cm. Hence they are labeled Texture 3 and Outline 3, and they are said to be in *conflict* with the Stereo 9 cue.

Different observers see curvature in these stimuli in different ways. Some people give emphasis to stereo, others to texture and outline, and yet others treat these cues equally. Technically, this is described as observers giving different *weights* to each cue in a process of *cue integration*.

One intriguing finding is that the way cue integration happens in these stimuli depends critically on the orientation of the ridge. Thus, for most observers, the texture and outline cue are more powerful in overcoming stereo for the *vertical* ridges than for the horizontal ones. Why this is so is controversial but it is clear that it is dangerous to generalise findings from cue combination studies that use just a few observers.

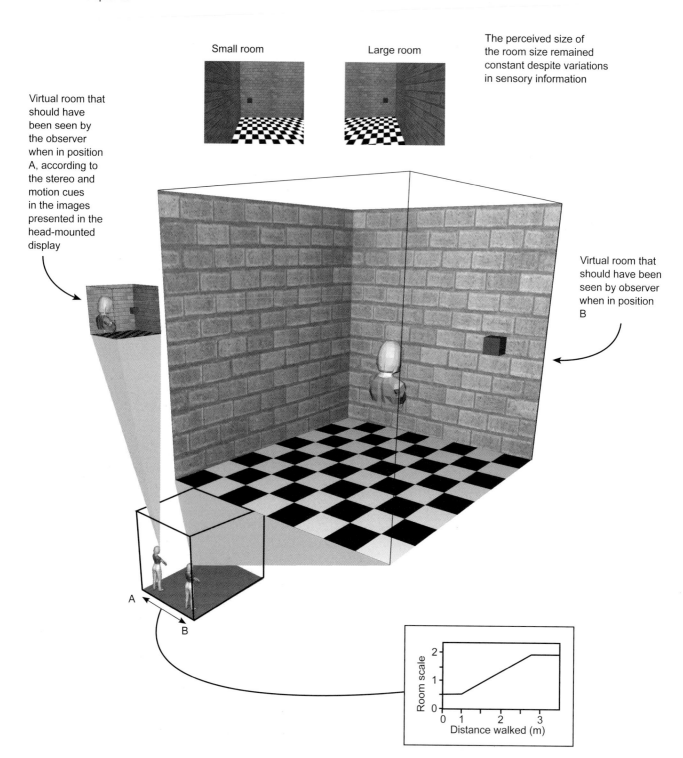

Small room

Large room

The perceived size of
the room size remained
constant despite variations
in sensory information

Virtual room that
should have
been seen by
the observer
when in position
A, according to
the stereo and
motion cues
in the images
presented in the
head-mounted
display

Virtual room that
should have been
seen by observer
when in position
B

A

B

Room scale

Distance walked (m)

20.13 Glennerster et al.'s use of computer graphics to study cue combination
The observer wore a head-mounted display that caused her to see a virtual room. As she walked from side to side (bottom
left) the images were adjusted in such a way that they were consistent with the room size increasing or decreasing (graph,
bottom right). Remarkably, the observer did not see a change in the size of the room, despite stereo and motion cues
showing that it had occurred. Courtesy Glennerster, Tcheang, Gilson, Fitzgibbon, and Parker (2006).

Pitting Stereo and Motion Against the Assumption of a Stable Visual World

The advent of relatively cheap computer graphics is having a considerable impact in studies of human vision generally and in particular in the field of cue combination. A remarkable experiment of this kind has recently been conducted by Andrew Glennerster, Lili Tcheang, Stuart Gilson, Andrew Fitzgibbon and Andrew Parker.

Observers wore a head-mounted computer-controlled binocular display that made it appear to them as though they were walking in a "virtual room," **20.13**. As they walked from left to right the computer adjusted the display creating images that would have arisen had the room gradually got bigger. Stereo and motion cues embedded in the display were sufficient, in principle, to allow the observers to see that room size had changed as they walked but this isn't what they saw. Instead, the room was perceived of as constant size. Hence the title Glennerster et al. gave to their paper: *Humans ignore motion and stereo cues in favour of a fictional stable world.* This is a remarkable finding, as in most circumstances the visual system regards stereo and motion cues as giving powerful cues about 3D scene structure.

Glennerster et al. interpreted their findings using a Bayesian framework. Specifically, they suggested that humans have an enormously strong prior in favor of objects in the world remaining the same size as they move around. This strong bias in favor of a stable world seemed to cause the observers' vision systems to recalibrate their stereo and motion modules to fit in with the assumption of a stable world.

The importance of this experiment for the present chapter is that it illustrates the broader context within which the combination of cues from different modules operates. The observers' perceptual systems were faced with a problem: should stereo and motion cues signaling that room size had altered be accepted at face value, or should these cues be evaluated using prior assumptions about the stability of the viewed world? The answer was clear: the stable world assumption was preferred and the information yielded by stereo and motion was interpreted in the light of that assumption.

Summary

In this chapter, we have explored an emerging framework for investigating how the brain might reconcile its many contradictory sensory inputs to achieve a rational compromise, based on the mathematical insights of the Reverend Bayes. His work, done in the 18th century, was introduced in Ch 13. You may find it surprising that Bayes' theorem does not appear in this chapter, given that the experiments described depend heavily on his work. The reason for this is that Bayes' theorem is buried fairly deeply in the mathematics which underpin the equations in this chapter. For the mathematically inclined, the presence of Bayes' theorem in cue combination can be traced via the texts given in Knill and Richards (1996).

Further reading

Ernst MO and Banks MS (2002) Humans integrate visual and haptic information in a statistically optimal fashion. *Nature* **415** 429–433. *Comment* One of the first Bayesian cue-combination papers to be published and relatively accessible.

Ernst MO and Bülthoff HH (2004) Merging the senses into a robust percept. *Trends in Cognitive Sciences* **8**(4) 162–169. *Comment* An introductory overview of this emerging research area.

Glennerster A, Tcheang L, Gilson SJ, Fitzgibbon AW, and Parker AJ (2006) Humans ignore motion and stereo cues in favor of a fictional stable world. *Current Biology* **16** 428–432.

Hillis JM, Watt SJ, Landy MS, and Banks MS (2004) Slant from texture and disparity cues: optimal cue combination. *Journal of Vision* **4** 967–992. *Comment* Technically demanding.

Knill DC and Richards W (1996) *Perception as Bayesian Inference.* (Eds) Cambridge University Press. *Comment* A comprehensive set of edited chapters but not for the faint hearted.

Knill DC and Saunders JA (2003) Do humans optimally integrate stereo and texture information for judgments of slant? *Vision Research* **43** 2539–2558. *Comment* Described in some detail above. Technically demanding but a clear account of cue combination.

21

Seeing in the Blocks World

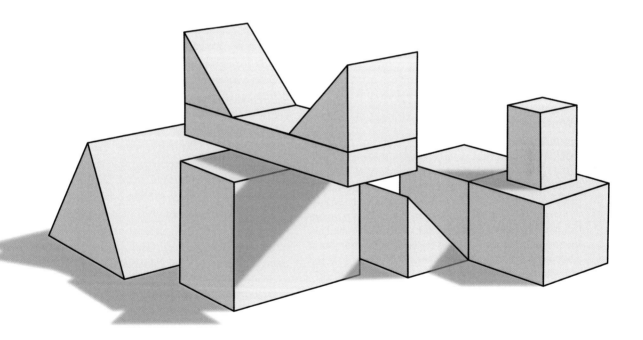

21.1 A scene from the toy blocks world

Ch 7 provided a broad review of the figure/ground problem. In this chapter, we pursue in detail a particular figure/ground task, that of segmenting images of blocks world scenes. We have two main goals. First, we want to illustrate some important general issues in understanding seeing, and particularly Marr's approach, which forms the theme of this book. Second, we introduce a particular type of algorithm, the *relaxation algorithm*, for implementing constraints. This has been widely used in computer vision and it has the advantage of lending itself to being implemented in nerve cells.

The Blocks World Problem

What do you see when you look at **21.1**? Your visual system tells you effortlessly that there is a heap of blocks of various shapes (wedges, cubes, etc.). We see the layout of these blocks as a scene with three-dimensional (3D) structure. We can easily report the existence of the arch, the triangular block partly obscuring a rectangular brick, the wedge resting on a cube, and so on. We see each block as a separate entity, a "thing," a *figure*, distinct from the other blocks around it, a *Gestalt*.

Why read this chapter?

In this chapter we describe research on a classic vision problem that was tackled in the early days (1960s and 1970s) of the emerging field of *artificial intelligence* (AI). This problem was to decide which edges in images of toy blocks belonged to each block. Hence, this chapter can be read as a continuation of Ch 7 on *Seeing Figure from Ground*. The blocks world domain was chosen by early AI researchers because it seemed hard enough to pose serious challenges, and yet sufficiently constrained to appear tractable. The research was characterised by a sequence of studies trying different approaches. Solutions that worked for images of complex blocks world scenes only gradually emerged, after various false turns. The main approaches are described and evaluated. The general lessons learnt in this quest are summarized. They remain instructive today, by providing a warning about hazards to be avoided in AI vision research. The relationship of this work to Marr's computational approach is reviewed. One conclusion is that some false turns were the result of not keeping clearly distinct the computational theory and algorithm levels of analysis.

Our visual system performs this *scene segmentation* task fluently, as it does for all manner of scenes—just look around wherever you are reading this book. But do not be misled by the ease with which your visual system does it. Scene segmentation is an extremely clever achievement of sophisticated visual processes. It remains a largely unsolved problem for complex natural scenes.

Research on the blocks world had as its goal finding out (making explicit) which parts of an image (lines, edges, line junctions) belong to which blocks in the scene. Choosing the relatively simple blocks world to study segmentation is an example of a common strategy in science: tackle a difficult problem by first picking a simpler case, hoping thereby to glean some general principles that will serve as a good foundation for more complex problems.

The trick in using this strategy profitably is to select a problem that offers the prospect of a reasonably early solution, while avoiding one that is trivially easy, or one that can be readily solved by uninteresting special purpose solutions. The blocks world was selected with this criterion in mind in vision research in the 1960s–1970s in the field of **artificial intelligence** (AI). AI can be roughly defined as studying how to get computers to do complex information processing tasks that are considered "intelligent" when humans do them.

It turned out that blocks world research did not prove a wholly good choice in this regard, because the domain was too simple to be extended directly to more complex cases. On the other hand, the story of blocks world research is one of successive elimination or revision of wrong turnings as their limitations were revealed. This process of refinement taught some very valuable lessons about how to proceed in general in investigating vision, and indeed other aspects of human cognition. These lessons are well worth describing and pondering.

What is the Input for Blocks World Research?

The natural starting point for blocks world segmentation is a description of the lines composing the image of the heaps of blocks. In short, an edge representation. Some of the early research extracted line descriptions from digitized images. However, making line information explicit from natural images is itself a complex vision processing task

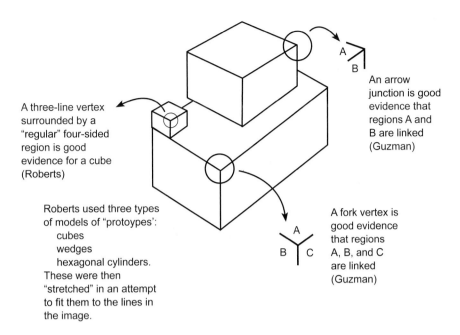

A three-line vertex surrounded by a "regular" four-sided region is good evidence for a cube (Roberts)

An arrow junction is good evidence that regions A and B are linked (Guzman)

Roberts used three types of models of "protoypes':
    cubes
    wedges
    hexagonal cylinders.
These were then "stretched" in an attempt to fit them to the lines in the image.

A fork vertex is good evidence that regions A, B, and C are linked (Guzman)

21.2 Segmentation by recognition (left-hand side) and by region linking (right–hand side).

(Chs 5 and 6). To get around that problem, hand-crafted line descriptions have sometimes been used in blocks world research. This artificial "ready made" input allows concentration on the segmentation task.

In what follows here, the assumption is made that line descriptions are available. This entails the assumption that segmentation should follow *after* line descriptions have been made explicit. This can be challenged, indeed, it has been. An alternative view is that object segmentation and line finding should go hand in hand at one and the same time.

Segmentation-by-recognition

The human visual system can search for figures against ground using its knowledge of the shapes of objects gained from previous experience. We saw this kind of processing at work in Ch 1 when *concept-driven processing* was introduced in connection with various camouflaged figures. Could this sort of *top-down processing* be going on when we segment the individual blocks in **21.1**? That is, could *specific object knowledge* about the shapes of wedges, cubes, etc., be at the heart of how human vision segments images of blocks world scenes?

Using object concepts was one of the first approaches employed in AI studies of blocks world segmentation. Roberts wrote an impressive computer vision program in 1965 that demonstrated a style of processing called *segmentation-by-recognition*.

The first stage of his program derived a line drawing of the scene from a photograph. Next, configurations of lines were detected as "good cues" to a possible prototype. For example, an image vertex surrounded by three regular polygons is a good cue for a cube, **21.2**. The selected prototype becomes a hypothesis that is tested for its fit to the image line data by exploring whether the prototype can be suitably "stretched" while preserving the geometrical integrity of the prototype. If the hypothesis survives the test, then it is accepted and the program goes on to deal with another part of the image. If it does not, then another prototype is considered and/or other cues selected.

All stages of Roberts's program were *model driven* (the models being the prototypes), except for the initial line-finding phase. And as far as the latter was concerned, some workers in the same tradition as Roberts even incorporated model testing into that phase as well (Falk; Shirai), in an

513

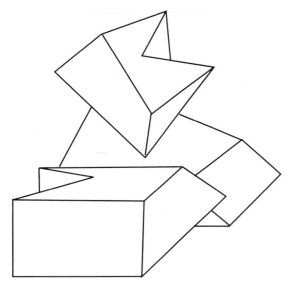

21.3 Some unusual blocks

attempt to overcome some of the (then surprising) difficulties that were experienced with extracting a good line description from a messy digital image.

Roberts's program thus implemented a time-honored view of perception. This is that vision is a process of detecting cues, selecting a hypothesis that is capable of making sense of them, drawing inferences from the hypothesis about further cues that should be present in the image, checking these inferences with a search for confirming and/or disconfirming cues, and repeating this sequence as many times as necessary.

This kind of theorizing about perception was espoused by the great 19th century German scientist Hermann von Helmholtz (1821–1894). He coined the phrase **unconscious inference** because we are not aware of these inferential processes. Segmented blocks pop up into our visual awareness and we have no ability to introspect on the unconscious processes involved. Ch 13 explains in detail modern conceptions of seeing as fundamentally inferential.

A question that may have occurred to you is: how did Roberts's program get its object knowledge in the first place? The answer is that Roberts himself provided his program with this knowledge for free; it did not have to learn it for itself. Obviously, humans learn a lot about the visual appearances of objects as we come into contact with them, both as children and as adults. The details of how we do this are largely mysterious.

Roberts' program achieved its goal. It was a very successful piece of research providing a clear demonstration of the power of using top-down knowledge. Nevertheless, his achievement left open the question: is concept driven processing essential for solving the blocks world segmentation task? Or could a solution be found that exploits only *general properties* about the blocks world being dealt with, rather than knowledge of the *specific shapes of objects* within that world? Attention thus turned to seeing how far a *bottom-up* approach could go in dealing with blocks world scenes.

A hint that this might be possible comes from unusual blocks of the sort shown in **21.3**. Presumably you have never before seen blocks quite like these and yet your visual system easily separates them one from another.

Segmentation by Image Region Linking

Guzman described in 1968 a program called SEE that tried to segment by seeking links between adjacent image regions. The underlying idea was to group together regions in the image that derived from each body in the scene according to evidence offered by different kinds of line junctions.

For example, Guzman reasoned that a fork (Y) junction is good evidence that all three of its bounding regions come from the same body and therefore that they are to be linked, **21.2**. An arrow junction is evidence that only two of the regions that comprise it are related. These rules of thumb are technically called **heuristics**.

SEE utilized many such heuristics to establish in a first pass over the line description all the evidence it could find about what image regions might be linked as coming from the same block. In a second stage, SEE then brought that evidence together, to provide an overall grouping of image regions. Each group corresponded to a constituent body in the scene. Again, the segmentation problem was solved. Or so it seemed.

In fact, problems abounded. To enable SEE to deal successfully with complicated scenes required the almost constant addition of new heuristics and modifications of old ones as counter-examples to its competence were discovered. The resulting messy patchwork of rules suggested a basic flaw in the approach. Why did SEE have so many problems?

The reason was that Guzman did not analyze sufficiently deeply *why* regions in the image associated with line junctions yielded good evidence about objects in the scene. What justifies the use of a Y junction as a clue that the three bounding regions come from the same block?

Attention thus turned to asking what the lines in the image depict about the corresponding entities in the scene. This new approach exploited advantages from maintaining a much clearer distinction between the image and scene domains.

Segmentation by Line Labeling

A major insight in bottom-up blocks world research was the realization that the segmentation task can be solved by working out what *types of scene edges* could have created each line in the image. This approach was independently proposed by Huffman and Clowes, it was developed by Mackworth, and it was brought to a magnificent climax by Waltz in his PhD thesis in 1975. We will describe only Waltz's work.

Waltz ignored, when starting out, edges from shadows. He considered first the very simple case of blocks whose vertices are made up of exactly three faces. He also restricted himself at first to blocks world images in which each line was the *projection* of two basic types of edge in the scene, each of which itself comes in two flavors, **21.4a**:

Boundary Edges: these delineate the surfaces belonging to different blocks. They are produced when one surface of a block lies in front of a surface from another block, or in front of the surface on which the pile of blocks lies. There are two sorts of boundary edges, symbolized by the labels → and ← using the convention that the obscuring surface is lying on the right-hand surface when facing in the direction of the arrow. Hence, which way the arrow points is a crucial matter as the two directions label a different relationship between the two scene surfaces forming the image line.

Interior Edges separate surfaces belong to the same block. A *concave* interior edge is formed by two surfaces of a block creating a concave depth variation from the observer's viewpoint. It is symbolized by

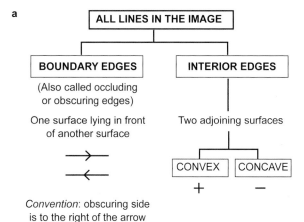

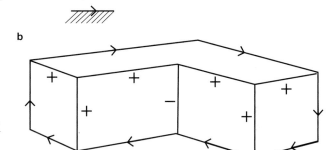

21.4 Line labels Used by Waltz
a Classification table.
b An example of the labels attached to a block.

the label - (*minus* sign). A *convex* interior edge is similar but here two surfaces of a block form a convex depth variation from the observer's viewpoint. It is symbolized by the label + (*plus* sign).

So there are two labels for boundary edges and two for interior edges, **21.4a**. The task now becomes that of making explicit, by attaching to each line in the image, one of the four symbols (+, -, ←, →) to show which one of these four types of scene entity produced each line. This is why this approach to the task of blocks world segmentation is called *line labeling*.

After scene edge labels have been correctly attached to image lines, it is easy to attain the desired segmentation. This is because sets of lines coming from individual blocks can be found using the boundary edges as a guide. You can see how this works by tracing the arrow symbols around the block in **21.4b**.

How then to begin the task of line labeling? The first step is to start with line junctions, as Guzman

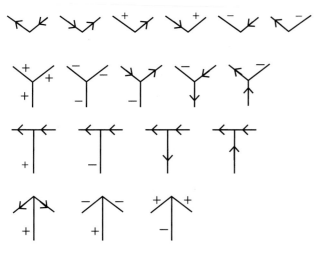

21.5 Junction dictionary

did but now in a very different way. There are four possible classes of junctions for the simple blocks we are presently considering: L, Y, T, and Arrow. Taking first an L junction, each arm can be one of the four line types, so the number of possible combinations is 4 × 4 = 16 different sorts of L. Y junctions have three lines, so for them we get 4 x 4 × 4 = 64 sorts of Ys. The Ts and the Arrows also produce 64 combinations, so the grand total is 16 + 64 + 64 + 64 = 208 line junction types (see Table 21.1 below).

It may seem an awful headache is in prospect deciding which junction type is appropriate for each line junction. The ambiguities seem hopelessly large. But it turns out that only 18 of the 208 junctions are physically possible. All the others cannot exist in the real world of rigid planar surfaces, as summarized in the table below.

Table 21.1 Combinatorial Explosion of Junction Types

Vertex Type	Number of Combinatorially Possible Junctions	Number of Physically Possible Junctions
L	16	6
Y (Fork)	64	5
T	64	4
Arrow	64	3
TOTALS	208	18

Technically, this is called *using a constraint to prune a combinatorial explosion*. Here the constraint is a physical one, derived from the nature of the blocks world under consideration.

The 18 physically possible junction configurations define a **junction dictionary**, **21.5**. The term dictionary is apt because each line junction type can be thought of as a legal "visual word" made up of 2 or 3 legal "visual letters" (the letters are the +, -, ←, → symbols attached to the lines forming each junction).

The line labeling approach also uses a second constraint, called the **consistency constraint**. This is that a scene edge has to be given the same line label at both ends. This is because it is physically impossible in the blocks world for it to change its character along its length. For example, a concave edge cannot change into a convex edge in the static world of rigid simple blocks made up of planar surfaces. It would be a completely different matter if blocks were to be allowed that were made of rubber so that their surfaces could be stretched and deformed from planarity.

Another piece of general blocks world knowledge embedded in Waltz's program was the assumption that the blocks will have been imaged with no problematic accidental alignments. This is called the **general viewpoint assumption** (recall the Ames Chair demonstrations in Ch 1). The top-down approach of Roberts also made this assumption. It is commonplace in vision theories for many different kinds of problems.

Waltz demonstrated that that the consistency constraint used in conjunction with the constraint afforded by the small junction dictionary permitted a unique line labeling interpretation for each image line. Having achieved this, tracing around the boundary labels distinguishes each block as a separated entity, **21.4b**. The segmentation task was then solved.

Relaxation Algorithms

You may at this point feel that you have got a reasonably good idea of the principles underlying Waltz's program but feel frustrated about not knowing the details of how it worked. How exactly did Waltz's program exploit the constraints afforded by his careful task analysis of the blocks world scene segmentation problem? This takes us into a different level of analysis, one concerned with defining the exact sequence of steps to be undertaken in solving an information processing task using the constraints identified in the task theory. Such a sequence is technically called an *algorithm*, a term we have met in previous chapters. The algorithm (or procedure) is rather like a cooking recipe that tells you how to prepare a given dish by carrying out a series of steps.

It might be helpful here to summarize the main take home message that we will be emphasizing from the blocks world research, to give you direction as to where we are going. First and foremost, begin with a detailed *task analysis* of the problem, with the intention of developing a *computational theory* cast in terms of valid constraints. Then implement the *constraints* in a suitable *algorithm*.

The algorithm used by Waltz was of a class known as a *relaxation algorithm*, also sometimes called a *cooperative algorithm*. This class has the following general formula when applied to a vision problem:

Step 1 Attach a list of all possible labels to each image part.

Step 2 Compare the possible labels for each part with those for related parts and eliminate mutually inconsistent labels, using the task theory to define inconsistency.

Step 3 Repeat Step 2 as many times as required until no further change in labels takes place. The technical term for repeating steps in this way is called iteration.

Relaxation algorithms are general purpose in character, which means that they can be used for implementing diverse theories. You just feed in labels suitable to the particular vision task under consideration, and define "mutually inconsistent" in Step 2 using the task theory. Hence, relaxation algorithms have been used in many different sorts of computer vision programs dealing with many different visual tasks. How they are used in any given case depends upon the knowledge being exploited to solve the task in question.

In the present line labeling example, the knowledge is not object specific knowledge as used by Roberts, but general knowledge applicable to *all* blocks in the range of scenes being considered. This knowledge is encoded in the junction dictionary (used in Step 1), and in the consistency constraint (enforced in Step 2).

Before showing in more detail how Waltz used a relaxation algorithm, it is worth noting that there are, broadly speaking, two types. Waltz used a *discrete* relaxation algorithm, as any given label was deemed to be present or absent. A second type uses estimates of the confidences in attaching labels to image parts (Marr and Poggio's stereo algorithm described in Ch 18 is an example).

Waltz's Discrete Relaxation Algorithm

Step 1 Find all the junctions in a line description of an image of the blocks world scene being dealt with. Label each one as either L, T, Y, or Arrow according to its structure. Consult the junction dictionary to find out the possible "legal" labelings for each line comprising each junction. When this first step is complete, there will be lists of possible labels at each line end, **21.6**.

Step 2 For each line, compare the lists of line labels at its two ends to find any labels that have no matching label at the other end. If any unmatched labels are found, eliminate the junction dictionary entities that provided the non-matching labels. This imposes the consistency constraint.

Figure **21.6** shows how this works for the junctions designated A and B. Junction A is an Arrow, junction B is an L. The three types of Arrow junction produce two possible labels, + and –, for the line between the junctions. The six types of L junction produce four possible labels. However, the + and – possibilities for this line from B lead to the elimination of all L junctions with arrow labels. This prunes away four of the six candidate L junctions. They simply could not exist. This sort of pruning is done for all junctions to eliminate all impossible junction types. You can get a good idea of how this works by following the sequence for junctions A, B, C, and D in **21.6**.

Step 3 Repeat Step 2 until no further eliminations result. If you are keen on learning jargon, this is described technically as iteratively cycling local neighborhood interactions to obtain a globally consistent line labeling interpretation of the image.

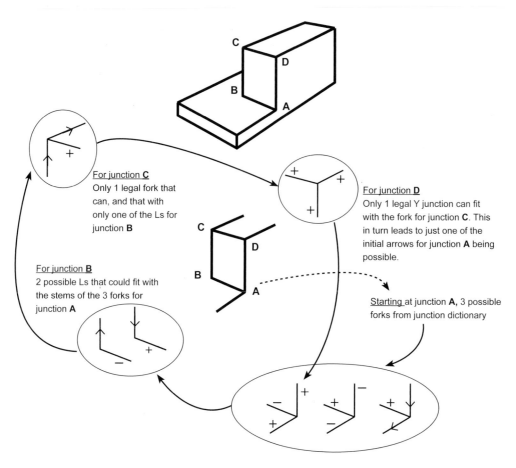

For junction **C**
Only 1 legal fork that can, and that with only one of the Ls for junction **B**

For junction **D**
Only 1 legal Y junction can fit with the fork for junction **C**. This in turn leads to just one of the initial arrows for junction **A** being possible.

For junction **B**
2 possible Ls that could fit with the stems of the 3 forks for junction **A**

Starting at junction **A**, 3 possible forks from junction dictionary

21.6 Line labeling relaxation algorithm at work
Redrawn from Winston (1984).

Step 4 Search the line labels for the boundary labels, to find sets that reveal the outlines of individual blocks.

If all has gone to plan, the segmentation problem is then solved.

Note that segmentation has been achieved here without any *specific* knowledge about the blocks present. The program ends by segmenting each block; it does not end by reporting whether each segmented block is a brick, a wedge, a cube, etc. How could it? It does not have embedded in it knowledge of any shapes. The only thing that it makes explicit is groups of lines belonging to each block. The end result is a set of groups of lines, each group being an unrecognized figure separated from its surround. But this is fine; it is the kind of output that Waltz set out to achieve.

Before leaving line labeling, it is helpful to note that a relaxation algorithm can be implemented using either *sequential* or *parallel processing*. In the former, each line junction is considered in turn, one after the other, as in **21.6**. In the latter, all line junctions are considered at every step.

Biological vision systems generally use parallel processing to make up for the poor operating speeds of their individual nerve cells. Computers have much faster components and can frequently perform well with sequential processing.

In Ch 18 we illustrated a relaxation algorithm implemented in processing elements that use excitation and inhibition to propagate constraints suitable for solving the stereo correspondence problem. These elements thus work in a roughly similar fashion to neurons.

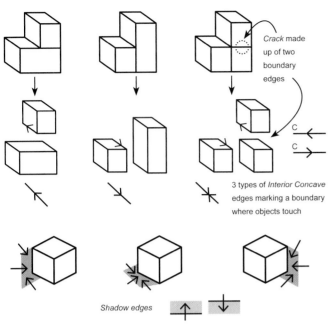

Crack made up of two boundary edges

3 types of *Interior Concave* edges marking a boundary where objects touch

Shadow edges

21.7 Adding new line labels

More Complex Blocks Worlds

So far we have considered only very simple blocks world scenes. The next research question was: could the line labeling approach generalize to more complex cases?

Waltz showed that it could, by going on to remove almost all restrictions on lighting and shadow conditions. This allowed blocks world scenes situations in which each image line could have any one of 99 possible scene edge interpretations, not just the four we have considered here.

Some of the new symbols used for labeling were the shadows and "cracks" shown in **21.7**. An updated classification that includes these new possible line labelings is shown in **21.8**. Waltz also lifted restrictions on the shapes of the blocks (though they still had to have planar surfaces), so greatly increasing the allowable junctions/vertices that could arise. To many people's surprise, his ***consistency filtering*** relaxation algorithm cut through the horrendous combinatorial explosion of possibilities to achieve segmentation. The reason was that some of the newly introduced junctions were relatively unambiguous, and they propagated their disambiguating influence through the entire network when consistency between neighboring vertices was imposed.

Problems with Line Labeling

Unfortunately, despite impressive achievements founded on a solid theoretical analysis of the task, deep problems with line labeling were discovered. Waltz's program produced on occasion legal labels for impossible objects. For example, in **21.9a** each junction is legal but the lines of this single block should all be labeled as boundary edges, and they are not. This is clearly unsatisfactory.

Another problem case is shown in **21.9b**, in which again legal labels have been attached. However, because the junction dictionary does not represent the concept of surface planarity it accepts this "impossible object," neglecting to note that surfaces A and B cannot be planar and hence that the object is illegal. Our visual systems tell us that the visible surfaces of the pyramid cannot be planar, because of the way the notch reveals that the lines joining the top to the base are not colinear. Blocks with curved surfaces were not allowed in the blocks world dealt with by Waltz's program. A case of a successful rejection of a block with non-planar surfaces is shown in **21.9c**.

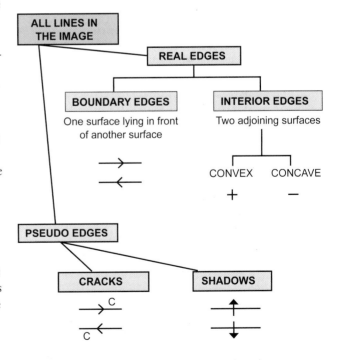

21.8 Waltz's extended set of line labels
Redrawn from Winston (1984)

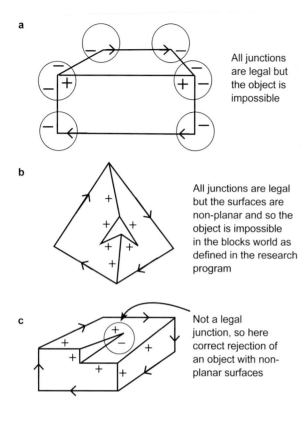

a — All junctions are legal but the object is impossible

b — All junctions are legal but the surfaces are non-planar and so the object is impossible in the blocks world as defined in the research program

c — Not a legal junction, so here correct rejection of an object with non-planar surfaces

21.9 Some limitations of the line labeling approach

It became apparent that the junction dictionary approach was fundamentally limited because, by confining itself only to labels representing edge types, it had no way of exploiting certain crucially important constraints about the relationships of *surfaces* in blocks world scenes.

The Gradient Space Representation

An important development in trying to overcome these problems was the use of the *gradient space* representation for describing surface slants. We met this concept in Ch 2 when considering seeing slant from texture gradients.

Gradient space is a convenient representation for thinking about surfaces. It is a 2D coordinate system (think of a conventional graph with two axes) in which planes in the 3D scene are represented as points in a graph. Thus, each pair of 2D coordinates in the gradient space graph, **21.10,** specifies the 3D orientation of a surface with respect to the observer's viewing position. The ver-

tical axis in gradient space (labelled as q) represents the size of the surface rotation of a plane about the horizontal axis in the scene. Similarly, the p axis in gradient space represents rotation of a plane about the vertical axis in the scene. Combinations of p and q rotations are represented in the gradient-space graph by points lying off the p and q axes. An example is shown of the point ($p=1$, $q=1$), which as p and q are the tangents of angles, represents a surface which is rotated by 45° around the vertical axis and 45° around horizontal.

Note that in the gradient space representation the actual depth of the surface plane from the viewer is not made explicit, nor is the spatial extent of the plane. However, gradient space can capture the concave/convex relationship between adjoining surfaces. We will not go into how it does so. We have introduced gradient space simply to illustrate a basic point needed to appreciate both the earlier approaches and work in other areas of vision.

This basic point is that the key thing to ask about a representation be it a model prototype (Roberts), a region label (Guzman), a line label (Waltz and others), or a point in gradient space is: what does it make *explicit*?

That is, what information does it convey in a form that can be used without more ado by subsequent processes?

If a representation does not make surface orientations explicit then it is clearly going to be inadequate in tasks where that information is crucial. The p,q gradient space representation moves on from Waltz's work in that it directly captures a critical aspect of blocks—the orientations in 3D space of their constituent surfaces.

Lessons from Blocks World Research

Blocks world research is now largely of historical interest—it is not a focus of current research. But that doesn't mean considering it is a waste of time. Far from it. It helped generate some fundamental lessons about how to go about studying vision in general, not just in the blocks world.

Lesson 1 Distinguish *Images* from *Scenes*

This sounds such an obvious point but it is in fact quite profound. One reason the line labeling approach made progress was by recognizing the

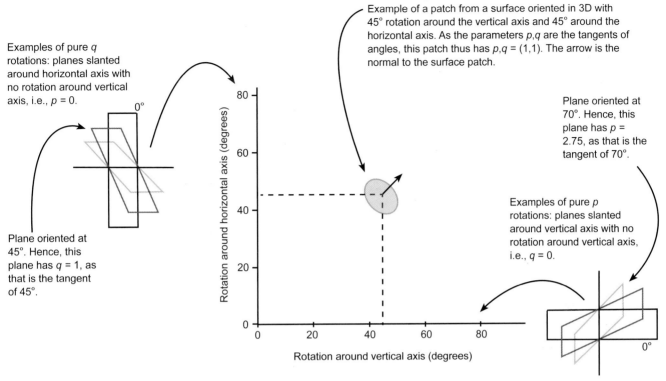

Example of a patch from a surface oriented in 3D with 45° rotation around the vertical axis and 45° around the horizontal axis. As the parameters *p,q* are the tangents of angles, this patch thus has *p,q* = (1,1). The arrow is the normal to the surface patch.

Examples of pure *q* rotations: planes slanted around horizontal axis with no rotation around vertical axis, i.e., *p* = 0.

Plane oriented at 70°. Hence, this plane has *p* = 2.75, as that is the tangent of 70°.

Plane oriented at 45°. Hence, this plane has *q* = 1, as that is the tangent of 45°.

Examples of pure *p* rotations: planes slanted around vertical axis with no rotation around vertical axis, i.e., *q* = 0.

21.10 Representing 3D surface orientation
Be careful to note that the gradient space parameters *p,q are* the tangents of the angles plotted in this graph—see various labels in the figure.

need to keep the image and scene domains sharply distinct, and to ask very clearly: what information is there in the image about entities in the scene? Guzman's approach ran into its ad hoc rules problem by reasoning about image region linking without sufficient scrutiny of rules about how the regions of blocks are configured in scenes.

As an aside here, we note that an image is a *visual projection* of a scene. The image provides input data for visual processes to work out what is in the scene. For this reason, vision has been characterized as ***image inversion.*** Cameras produce images of scenes, whereas vision produces scene descriptions from images (Ch 1).

Lesson 2 Define Carefully the Scene Description Task to be Solved: Task Analysis & Task Theory

What is it that is being *described* (made explicit) in a *representation* using what vocabulary of *symbols*? A clear task analysis is required, whose objective is to specify a computational theory of the task. The latter is a principled method for solving the task using appropriate knowledge. The computational theory as specifies constraints for solving the task in hand, and justifying why they are good ones for that task.

The full task theory for line labeling in fact involved a good deal of math, proving what junctions were and were not allowable. That level of detail would be inappropriate in this book, but it is worth noting that math is a typical feature of a fully specified vision task theory.

This does not mean that everyone working on vision has to be a mathematician, far from it. But it does mean that studying vision is nowadays an inter-disciplinary activity and that collaborations between mathematicians, computer vision scientists, psychologists, and neuroscientists are likely to prove the most effective research teams.

Lesson 3 Visual Processes Exploit Knowledge

Knowledge used in visual processes can be classified into two broad kinds: *specific* knowledge about particular objects or particular scenes, or *general* knowledge applicable to a class of objects or a class of scenes.

Using the former is called *concept driven* or *top-down* processing, whereas the latter can be embedded in what are often called *data driven* or *bottom-up* processes. Roberts's program illustrated the former, by showing that *specific* object knowledge can be effective.

Waltz's program illustrated the latter: he showed the benefits of that can flow from using *general* knowledge about blocks made up of planar surfaces, and *encoding* this knowledge in a dictionary of junction types.

An important lesson from the blocks world story was: do not resort to top-down knowledge too quickly when confronting a vision problem. Much can be achieved from data driven processing, given a thorough task analysis that reveals good *general constraints* for the scenes in question.

Lesson 4 Distinguish Algorithms from Task Theories

Having completed the task analysis, the resulting computational theory needs to be implemented in an *algorithm* of some sort. Often a variety of different kinds of algorithm is available to implement the same task theory. We have not illustrated this point in this chapter, except to remark in passing that relaxation algorithms can be applied sequentially or in parallel.

It is important to distinguish the computational theory from its implementation. If a program fails, is the problem located in an inadequate task analysis, or in shortcomings of the algorithm used to implement it? This is a key point that helps clarify debates about vision and indeed other cognitive processes.

This point is so important that it deserves special emphasis: *computational theories and algorithms for implementing task theories are not one and the same thing.*

The need for a distinction of this kind has been recognized by many vision researchers but it has become well known from the work of David Marr.

We reviewed his approach in Ch 2 where we set it out as the underlying theme of this book. Indeed, the main purpose of this chapter is to provide a further detailed example of this approach at work.

We must, however, immediately repeat a cautionary remark made elsewhere in this book: some vision researchers have argued that, at least in some cases, it is difficult to distinguish the computational theory from algorithm implementing the theory. We discuss this issue again in our final chapter, *Seeing Summarized.*

Unfortunately, the words computational and algorithm confuse many people. These terms encourage the erroneous conclusion that this approach to vision is intrinsically bound up with computers. That is wrong, completely wrong. One aim of this book is to show that we need computational theories and algorithms when considering how human visual brain processes work, not just when building computer vision systems.

Lesson 5 Avoid Procedural Hacking

When developing a computer vision program, it is often tempting to start without the benefit of a clear task analysis. Plunging ahead in this way can in fact be useful as experimental results can prompt theoretical ideas and thus help to articulate a good computational theory. Also, a pilot system might have the merit of being an *existence proof* that an artificial vision system can be built. The potential downside, however, is a practice known as *hacking*, which in this context does not mean finding a way to break computer security systems.

Hacking here means developing a program in a piecemeal fashion without any real concern for theory development, adding a tweak here and tweak there to patch around problems revealed by failures. Even if hacking finally produces a program showing reasonably decent performance, there remains the worry that the reasons why the algorithm ends up more or less working are unknown. What knowledge is the program exploiting exactly? Why do the ad hoc heuristics work? Unless the program is based on a well specified task theory, it remains mysterious. This was the deep problem with Guzman's region labeling approach, and why ultimately it was rejected. His work was pioneering and his program SEE could deal with some remarkably complex blocks world scenes. It was a very useful step forward. It is, however, no longer thought to be a good way of dealing with blocks world scenes.

In the worst case, hacking could end up inventing a new "seeing system" that itself needs research to see why it is working. This is not exactly the goal of the enterprise. So a key take-home message is: remember always that *to understand vision requires the articulation of good vision task theories.*

However, it may well be a good approach when dealing with the kind of expertise typical of certain other areas of human cognition, such as some kinds of thinking, like medical diagnosis, or playing chess. In such cases, a good task theory may not exist; indeed, it may not even be attainable in principle (will we ever have a good theory of chess?). In such cases, human cognition seems to rely on pattern recognizing a set of features (e.g., medical symptoms, a given chess board position) that are indicative of particular entities (e.g., diseases, or in chess the moment to launch an attack) and using these to trigger actions that are known to work, at least some of the time.

This form of expertise is naturally encoded as a set of rules: IF A THEN DO X, IF B THEN DO Y, and so on. Computer-based medical expert systems commonly take this format. They have to grapple with the problems thrown up when different rules compete, as they often do (this is the *conflict resolution problem).*

It is possible that vision is a special case for which good task theories can be sought. Marr called such cases ***Type 1 problems.*** For other problems, task theories may not in principle be attainable. Marr called these ***Type 2 problems***, giving as an example protein folding, in which a mass of interactions between the charges on different parts of long molecule determine its final folded-up shape.

Marr thought that progress was being hampered in areas of cognition such as memory amd language because the research is not seeking Type 1 theories. Reading the Epilogue to his book *Vision* provides some stringent (and entertaining) admonishments to those who build rule-based systems; see for example pp. 347–348 of Marr's book.

Marr thought that the underlying reason why seeing is such a fluent and effortless process for us is that human vision has "discovered" via evolution that seeing is a Type 1 problem. Accordingly, it implements a wide variety of good constraints for solving vision problems.

So the moral to be drawn from the Guzman story is: concentrate hard on task analysis before going far in "hacking up" a set of rules that seem to work (to some degree) for no very clear reasons. Or at any rate, always be mindful when hacking up such rules that they are no more than experimental stepping stones towards the goal of a proper task theory.

Lesson 6 Distinguish Task Theories and Algorithms from Brainware

Where does the brain fit into this account? Any complete theory of human vision must include an account of structures found in the eye and the brain. So you may be thinking it a bit strange that we have got this far in discussing the blocks world segmentation task without once having mentioned the stuff inside our heads that actually does the blocks world segmentation. Is there anything that can be learnt from the brainware?

In fact, we know of no neurophysiological work on the brain mechanisms involved in blocks world segmentation. However, there are some general points worth making at this juncture.

Let's suppose that a neuroscientist discovered a cell that is active when a given block is segmented. Would you be satisfied with that discovery as an explanation of segmenting blocks world scenes?

We hope not, having read this and preceding chapters. Surely we could only claim to understand such brain processes if we could say what task theory and what task algorithm that cell and its associated cells were implementing. This is an important point.

A brain cell (or a circuit made up of cells) whose activity could be demonstrated to be a *neural correlate* of a given block being "seen as segmented" would be an impressive and interesting achievement. But it would immediately prompt the question: *How* are those cells "solving" the task?

So we might enquire: are they encoding a junction dictionary? Or a high level block concept as used by Roberts? Or perhaps something else altogether, such as a point in gradient space? Can the details of their operation suggest that they mediate a relaxation algorithm, with different cells coding different line labels and with activity in some cells getting switched off when the labels they represent are eliminated connections implementing the consistency constraint? Or might such cells implement a different sort of algorithm? And, if so, what one?

The underlying point here is this: *we need the computational theory and algorithm levels of understanding to help us understand how the brain works.*

It is a pity that the blocks world story cannot be rounded off by describing brain cells implementing a blocks world theory. Even so, the story of blocks world research helps illustrate the claim that unless neuroscientists study the brain guided by task theories they will not know the right things to look out for.

The issue just discussed is customarily cast in relation to so-called **grandmother cells**. These are cells that become active when you see a complex and highly specific object, such as your grandmother. There is now some evidence that cells of this sort may exist. There are lots of facets to this idea (Ch 11) but first and foremost is the one being argued here: simply discovering a grandmother cell would not tell us how it works, how it becomes active, the principles underlying its design. Its discovery would not allow us to build a computer system able to recognize granny.

One aspect of this debate worth noting is the following: the task theory/algorithm/hardware distinctions are very clear in computer science, where the irrelevance of the hardware for many aspects of theory and algorithm development is clearly evident. But is this really true of the brain? Neuroscientists often seem to do their research with the working hypothesis that the algorithm quite simply *is* the brainware. Have they got a point, or are they simply deluded?

This may be a case of only time will tell. For the present, we agree with Marr that keeping the computational theory/algorithm/brainware domains separated as distinct areas of analysis is the right way to proceed.

Lesson 7 Help from Studies of Human Vision

Blocks works research has made little or no use of studies of human vision, beyond simply using human seeing as an existence proof that the task was solvable. Could detailed guidance have been gained from studies of human vision? The answer in this particular case is no, not at present. We simply have no significant clues from studies of human vision as to whether human vision uses, say, the kind of knowledge embedded in the junction dictionary and the consistency constraint when segmenting blocks world scenes. But this shortcoming does not prevent this problem being a good vehicle for introducing the key ideas of task theory and algorithm.

However, there are some things about human vision that touch in a general way on this question. For example, Ch 1 used an impossible object to illustrate that human vision is prepared to perceive correctly the structure of various local 3D features at the price of allowing an absurd, indeed physically impossible, overall configuration. In other words, the consistency constraint used by Waltz does not seem to be imposed by human vision in dealing with all scenes, even if it uses this constraint wherever it can.

A Last Example on Levels of Analysis

Many people find the argument being made about different levels of understanding elusive, difficult to grasp. The following example may help.

Suppose you discovered an archaeological relic composed of various pieces of clockwork. It would be a fairly simple task to describe the properties of the various parts of the relic in terms of which gear turned which, what gear ratios were embedded in the system, and so on. But your understanding of this piece of hardware would not be complete until you knew the design principles underlying the "thing" you had discovered.

Is it a clock? Or could it be a calculator of some sort? Or the control mechanism of an industrial device? Once that information was available, your appreciation of the logic underlying the relic's design would be dramatically transformed—it would lead you to appreciate many aspects of the mechanism which had hitherto gone unappreciated, perhaps even completely unnoticed. Indeed, it would lead you to *see* things in the structure that you had overlooked before (compare the Dalmatian dog and other hidden figure illustrations in Ch 1).

This is by no means a far-fetched scenario. The *Antikythera mechanism* is an ancient Greek device dated around 200–100 BC. It is a set of over 30 precisely shaped gear wheels, dials, and pointers held in a wooden case. It was discovered in a shipwreck around 1900 but it was only in 1965 that it was re-found in a cigar case in the National Archaeological Museum in Athens.

Trying to work out the Antikythera mechanism's function has proved a big challenge, with two rival explanations emerging.

One is that it is a kind of planetarium; the other that it is an instrument for predicting eclipses (Marchant, 2008).

This story illustrates nicely the general point we are making here: that to fully understand a device you need to know about the structure of the task which it was designed to solve as well as the structure of the mechanism itself. Simply describing the arrangement of the gear wheels is not enough.

The same seems to be true of visual mechanisms discovered by neurophysiologists: the properties of neurons can now be described in considerable detail, but to fully understand their design needs a deep understanding of the tasks they are performing.

You might want to ask at this point: where is the algorithm level in this relic example? An algorithm specifies in detail a sequence of operations to achieve a given end. But a device made up of gear wheels does not seem to be built to execute a sequence of steps, as in a computer program implementing an algorithm.

This is an important point because it illustrates how difficult it can be to disentangle algorithms from hardware. Suppose the device that is being studied is an ancient arithmetical calculator of some kind. The task theory level of understanding would be the fundamental logic of counting, of using numbers. Number systems of various kinds exist. We use Arabic notation with a base of ten. The Romans had a quite different system. The distinction between the two is at the algorithm level. So to understand the ancient calculator would entail that we knew whether it was implementing Arabic, Roman, or some other number system. The hardware level would be to do with gear wheels, abacus beads, or whatever else was found in the device.

Marr used this example of counting machines to illustrate his contention that in trying to understand an information-processing device it is necessary to keep clearly distinct the issues to be faced at three different levels: task theory, algorithm, and brainware/hardware. These levels need to be carefully distinguished while at the same time recognizing that they are complementary. Each one has its own preoccupations and concerns and all are important for a full understanding of what is going on.

They are connected because, for example, choosing an algorithm might be influenced by what hardware is available. But each level, Marr claimed, has its own concerns and they need to be distinguished to achieve a clear understanding of the brain.

Marr further claimed that many investigations of biological vision systems, and the creation of many man-made ones, have failed to recognize the importance of these *levels of analysis* distinctions, with the result that progress has been slowed. Marr has not been alone in making this claim (see Chs 23 for antecedents).

Blocks World Segmentation: An Overview

Research on the blocks world contributed to the clarification of some important issues in image understanding. It helped shape our present conceptions of low-level vision. It explored in detail the scope of certain representations and how to use consistency constraints within those representational schemes for filtering out incorrect interpretations.

The low-level representations it developed were limited in scope. They are related to what Marr called the *2.5D sketch*, and Barrow and Tenenbaum called *intrinsic images*. These low-level (i.e., non-object based) representations are concerned with *viewer-centred representations* making explicit the depth relationships of visible surfaces.

But the blocks world research described here dealt with just one depth cue (shape-from-contour), whereas the 2.5D sketch integrates the outputs from many other depth cues, for example cues such as shape-from-stereo, -motion, and -texture.

As new work explored those cues the blocks world was left behind. It had proved a stepping stone but not a destination. It was a fine microworld as far as it went for exploring a particular set of clues implicit in line images. However, it was too limited in scope for it to serve as a useful domain for exploring vision in general.

Another limitation of much blocks world research, including that of Waltz, was that it chose to concentrate on the segmentation problem by ignoring the problem of how to get good line and good line junction descriptions (Roberts was an exception.). This was a justifiable interim research strategy but it needs to be borne in mind that Waltz's program would be hopelessly out of its depth if it had anything other than perfect line descriptions as its input.

Further Reading

Marchant J (2008) *Decoding the Heavens: Solving the Mystery of the World's First Computer.* William Heinemann. *Comment* Describes the quest to understand the Antikythera mechanism.

Marr D (1982) *Vision.* Freeman: San Francisco. *Comment* Marr's highly influential book.

Mayhew JEW and Frisby J P (1984) Computer Vision. In *Artificial Intelligence: Tools, Techniques, and Applications.* Eds O'Shea T and Eisenstadt M, pp. 3501–3521. New York: Harper & Row. *Comment* This chapter is largely based on this source. If you want to pursue this topic in more detail it gives full references and suggestions for further reading.

Winston PH (1984) *Artificial Intelligence.* 2nd Edition. Addison-Wesley. *Comment* Excellent on classic work on the line labeling approach to the blocks world.

22

Seeing and Consciousness

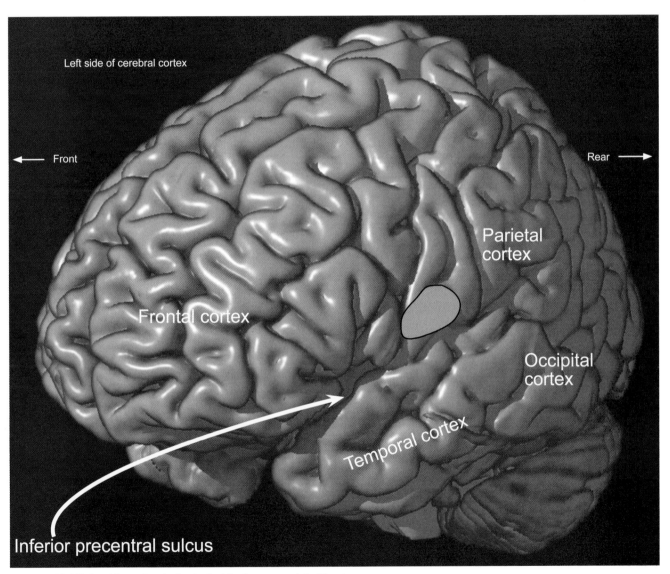

Left side of cerebral cortex

← Front

Rear →

Parietal cortex

Frontal cortex

Occipital cortex

Temporal cortex

Inferior precentral sulcus

22.1 Using binocular rivalry to find brain areas related to awareness
If each eye is presented with a different image (e.g., a dog to one eye, and a house to the other) then there arises a sequence of alternating perceptions, with awareness swapping between percepts based on one or other image. This phenomenon is called **binocular rivalry** and it has been used as a probe in studies trying to track down brain sites associated with awareness. The grey patch in this figure highlights a region of fronto-parietal cortex that has been linked with rivalry alternations. This region is identified only approximately in this figure. See Lumer, Friston, and Rees (1998) for brain images from a functional magnetic resonance imaging (fMRI) study probing brain sites associated with rivalry.

The most obvious thing about our visual world is that we are conscious of it. Open your eyes and there it is. The scene in front of you enters your awareness in all its glory. But what has the science of seeing got to say about conscious visual experience? The answer many would give to this question is blunt: nothing. They think that a famous quotation from Stuart Sutherland, written by him in 1989, still sums up the situation pretty well:

> Consciousness is a fascinating but elusive phenomenon; it is impossible to specify what it is, what it does, or why it evolved. Nothing worth reading has been written about it.

Others think this is a bit unfair. We suggest that it would be kinder to the many scientists and philosophers who have tackled this ancient issue to say, with apologies to Ford Prefect of *Hitchhiker's Guide to the Galaxy* fame: almost nothing.

Certainly not everyone agrees with Sutherland, by any means. There has been a massive recent resurgence over the past decade or so in the study of consciousness, which has led to new journals, new professional associations, and many conferences. This effort has produced at least one benefit: we can take cold comfort from being a little clearer *why* consciousness is such a baffling topic.

Indeed, our main goal in this chapter is not to provide a thorough review of current work on consciousness. Rather, it is to help the reader better understand why consciousness is so baffling, and to provide some pointers into the extensive literature.

Curiously, although perceptual experience is the stock-in-trade of every psychologist who studies perception (except perhaps diehard behaviorists, who would insist that they do not study conscious experience, only perceptual behavior), few psychologists bother much about its conscious aspect. They just get on and study the various phenomena of interest to them. They take the pragmatic view

Why read this chapter?

Consciousness remains the fascinating but baffling problem that it has always been, despite much effort in recent years to use modern imaging methods to explore brain mechanisms that might underpin consciousness. Various accounts of consciousness are reviewed and associated philosophical difficulties are described.

that consciousness remains a great mystery, despite advances in our knowledge of perceptual mechanisms, but precious little can be said about it sensibly so far in terms of scientific theories. So they ignore it, as we have done throughout this book.

Consciousness and Brain Cells

In thinking about this exceedingly thorny topic, the starting point has to be that somehow the whole of our perceived visual world is tucked away in our skulls as an inner representation which stands for the outside world, **22.1**. At first, it is difficult and unnatural to disentangle the "perception of a scene" from the "scene itself" (Ch 1). And it can come as a bit of a shock to the newcomer to the field to realize deeply for the first time how amazing it is that the offal between our ears somehow supports something so astounding as visual awareness.

The only available candidates for generating awareness, brain cells, seem in our present state of knowledge quite insufficient for the task. The "inner screen" theory described in Ch 1 posits a direct relationship between conscious visual experiences on the one hand and activity in certain brain cells on the other. That is, activity in certain cells is somehow accompanied by conscious experience.

Proposing this kind of parallelism between brain-cell activity and visual experience is characteristic of many theories of perceptual brain mechanisms. But is there more to it than this? Can the richness of visual experience really be identified with activity in a few million, or even a few trillion, brain cells? Are brain cells the right kind of entities to provide, somehow, conscious perceptual experience?

Dualism

In facing this question, the "ordinary person" who was pressed in Ch 1 into speculating about perceptual mechanisms and who came up with the "inner screen" theory in reply, might wish to withdraw to what seems safer ground, and say that activity in brain cells cannot provide a completely satisfactory account of perceptual experience.

To do so would be to join a long tradition of philosophical thought which claims that consciousness cannot be wholly and completely identified with the activity of matter, even brain matter, and

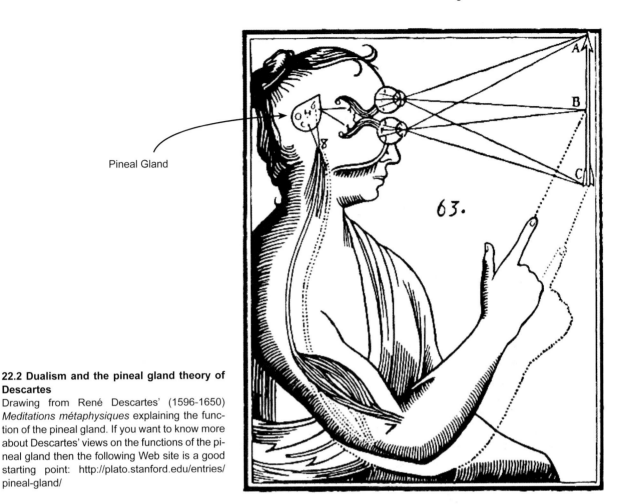

Pineal Gland

22.2 Dualism and the pineal gland theory of Descartes
Drawing from René Descartes' (1596-1650) *Meditations métaphysiques* explaining the function of the pineal gland. If you want to know more about Descartes' views on the functions of the pineal gland then the following Web site is a good starting point: http://plato.stanford.edu/entries/pineal-gland/

that it must be carried by a different "substance" of some kind.

This viewpoint, whose origins lie deep in antiquity but which is usually associated with the great French philosopher René Descartes (1596-1650), is called *dualism* because it proposes two sorts of substance in the universe, mind and matter. These substances are distinct, it is claimed, although somehow they can relate to one another.

But does dualism offer only a name for the dilemma—the apparent difference of mind and matter—rather than an explanation?

When asked: "Who created the Earth?", the answer "Slartibartfast" (who did the job in *Hitch-hiker's Guide to the Galaxy*) immediately invites the question: "And who created Slartibartfast?"

Does a similar logical difficulty obtain with dualism? It asserts a mind/matter difference, but if

that is all it can offer then it has not taken us very far. If mind is not matter, then what is it?

The dualist may have no answer to that question but would nevertheless insist that it is wrong to equate activity in certain brain cells simply and straightforwardly with the conscious experience of the scene observed. The dualist might hold that activity in such cells is *necessary* for the conscious experience, a precursor if you like, but consciousness itself must be quite different from brain matter, even though the unspecified "mind substance" must somehow co-exist with brain cells.

Descartes recognized the need for mind and matter to "speak" to one another and proposed that the site of interaction was a small structure lying in the middle of the brain, called the *pineal gland*, **22.2**. He chose this structure for this important role because it is singular, as he thought

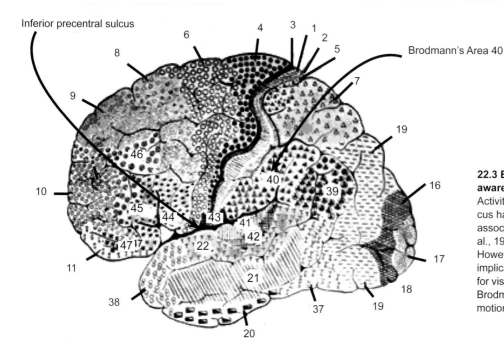

Inferior precentral sulcus

Brodmann's Area 40

22.3 Brain areas associated with visual awarenes
Activity around the inferior precentral sulcus has been identified in many studies as associated with visual awareness (Lumer et al., 1998; Stoerig and Cowey, 2007, 2009). However, many other sites have also been implicated as contributing to or necessary for visual awareness (Gaillard et al., 2009). Brodmann's Area 40 has been linked to the motion aftereffect, as described in the text.

consciousness to be. (Most brain structures are replicated in the brain's two halves. Just as we have left and right legs, so too we have left and right cerebral hemispheres: Chs 1 and 9.) No one now takes this pineal gland theory seriously, but dualism itself is still the usual starting point for discussions of consciousness. This is so even though it is nowadays a view espoused by few philosophers and brain scientists.

Identity Theory

If pressed, most contemporary psychologists and neurophysiologists would agree that they either explicitly or implicitly adopt one or other version of the ***identity theory*** as a working hypothesis. That is, they believe that any given conscious experience is an attribute of activity in one or more brain cells, that there is no need to invoke a special and independent substance to explain mind. This view is sometimes described as rejecting the idea of a ***ghost in the machine***.

Nevertheless, even if it turns out to be correct to assert an identity between conscious experience and some sort of brain cell activity, this leaves many questions unanswered. Is consciousness

associated with all brain cell activity, or just some? Evidence on the latter will be described shortly. It leads naturally to the follow-up question: why is experience related only to some brain cells and not others? What is special about the ones identified with experience?

Another follow-up question is: would finding a special substance possessed only by those brain cells, whose activity is the correlate of consciousness, solve the real mystery—which is that consciousness seems to us so very different from material things?

For example, when we perceive redness, which is an instance of perceived object properties that philosophers call ***qualia***, would it help much to find out that redness is the result of only a certain type of brain cell firing?

To be sure, this would be a very interesting result, a good launching pad for further work perhaps. Even so, it is not obvious that it would lead to a resolution of the matter vs. mind stumbling block.

For example, suppose we were able to say that activity of type X in certain cells is the correlate of the experience of redness, and that of type Y

in those or other cells is the correlate of blueness. That would still leave us bemused about what exactly consciousness is and how it is "caused" by these putative attributes of cells. The apparent arbitrariness at the core of the brain cell identity theory of consciousness (sometimes called the **explanatory gap**) would still remain. It would simply be shifted to a new level of anatomical detail.

Hazard of Essentialism

Whichever way things turns out on this millennia-old question, anyone thinking about the mystery of consciousness from a scientific standpoint needs to be on guard against a conception of science that the philosopher Karl Popper termed **essentialism**. Stanovich neatly summed up this hazard in 1995 as follows:

> Do physicists really know what gravity is? I mean *really*. What is the real meaning of the term *gravity*? What is the underlying essence of it? ... The proper answer to [these] questions is that physicists *don't* know what gravity in this sense is.

And this is so despite physicists having sophisticated equations describing with great accuracy phenomena arising from the gravitational force attracting bodies one to another.

Richard Feynman, the Nobel Prize–winning physicist, was well aware of this hazard. It is worth quoting at length from his sobering account, written in 1963, of the concept of energy:

> There is a fact, or if you wish, a *law*, governing all natural phenomena that are known to date. ... The law is called the *conservation of energy*. It states that there is a certain quantity, which we call energy, that does not change in the manifold changes which nature undergoes. This is a most abstract idea, because it is a mathematical principle; it says there is a numerical quantity which does not change when something happens. It is not a description of a mechanism, or anything concrete; it is just a strange fact that we can calculate some number and when we finish watching nature go through her tricks [such as when a flow of electricity generates heat] and calculate the number again, it is the same ... This number is called energy and it has a large number of different forms, and there is a formula for each one: gravitational energy, kinetic energy, heat energy, elastic energy, electrical energy, chemical energy, radiant energy, nuclear energy, mass energy ... It is important to realize that in physics today we have no knowledge of what energy *is*. We do not have a picture that energy comes in little blobs of a definite amount. It is not that way ... It is an abstract thing in that it does not tell us the mechanism or the reasons for the various formulas.

Feynman's point is that physicists don't "really" know what energy is. One could add they don't even seek an answer to that question (at least, so it seems to us). They get on very well without that "knowledge." The achievements of physicists are often regarded as the very pinnacle of modern science, not without justification. But if they can be content about not knowing what energy is then perhaps we should be content with a similar situation regarding consciousness. The "essence" of consciousness might ever elude us; indeed, expecting one day to discover its "essence" might be a mistake because there may be no "essence" to be found, any more than there may not be an "essence" of energy.

Picking a Tractable Sub-Problem

An antidote to the hazard of essentialism is Peter Medawar's famous dictum, which he proposed in 1967: *science is the art of the soluble*. One such solvable problem may be finding out which parts of the brain are specially related to consciousness, with the latter being defined for this purpose as synonymous with awareness.

Taylor concluded in 2000, from various lines of evidence, that a brain region called the *inferior parietal lobe* is "the essential site for consciousness." One intriguing experiment that led him to this conclusion was his brain imaging study providing evidence that a region known as Brodmann's Area 40 in the inferior parietal lobe is active during awareness of the *motion aftereffect*, Ch 8.

This illusion is the awareness of visual motion that follows prolonged exposure to a stimulus moving in just one direction. That is, the motion aftereffect is an illusory movement of stationary objects in a direction opposite to that seen during the prolonged exposure.

This effect can be seen after gazing at a waterfall for 30 seconds or so, after which if gaze is transferred to, say, a nearby set of stationary rocks then these appear to move upward.

Taylor found that a network of different areas seems to be involved in the motion aftereffect. However, he interpreted his evidence overall as suggesting that Brodmann's area 40 in the inferior parietal lobe on the right hand side of the brain may be the active site when just the motion aftereffect is experienced, with scene, head and eyes all held stationary, **22.3**.

Taylor's work on the brain site for the motion aftereffect is but one of a large number of studies pursuing the possibly tractable problem of finding the neural correlate of consciousness. Francis Crick, famous Nobel Prize winner for his co-discovery with James Watson of the DNA genetic code, has been particularly active in promoting this line of enquiry, along with his colleague Kristof Koch (Crick, 1994; Crick and Koch, 2003).

One approach to finding brain sites specifically associated with awareness use the phenomenon of ***binocular rivalry***. The study by Lumer et al. (1998) of this kind is explained briefly in the legend to **22.1**. They found a particular region in fronto-parietal cortex linked to awareness. In contrast, Gaillard et al., (2009, p.472) concluded from intracranial electroencephalogram recordings taken during conscious and nonconscious processing of briefly flashed words that there existed a "distributed state of conscious processing."

The next few years may reveal whether seeking brain sites linked to consciousness proves to be a tractable problem. Many are doubtful. For example, a skeptical view of the value of studying brain mechanisms underlying consciousness is held by the neurophysiologist David Hubel:

> GH Hardy is supposed to have said that a mathematician is someone who not only does not know what he is talking about, but also does not care. Those who discuss in depth subjects such as the physiology of the mind probably care, but I cannot see how they could possibly know.

[Hardy is here referring to pure mathematicians who create abstract systems of symbols which are not intended to map on to any physical objects.]

Seeking neurophysiological correlates of consciousness is but one active strand in the resurgence of interest in consciousness over the last decade. There is now a professional society for its study, that brings together philosophers, psychologists and neuroscientists interested in consciousness.

Blindsight

One phenomenon that has become famous for its bearing on the question of the neural correlates of awareness is a condition dubbed by Larry Weiskrantz and Elizabeth Warrington in 1973 as ***blindsight*** (see *Furher Reading*). Patients who suffer a serious brain injury to the *striate cortex*, a region at the back of the brain which is the receiving site for many fibers in the optic nerve (Ch 1), are usually rendered blind. Even so, when these patients are induced to point to a spot of light that is lit up somewhere in front of them, then they often point in the right place. They do so even though they protest that they find this a strange thing for them to be asked to do as they say they can't see anything.

Blindsight seems to be possible because some of the fibers from the eyes go to brain regions other than the striate cortex, and these other regions use this retinal information for such things as directing gaze. The phenomenon of blindsight may reflect the continuing operation of these undamaged parts of the brain, which thus appear to work without any associated visual awareness. If this is so, then it is a case of an image-processing brain region whose activity is *not* a neural correlate of consciousness.

Alan Cowey, an eminent neurophysiologist, has done a lot of fine work over many years on blindsight. A sample of his papers are cited in *Further Reading* if you want to find out about recent developments in this field. One important point he makes is that blindsight is not a case of "normal vision without awareness." He has demonstrated that the visual systems of blindsight patients are in a very sorry state and are far from operating normally. For example, their optic tracts show deterioration and they lose many of their retinal ganglion cells of the P type (see Chs 6, 8, and 9). Another caution that Cowey and others urges is that blindsight has nothing to say about the philosophical problems of qualia.

One complexity to which Cowey has drawn attention is that there appear to be two types of blindsight: no awareness at all and awareness of "something has happened but no idea what." Also, blindsight patients can in some circumstances experience bright flashes (called *phosphenes*) induced by transcranial magnetic stimulation. This shows that they have not completely lost the ability for visual experiences.

The upshot of these various considerations is: blindsight is a complex phenomenon mediated by a far from normal visual system and its interpretation is accordingly also complex.

Functions of Visual Awareness

Blindsight brings to the fore a deceptively simple question about awareness: why do we have it? If some brain cells can use visual information for various tasks without awareness, why are we visually conscious at all? Why are we not visual zombies? That is, why aren't we biological machines that can processes images without any awareness being associated with the outcomes—just as seems to happen in patients with blindsight, at least for the simple task of pointing to a target?

Richard Gregory has suggested that the function of visual awareness could be to serve as a code for the "visual here and now" (Gregory, 2009). That is, visual awareness may be the brain's way of coding the visual world that exists at any given moment, as distinct from memories of past events. Of course, as Gregory is well aware, this does not give a scientific lever on the deep mystery of qualia, but it is an intriguing idea about the function of our awareness of the visual world.

Others have suggested that consciousness is associated with long-term planning aimed at achieving long-term goals. One line of evidence that has prompted this idea is that action can precede consciousness of that action. This is strange indeed: it suggests a monitoring function for consciousness rather than consciousness being (always) the precursor of actions. (For a discussion of the problem of correlation and causality in interpreting neurophysiological data, see p. 350.)

Another strange finding that may have a bearing on this issue arises from brain scans taken of patients in deep coma. If they are asked, despite being apparently completely unconscious, to imagine, say, playing tennis, then their brain scans reveal areas that "light up" in a way similar to what happens in the brains of normal conscious people when asked to imagine playing tennis. This is surprising indeed. Is it that the coma patients are after

all "conscious" but just cannot talk to us? Or is this a case of the brain engaging in processing without associated consciousness, much as happens, say, for brain sites controlling heart rate?

Consciousness and Attention

Imagine yourself being asked to view a video of a basketball match and count the number of passes made by one of the teams. This task demands careful attention to what is going on. Then imagine being asked afterwards if you had seen a gorilla among the players. Typically people answer this question by saying something like: "A gorilla? No, of course not. It was a normal basketball match." They are then astonished when, given an opportunity to view the basketball video again, they see the gorilla.

This phenomenon might have a parallel in road traffic accidents in which cyclists are knocked down by cars. Car drivers who have done this sometimes say: "I was concentrating hard on my driving but I just didn't see the cyclist." If cyclists are not a commonplace event on the road in question then car drivers are not expecting them and hence they are less likely to enter awareness.

The unseen gorilla demonstration highlights the fact that what enters our visual awareness is governed by a selective process which is given the name **attention**. William James, the famous psychologist, famously wrote in 1890 in his great work *Principles of Psychology*:

> Everyone knows what attention is. It is the taking possession by the mind, in clear and vivid form, of one out of what seem several simultaneously possible objects or trains of thought. Focalization, concentration, of consciousness are of its essence. It implies withdrawal from some things in order to deal effectively with others, and is a condition which has a real opposite in the confused, dazed, scatterbrained state which in French is called distraction, and *Zerstreutheit* in German.

Indeed. And yet, a century after these words were written, we have little understanding of the brain processes involved.

See Stoerig (2006) and Stoerig and Cowey (2006, 2007) for reviews of recent neurophysiological studies exploring attention and awareness in a blindsight patient.

Perceptual World of the Bat

Some philosophical discussions of the problem of consciousness have taken as their starting point the following thought experiment, which is quite good fun to muse about if you have a philosophical bent.

Consider what the perceptual world of a bat might be like as it flies around at night relying on its echo location to detect obstacles in the darkness. Assume for the sake of this experiment that bats are conscious creatures, so they have an awareness of their surroundings. What might this awareness be like?

This scenario has a famous role in discussions of the qualia problem, following its introduction by Nagel in 1979 to discuss certain philosophical issues. He thought that anyone who has seen a bat flying around in a room is instantly struck by its fundamentally alien nature, so much so, he argued, that we could not possibly know what it is like to be a bat. We will not go into that debate here; if you are interested see Nagel (1979).

For our present purposes we ask you to take a different tack: try entertaining the idea that the bat's perceptual world might be similar in some important respects to our own *visually* perceived world. Of course, it would be surprising if it were not hugely different in lots of ways. Sonar (the location of objects by echoes) might make available object attributes that vision fails to do nearly as well, e.g., whether a surface is hard or soft, but be incapable giving evidence about surface colors.

Even so, might bats experience a 3D world akin to our own visual world in some ways, a world of objects to avoid, their spatial layouts and some of their surface properties? And all this while flying around in the dark. Might they even use qualia akin to our color perceptions to code for the quite different object property of hard vs. soft surfaces?

Presumably we will never know. But the point of this thought experiment is to ponder that the bat's perceptual world *could* be like our own visual world in some fundamental ways. There appears to be no obvious logical objection to this idea despite their input devices, their ears, not initiating the perception of their world with anything akin to retinal images. This intriguing proposition may help convince you that the representation that constitutes our visual world may, in a deep sense, be only arbitrarily related to "eyes as cameras."

Seeing, Consciousness, and Computers

What is the prospect of vision research culminating in the building of a seeing machine able match human visual performance? And would such a machine be a zombie or have visual awareness?

Some regard the task of building a seeing machine able to match human vision as so difficult as to be centuries away, perhaps impossible altogether. For them, neural tissue, with its tiny components and richly interconnected networks, is the only material up to the job. But it needs to be remembered that we have not yet had much more than five decades of developing the modern electronic computer, and even less of what could properly be called an attempt to build a machine-vision system.

To be sure, progress has not been as rapid as originally hoped. It was thought by some of the pioneers of artificial intelligence working in the 1960s that the "vision problem" would be solved after a few months of research. We know better now although there are some distinguished computer scientists who anticipate a solution to the "vision problem" is not that far off. We are not so optimistic. In any event, everyone working in the field recognises that the problems of mimicking all aspects of mind, of creating artificial intelligence, have turned out to be much more difficult than at first suspected by many enthusiasts, and this bitter experience has substantially moderated the optimism of many.

Yet we are constantly witnessing the advent of new computer technology and perhaps something will turn up that makes a big difference. One hardware development that has offered hope to some has been the advent of the "distributed processor." This is a computing system composed of many more or less independent processing sub-units, each able to get on with its own job, but also able to "talk" to one another and to other parts of the machine. This looks to them like the true dawn of genuinely **parallel** processing, hitherto the province par excellence only of biological brains.

In the simplest desktop computers, there is just one "central processor," a single device through which all parts of a computation have to proceed, one by one, in so-called *serial* mode: any particular piece of computation has to wait upon the completion of the one before it, or else interrupt it.

A serial device can, awkwardly and relatively slowly, perform certain computations that ideally require parallel processing, for example by using *iteration*. This is a computing procedure in which a solution is gradually approached in a sequence of steps in which a collection of computations is performed by a single processor. An example is the Marr/Poggio stereo algorithm described in Ch 13. Iteration of this type can be slow and clumsy but now there are parallel-processing computers composed of myriad micro-processors, each one "doing its own thing" at any given moment.

The distributed processor computer architecture is reminiscent of the hypercolumns of the visual cortex. Each hypercolumn "looks" at its own particular part of the retinal image and processes whatever information it finds there, thus contributing its bit to the feature description of the entire scene. All hypercolumns work in parallel, although they probably "speak" to one another, as neighbors, during figure/ground segmentation, just as distributed microprocessors can be made to do. Hypercolumns feed other brain sites (other microprocessors), and so on, in one extensive parallel-processing network which culminates in the symbolic scene description which constitutes sight.

Thus, with the development of the distributed processing computer, we are in a better position to implement seeing theories able to process natural images in "real time" (computer jargon for dealing with inputs as they actually come about in a real-world setting, rather than on the basis of a memory of the inputs stored at the time of their original production and then used subsequently). This parallel computer machinery is also valuable for those interested in simulating parallel-processing networks of the kind found in biological vision.

We turn now to the second question posed at the beginning of this section. Suppose a seeing machine was capable of matching human vision, in the sense that if confronted with a natural scene it could print out on its typewriter, or "speak" on some other output device, an explicit description of what the scene contains. Would such a machine have conscious visual experience?

Presumably we will never know. After all, how do you know that another person is conscious? You may think this is a reasonable belief, because other people say they are and because they are built rather like you. Those factors may make you content to extrapolate from knowledge of your own conscious experience and accept that others also share this attribute. But such an argument would not satisfy a determined skeptic, as debate between philosophers throughout the ages on this time-honored question testifies.

It seems therefore that we will have to remain as agnostic about machine consciousness as we are about human consciousness. All we can be sure of, given a perceptual ability which matches that of humans, is that machines will *appear* to be conscious. Already computer programmers are typically anthropomorphic in their everyday dealings with their machines. They use phrases like "It's thinking hard right now," "It got confused then," "It's suffering an illusion," "It thought I wanted x but in fact I wanted y," etc. This is the natural way to talk about clever machines, just as it is the natural way we talk about clever animals.

Perhaps anthropomorphism will become much more marked, even rather spooky ("Is there somebody in there?") when we start to deal with truly high-class seeing machines. Indeed, we suspect that clever perceiving machines will prove far more impressive than, say, clever chess-playing ones. (Although it is probable that the development of the latter will hinge upon the former, in that to play good chess requires "seeing" in the positions of the pieces certain possibilities, a form of thinking closer to vision than perhaps we normally acknowledge, despite the frequent use of the word "seeing" to refer to understanding, for example : "Ah! I see what you mean.")

In fact, as the pursuit of artificial intelligence proceeds, perhaps we will have to adjust our notions about the nature of man, just as the Victorians had to adjust theirs in the face of Darwin's theory of evolution. "Man as an animal? Rubbish!" was the all too common, but also very understandable, reaction to Darwin's ideas.

Today a similar response is: "Man as a machine? Ridiculous!", quickly followed by remarks revealing some sadly ignorant claims: "Machines can't think," "Computers are no more than large, electronic arithmetic calculators," "Machines do only what they are told to do," and so on. On the one hand, these claims raise large philosophical, such as *What is thinking?* And on the other, they betray a lack of awareness about the capabilities of present-day sophisticated computers. Machines are not nearly so limited as these critics claim, and this fact is not widely recognised.

Even science fiction and the television space sagas have not yet convinced many people that it is sensible and proper to consider man as a type of machine, albeit a very special kind of machine (just as man is a very special kind of animal). Perhaps this is because a common science fiction theme is that "Man is more than a robot," Often, at the last moment, the plot is resolved not by the appearance of a *deus ex machina* but, so to speak, by a *homo ex machina*. Man's body is machine-like, yes, but his intellect—no! His mind has some special non-machine-like quality which saves the day.

(*Deus ex machina* is literally translated as "god from the machine". Nowadays it refers to a plot device in which a person or thing appears "out of the blue" to help a character to overcome a seemingly insolvable difficulty.)

This was true of *Star Wars*, the enormously successful space classic, where the hero (human, of course) won the day by switching off his computer and then beating the enemy with his own bare hands—or, perhaps one should say, his own bare mind. Indeed, the dramatic success of this film might in part be ascribed to the fact that it panders to man's desire to see himself as a cut above computers and robots, just as he craved to see himself as a cut above animals. Will this view fade as people become aware that distinction between humans and machines may be more apparent than real?

Machines need not be rigid slaves bound for ever to follow their instructions to the letter, if this means showing no creativity, no learning, no adaptability, no perceiving. Already we know how to instruct computers to exhibit these qualities in certain situations. But we suspect the full consequences of artificial intelligence will not become manifest to non-experts until the problem of seeing is solved. Only when the computer is given its own eyes, and its own capacity for explicit symbolic scene description, will it reveal its true potential for life-like action.

Meanwhile, our view is that extreme caution is the order of the day. As noted earlier, some pioneers of artificial intelligence thought that solving the vision problem would take only a few months. Harsh experience has dramatically tempered that naïve optimism. In fact, despite spending our whole working lives studying seeing, we are fully aware of how woefully ignorant we are about exactly how it works (see our final chapter).

Summary

To sum up this brief survey on consciousness and seeing, visual awareness remains as mysterious now as it has ever been. On the other hand, we may see progress soon on some tractable questions about the neural correlates of consciousness. New methodologies, such as brain imaging, may provide the tools needed to find these elusive neural correlates. Meanwhile, the unresolved mysteries of consciousness do not stop visual scientists studying vision to good effect, as we hope this book has demonstrated.

We end this chapter with a famous quotation of Ludwig Wittgenstein, written in his *Tractatus Logico Philosophicus* in 1922. Although there are many who would disagree, it seems to us that it fits well the problem of qualia: "Whereof one cannot speak, thereof one must be silent."

Meanwhile, to lighten things a little: *Consciousness: that annoying time in between naps.*

Further Reading

Blackmore S (2003) *Consciousness : An Introduction.* UK Edition Published by Hodder and Stoughton London. *Comment* An excellent starting point for getting into the large literature on consciousness studies.

Blackmore S (2003) *Conversations on Consciousness* Oxford University Press. *Comment* A compendium of 20 interviews conducted with major figures in the field of consciousness studies. A useful way to become acquainted with the diverse outlooks of contemporary leaders in the field.

Blakemore C and Greenfield S (1987) *Mindwaves: Thoughts on Intelligence, Identity and Consciousness.* Blackwell Publishing Limited.

Chalmers D (Ed) and Bourget D (2009) *Online Papers on Consciousness. Comment* A directory of 5326 free online papers on consciousness in philosophy and in science, and of related topics in the philosophy of mind. http://consc.net/online

Crick F (1994) *The Astonishing Hypothesis* Simon and Schuster. *Comment* Sets out Crick's approach to pursuing the neurophysiological correlates of consciousness.

Crick F and Koch C (2003) A framework for consciousness. *Nature Neuroscience* **6** 119-126.

Feynman RP (1998) *Six Easy Pieces* Penguin Books. *Comment* A selection of lecture notes from their first publication in 1963 by California Institute of Technology. The quotation in the text is composed from pages 69–72 in Ch 4 on the Conservation of Energy.

Gaillard R, Dehaaene S, Adam C, Clémenenceau, S, Hasboun D, Baulac M, Cohen L, and Naccache L (2009) Converging intracranial markers of conscious access. *PLOS Biology* **7**(3) 472–491.

Gregory RL (2009) *Seeing Seeing Through Illusions: Making Sense of the Senses.* Oxford University Press, Oxford. *Comment* Among much else, Gregory proposes that the functional role of visual awareness is to encode the "here and now."

Gregory R (2005) Musing on Blakemore's Paradox *Perception* **34**(9) 1043–1167. *Comment* A short lively discussion of this paradox, which is: if every

conscious mental event is given by some corresponding physical neural event, there is nothing that awareness could add to physical neural causes. So why, if of no causal use for survival, should consciousness have evolved by natural selection? What would consciousness add to the survival skills of a zombie?

Humphreys N (2006) *Seeing Red* Belknap Press of Harvard University. *Comment* Nick Humphreys has fearlessly pursued the study of consciousness over many years, in the face of much skepticism about the worth of so doing. This engaging short book summarizes his unusual outlook.

Lumer ED Friston KJ and Rees G (1998) Neural correlates of perceptual rivalry in the human brain *Science* **280** 1931–34. *Comment* The source of 2.1.

Medawar P (1967) *The Art of the Soluble.* London: Methuen.

Nagel T (1979) What is it Like to be a Bat? *Mortal Questions* Cambridge University Press: UK.

Popper K (1945) *The Open Society and its Enemies.* Routledge: London.

Rees G (2005) *Decoding Consciousness* The *Francis Crick Lecture* presented at the Royal Society of London, 5 December 2007. *Comment* Discusses recent brain studies linked to consciousness. See http://royalsociety.org/page.asp?id=8303

Schurger A, Cowey A, Cohen JD, Treisman A and Tallon-Baudry C (2008) Distinct and independent correlates of attention and awareness in a hemi-anopic patient. *Neuropsychologia* **46** 2189–2197. *Comment* Reports an experiment exploring the intimately related concepts of attention and awareness in which the subject is the blindsight patient, referred to with the initials GY.

Stanovich K (2009) *How to Think Straight about Psychology.* 9th Ed. Allyn and Bacon.

Stoerig P (2006) Blindsight, conscious vision, and the role of the primary visual cortex. *Progress in Brain Research* **155** 217–234. *Comment* An excellent and detailed review of blindsight research.

Stoerig P and Cowey A (2007) Blindsight: Quick Guide. *Current Biology* **27**(19) 822–824. *Comment* An excellent starting point for getting into the blindsight literature.

Stoerig P and Cowey A (2009) Blindsight. In: T Baynes, A Cleermans and P Wilken (Eds) *Oxford Companion to Consciousness*. Oxford University Press, Oxford. Pages 112–116.

Sutherland S (1989) *International Dictionary of Psychology*. London: Crossroad.

Taylor JG (2003) Neural models of consciousness. In Arbib MA (Ed) *The Handbook of Brain Theory and Neural Networks*. 2nd Ed. Pages 263–267. *Comment* A short review.

Taylor JG, Scmitz N, Ziemons K, Greoss-Ruyken ML, Mueller-Gartner HW and Shah NJ (2000) The network of areas involved in the motion after-effect. *Neuroimage* **11** 257–270.

Weiskrantz L (1986) *Blindsight*. Oxford: Clarendon. *Comment* A classic work summarizing early research on blindsight.

Wittgenstein L (1922) *Tractatus Logico-Philosophicus*. Logisch-Philosophische Abhandlung.

Seeing Summarized

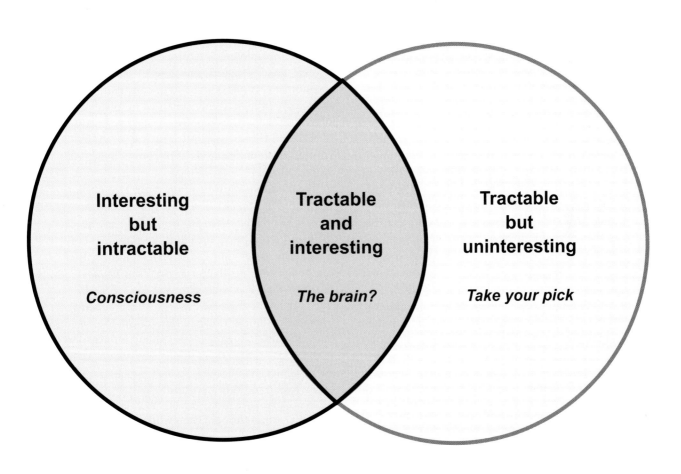

23.1 Space of all possible problems
These three classes are discussed *inter alia* in this chapter.

There is a story recounted by Nick Humphrey about an Oxford philosopher who gave a course of lectures on *What do we see?* He started off with the idea that we see colors, but he gave that up in the third week and instead argued that we see things. But that did not work out well either, and by the end of the course he admitted ruefully: "I'm darned if I know what we do see."

Has this book done any better? It began with a not entirely dissimilar question to the one chosen by this hapless philosopher. Our question was: what goes on inside our heads when we see? An imaginary "ordinary person" produced his photographic or "inner screen" theory in reply. We hope the general reader or the beginning student for whom this book is intended could now have a better shot at it, perhaps answering along the following lines:

Seeing is a matter of building up a *representation of the scene* observed. The job of this representation is to make a variety of scene attributes *explicit*, by which is meant immediately available for subsequent processes (i.e., no need for further work), be they further visual analyses or processes guiding actions.

The photographic analogy is no good because a photograph simply replaces one image with another, leaving all the information about the scene implicit within the new image. That is hopeless. When we see things, we are engaged in a process of recognition of features, objects, colors, movements, and other attributes of a scene. Seeing enables us to point to things, to pick them up, to talk about them, in a word to *act* in relation to them. Seeing does so because it makes explicit what the visual scene contains in a description cast in a language of *symbols*.

Seeing must be a symbolic process because the world itself obviously does not exist inside our heads, and so our "internal" visual world must be a collection of symbols standing for the scene and its various attributes. Uncertainty surrounds the nature of the brain's symbols for seeing. Various theories exist and little that is definite can be said at this stage.

What we can be sure about is that arriving at an explicit scene description is not a straightforward business. Each eye's *retinal image* contains useful data about scene but it is generally *ambiguous*. Various *measurements* are taken from these images, and these are *interpreted* to give the required representations of attributes of the scene. This process of

interpretation is critically dependent on exploiting *constraints* that resolve the ambiguities.

We will now develop various issues arising from this summary.

Constraints for Interpreting Image Data

Sometimes useful constraints can be found from geometrical analyses of image formation. Our examination of stereo geometry in Chs 18 and 19 provides an example of this kind. Another example is analyzing the nature of texture cues available in images for obtaining shape from texture, Ch 2.

Other constraints are generally valid assumptions about the nature of the objects and surfaces in our visual world. An example of this kind underpins the use of using various figural grouping principles when linking edge segments. Those principles are founded on the constraint that our world is generally made up of extended objects rather than isolated points. Hence figural grouping can legitimately exploit various properties about object borders, textures, colors, and so on.

Another example of this sort of constraint is the surface smoothness assumption used in solving the stereo correspondence problem. Good constraints of this kind have presumably been discovered by natural selection during the evolution of biological vision systems.

The constraints just mentioned can be embedded (exploited) in *low-level* processes running in a *bottom-up* or *data driven fashion* without any need for *high-level* information about the particular objects in the scene. That said, high level information can be valuable, in which case the terms *top-down* or *conceptually driven* visual processing are used. The human visual system has many fibers sending information up the visual pathways from the eyes but it has also has many fibers sending information in the opposite direction (at least as far down as the lateral geniculate nucleus). Perhaps the latter fibers play a role in top-down processing.

Marr's Computational Framework

What sort of answer will ultimately come to be seen as a "full" explanation of seeing? We simply do not know, despite having spent most of our academic lives studying seeing.

That said, in our opinion the current best bet for a framework for guiding vision research has been provided by David Marr. His *computational approach* has been the linking theme throughout this book. It is worth reiterating here its main features. Marr argued that it is important to distinguish three different **levels** when analyzing complex information processing systems:

Computational theory

What is the nature of the problem to solved, what is the goal of the computation, why is it appropriate, and what is the logic of the strategy by which it can be carried out? Identifying constraints for solving the problem is a key feature of this level.

Representation and algorithm

How can the computational theory be implemented? In particular, what is the representation for the input and output, and what is the algorithm for the transformation from input to output? In other words, how are the constraints to be put to work in an algorithm?

Hardware implementation

How can the representation and the algorithm be realized physically? It may help to add that an algorithm is a specification of a sequence of steps that must be carried out.

It may help to add that an algorithm is a definite method, a specification of a sequence of steps that must be carried out. However, the steps specified may well not be realized in the hardware in a serial manner (see later comments).

Similar conceptual frameworks to Marr's existed before he began his research, for example in the field of **cybernetics** (usually defined as the interdisciplinary study of the structure of regulatory systems, and hence closely related to control theory and systems theory).

Perhaps the reason that Marr's name is particularly associated with this approach is because he and his colleagues in the Massachusetts Institute of Technology published a stream of studies in the 1970s and 1980s showing the approach at work in interpreting *natural* images. They were able to do this by exploiting powerful computing resources of a standard not generally available at that time.

The term *computational theory* is an unfortunate label for the top level of analysis because some have read it as a claim that the brain works in a similar way to a digital computer. Nothing could be further from the truth. The brain may be implementing the same constraints to solve a task as a suitably programmed digital computer (i.e., using the same

computational level theory) but its operations at the hardware level may well be radically different, and usually are. Marr chose the name *computational theory* because he regarded seeing as an *information processing problem* and that term leads naturally to the label *computational*.

Also, be careful to avoid the temptation of concluding that Marr's algorithm level implies that the brain performs visual processing by executing a series of steps. Any one who has programmed a digital computer knows that programming languages are designed to allow the user to perform a sequence of operations. But many regard this as a most unlikely model for how the neurons of the brain work.

For Marr, the algorithm level exists to clarify different possible ways of exploiting the constraints identified at the computational level. Specifying different possible algorithms for exploiting a given set of constraints is an altogether different matter from implementing a given algorithm using neurons.

Consider, for example, the interpretation of bipolar cells in the retina as implementing a particular algorithm for measuring image intensity gradients. These bipolar cells are not thereby held to be executing as a sequence of steps in a *serial* manner the operations specified in the Marr/Hildreth algorithm in Ch 5.

Rather, they are regarded as a particular species of "hardware" that implements the algorithm using a receptive field comprised of excitatory and inhibitory regions working in a *parallel* fashion. The conclusion here is: beware of confusing the different types of concerns dealt with at each of Marr's levels.

We need to state a further cautionary note at this point. Marr distinguished between two sorts of problems. He called the first *Type 1*: these are problems for which a theoretical account is attainable in principle, and which are therefore amenable to his kind of three-levels analysis.

The second, *Type 2*, are problems for which all that can be achieved is the description of the interplay of many interacting factors. Marr's example was the way a protein molecule folds upon itself. He speculated that one could do no better than describe the interplay of the various intra-molecular forces that determine the molecule's final shape. Whether this remains a good example given recent developments in predicting protein shapes from knowing their constituent parts is a moot point. However, it remains possible that some problems may be of Type 2, and therefore simply not amenable to theoretical analysis.

Whether the general problem of brain function, and the specific problem of seeing, are Type 2 problems remains to be seen. As far as we are aware, there is at present no way of deciding in advance whether a problem is of Type 1 or Type 2. Marr argued that the speed and fluency of seeing suggests that it is founded upon constraints that can be discovered by computational analyses. Hence, he thought it was likely to be a Type 1 problem. This still seems a reasonable point of view, insofar as we have been able to select for this book at least some sub-problems within vision that have yielded to a Type 1 analysis. On the other hand, it is not inconceivable that some Type 2 problems could be solved quickly by the brain using its massively parallel neural networks.

We now turn to summarizing different approaches to studying seeing. After that, we discuss how Marr's approach has fared since his untimely death in 1980 aged 35 years from leukemia.

Computer Vision Approach

Trying to build a computer vision system has advantage that it forces attention to the fundamental processing requirements of a given visual processing task. Usually the "first shots" tend to misfire badly. This throws the builders back to thinking hard about what *exactly* the problem is that has to be solved, and then what computational strategies might solve it.

It is at this point that knowledge of the psychology and physiology of seeing can be valuable for suggesting ideas. But it is also true that many applied workers in the area, trying to invent machine vision for industrial applications (for which they can control rather precisely the visual inputs), can adopt *ad hoc* solutions.

Examples are the specially designed numbers on cheques, or the pattern recognition methods used in iris recognition systems. These are adequate for their particular problem, but they may cast little light on what goes on inside biological visual systems. The latter have to make do with *natural* images of natural scenes, given to them by some-

times low-quality eyes, and so they have developed a sophisticated range of strategies to cope with their problematic inputs.

It is thus important not to confuse Marr's computational framework with "computer vision." Marr was highly critical of some studies in the emerging field of artificial intelligence that used *ad hoc* approaches when trying to develop computer vision systems (see Ch 21 on some of the early *blocks world* studies). The rationale underlying some of those systems was far from clear. That is, in Marr's opinion they lacked a decent computational theory, and this was why the systems he was criticising worked so poorly.

Psychophysical Approach

The psychophysicist provides methodologies for studying the input-output (retinal-images-to-perceptions) performance of the best visual systems presently known, those of human and other animals.

Often, the phenomena chosen for study are illusions, each illusion being regarded as the result of a misapplication of an interpretative stratagem (low-level or high-level) to a scene. As Richard Gregory has argued, illusions thus provide clues about the constraints which the visual system has found it profitable to employ in general for interpreting retinal images. Gregory often uses the terms *strategies* in a way analogous to Marr's use of the term constraints.

The value of psychophysics is thus obvious. However, there is a danger that the delights, indeed the beauty, of many visual phenomena can lead to studies which examine them only for their own sake. Doing so misses the opportunities they provide for developing computational theories about how vision tasks can be solved.

Another hazard facing users of the psychophysical approach is to regard a visual phenomenon as "explained" if it can be linked to the operation of certain neurons.

Such links are fine as far as they go. For example, the use of various aftereffects for probing possible neural mechanisms is one of the jewels in the crown of vision research. However, such links are not in themselves sufficient for a complete explanation of visual phenomena, as they lack the insights conveyed by computational theories as to *why* the neural mechanisms might exist.

To illustrate the latter point, when spatial frequency tuned channels were first being investigated a speaker at a conference was asked what they were for. He replied, jokingly: "*No idea really; perhaps they are detectors of radiators or picket fences.*" A well-argued account of their functional role as image intensity gradient measuring devices awaited Marr and Hildreth's work in 1976 on the computation of the raw primal sketch.

Neurophysiological and Neuroanatomical Approaches

These approaches utilize techniques for studying the constituent parts of biological visual systems directly, for example, using microelectrode recordings allied to neuroanatomical investigations. These kinds of studies provide information on the biological "nuts and bolts" of seeing, as it were, and consequently great strides forward have been taken in neuroscience over the past five decades.

But again we emphasize that it is not enough to study the components of the brain: we need also to understand the tasks that neurons in the brain are carrying out. It is worth repeating Marr's famous quotation, stated in our first chapter:

> Trying to understand perception by studying only neurons is like trying to understand bird flight by studying only feathers: it just cannot be done. In order to study bird flight we have to understand aerodynamics: only then do the structure of the feathers and the different shapes of bird wings make sense.

So it is for neurons: we need to understand their properties in terms of the tasks they are performing, and that entails devising good computational theories.

Brain Imaging

The last two decades have seen the emergence of the massively burgeoning research field exploiting techniques for **brain imaging**. It is too early to judge whether this kind of work (in which both authors of this book have engaged) will lead to truly deep insights about the brain. It is easier right now to see some of its limitations and hazards.

In particular, when a given brain region is shown to "light up" under a given set of stimu-

lus or task conditions, we are not much further forward in understanding how the brain works. Knowing where a task is executed in the brain is no substitute for asking *what is the nature of the task and how it can be solved in principle.*

Moreover, in the case of brain imaging studies of visual mechanisms, it seems to us that rather little of fundamental importance has been added so far to the brain mapping done with conventional microelectrodes. Perhaps this situation will change as brain imaging techniques advance in terms of their spatial and temporal resolution. For us, the jury is still out on brain imaging.

Modeling Approach

If we really understand how a system works then we should be able to build a working model of that system. This means implementing it in a suitable man-made computing system. When this is done, it becomes evident that it is not sufficient, for example, to declare that the firing of one cell inhibits another cell. We need to know precisely how much the firing of one cell inhibits the firing of another. Such knowledge can only be expressed mathematically. In other words, the computational approach demands a mathematical account of the functional relationships between the different components of a system.

Hence, the ideal end point of the computational approach is the building of working models. This implies a particularly rigorous form of understanding, one that is not tested adequately by asking another theorist if a theory makes sense. Rather, it is necessary to ask if the physical world if it can support a working model which instantiates the theory. In other words, if we claim to understand how seeing works, then we should not explain the theory only in words. We also need to write down the equations which give a precise account of the theory, and then build a model based on those equations.

But, as usual, there is an accompanying hazard, in this case one that is sometimes called informally *growing mathematical hair.* That is, it is important to avoid writing down equations without grounding those equations in a very clear task analysis that identifies the constraints being exploited. Without this essential step, it is too easy to invent equations whose value is difficult to discern. This is why it is good methodology to devise computational theories underpinned by the insights arising from experiments in psychophysics, neuroscience and computer vision.

In fact, the diverse contents of this book illustrates that the problem of seeing is best tackled by a combined assault using psychophysical, neurophysiological, and computational methods in unison. This does not mean that it is necessary to be a jack of all trades to study vision, but it does imply that a familiarity with all three literatures is important, and that a team approach is probably the best way forward.

Image Statistics Approach

There has been a recent emergence of studies aimed at exploiting the image statistics of natural images to model various visual phenomena. This work can be viewed as a modern version of a long-standing way of thinking about perception, called the ***empiricist approach.*** For example, Purves and colleagues have recently published several studies aimed at using the statistics of images (e.g., distributions of pixel intensities, or of orientations of edge segments) in simulations of various visual effects, such as contrast illusions.

The next decade should decide whether this approach proves to be a major new development in how to study seeing. It is certainly interesting and it will be worth keeping a look-out for results. Our concern at present is that even if a successful simulation is achieved it can remain unclear what is really going on at the computational theory level. In other words, it seems a hazard with this approach that it becomes an exercise in algorithm development, leaving open the question: if the algorithm works, *why* does it work? What exactly are the constraints being picked up in the image statistics? We do not regard it as very satisfying simply to state that regularities in image statistics capture the "way the world behaves." That is fine as far as it goes, but it leaves open the question: what are the underlying causes of the regularities?

In the case of the image statistics of oriented edges (Ch 4) it is easy to see the underlying causes of the peaks that occur for vertical and horizontal edges. However, we asked in Ch 16 when Purves et al. used image statistics to obtain lightness values: are those statistics determined by illumination

gradients being typically shallow compared with reflectance boundaries, as exploited in the Land-Horn theory, or could there by some other reason? This strikes us as an interesting question to ask. It is left unanswered simply by measuring and using image pixel statistics, and stopping analysis there.

We are reminded here of Marr's remarks about trying to understand flight in the animal kingdom. A statistical approach might discover that flight is statistically closely associated across a wide range of species with having wings. But leaving matters there would leave untouched the question of the hugely different aerodynamics underlying the flights of birds and bees.

The next decade should decide whether the image statistics approach proves to be a major new development in how to study seeing. It is certainly interesting and it will be worth keeping a look-out for results. Anderson and Kim (2009) report a challenge to its adequacy for the perception of gloss and lightness.

Multiple Representations

The initial step in building an explicit scene description is encoding retinal images as *gray-level descriptions* in the receptors of the retina (a description which is perhaps the closest our visual system ever gets to a photographic type of representation). Although the term gray-level is a commonplace one for receptor activities, in fact receptors vary in their sensitivities to light of different wavelengths, hence beginning the processes that culminate in color perception.

Pixels values (roughly equivalent to receptor activities) make explicit only point-by-point image intensities. They provide data for building numerous different kinds of representations. These include the lightnesses of surfaces (a task requiring a solution to the problem of discounting the illuminant), the slants of surfaces, features such as edges and corners, and objects. Hence, seeing can be thought of as the task of creating multiple representations of scenes.

Active Vision

Despite the reference just made to top-down processing, the idea of moving from low-level to high-level representations was a cornerstone of Marr's research. He stated in the opening paragraph of his highly influential book that vision is the process of discovering from images what is present in the world and where it is. This was expressed more succinctly in his slogan: *Vision is knowing what is where by looking.*

But that approach has been challenged over the past two decades by a different paradigm, dubbed *active vision* because its key idea is to build seeing systems geared to specific sorts of visually guided actions. Andrew Blake and Alan Yuille summed up this approach in 1992 as follows:

> … an active vision system is … selfish. It picks out properties of the images which it needs to perform its assigned task, and ignores the rest …. Work in this field has emphasized the simplifications that arise for an observer that is parsimonious, opportunistic and above all mobile.

Active vision research in the world of computer vision has developed, for example, systems for automatic surveillance. In these, the *action* might be guiding a video camera rig to track a potential suspect, with the visual processes built to do that having no idea of what the tracked object is. This is thus an example of a vision module designed to perform a certain action (camera control) without doing anything else (such as recognizing the tracked object).

There has also been a vigorous program of research in biological vision guided by the ideas of active vision (Findlay and Gilchrist, 2003, is a good starting point for searching that literature). One major claim has been that there exist separate visual pathways in the brain (the "dorsal and ventral streams") for mediating visually guided actions, such as posting a letter (dorsal stream), and for object recognition (ventral stream). This research has been highly influential although recent work has reported a large degree of overlap between the functions of these highly interconnected yet anatomically distinct systems (Farivar, 2009).

The active vision paradigm illustrates a key question that arises in any attempt to construct a computational theory of seeing: "What is seeing for?" The answer to this question has shifted over time, reflecting the changing emphasis within the research field of vision. The answer given by enthusiasts of active vision is clear: vision is for guiding actions. See *Further Reading*.

But it would be a mistake to think that vision is *only* for guiding actions. If we have learned anything over the last 50 years of vision research, it is that seeing is not for any single thing. It includes recognizing objects as well as guiding hands to grasp objects and to catch balls, or guiding feet over rough terrain. In short, seeing is good for many things, including the questions of *seeing what?* and *seeing where?* Indeed, it is self-evident that often there is little point in knowing what a thing is without knowing where it is, and vice versa.

Learning and Seeing

We have not referred much to the literature on the role of learning in seeing, confining our remarks to the effects of early visual environments on the orientation tuning of cells in the striate cortex (pp. 85-88). However, recently there has been considerable attention given to the role of learning in active vision systems. For example, such systems can have learning algorithms embedded in them for the control of eye movements involved in choosing directions of gaze (e.g., Ballard and Hayhoe, 2009).

Learning could also be involved in acquiring image statistics (see remarks above on the latter).

Another example of the use of learning in seeing has been in building models capable of learning the surface smoothness constraint for solving the stereo correspondence problem (e.g., Stone, 1996).

Learning is also involved in diverse other visual tasks, such as recognizing objects. A good starting point for exploring the literature on perceptual learning is the set of articles edited by Manfred Fahle and Tomaso Poggio (2002).

We suspect that learning will be a growth area in the study of seeing over the next couple of decades.

Bayesian Theories

The past two decades have seen an explosion of studies on seeing using Bayes' theorem. Despite the long historical pedigree of this theorem (devised in the eighteenth century), its potential for application within vision has only recently become widely accepted. It offers a principled way to implement constraints, which are put to work as the *priors* in Bayesian theories. Bayesian approaches of this kind can be found both when computational theories are being specified and also when algorithms are being developed. The scope for exploiting Bayes' theorem is thus very wide.

As we have observed in this book, whatever else it may be good for, Bayes' theorem is ideally suited to solving problems in which different, and often contradictory, sources of information must be reconciled to yield a sensible conclusion. This is a hard problem but the recent innovation making this problem tractable is the application of Bayesian principles. An example is the *cue combination problem* that has to be resolved when bringing together information provided by different cues to 3D surface slant (Ch 20). Thus, Bayes' theorem may not only be used to solve individual problems in vision; it may also provide a framework for reconciling apparently disparate sub-fields within vision.

Another valuable aspect of the Bayesian approach is that it provides a way of quantifying the importance of various constraints, and it also helps refine the notions of *seeing as unconscious inference* and *perception as hypothesis* (Ch 13).

How does the Bayesian approach relate to Marr's framework? We touched on this question in Ch 16 (p. 395) when we asked: is Jaramillo and Pearlmutter's Bayesian account of lightness perception a computational theory in Marr's sense of the term? That is, does it identify key constraints arising from the analysis of a particular task, and show that exploiting those constraints solves the task?

We concluded that the first of the constraints used by Jaramillo and Pearlmutter (that image noise is gaussian) satisfies this requirement. However, the second constraint they exploited (that image grey levels are gaussian) is not consistent with typical image gray levels.

Thus, their "gaussian image" assumption is used as a constraint, but this assumption is not consistent with a computational analysis of the problem. The assumption "works," insofar as it yields various lightness illusions, which was the goal of the exercise. However, this type of model would be more convincing if the illusions were obtained for models which assume non-gaussian distributions, especially if those non-gaussian distributions coincided with the particular distributions which characterize typical images. To our knowledge, this hypothesis has not been tested, so the jury is still out with regard to this issue

We reiterate this analysis here to give a concrete example of how it can be tempting to use Bayes' rule as an "algorithm engine" that "works" (in a specific case), and lose sight of the bigger picture, which is: exactly what constraints are being fed into this engine and are they defendable in terms of a full analysis of the vision problem being addressed?

To summarize our view on this important contemporary issue (important because of the current prevalence of Bayesian models of visual processes): render unto Bayes the things that are Bayes'. The Bayesian framework is an excellent, indeed the "proper", mathematical framework for implementing constraints in many circumstances. But we would add: as always, don't be seduced into using mathematically sophisticated algorithms without constantly asking—what constraints are being exploited and can they be justified from a computational analysis of the problem?

Philosophical Issues

Consider the following statement from Marr's book (p. 68):

> the true heart of visual perception is the inference from the structure of an image about the structure of the real world outside. The theory of vision is exactly the theory of how to do this, and its central concern is with the physical constraints and assumptions that make this possible.

Readers will recognize this as precisely the stance we have adopted in this book. It is one we suspect is widely shared nowadays by many vision scientists. Yet is it defendable philosophically?

Jan Koenderink, a highly distinguished physicist and psychophysicist, thinks that it is fundamentally flawed. His reasons are interesting.

Koenderink notes that Marr's statement espouses a version of the philosophical position known as *realism*. This holds that there is a physical world "out there" whose existence is independent of our conceptual schemes, linguistic practices, beliefs, etc. Philosophers who profess realism also typically believe that truth amounts to a correspondence between belief and reality.

Koenderink adopts the philosophical stance of realism (as we do ourselves) but he regards Marr's statement as going a step further and implying acceptance also of a version of the philosophical position of *objectivism*. There are various ways in which this term is used in philosophy, but here it refers to the claim that not only is there a real world but that it is also possible to know its properties.

Objectivism is a controversial hot topic in contemporary philosophy, and particularly in the philosophy of science. (It might help some readers to note that we are not here referring to objectivism as defined by Ayn Rand.)

Hilary Putnam, cited approvingly by Koenderink, is a respected American philosopher who is a well known critic of objectivism. Putnam draws attention *inter alia* to the fact that modern-day physics has still not resolved the question of whether a photon is a wave or a particle, as originally posed. Physicists now regard it as either one or the other depending on which view is most suitable for explaining a given phenomenon.

Such flexibility casts doubt, Putnam argues, on believing in an attainable "truly objective reality" about the nature of a photon. Interested readers can explore this issue by a suitable Web search but they should be ready to risk becoming bogged down in a philosophical quagmire of hotly contested points of view. We simply ask: in what sense does Marr's statement entail acceptance of an objectivist stance?

We read Marr as a pragmatist in this regard. When he refers to "the structure of the real world outside" we think he is using this term in the ordinary sense. That is, to refer to what you would find if you used simple measuring instruments such as rulers and photometers to measure the properties of objects. That seems good enough to us for justifying our acceptance of his statement as a basis for research into seeing.

That said, Koenderink develops his point in an interesting way. He points out that various aspects of our visual perceptions have no clear correlates in the physical world. Consider, for example, looking at a tree in winter with no foliage. We readily discern the overall shape of the tree from the boundary of where its twigs and branches end. We may even be able to use its shape to decide whether the

tree is an oak, an elm, or whatever. By definition, our representation of this boundary is a mental construct. Can it be said to represent a boundary that is physically present in the "real world?"

The problem is that the boundary does not exist as a continuous physical structure, but is an inference of a shape describing the endpoints of a set of elements (twig endings). This inference is an example of the application of the figural grouping principles described in Ch 7.

We note, to avoid possible misunderstandings, that this tree-in-winter example is our own, not Koenderink's, but we think it illustrates correctly his argument, which is: Marr must be mistaken in claiming that *the true heart of visual perception is the inference from the structure of an image about the structure of the real world outside* because here we have a case in which there is no structure in the world that directly matches our perception.

Some readers with a philosophical bent might find this a really interesting debate. Others might wonder whether it is philosophical hair-splitting of a very high order. The latter group might suggest that if one took suitable measuring instruments to the tree and logged the density of twigs in the tree and the space around it, there would emerge a pretty clear demarcation boundary in those physical measurements. Is this not a species of "physical reality?" Human vision makes that density boundary explicit.

We end this discussion by remarking that what is going on in our tree boundary example is perfectly transparent (please forgive the pun). It makes sense to assert that the tree has an outline boundary that is rooted in the physical structure of the tree, and that visual processes make that boundary explicit. There is no mystery here akin to the photon wave/particle duality problem in physics. It is the business of philosophers to worry about such issues as the nature of objectivism but pragmatic vision scientists, like pragmatic physicists, just get on and do what they regard as "sensible" things in tackling tractable problems.

How Has Marr's Approach Fared?

Marr's widely-cited book was completed by colleagues and published posthumously two years after he died. It is interesting to ask: how has his work fared in the three decades since his death?

First, Marr's specific theories of particular vision tasks have been largely overtaken by subsequent work. This means that his book is now mainly (but by no means wholly) of only historical interest as far as theories of various specific vision tasks are concerned (while remaining of great value in explaining his approach to vision). It is a great pity, indeed we would say tragic, that leukemia denied him the opportunity to follow up his early work toward his goal of establishing lasting theories of vision tasks. That said, the three-levels framework expounded in his book survives as an enduring contribution to vision research, which is why we have used it as the theme of this book.

It is a curious fact that the authors of many textbooks on vision, and also on cognitive psychology, introduce early on Marr's three-levels approach (computational theory, algorithm, hardware), but then largely ignore it in the way they deal with the various topics that they choose to include in their book. (Quinlan and Dyson's excellent textbook is an exception.)

We have tried to do the opposite: we also set out Marr's framework early on but then tried to use it systematically when tackling a wide range of vision problems. Why haven't other authors of texts on biological vision done the same?

One reason may be that many vision scientists simply have a different conception from Marr as to what constitutes a satisfactory theory of perception. Some take the view that finding a link between a given visual phenomenon and a neurophysiological process amounts to *Job Done*. We disagree, as did Marr (and as do others, including significant predecessors with a similar general outlook; see Ch 1, p. 11), and we explained why when discussing above the pros and cons of the psychophysical and neurophysiological approaches.

Another reason for widespread neglect in using Marr's approach could be that it requires skills that some scientists of seeing just don't have. Specifically, it typically utilizes math somewhere along the line. That said, many constraints are easy to describe in words, such as the surface smoothness constraint utilized in the solution of the stereo correspondence problem.

Moreover, constraints have been identified from psychophysical studies without need of mathematics (figural grouping principles are a case in point),

even if they have subsequently been rendered more precise by mathematical formalisms. The importance of math means that for many the computational approach can at best be pursued in a multi-disciplinary manner.

Sophisticated Ignorance

We have explained in this book that answers to many questions about seeing are simply not available. The history of science shows that if a science lacks answers then the problem can often be traced back to a lack of well-defined questions. We have emphasized the computational framework because we think it is always valuable to bear in mind the key question: *what is the vision task to be solved?*

This emphasis has the potential for precision in formulating well defined questions. For example, it is not enough to ask interesting but vague questions, such as "how does the brain work?" Instead, we need to ask precise questions for which answers can be sought that are themselves sufficiently precise that they can be tested to destruction. As noted by EE Cummings (1938), "Always the beautiful answer who asks a more beautiful question."

If you were new to the field of seeing when you started reading this book, then we hope you have learned a little about the subject. Looking back on our own careers in studying seeing, we now realize that at the outset our ignorance could be described as quite narrow, inasmuch as we had a very limited idea of the range of questions that had to be answered. Moreover, our prior knowledge on those questions was extremely limited. We hope now, having explored various aspects of seeing in this book, that the ignorance you now have is akin to our own, not just smaller than it once was, but qualitatively different.

The science of vision has a very long way to go before it can be said to offer a good understanding of how seeing works. We hope you now share with us, obviously not a neat set of answers, but a sophisticated kind of ignorance that is wide and rich with questions.

Figure **23.1** defines three fundamental types of problem. First, there are problems that are interesting but currently ***intractable***, such as consciousness, time travel, faster-than-light travel, immortality. Attempted solutions to these sorts of problems presently fail the maxim laid down by Peter

Medawar: *Science is the art of the soluble.*

Then there are problems that are **tractable** (i.e soluble) but uninteresting (at least, when considered in the larger scheme of things), such as making paint dry faster.

Finally, there are problems that are both *interesting* and *tractable*. Historically, problems often become tractable by some technological innovation which lifts them out of the intractable class. For example, the discovery of DNA depended critically on the development of X-ray diffraction. We think, or at any rate we hope, that vision sits in the class of tractable and interesting problems. However, unlike DNA, vision probably does not require the development of new technology. Instead, it requires theoretical developments.

In a recent Royal Society lecture, Daniel Wolpert described how the number of pages in a classic neuroscience text increases every time it is republished, which happens every few years. He protested that this is not really how science should progress. What is required, he argued, is a new grand over-arching theory which would take the current mass of more or less related facts and compress them down into a coherent whole.

Does this argument apply to seeing? We are confronted with mountains of data, mountains which grow with each passing year. Should we aim for a grand unifying theory of vision able to explain these mountains of data succinctly?

Sadly, to our minds, too much research within vision consists of studying engaging (and often dramatic) phenomena, with too little concern for the question *What are the vision tasks?* that these phenomena illuminate. Such work is often fascinating, because the phenomena are often astounding, but our worry is that it acts as a distraction. The reason is that studying the phenomena for their own sake risks studying mechanisms rather than the structure of visual tasks. Marr's coruscating critique of much research being done in his day in the fields of artificial intelligence, memory, and vision deserves revisiting in this regard, as it seems to us as relevant now as when it was in the 1970s. For example, he writes in the Epilogue of his book (p. 347):

Mechanism-based approaches are genuinely dangerous. The problem is that the goal of these studies is mimicry rather than true understanding…If we believe that the aim of information processing studies is to formulate and understand particular information-processing problems, then the structure of those problems is central, not the mechanisms through which those solutions are implemented.

We strongly recommend that the reader reads the whole of this Epilogue, as we think it still has much to say about where vision research should be heading.

So the next time you gaze, along with us, eyes wide open in amazement, at some astounding visual phenomenon, ponder on the thought that behind it somewhere lies unseen a beautiful theory about a visual task.

Perhaps a grand over-arching theory of seeing is not in principle realizable, for reasons which we don't yet understand. But if one is attainable then our best bet is that its seeds are most likely to be found using Marr's computational framework.

Summing Up

Little more need be said by way of summary in this final chapter, but a last reminder should be added that the literature of visual science is vast and that necessarily only a small sample has been dealt with here. All chosen topics have been simplified to communicate essentials, but we hope not to the point of oversimplification, and thus not to the point of serious distortion. Readers who find their appetite whetted for further information about seeing can follow up references given in the sections entitled *Further Reading*, and/or conduct Web searches.

Reflections on Scientific Revisions

Various topics in this book provide examples of how a scientific theory begins life as a highly provisional idea. The theory is then disrupted by inconvenient facts. However, these data stimulate a new theory which, because it is closer to the truth, is often more elegant than the theory it replaced. The scientific process is thus theory→revision→new theory→revision, and so on.

This sequence reflects the fact that science is an intellectual enterprise which is designed to incorporate both old and new findings by creating increasingly elegant and general theories. Of course, individual scientists can be as stubborn about *not* changing their minds as anyone else. They might so regret the death of a nice theory that they challenge "the facts" which contradict their beloved theory. And of course they might sometimes be right so to do. But, in the end, evidence rules the day within the scientific community. If one scientist remains in denial about the need for changes to a theory, you can be sure others will be vociferous in their objections to the theory.

However, it is important not to exaggerate the provisional character of scientific theorizing because some theories become accepted as facts. A case in point is Harvey's theory, advanced in 1628, that the heart is a pump for squeezing blood around the body through the arteries and back to the heart via the veins. This was a major and controversial revision to existing ideas in its day. Nowadays you would risk being thought mad to reject it. Another example illustrating the same general point is the "theory" that the earth circles the sun, rather than the other way around.

We have chosen in this book to promote the claims of the computational approach to biological vision as being a good, indeed we presently think the "correct," way to study seeing. But who knows what the future will bring? We, like all scientists, reserve the right to change our minds when new evidence is found. Science is not a faith system.

Or at any rate, the only way "faith" comes into it is the belief that evidence is crucial to progress. But even that single article of "faith" is based on c.400 years of history showing that scientific progress is achieved by using evidence. Evidence is a wedge that separates rational thought from superstition, it is the only means we have to interrogate nature. And, as Richard Feynman famously remarked: *nature cannot be fooled.*

It could be that future of seeing studies will be radically different from our present best bet. It is worth ending by repeating what we said earlier:

> What sort of answer will ultimately come to be seen as a "full" explanation of seeing? We simply do not know, despite having spent most of our academic lives studying seeing.

Also, it needs to be added that some have doubted whether a "grand theory" will ever be achieved, wondering whether its complexities will defy human understanding. Moreover, it might even be that thinking in terms of a "grand theory" is a mistake: perhaps seeing will turn out to be no more than the operation of myriad different modules. At this point in the history of vision, we are confronted with oceans of data, in which can be found small islands, corresponding to theories which explain just the data lapping at their shores. Only time will tell whether or not these apparently disparate islands in fact comprise an archipelago, a "grand theory" of seeing.

Further Reading

Anderson BL and Kim J (2009) Image statistics do not explain the perception of gloss and lightness. *Journal of Vision* **9**(11).

Ballard DH and Hayhoe MM (2009) Modeling the role of task in the control of gaze. *Visual Cognition*, **17** 1185-1204. *Comment* Dana Ballard is a highly respected computer vision scientist who has been active in recent years in exploring the use of reinforcement learning algorithms in the development of active vision systems. His Web site is a good starting point if you want to investigate his work and the associated literature: http://www.cs.utexas.edu/~dana/#publications.

Blake A and Yuille A (1992) *Active Vision*. The MIT Presss. *Comment* The quotation from these authors is on page xv of their *Introduction* to this edited collection of diverse papers.

Fahle M and Poggio T (2002) Perceptual Learning. The MIT Press. *Comment* A set of articles giving a broad overview of the field.

Farivar R (2009) Dorsal-ventral integration in object recognition. *Brain Research Reviews* **61** 144–153. *Comment* Discusses current views on the "dorsal/ventral streams hypothesis."

Findlay JM and Gilchrist ID (2003) *Active Vision: The Psychology of Looking and Seeing*. Oxford Psychology Series. *Comment* A review of recent psychological work using the active vision paradigm.

Koenderink JJ (2002) *Marr's Vision in Retrospect*. Paper presented to the *Marr Symposium*, European Conference on Visual Perception, Glasgow.

Medawar P (1967) *The Art of the Soluble.* London: Methuen.

Quinlan Q and Dyson B (2008) *Cognitive Psychology.* Pearson Education Ltd. *Comment* A fine textbook which emphasizes Marr's approach. It is a good complement to this book in that it has little on neurophysiology and much on cognitive process such as memory and attention.

Stone JV (1996) Learning perceptually salient visual parameters using spatiotemporal smoothness constraints. *Neural Computation* **8**(7) 1463-1492.

Wolpert D (2005) The Puppet Master: How the brain controls the body. *Francis Crick Lecture* presented at the Royal Society of London, Thursday 8 December 2005.

Index